Cambridge International AS & A Level Biology

STUDENT'S BOOK

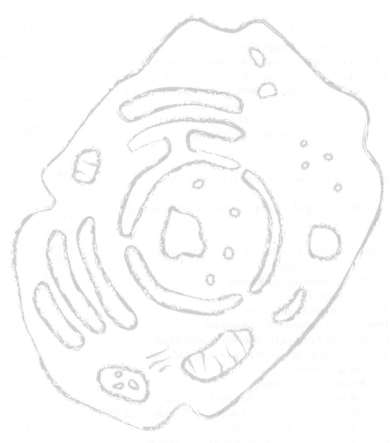

David Martindill, Michael Smyth
and Mike Smith

William Collins' dream of knowledge for all began with the publication of his first book in 1819.
A self-educated mill worker, he not only enriched millions of lives, but also founded a flourishing publishing house.
Today, staying true to this spirit, Collins books are packed with inspiration, innovation and practical expertise.
They place you at the centre of a world of possibility and give you exactly what you need to explore it.

Collins. Freedom to teach.

Published by Collins
An imprint of HarperCollins*Publishers*
The News Building
1 London Bridge Street
London
SE1 9GF

HarperCollins*Publishers* Macken House
39/40 Mayor Street Upper
Dublin 1
D01 C9W8
Ireland

Browse the complete Collins catalogue at
www.collins.co.uk

British Library Cataloguing-in-Publication Data

A catalogue record for this publication is available from the
British Library.

Authors: David Martindill, Michael Smyth, Mike Smith
Publisher: Elaine Higgleton
Product manager: Joanna Ramsay
Content editor: Tina Pietron
Project manager: Susan Lyons
Development editor: Rebecca Ramsden
Copyeditors: Amanda Harman and Naomi Mackay
Typesetter, illustrator: Jouve India Private Ltd
Indexer: Jane Henley
Proofreader: David Hemsley
Artwork permissions researcher: Rachel Thorne
Maps created or checked by © Collins Bartholomew Ltd 2019
Cover designer: Gordon MacGilp
Cover illustration: Mike Smith and Ann Paganuzzi
Production controller: Lyndsey Rogers
Printed by Ashford Colour Press Ltd
With thanks to David Styles for his help on this project.

MIX
Paper | Supporting
responsible forestry
FSC™ C007454

This book contains FSC™ certified paper and other controlled
sources to ensure responsible forest management.

For more information visit: www.harpercollins.co.uk/green

This book is produced from independently certified FSC™
paper to ensure responsible forest management.

For more information visit: www.harpercollins.co.uk/green

Cambridge International copyright material in this publication
is reproduced under licence and remains the intellectual
property of Cambridge Assessment International Education.

Exam-style questions have been written by the authors.
In examinations, the way marks are awarded may be different.
References to assessment and/or assessment preparation are
the publisher's interpretation of the syllabus requirements and
may not fully reflect the approach of Cambridge Assessment
International Education.

Cambridge International recommends that teachers consider
using a range of teaching and learning resources in preparing
learners for assessment, based on their own professional
judgement of their students' needs.

The publishers gratefully acknowledge the permission granted
to reproduce the copyright material in this book. Every effort
has been made to trace copyright holders and to obtain their
permission for the use of copyright material. The publishers
will gladly receive any information enabling them to rectify
any error or omission at the first opportunity.

Contents

Contents

Getting the best from this book

Welcome to *Collins Cambridge International AS & A Level Biology Student's Book*. This textbook has been designed to equip you with the knowledge and understanding required for the Cambridge International AS & A Level Biology 9700 syllabus.

SAFETY NOTE

This book is a textbook and not a laboratory or practical manual. As such, you should not interpret any information in this book that relates to practical work as including comprehensive safety instructions. Your teachers will provide full guidance for practical work and cover rules that are specific to your school.

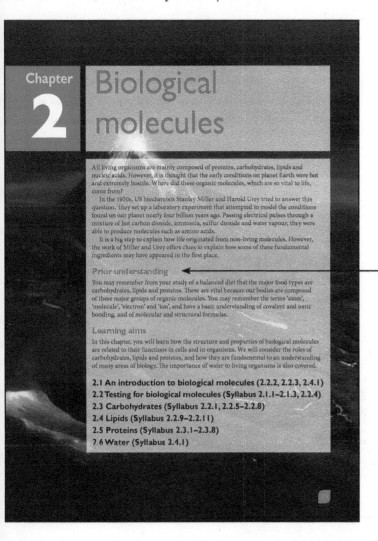

Remind yourself of relevant knowledge and understanding you already have before starting a new topic.

Check your understanding regularly throughout. Use the 'Clues' if you need a pointer or are unsure as to how to arrive at the answer.

6. Draw a table to compare the structure and properties of starch and cellulose. Use Table 2.9 to help you. You should include the following features:
   - the sugar units from which they are made
   - the type of glycosidic bonds that join them together
   - whether or not they are made of one type of molecule
   - whether or not they branch
   - the ease with which they can be digested
   - where they are found
   - their functions.

7. Discuss how amylopectin and glycogen are readily hydrolysed to release glucose for respiration.

8. Trehalose is a disaccharide formed by an α(1,1) link between two α-glucose monomers. Use this information to draw the full structural formula of trehalose.

Tip

Starch and glycogen are storage polysaccharides, whereas cellulose is a structural polysaccharide

Challenge yourself with questions requiring deeper engagement and thinking skills.

Develop your skills in evaluating practical planning, design, methods and data quality.

Experimental skills 2.2: Investigating dietary starch

In most plant cells, starch is stored in grains called **granules**. These are packed into the cytoplasm as energy stores, and they can be hydrolysed to release glucose molecules when required.

QUESTIONS

P1. Explain the advantages of storing starch rather than glucose in plant cells.

The percentage of amylose and amylopectin in starch granules varies depending on the species

of plant. In some plant tissues, such as in mung beans, the starch consists of more amylose than in others, such as in potatoes. The relative proportions of amylose and amylopectin can be expressed as an amylose : amylopectin ratio.

Five starch granules were measured in a cell taken from seven different plant species and a mean for each was calculated. Table 2.10 shows the results.

Consolidate and reflect upon your learning.

Key ideas

→ Proteins are polymers made from amino acids joined by peptide bonds.
→ There are 20 different amino acids found in living organisms. Each has a central carbon atom, to which a carboxyl group and an amine group are joined. The fourth bond on the carbon atom holds an R group or side chain, which varies between different amino acids.
→ A long chain of amino acids is known as a polypeptide. A protein may be made up of just one or several polypeptides.
→ Hydrogen bonds, hydrophobic interactions, ionic bonds and disulfide bonds help to hold a protein molecule in shape.
→ The sequence of amino acids in a polypeptide or protein is known as its primary structure. Regular coiling or folding forms the secondary structure, and further folding forms the tertiary structure. The association of more than one polypeptide forms the quaternary structure.
→ The shape of a protein is strongly related to its function. Globular proteins are usually soluble and often have roles in metabolism. Fibrous proteins are insoluble and generally have structural roles.

2.6 Water

Life on Earth probably evolved in the water, and the majority of the total mass of living organisms is water. Simply put, without water, there would be no life on Earth. It is a special molecule with extraordinary properties.

To understand the importance of water to life on Earth, we need to understand the structure of its molecules and how they behave with other molecules. A water molecule is made up of two hydrogen atoms joined by covalent bonds to a single oxygen atom, so its formula is H_2O.

As we have encountered throughout this chapter, all atoms contain electrons, and an electron has a small negative charge. In many molecules, including water, the electrons in the chemical bond are not shared equally between the oxygen and the hydrogen atoms. This results in the oxygen atom having a tiny negative charge, and each hydrogen atom having a tiny positive charge. The water molecule is said to be polar. The molecule can also be said to have a **dipole**. As you read earlier, although its overall charge is zero, it has a small positive charge in some places (near the two hydrogen atoms), and a small negative charge in another (near the oxygen atom). These small charges are written δ– and δ+, as shown in Figure 2.45a.

Negative charges and positive charges are attracted to one another. This means that each water molecule is attracted to its neighbours. The negative charge on one water molecule is attracted to the positive charge on another water molecule nearby. These attractions are called hydrogen bonds (Figure 2.45b). Although they are important, hydrogen bonds are much weaker than the strong covalent bonds that hold the oxygen and hydrogen atoms firmly together.

Avoid common errors and benefit from tips on how to remember key facts.

Tip

Other molecules are also polar, especially those that have –OH, –C=O or –NH groups. Water can form hydrogen bonds with these, and these groups can form hydrogen bonds with each other.

See and understand how different topics link together throughout the book.

Link

The mechanism by which oxygen is loaded to and unloaded from haemoglobin is explored in greater detail in Chapter 8.

- **High solubility:** the four polypeptide chains are coiled up so that the hydrophilic R groups on the residues of each are on the outside of the molecule. They can therefore form hydrogen bonds with water molecules. Hydrophobic R groups are mostly found inside the molecule, which means that they do not interact with the aqueous environment.
- **Ability to bind oxygen:** the iron (Fe) ion, found in the haem group of each polypeptide chain, can combine with oxygen. Because there are four haem groups, one haemoglobin molecule can combine with four oxygen molecules (eight oxygen atoms).
- **Ability to release oxygen:** the shape of the haemoglobin molecule is able to change to better enable it to pick up oxygen when the oxygen concentration in its surroundings is high, and to release oxygen more readily when it is low.

Develop strategies for answering questions, including mathematical skills.

Worked example

With reference to named examples, explain the differences between globular proteins and fibrous proteins.

Answer

Globular proteins, which includes all enzymes and other proteins with physiological roles such as haemoglobin, are roughly spherical in shape and are soluble in water. This is because they have amino acid residues with hydrophobic R groups positioned on the inside of the protein, and amino acid residues with hydrophilic R groups positioned on the outside of the protein that interact with the aqueous environment. Fibrous proteins, which include proteins such as keratin and collagen that have structural functions, are insoluble in water. This is because they have amino acid residues with hydrophobic R groups positioned on the outside of the protein.

Apply your new knowledge and practical or mathematical skills in different contexts and sharpen your analysis and evaluation skills.

Assignment 2.1: Scurvy – the 'scourge of the sea'

Use the information in the passage below, and your own knowledge, to answer the questions that follow.

Fruit and vegetables are a vital part of a balanced diet. One reason for this is that many are rich sources of vitamin C. The body needs this molecule to keep our bones, skin, tissues and blood vessels healthy. This is because vitamin C is required for the conversion of proline, an amino acid, into hydroxyproline, an amino acid derivative. Hydroxyproline is a key component of collagen. Early symptoms of scurvy, a disease caused by a lack of vitamin C, include skin that bruises easily and bleeding gums (Figure 2.42).

Scurvy used to be common in many parts of the world. Famously, it often affected sailors, who could sometimes spend many months on ships at sea without fresh fruit and vegetables to eat. In 1752, the British doctor James Lind

Figure 2.42 An early symptom of scurvy is bleeding gums. The red colour of blood is due to the high concentration of the haemoglobin protein in red blood cells.

Strengthen your understanding of topics and the key concepts by drawing out the mini mind maps and expanding them.

CHAPTER OVERVIEW

Review the mini mind map to link new learning with your prior learning, and to highlight which of the key concepts underpin the chapter.

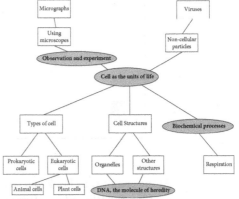

Try copying this mini mind map and expanding upon it. Use your notes to help you explore how the essential ideas, theories and principles can be linked further together.

Check that you have learned the key points in the syllabus for this area and revisit the relevant topics if you want to consolidate your learning.

WHAT YOU HAVE LEARNED

Now that you have finished this chapter you should be able to:

- make temporary preparations of cellular material suitable for viewing with a light microscope
- draw cells from microscope slides and photomicrographs
- calculate magnifications of images and actual sizes of specimens using drawings, photomicrographs and electron micrographs (scanning and transmission)
- use an eyepiece graticule and stage micrometer scale to make measurements and use the appropriate units, millimetre (mm), micrometre (μm) and nanometre (nm)
- define resolution and magnification and explain the differences between these terms,

Check and embed your
learning with review
questions.

CHAPTER REVIEW

1. Using the terms 'specific' and 'complementary' in your answer, define the term 'active site'.

2. When clear egg 'white' is boiled during cooking, soluble proteins form an insoluble precipitate. This is due to the hydrophobic R groups in the centre of the molecule becoming exposed to the water.

   a. Explain why heating egg 'white' causes the hydrophobic R groups in the centre of the molecule to become exposed to the water.

   b. Predict, with an explanation, what you would see if an egg was cracked into a beaker of vinegar (vinegar has a pH of about 3–4).

3. Draw a table to compare and contrast how denaturation of enzymes occurs as a result of extremes of temperature and extremes of pH.

4. Match each word with its correct definition.

(a) Enzyme	(i)	Region of an enzyme molecule that binds to a protein during a reaction.
(b) Substrate	(ii)	Molecule produced by a reaction.
(c) Product	(iii)	Molecule that an enzyme acts upon.
(d) Active site	(iv)	Molecule that inhibits enzyme reactions by altering the shape of the enzyme molecule without binding to the active site.
(e) Competitive inhibitor	(v)	Catalyst (usually a protein) produced by cells, which controls the rate of chemical reactions in cells.
(f) Non-competitive inhibitor	(vi)	Molecule that binds to the active site in an enzyme but no reaction takes place.

5. Inhibitors can either be competitive or non-competitive.

   a. Identify the type of inhibition (competitive or non-competitive) that happens in the following biochemical scenarios.

      i. Ethylene glycol is an ingredient in antifreeze, which is added to vehicle engines to prevent them freezing in cold weather. Pets may drink antifreeze if it is spilt, and the conversion of ethylene glycol to toxic products by alcohol dehydrogenase can cause kidney failure. Affected animals can be treated by providing them with ethanol to drink, which has a similar shape to ethylene glycol.

      ii. During a part of photosynthesis called the light independent reaction, oxygen will bind with the enzyme ribulose bisphosphate carboxylase to prevent it catalyzing a reaction between carbon dioxide, which has a similar structure to oxygen, and a larger organic molecule.

      iii. Aspirin binds to cyclooxygenase, an enzyme involved in the synthesis of products called prostaglandins. This reduces inflammation. Arsenic reduces the V_{max} of this reaction but does not have an effect on the value of K_m.

   b. Outline an experiment that a researcher should carry out to determine whether thiourea is a competitive or non-competitive inhibitor of an enzyme.

6. The graph in Figure 3.34 shows the effect of substrate concentration on the initial rate of an enzyme-catalysed reaction.

   a. The graph is labelled with the letters X and Y. Which of the following descriptions (i) to (v) match part X, Y, or X and Y on the graph?

At the end of your book you will also find:
- clues for selected questions if you are unsure how to arrive at the answer
- practice exam-style questions to embed your learning
- a complete glossary so that you can check and consolidate your understanding of key terms
- information to help you develop and refine the skills you will use during your course and in your examinations
- complete answers for questions in the Student's Book.

1.1 The microscope

cell studies

Chapter

1

Cell structure

Scientists working for the Deep Carbon Observatory project have investigated life far below the Earth's surface. They have shown that approximately 70% of the Earth's bacteria and other microorganisms live underground. The scientists used microscopes to study samples from mines and boreholes, finding microorganisms living up to 5 km deep. Many of these organisms are very different from their surface-dwelling relatives, with a genetic diversity at least equal to, if not greater than, that found above the surface.

The ability of these organisms to survive in such extreme conditions will change the way scientists search for life elsewhere in the universe. Until now, attention has been focused on places where water may be found. Now scientists will need to consider the possibility of life surviving far below rocky surfaces too.

Prior understanding

In your previous studies you may have used microscopes or hand lenses to study biological specimens. You may have made drawings of specimens and calculated their magnification. You may be familiar with some different types of cell, such as animal and plant cells, as well as the functions of some of the structures inside cells.

Learning aims

In this chapter, you will learn how to use microscopy to study cells, including the function of an eyepiece graticule and stage micrometer scale, and distinguish between resolution and magnification. You will learn about two basic types of cell – prokaryotic and eukaryotic – as well as the key features of non-cellular viruses. You will also learn about the functions of cell structures, including the role of ATP in energy transfer.

1.1 The microscope in cell studies (Syllabus 1.1.1–1.1.5)

1.2 Cells as the basic units of living organisms (Syllabus 1.2.1–1.2.7)

1.1 The microscope in cell studies

Although lenses have been used as magnifying glasses for at least a thousand years, the first microscopes were built around the start of the 17th century. Some used a single lens but others were compound microscopes that used two lenses: an objective lens and an eyepiece lens, very similar to many microscopes today.

Later in the 17th century, scientists used microscopes to make the first detailed observations of biological structures. In Italy, Marcello Malpighi studied the structure of animal and plant tissues and was the first person to see blood capillaries. Cells were discovered and named by Robert Hooke in England, who thought that what he saw in a thin slice of cork resembled the small rooms, or **cells**, that monks lived in. Antonie van Leeuwenhoek, a Dutch scientist, discovered bacteria, single-celled organisms called protists and sperm cells, as well as making many other detailed observations, for example of red blood cells and muscle fibres.

In the 20th century, **electron microscopes** were developed. These use a beam of electrons, rather than light, to produce an image, which shows much more detail than is possible with a light microscope. For example, the structure of viruses, which are much smaller than cells, was first seen using an electron microscope (although their existence had been known much earlier). A **transmission electron microscope (TEM)** passes electrons through a specimen to produce an image. A **scanning electron microscope (SEM)** detects electrons that have been reflected from the surface of a specimen to produce a 'three-dimensional' image.

Images produced using a microscope are called **micrographs**: images produced by a **light microscope** are called **photomicrographs** (Figure 1.1a); images produced by an electron microscope are called **electron micrographs** (Figures 1.1b, c).

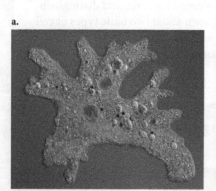
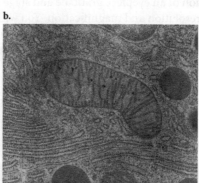
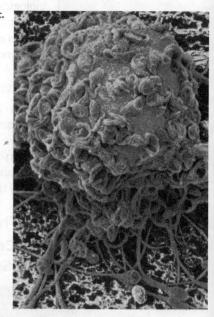

Figure 1.1 Micrographs of: *Amoeba*, a single-celled organism, taken with a light microscope (**a**); a mitochondrion, taken with a TEM (**b**); a human egg cell covered with sperm cells, taken with a SEM (**c**).

1. Explain the difference between a photomicrograph and a photograph.
2. Look at Figure 1.1c. Explain how you can tell that the image has been taken with a SEM and not a TEM.

MAGNIFICATION AND RESOLUTION

Microscopes produce images larger than the original objects. This is called **magnification**. Although in theory there is no limit to how many times you can magnify an image, at some point you will stop seeing any more detail. The ability to see detail is a measure of the microscope's **resolution** (or resolving power). Think about a digital photograph: if you keep enlarging it, the magnification increases. However, when you get to the point where you can see the separate pixels, then greater magnification will not lead to greater resolution because you will not be able to see any more detail.

Magnification

Magnification is the number of times larger an image is than the actual object. For example, if the image of a cell is 50 mm long and the actual cell is 0.1 mm long, then the magnification is ×500.

Magnification can be calculated using the equation:

$$\text{magnification} = \frac{\text{size of image}}{\text{size of object}}$$

The following units are used when measuring cells and cell structures: millimetres (mm), micrometres (μm) and nanometres (nm).

- One **millimetre** is one-thousandth of a metre.
- One **micrometre** is one-thousandth of a millimetre and one-millionth of a metre.
- One **nanometre** is one-thousandth of a micrometre, one-millionth of a millimetre and one-billionth of a metre.

This is summarised in Table 1.1.

Unit	Symbol	Comparison with 1 metre	
millimetre	mm	$1\text{ mm} = 1 \times 10^{-3}\text{ m}$	$1\text{ m} = 1 \times 10^{3}\text{ mm}$
micrometre	μm	$1\text{ μm} = 1 \times 10^{-6}\text{ m}$	$1\text{ m} = 1 \times 10^{6}\text{ μm}$
nanometre	nm	$1\text{ nm} = 1 \times 10^{-9}\text{ m}$	$1\text{ m} = 1 \times 10^{9}\text{ nm}$

Table 1.1 Units used when making microscope observations of biological specimens.

Worked example

a. Convert 5000 nanometres to micrometres.

Answer
1000 nm = 1 μm, so 5000 nm = 5 μm.

b. Convert 72 millimetres to nanometres. Give your answer in standard form.

Answer
$1\text{ mm} = 1 \times 10^{6}\text{ nm}$, so $72\text{ mm} = 72 \times 10^{6}\text{ nm}$. In standard form this is $7.2 \times 10^{7}\text{ nm}$.

3. The length of a sperm cell is measured as 50 μm. Give the length, in standard form, in
   a. mm   b. nm.
4. The diameter of the nucleus in a human cell is approximately 6 μm. An electron micrograph is made of the nucleus at magnification ×4000.
   a. Give the actual diameter of the nucleus in nanometres.
   b. Calculate the diameter of the image of the nucleus. Give your answer in millimetres.

Human cells range in diameter from about 7 μm (red blood cell) to about 100 μm (egg cell). Bacteria cells, which are usually smaller than human cells, have diameters between 0.2 μm and 10 μm.

Tip

Standard form is used to write very large or very small numbers in a shorter format. For example, 5 000 000 in standard form is 5×10^{6} and 0.000005 is 5×10^{-6}.

Tip

Always try to show your working in calculations, even if you are not specifically asked for it.

Tip

When writing in standard form, the first part of the number must be at least 1.0 but smaller than 10.

Tip

You may need to give an answer in different units from those in a question. For magnification calculations, make sure both the image and object measurements are in the same unit.

You can work out the magnification of a compound light microscope (see Figure 1.2) by multiplying together the magnifications of the eyepiece lens and the objective lens. (The magnification is usually shown on the side of a lens.) For example, if an eyepiece lens has a magnification of ×10 and the objective lens has a magnification of ×40, then the total magnification = 10 × 40 = ×400.

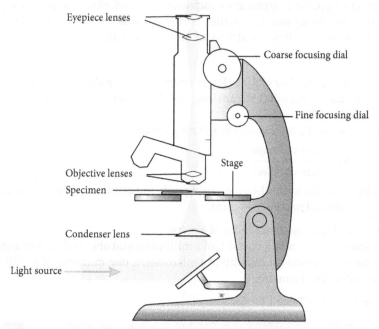

Figure 1.2 A compound light microscope. Note that often the eyepiece and objective lenses are each composed of more than one lens.

Resolution

Resolution is the ability to distinguish between two objects that are close together – to be able to see them as separate objects. The reason that electron microscopes (see Figure 1.3) produce images with a greater resolution than light microscopes is to do with wavelength. Electrons have a much smaller wavelength than the wavelength of light, which allows for greater resolution of the image. Light microscopes have a resolution limit between 200 nm and 250 nm. The limit of resolution of a SEM is about 1 nm and that of a TEM about 0.5 nm. This means that the maximum useful magnification of a light microscope is approximately ×1500, whereas for electron microscopes magnification can be up to ×500 000 for a TEM (and about half that for a SEM). As a comparison, Figure 1.4 shows chloroplasts viewed with both a light microscope and a TEM. Only the latter clearly shows the internal structure of a chloroplast.

Comparison of light microscopes and electron microscopes

Although electron microscopes allow much greater magnification and resolution than light microscopes, they do have some disadvantages. The images produced by electron microscopes do not show colour (because they are formed by the action of electrons, not light). If you see a coloured electron micrograph this is because the colour has been added afterwards, for example by a computer. These images are known as false-colour images.

The other main disadvantage is that electron microscopes cannot be used to view living tissue or living organisms. This is because the prepared specimens have to be placed in a vacuum inside the microscope.

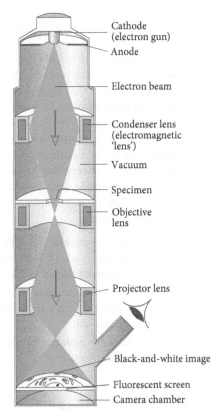

Cathode
(electron gun)

Anode

Electron beam

Condenser lens
(electromagnetic
'lens')

Vacuum

Specimen

Objective
lens

Projector lens

Black-and-white image

Fluorescent screen

Camera chamber

a.

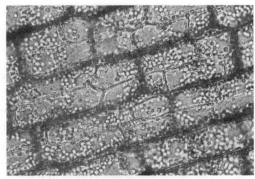

b.

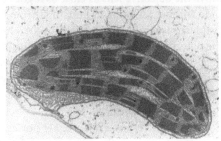

Figure 1.4 (a) Micrographs of plant cells, taken with a light microscope – the green structures inside the cells are chloroplasts; **(b)** micrograph of a single chloroplast, taken with a transmission electron microscope.

Figure 1.3 A transmission electron microscope. (Note that you do not need to learn the structure of an electron microscope.)

Electron microscopes are also much larger, more expensive, and more complicated to maintain than light microscopes.

5. Explain the difference between resolution and magnification.
6. Give one advantage and one disadvantage of an image produced by an electron microscope compared with one produced by a light microscope.

Experimental skills 1.1: Preparing and observing slides

Preparing slides of biological material for viewing under a light microscope is usually quite straightforward. For looking at individual cells, the specimen should be no more than one cell thick. This can be obtained by peeling off a suitable layer of tissue, for example from the inside of a piece of onion. Another way is to take a sample of tissue and squash it so the cells separate from each other. Stains are often added to the specimen to make the cell structures more visible.

Preparing an onion cell slide
A student carried out the following steps to prepare and then observe a slide of onion cells:

1. Peel a layer of tissue from the inside of a piece of onion. The layer should be one cell thick.
2. Trim the piece of tissue so that it is no wider than a coverslip (approximately 10 mm × 10 mm). Place

the tissue on a microscope slide, making sure the tissue is flat and not folded over.
3. Add a drop of orange-brown iodine–potassium iodide solution as a stain.
4. Carefully lower a coverslip; place one side down first to avoid trapping air bubbles.
5. Gently press down on the coverslip with tissue paper to absorb any excess liquid.

QUESTIONS
P1. Suggest reasons for the following steps in preparing a microscope slide:
   a. making sure that the specimen tissue is not folded over
   b. using a stain
   c. using a coverslip.

Observing onion cells

The student observed the onion cell slide under a light microscope, carrying out the following steps:

1. Place the slide on the microscope stage, securing it with the clips.
2. Move the objective lens with the lowest power (magnification), so it is above the specimen.
3. Adjust the mirror (or light source) to ensure the specimen is suitably bright when observed.
4. Move the objective lens as close as it goes to the slide without touching it. Looking through the eyepiece, use the coarse focusing dial to move the objective lens away from the slide until the specimen comes into focus. Use the fine focusing dial to make the image as sharp as possible. Move the slide around on the stage to observe different parts of the specimen.
5. Use the objective lenses with higher powers, in the same way, to observe the cells in more detail, before making biological drawings of several cells.

Biological drawings should identify clearly and simply the main features of what is being observed:

- use a pencil, not a pen
- use continuous lines, not feathery or broken
- do not use shading
- just draw what you have been asked to
- ensure all contents are drawn, in the correct proportions, in the correct positions
- include correct labels, with lines drawn with a ruler, and without arrowheads
- use as much of the available space as possible
- add a title and show the magnification where possible.

P2. Suggest reasons for the following steps in observing a microscope slide:
  a. focusing by starting with the objective lens close to the slide and then moving it away
  b. focusing first with the coarse focusing dial and then with the fine focusing dial
  c. making your first observations with a low power objective lens before moving to a higher power lens.

P3. Figure 1.5 shows an onion cell slide observed using a light microscope.

Figure 1.5 Onion cells observed with a light microscope.

Produce an accurate biological drawing of one of the onion cells shown in Figure 1.5.

Using an eyepiece graticule and stage micrometer scale

An **eyepiece graticule** fits inside the eyepiece and shows a scale when you look through it. If you know the distance between each division on the scale then you can use the graticule to make measurements of the specimen you observe. You can work out this distance using a **stage micrometer scale** – this is a microscope slide with another scale etched on to it. Working out the distance between the divisions on the graticule scale is an example of **calibration**.

The following steps explain how to calibrate an eyepiece graticule:

1. Turn the eyepiece or the slide to line up the scale on the graticule with the scale on the stage micrometer, as shown in Figure 1.6.
2. Count the number of divisions on the graticule that match up with a set distance on the stage micrometer scale.
3. Calculate the distance between each division on the eyepiece graticule.

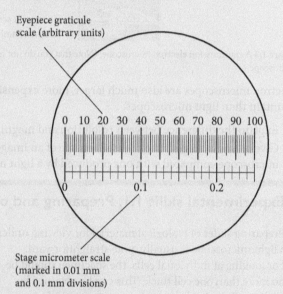

Eyepiece graticule scale (arbitrary units)

Stage micrometer scale (marked in 0.01 mm and 0.1 mm divisions)

Figure 1.6 Calibrating an eyepiece graticule.

P4. Look at Figure 1.6.
  a. State how many eyepiece graticule divisions correspond with 0.1 mm on the stage micrometer scale.
  b. Calculate the distance between the divisions on the eyepiece graticule. Give your answer in millimetres in standard form.
  c. A student used the eyepiece graticule to observe a cell. The diameter of the cell measured 12 divisions on the graticule scale. Calculate the actual diameter

of the cell. Give your answer in micrometres.

d. The student then used a different objective lens with a higher magnification to observe the cell nucleus in more detail. The student measured the diameter of the nucleus using the same eyepiece graticule. Explain what else the student needs to do to make the measurement accurate.

Using a grid to estimate the number of cells

The number of cells on a slide, or in the field of view, can be estimated, for example by using a grid, which can be fitted inside the eyepiece or be on a slide itself. By counting the number of cells in several sections of the grid, an estimate can be made of the total number of cells.

The following steps explain how to do this:

1. Count the number of cells in several sections of the grid.
2. Calculate the mean number of cells per section.
3. Multiply the mean number by how many times bigger the whole sample is than a single section.

P5. Figure 1.7 shows red blood cells observed using a light microscope and a grid. Each square on the grid measures 1 mm by 1 mm.

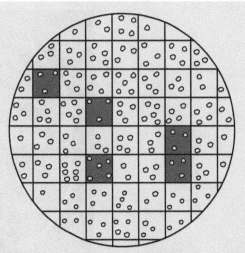

Figure 1.7 Red blood cells viewed through a light microscope.

a. The total area of the blood sample on the whole slide is 4 cm². Use information from the shaded squares in Figure 1.7 to estimate the total number of red blood cells on the slide.

b. Suggest how the accuracy of this estimate could be improved.

> **Link**
>
> In Chapter 18, quadrats are used to estimate population sizes during ecological fieldwork. Using a microscope grid follows the same principle of using samples to estimate a total number.

Assignment 1.1: Observing human cheek cells

A student used a cotton bud to wipe the inside of their cheek. They wiped the cotton bud across a clean microscope slide. They put one drop of methylene blue stain (a dark blue stain) on the smear and placed a coverslip over it. They observed the slide with a light microscope. Figure 1.8 shows one of the cheek cells as seen under the microscope.

QUESTIONS

A1. Look at Figure 1.8.
    a. The cell diameter shown measures 68 mm on the photomicrograph. The magnification is ×12 000. Calculate the actual diameter of the cell in micrometres. Give your answer to two significant figures.
    b. Describe how you could calculate a more accurate value for the average diameter of the cell.

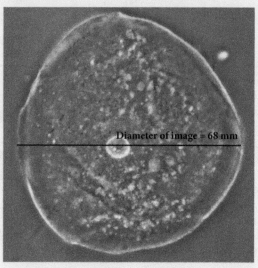

Figure 1.8 Photomicrograph of a human cheek cell.

A2. Three students made biological drawings of the cheek cell, as shown in Figure 1.9.

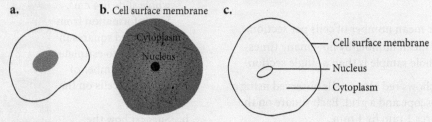

Figure 1.9 Biological drawings of a human cheek cell.

a. Explain which diagram shows the best biological drawing of the cheek cell.
b. State how the diagram you have chosen in part (a) could be improved further.

Key ideas

→ Light microscopes produce magnified images, allowing the study of cells and their components.

→ Electron microscopes produce images of greater resolution and magnification than light microscopes, allowing the detailed study of cell organelles.

→ A transmission electron microscope (TEM) produces images by passing electrons through an object, whereas a scanning electron microscope (SEM) produces images from electrons reflected from the surface of an object.

→ Magnification is how many times bigger an image is than the original object.

→ Resolution is the shortest distance between two separate objects that can still be seen as separate.

→ Eyepiece graticules and stage micrometer scales can be used to make measurements of objects seen through microscopes.

1.2 Cells as the basic units of living organisms

All living organisms are made up of cells. Some organisms consist of a single cell, for example *Amoeba*, whereas an adult human may be made up of over 37 trillion cells (one trillion = 10^{12}), consisting of many different types (see Figure 1.10). Current estimates are that there are around 1.3 million different species on the planet, but there are only two basic types of cell: **eukaryotic cells** and **prokaryotic cells**. Organisms with eukaryotic cells include all animals, plants and fungi. Organisms with prokaryotic cells include bacteria. Prokaryotic cells are smaller and lack many of the features of eukaryotic cells, such as a nucleus. Even smaller than prokaryotic cells are **viruses**, which do not have a cellular structure and are not considered as living because they do not carry out all the functions of living organisms: for example they do not respire.

7. Look at Figure 1.10a. Explain whether *Stentor* is a prokaryotic or a eukaryotic cell.

STRUCTURE OF EUKARYOTIC CELLS

Animals, plants, fungi and many single-celled organisms are composed of eukaryotic cells. Although there are many different types of eukaryotic cell, they have many features in common.

Cell surface membrane

Figure 1.11 shows the structure of a typical animal cell. It has many of the features found in other types of eukaryotic cells too. Figure 1.12 shows part of an animal cell, as seen using an electron microscope.

Link

Scientists used to classify living things into two groups: those with eukaryotic cells and those with prokaryotic cells. Chapter 18 explains why organisms are now classified into three main groups.

a.

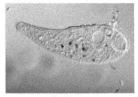

b.

Figure 1.10 A micrograph of *Stentor*, a single-celled organism (**a**). A human is composed of trillions of cells – a multicellular organism (**b**).

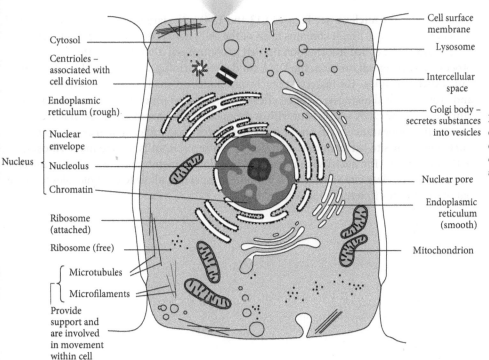

Figure 1.11 The structure of a typical animal cell, as seen using an electron microscope.

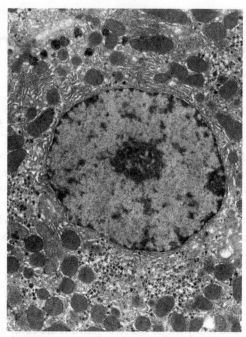

Figure 1.12 An electron micrograph of part of an animal cell.

Link

In Chapter 4, you will see how the structure of the cell surface membrane controls the movement of substances into and out of cells.

The animal cell is surrounded by a **cell surface membrane** (often just known as a cell membrane). As well as containing all the other cell structures, and separating the cell from its surroundings, the membrane also controls the substances that enter and leave the cell. Inside the cell are many smaller structures called **organelles** within a semi-liquid substance called **cytosol**. The cytosol and the organelles (apart from the nucleus) make up the **cytoplasm**.

Nucleus

In many cells, the **nucleus** (see Figures 1.13 and 1.14) is the largest organelle, with a typical diameter of about 10 μm. It contains most of the cell's deoxyribonucleic acid (**DNA**), which carries the coded information needed to control cell division and other cell processes. During cell division, the DNA condenses, forming short thread-like structures called **chromosomes**. The material that chromosomes are made from, a mixture of DNA and proteins called histones, is called **chromatin**. At other times, the chromosomes 'unravel' and the DNA spreads out within the nucleus; the chromatin in the nucleus then has a grainy appearance, with one or more darker regions visible, known as nucleoli. A **nucleolus** is the site of ribonucleic acid (**RNA**) synthesis. RNA is used to make **ribosomes** in the cytoplasm. The nucleus is surrounded by a double membrane known as the **nuclear envelope** (or nuclear envelope), which has pores (gaps) that allow RNA molecules to move out of the nucleus into the cytoplasm.

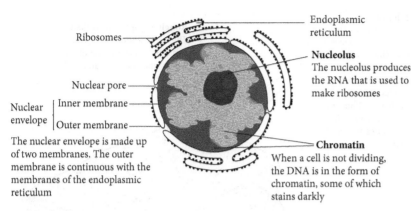

Ribosomes

Endoplasmic reticulum

Nucleolus
The nucleolus produces the RNA that is used to make ribosomes

Nuclear pore

Nuclear envelope { Inner membrane

Outer membrane

The nuclear envelope is made up of two membranes. The outer membrane is continuous with the membranes of the endoplasmic reticulum

Chromatin
When a cell is not dividing, the DNA is in the form of chromatin, some of which stains darkly

Figure 1.13 A nucleus.

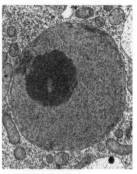

Figure 1.14 False-colour electron micrograph of a nucleus. The nucleus is shown in light orange and the nucleolus as a darker brown area. You can see the double-layered nuclear envelope as two dark lines. The green organelles are mitochondria.

8. Suggest why it is important to control what enters and leaves a cell.

9. During cell division, the DNA in the nucleus condenses, forming compact chromosomes. In between divisions, much of the DNA 'unravels'. Suggest an explanation for these changes.

Endoplasmic reticulum and ribosomes

Endoplasmic reticulum (ER; see Figures 1.15 and 1.16) is a network of membranes that form channels throughout the cytoplasm. Some of these channels join up with the nuclear envelope and some may join with the cell surface membrane. There are two types of ER:

- **Rough endoplasmic reticulum** (rough ER, or RER) has ribosomes attached to its outer surface, giving it a 'bumpy' appearance – hence its name. Ribosomes are very small organelles formed of RNA and are where proteins are synthesised. The proteins move into the channels formed by the membranes, through which they travel to other parts of the cell.
- **Smooth endoplasmic reticulum** (smooth ER, or SER) has no ribosomes associated with it. It is involved in several functions, including the synthesis of hormones and lipids.

Although most ribosomes are attached to the rough ER, some are found free in the cytoplasm. Some ribosomes are also located inside mitochondria and chloroplasts. Those found in the cytoplasm, including those on rough ER, are sometimes described as **80S ribosomes**. (The S refers to how quickly they form sediment in a process called centrifugation.) Those found in mitochondria and chloroplasts are smaller **70S ribosomes**.

Link

There is more about the structure of DNA, chromosomes and RNA, and how they are involved in cell division and protein synthesis, in Chapters 5, 6 and 16.

Link

In Chapter 6, you will see how ribosomes, and other types of RNA, are involved in producing proteins.

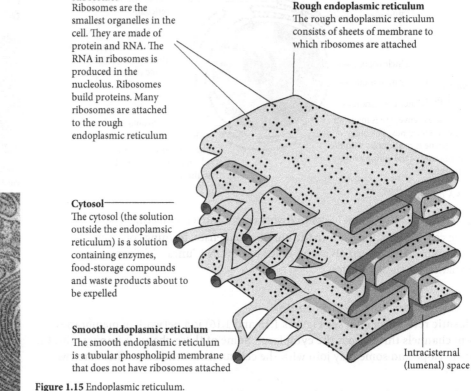

Ribosomes
Ribosomes are the smallest organelles in the cell. They are made of protein and RNA. The RNA in ribosomes is produced in the nucleolus. Ribosomes build proteins. Many ribosomes are attached to the rough endoplasmic reticulum

Rough endoplasmic reticulum
The rough endoplasmic reticulum consists of sheets of membrane to which ribosomes are attached

Cytosol
The cytosol (the solution outside the endoplamsic reticulum) is a solution containing enzymes, food-storage compounds and waste products about to be expelled

Smooth endoplasmic reticulum
The smooth endoplasmic reticulum is a tubular phospholipid membrane that does not have ribosomes attached

Intracisternal (lumenal) space

Figure 1.15 Endoplasmic reticulum.

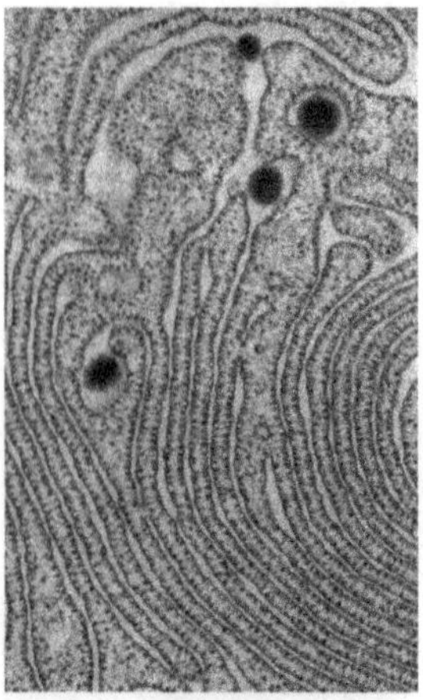

Figure 1.16 Electron micrograph of endoplasmic reticulum.

10. Look at Figure 1.16. What type of ER does it show? Explain your answer.

★ 11. Suggest why some substances are transported through the ER rather than just diffusing throughout a cell.

Golgi body

The **Golgi body** (also known as the Golgi apparatus or Golgi complex; see Figures 1.17 and 1.18) is an organelle that stores substances such as proteins before they are transported elsewhere in the cell. The substances may also be modified, for example by adding short carbohydrate chains to proteins to form glycoproteins. The Golgi body is formed from a group of flattened spaces or **vesicles**. The processed substances are packaged into their own vesicles, which leave the Golgi body and either move to other parts of the cell or fuse with the cell surface membrane to release their contents outside the cell (a process known as **exocytosis**).

Link

In Chapter 4, you will find out more about how exocytosis moves substances out of cells.

12. Which process occurs in both the Golgi body and the endoplasmic reticulum?
    A. Movement of substances between cells
    B. Production of new DNA
    C. Production of new substances
    D. Protein synthesis

★ 13. Suggest one type of cell that will contain a large number of Golgi bodies.

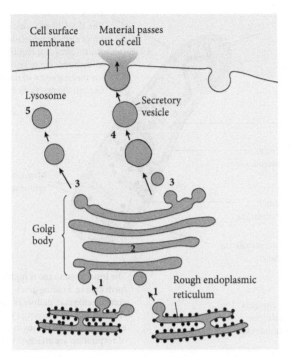

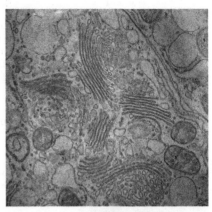

Figure 1.18 Electron micrograph of the Golgi body.

1 Vesicles containing proteins 'bud off' from the RER and fuse with the Golgi body.
2 Inside the Golgi body, the proteins are modified ready for use elsewhere in the cell, or for export.
3 Vesicles containing the modified proteins 'bud off' from the Golgi body.
4 Some of these vesicles move to the cell surface membrane, where they fuse with it and empty their contents outside of the cell.
5 Some of the vesicles are lysosomes, containing hydrolase enzymes that can hydrolyse material inside the cell.

Figure 1.17 The Golgi body.

Mitochondria

Mitochondria (singular: mitochondrion; see Figures 1.19 and 1.20) are spherical or elongated ('sausage-shaped') organelles found in most eukaryotic cells. They can be relatively large compared with other organelles, being around 0.5–1.5 μm wide and 3–10 μm long. They are the site of **aerobic respiration**. **Respiration** is a process that breaks down substances such as glucose to release energy used to make adenosine triphosphate (**ATP**). Aerobic respiration uses oxygen to make more ATP per glucose molecule than can be released in **anaerobic respiration** (respiration in the absence of oxygen). It is ATP that directly provides the **energy** that cells transfer for energy-requiring processes, such as active transport (see Chapter 4) and muscle contraction (see Chapter 15).

Mitochondria have two membranes: an outer membrane and a highly folded inner membrane (the folds are known as **cristae**). The cristae hold many of the enzymes involved in respiration, although there are other enzymes involved found in the fluid-filled internal space (known as the **matrix**).

Mitochondria contain 70S ribosomes and their own small circular molecules of DNA.

Link

Many of the stages of respiration occur in the mitochondria: in the matrix and on the inner membrane. You will learn about this in Chapter 12.

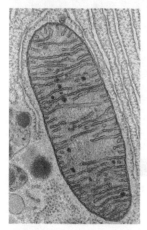

Figure 1.20 Electron micrograph of a mitochondrion.

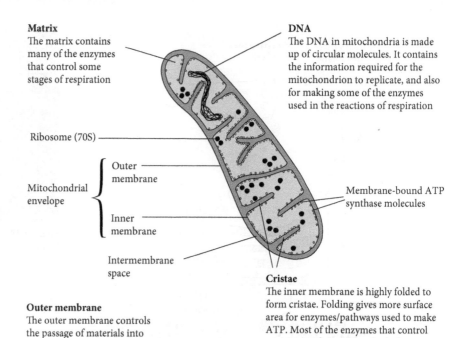

Matrix
The matrix contains many of the enzymes that control some stages of respiration

DNA
The DNA in mitochondria is made up of circular molecules. It contains the information required for the mitochondrion to replicate, and also for making some of the enzymes used in the reactions of respiration

Ribosome (70S)

Mitochondrial envelope

Outer membrane

Inner membrane

Intermembrane space

Membrane-bound ATP synthase molecules

Cristae
The inner membrane is highly folded to form cristae. Folding gives more surface area for enzymes/pathways used to make ATP. Most of the enzymes that control energy transfer reactions in the later stages of respiration are attached to the cristae

Outer membrane
The outer membrane controls the passage of materials into and out of the mitochondrion

Figure 1.19 A mitochondrion.

Lysosomes

Lysosomes (look back at Figure 1.11) are vesicles produced by the Golgi body, containing digestive enzymes called **lysozymes**. Lysosomes are used within cells to break down old or unneeded organelles, and they are used by white blood cells called **phagocytes** to digest any engulfed **pathogens**. Lysosomes are also used to destroy unwanted cells and tissues, for example to reduce the muscle of the uterus after birth, and to reduce milk-producing tissue after weaning.

14. Suggest one type of cell that will contain large numbers of mitochondria. Explain your choice.

15. Suggest why it is important that lysozymes inside a cell remain inside lysosomes.

Centrioles, microtubules and cilia

Centrioles are small organelles found in all animal cells but absent in many plants. They are the site of formation of fine protein filaments known as **microtubules**, or microfilaments. Microtubules help form a supportive network inside a cell, supporting its shape or helping it to move. They also organise and move organelles within a cell, and they form the **spindle** (spindle fibres) during cell division. Some microtubules form external structures called **cilia** that are involved with movement: some single-celled organisms use cilia to move (see Figure 1.21); they are also found inside multicellular animals, for example on ciliated epithelial cells, which move to expel mucus from the lungs.

Figure 1.21 The cilia (the tiny 'hairs' around the outside) of *Paramecium* move, propelling it through water.

Link

In Chapter 11, you will learn about the essential role that phagocytes, and other types of white blood cell, have in the body's responses to infection.

Microvilli

Microvilli are small finger-like projections found on the surface of some cells to increase surface area. For example, they are found on the surface of the epithelium cells covering the villi in the small intestine, where they increase surface area for the absorption of digested food (see Figure 1.22).

16. Smoking tobacco damages the ciliated epithelial cells in our lungs. Suggest why this leads to the heavy cough of many smokers.

17. Microvillus inclusion disease is a health condition in which microvilli in the small intestine do not form properly. The condition can often be fatal. Suggest why.

Link

You will find out how the spindle is involved in moving chromosomes during cell division in Chapters 5 and 16.

Link

In Chapter 9, you will see how different cells, including ciliated epithelial cells, help maintain the health of the human gas exchange system.

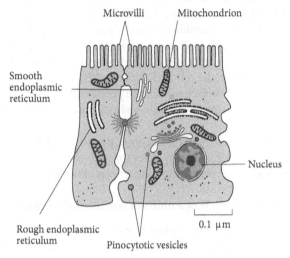

Figure 1.22 An epithelial cell on the surface of a villus.

Plant cells - cell walls and plasmodesmata

Figure 1.23 shows the structure of a typical plant cell. It has many of the features found in animal cells, but in addition, plant cells also contain a cell wall, chloroplasts, plasmodesmata and a large permanent vacuole. Figure 1.24 shows a plant cell, as seen using an electron microscope.

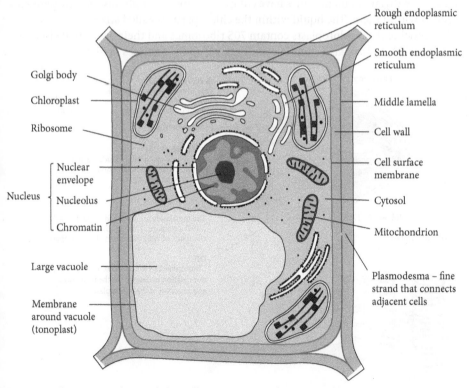

Figure 1.23 The structure of a typical plant cell, as seen using an electron microscope.

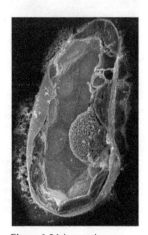

Figure 1.24 A scanning electron micrograph of a plant cell.

Link

You will learn about the structure and function of cellulose and other polysaccharides in Chapter 2.

Link

How cells absorb water is explained in Chapter 4. In Chapter 7, you will see how water and other substances are transported from one part of a plant to another.

Figure 1.25 Electron micrograph of cellulose microfibrils in the cell wall of an alga.

Link

In Chapter 13, you will see how the different parts of a chloroplast are involved in the different stages of photosynthesis.

Outside the cell surface membrane, the plant cell is surrounded by a **cell wall**. Cell walls are found on the outside of the cells of plants, algae and fungi, but never on animal cells. Plant and algal cell walls are made of a substance called **cellulose**, which is a polysaccharide (see Chapter 2). Cellulose is a long molecule, and in the cell wall, parallel cellulose molecules are packed together in bundles called **microfibrils**. The cell wall is made of many layers of these microfibrils embedded in a substance called **pectin**. In different layers, the microfibrils run in different directions, adding strength to the structure (see Figure 1.25). Unlike the cell surface membrane, the cell wall is completely permeable and does not control in any way the substances that enter or leave a cell. Its function is to provide the cell with support. When a plant cell absorbs water, its volume increases and the cell surface membrane presses against the cell wall, making the whole structure rigid. This is why even non-woody plants can support themselves to stay upright.

Joining adjacent plant cells together is a layer of pectin called the **middle lamella**. **Plasmodesmata** (see Figure 1.23) are tiny holes through the middle lamella between adjacent cells through which cytoplasmic connections allow substances to move from one cell to another.

The cell walls of fungi are made of a different polymer called **chitin**.

18. Suggest why dehydration causes a plant to wilt (droop).
19. Suggest how large animals support their body if they do not have cell walls.

Chloroplasts

Chloroplasts (see Figures 1.26 and 1.27) are organelles found in plant and algal cells and are the site of **photosynthesis**. The green pigment **chlorophyll** in chloroplasts absorbs light, and the subsequent energy transfer is used to build up glucose and other complex organic molecules.

Chloroplasts have two outer membranes (forming the **chloroplast envelope**) as well as a series of internal membranes forming **thylakoids** that stack to form **grana** (singular: granum). The thylakoid membranes have chlorophyll and other photosynthetic pigments embedded within them. The liquid within the chloroplasts is called **stroma**.

Like mitochondria, chloroplasts contain 70S ribosomes and their own small circular molecules of DNA. They also contain starch grains, and sometimes lipid droplets.

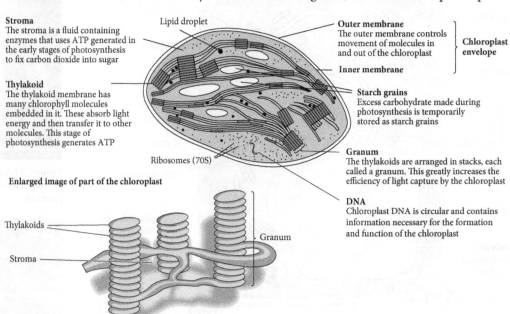

Stroma
The stroma is a fluid containing enzymes that uses ATP generated in the early stages of photosynthesis to fix carbon dioxide into sugar

Thylakoid
The thylakoid membrane has many chlorophyll molecules embedded in it. These absorb light energy and then transfer it to other molecules. This stage of photosynthesis generates ATP

Lipid droplet

Ribosomes (70S)

Enlarged image of part of the chloroplast

Outer membrane
The outer membrane controls movement of molecules in and out of the chloroplast

Inner membrane

} **Chloroplast envelope**

Starch grains
Excess carbohydrate made during photosynthesis is temporarily stored as starch grains

Granum
The thylakoids are arranged in stacks, each called a granum. This greatly increases the efficiency of light capture by the chloroplast

DNA
Chloroplast DNA is circular and contains information necessary for the formation and function of the chloroplast

Thylakoids

Stroma

Granum

Figure 1.26 The structure of a chloroplast.

Vacuole

All cells contain some liquid-filled spaces called **vacuoles**. Many are small and short-lived – for example vesicles. However, plant cells usually have a large permanent vacuole, which can take up much of the space inside the cell (see Figure 1.23). The vacuole is surrounded by a single membrane called the **tonoplast** and contains a fluid called **cell sap**. The main functions of plant vacuoles are to help maintain cell rigidity by absorbing water so the cell contents press against the cell wall, and to store substances such as sucrose and amino acids, as well as waste products that may be harmful to the rest of the cell.

20. Not all plant cells contain chloroplasts. Suggest why and give an example of a plant cell without chloroplasts.
21. Some of the waste products stored in plant cell vacuoles have an unpleasant taste to animals. Suggest how storing these products may benefit a plant.

Figure 1.27 Electron micrograph of a chloroplast.

Comparison of the structure of typical plant and animal cells

Plant and animal cells have many structural features in common, although there are differences. These similarities and differences can be seen in Figures 1.11 and 1.23.

Worked example

a. Compare the structures of a typical plant cell and animal cell.

Answer

Plant and animal cells have many structures in common: cell surface membrane; nucleus, including the nuclear envelope and nucleolus; rough and smooth endoplasmic reticulum; Golgi bodies; mitochondria; ribosomes; lysosomes; centrioles and microtubules. There are some features found in plant cells that are not found in animal cells: cell wall; chloroplasts; plasmodesmata; large permanent vacuole with tonoplast.

b. Suggest reasons for the similarities and differences between a typical plant and animal cell.

Answer

Plant and animal cells have many features in common because they need them for the same reasons: for example, a cell surface membrane to control movement of substances in and out of the cell, a nucleus containing genetic information in DNA, and mitochondria for respiration. However, whereas animals take in the organic substances they need as food from their environment, plants build up their organic substances by photosynthesis, and so need chloroplasts for this. Plants also need cell walls and large vacuoles to make their cells rigid, which is necessary for the support of the plant; animals gain support either through having a firm skeleton, by living in water, or by being very small and not needing to support their body weight against gravity.

> **Link**
>
> In Chapter 4, you will learn about how plant cells, and their vacuoles, absorb water by osmosis.

STRUCTURE OF PROKARYOTIC CELLS

Prokaryotic cells are much smaller than eukaryotic cells, and many of the structures commonly present in eukaryotic cells are absent. Prokaryotic cells are thought to have evolved much earlier and to have later given rise to eukaryotic cells. Many scientists think that eukaryotic cells formed when prokaryotic cells joined together to form larger cells. This is called the 'endosymbiotic' theory. The theory was first suggested at the start of the 20th century by Russian botanist Konstantin Mereschkowski, and later developed by other scientists, most notably, in the second half of the century, by US evolutionary biologist Lynn Margulis. Some of the evidence for the theory comes from the similarities in structure found between prokaryotic cells and mitochondria and chloroplasts, which, it is suggested, are descended from once separate, free-living prokaryotic cells.

Figure 1.28 shows the structure of a typical bacterium as an example of a prokaryotic cell.

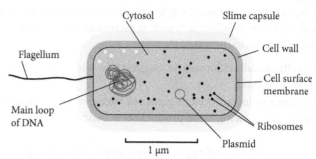

Figure 1.28 The structure of a typical bacterium, a single-celled prokaryotic organism.

Key features of typical bacterial cells include the following:

- They are **unicellular**. Although some do form strands joined end to end, they do not form complex **multicellular** organisms like many eukaryotic cells.
- They are small, generally 1–5 μm in diameter, compared with eukaryotic cells, which are typically 10–100 μm in diameter.
- They have cell walls composed of **peptidoglycan**, a polymer made of polysaccharide and peptide chains (see Chapter 2). By contrast, eukaryotic cells have cell walls made of cellulose (plants and algae) or chitin (fungi), or have no cell wall at all (animals).
- Their DNA is circular and not associated with histone proteins. Eukaryotic DNA is associated with histones to form chromatin and chromosomes (although mitochondria and chloroplasts do have circular DNA). In addition to a main DNA loop, bacteria also contain much smaller loops of DNA called **plasmids**, although, in total, a prokaryotic cell contains far less DNA than a eukaryotic cell.
- They contain 70S ribosomes, unlike the 80S ribosomes found in eukaryotic cells (although mitochondria and chloroplasts have 70S ribosomes).
- They do not contain organelles surrounded by a double membrane, such as nuclei, mitochondria or chloroplasts.
- They do not have endoplasmic reticulum, Golgi bodies, lysosomes or microtubules.
- Some bacterial cells have **flagella** (singular: flagellum), which help them to 'swim'.
- Many bacterial cells have an extra layer outside the cell wall called a **capsule** (or slime capsule) made of polysaccharide (see Chapter 2). Capsules provide protection against attack from other cells, as well as from physical harm such as desiccation (drying out).

Worked example

---

a. Compare the outer structures of a typical bacterial cell and animal cell.

Answer

Both types of cell are surrounded by a cell surface membrane; however, a bacterial cell also has a peptidoglycan cell wall. Many bacterial cells also have a capsule layer outside the cell wall.

b. Bacterial cells, like all prokaryotic cells, do not have mitochondria. However, they still need to respire to produce ATP. Suggest where respiration occurs in a bacterial cell.

Answer

In bacteria, all stages of respiration occur in the cytoplasm.

---

22. Describe how the DNA in prokaryotic cells is different from that in eukaryotic cells.

Link

In Chapter 10, you will see examples of bacteria that cause infectious disease (pathogens). Chapter 19 explains how bacterial plasmids are used in genetic technology.

Tip

In questions like this, do not just refer to 'it' in your answer. Make sure you are clear which type of cell you are referring to.

VIRUSES

Viruses are tiny particles that are much smaller than prokaryotic cells, being around 20–400 nm in diameter. They are non-cellular structures composed of a **nucleic acid** core (either DNA or RNA; see Chapter 6) surrounded by a protein coat called a **capsid**. Some viruses also have an outer envelope made of phospholipids (see Chapter 2).

Figure 1.29 shows the structure of a typical virus.

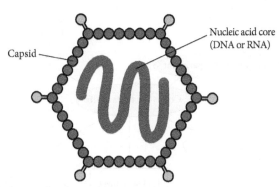

Capsid

Nucleic acid core
(DNA or RNA)

Figure 1.29 The structure of a virus particle.

Although viruses are associated with living organisms, they are not considered to be alive themselves. They are not made of cells, and they do not demonstrate the seven characteristics of living organisms (movement, respiration, sensitivity, growth, reproduction, excretion, nutrition). The only characteristic they do demonstrate is reproduction, but even then, they need living cells to be able to reproduce. When a virus invades a cell, the cell replicates the viral DNA or RNA, makes new protein coats and assembles new viruses. The viruses then burst out of the cell, killing it.

23. How many times greater in size is the diameter of a prokaryotic cell of 5 μm compared with that of a virus with a diameter of 25 nm?

Key ideas

→ Eukaryotic cells are larger and more complex than prokaryotic cells, containing many structures not present in prokaryotic cells: nucleus, endoplasmic reticulum, Golgi bodies, mitochondria, chloroplasts, lysosomes, centrioles, microtubules.
→ Although eukaryotic and prokaryotic cells may have some features in common, there are still differences in these features: in the size of the cells, in the size of ribosomes, in the composition of cell walls.
→ Prokaryotic cells have some features not found in eukaryotic cells, such as plasmids.
→ Animal and plant cells share many cell structures, but only plant cells have a cell wall, chloroplasts and a large permanent vacuole.
→ Cells use ATP from respiration for energy-requiring processes.
→ Viruses are non-living, non-cellular particles that can only reproduce by attacking living cells.

Link

Chapter 6 explains how DNA is replicated and how proteins are synthesised. In Chapter 10, you will see the factors involved in the control of one type of virus: HIV.

Tip

In calculations like this, convert all the values to the same unit. Usually it is easier to convert to the smaller unit.

CHAPTER OVERVIEW

Review the mini mind map to link new learning with your prior learning, and to highlight which of the key concepts underpin the chapter.

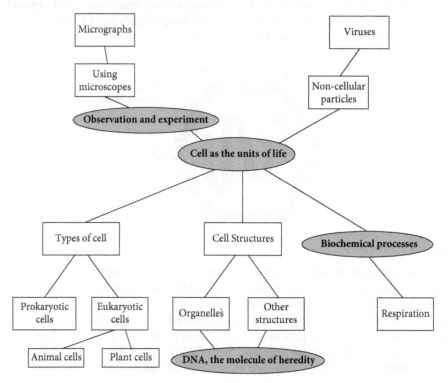

Try copying this mini mind map and expanding upon it. Use your notes to help you explore how the essential ideas, theories and principles can be linked further together.

WHAT YOU HAVE LEARNED

Now that you have finished this chapter you should be able to:

- make temporary preparations of cellular material suitable for viewing with a light microscope
- draw cells from microscope slides and photomicrographs
- calculate magnifications of images and actual sizes of specimens using drawings, photomicrographs and electron micrographs (scanning and transmission)
- use an eyepiece graticule and stage micrometer scale to make measurements and use the appropriate units, millimetre (mm), micrometre (µm) and nanometre (nm)
- define resolution and magnification and explain the differences between these terms, with reference to light microscopy and electron microscopy
- recognise organelles and other cell structures found in eukaryotic cells and outline their structures and functions, limited to:
  - cell surface membrane
  - nucleus, nuclear envelope and nucleolus
  - rough endoplasmic reticulum
  - smooth endoplasmic reticulum
  - Golgi body (Golgi apparatus or Golgi complex)

- mitochondria (including the presence of small circular DNA)
- ribosomes (80S in the cytoplasm and 70S in chloroplasts and mitochondria)
- lysosomes
- centrioles and microtubules
- cilia
- microvilli
- chloroplasts (including the presence of small circular DNA)
- cell wall
- plasmodesmata
- large permanent vacuole and tonoplast of plant cells
- describe and interpret photomicrographs, electron micrographs and drawings of typical plant and animal cells
- compare the structure of typical plant and animal cells
- state that cells use ATP from respiration for energy-requiring processes
- outline key structural features of a prokaryotic cell as found in a typical bacterium, including:
  - unicellular
  - generally 1–5 µm diameter
  - peptidoglycan cell walls
  - circular DNA
  - 70S ribosomes
  - absence of organelles surrounded by double membranes
- compare the structure of a prokaryotic cell as found in a typical bacterium with the structures of typical eukaryotic cells in plants and animals
- state that viruses are non-cellular structures with a nucleic acid core (either DNA or RNA) and a capsid made of protein, and that some viruses have an outer envelope made of phospholipids.

CHAPTER REVIEW

1. Which cell structure is not found in prokaryotic cells?
   A. 70S ribosome
   B. 80S ribosome
   C. Cell wall
   D. Plasmid
2. Where is RNA made?
   A. Centriole
   B. Golgi body
   C. Lysosome
   D. Nucleolus
3. A photomicrograph image of a cell has a diameter of 15 mm. The actual cell has a diameter of 75 µm. Calculate the magnification of the image.
4. A student uses a light microscope with an eyepiece lens of magnification ×10 and an objective lens of magnification ×25. Calculate the total magnification of the image produced by the microscope.
5. Figure 1.30 shows an electron micrograph of a leaf.
   State precisely what type of microscope was used to produce the micrograph in Figure 1.30. Give a reason for your answer.

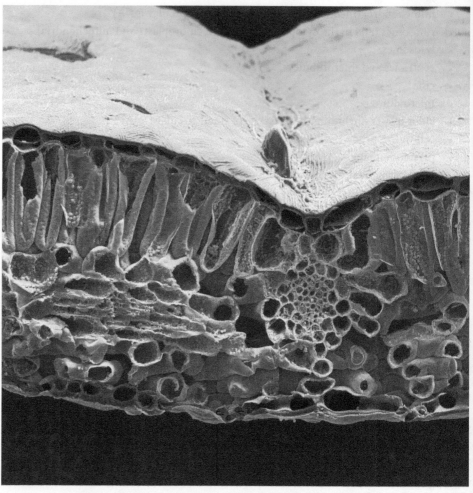

Figure 1.30 Electron micrograph of a section through a leaf.

6. Give three structures found in plant cells that are not found in animal cells.
7. Explain where most ATP is produced in a eukaryotic cell.
8. Describe the function of a lysosome.
9. Give three features of a prokaryotic cell that distinguish it from a eukaryotic cell.
10. Viruses are not regarded as living organisms. Give two reasons why.

Chapter 2

Biological molecules

All living organisms are mainly composed of proteins, carbohydrates, lipids and nucleic acids. However, it is thought that the early conditions on planet Earth were hot and extremely hostile. Where did these organic molecules, which are so vital to life, come from?

In the 1950s, US biochemists Stanley Miller and Harold Urey tried to answer this question. They set up a laboratory experiment that attempted to model the conditions found on our planet nearly four billion years ago. Passing electrical pulses through a mixture of hot carbon dioxide, ammonia, sulfur dioxide and water vapour, they were able to produce molecules such as amino acids.

It is a big step to explain how life originated from non-living molecules. However, the work of Miller and Urey offers clues to explain how some of these fundamental ingredients may have appeared in the first place.

Prior understanding

You may remember from your study of a balanced diet that the major food types are carbohydrates, lipids and proteins. These are vital because our bodies are composed of these major groups of organic molecules. You may remember the terms 'atom', 'molecule', 'electron' and 'ion', and have a basic understanding of covalent and ionic bonding, and of molecular and structural formulae.

Learning aims

In this chapter, you will learn how the structure and properties of biological molecules are related to their functions in cells and in organisms. We will consider the roles of carbohydrates, lipids and proteins, and how they are fundamental to an understanding of many areas of biology. The importance of water to living organisms is also covered.

2.1 An introduction to biological molecules (2.2.2, 2.2.3, 2.4.1)

2.2 Testing for biological molecules (Syllabus 2.1.1–2.1.3, 2.2.4)

2.3 Carbohydrates (Syllabus 2.2.1, 2.2.5–2.2.8)

2.4 Lipids (Syllabus 2.2.9–2.2.11)

2.5 Proteins (Syllabus 2.3.1–2.3.8)

2.6 Water (Syllabus 2.4.1)

2.1 An introduction to biological molecules

As we saw at the start of this chapter, the emergence of organic molecules was probably a precursor to life on Earth. **Molecules** are the fundamental units of life. These are defined as particles with two or more **atoms** combined. Joining different atoms together in different quantities, or in different ways, produces molecules with a range of different properties. **Molecular biology** is the study of these molecules, and the ways in which they react with other particles is **biochemistry**.

In organisms, most of the atoms are of just six different kinds – hydrogen, oxygen, carbon, nitrogen, sulfur and phosphorus – with other trace elements being required in smaller quantities. As we will see during the course of this chapter, the element carbon forms a central 'skeleton' in all biological macromolecules.

MONOMERS, POLYMERS AND MACROMOLECULES

Although there is a vast number of different biomolecules, they can be grouped into different classes. These are monomers, polymers and macromolecules.

- A **monomer** is a molecule that is used as the building block for the synthesis of a polymer. Monomers are joined together by covalent bonds in **condensation reactions** to make polymers.
- A **polymer** is a giant molecule made from many similar repeating subunits, called **residues**, joined together by strong covalent bonds in a chain. Polymers can be broken into smaller subunits, often monomers, in **hydrolysis** reactions.
- A **macromolecule** is a giant molecule consisting of thousands of atoms, built from a small number of simple molecules. In this chapter, we look at the macromolecules **carbohydrates**, **lipids** and **proteins**, and describe the characteristics that make them the chemicals of life.

Table 2.1 summarises the four main biological macromolecules and gives further details about them.

Macromolecule	Monomer	Polymer	Covalent bond
carbohydrate	sugar	starch, cellulose, glycogen	glycosidic
protein	amino acid	polypeptide	peptide
lipid	N/A	N/A	ester
nucleic acid	nucleotide	DNA, RNA	phosphodiester

Table 2.1 Carbohydrates, proteins, lipids and nucleic acids are the four most common macromolecules in living organisms. Lipids are unusual in that they do not consist of a polymer as they do not consist of repeating residues.

HYDROPHILIC AND HYDROPHOBIC MOLECULES

The properties of biological molecules are numerous and diverse. However, during this chapter, we will encounter one clear difference between them. This property is the degree to which the molecule is described as **hydrophilic** or **hydrophobic** – meaning how well it mixes with the **polar** solvent, water. There are several alternative terms that describe this property, which is due to the structure of the molecule itself. These are summarised in Table 2.2.

Molecules that mix with water are:	Molecules that do not mix with water are:
hydrophilic	hydrophobic
polar	non-polar
charged	uncharged
water-soluble	water-insoluble
lipid-insoluble	lipid-soluble

Table 2.2 One property that can differ between biological molecules is their ability to mix with water. The terms on the left describe molecules that readily mix with water; those listed on the right are used to describe molecules that do not mix well with water.

HYDROGEN BONDING

All atoms contain electrons, and an electron has a small negative charge. In many molecules, the electrons in the chemical bond are not shared equally between the oxygen and the hydrogen atoms. In water, for example, this results in the oxygen atom having a tiny negative charge, and each hydrogen atom having a tiny positive charge. These molecules are said to be polar. Although the overall charge is zero, a polar molecule has a small positive charge in some places, and a small negative charge in others. These small charges are written δ– and δ+. The symbol δ is the Greek letter delta, so you can say 'delta minus' and 'delta plus'.

Negative charges and positive charges are attracted to one another. This means that some molecules are attracted to their neighbours. The negative charge on one molecule is attracted to the positive charge on another molecule. These attractions are called **hydrogen bonds**, and you will encounter them throughout this chapter. Although they are important, hydrogen bonds are much weaker than the strong **covalent bonds** that hold the atoms in molecules firmly together.

2.2 Testing for biological molecules

For a number of reasons, scientists often need to test for the presence of key biological molecules in a variety of animal or plant materials. For example, it may be important to identify the contents of mixtures of molecules in foods, or to work out the activity of digestive enzymes.

IODINE TEST FOR STARCH

To test for starch, add an orange–brown solution of iodine dissolved in **potassium iodide solution** (often abbreviated to **iodine–potassium iodide solution** or simply 'iodine solution') to the substance. A blue–black colour indicates the presence of starch (Figure 2.1).

> **Tip**
> Stating 'no change' to record a negative result is insufficient. You should indicate what colour remains. For example, 'iodine–potassium iodide solution remains orange–brown' is an accurate observation of a negative starch test.

Figure 2.1 When iodine–potassium iodide solution encounters starch, its colour changes from orange–brown to blue–black.

TESTS FOR SUGARS

Tests for sugars can distinguish two groups of sugars: the **reducing sugars** and the **non-reducing sugars**. The term 'reducing sugar' indicates that some sugars can reduce, or donate electrons to, other molecules.

The **Benedict's test** for a reducing sugar involves adding a solution containing copper sulfate called **Benedict's solution** (sometimes called **Benedict's reagent**) to the sample and heating the sample to at least 80 °C in a water bath. If a reducing sugar is present, the Cu^{2+} ions in copper sulfate are reduced by the sugar to Cu^+ ions, resulting in a colour change.

The colour seen in a positive Benedict's test depends on the concentration of reducing sugar present in the sample (Figure 2.2). As you will see in the Experimental skills section, this can be used to estimate the concentration of reducing sugars in a sample.

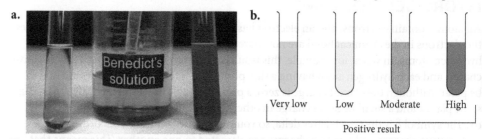

Figure 2.2 The colour change of a positive Benedict's test is caused by the precipitation of copper (I) oxide (a). The more reducing sugar present in the original sample, the more precipitate there is. Lower concentrations of reducing sugar will give a colour change from blue to green or yellow, whereas higher concentrations will give a colour change to orange or red (b).

Glucose, **fructose**, **galactose**, **maltose** and **lactose** are all reducing sugars, but **sucrose** is not – it is a non-reducing sugar. It will not give a positive test result with Benedict's solution. However, as we will see later, sucrose consists of a molecule of glucose and fructose joined together. After sucrose is boiled with dilute acid to hydrolyse (split) it into glucose and fructose, a positive result is observed (Figure 2.3). This is called **acid hydrolysis**.

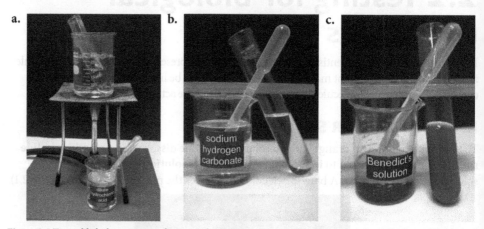

Figure 2.3 To establish the presence of a non-reducing sugar in a sample, the mixture should be heated strongly, with acid (a). Then, the sample is neutralised using sodium hydrogencarbonate (b). The Benedict's test will then give a positive result if the original sample contains a non-reducing sugar (c).

EMULSION TEST FOR LIPIDS

To test for the presence of a lipid in a sample, the sample is first crushed (if it is a solid) and then shaken with ethanol in a clean test-tube. Lipids are soluble in ethanol, so if there is any lipid in the sample, some of it will dissolve. The second step is to gently pour the ethanol into another test-tube containing water. If there is any lipid dissolved in the ethanol, it will form little droplets in the water. The mixture of these microscopic lipid droplets and water is known as an **emulsion**. The lipid droplets reduce the amount of light passing through, so the water becomes cloudy (milky white) rather than being transparent (Figure 2.4).

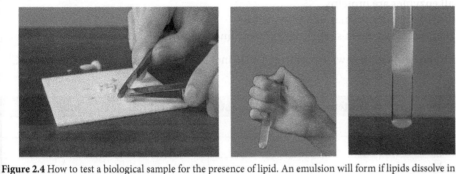

Figure 2.4 How to test a biological sample for the presence of lipid. An emulsion will form if lipids dissolve in ethanol when mixed with water.

Figure 2.5 The biuret test for proteins. Biuret solution is pale blue but turns pale purple in the presence of protein.

BIURET TEST FOR PROTEINS

To test a sample for the presence of protein, biuret solution is added to the biological sample and mixed gently. Unlike the Benedict's test, there is no need to heat. As shown in Figure 2.5, if protein is present, blue biuret solution will become pale purple (sometimes called lilac, or mauve).

The biochemical tests that can be carried out for biological molecules, with the positive results listed, are summarised in Table 2.3. Note that an **observation** is very different from a **conclusion**. An observation is a description of what is seen, whereas a conclusion is a statement that can be inferred from an observation. For example, 'Benedict's solution changes from blue to red' is an observation; 'the sample contains a high concentration of reducing sugar' is a conclusion.

> **Tip**
> Remember to state the original colour of the test solution. Get into the habit of stating the colour beforehand; for example, 'blue biuret solution' or 'orange–brown iodine–potassium iodide solution'.

Biological molecule	Reagent(s) used	How test is carried out	Positive result
reducing sugar (e.g. glucose)	blue Benedict's solution	add Benedict's solution to sample in a test-tube; heat in a water bath set at 80° C	colour change from blue to green, yellow, orange or red
non-reducing sugar (e.g. sucrose)	hydrochloric acid and blue Benedict's solution	once the reducing sugar test has proved negative, boil with dilute acid; add sodium hydrogencarbonate to neutralise; carry out reducing sugar test, as described above	colour change from blue to green, yellow, orange or red
starch	orange–brown iodine–potassium iodide solution	add a few drops of iodine–potassium iodide solution to the sample	colour change from orange–brown to blue–black
lipid	ethanol distilled water	shake the sample with ethanol in a test-tube tube; allow to settle; pour clear liquid into water in another test-tube	cloudy white emulsion
protein	blue biuret solution	add biuret solution to sample in a test-tube	colour change from blue to pale purple

Table 2.3 The reagents, procedures and positive results of biochemical tests for biological molecules.

Worked example

An investigation was carried out into the activity of two enzymes, sucrase and amylase. Four reaction mixtures were prepared, as shown in Table 2.4.
To answer this question you will need to know that:

- all enzymes are proteins
- amylase breaks down starch into maltose
- sucrase breaks down sucrose into glucose and fructose.

Three biochemical tests were performed after incubating each reaction mixture at 37 °C for 1 hour.

Reaction mixture contents	Benedict's test	Biuret test	Iodine test
sucrose + sucrase			
sucrose + amylase			
starch + sucrase			
starch + amylase			

Table 2.4

Complete the table by indicating which tests would give positive (✔) and negative (✘) results.

Answer

Reaction mixture contents	Benedict's test	Biuret test	Iodine test
sucrose + sucrase	✔	✔	✘
sucrose + amylase	✘	✔	✘
starch + sucrase	✘	✔	✔
starch + amylase	✔	✔	✔

- All four reaction mixtures would give a positive result for the biuret test, as they contain enzymes, which are proteins.
- The first and last reaction mixtures would give a positive result for the Benedict's test, because they contain reducing sugars which are produced as a result of the hydrolysis of sucrose or starch, respectively.
- Only the reaction mixtures that contain starch would give a positive result for the iodine test.

1. A student made the following observations when they tested a sample for three biological molecules:
   - orange–brown colour when iodine–potassium iodide solution is added
   - pale purple colour when biuret solution is added
   - cloudy emulsion when shaken with ethanol and mixed with water.

   Identify which types of macromolecule are present in the sample.

2. Boiling a disaccharide with acid hydrolyses it into monosaccharides. Explain why the Benedict's test gives a positive result after, but not before, sucrose has been heated strongly with acid. Use the information in the Worked Example to help you.

★ 3. A scientist investigated the glucose content of a range of soft drinks to compare which had the greatest concentration of this sugar. Based on the results of the investigation, the scientist ranked the fruit juices in order, with fruit juice 1 having the highest glucose concentration, and fruit juice 5 having the lowest glucose concentration. The results are shown in Table 2.5.

Fruit juice brand	Observation with Benedict's test	Rank
A	changed from blue to orange	2
B	changed from blue to red	1
C	changed from blue to yellow	3
D	no change – remained blue	5
E	changed from blue to green	4

Table 2.5 The results of an investigation into the glucose concentration of five samples of fruit juice.

a. Suggest how the scientist was able to rank the fruit juices in order of reducing sugar concentration.

b. Describe **three** factors that must be standardised in order to make a valid comparison of the glucose concentrations of the soft drinks.

c. Suggest why estimating the concentration of glucose in soft drinks might give an underestimate of their sugar content.

Experimental skills 2.1: Diagnosing diabetes

Diabetes is a disorder in which a person cannot properly control the concentration of glucose in their bloodstream. One symptom of diabetes is the passing of urine containing glucose (Figure 2.6). Glucose is not found in the urine of a healthy person.

A scientist wanted to use Benedict's solution to develop a quick test to diagnose diabetes.

The scientist made **serial dilutions** of a 1% glucose solution. This reduces the concentration by half between each successive dilution. The scientist wanted to make 10 cm³ of each concentration.

Figure 2.6 The first treatment for diabetes was recorded in Ancient Egypt on the Apers Papyrus. At around the same time, Indian physicians described how the disease led to *madhumeha*, or 'honey urine', noting that it would attract ants.

QUESTIONS

P1. With reference to Figure 2.7, show how the scientist should produce three more solutions of glucose. The values for volume that you choose should produce solutions of the concentrations shown.

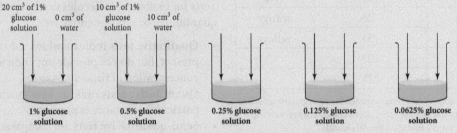

Figure 2.7 The preparation of serial dilutions using a 1% glucose solution and distilled water.

To measure the volumes of glucose solution and distilled water in their serial dilutions, the scientist used a 10 cm³ syringe, shown in Figure 2.8.

Figure 2.8 The 10 cm³ syringe used in this investigation.

All pieces of equipment used for measuring volumes of liquids and gases have a **percentage error**. This value gives an indication of the accuracy of the quantity that can be measured using this item. The value for percentage error can be calculated using the following equation:

$$\text{percentage error} = \frac{\text{uncertainty}}{\text{value being measured}} \times 100$$

where uncertainty is equal to ± half the smallest division on the measuring instrument.

P2. Describe and explain **one** source of error that could be encountered when measuring volumes of liquid using a syringe.

P3. Calculate the percentage error when using this 10 cm³ syringe with 0.2 cm³ graduations. Give your answer to three significant figures.

The scientist put 5 cm³ of the 1% glucose solution into a test-tube and added an equal volume of Benedict's solution. The contents of the test-tube were then observed while it was heated in a water bath set at 85 °C for 60 seconds. The scientist recorded two pieces of data for each solution:

- the time taken for the first appearance of a colour change, in order to produce a **calibration curve**
- the colour after heating for 60 seconds, in order to produce a series of **colour standards**.

Table 2.6 shows the scientist's results.

Concentration of glucose solution/%	Time taken for the first appearance of a colour change/ seconds	Colour of Benedict's solution at 60 seconds
1.0000	12	red
0.5000	26	orange
0.2500	55	yellow
0.1250	107	
0.0625	221	

Table 2.6 Two pieces of data were recorded for each glucose solution: the time taken for the first appearance of a colour change, and the colour of the Benedict's solution after 60 seconds.

P4. a. Complete the table to predict the colour of Benedict's solution at 60 seconds for the solutions of glucose of concentrations 0.125% and 0.0625%.

b. Identify a hazard when using the apparatus and reagents, and describe how the risk could be minimised.

c. Explain why it was important for the scientist to wash the test-tube thoroughly with distilled water between each experiment.

d. Suggest an appropriate control and why this should be included in the investigation.

The scientist decided to plot a calibration curve using this data to determine the concentration of glucose in a urine sample (Figure 2.9).

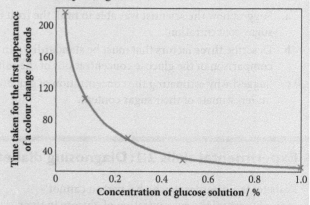

Figure 2.9 A calibration curve for glucose concentration, based on the time taken for Benedict's solution to begin to change colour.

P5. a. Give **two** ways in which the presentation of this graph could be improved.

b. Describe in detail how the glucose concentration in a sample of urine could be estimated using this calibration curve.

P6. The scientist also attempted to estimate the concentration of glucose in a patient's urine using the colour standards by recording the colour of the Benedict's solution after heating for 60 seconds. Suggest **two** reasons that may cause an inaccurate measurement of the concentration of glucose in urine using this method.

Tests for biological molecules can be qualitative, quantitative or semi-quantitative:

- **Qualitative tests** indicate which substances are present, but do not provide any indication of the concentration of these substances.
- **Quantitative tests** indicate how much of a particular substance is present.
- **Semi-quantitative tests** are comparative: they illustrate whether a sample has more or less of a particular substance than another sample, and they can sometimes be used to give a rough estimate of their concentration.

P7. Outline why both glucose tests undertaken by the scientist in this investigation are semi-quantitative in nature.

Key ideas

→ Living organisms are made up of many types of macromolecule, most of which contain atoms of carbon, hydrogen, oxygen and nitrogen.

→ Many biological molecules are polymers, made up of long chains of smaller molecules, called monomers, joined together.

→ Iodine dissolved in potassium iodide solution is used to test for starch. This orange–brown solution turns blue–black in its presence.

→ Blue Benedict's solution can be used to test for reducing sugars. Non-reducing sugars do not give a positive result with Benedict's solution, but they will do if they are first hydrolysed using hot acid. The positive result is a colour change, from green through to yellow, orange and red, depending on the concentration of reducing sugars in the sample.

→ The emulsion test can be used to test for the presence of lipids.

→ Blue biuret solution is used to test for proteins. It turns pale purple in their presence.

2.3 Carbohydrates

All carbohydrates contain the elements carbon, hydrogen and oxygen. They are always present in the same ratio, giving carbohydrates the general formula CH_2O_n.

There are three basic types of carbohydrate molecule: monosaccharides, disaccharides and polysaccharides.

- **Monosaccharides**, the monomers, are single sugars.
- **Disaccharides**, the dimers, are double sugars (made from two monosaccharides).
- **Polysaccharides**, the polymers, are multiple sugars (polymers of many monosaccharides).

Figure 2.10 shows the relationship between monosaccharides, disaccharides and polysaccharides. Also listed are some examples of these carbohydrates that we will encounter in this section.

Tip

A monomer is a single molecule. A dimer is formed when two monomers link together. A polymer is formed when many monomers link together.

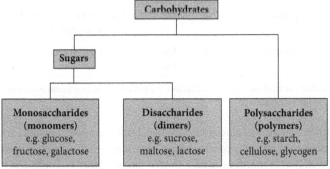

Figure 2.10 Carbohydrates consist of the sugars, which can be called monosaccharide monomers and disaccharide dimers, and the polysaccharide polymers.

MONOSACCHARIDES

Monosaccharides are simple sugars – small, hydrophilic molecules that taste sweet. These are the monomers from which all larger carbohydrates are made. Common monosaccharides include glucose, galactose and fructose.

Tip

Sugars usually have names ending in '-ose'.

Glucose, the main source of energy for cells, is a hexose (six-carbon) sugar that has the **molecular formula** $C_6H_{12}O_6$. All other hexose sugars, such as fructose and galactose, have the same molecular formula. However, their atoms are in different positions, so they have different **structural formulae**. This means that they are **isomers** of glucose, and have different properties.

Even glucose itself does not have just one structure. It has two closely related isomers, called **alpha-glucose (α-glucose)** and **beta-glucose (β-glucose)**. These are shown in Figure 2.11. Five of the carbon atoms (numbered 1–5 in the diagram) are part of a ring. The difference between these two isomers is due to the orientation (position) of the hydrogen and hydroxyl (−OH) groups on carbon atom 1.

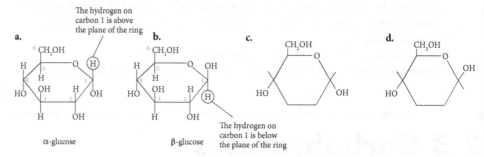

Figure 2.11 The structures of α-glucose and β-glucose. The full structural formulae of the sugars are shown for α-glucose (a) and β-glucose (b). The ring forms of the two molecules are shown in (c) and (d).

DISACCHARIDES

Disaccharides are sometimes referred to as 'double sugars'. This is because they consist of molecules made of two monosaccharides joined together. Like monosaccharides, all disaccharides are hydrophilic and sweet. Table 2.7 shows the monosaccharides that are joined together to make three different disaccharides.

Disaccharide	Monosaccharide units	Uses in living organisms
maltose	α-glucose + α-glucose	found in germinating seeds when starch stores are broken down by amylase
sucrose	α-glucose + fructose	the main transport sugar in the phloem vessels of plants
lactose	α-glucose + galactose	found in milk; the source of energy for infant mammals

Table 2.7 The disaccharides maltose, sucrose and lactose are each made of two monosaccharides. They have a variety of uses in living organisms.

Figure 2.12 shows how two α-glucose molecules, both monosaccharides, are joined together. The process is called a condensation reaction, because a water molecule is formed. The bond that is made, around a shared oxygen atom between the two original monosaccharide molecules, is called a **glycosidic bond**. In this reaction, the disaccharide produced is called maltose. The monosaccharides, now joined, are referred to as **residues**.

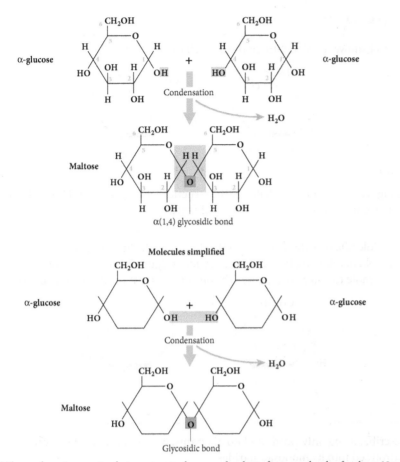

Figure 2.12 The condensation reaction between two α-glucose molecules to form a molecule of maltose. Note that in this example the glycosidic bond is formed between carbon atoms 1 and 4 of the neighbouring glucose molecules.

Disaccharides can also be broken down into their two component monosaccharides, in a reaction called hydrolysis. This is shown in Figure 2.13 for two other disaccharides, sucrose and lactose. In all cases of hydrolysis, a water molecule must be used: the term 'hydrolysis' literally means 'splitting with water'. Enzymes are usually required to catalyse the hydrolysis reaction and most of the enzymes involved in digestion fall into this category.

> **Tip**
> Remember that the molecular formula of a disaccharide will not be exactly two monosaccharides joined together. A molecule of water is lost. For example, maltose has the molecular formula $C_{12}H_{22}O_{11}$.

> **Link**
> Enzymes, proteins that catalyse specific biochemical reactions, will be explored in greater detail in Chapter 3.

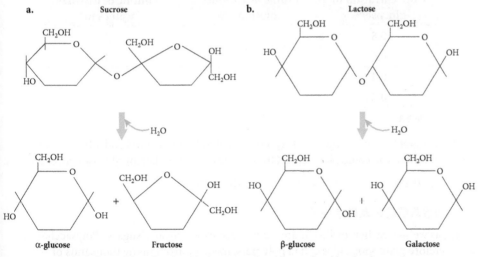

Figure 2.13 The hydrolysis of disaccharides, which requires the addition of a molecule of water, into two monosaccharides. Sucrose (a) is hydrolysed to give α-glucose and fructose; lactose (b) is hydrolysed into β-glucose and galactose.

Tip

Hydrolysis is
the opposite of
condensation. The
laboratory test for
non-reducing sugars,
such as sucrose and
lactose, relies on
their hydrolysis into
reducing sugars,
which then give a
positive result with
Benedict's solution.

Worked example

Figure 2.14 shows the structure of the disaccharide sucrose.

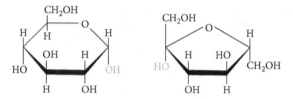

Figure 2.14

Using Figure 2.14, your knowledge of the structure of α-glucose and how a hydrolysis reaction occurs, determine the structure of fructose.

Answer

When a molecule of water reacts with the disaccharide during hydrolysis, the glycosidic bond is broken, releasing two monosaccharides (Figure 2.15). The one on the left should be recognisable to you as α-glucose. The one on the right must therefore be fructose.

Figure 2.15

Tip

You will not be
expected to recall the
structure of sugars
other than α-glucose
and β-glucose. This
question requires
you to apply your
knowledge.

4. Describe, using only words and no diagrams, how a molecule of sucrose can be hydrolysed into its monosaccharides.

5. The techniques of serial dilution and simple (proportional) dilution, which you encountered earlier, can be used to prepare solutions of different concentrations. The concentration of the original solution, called a standard solution, determines the volume of distilled water that needs to be added to produce a solution of a given concentration.

   a. Copy and complete Table 2.8 to show how to make a range of concentrations when given a 1% standard solution of glucose. The first two have been done for you.

Tip

Once a monomer has
been incorporated
into a polymer, it
is referred to as a
residue.

Concentration of glucose/%	Volume of 1% glucose solution/cm³	Volume of distilled water/cm³
0.8	8	2
0.6	6	4
0.4		
0.2		

Table 2.8

   b. Suggest how, using the simple (proportional) dilution method and a 1% standard solution, you could make up 1000 cm³ (1 dm³) of the following glucose concentrations:

   i. 0.2%   ii. 0.05%   iii. 0.001%.

POLYSACCHARIDES

'Poly', as we saw earlier, means 'many' and 'saccharides' means 'sugars'. Polysaccharides are therefore giant molecules called polymers, made up from many thousands of monosaccharide residues. Like disaccharides, they can also be broken down into smaller units by breaking the glycosidic bonds between their residues in hydrolysis reactions.

Common polysaccharides include starch, cellulose and **glycogen**. Although they have remarkably different properties, as we will see in this next section, all polysaccharides are relatively insoluble in water. This is not because they are hydrophobic. It is because their molecules are so large that they cannot spread out between the water molecules as smaller molecules do.

Starch

Starch is a carbohydrate that may be familiar to you. It is in many of the foods we eat, such as rice and potatoes, and foodstuffs made from plant materials such as pasta and bread (Figure 2.16).

Starch is actually a mixture of two different polysaccharides, **amylose** and **amylopectin**, found together in starch granules inside plant cells. These two components have different molecular structures, which give them different properties. However, they both consist of only one type of monomer: α-glucose.

To fully understand the structure of starch, it is important to understand the difference between the two types of glycosidic bond found between α-glucose residues.

- Most glycosidic bonds in amylose and amylopectin are between carbon 1 on one α-glucose and carbon 4 on the next, so they are called $\alpha(1,4)$ glycosidic bonds.
- Although not as common, there are also some $\alpha(1,6)$ glycosidic bonds in amylopectin only, which form branches in the chain.

Amylose is an unbranched polymer in which residues are joined by $\alpha(1,4)$ glycosidic bonds. This forms a tightly packed spiral molecule that can be packed into small spaces, stabilised by hydrogen bonding between the glucose units. The glucose chains of amylopectin have $\alpha(1,4)$ glycosidic bonds and $\alpha(1,6)$ glycosidic bonds. This allows branching of the chain, providing a large number of 'ends' to which enzymes can attach and detach glucose units. Part of a starch molecule is shown in Figure 2.17.

Figure 2.16 Starch is a common carbohydrate in many types of food.

Link

The insolubility of starch is important. As we will see in Chapter 4, starch does not have a significant effect on water movement into and out of plant cells by osmosis.

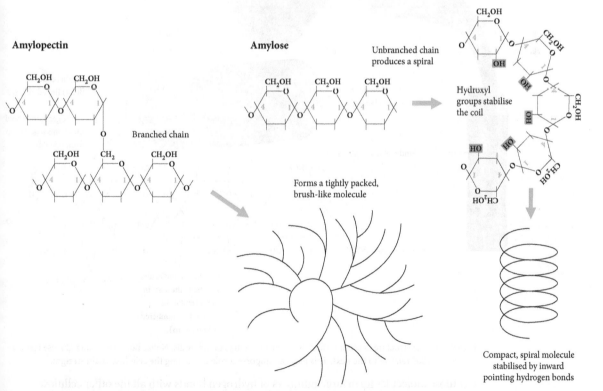

Figure 2.17 Starch is a mixture of two distinct macromolecules, both composed of α-glucose residues joined by glycosidic bonds. Amylose is an unbranched polysaccharide that produces a spiral, stabilised by hydrogen bonding between the hydroxyl (–OH) groups between different residues. Amylopectin is a branched polysaccharide.

Glycogen

Animals and fungi do not use starch as a storage carbohydrate; instead, they use a carbohydrate called glycogen. Glycogen molecules are very similar to amylopectin molecules – they are made up of α-glucose molecules joined by α(1,4) and α(1,6) glycosidic bonds – but they have far more branches.

The structure of glycogen suits its function as a storage molecule in animals. Owing to its highly branched nature, it can be built up and broken down very quickly, allowing the animal to store energy efficiently and release it on demand. The huge number of 'ends' increases the rate at which enzymes can hydrolyse (break down) the molecule to remove glucose. In humans, glycogen is stored in large amounts in the liver and the muscles and is usually hydrolysed during prolonged exercise, when the immediate supply of glucose in cells is used up.

Cellulose

Cellulose is the most common polysaccharide on Earth. This is because it is the main ingredient in plant cell walls. It also has a number of practical uses, including the production of paper. The papyrus shown in Figure 2.6 consists mainly of cellulose.

Like amylose, cellulose is made by joining glucose molecules by glycosidic bonds between carbons 1 and 4. However, the key difference is that cellulose molecules are made up of long chains of β-glucose molecules (Figure 2.18). Owing to the slight difference in the structure of this isomer (see Figure 2.15), when β-glucose molecules join to each other, each one is upside down (inverted 180°) relative to one another. This limits hydrogen bonding within the chain and prevents the formation of a spiral, like amylose. Instead, this produces long, straight, unbranched molecules that lie side by side.

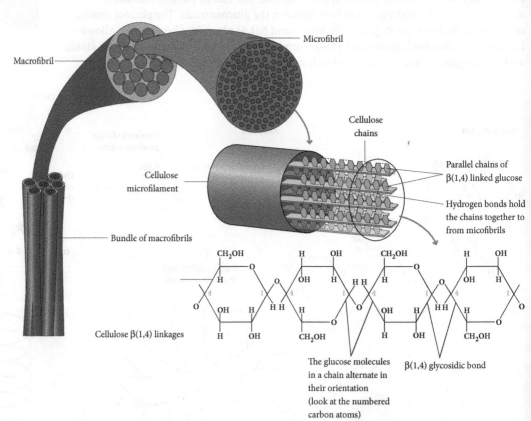

Figure 2.18 The structure of cellulose and its arrangement in plant cell walls. Notice how every other glucose residue is inverted 180° relative to the next. This plays an important role in keeping the cellulose chain straight.

Cellulose molecules form large numbers of hydrogen bonds with all the other cellulose molecules lying alongside them, forming groups of parallel molecules called microfibrils. Although an individual hydrogen bond is very weak, the large numbers of them mean that the microfibrils have great strength. Microfibrils are further packaged into **macrofibrils**, which are very strong. This makes cellulose a suitable material for plant cell walls, because it is strong and flexible. It will not break easily if the plant cell swells as it absorbs water.

Table 2.9 summarises the similarities and differences between the polysaccharides.

Polysaccharide	Name of monomer	Glycosidic bonds joining monomers	Description	Function
amylose	α-glucose	α(1,4)	helical, compact	long-term energy store in plants
amylopectin	α-glucose	α(1,4) and α(1,6)	branched, compact	short-term energy store in plants
glycogen	α-glucose	α(1,4) and α(1,6)	branched, compact	energy store in animals and fungi
cellulose	β-glucose	β(1,4)	straight-chained	high tensile strength in cell walls

Table 2.9 The structure and function of common polysaccharides. Starch is a mixture of both amylose and amylopectin.

6. Draw a table to compare the structure and properties of starch and cellulose. Use Table 2.9 to help you. You should include the following features:
   - the sugar units from which they are made
   - the type of glycosidic bonds that join them together
   - whether or not they are made of one type of molecule
   - whether or not they branch
   - the ease with which they can be digested
   - where they are found
   - their functions.

Tip

Starch and glycogen are storage polysaccharides, whereas cellulose is a structural polysaccharide

7. Discuss how amylopectin and glycogen are readily hydrolysed to release glucose for respiration.

8. Trehalose is a disaccharide formed by an α(1,1) link between two α-glucose monomers. Use this information to draw the full structural formula of trehalose.

Experimental skills 2.2: Investigating dietary starch

In most plant cells, starch is stored in grains called **granules**. These are packed into the cytoplasm as energy stores, and they can be hydrolysed to release glucose molecules when required.

QUESTIONS

P1. Explain the advantages of storing starch rather than glucose in plant cells.

The percentage of amylose and amylopectin in starch granules varies depending on the species of plant. In some plant tissues, such as in mung beans, the starch consists of more amylose than in others, such as in potatoes. The relative proportions of amylose and amylopectin can be expressed as an amylose : amylopectin ratio.

Five starch granules were measured in a cell taken from seven different plant species and a mean for each was calculated. Table 2.10 shows the results.

Plant tissue	% Amylose	% Amylopectin	Amylose : amylopectin ratio	Mean starch granule diameter/μm
wheat	26	74	0.351	22
sweet potato	23	77	0.299	20
maize	24	76	0.316	20
tapioca	18	82	0.220	18
potato	17	83	0.205	15
mung bean	31	69		10
rice	21	79		8

Table 2.10 The mean starch granule diameter and amylose : amylopectin ratios of starch extracted from different plant tissues.

P2. Calculate the two missing amylose : amylopectin ratios in Table 2.10.

P3. Suggest which plant tissue would be able to hydrolyse its starch energy stores most rapidly. Explain your choice.

★ **P4.** Suggest an explanation for these findings of the investigation into the relationship between the diameter of the starch granules and the amylose : amylopectin ratio.

To estimate the amylose : amylopectin ratio of a plant tissue, a procedure using different enzymes that digest starch can be used. Amylases are enzymes that break down starch. In an investigation, extracts containing amylases were obtained from three species of fungus. The first two, extract 1 and extract 2, catalyse the hydrolysis of α(1,4) glycosidic bonds. A third, extract 3, catalyses the hydrolysis of α(1,6) glycosidic bonds. Figure 2.19 shows the approximate positions in amylopectin and amylose where the amylase enzymes in these three fungal extracts hydrolyse the glycosidic bonds.

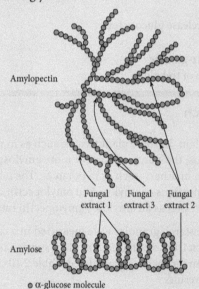

Figure 2.19 The approximate positions in amylopectin and amylose where the amylase enzymes in three different fungal extracts hydrolyse glycosidic bonds.

After a sample of starch has been digested, a procedure called **chromatography** can be used to give an indication of its amylose : amylopectin ratio. Chromatography involves separating a mixture of molecules according to their size or density. The procedure consists of adding a sample of a substance at the origin on a chromatogram (usually a sheet of plastic coated with a special gel), and then keeping the sample partially immersed in a solvent. As the solvent moves along the gel, the contents of the mixture are separated according to their density. The denser the molecules, the shorter the distance they will move (Figure 2.20). Special dyes can be used to make the position of the molecules visible on the chromatogram.

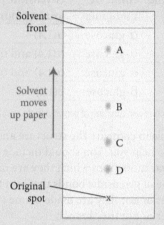

Figure 2.20 Chromatography is a laboratory procedure used to separate a mixture of molecules according to their density. In this example, a mixture of four molecules was applied to the original spot (origin). As the solvent moved up the paper, the molecules separated. Here, molecule A has the lowest density and molecule D the highest.

A scientist investigated the relative amylose : amylopectin ratios of three plant tissues by incubating 1% starch suspension from each separately with fungal extract 3. When hydrolysis was complete, the **products** of hydrolysis were removed and were separated by chromatography. Figure 2.21 shows the chromatograms produced using each of the mixtures after incubation.

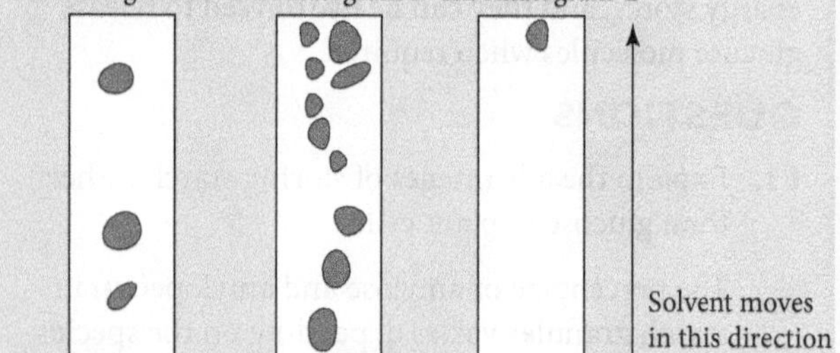

Chromatogram **A** Chromatogram **B** Chromatogram **C**

Solvent moves in this direction

Origin

Figure 2.21 The results of an investigation into the relative amylose : amylopectin ratios of three plant tissues. A chromatogram was prepared for each plant tissue, which had a starch solution incubated with amylase found in fungal extract 3.

P5. Identify **three** variables that must be standardised to make a valid comparison of these results. Describe how **one** of these variables could be standardised.

P6. Suggest how the scientist was able to determine that the hydrolysis of starch was complete.

P7. Describe in detail how the results of this investigation could be used to estimate the relative amylose : amylopectin ratios of these three plant tissues.

P8. Predict, with reasons, what you would expect to see on the chromatogram if a:

a. 2% cellulose suspension was used

b. 2% glycogen suspension was used.

P9. Explain why the amylases in fungal extract 1 cannot be used instead of fungal extract 3 to estimate the amylose : amylopectin ratio of starches (refer to Figure 2.19 to help you answer this question).

A Level analysis

Pearson's linear correlation coefficient (r)

The scatter graph in Figure 2.22 shows the results of an investigation into the relationship between the diameter of starch granules and the amylose : amylopectin ratio of the starch granule (Table 2.10).

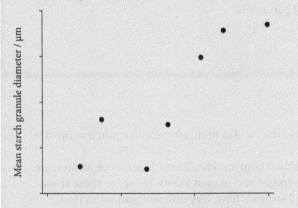

Amylose : amylopectin ratio/no units

Figure 2.22 A scatter graph showing the relationship between the amylose : amylopectin ratio of starch from a number of plant tissues, and the diameter of starch granules found in those tissues.

Based on these findings, a **statistical test** was used to find the significance (strength) of the relationship between the diameter of starch granules and the amylose : amylopectin ratio. You will encounter a number of statistical tests in your course.

Pearson's linear correlation can be used to test for a **correlation** between two sets of data that are normally distributed. This provides a value called Pearson's linear correlation coefficient (r), calculated using the following equation:

$$r = \frac{\sum xy - n\overline{x}\,\overline{y}}{(n-1)s_x s_y}$$

s = standard deviation

x, y = observations

$\overline{x}, \overline{y}$ = means

Correlations can be positive, which means that as one variable increases, so does the other; or negative, where as one increases, the other decreases. In effect, this statistical test gives an indication as to how close the data is to a trend line.

The number, or coefficient, obtained by performing Pearson's linear correlation test is on a scale between +1 and −1, where +1 indicates a total positive correlation, −1 a total negative correlation and 0 no correlation.

P10. a. Explain why this set of data allows for Pearson's linear correlation to be used to test for a correlation between these two variables.

b. Explain why a correlation between two variables may not necessarily imply a causative relationship.

c. Pearson's coefficient of correlation (r) was found to be 0.775 for the scatter graph in Figure 2.22. State what this value suggests about the relationship between the amylose : amylopectin ratio of starch and the diameter of starch granules.

Table 2.11 shows part of a table of critical values for Pearson's linear correlation test (r).

Number of pairs of data (n)	Probability level (p)			
	0.10	0.05	0.02	0.01
1	0.988	0.997	0.9995	0.9999
2	0.900	0.950	0.980	0.990
3	0.805	0.878	0.934	0.959
4	0.729	0.811	0.882	0.917
5	0.669	0.754	0.833	0.874
6	0.622	0.707	0.789	0.834
7	0.582	0.665	0.750	0.798
8	0.549	0.632	0.716	0.765
9	0.521	0.602	0.685	0.735
10	0.497	0.576	0.658	0.708

Table 2.11

d. Describe how the student calculated the degrees of freedom.

The y-axis of the scatter graph is labelled "Mean starch granule diameter / μm".

e. Describe how the student used the probability table to find out if the value for $r = 0.775$ is significant.

f. Suggest **two** reasons why the method used for the investigation into the effect of relationship between the amylose : amylopectin ratio of starch and the diameter of starch granules may have given results that are not reliable.

Key ideas

→ Carbohydrates are made up of carbon, hydrogen and oxygen.

→ Monosaccharides are single-unit sugars. Glucose, fructose and galactose are monosaccharides.

→ Monosaccharides, such as glucose, can exist in two different forms called isomers. The two isomers in glucose are called α-glucose and β-glucose.

→ Disaccharides are made of two monosaccharide molecules joined by a glycosidic bond. Maltose, sucrose and lactose are disaccharides.

→ A condensation reaction joins together two monosaccharides. The reaction that breaks them apart is called hydrolysis.

→ Polysaccharides are polymers of monosaccharides. Starch, glycogen and cellulose are polysaccharides.

→ Starch and glycogen have compact, coiled and branched molecules made of long chains of α-glucose monomers. They are used as energy stores in plants and animals, respectively.

→ Cellulose are straight, unbranched molecules made of long chains of β-glucose monomers. Cellulose molecules associate with each other to form microfibrils. They are found in cell walls of plant cells.

2.4 Lipids

Lipids can be fats or oils. Fats tend to be solid at room temperature, whereas oils are liquid.

Most lipids belong to a family called **triglycerides**. These, like carbohydrates, are molecules that contain carbon, hydrogen and oxygen atoms. However, these atoms are arranged in a different way and so lipids have very different properties from carbohydrates. Some lipids, as we shall see later, can contain another element – phosphorus. These are the **phospholipids**.

TRIGLYCERIDES

Most lipids are triglycerides. These are hydrophobic, or non-polar, in nature. If mixed with water, triglyceride molecules group together to form droplets and form a separate layer to the water.

Triglycerides are made up of one molecule of **glycerol** (a type of alcohol), to which three **fatty acids** have been joined. The 'tri' portion of the triglyceride name refers to the fatty acids – carboxylic acids with at least four, but usually more, carbon atoms in their hydrocarbon chain. Figure 2.23 shows how these four molecules join together, in three separate condensation reactions, to form a triglyceride. **Ester bonds** form between these molecules.

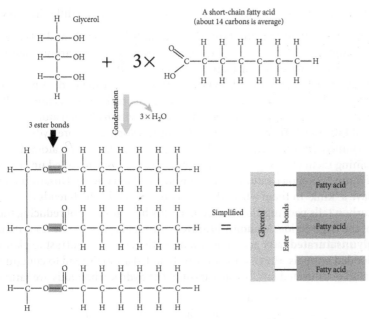

Figure 2.23 The formation of a triglyceride from one glycerol molecule and three fatty acids.

Triglycerides have a number of functions in living organisms. In all cases, the structure of the molecule is important to its functions. These include:

- **Energy stores:** many animals store energy in the form of triglycerides, which have a great advantage over carbohydrates in this respect. Because they contain more C–H bonds than carbohydrates, they can yield more energy during respiration, per unit mass, when they are oxidised. In plants, triglycerides are often found in the cotyledons of seeds, ready to be used to release energy during germination. In animals, they are found in **adipose tissue**, which often lies just beneath the skin. Each adipocyte (fat cell) contains an oil droplet, which can almost completely fill the cell (Figure 2.24).
- **Thermal insulation:** another benefit of storing triglycerides in cells under the skin is to insulate the animal against heat loss. Blubber is an especially thick layer of adipose tissue found under the skin of some marine animals such as whales and seals. This is particularly effective at reducing the rate of heat loss from animals to cold water.
- **Buoyancy:** the density of triglycerides is low. Blubber therefore acts as a superb buoyancy aid, keeping aquatic mammals afloat in the water.
- **Waterproofing:** oily secretions from the skin in mammals, and from the preen gland in birds, act as a water repellent. The hydrophobic properties of the triglycerides prevent fur and hair from becoming waterlogged when wet.

> **Link**
>
> Waxy suberin is a lipid in the Casparian strip of endodermal cells in plant roots forces water into the symplast pathway. In Chapter 7, you will see that this is crucial to plant transport.

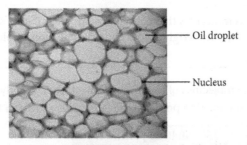

Figure 2.24 A microscope image of adipocytes (fat cells) found under the skin in animals. The large oil droplets completely fill the cell, which can sometimes squeeze the nucleus up against the cell surface membrane.

Because the glycerol molecule is the same in all triglycerides, the properties of different triglycerides depend on the structure of the fatty acids. Fatty acids are a diverse family of

> **Tip**
>
> When a molecule is oxidised, it often means that it reacts with oxygen. When carbohydrates and lipids are oxidised, ATP, carbon dioxide and water are the products.

molecules that share a —COOH (carboxylic acid) group. They can differ in two ways:

- **The length of the hydrocarbon tails:** this is usually between 14 and 28 carbon atoms in length. These can often be too long to draw, so hydrocarbon chains can be represented by the letter R and a fatty acid as R–COOH.
- **The saturation of the hydrocarbon tails:** carbon atoms can bond with up to four atoms to form long chains. A fatty acid in which the carbon atoms use up all their four bonds, two with other carbon atoms and two with hydrogen atoms, is called a **saturated fatty acid**. There are no spare bonds to which any more hydrogen atoms could be joined. In some fatty acids, however, some of the carbon atoms have a double bond joining them to a neighbouring carbon atom. These are called **unsaturated fatty acids**. Figure 2.25 shows saturated and unsaturated fatty acids. Unsaturated fatty acids have a 'kink' in them where there is a double bond, which tends to prevent the hydrocarbon tails packing closely together. This has the effect of reducing the melting point of the triglyceride. **Monounsaturated** fatty acids possess one C=C double bond and **polyunsaturated** fatty acids contain more than one. In contrast to plants and cold-blooded animals, the tissues of warm-blooded animals tend to contain a higher proportion of saturated fats, because most of the hydrocarbon tails are saturated.

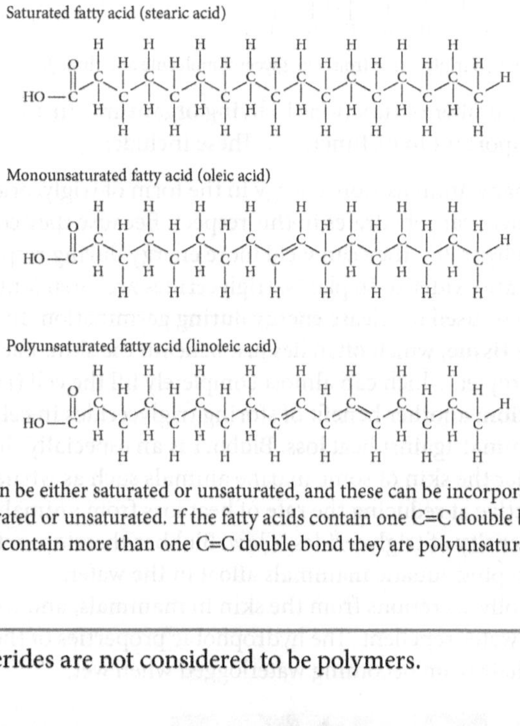

Figure 2.25 Fatty acids can be either saturated or unsaturated, and these can be incorporated into triglycerides that are described as saturated or unsaturated. If the fatty acids contain one C=C double bond then they are monounsaturated. If they contain more than one C=C double bond they are polyunsaturated.

Worked example

Explain why triglycerides are not considered to be polymers.

Answer

Polymers are defined as molecules that consist of a large number of smaller molecules that have been joined together. Triglycerides consist of just four molecules – a glycerol and three fatty acids.

9. Explain why three molecules of water are released when a triglyceride is formed.
10. Table 2.12 shows the melting points of the fatty acids shown in Figure 2.25.

Fatty acid	Melting point/ °C
linoleic acid	−5
oleic acid	13
stearic acid	69

Table 2.12 Melting points of three fatty acids.

a. Using the information in Table 2.11 and Figure 2.25, suggest reasons for the different melting points of these three fatty acids.

b. Describe in detail what is meant by the term 'polyunsaturated triglyceride'.

11. An investigation was conducted into the energy content of different biological molecules. Identical masses of pure, dried carbohydrate and lipid were burned under identical volumes of water, as shown in Figure 2.26. The temperature rise was then recorded, and calculations were performed to give a value for energy content. Table 2.13 shows typical energy values of samples of carbohydrate and lipid.

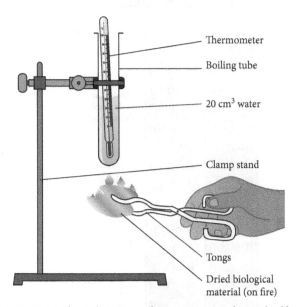

Figure 2.26 Simple equipment can be used to estimate the energy content of a sample of food in the laboratory.

Respiratory substrate	Energy value/$KJ\,g^{-1}$
carbohydrate	15.8
lipid	39.4

Table 2.13 Typical energy content of samples of carbohydrate and lipid.

a. Calculate how many times more energy can be released by a sample of lipid than an equal mass of carbohydrate. Give your answer to one decimal place.

b. In animals, triglycerides are usually stored for long-term energy requirements and are not as useful for short-term release. Suggest why.

c. Describe and explain one improvement that could be made to the method shown in Figure 2.26 that would provide more accurate data.

☆ d. A student referred to this experimental study and claimed that all lipids are a better source of energy than carbohydrates. Using your knowledge of both molecules and how they are stored, evaluate this statement.

PHOSPHOLIPIDS

Phospholipids are a special type of lipid with an unusual property. They are **amphipathic**, which means that one part of the molecule is hydrophilic, and another part is hydrophobic. Figure 2.27 shows the structure of a phospholipid. In place of one of the three fatty acid hydrocarbon tails, a polar, negatively charged phosphate head is attached to the glycerol molecule. Two **non-polar**, uncharged fatty acid tails are also attached to the glycerol. The phosphate group of a phospholipid is often referred to as the 'head', whereas the two hydrocarbon chains are called 'tails'.

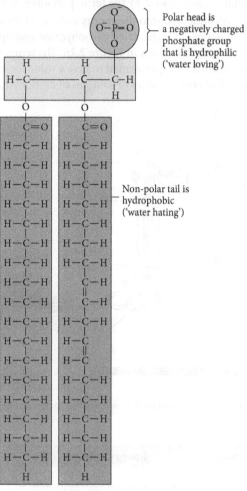

Polar head is a negatively charged phosphate group that is hydrophilic ('water loving')

Non-polar tail is hydrophobic ('water hating')

Link

In Chapter 4, you will explore phospholipids in greater detail, and see how these molecules form the basis of the surface membrane of all cells.

Figure 2.27 The structure of a phospholipid. The phosphate head is polar and hydrophilic. The fatty acid tails are non-polar and hydrophobic.

The amphipathic nature of the phospholipids explains the behaviour of the molecules when they come into contact with water or an aqueous solution. The phosphate heads are hydrophilic, and they are attracted to the water. However, the hydrocarbon tails are attracted to each other and not water – they are hydrophobic. This results in the spontaneous assembly of a phospholipid bilayer, as shown in Figure 2.28.

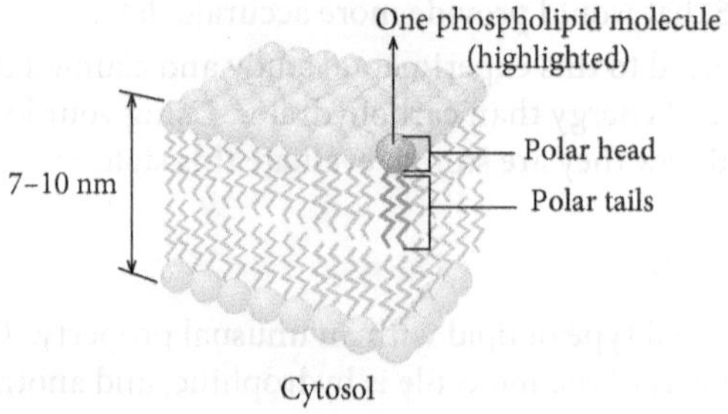

Extracellular environment

One phospholipid molecule (highlighted)

Polar head

Polar tails

7–10 nm

Cytosol

Figure 2.28 The structure of the phospholipid bilayer. The phosphate heads are polar and hydrophilic. The fatty acid tails are non-polar and hydrophobic. The phospholipids in a membrane are arranged tail to tail, forming a bilayer.

Key ideas

→ Non-polar triglycerides and amphipathic phospholipids are two groups of lipid.
→ Triglycerides are formed when three fatty acids form ester bonds with a glycerol molecule, in a condensation reaction.
→ Fatty acids in which all the carbon–carbon bonds are single are known as saturated fatty acids. If at least one carbon–carbon bond is a double bond, the fatty acid is unsaturated.
→ In phospholipids, a negatively charged phosphate group takes the place of one of the fatty acids.
→ Phospholipids are found in the cell surface membrane of all organisms.

2.5 Proteins

Proteins are an extremely important class of macromolecule in living organisms. More than 50% of the dry mass of most cells is protein, and it is estimated that in humans there may be millions, or even billions, of different proteins. The variety of proteins, and their functions in living organisms, is phenomenal. As shown in Figure 2.29, proteins serve multiple purposes in mammals, while in other organisms their functions can be even more diverse, from functioning as antifreeze in Arctic insects to toxins produced by pathogens.

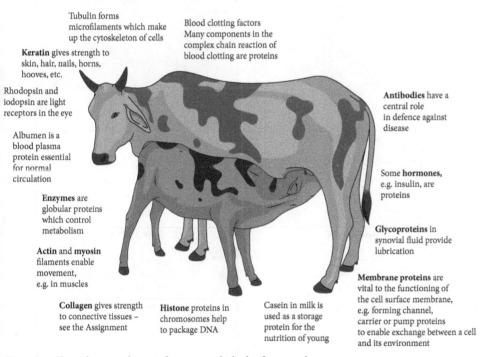

Tubulin forms microfilaments which make up the cytoskeleton of cells

Blood clotting factors
Many components in the complex chain reaction of blood clotting are proteins

Keratin gives strength to skin, hair, nails, horns, hooves, etc.

Rhodopsin and iodopsin are light receptors in the eye

Albumen is a blood plasma protein essential for normal circulation

Enzymes are globular proteins which control metabolism

Actin and **myosin** filaments enable movement, e.g. in muscles

Collagen gives strength to connective tissues – see the Assignment

Histone proteins in chromosomes help to package DNA

Casein in milk is used as a storage protein for the nutrition of young

Antibodies have a central role in defence against disease

Some **hormones,** e.g. insulin, are proteins

Glycoproteins in synovial fluid provide lubrication

Membrane proteins are vital to the functioning of the cell surface membrane, e.g. forming channel, carrier or pump proteins to enable exchange between a cell and its environment

Figure 2.29 The wide range of protein functions in the body of an animal.

Like carbohydrates and lipids, the shape of a protein is the key to its role in the cell. However, unlike these other macromolecules, proteins have an enormous variety of shapes and sizes.

At their simplest level, all proteins are made from the same monomers. These are called **amino acids**. In addition to the elements carbon, hydrogen and oxygen, amino acids always contain nitrogen and sometimes sulfur. Figure 2.30 shows the simple structure of an amino acid. As its name suggests, the monomer contains an **amine group** and a **carboxyl group**. Both of these chemical groups are attached to a central carbon atom, called the alpha-carbon (α-carbon). A hydrogen atom is also attached to the α-carbon.

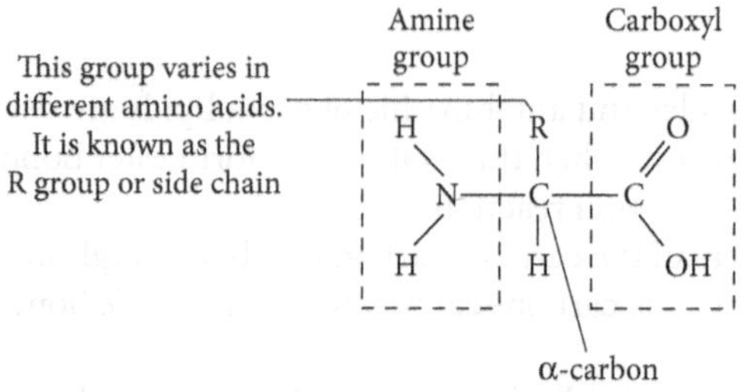

This group varies in different amino acids. It is known as the R group or side chain

Amine group

Carboxyl group

α-carbon

Figure 2.30 The simple structure of an amino acid. The 'R' group varies between different amino acids, of which there are 20. Note that when an amino acid is dissolved in water, the amine group, $-NH_2$, accepts hydrogen ions, to become NH_3^+. The carboxyl group, $-COOH$, breaks apart to produce hydrogen ions, H^+, and becomes $-COO^-$.

Link

In Chapter 4, you will see how membrane proteins are enriched in amino acids with R groups that interact with the hydrophilic or hydrophobic sections of the cell surface membrane.

There are 20 amino acids in the proteins of living organisms, which differ in the identity of the 'R' group found attached to the α-carbon. The simplest and smallest R group is a hydrogen atom (in the amino acid called glycine), but all the other R groups contain carbon and can be very complex (Figure 2.31). The different side chains give each amino acid different properties, and this affects the structure and hence function of the proteins they form.

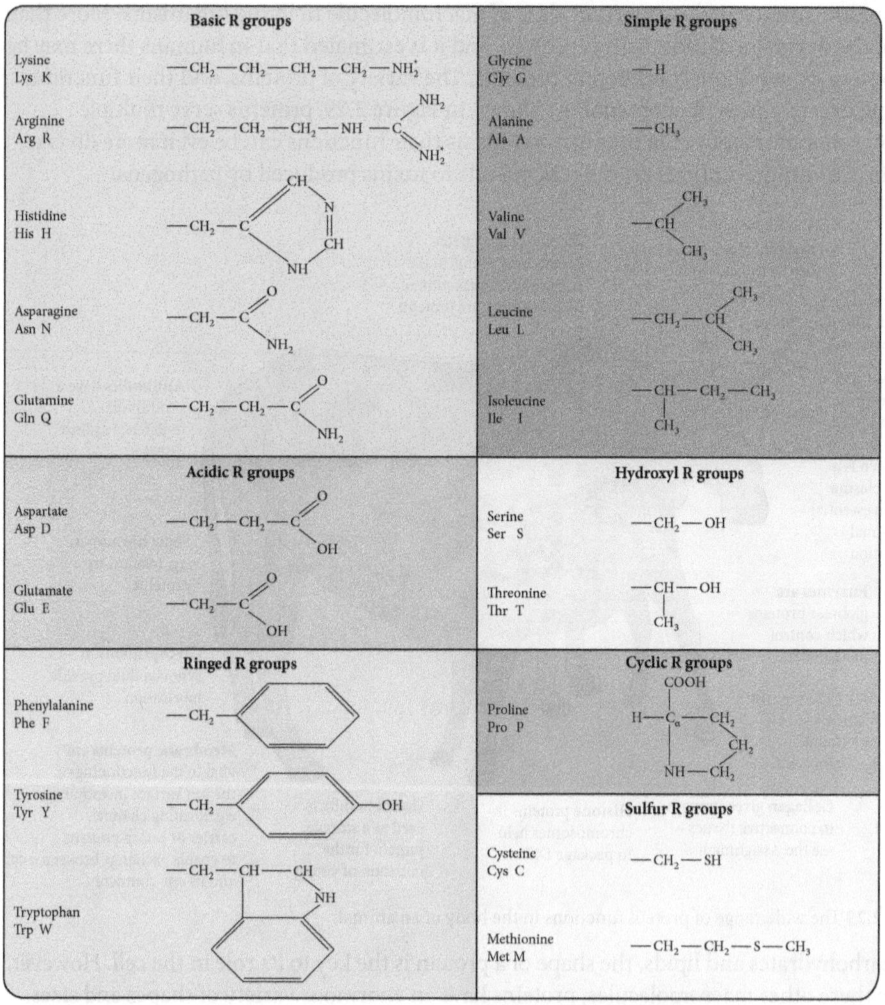

Figure 2.31 The different R groups of the 20 amino acids, as well as their three-letter and single-letter shortened names, that make up all proteins. **You do not need to learn these structures, or the names/abbreviations of the amino acids.**

Proteins are polymers made of many amino acids joined together in long chains. Small proteins can have 50–100 amino acid residues, whereas larger proteins can have several tens of thousands. In all cases, these are joined together by chemical bonds into a **polypeptide chain**. In this respect, they are similar to polysaccharides. However, unlike starch and cellulose, which are made from only one type of monomer (glucose), proteins are made from 20 different naturally occurring amino acid monomer units. These can be assembled in any order, making an endless variety of protein structures possible. It is like having an alphabet of 20 letters to make up words hundreds or even thousands of letters long. To take this idea further, consider Table 2.14. A typical polypeptide consists of far more than 10 amino acids, but even a polypeptide of this length could have over 10 trillion possible amino acid sequences!

Number of amino acids in the polypeptide	Number of possible amino acid sequences
1	$20^1 = 20$
2	$20^2 = 400$
5	$20^5 = 3.2$ million
10	$20^{10} = 10.24$ trillion

Table 2.14 The diversity of amino acid sequences of polypeptides. Most proteins have much more than 10 amino acid residues, resulting in an almost infinite possible number of amino acid sequences.

The structure of proteins is clearly more complex than the structure of polysaccharides and lipids. Therefore, biochemists refer to different levels of structure – primary, secondary, tertiary and quaternary – in order to describe how a series of amino acids in a polypeptide chain becomes a protein.

THE PRIMARY STRUCTURE OF PROTEINS

Amino acids can be joined together by **peptide bonds**. Joining two amino acids together requires a condensation reaction, with the loss of a water molecule. The product is a molecule called a **dipeptide**. This reaction is shown in Figure 2.32.

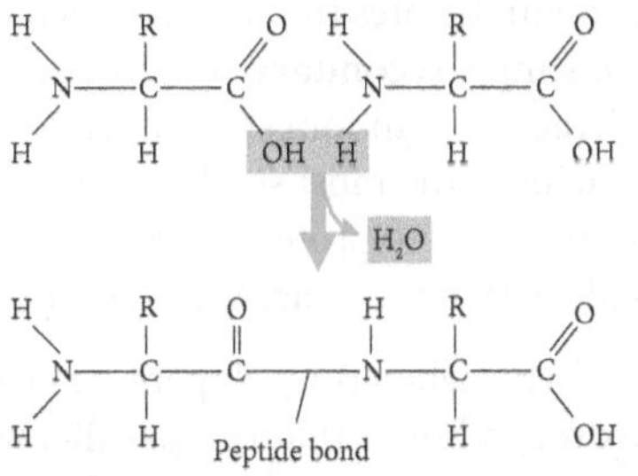

Figure 2.32 Two amino acids with any R groups can be joined in a condensation reaction, with the removal of water, to form a dipeptide. The bond between the two residues is called a peptide bond.

The addition of another amino acid in the same way to either end of the dipeptide would produce a tripeptide, and so on, until a polypeptide is formed. The **primary structure** of a protein is therefore not really a structure at all. It is simply a description of the sequence of amino acid residues in the polypeptide chain. However, this is fundamental to the structure of the protein as it determines how it takes shape. Figure 2.33 shows part of the amino acid sequence of human insulin, a relatively small protein of 51 amino acids.

Glu	Ala	Leu	Tyr	Leu	Val	Cys	Gly	Glu	Arg	Gly	Phe

Figure 2.33 A section of the primary structure of the human insulin protein. The three-letter codes represent different amino acids, as listed in Figure 2.31.

Tip

No matter how many amino acids there are in the polypeptide chain, if there is a –COOH group at one end, there will be an –NH$_2$ group at the other end.

Tip

When you draw a biological molecule, it is useful to remember the C-4, N-3, O-2, H-1 rule – check that you have drawn the appropriate number of bonds for each atom.

Link

The primary structure of any protein is determined by at least one gene. As you will see in Chapter 6, changing just one amino acid can significantly alter a protein's function.

Worked example

Figure 2.34 shows the structure of a dipeptide consisting of the amino acids serine (first residue) and cysteine (second residue).

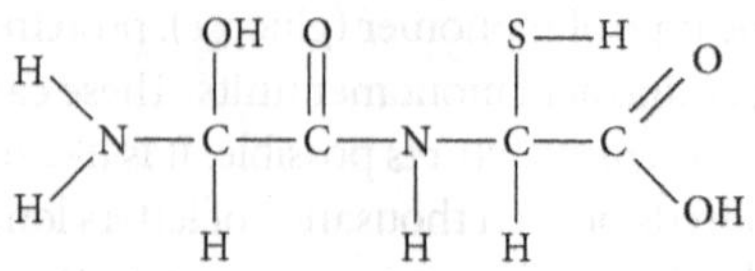

Figure 2.34

Using your knowledge of amino acids and how a hydrolysis reaction occurs, use Figure 2.34 to show how this dipeptide can be broken down into its two amino acid monomers.

Answer

When a molecule of water reacts to the dipeptide during hydrolysis, the peptide bond is broken, releasing two amino acids as shown in Figure 2.35. You can check the answer by referring to the amino acid structures in Figure 2.31.

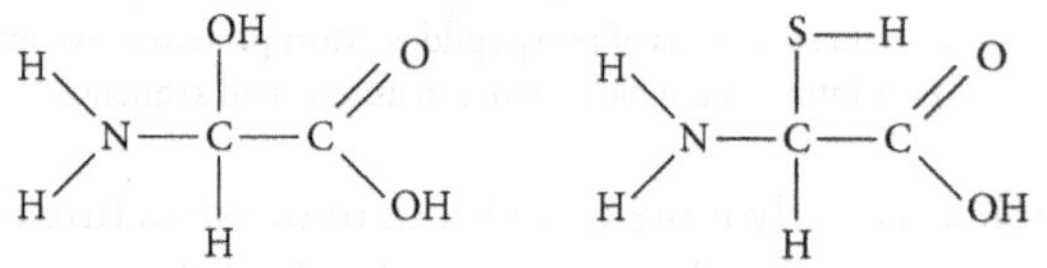

Figure 2.35

12. Explain in words, without drawing a diagram, how a peptide bond is formed between two amino acids.
13. Draw a detailed diagram to show how the amino acids glycine and alanine join together in a condensation reaction. Refer to Figure 2.31 to help you.

THE SECONDARY STRUCTURE OF PROTEINS

The primary structure of a protein dictates the way in which the linear chain of amino acids twists and turns to form simple **secondary structures**.

Secondary structures are irregular, random structures. Residues in the polypeptide chain interact with others to achieve the most stable arrangement of hydrogen bonds. These exist in different places in different proteins, producing an almost infinite variety of molecular shapes. Two secondary structures are very common:

- The **alpha-helix (α-helix)** forms when the polypeptide chain coils round in a tight spiral, held in place by hydrogen bonds that run parallel with the long helical axis. Hydrogen bonds form between the hydrogen atom of the amine group of one amino acid, and the oxygen atom of a carboxyl group of another amino acid further along the chain, as shown in Figure 2.36a.
- The **beta-pleated (β-pleated) sheet** is another secondary structure found in most proteins. It consists of a flat structure in which two or more amino acid chains run parallel to each other, in a zig-zag pattern, linked by hydrogen bonds (Figure 2.36b).

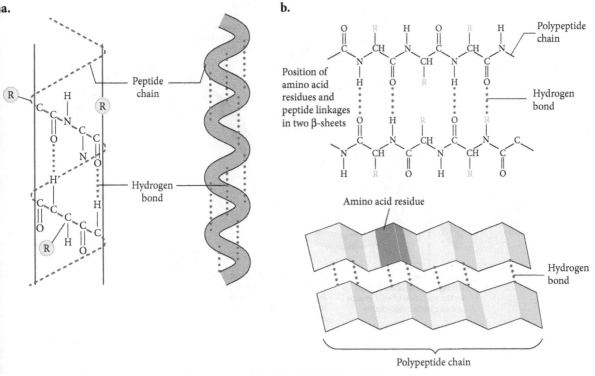

Figure 2.36 The two secondary structures of proteins. The alpha-helix (α-helix) (a) and the beta-pleated (β-pleated) sheet (b).

THE TERTIARY STRUCTURE OF PROTEINS

So far, we have seen how a chain of amino acids – a polypeptide – has taken on just two simple structures, α-helices and **β-pleated sheets**. However, the folding of the polypeptide chain is much more complex than this. Each type of protein has its own particular and unique **tertiary structure**, and this is crucial to its function.

To take on a tertiary structure, the secondary structures of a polypeptide chain usually fold again and interact with each other, forming all kinds of complex three-dimensional (3-D) shapes. These shapes are held in place by four different types of chemical bond. These usually form between the R groups of the amino acids. The bonds are shown in Figure 2.37, and their properties are summarised in Table 2.15.

Tip

Do not confuse the alpha-helix of proteins with the double helix of DNA – helices are found in a wide variety of biological molecules.

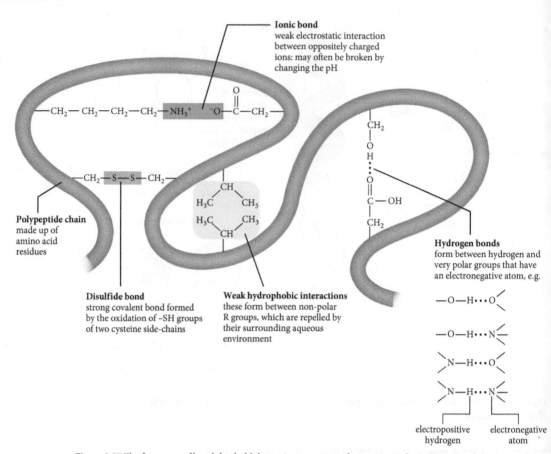

Figure 2.37 The four types of bond that hold the tertiary structure of a protein, its three-dimensional shape, together.

Type of bond	Formed between
hydrogen bonds	H atom and an electronegative atom, usually O or N
ionic bonds	oppositely charged ions
van der Waals forces	non-specific, between nearby atoms
disulfide bonds	two SH-containing cysteine residues
hydrophobic interactions	R group of some amino acids and surrounding water

Table 2.15 The bonds found in the tertiary structure of many proteins. These help to hold the three-dimensional form of the protein in a precise shape. Note that the only strong bond is the disulfide bond, which is covalent.

Link

The loss of a tertiary structure of a protein causes a loss of its function. In Chapter 3, we will see how this affects the function of enzymes, which are proteins.

The tertiary structure of a protein has a very important role to play in the function of that protein. The large number of **disulfide bonds** in keratin, for example, makes body structures such as hair, nails and even horns very tough (Figure 2.38).

Figure 2.38 The oryx of East Africa and the Arabian Peninsula. Like all mammals, keratin is found in their hair, wool, nails and the surface layers of the skin. Keratin forms long molecules that coil and associate with each other via disulfide bonds to form filaments. The protein can be strong enough to make sword-like horns, strong enough to kill a competitor.

> **Tip**
>
> When amino acids join together, a linear (straight-chain) 'polypeptide' is formed. Only after the molecule takes on a functional, three-dimensional shape can we use the term 'protein' to refer to it.

THE QUATERNARY STRUCTURE OF PROTEINS

If a protein consists of only one polypeptide chain, the tertiary structure is the final, functional, shape of the molecule. Some proteins, however, consist of more than one polypeptide. These are said to have a **quaternary level of structure**.

Many proteins are made up of more than one polypeptide chain. **Haemoglobin** is a good example. It contains four polypeptide chains: two of one type (called α-chains) and two of another type (called β-chains). We will explore haemoglobin more in the next section of this chapter. Insulin, the protein hormone involved in regulating blood glucose concentration, is shown in Figure 2.39.

> **Tip**
>
> Quaternary structures do not mean that the protein has four polypeptides; other examples of proteins with quaternary structure include collagen (three polypeptides) and insulin (two polypeptides).

a.

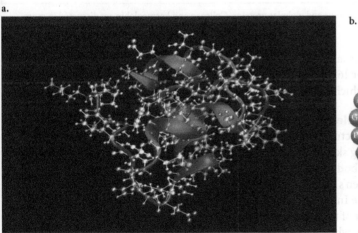

b.

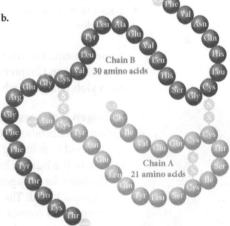

Figure 2.39 Biochemists use computer software to work out the shape of proteins from their amino acid sequences. Here is a computer-generated model of an insulin molecule, which regulates blood glucose concentration (**a**). It has two polypeptide chains, chain A and chain B, shown in blue and red. The individual atoms of the amino acids are shown as balls joined by sticks. Three purple straws represent disulfide bonds helping to hold the two polypeptides in a precise shape. A two-dimensional representation of the primary structure is given in (**b**), with the three disulfide bonds shown.

The different polypeptides in a protein with quaternary structure are held together by the same bonds illustrated and described in Figure 2.37 and Table 2.15.

Tip

The function of a protein depends on its tertiary structure. This is determined by its primary structure, which determines where the four types of bonds can form.

FIBROUS AND GLOBULAR PROTEINS

Every protein belongs to one of two categories. These differ on the basis of their final three-dimensional structure. **Fibrous proteins** contain polypeptides that bind together to form long, narrow fibres or sheets. **Globular proteins** are compact and approximately spherical, or 'ball-shaped'. Other features of globular and fibrous proteins are compared in Table 2.16.

Fibrous proteins	Globular proteins
polypeptides bind together to form long fibres or sheets	spherical / ball-shaped
insoluble in water, because hydrophobic R groups point outwards to interact with the surrounding water molecules	soluble in water, because hydrophobic R groups point inwards to face away from the surrounding water molecules
usually consist of proteins with only secondary structure, and only rarely a tertiary and quaternary structure	consist of proteins with a tertiary structure and sometimes a quaternary structure
physically tough and have a structural function	function in metabolic reactions (have physiological roles)
repetitive amino acid sequence with hydrophobic R-groups on the outside of the molecule	irregular amino acid sequence with hydrophilic R-groups on the outside of the molecule
less sensitive to changes in pH and temperature	more sensitive to changes in pH and temperature
examples include collagen (bone), keratin (hair), tubulin (cytoskeleton), elastin (lungs) and actin (muscle)	examples include haemoglobin, membrane proteins, receptors and all enzymes

Table 2.16 The properties of fibrous and globular proteins. Collagen and haemoglobin are discussed in more detail below.

Some proteins are unusual in that they can have both fibrous and globular sections: as you will see in Chapter 16, the muscle protein myosin, for example, has a long fibrous tail but a globular head, which acts as an enzyme.

Link

Collagen is found in the walls of arteries, as you will see in Chapter 8. This helps these blood vessels resist tearing as the blood pressure rises with each heartbeat.

Collagen – an example of a fibrous protein
Collagen, a fibrous structural protein, gives strength to a number of materials in the human body, including skin, bone, teeth, tendons and cartilage. About a quarter of all protein in the human body is collagen.

The basis for collagen's high tensile strength, or ability to stretch without breaking, is how it is packaged. The Indian physicist G.N. Ramachandran showed in the 1950s that **collagen fibres** consist of three **collagen polypeptides**. Each of these contains at least a thousand amino acids, which have been intertwined to form a triple helix held together by hydrogen bonds. Often the fibres are packed one on top of each other together to make near-rigid rods, as shown in Figure 2.40. If the process of collagen formation is faulty, as you will see in the Assignment, serious disorders can result.

a.

Collagen fibres

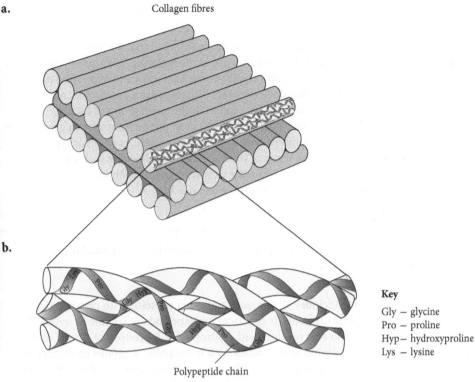

b.

Key
Gly — glycine
Pro — proline
Hyp— hydroxyproline
Lys — lysine

Polypeptide chain

Figure 2.40 The structure of collagen. The primary structure of the collagen polypeptide is very regular, consisting of a repeating sequence of glycine and two other amino acids, usually proline and hydroxyproline. This arrangement prevents the polypeptide from taking on the usual α-helix and β-pleated sheet. Every third amino acid is glycine. The R group of glycine is just a single hydrogen atom, whose small size allows the three polypeptide chains in a molecule to pack together very tightly into a compact rope-like structure.

Haemoglobin – an example of a globular protein

Haemoglobin is an example of a globular protein. The function of haemoglobin is the transport of oxygen from the lungs to respiring tissues. It is found inside red blood cells.

Haemoglobin has a roughly spherical shape, like all globular proteins, but it is unusual in that it has a quaternary structure. As we saw earlier, it consists of four polypeptide chains, two of one type (called **α-chains**) and two of another (called **β-chains**). Each of these polypeptide chains is attached to a non-protein **prosthetic group**, called a **haem group**, which contains an iron (Fe) ion (Figure 2.41).

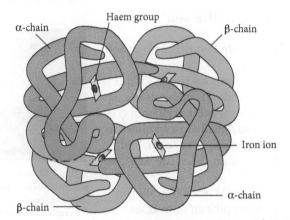

Haem group
α-chain
β-chain
Iron ion
α-chain
β-chain

Figure 2.41 The quaternary structure of the haemoglobin protein. The α-chains and β-chains are sometimes called α-globin and β-globin polypeptides.

The structure of haemoglobin gives it some very important properties, which are vital for its function:

- **High solubility:** the four polypeptide chains are coiled up so that the hydrophilic R groups on the residues of each are on the outside of the molecule. They can therefore form hydrogen bonds with water molecules. Hydrophobic R groups are mostly found inside the molecule, which means that they do not interact with the aqueous environment.
- **Ability to bind oxygen:** the iron (Fe) ion, found in the haem group of each polypeptide chain, can combine with oxygen. Because there are four haem groups, one haemoglobin molecule can combine with four oxygen molecules (eight oxygen atoms).
- **Ability to release oxygen:** the shape of the haemoglobin molecule is able to change to better enable it to pick up oxygen when the oxygen concentration in its surroundings is high, and to release oxygen more readily when it is low.

Link

The mechanism by which oxygen is loaded to and unloaded from haemoglobin is explored in greater detail in Chapter 8.

Worked example

With reference to named examples, explain the differences between globular proteins and fibrous proteins.

Answer

Globular proteins, which includes all enzymes and other proteins with physiological roles such as haemoglobin, are roughly spherical in shape and are soluble in water. This is because they have amino acid residues with hydrophobic R groups positioned on the inside of the protein, and amino acid residues with hydrophilic R groups positioned on the outside of the protein that interact with the aqueous environment. Fibrous proteins, which include proteins such as keratin and collagen that have structural functions, are insoluble in water. This is because they have amino acid residues with hydrophobic R groups positioned on the outside of the protein.

14. Compare what is meant by the terms 'polypeptide' and 'protein'.
15. Explain how the tertiary structure of all enzymes results in them being globular.
16. **a.** Explain in detail how the primary structure of a protein determines its function.
    b. With reference to Figure 2.31, identify which amino acids could form disulfide bonds.
    c. Suggest why proteins have a much greater variety of purposes than carbohydrates in living organisms.
17. A substance called β-mercaptoethanol breaks disulfide bonds. Suggest reasons to explain why β-mercaptoethanol will not affect the function of all proteins.

Assignment 2.1: Scurvy − the 'scourge of the sea'

Use the information in the passage below, and your own knowledge, to answer the questions that follow.

Fruit and vegetables are a vital part of a balanced diet. One reason for this is that many are rich sources of vitamin C. The body needs this molecule to keep our bones, skin, tissues and blood vessels healthy. This is because vitamin C is required for the conversion of proline, an amino acid, into hydroxyproline, an amino acid derivative. Hydroxyproline is a key component of collagen. Early symptoms of scurvy, a disease caused by a lack of vitamin C, include skin that bruises easily and bleeding gums (Figure 2.42).

Scurvy used to be common in many parts of the world. Famously, it often affected sailors, who could sometimes spend many months on ships at sea without fresh fruit and vegetables to eat. In 1752, the British doctor James Lind

Figure 2.42 An early symptom of scurvy is bleeding gums. The red colour of blood is due to the high concentration of the haemoglobin protein in red blood cells.

found that eating citrus fruit cured many of the symptoms of scurvy sufferers. However, scurvy rates remained high in Britain and many countries into the 20th century, partly due to the high cost of fresh fruit and vegetables.

QUESTIONS

A1. Haemoglobin and collagen are very common proteins in the human body.
   a. Compare the chemical elements found in haemoglobin and collagen.
   b. Explain why collagen is classified as a fibrous protein.
   c. List the similarities **and** differences between the tertiary structures of collagen and the structure of haemoglobin.

A2. Figure 2.43 shows the structures of the amino acid proline and hydroxyproline, the amino acid derivative that is made by the body.

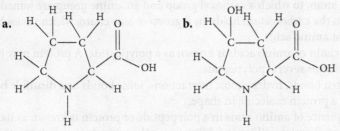

Figure 2.43 The amino acid proline (a) and the amino acid derivative hydroxyproline (b).

> **Tip**
> Although there are 20 amino acids found in most human proteins, a small number of amino acid derivatives can be made by chemically modifying them.

Proline is an unusual amino acid because the R group has formed a bond with the amine group.

   a. Draw a diagram to show how proline and hydroxyproline are joined together to form a dipeptide. You should indicate the position of the peptide bond.
   b. Explain why the formation of a dipeptide is described as a condensation reaction.

A3. Refer to the structure of collagen shown in Figure 2.40 earlier in this chapter. Explain in detail why a lack of vitamin C results in skin that bruises easily and bleeding gums.

A4. A pharmaceutical company suggested that the introduction of vitamin C to formula milk fed to babies explains why scurvy rates in England reduced during a period in the 20th century. Figure 2.44 shows a graph that the company published which they argued supports this claim. Table 2.17 shows the population of England at various years during this time period.

> **Tip**
> To find the percentage change in samples of data over time, work out the difference between the two values, divide the difference by the initial value, then multiply by 100.

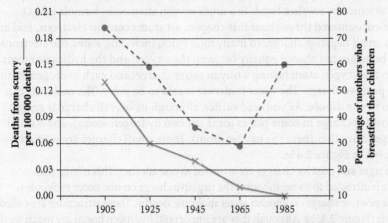

Figure 2.44 The number of deaths from scurvy during the 20th century in England and the proportion of English women who breastfed their children during the same period.

Year	Population
1911	3.36×10^7
1931	3.74×10^7
1939	3.81×10^7
1951	3.87×10^7
1961	4.12×10^7
1971	4.35×10^7
1981	4.60×10^7
1991	4.82×10^7

Table 2.17 The population of England during the 20th century.

a. Calculate the percentage change in the population of England from 1911 to 1991. Give your answer to one decimal place.

b. Calculate the percentage change in the percentage of mothers who breastfed their children during 1905 to 1965 shown in this study. Give your answer to one decimal place.

★ c. Using all of the information provided, evaluate the claim made by the pharmaceutical company that the introduction of vitamin C to formula milk fed to babies explains why scurvy rates in England reduced during this time period.

Key ideas

→ Proteins are polymers made from amino acids joined by peptide bonds.

→ There are 20 different amino acids found in living organisms. Each has a central carbon atom, to which a carboxyl group and an amine group are joined. The fourth bond on the carbon atom holds an R group or side chain, which varies between different amino acids.

→ A long chain of amino acids is known as a polypeptide. A protein may be made up of just one or several polypeptides.

→ Hydrogen bonds, hydrophobic interactions, ionic bonds and disulfide bonds help to hold a protein molecule in shape.

→ The sequence of amino acids in a polypeptide or protein is known as its primary structure. Regular coiling or folding forms the secondary structure, and further folding forms the tertiary structure. The association of more than one polypeptide forms the quaternary structure.

→ The shape of a protein is strongly related to its function. Globular proteins are usually soluble and often have roles in metabolism. Fibrous proteins are insoluble and generally have structural roles.

2.6 Water

Life on Earth probably evolved in the water, and the majority of the total mass of living organisms is water. Simply put, without water, there would be no life on Earth. It is a special molecule with extraordinary properties.

To understand the importance of water to life on Earth, we need to understand the structure of its molecules and how they behave with other molecules. A water molecule is made up of two hydrogen atoms joined by covalent bonds to a single oxygen atom, so its formula is H_2O.

As we have encountered throughout this chapter, all atoms contain electrons, and an electron has a small negative charge. In many molecules, including water, the electrons in the chemical bond are not shared equally between the oxygen and the hydrogen atoms. This results in the oxygen atom having a tiny negative charge, and each hydrogen atom having a tiny positive charge. The water molecule is said to be polar. The molecule can also be said to have a **dipole**. As you read earlier, although its overall charge is zero, it has a small positive charge in some places (near the two hydrogen atoms), and a small negative charge in another (near the oxygen atom). These small charges are written $\delta-$ and $\delta+$, as shown in Figure 2.45a.

Negative charges and positive charges are attracted to one another. This means that each water molecule is attracted to its neighbours. The negative charge on one water molecule is attracted to the positive charge on another water molecule nearby. These attractions are called hydrogen bonds (Figure 2.45b). Although they are important, hydrogen bonds are much weaker than the strong covalent bonds that hold the oxygen and hydrogen atoms firmly together.

Tip

Other molecules are also polar, especially those that have −OH, −C=O or −NH groups. Water can form hydrogen bonds with these, and these groups can form hydrogen bonds with each other.

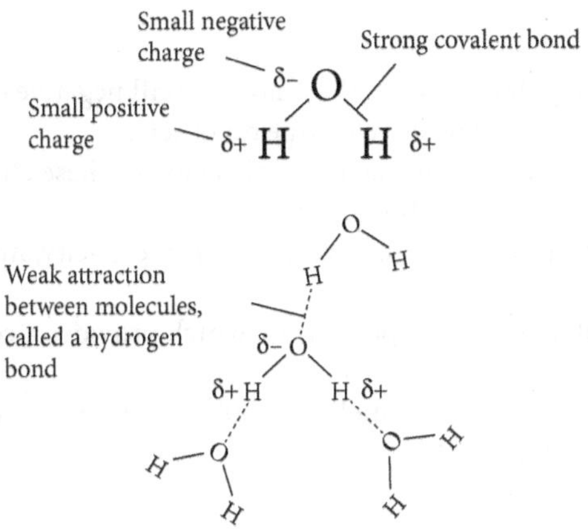

Link

In Chapter 7, you will see how the hydrogen bonds between water molecules promote cohesion and adhesion, which enables water to move up plant stems against the force of gravity.

Figure 2.45 Water molecules are polar and have the ability to form hydrogen bonds with each other and other molecules.

The hydrogen bonds between molecules of water are responsible for a wide range of its properties, summarised in Table 2.17.

Property	Explanation	Relevance to living organisms
Excellent solvent action	Water dissolves solutes – it is an excellent solvent. This is because of its polar nature. The δ– and δ+ charges on the water molecules mean that charged particles (for example, the ions in sodium chloride) and the polar regions of some larger molecules (e.g. glucose) are surrounded by a 'shell' of water molecules. This causes solids to spread out among the water molecules and dissolve into a solution.	Solute particles, when dissolved, are free to move around and to react with other particles that are also in solution. Metabolic reactions only take place because the reactants are aqueous, because they are free to collide and react with other particles. Water's solvent properties are also important in transporting substances around and out of organisms, whether that be the blood plasma or urine in animals, or the solutions carried by the xylem and phloem vessels of plants.
High specific heat capacity	Water has a high **specific heat capacity**. This is the amount of heat energy that has to be added to a given mass of a substance to raise its temperature by 1 °C. Significant heat energy is required to increase the temperature of water to break the hydrogen bonds between water molecules, separating them from each other.	Water helps to keep temperatures stable. Because most organisms are mostly made of water, the water in and around cells absorbs a lot of heat energy without its temperature rising very much. This is also very helpful to aquatic organisms. Large bodies of water, such as the sea or a lake, do not change temperature as quickly or as greatly as the air.
High latent heat of vaporisation	The energy that is lost from liquid water as it evaporates, to form a gas, is called the **latent heat of vaporisation**. This value is high for water, so the evaporation of small quantities of water has a large cooling effect.	Water helps to keep organisms cool. Sweat, which is mostly water, lies on the skin surface and evaporates. As it evaporates, the skin surface is cooled. Plants, too, are cooled down when water evaporates from their leaves during transpiration.

Table 2.18 The importance of hydrogen bonds to the properties of water. The ability of water molecules to attract each other explains many of the properties that make it vital to living organisms.

18. a. List how the following properties of water are vital to living organisms.

    i. It is an excellent solvent.

    ii. It has a high specific heat capacity.

    iii. It has a high latent heat of vaporisation.

  ⭐ **b.** Explain how hydrogen bonding between water molecules offers an explanation for each of these properties.

Key ideas

→ Water molecules are polar – that is, they have a small negative charge on some parts of the molecule and a small positive charge on others.

→ Water molecules are attracted to each other because of these charges. The weak attractions are called hydrogen bonds.

→ Hydrogen bonding causes water to have a high heat capacity, minimising temperature changes.

→ Water has a high latent heat of vaporisation, which provides a cooling effect when it evaporates.

→ Water is an excellent solvent, in which metabolic reactions can take place and substances can be transported.

CHAPTER OVERVIEW

Review the mini mind map to link new learning with your prior learning, and to highlight which of the key concepts underpin the chapter.

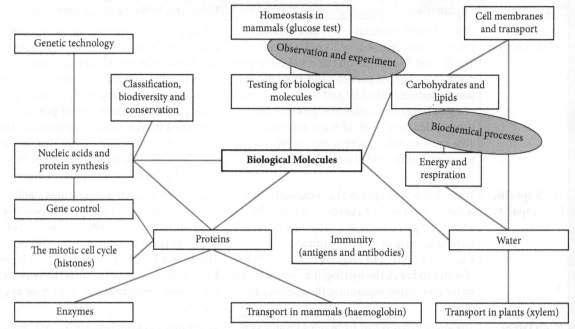

Try copying this mini mind map and expanding upon it. Use your notes from other chapters to help you explore how the essential ideas, theories and principles can be linked further together.

WHAT YOU HAVE LEARNED

Now that you have finished this chapter you should be able to:

- describe and carry out the Benedict's test for reducing sugars, the iodine test for starch, the emulsion test for lipids and the biuret test for proteins
- describe and carry out a semi-quantitative Benedict's test on a reducing sugar solution by standardising the test and using the results (time to first colour change or comparison to colour standards) to estimate the concentration
- describe and carry out a test to identify the presence of non-reducing sugars, using acid hydrolysis and Benedict's solution

- describe and draw the ring forms of α-glucose and β-glucose
- define the terms 'monomer', 'polymer', 'macromolecule', 'monosaccharide', 'disaccharide' and 'polysaccharide'
- state the role of covalent bonds in joining smaller molecules together to form polymers
- state that glucose, fructose and maltose are reducing sugars and that sucrose and lactose are non-reducing sugars
- describe the formation of a glycosidic bond by condensation, with reference both to polysaccharides and to disaccharides, including sucrose
- describe the breakage of a glycosidic bond in polysaccharides and disaccharides by hydrolysis, with reference to the non-reducing sugar test
- describe the molecular structure of the polysaccharides starch (amylose and amylopectin) and glycogen and relate their structures to their functions in living organisms
- describe the molecular structure of the polysaccharide cellulose and outline how the arrangement of cellulose molecules contributes to the function of plant cell walls
- state that triglycerides are non-polar hydrophobic molecules and describe the molecular structure of triglycerides with reference to fatty acids (saturated and unsaturated), glycerol and the formation of ester bonds
- relate the molecular structure of triglycerides to their functions in living organisms
- describe the molecular structure of phospholipids with reference to their hydrophilic phosphate heads and hydrophobic fatty acid tails
- describe and draw the general structure of an amino acid and the formation and breakage of a peptide bond
- explain the meaning of the terms 'primary structure', 'secondary structure', 'tertiary structure' and 'quaternary structure' of proteins
- describe the types of interaction that hold these molecules in shape:
  - hydrophobic interactions
  - hydrogen bonding
  - ionic bonding
  - covalent bonding, including disulfide bonds
- state that globular proteins are generally soluble and have physiological roles and fibrous proteins are generally insoluble and have structural roles
- describe the structure of a molecule of haemoglobin as an example of a globular protein, including the formation of its quaternary structure from two alpha (α) chains, and two beta (β) chains and a haem group
- relate the structure of haemoglobin to its function, including the importance of iron in the haem group
- describe the structure of a molecule of collagen as an example of a fibrous protein, and the arrangement of collagen molecules to form collagen fibres
- relate the structures of collagen molecules and collagen fibres to their function
- explain how hydrogen bonding occurs between water molecules and relate the properties of water to its roles in living organisms, limited to solvent action, high specific heat capacity and latent heat of vaporisation.

CHAPTER REVIEW

1. Explain why a sample of food might give a negative result with the Benedict's test before acid hydrolysis, but a positive result after.
2. Draw a table to list the similarities and differences between monosaccharides and amino acids.
3. Chemical bonds are found in all biological molecules. Complete the table below by adding a tick (✔) or a cross (✘) to describe the types of bond found in different biological molecules.

Chemical bonds found in:	hydrogen bond	disulfide bond	peptide bond	α(1,4) glycosidic bond
Primary structure of a protein				
Tertiary structure of a protein				
Amylopectin				
Cellulose				

4. Most animals can digest starch, but only a few can digest cellulose. Suggest why.

5. Human hair is composed of keratin. With reference to the molecular structure of this protein, suggest why hair does not dissolve when it is washed.

6. By referring to examples of proteins in your answer, evaluate the following statement: 'In order to have a quaternary structure, a protein must first have a tertiary structure.'

7. Read the following passage and answer the questions that follow.

There are an estimated 450 million people with diabetes around the world. Of these, about 10 per cent have type 1 diabetes. As a result, about 4.5 million people need a daily supply of insulin.

Insulin is a hormone consisting of two short polypeptide chains attached together. However, it is still a complex molecule and far too big to be made in the laboratory. Insulin extracted from the pancreases of animals can be effective but most is now made by genetic engineering. A gene which codes for insulin is put into a microorganism such as a bacterium, which will then use its cellular machinery to synthesise the protein.

Recent advances have allowed scientists to produce a variety of different insulins: rapid-acting, short-acting, intermediate-acting and long-acting. None of these are the same molecule that is made by the human body. They are synthetic analogues, molecules that have the same basic mode of action as insulin but can be adapted to work on different timescales.

a. Explain how a molecule of insulin shows the four levels of organisation of protein molecules.

b. Identify the evidence in the passage that insulin is a globular protein?

c. Suggest why synthetic analogues of insulin must be very similar in structure to the insulin hormone in order to function normally.

Chapter 3

Enzymes

Plastic is a very useful material. It is insoluble, unreactive and durable. However, plastics take a very long time to biodegrade and can have devastating effects on the environment.

In 2016, scientists working in a Japanese waste disposal facility made a significant discovery. They found a bacterium that produces an enzyme able to break down the substance polyethylene terephthalate (PET), used in the manufacture of plastic bottles. The bacteria appeared to absorb the products of PET digestion as a food source. The hope is to further modify the enzyme so that it can break down discarded waste within hours, rather than days.

Some of the most potentially important discoveries in the history of science have been made by chance. Might this development offer a solution to tackle the global problem of plastic pollution?

Prior understanding

You may know that enzymes are proteins that act as biological catalysts. The sequence of amino acids comprises the primary structure of a protein, which determines their secondary, tertiary and quaternary structure. You may also remember the types of chemical bonds that hold these molecules in their characteristic shapes.

Learning aims

In this chapter, you will learn how enzymes control the rate of specific chemical reactions inside and outside cells. You will learn how, and why, enzymes bind to other molecules, and the effect that these interactions have on enzyme activity. You will see how the rate of enzyme-catalysed reactions can be measured in the laboratory, and how biotechnologists use enzymes in an immobilised state to produce useful products.

3.1 Mode of action of enzymes (Syllabus 3.1.1–3.1.4)

3.2 Factors that affect enzyme action (Syllabus 3.2.1–3.2.4)

3.1 Mode of action of enzymes

When asked to define 'enzyme', many will describe their role in the digestion of food. These extracellular enzymes function in the alimentary canal, outside of the cells that produced them. Salivary amylase, for example, begins the digestion of starch in the mouth, and protease enzymes in the stomach hydrolyse proteins. Some organisms secrete extracellular enzymes into their surroundings and absorb the molecules as their nutrients, including the plastic-digesting bacteria mentioned at the start of this chapter. However, enzymes are also found inside cells. As you saw in Chapter 1, lysosomes are vesicles that contain hydrolytic enzymes that digest debris and worn-out organelles in the cytoplasm. These are intracellular enzymes.

Enzymes were discovered by two brothers, Hans and Eduard Buchner, who were awarded the Nobel Prize for this discovery in 1907. They were successful in demonstrating that an extract of yeast cells could digest sucrose, a substrate, into alcohol and carbon dioxide, two products. The term 'enzyme', which means 'in yeast', was chosen to name these new-found substances that came from cells. Most enzymes have names that end in '–ase' (e.g. amylase and lipase).

It is not just yeast that has enzymes. Nor are they a group of substances that just break substrates into products. Figure 3.1 shows a small number of the countless chemical reactions, known collectively as metabolism, that occur in a cell. Some of these reactions release energy, some synthesise new substances and others break down waste products. Nearly all of them depend on the products of other reactions.

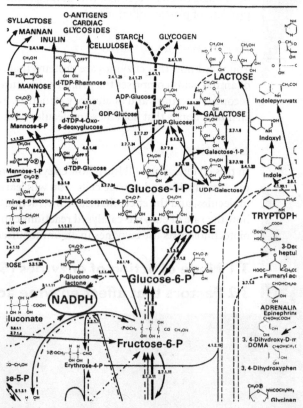

Figure 3.1 The cell is a busy place. The term 'metabolism' is given to the complex network of interconnected chemical reactions within a cell; a few of them are shown here. Each number corresponds to an enzyme that catalyses that reaction.

THE EFFECT OF ENZYMES ON CHEMICAL REACTIONS

One enzyme that may be familiar to you from previous work is **catalase**. Its substrate is hydrogen peroxide (H_2O_2), a colourless liquid that looks very similar to water. A flask of hydrogen peroxide gradually decomposes to form water and oxygen, but the process takes several months. The equation for this reaction is:

$$2H_2O_2 \rightarrow 2H_2O + O_2$$

However, in the presence of a few drops of catalase solution, the hydrogen peroxide begins to bubble vigorously. This produces a column of froth, which rises up the flask, as shown in Figure 3.2a. There are some living organisms that are able to use this chemical reaction as a useful defence strategy (Figure 3.2b).

a.

b.

Figure 3.2 The addition of plant tissue, which contains the enzyme catalase, to a flask of hydrogen peroxide increases the rate of decomposition, which causes a rapid release of heat, water and oxygen (a). The bombardier beetle is able to spray an attacking predator with the products of this reaction (b).

Catalase is an intracellular enzyme found in all living cells. A tissue sample from any organism will have a similar effect to a solution of pure catalase. The reason for this spectacular increase in reaction rate is that the enzyme increases the rate of the reaction. In this regard, enzymes are often described as biological **catalysts**. The substrate for catalase is hydrogen peroxide, while water and oxygen are the products of the reaction that the enzyme catalyses.

Most of the reactions that occur in living cells would occur so slowly without enzymes that they would virtually not happen at all. However, it is more accurate to say that enzymes control the rates of the interconnected network of reactions in a cell, to ensure that they happen at appropriate rates. This means that substrates are broken down, and products are made, in just the right amounts and at just the right moments.

A single catalase molecule can break down millions of molecules of hydrogen peroxide in one second. This is good news for cells: hydrogen peroxide, a by-product of respiration, is a powerful oxidising agent, which can damage molecules including DNA.

1. Distinguish between intracellular and extracellular enzymes.
2. A small change in the primary structure of an enzyme molecule can prevent it from working. Use your knowledge of protein structure to explain why.

Experimental skills 3.1: Comparing catalase concentrations

Two students conducted separate investigations into the catalase concentrations of a number of tissues taken from different plants.

Both students obtained an equal mass of samples from four plant tissues. After cutting the fresh tissue into small pieces with a knife, they blended them into a pulp.

Both students used hydrogen peroxide as a reactant in the investigation, with equal volumes in each experiment. Hydrogen peroxide decomposes into oxygen in an extremely exothermic reaction (a lot of heat is given out). They planned to measure the volume of oxygen released in 30 seconds, but they used different methods to collect the oxygen released. The first student used a gas syringe, as shown in Figure 3.3a. The second student used a method of gas collection that involved the downward displacement of water from an inverted measuring cylinder (Figure 3.3b).

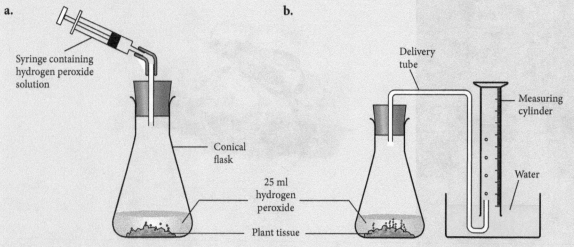

Figure 3.3 The rate of some enzyme-catalysed reactions can be measured by collecting the gas released. Here, oxygen is collected in a gas syringe (a) or by collection in an in an inverted measuring cylinder by the downward displacement of water (b).

QUESTIONS

P1. Identify the independent and dependent variables in this investigation.

P2. State the hazard with the greatest level of risk when using the apparatus and reagents in this investigation.

P3. The students each recorded their results on a scrap of paper, as shown in Figure 3.4. Draw a table for each student to present their data.

Figure 3.4 The results obtained by the students (the volume of oxygen collected in 30 seconds).

P4. Explain why a bar chart, rather than a line graph, would be the appropriate choice in this investigation to show the collected data.

P5. The teacher claimed that the students' results are likely to have a low reliability. Explain what this means.

P6. The students used the same mass of plant tissue and volume of hydrogen peroxide in their investigations. Assuming that all other variables were standardised and kept the same, suggest a reason for the difference in results between the two students.

P7. An alternative method to follow the progress of this enzyme-catalysed reaction would be to place the conical flask containing the hydrogen peroxide and plant tissue onto a top-pan balance. As the reaction progresses, there will be a change in mass, as shown in Figure 3.5. In this case, the teacher suggested that the students should measure the volume of gas released every 5 seconds and plot a line graph for each plant tissue.

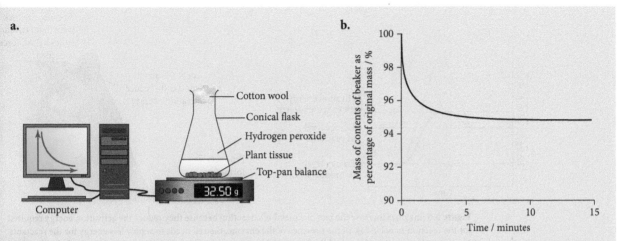

Figure 3.5 The rate of some enzyme-catalysed reactions can be found by measuring the mass of the reactant that remains by placing the reaction vessel onto a top-pan balance (a) and recording the loss of mass at regular time intervals (b).

a. Explain why the mass of the contents of the conical flask would reduce over time.

b. Give two reasons why this method would give more accurate results than the methods used by the students in their studies.

c. Describe the shape of the graph.

d. Explain why the mass of contents as a percentage of original mass was calculated, rather than total mass.

e. Explain how data-logging equipment using a computer, rather than manual reading of the top-pan balance scale, would improve the accuracy of the collected data.

So, enzymes are biological catalysts that increase the rate of metabolic reactions. But how do they do this? For any chemical reaction to take place, bonds must be broken in the reactant molecules before new ones can form. The energy needed to break these bonds, and so start the reaction, is called the **activation energy**. Usually, once the reaction starts, no further input of energy is required, and it continues until all reactants are gone. It is a bit like striking a match. Once an initial 'burst' of energy is provided, the match head burns by itself unaided and very quickly, until the chemicals on the head are used up.

Figure 3.6 shows the energy changes that take place during a typical chemical reaction. Before the reaction, the energy of the molecules is steady. Then something happens to raise the energy level – perhaps one molecule is hit by another very hard, or perhaps they are heated up. If the molecules reach the activation energy, the reaction then takes place: the bonds in the reactants are broken, which allows new bonds to form between products. The product molecules usually have a lower energy level. Hence the activation energy is significantly lowered in the presence of enzymes; this allows reactions to take place at the relatively low temperatures normally found in living organisms.

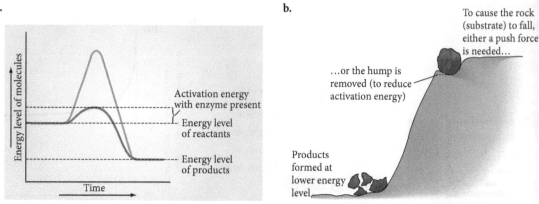

Figure 3.6 Enzymes increase the rate of a metabolic reaction because they reduce the activation energy required for the reaction to begin (a). In the presence of the enzyme, the cell needs to supply less energy for the reactants to react and become products (red line), than if the enzyme was absent (grey line). An analogy of what is going on is a rock (the substrate) perched on a slope, prevented from rolling down by a small hump (the activation energy) in front of it (b). The rock can be pushed over the hump. Alternatively, the hump can be dug away (the activation energy can be lowered), allowing the rock to roll and shatter at a lower level (into the products). In both cases, the energy content of the substrate and products is not changed; the pathway follows a lower energy course.

MECHANISMS OF ENZYME ACTION

As you saw in Chapter 2, almost all enzymes are proteins. They consist of at least one polypeptide chain that has folded to form a three-dimensional tertiary structure, roughly spherical, and globular in nature. Like all globular proteins, enzyme molecules have hydrophilic R groups (amino acid side chains) on the outside of the molecule, to ensure that they are soluble in water.

The shape of an enzyme is crucial to its function. Of greatest importance is the **active site** of an enzyme. This is a 'dent' or 'pocket', the shape of which is very specific and is **complementary** to the shape of the substrate, as shown in Figure 3.7. This is lined with R groups of around 5 to 10 very specific amino acids within the polypeptide chain.

> **Tip**
>
> Not all enzymes break down their substrates. For example, DNA polymerase catalyses the synthesis of larger molecules from smaller molecules.

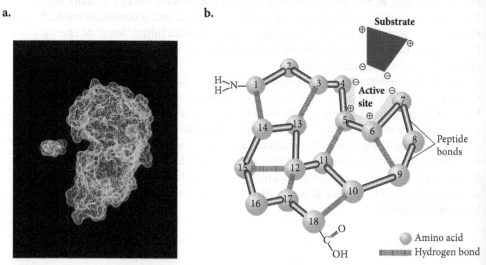

Figure 3.7 (a) The three-dimensional shape of ribonuclease A, an enzyme that helps break up a substrate called mRNA in bacteria. The active site, the 'pocket' into which the substrate fits, is clearly visible. (b) This schematic diagram shows the three-dimensional structure of an enzyme molecule. Hydrogen bonds are shown in red. The shape and the electrical charges on the substrate closely match those of the active site.

There are two hypotheses that have been used to help explain how enzymes reduce the activation energy of a metabolic reaction. These are called the **lock-and-key hypothesis** and the **induced-fit hypothesis**. Both are dependent on the specificity of the active site and its ability to attach to the substrate. Both models agree that once the substrate(s) are converted into product(s), these leave the active site of the enzyme. The enzyme remains chemically unchanged and is now ready to interact with the next substrate molecule.

The lock-and-key hypothesis of enzyme activity was first proposed in 1894. It suggests that the substrate of the enzyme has a shape that is complementary to the active site. That is, the size, shape and chemical nature of the active site correspond perfectly with those of the substrate molecule. Therefore, the molecules fit together like a key (the substrate) into a lock (the enzyme's active site). This is because the active site where the substrate molecule binds has a precise shape and distinctive chemical properties, namely the presence of particular chemical groups and bonds. This explains why different catalases from different organisms all catalyse the breakdown of hydrogen peroxide: their active sites are of a very similar shape to this substrate.

In a solution containing both enzyme and substrate molecules, both kinds of molecules are in constant motion. They often collide with one another. When a substrate molecule hits the active site on an enzyme molecule it interacts with the R groups of the amino acids lining this site. The substrate is held in place by temporary bonds that form between the substrate and some of the R groups of the enzyme's amino acids. The substrate slots into the active site and forms temporary associations with the R groups, including ionic bonds and hydrogen bonds. The temporary combination of enzyme and substrate is called an **enzyme–substrate complex**. This exists for the briefest of instants.

Enzymes do this by reducing the activation energy required for the reaction to take place. How? This very much depends on the type of reaction that the enzyme catalyses. If a bond in the substrate is to be broken, that bond might be weakened by the enzyme, making it more likely to break. Or, if a bond is to be made between two molecules, the two molecules can be held in exactly the right position and orientation and brought together, making the bond more likely to form. The enzyme can also make the local conditions inside the active site quite different from those outside (such as pH, water concentration, charge), so that the reaction is more likely to happen.

The lock-and-key model would suggest that the shape of the active site is fixed and that one enzyme can bind to only one substrate. However, experimental evidence in the early 20th century suggested that some enzymes can catalyse numerous reactions, involving a number of similar substrates. This led to the induced-fit hypothesis, represented in Figure 3.8, which was proposed in 1958. In this model, the active site is thought to be slightly flexible, rather than being rigid and of a fixed shape. The arrival of the substrate causes (induces) a small and temporary **conformational change** in the shape of the enzyme's active site that makes it exactly complementary in shape to that of the substrate molecule. This allows the substrate to bind with it to form an enzyme–substrate complex. The slight modifications to the tertiary structure are due to the R group of the amino acids at the active site of the enzyme, as the enzyme interacts with the substrate.

Tip

The active site of an enzyme and the substrate are specific shapes, but are not the same shape. Neither do they 'match'. They are said to be complementary.

Tip

The best analogy for the induced-fit hypothesis is how a glove can fit hands of many sizes, and 'moulds' around them. But the glove could never fit anybody's foot!

Link

Conformational changes, caused by binding to another molecule, are not limited to enzymes. You will see in Chapter 4 that this also happens to carrier proteins in the cell surface membrane.

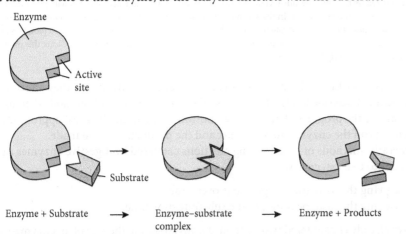

Figure 3.8 The induced-fit hypothesis. Before the substrate binds, the enzyme's active site appears to be a fixed shape. However, when the substrate binds, the active site is induced to change slightly by molecular interactions between the two molecules, and 'moulds' around the substrate to form an enzyme–substrate complex. As the products are released from the active site, the active site takes on its original structure.

INVESTIGATING THE PROGRESS OF AN ENZYME-CATALYSED REACTION

Earlier in this chapter, we learned that a number of methods can be used to investigate the progress of a reaction catalysed by the enzyme catalase. In most investigations, you can take measurements every 30 seconds, but you may need to have a partner to help. In addition, for reasons we saw earlier, initial rates should be calculated when comparing enzyme activity under different conditions.

In many investigations, it is necessary to convert a time value into a value for the rate of reaction. Consider the relationship between these two quantities. If an enzyme-catalysed reaction has a high rate, then a change (for example, the collection of a volume of gas) will happen in a shorter period of time. The opposite would be true for a reaction with a low rate. There are two ways to convert time values into rate values:

- plotting a graph of the change in mass or volume over time, and calculating the gradient of the line at any particular time value
- calculating values for rate of reaction by dividing 1 by the time value, and plotting these values on a graph.

Both options give a quantitative value with which to compare rates of reaction. The larger the value, the quicker the reaction. Worked examples are shown in Figure 3.9.

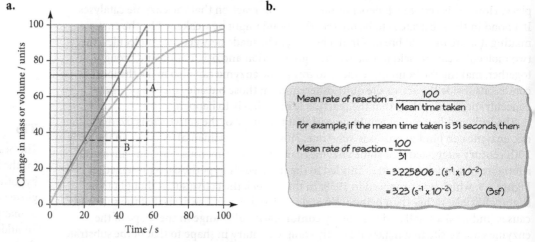

Figure 3.9 Two mathematical methods are commonly used to convert values of time to rate of change. In the first method (a), the gradient of the plotted line is calculated by drawing a tangent at the maximum slope of the plotted curved line, which forms a triangle whose hypotenuse is parallel to the required value on the *x*-axis. Dividing the distance covered by the opposite side (A) by the distance covered by the adjacent side (B) will give a value that is proportional to the rate of reaction. In the second method (b), data can be processed by calculating the reciprocals of their mean times as a measure of rate of reaction. To generate larger values that are easier to plot, you can multiply the reciprocal by 100 or even 1000 (100/time or 1000/time), provided that you indicate this on the axis with a label '×100' or '×1000'.

Much of what we know about enzymes comes from our study of how they affect the rate of biochemical reactions in the laboratory. There are a number of methods that are used to investigate the progress of an enzyme-controlled reaction. The most appropriate one to use depends on the enzyme, its substrate, and the products that are made.

Generally, methods of investigating reactions catalysed by digestive enzymes can be grouped into two categories:

- measuring the formation of products over time
- measuring the disappearance of the substrate over time.

We have already encountered two methods for investigating the rate of an enzyme-catalysed reaction (Figures 3.3 and 3.5). In both of these cases, the measurement depended on the release of gas (oxygen) from the reaction mixture. Its loss could be measured by collecting the gas produced, or by measuring the mass of substrate lost. However, not all enzyme-catalysed reactions release a gas as a product. Therefore, other indirect methods can be used to track the rate of reaction. Throughout the rest of this chapter, you will see how these methods have been used to investigate how enzymes catalyse a variety of reactions, and how a number of factors – temperature, pH and chemical inhibitors – affect enzyme activity.

3. Hexokinase is an enzyme that transfers a phosphate group to hexose (six-carbon) sugars, including glucose, fructose and galactose.

   a. With reference to activation energy, explain how hexokinase catalyses these reactions.

   b. Describe the difference between the lock-and-key hypothesis and the induced-fit hypothesis of enzyme action.

   c. Using the information above, explain why the action of hexokinase supports the induced-fit hypothesis.

4. Scientists recently discovered an enzyme in bacteria that digests plastic. It is hoped that one day this enzyme could be used to break down waste plastic to prevent its burial in landfill sites or disposal at sea.

   a. Is this likely to be an extracellular or intracellular enzyme? Explain your choice.

   b. The scientists have suggested that it may be possible to engineer bacteria that live in high-temperature environments to produce this enzyme. Suggest why this could be an advantage.

   ☆ c. Discuss the possible advantages and disadvantages of breaking down waste plastic with enzymes.

Key ideas

→ Enzymes are globular proteins that catalyse metabolic reactions.

→ Enzymes lower the activation energy of a reaction, making it possible for the reaction to happen at normal temperatures in a given organism.

→ For enzyme-catalysed reactions that involve the release of a gas, the rate of the reaction can be measured in the laboratory by measuring the volume of gas released, or loss in mass of the solution, in a given time.

→ An enzyme molecule has an active site into which its substrate molecule fits. This is called the lock-and-key hypothesis.

→ Enzymes can act only on their specific substrate, because only that substrate has the correct shape to bind perfectly with the enzyme's active site.

→ The arrival of the substrate molecule slightly changes the shape of the active site, so that the enzyme and substrate can bind together. This is called the induced-fit model.

→ The temporary combination of the enzyme and substrate is called an enzyme–substrate complex.

3.2 Factors that affect enzyme action

As you saw earlier, the properties of proteins help to explain how enzymes work by lowering the activation energy of a metabolic reaction. This can also explain how and why they respond to changes in environmental conditions and other molecules both inside and outside of cells.

A number of factors affect the ability of enzymes to catalyse metabolic reactions, including:

- temperature
- pH
- substrate concentration
- enzyme concentration
- the presence of inhibitors.

THE EFFECT OF CHANGES IN TEMPERATURE ON THE RATE OF ENZYME-CATALYSED REACTIONS

In many animals, maintaining a constant body temperature can be a matter of life and death. In humans, a fall in the core temperature of the body by a few degrees Celsius can be life-threatening. The same is true if the temperature rises above 40 °C. Temperature can upset the delicate balance of enzyme-catalysed reactions in cells.

Temperature is a key factor that affects the rate of enzyme-catalysed reactions. However, there are many reasons for this relationship, which can be investigated in the laboratory.

One investigation into the effect of temperature is the hydrolysis of a milk protein called casein by a protease enzyme. Proteases such as **trypsin** catalyse the hydrolysis of casein, which gives milk its white colour. This reaction is characterised by the decolourisation of the milk as the casein is hydrolysed into colourless peptides. This gives a way to estimate the rate of the reaction in a school laboratory (Figure 3.10). Carrying out this investigation at a range of different temperatures can illustrate the effect of temperature on an enzyme.

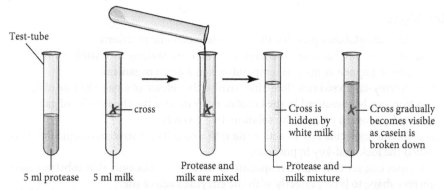

Figure 3.10 One method used to measure the rate of an enzyme-catalysed reaction. If the reaction involves the decolourisation of a substrate, the time taken for a cross drawn on the side of the reaction tube to appear can be measured. The shorter the time, the faster the reaction.

A sufficient number and range of temperatures should be chosen at which the reaction rate is measured. These values of the **independent variable** must be carefully chosen, as it will have an impact on the accuracy of a line graph that is drawn later. Related to this, apparatus is required that will maintain the reaction mixture at a particular and fixed temperature. Options include:

- a simple water bath, consisting of a beaker of water heated by a Bunsen burner on a tripod and gauze or an electrical hotplate; small volumes of cold or hot water can be added to finely adjust the temperature to keep it constant
- a thermostatically controlled water bath, which can maintain a constant temperature with much less variation than a manual water bath.

If investigating the effects of temperature, a value should be chosen for at least one of the water baths that is likely to be close to the temperature at which the enzyme normally works in nature. This is called the **optimum temperature** for the enzyme. In the case of human enzymes, around 37 °C should be used. Plant enzymes often have lower optimum temperatures, and enzymes from bacteria or fungi (the source of many commercially available enzymes) are often much higher.

In any investigation, it is important to ensure that the independent variable is the only factor that is changed. All other factors that may have an effect on the rate of reaction must be **controlled** (standardised), namely kept constant. For reasons we will see later in this chapter, these factors include pH, substrate concentration and enzyme concentration. pH can be kept constant by using a buffer solution, added to the reaction vessel, that maintains the pH at a given value. This is particularly important if the reaction may lead to a pH change. Standardising variables other than the independent variable ensures that we can make a **valid** comparison of the data collected.

A **control experiment** is also required in investigations involving enzymes. The purpose of a control experiment is to remove the effect of the independent variable, to ensure that any change in the rate of reaction is owing to the factor being investigated. In most investigations using enzymes, an experiment is carried out in which no enzyme is added. Even better, for reasons we will see later, it is better to use enzyme that has been boiled.

It is also important in an investigation to ensure that the collected data is **reliable**. This involves repeating the experiment for each value of the independent variable in order to calculate a mean value. Normally three to five trials are conducted for each value in an enzyme investigation. The benefit of this is that any **anomalous data** – measurements that are clearly different from those in other trials – stands out. Reliable data is said to be **replicable** – if another person were to undertake the investigation, they would obtain very similar results.

Finally, a good investigation involving enzymes must take steps to attempt to standardise the end-point to ensure that the measurements of the **dependent variable** are **accurate**. In this investigation, there would be significant **subjectivity** of the end-point judgements – what one person may judge to be the time taken for the cross to appear may not be the same for another person. Being careful to draw the cross on the test-tube rather than on a piece of card helps to ensure the cross is the same distance from the solution each time. Similarly, looking at the cross from the same angle and distance, and ensuring that the light is the same for each judgement, perhaps by illuminating the test-tubes the same way each time with a bench lamp, would improve the accuracy of the measurements.

Alternatively, to improve the accuracy of measuring the disappearance of a substrate over time, a **colorimeter** can be used (Figure 3.11). This piece of equipment provides a quantitative reading of the proportion of light absorbed by a liquid. The greater the absorbance, the more concentrated the pigment in the solution. The liquid is poured into a special rectangular tube called a cuvette and light is shone through. The light transmitted is read by a light-sensitive meter on the other side. Pure water, for example, would have an absorbance of zero and a transmission of 100%. The colorimeter reading is standardised ('zeroed') with distilled water before every reading. A colorimeter can be used, for example, to measure how much starch is left in a solution to which amylase and iodine–potassium iodide solution has been added. Over time, the blue–black colour will change to orange–brown.

Tip

Standardised variables are factors that must be kept constant in an investigation to ensure that conclusions drawn are valid.

Tip

A control experiment is carried out to ensure that the effect of changes in the independent variable (the factor that is changed) are responsible for changes that occur to the dependent variable.

Figure 3.11 A colorimeter and cuvettes.

The results of a study such as this are shown in Table 3.1 and are plotted on a line graph in Figure 3.12. Values for the rate of reaction, with units $s^{-1} \times 10^{-3}$, were calculated as described in Figure 3.9.

Temperature / °C	Time taken for cross to appear / s				Mean rate of reaction / $s^{-1} \times 10^{-3}$
	Trial 1	Trial 2	Trial 3	Mean	
5	88	96	130	104.7	9.6
25	54	51	52	52.3	19.1
45	22	19	20	20.3	49.2
65	163	160	154	159.0	6.0
85	NR	NR	NR	NR	0.0

Table 3.1 The results of an investigation into the effect of temperature on the rate of casein hydrolysis by a protease. Notice how the same number of decimal places are provided for each data point for the mean time and mean rate of reaction, even for whole numbers. NR = no reaction.

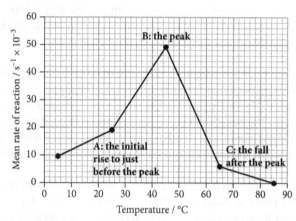

Figure 3.12 The results of an investigation into the effect of temperature on the rate of casein hydrolysis by a protease.

Tip

You need to recognise when it is appropriate to join the points on a graph with straight rules lines, and when it is appropriate to use a line (straight or curved) of best fit.

Tip

This graph shows temperature, not time. Do not use 'time-related' words when describing it. Refer to steep changes or gradients of the line, not words such as 'quick' or 'gradual'.

Tip

Enzymes are not 'killed' by high temperatures – they were never 'alive'. Neither are they 'destroyed'. The protein that was the enzyme still exists, but in an unfolded, inactive form.

This investigation shows clearly that a change in temperature has an effect on the rate of an enzyme-catalysed reaction using molecules dissolved in solution. But why? The descriptions and explanations below provide an analysis.

A. As the temperature of this solution increases, the kinetic energy of the molecules increases too. They move around with greater speed and are more likely to collide with enough activation energy and to form enzyme–substrate complexes. This is because the enzyme and substrate molecules both have more kinetic energy so collide more often, and also because more molecules have sufficient energy to overcome the (greatly reduced) activation energy. At temperatures before the peak in the graph, there is a steep increase in the rate of the reaction.

B. As the temperature increases, the tertiary structure of the enzyme begins to change as the atoms and bonds within its structure gain kinetic energy. At a particular temperature, called the optimum temperature, the tertiary structure of the enzyme is such that the active site is perfectly suited to the shape of the substrate. It is at this temperature that the likelihood of an enzyme–substrate complex forming is highest and therefore the highest rate of reaction is recorded.

C. At temperatures higher than the optimum temperature of an enzyme, there is usually a steep decrease in the rate of reaction. The enzyme molecules gain energy and vibrate so rapidly that the weak hydrogen bonds and ionic forces of attraction that help to maintain the tertiary structure of the protein molecule break and the molecule can become irreversibly distorted. Since the shape of the active site in an enzyme molecule is crucial for it to work, any change in shape will inactivate the enzyme and prevent it from binding to its substrate. Once broken, the hydrogen bonds generally do not re-form in their original positions. The enzyme is said to be **denatured**. This happens for most enzymes at temperatures above 45 °C.

Remember from Figure 3.1 that metabolic reactions are not isolated. The change in activity of one enzyme can be affected by other reactions. The clear effect of a change in temperature on enzyme activity suggests an explanation for the importance of maintaining a physiological temperature in many organisms, including humans.

Worked example

Sucrase digests sucrose to the reducing sugars glucose and fructose. Suggest how researchers could estimate the rate of activity of sucrase.

Answer
The researchers should take samples of the reaction mixture at set time intervals and perform the Benedict's test on them. They should note the colour change that occurs (from blue to green/yellow/orange/red), using a colorimeter/colour standards.

Tip
The rate is not zero at 0 °C, so enzymes still work in the fridge (and food still spoils over time), but they work slowly.

5. A student says that 'all bonds in a protein are broken when it is denatured'. Evaluate the accuracy of this statement.

6. An experiment was carried out to investigate the effect of temperature on the hydrolysis of starch. The presence of starch can be tested for using iodine–potassium iodide solution. Iodine–potassium iodide solution is orange–brown and turns blue–black in the presence of starch. The experiment measured the time taken for the starch in a solution to be completely hydrolysed. The results are shown in Table 3.2.

| Time / | Iodine test results | | | | |
minutes	5 °C	25 °C	40 °C	55 °C	70 °C
0	blue–black	blue–black	blue–black	blue–black	blue–black
1	blue–black	blue–black	orange–brown	blue–black	blue–black
2	blue–black	blue–black		blue–black	blue–black
3	blue–black	blue–black		blue–black	blue–black
4	blue–black	orange–brown		blue–black	blue–black
5	blue–black			blue–black	blue–black
6	blue–black			orange–brown	blue–black
7	blue–black				blue–black
8	orange–brown				

Table 3.2 The results of an investigation into the effect of temperature on the rate of starch hydrolysis catalysed by amylase.

a. Explain why the enzyme and substrate solutions should be incubated in the water bath for 5 minutes before being mixed together.

b. Explain why an identical volume of iodine–potassium iodide solution reagent was added to each sample.

☆ c. Iodine–potassium iodide solution can interfere with the formation of enzyme–substrate complexes. Suggest why the iodine–potassium iodide solution is carried out on samples taken from the reaction mixture, rather than directly on the reaction mixture.

☆ d. Convert the data in Table 3.2 into a rate of reaction for each temperature. Record the data in a new table.

e. Describe one way in which you could change the method to determine a more accurate value for the optimum temperature.

f. Without using the term 'denature' in your answer, explain the effect of incubating the enzyme at a temperature of 70 °C in this study on the rate of the catalysed reaction.

g. Describe why amylase, like most enzymes, will not significantly catalyse reactions at very low temperatures.

7. In an investigation into the effect of temperature on the rate of the activity of a protease enzyme, a scientist collected the data shown in the graph in Figure 3.13.

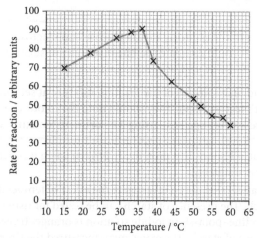

Figure 3.13 The effect of temperature on the rate of a reaction catalysed by a protease enzyme.

a. The effect of temperature on the rate of a biochemical reaction can be described by the term 'temperature coefficient', Q_{10}. This indicates the factor by which the rate of reaction is increased by increasing the temperature by 10 °C. It can be found using the following formula:

$$Q_{10} = (\text{rate of reaction at } T \,°C + 10 \,°C) / \text{rate of reaction at } T \,°C$$

where T = temperature.

i. Calculate the Q_{10} values of the enzyme used in this study between 20 °C and 30 °C, and between 25 °C and 35 °C.

ii. Explain any difference in the values you calculated in part (i).

b. Describe in detail the effect of temperature on the percentage activity of the protease.

☆ c. Explain why the rate of reaction does not fall to zero after the optimum temperature.

THE EFFECT OF CHANGES IN PH ON THE RATE OF ENZYME-CATALYSED REACTIONS

Enzymes have an **optimum pH**, as well as an optimum temperature. Changing pH either side of the optimum affects enzyme activity, causing changes in the shape of the active site and therefore the rate of the catalysed reaction.

Most enzymes are denatured when placed into solutions that are very different from their optimum. This is because the hydrogen ions (H^+) in an acid or the hydroxyl ions (OH^-) in an alkali are attracted to the negative and positive charges on the R groups of the amino acids in the polypeptide chains that make up the enzyme. This can disrupt the hydrogen bonds and ionic bonds that maintain the enzyme molecule's three-dimensional shape, distorting the shape of the active site of the enzyme, and reducing the likelihood of enzyme–substrate complexes forming. However, unlike the effect of temperature, the denaturing of enzymes by extreme pH values can be reversible: returning an enzyme to its optimum pH can restore its function.

The pH of an enzyme solution can be varied in the school laboratory by using **pH buffers**. These are solutions that have a specific pH and maintain that pH even during a chemical reaction that may produce acidic or alkaline products. Indicator solutions, or an electronic pH meter attached to a probe, can be used to check the pH values. One investigation commonly used to investigate the effect of pH on enzyme activity uses a spotting tile and determines the end-point of a colour change. This is used with a range of enzyme-catalysed reactions, but commonly in the school laboratory with the hydrolysis of starch into maltose by amylase. A small volume of iodine dissolved in potassium iodide solution is placed into the wells in the spotting tile and turns blue–black when starch is added (Figure 3.14). Successive samples are taken from the reaction mixture at regular time intervals (usually every 30 seconds), often by dipping into the solution with the end of a glass rod. Initially there is a high concentration of starch, which turns the iodine-potassium iodide solution blue–black. If this colour lightens, starch is breaking down. When there is no starch, the iodine–potassium iodide solution remains orange–brown. The time taken to reach the end-point of the experiment, when all the blue–black colour has gone, is then measured. This shows the loss of starch from the reaction mixture over time. The results of one experiment using this method are shown in Figure 3.14.

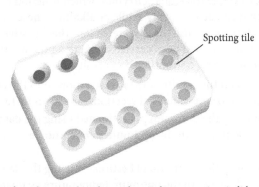

Spotting tile

Figure 3.14 The digestion of starch into maltose by amylase. Iodine–potassium iodide solution turns blue–black in the presence of starch. When iodine no longer changes colour, there is no starch present. Samples are taken at each 30 seconds and applied to iodine–potassium iodide solution that has already been placed into the wells.

The results of a study such as this are shown in Table 3.3 and are plotted on a line graph in Figure 3.15. Values for the rate of reaction, with units $s^{-1} \times 10^{-3}$, were calculated as described in Figure 3.9.

pH / no units	Time taken to detect no starch / s						Mean rate of reaction / $s^{-1} \times 10^{-3}$
	Trial 1	Trial 2	Trial 3	Trial 4	Trial 5	Mean	
2	NR	NR	NR	NR	NR	N/A	0.0
4	210	240	240	240	210	228	4.4
6	60	90	60	90	90	78	12.8
8	30	30	30	30	30	30	33.3
10	60	90	90	60	90	30	12.8
12	150	150	180	180	180	78	6.0
14	330	360	360	390	390	366	2.7

Table 3.3 The results of an investigation into the effect of pH on the rate of starch hydrolysis by amylase. Note how the same number of decimal places are provided for each data point for the mean rate of reaction, even for whole numbers.

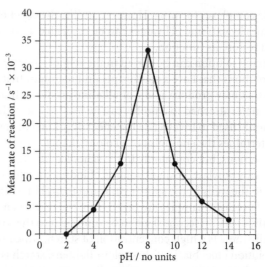

Figure 3.15 The results of an investigation into the effect of pH on the rate of starch hydrolysis by amylase.

Almost all intracellular enzymes in humans have an optimum pH of around 7.35. This is the pH value of the cytoplasm of most of our cells, which have special processes to keep pH constant. However, extracellular enzymes, which function outside of cells, have optimum pH values that can be extremely acidic or alkaline. These include the hydrolytic enzymes found inside lysosomes and those secreted into the digestive tract of animals. If the investigation in Figure 3.15 had been carried out with a protease from the stomach, the results would have shown an optimum pH of about 1–2.

8. This question focuses on the results of the investigations into the effect of temperature and pH on the rate of enzyme-catalysed reactions, as shown in Tables 3.1 and 3.3.

    a. Give one reason why the optimum pH estimated using the data in Table 3.3 would be more accurate than the optimum temperature estimated using the data in Table 3.1.

    b. Give one reason why the optimum pH estimated using the data in Table 3.3 would be more reliable than the optimum temperature estimated using the data in Table 3.1.

    c. Identify any data in both Tables 3.1 and 3.3 that appears to be anomalous. Give reasons for your choice and explain how the investigator should treat this anomalous value.

THE EFFECT OF CHANGING SUBSTRATE AND ENZYME CONCENTRATION ON THE RATE OF ENZYME-CATALYSED REACTIONS

The concentrations of both the enzyme and the substrate affect the frequency with which enzyme–substrate complexes form. This is because, in a given volume, a more concentrated solution contains more particles than a dilute solution. Collisions between molecules are therefore more likely, and so is the likelihood of formation of enzyme–substrate complexes.

The effect of enzyme concentration on the rate of an enzyme-catalysed reaction can be investigated by a method that uses the digestion of lipids into fatty acids and glycerol. The formation of fatty acids as a product in this reaction causes the pH of the solution to fall. The faster the drop in pH, the greater the rate of reaction. This can be detected using a coloured indicator such as cresol red dye, as shown in Figure 3.16.

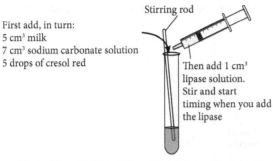

Stirring rod

First add, in turn:
5 cm³ milk
7 cm³ sodium carbonate solution
5 drops of cresol red

Then add 1 cm³ lipase solution. Stir and start timing when you add the lipase

Figure 3.16 The digestion of fats (triglycerides) into fatty acids and glycerol by **lipase**. Sodium carbonate is added to increase the pH to alkaline levels, which results in the cresol red dye turning the solution red in colour. As the lipids in the milk are digested by the lipase, fatty acids are produced, which cause a drop in pH. The stopwatch is stopped when the solution becomes orange.

Enzyme solutions of different concentrations can be prepared by either serial dilution or **proportional (simple) dilution** methods. The key differences between these two procedures are summarised in Table 3.4.

	Serial dilution	Proportional (simple) dilution
Overview of method	1 cm³ of the stock solution volume is added to 9 cm³ distilled water. This new solution is mixed before 1 cm³ is added to another 9 cm³ of distilled water. This process is continued to produce a range of solutions of concentrations that are one-tenth of the concentration of the previous one.	A specific volume of the stock solution is added to a specific volume of distilled water and mixed well in order to make a solution of a specific concentration. The formula $V_1 C_1 = V_2 C_2$ may be used, where V denotes volume, C denotes concentration, and the numbers 1 and 2 denote initial and final values.
Example (assuming a 1 g dm⁻³ solution is used)	• Transfer 1 cm³ of a 1% solution to a clean test-tube and add 9 cm³ of distilled water to make a 0.1% solution. • Transfer 1 cm³ of the 0.1% solution to a clean test-tube and add 9 cm³ of distilled water to make a 0.01% solution. • Transfer 1 cm³ of the 0.01% solution to a clean test-tube and add 9 cm³ of distilled water to make a 0.001% solution. • Transfer 1 cm³ of the 0.001% solution to a clean test-tube and add 9 cm³ of distilled water to make a 0.0001% solution.	• Transfer 8 cm³ of the 1% solution to a clean test-tube and add 2 cm³ of water to make a 0.8% solution. • Transfer 6 cm³ of the 1% solution to a clean test-tube and add 4 cm³ of water to make a 0.6% solution. • Transfer 4 cm³ of the 1% solution to a clean test-tube and add 6 cm³ of water to make a 0.4% solution. • Transfer 2 cm³ of the 1% solution to a clean test-tube and add 8 cm³ of water to make a 0.2% solution.
Applications	Better to use when a broader range of concentrations of a solution are to be investigated.	Better to use when very specific concentrations of a solution are required, or if a narrow range of concentrations of a solution are to be investigated.

Table 3.4 A comparison of the serial and proportional (simple) dilution methods. Assume that the stock solution (to be diluted) has a concentration of 1%.

The results of an investigation into the effect of lipase concentration using this method are shown in Figure 3.17.

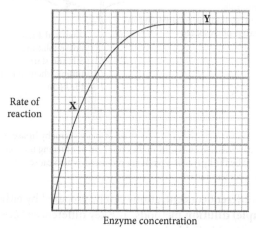

Figure 3.17 The results of an investigation into the effect of enzyme concentration on the rate of an enzyme-catalysed reaction. Labels X and Y correspond to the text in the paragraph below.

The shape of a graph showing the effect of enzyme concentration on the rate of reaction is very characteristic. It rises with a relatively constant gradient (X), but then levels off and reaches a plateau (Y), after which no further increases in enzyme concentration have any effect on the rate of reaction. How can it be explained?

As we increase the enzyme concentration in a solution of its substrate, the chance of an enzyme molecule colliding with a substrate molecule increases. Therefore, the likelihood of enzyme–substrate formation increases, with a similar effect on the rate of reaction (X). However, at high enzyme concentrations, there may not be enough substrate molecules to keep them all busy all the time, so the rate levels off (Y).

A similar graph would be seen if we were to keep the enzyme concentration constant and increase the substrate concentration (Figure 3.18). However, despite the similar shape of the graphs, the relationship has a slightly different explanation. Eventually, we will reach a point where all the active sites are fully occupied by substrate molecules at any given time. The enzyme molecules are all engaging in catalysis as quickly as they can, receiving substrates, turning them into products and releasing them (Figure 3.19). No matter how much more substrate is available the enzyme cannot work any faster. This explains why this curve levels off at high substrate concentrations. The enzyme is said to have become saturated with substrate.

Link

In both graphs, the plateau can be explained by a factor other than the independent variable acting as a **limiting factor**. You will encounter limiting factors again in Chapter 13.

Tip

In the environment of a cell, where enzymes are found in much smaller numbers than the substrates they digest, substrate concentration is rarely a limiting factor.

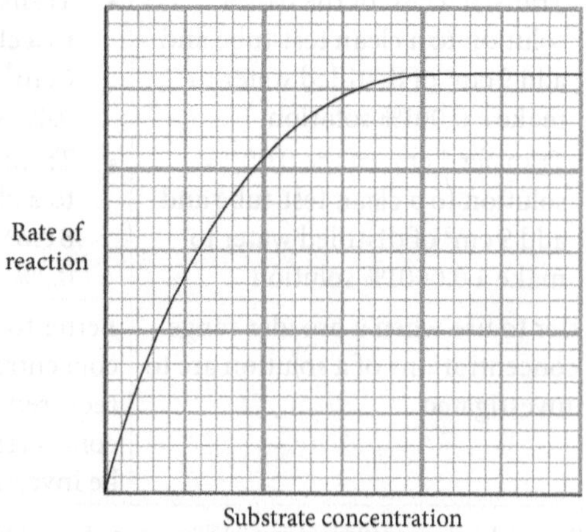

Figure 3.18 The results of an investigation into the effect of substrate concentration on the rate of an enzyme-catalysed reaction.

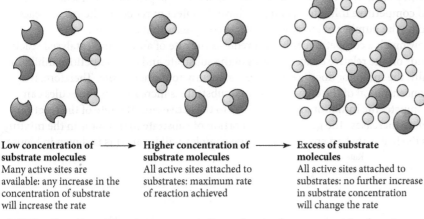

Low concentration of substrate molecules
Many active sites are available: any increase in the concentration of substrate will increase the rate

→ **Higher concentration of substrate molecules**
All active sites attached to substrates: maximum rate of reaction achieved

→ **Excess of substrate molecules**
All active sites attached to substrates: no further increase in substrate concentration will change the rate

Figure 3.19 The effect of increasing substrate concentration on the rate of an enzyme-catalysed reaction. If an excess of substrate is added, the substrate molecules are effectively 'queuing up' to access the active site of an enzyme.

THE EFFECT OF INHIBITORS ON THE RATE OF ENZYME-CATALYSED REACTIONS

Some chemical substances are able to reduce the activity of enzymes. These substances therefore have an effect on the rate of enzyme-catalysed reactions. They are called inhibitors and can be grouped into two categories:

- **competitive inhibitors**
- **non-competitive inhibitors**.

The mechanism of action of these two types of inhibitor is summarised in Figure 3.20.

a. Competitive inhibition

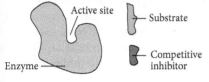

Active site

Enzyme

Substrate

Competitive inhibitor

The usual substrate molecule can form a complex with the enzyme

A competitive inhibitor molecule can also form a complex with the enzyme, preventing the usual reaction from taking place. The substrate molecule and the inhibitor molecule compete for the active site

b. Non-competitive inhibition

The substrate can form a complex with the enzyme at the active site

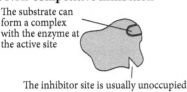

The inhibitor site is usually unoccupied

The substrate molecule is unable to form a complex with the enzyme

An inhibitor molecule attaches to the enzyme, changing the enzyme's shape and affecting the active site

Figure 3.20 (a) The effect of a competitive inhibitor on the rate of reaction. (b) The effect of a non-competitive inhibitor on the rate of reaction.

As you saw earlier when we encountered the induced-fit hypothesis, although enzymes have substrate specificity, it is possible for more than one substrate to bind to an enzyme's active site. However, some molecules that are not substrates for the enzyme have a similar size and shape complementary to the active site of an enzyme. If they are present in the same solution as the enzyme and its substrate, then they will compete for the active site – they bind to it, but no reaction takes place and no products are formed. These are called competitive inhibitors, and they have the effect of reducing the rate of reaction by reducing the number of enzyme–substrate complexes that can form.

How much a competitive inhibitor reduces the rate of an enzyme-catalysed reaction depends on the relative concentrations of the substrate and the competitive inhibitor. Usually, the inhibitor does not attach permanently to the active site. Therefore, if the concentration of the enzyme's normal substrate is increased, its molecules can 'outcompete' the inhibitor and gain access to the active site. The rate of the reaction therefore increases. The greater the proportion of substrate to inhibitor in the mixture, the more likely it is that an enzyme–substrate complex will form. This is shown in Figure 3.21.

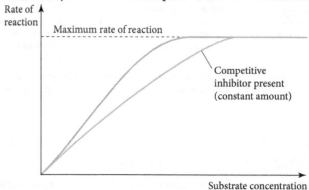

Figure 3.21 The effect of a competitive inhibitor on the rate of reaction. Note that the effect of the inhibitor can be overcome by adding more substrate.

Some substances inhibit enzyme reactions in a different way. They do not attach to the active site but to a different part of the enzyme, away from the active site. This is called an **allosteric site**. This alters the shape of the enzyme and thus the active site, reducing the likelihood of enzyme–substrate formation and therefore the reaction rate (Figure 3.22). These substances are called non-competitive inhibitors, because there is no **competition** between them and the substrate for the active site. The shape of a non-competitive inhibitor molecule may be completely different from that of the usual substrate. It is often a much smaller molecule, and it can even be an atom or ion. Note that, unlike the condition with competitive inhibitors shown in Figure 3.21, excess substrate will not overcome non-competitive inhibition. This is because there is no competition for active sites between the substrate and the inhibitor.

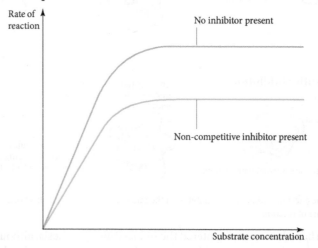

Figure 3.22 The effect of a non-competitive inhibitor on the rate of reaction. Note that the effect of the inhibitor cannot be overcome by adding more substrate.

Some non-competitive inhibitors can affect many different enzymes. This is why heavy metal ions such as mercury, lead and arsenic are so poisonous. They are non-competitive inhibitors of many different enzymes, and so they can prevent many of the body's metabolic reactions taking place. Other non-competitive inhibitors are more specific, and some even have useful applications: for example, the nicotinamide family of insecticides and some medicinal drugs including statins, which reduce blood cholesterol levels, and some treatments for cancer.

THE AFFINITY OF ENZYMES FOR THEIR SUBSTRATES

So far in this chapter, we have quantitatively investigated the activity of enzymes by assessing the rate at which they catalyse a given reaction. However, for the purpose of comparing enzymes used in industry and medicine, further comparison of enzymes is required.

In 1913, German biochemist Leonor Michaelis and Canadian physician Maud Menten introduced a mathematical model to describe the **affinity** of an enzyme. 'Affinity' is a term that describes the attraction between an enzyme and its substrate: that is, the likelihood of an enzyme–substrate complex forming. The two scientists carried out experiments to assess the rate of reactions catalysed by different enzymes at a range of substrate concentrations.

In their experiments, Michaelis and Menten used the term 'velocity' (symbol V), which means speed, to describe the term rate. Therefore, **maximum velocity** (V_{max}) indicates the initial rate of reaction at which all enzyme molecules are saturated with their substrate. This can be found by identifying the concentration of substrate at which the graph forms a plateau. They proposed the use of a constant, known as the **Michaelis–Menten constant**, or K_m. This is defined as the substrate concentration that causes an initial rate of reaction that is half of the V_{max}:

$$K_m \rightarrow \tfrac{1}{2}V_{max}$$

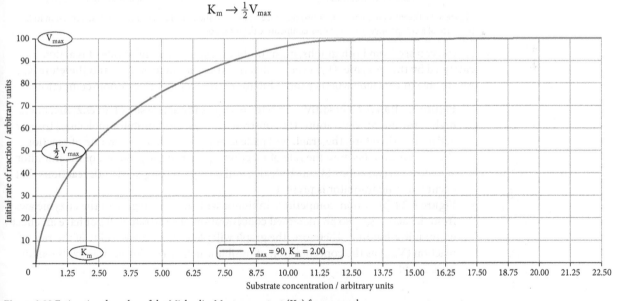

Figure 3.23 Estimating the value of the Michaelis–Menten constant (K_m) from a graph.

When comparing reaction rates, it is best to look at the very start of the reaction. This is because, as the reaction progresses, the concentration of substrate in the reaction reduces, as it is converted into products. This also applies to investigations into factors that affect enzyme activity; this ensures that any difference in rate is due to the factor, and not substrate concentration.

As you can see from Figure 3.23, the value of K_m is therefore an indicator of the affinity of an enzyme for its substrate. The smaller the value of K_m the higher the affinity of the enzyme for its substrate, and so the lower the concentration of substrate required to achieve a given rate. K_m can be experimentally determined at specified pH and temperature and has units that are the same as substrate concentration, which are usually mol dm^{-3}.

THE EFFECT OF INHIBITORS ON THE VALUE OF K_m

It is not possible to directly observe how molecules interact. So how do researchers determine whether an inhibitor is competitive or non-competitive?

One way is to carry out experiments to investigate the initial rate of reaction and the value of K_m, using different substrate concentrations. Figure 3.24 shows the results of an investigation into the effects of two different inhibitors of the same enzyme.

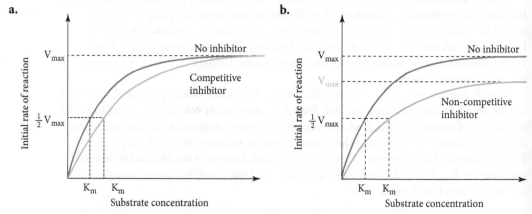

Figure 3.24 The effect of increasing substrate concentration on the rate of an enzyme-catalysed reaction in the presence of a competitive (a) and a non-competitive (b) inhibitor.

As you can see from both graphs, both inhibitors reduce the initial rate of reaction catalysed by this enzyme. However, by determining the value of V_{max}, and therefore K_m, we are able to conclude whether the inhibitor used is competitive or non-competitive.

- **Figure 3.24(a):** A competitive inhibitor occupies the active site of an enzyme, but at high substrate concentrations the substrate will displace inhibitor molecules from the active site. Therefore, the graph will plateau at the same maximum rate of reaction (V_{max}). However, because the rate of reaction is lower in the presence of the inhibitor at lower concentrations of substrate, K_m will be higher compared with reactions that occur when no inhibitor is present.
- **Figure 3.24(b):** A non-competitive inhibitor will bind to a region of the enzyme away from the active site, so an increased concentration of substrate will not have an effect on inhibitor binding. Therefore, the graph will plateau at a lower value of the rate of reaction (V_{max}) as the concentration of inhibitor increases. However, K_m will remain unchanged.

Worked example

Urease is an enzyme that catalyses the breakdown of urea into carbon dioxide, water and ammonia:

$$\text{urea} \xrightarrow{\text{urease}} \text{carbon dioxide} + \text{water} + \text{ammonia}$$

A researcher investigated the effect of an inhibitor of urease, called thiourea. An initial experiment was conducted in the absence of inhibitor. The results are shown in Figure 3.25.

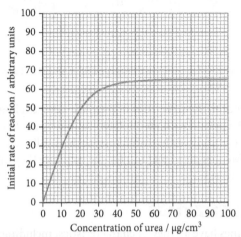

Figure 3.25 The effect of the concentration of urea on the initial rate of breakdown of urea catalysed by urease.

a. Define the term 'Michaelis–Menten constant' (K_m) and estimate its value from the graph.

Answer

The value of substrate concentration that results in an initial rate of reaction that is half the value of the maximum rate of reaction (V_{max}). Reading this value from the graph, $\frac{1}{2}V_{max} = 12 \, \mu g/cm^3$.

b. Ammonia dissolves in water to form ammonium hydroxide, which is an alkaline substance. Suggest how the researcher could measure the initial rate of the reaction in this investigation.

Answer

The increase in pH caused by production of ammonia could be measured using a pH indicator or pH meter.

---

9. Explain why it is important to determine the initial rate of reaction when investigating the effect of a competitive inhibitor on an enzyme.

10. Extracellular enzymes are used by many organisms as defence mechanisms against pathogens or toxic substances produced by other organisms. Two examples are lysozyme and beta-lactamase.

- Lysozyme is a human enzyme found in secretions of the tear ducts. It hydrolyses chemical bonds found between polysaccharides in the cell walls of bacteria, causing them to burst.
- Beta-lactamase is an enzyme secreted by bacteria that are resistant to the antibiotic penicillin. It breaks down the penicillin molecule before the antibiotic can harm the cell.

a. Lysozyme has no effect on plant cell walls. Using your knowledge of enzyme specificity, explain what this suggests about plant and bacterial cell walls.

b. The value of K_m for human lysozyme is 6 mmol dm^{-3}, whereas the value for penicillinase in bacteria is 50 mmol dm^{-3}. Explain what this difference in the K_m values reveals about these two enzymes.

☆**c.** Bacteria that are resistant to penicillin can cause life-threatening illnesses in humans. One treatment is the competitive inhibitor sulbactam. The structures of penicillin and sulbactam are shown in Figure 3.26.

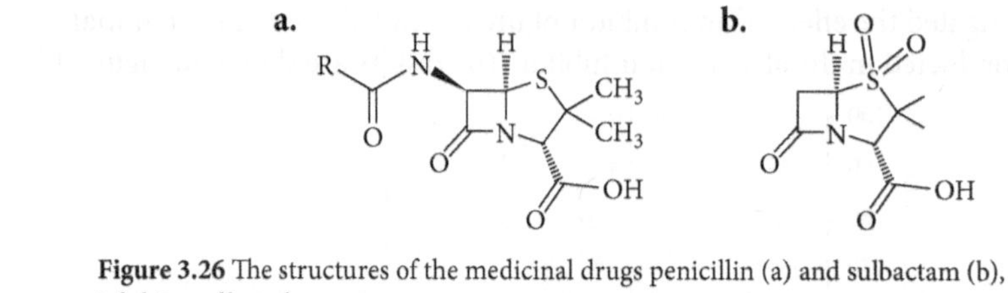

a. **b.**

Figure 3.26 The structures of the medicinal drugs penicillin (a) and sulbactam (b), a competitive inhibitor of beta-lactamase.

🔍 **i.** With reference to the structures of the molecules, explain how sulbactam can help to treat a person with an infection caused by penicillin-resistant bacteria.

ii. The presence of competitive inhibitors, such as sulbactam, increases the Michaelis–Menten constant (K_m) for the enzymes they inhibit. Explain why the K_m value increases.

IMMOBILISING ENZYMES

As we have seen, enzymes have a range of characteristics, including:

- they are highly specific, catalysing changes in one particular compound
- they are efficient – only a tiny quantity of enzyme is required to catalyse the production of a large quantity of product
- they are effective at body temperatures – only a limited input of energy as heat is required.

These characteristics, among others, make them very valuable substances, not only in living organisms, but also for human use. Enzymes from yeast have for centuries been used to brew alcoholic drinks, for example. Enzymes are also commonly added to detergents such as washing powders to digest food stains. However, these proteins can be very expensive.

Enzymes can be used as solutes in solution, or they can be **immobilised** in a **bioreactor** (reaction column). This involves attaching them to an insoluble support, before passing a liquid containing the enzyme's substrate over them. As the substrate runs over the surface of the immobilisation material, the enzymes catalyse a reaction that converts the substrate into product. The product continues to trickle down the column, emerging from the bottom, where it can be collected and purified. There are a number of methods of immobilising enzymes, three of which are shown in Figure 3.27. The most common method used in school laboratories involves encapsulation using **sodium alginate** and calcium chloride, which will be considered in the Experimental skills section later in this chapter.

> **Link**
>
> As you will see in Chapter 4, some digestive enzymes are naturally immobilised in the surface membranes of cells lining the alimentary canal of animals. In humans, maltase is fixed in the cell surface membranes of cells of the intestinal epithelium.

a. Cross-linkage

b. Encapsulation

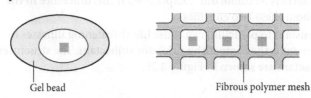

C. Adsorption

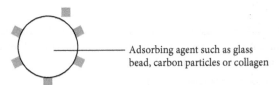

— Adsorbing agent such as glass
bead, carbon particles or collagen

Figure 3.27 There are a variety of methods of immobilising enzymes, three of which are shown. Enzymes can be chemically cross-linked, using an agent such as glutaraldehyde (a). They can be entrapped or encapsulated in gel beads or a mesh (b). They can also be attached to the surface of glass beads in a process called adsorption (c).

The first commercial use of immobilised enzymes took place in the 1970s, when immobilised glucose isomerase was used to make high-fructose syrups from starch. Today, immobilised enzymes are widely used in the food industry and the pharmaceutical industry. The main advantage of immobilising enzymes is that the protein remains unchanged after the reaction it catalyses, so it can be reused. In addition, less enzyme is wasted because it does not need to be removed from the products. This is one of a number of advantages of using immobilised enzymes instead of free enzymes in industrial processes, listed in Table 3.5.

Advantages	Disadvantages
Less contamination of products. It does not need to be recovered or purified from the products, which means it can be reused in a continuous procedure with less downstream processing. This reduces wastage.	The material used to immobilise the enzyme may affect the affinity of the enzyme for the substrate by obstructing its active site (K_m is generally increased).
Remains functional in a wide range of temperature and pH conditions. This is due to immobilisation material stabilising its tertiary structure. It is less likely to be denatured and so has a longer shelf-life.	The reduced movement of the enzyme reduces the number of collisions between enzymes and substrate molecules, reducing the rate of reaction.

Table 3.5 The advantages and disadvantages of using immobilised enzymes rather than free enzymes in industrial processes.

Key ideas

→ An enzyme has an optimum temperature at which its activity is greatest. At lower temperatures, the enzyme and substrate molecules have lower kinetic energy, and so do not collide as frequently. At higher temperatures, hydrogen bonds within the enzyme molecule break, so the enzyme loses its shape and is said to be denatured.

→ An enzyme has an optimum pH at which its activity is greatest. At a higher or lower pH, the hydrogen and ionic bonds within the enzyme are disrupted, so the active site loses its shape.

→ Increasing either the substrate concentration or the enzyme concentration increases the frequency with which substrate and enzyme molecules collide, and therefore increases the rate of reaction.

→ Inhibitors are substances that reduce enzyme activity.

→ Competitive inhibitors fit into the active site and prevent the normal substrate from binding. If the concentration of the substrate is increased, then there is a better chance that the substrate molecule will get into the active site before an inhibitor, and the reaction rate increases.

→ Non-competitive inhibitors bind with the enzyme at a position other than its active site. This causes the enzyme to change its shape so the active site can no longer bind with the substrate. Increasing the substrate concentration therefore has no effect on the reaction rate.

→ The use of immobilised enzymes in industry has a number of advantages compared with the use of free enzymes.

3 Enzymes

Worked example

Table 3.6 shows the results of an investigation, conducted by a food manufacturer, into the activity of an enzyme called glucose isomerase. This enzyme is used to convert glucose into fructose in the production of food and drink.

Temperature / °C	Enzyme activity / %	
	Immobilised glucose isomerase	Non-immobilised glucose isomerase
35	88	90
40	95	92
45	92	84
50	90	62
55	85	45
60	77	23
65	57	10
70	35	3

Table 3.6

Glucose isomerase is not an expensive enzyme to buy and the manufacturer is not concerned about enzyme wastage. Using the information in the table, and your own knowledge, give two reasons that the manufacturers may also be keen to use non-immobilised enzyme. You may wish to draw a graph to help you answer this question.

Answer

The data from Table 3.6 is plotted as shown in Figure 3.28. Non-immobilised enzymes are likely to have a lower optimum temperature than immobilised enzymes, which will save money as the reaction mixture will require less heating. Immobilising enzymes can reduce their affinity for their substrate, so using non-immobilised enzymes could make the catalysis of the substrate more efficient (Figure 3.28).

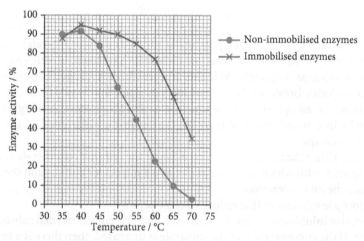

Figure 3.28

11. Evaluate the advantages and disadvantages of the methods of enzyme immobilisation shown in Figure 3.27.
12. In industry, clear apple juice is produced by incubating apple pulp with the enzyme pectinase. Pectin has a high water-binding capacity and reduces the yield of pressed

juice. Figure 3.29 shows the results of an investigation that compared the volume of clear apple juice produced using enzymes that were applied to the reaction in three different ways. The first used non-immobilised (free) enzymes in solution. The second used enzymes bound to a gel membrane surface (adsorption). The third used enzymes immobilised inside beads (encapsulation).

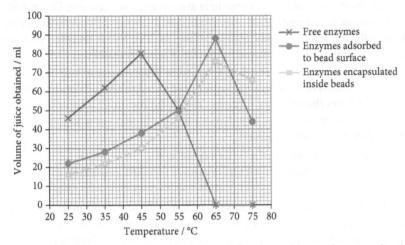

Figure 3.29 The results of an investigation that compared the volume of clear apple juice produced using three different enzyme set-ups.

Compare the different methods of making clear apple juice and give explanations for the differences between them.

Assignment 3.1: Surviving in space

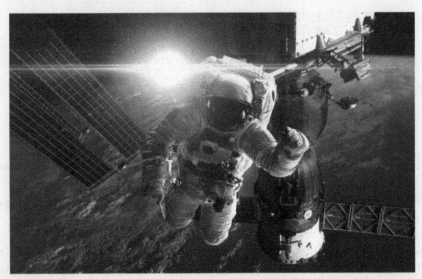

Figure 3.30 The human body has not evolved to live in microgravity, so researchers face unique challenges in space.

'Life on the International Space Station (ISS) is not too different to life on Earth,' says Gennady Padalka, a researcher who holds the record for the longest time in space, at 879 days. Most would disagree. From what food to eat, to how to dispose of body waste, there is little in the life of a space researcher that seems familiar to most of us (Figure 3.30).

QUESTIONS

A1. Most foods naturally contain enzymes (or microorganisms that release enzymes) that cause food to spoil after a few days or weeks. Explain why food taken into space by researchers is ultra-heated before it is taken on a space mission.

A2. Some reports suggest that the immune systems of space researchers are less effective in space. One possible reason for this is an increased activity of an enzyme called 5-lipoxygenase. This enzyme's overactivity can reduce the ability of white blood cells to defend the body against pathogens. Using your knowledge of the way in which enzymes work, suggest a treatment for the reduced immunity of space researchers.

A3. Immobilised enzymes can be used in bioreactors that attach to space suits. The bioreactors use urease enzyme that has been attached to powdered charcoal or small gel beads produced using calcium chloride. The urease catalyses the breakdown of urea, which has the chemical formula $CO(NH_2)_2$, forming carbon dioxide and ammonia. These products react to form ions, which are then removed by the bioreactor with the other components of urine, meaning that only water is left behind.

    a. Describe the advantages of installing a urease bioreactor in a space suit.

    b. Suggest why it is possible to use a urease enzyme from any organism (not just human) in a space suit bioreactor.

In the development of a urease bioreactor, a bioengineer found that immobilising urease in beads using manganese chloride rather than calcium chloride increased the rate of urea hydrolysis. This is because the manganese ion (Mn^{2+}) can bind to some enzymes and have a positive effect on catalytic activity. The bioengineer conducted a series of investigations to find how the Michaelis–Menten constant (K_m) of urease varies with the concentration of Mn^{2+} used during the encapsulation process (Figure 3.31).

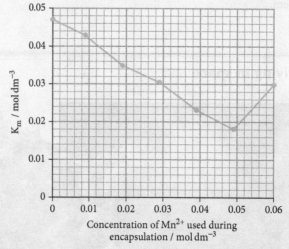

Figure 3.31 The results of an investigation into the effect of manganese ion concentration and the Michaelis–Menten constant (K_m) of immobilised urease.

    c. Of the five concentrations of manganese ions, identify which one should be used during encapsulation in order to maximise the activity of urease.

   ⋆**d.** Suggest explanations for the results of this investigation.

    e. The scientist was concerned that there may be some degradation of the beads over time, resulting in the contamination of the products of the reaction with enzyme. Describe a simple laboratory test that could be carried out on the products of the reaction column to test this.

A new procedure that can be used to immobilise enzymes is to trap them in a compartment behind a selectively permeable membrane that allows reactants in and products out, but does not allow the enzymes out.

   ⋆**f.** Describe an advantage of this method compared with immobilising the enzymes in gel beads.

    g. Suggest why this method could not be used with enzymes that make macromolecules.

Experimental skills 3.2: Investigating the effect of bead diameter on the efficiency of an enzyme-catalysed reaction

Immobilised enzymes are widely used in industry to maximise the efficiency of biochemical reactions. As we saw in Figure 3.27, there are a variety of methods that can be used to immobilise enzymes for use in industrial processes. In school, enzymes can be easily immobilised by encapsulation inside gel beads.

In this experiment, you will see how the enzyme lactase can be 'trapped' inside alginate beads, and that its ability to hydrolyse its substrate, lactose, is not affected by immobilisation. Lactase catalyses a hydrolysis reaction to break down lactose into glucose and galactose. This illustrates how a continuous process could be used to make food and drinks for people who are lactose-intolerant.

The following investigation will consider how bead diameter affects the yield of lactose-free milk. You will obtain qualitative results from observations of colour changes using a number scale for intensity of colour with a key, and denaturing an enzyme by boiling, to act as a control.

Figure 3.32 illustrates how gel beads can be made in the laboratory. When the solution of sodium alginate is added drop by drop to a solution of calcium chloride, spherical beads of calcium alginate form. This is because calcium ions cross-link the alginate strands. The effect of this is that the mixture of reacting solutions becomes semi-solid in nature.

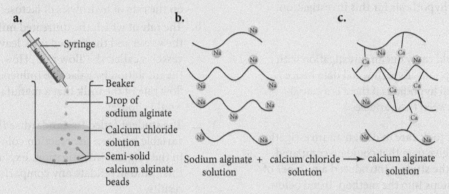

Figure 3.32 Gel beads can be made in the school laboratory using two solutions, sodium alginate and calcium chloride (a). Sodium alginate is soluble (b) but the addition of calcium ions produces calcium alginate that precipitates from the solution in the form of a semi-solid (c).

Beads containing an enzyme can be made by mixing solutions of enzyme and sodium alginate together first. This mixture can be drawn up into a syringe and gently dropped into a beaker containing a solution of calcium chloride. This produces small spheres containing enzyme molecules mixed with the insoluble calcium alginate. Therefore, the enzyme is encapsulated throughout the bead. Beads containing an enzyme, such as lactase, can then be used in a continuous process to digest a substrate, as shown in Figure 3.33.

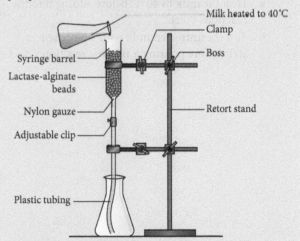

Figure 3.33 The use of beads containing immobilised lactase in a reaction column. The adjustable clip is tightened before the milk is poured into the reaction column. After 5 minutes, the clip is loosened, and the milk collects in the beaker.

QUESTIONS

A student wishes to carry out an investigation to find the diameter of lactase beads that produces the greatest yield of lactose-free milk when used in a reaction column. To do this, the student produced a number of samples of lactase-alginate beads of different diameters of the same total mass. Lactase catalyses a hydrolysis reaction to break down lactose into glucose and galactose. The student intended to use a glucose test stick to measure the concentration of glucose in the milk collected from the bottom of the reaction vessel. Glucose test sticks give a semi-quantitative measure of the concentration of glucose in a solution, by providing a colour that can be compared with a colour chart.

P1. Identify the independent and dependent variables in this investigation.

P2. State a null hypothesis for this investigation.

Tip

A scientist should carry out an investigation with no prior assumptions. Experimental data hence invalidates a null hypothesis if there is a causal relationship between two variables.

P3. The student took care to design an investigation that would give data that could be compared. To do this, the student introduced a number of important steps into the method, listed below. Explain why each step was taken.

    a. Heat the milk to 40 °C before adding it to the reaction vessel.

    b. Test the untreated milk with a glucose test strip before pouring it into the reaction vessel.

    c. Allow at least 5 minutes before the adjustable clip is untightened, to allow the milk to flow through the plastic tubing and into the beaker.

    d. Add the enzyme to the sodium alginate rather than the calcium chloride when the beads are made.

P4. Suggest a suitable control experiment for this investigation.

P5. This investigation used a semi-quantitative measurement of glucose concentration. Describe how a quantitative measure of glucose concentration could be obtained.

P6. Galactose has a similar tertiary structure to lactose and is an inhibitor of lactase.

    a. What type of inhibitor is galactose? Explain your answer. What impact will galactose have on the rate of hydrolysis of lactose?

    b. The rate at which the untreated milk enters the vessel and the treated milk leaves the vessel is called the 'flow rate'. How might the inhibition by galactose influence the flow rate of the milk that a manufacturer would choose?

    c. The student failed to standardise this variable between the reaction columns used in the original investigation. Explain why this could invalidate any comparison of the results.

P7. An alternative to immobilising lactase in gel beads is to immobilise *Kluyveromyces lactis*, a species of yeast that grows naturally in milk and produces lactase. Evaluate the advantages and disadvantages of this method.

CHAPTER OVERVIEW

Review the mini mind map to link new learning with your prior learning, and to highlight which of the key concepts underpin the chapter.

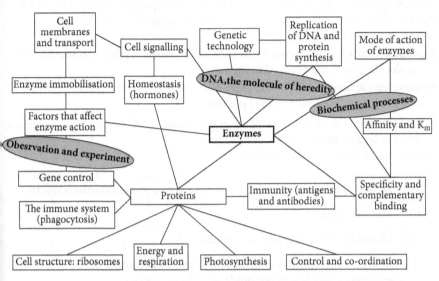

Try copying this mini mind map and expanding upon it. Use your notes from other chapters to help you explore how the essential ideas, theories and principles can be linked further together.

WHAT YOU HAVE LEARNED

Now that you have finished this chapter you should be able to:

- state that enzymes are globular proteins that catalyse reactions inside cells (intracellular enzymes) or are secreted to catalyse reactions outside cells (extracellular enzymes)
- explain the mode of action of enzymes in terms of an active site, enzyme–substrate complex, lowering of activation energy and enzyme specificity, including the lock-and-key hypothesis and the induced-fit hypothesis
- investigate the progress of enzyme-catalysed reactions by measuring rates of formation of products using catalase and rates of disappearance of substrate using amylase
- outline the use of a colorimeter for measuring the progress of enzyme-catalysed reactions that involve colour changes
- investigate and explain the effects of the following factors on the rate of enzyme-catalysed reactions:

  - temperature
  - pH (using buffer solutions)
  - enzyme concentration
  - substrate concentration
  - inhibitor concentration

- explain that the maximum rate of reaction (V_{max}) is used to derive the Michaelis–Menten constant (K_m), which is used to compare the affinity of different enzymes for their substrates
- explain the effects of reversible inhibitors, both competitive and non-competitive, on enzyme activity
- investigate the difference in activity between an enzyme immobilised in alginate and the same enzyme free in solution, and state the advantages of using immobilised enzymes.

CHAPTER REVIEW

1. Using the terms 'specific' and 'complementary' in your answer, define the term 'active site'.

2. When clear egg 'white' is boiled during cooking, soluble proteins form an insoluble precipitate. This is due to the hydrophobic R groups in the centre of the molecule becoming exposed to the water.

    a. Explain why heating egg 'white' causes the hydrophobic R groups in the centre of the molecule to become exposed to the water.

    b. Predict, with an explanation, what you would see if an egg was cracked into a beaker of vinegar (vinegar has a pH of about 3–4).

3. Draw a table to compare and contrast how denaturation of enzymes occurs as a result of extremes of temperature and extremes of pH.

4. Match each word with its correct definition.

(a) Enzyme	(i)	Region of an enzyme molecule that binds to a protein during a reaction.
(b) Substrate	(ii)	Molecule produced by a reaction.
(c) Product	(iii)	Molecule that an enzyme acts upon.
(d) Active site	(iv)	Molecule that inhibits enzyme reactions by altering the shape of the enzyme molecule without binding to the active site.
(e) Competitive inhibitor	(v)	Catalyst (usually a protein) produced by cells, which controls the rate of chemical reactions in cells.
(f) Non-competitive inhibitor	(vi)	Molecule that binds to the active site in an enzyme but no reaction takes place.

5. Inhibitors can either be competitive or non-competitive.

    a. Identify the type of inhibition (competitive or non-competitive) that happens in the following biochemical scenarios.

    i. Ethylene glycol is an ingredient in antifreeze, which is added to vehicle engines to prevent them freezing in cold weather. Pets may drink antifreeze if it is spilt, and the conversion of ethylene glycol to toxic products by alcohol dehydrogenase can cause kidney failure. Affected animals can be treated by providing them with ethanol to drink, which has a similar shape to ethylene glycol.

    ii. During a part of photosynthesis called the light independent reaction, oxygen will bind with the enzyme ribulose bisphosphate carboxylase to prevent it catalyzing a reaction between carbon dioxide, which has a similar structure to oxygen, and a larger organic molecule.

    iii. Aspirin binds to cyclooxygenase, an enzyme involved in the synthesis of products called prostaglandins. This reduces inflammation. Arsenic reduces the V_{max} of this reaction but does not have an effect on the value of K_m.

    b. Outline an experiment that a researcher should carry out to determine whether thiourea is a competitive or non-competitive inhibitor of an enzyme.

6. The graph in Figure 3.34 shows the effect of substrate concentration on the initial rate of an enzyme-catalysed reaction.

    a. The graph is labelled with the letters X and Y. Which of the following descriptions (i) to (v) match part X, Y, or X and Y on the graph?

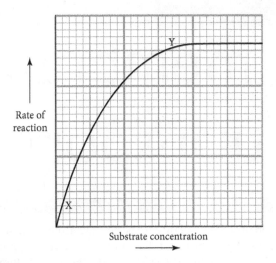

Figure 3.34

    i. At this point, there is an excess of substrate in the mixture.

    ii. At this point, the enzyme's active sites are saturated with substrate.

    iii. At this point, adding more enzyme could increase the rate of reaction.

    iv. At this point, adding more substrate could increase the rate of reaction.

    v. At this point, increasing the temperature could increase the rate of reaction.

  b. Describe and explain the change in the shape of this curve if the experiment were to be repeated with the following inhibitors. You should refer to the values of V_{max} and K_m in your answers.

    i. a competitive inhibitor

    ii. a non-competitive inhibitor

7. One product of lipid hydrolysis, a reaction catalysed by lipase, is fatty acids. Fatty acids reduce the pH of the solution and this can be made visible with the use of a colour indicator.

You have been asked to conduct an investigation to estimate the temperature at which the enzyme lipase is denatured within 5 minutes.

Complete the table to show how you would enhance the validity of the conclusions drawn from this investigation, and the accuracy and reliability of the estimate you would make.

I would enhance the validity of conclusions by...	I would enhance the accuracy of my estimate by...	I would improve the reliability of the result by...

Cell membranes and transport

Every year, over 100 000 organ transplants are performed around the world. There are many challenges. The donor may be in a different country from the patient, thousands of miles away. It can be a race against time to transport the organ before the cells within it die.

Recently, bioengineers have developed special containers, called Organ Care Systems. These containers store organs during transport and provide the organ with an artificial blood supply. A donor heart can even continue to beat long after it has been removed from a donor's body.

Crucial to the success of a transplant is the solution in which the organ is immersed. A solution that is too dilute will cause the cells in the organ to swell and burst; too concentrated, and the organ will dehydrate.

Prior understanding

From your work in Chapter 1, you may be able to outline the structure and function of the cell surface membrane in cells. From Chapter 2, you may recall the key molecular components of the membrane, and the difference between hydrophilic and hydrophobic (polar and non-polar) substances. You have probably learned that diffusion is the result of the random motion of particles, and that osmosis is the diffusion of water across partially permeable membranes.

Learning aims

In this chapter, you will learn how phospholipid molecules associate with other molecules to form cell surface membranes. You will also see how the fluid and mosaic nature of this structure explains its properties and functions. This includes how it controls the entry and exit of specific substances into and out of a cell.

4.1 Fluid mosaic membranes (Syllabus 4.1.1–4.1.4)

4.2 Movement of substances into and out of cells (Syllabus 4.2.1–4.2.6)

4.1 Fluid mosaic membranes

A cell would not be a cell without its membranes. Cell surface membranes act as a barrier to separate the outside environment of all cells from their contents, the cytosol, and the metabolic reactions that happen in the cytoplasm.

Figure 4.1 shows an electron micrograph of the cell surface membrane. It is an extremely thin structure, approximately 7 nanometres (nm) in width, yet it has the strength to keep all of the cell's contents in place. Many of the organelles of the eukaryotic cell, including the Golgi body, chloroplasts and mitochondria, are also surrounded by membranes with a similar structure to that of the cell surface membrane.

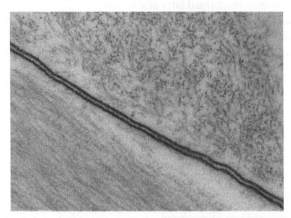

Figure 4.1 The cell surface membrane. This electron micrograph shows two dark lines, very close together. The inside of the cell is the upper half. Magnification ×370 000.

The structure and properties of cell surface membranes are quite unique. The membrane is a mixture of at least four different types of molecule, each with its own important functions. A cell surface membrane contains phospholipids, **cholesterol**, **proteins**, glycoproteins and glycolipids. A simplified diagram of its structure is shown in Figure 4.2.

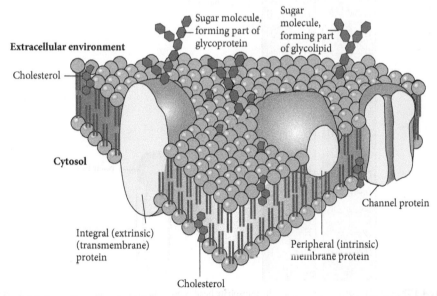

Figure 4.2 A three-dimensional artist's impression of the cell surface membrane in animal cells. This is the fluid mosaic model of the structure of a cell surface membrane. The membrane has been described as 'protein icebergs floating in a phospholipid sea'.

Figure 4.3 The different molecules found in the cell surface membrane fit together like tiles in a pattern, called a mosaic. However, unlike the mosaic tiles in this work of art, the molecules that make up a membrane are not all fixed in place. The motion of the molecules causes the pattern to change over time – the structure is also fluid.

The cell surface membrane has two key properties. First, the structure behaves like a fluid more than it behaves like a solid. The different molecules that make up the flexible membrane can move in relation to each other. As a result of such lateral (sideways) movement, the scattered pattern, or mosaic, of different molecules in the cell surface membrane is constantly in motion (Figure 4.3). The cell surface membrane is therefore a very dynamic, ever-changing structure. This is called the **fluid mosaic model**.

PHOSPHOLIPIDS

The most common molecules in the cell surface membrane are phospholipids. Phospholipids are described as **amphipathic**, which means that different parts of the molecule have different properties. A polar, negatively charged phosphate head is attached via a glycerol molecule to two non-polar, uncharged fatty acids.

Figure 4.4 summarises the structure of a phospholipid. The phosphate group of a phospholipid is often referred to as the 'head', whereas the two hydrocarbon chains are the 'tails'. The head and the tails are often represented by drawing a circle and two lines, respectively.

Link

In Chapter 2, you studied the structure of lipids, or triglycerides. A phospholipid is a triglyceride with a phosphate group in place of one of the hydrocarbon tails on the glycerol.

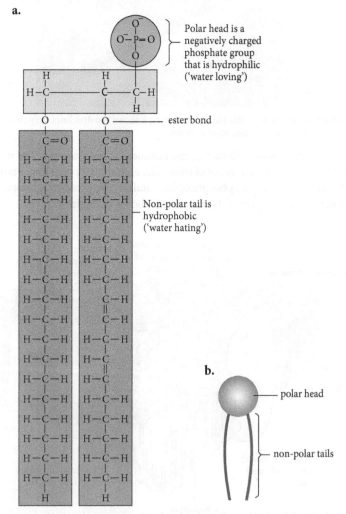

Figure 4.4 The structure of a phospholipid. The phosphate head is polar and hydrophilic. The fatty acid tails are non-polar and hydrophobic. The full molecular structure (a) and schematic representation (b) are shown.

The amphipathic nature of phospholipids explains their behaviour of the molecules when they come into contact with water or an aqueous solution. The phosphate heads are hydrophilic, so they are attracted to the water. However, the hydrocarbon tails are attracted to each other and not water – they are hydrophobic. This causes the phospholipids to spontaneously assemble into a double layer (a bilayer), with the phosphate heads pointing outwards, towards the water, and the fatty acid tails pointing in towards each other. This structure is called a **phospholipid bilayer** and is shown in Figure 4.5. The two dark lines visible in Figure 4.1 represent the two layers of phosphate heads that face the inner and outer layers of the membrane.

> **Tip**
> The term 'phobia' is defined as a fear of something. For example, arachnophobia is a fear of spiders. The aversion to water shown by the phospholipid tails is known as hydrophobia.

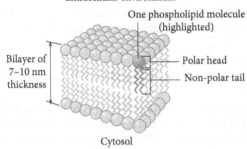

Figure 4.5 The structure of the phospholipid bilayer. The phosphate heads are polar and hydrophilic. The fatty acid tails are non-polar and hydrophobic. The phospholipids in a membrane are arranged tail to tail, forming a bilayer.

If a small number of phospholipids are shaken with water, they form stable ball-like structures called **micelles**, in which the hydrocarbon tails are found inside the ball. Sometimes, a small volume of water is 'locked up' inside a micelle. This is called a liposome. A liposome is shown in Figure 4.6. It is similar in structure to vesicles that form from the Golgi body, found in eukaryotic cells. In recent years, artificial **liposomes** have been produced in the laboratory as a means of delivering medicinal drugs and working copies of genes into cells in the body.

1. Describe the structure of a phospholipid molecule.
2. Explain why a small number of phospholipids will arrange themselves into a micelle if mixed with water. Use the term 'amphipathic' in your answer.
3. Suggest how the structure of a liposome enables it to:
   a. act as a vesicle to carry out its functions in a eukaryotic cell
   ☆ b. deliver medicinal drugs to body cells.
4. Some phospholipids are unsaturated. This means that they have a greater number of carbon-to-carbon double bonds in their fatty acid tails. This has the effect of introducing 'kinks' (bends) into the tails that prevent the closepacking of the phospholipids.

Suggest why some organisms have been found to vary the balance between saturated and unsaturated fatty acids and the amount of cholesterol in their membranes as their environmental temperature changes.

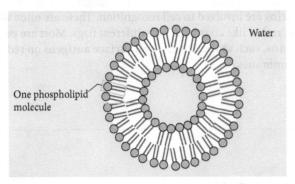

Figure 4.6 The structure of a liposome, a membrane-bound compartment that forms spontaneously when a small number of phospholipids are mixed with water.

Link

In Chapter 2 you saw that some of the variable R groups of amino acids are classed as hydrophobic or hydrophilic.

Link

In Chapters 12 and 13 you will see why chloroplast and mitochondrial membranes contain such a high percentage of protein. These 'electron transport chains' are vital for photosynthesis and respiration.

MEMBRANE PROTEINS

Although the most common molecule is the phospholipid, cell surface membranes also contain a range of other molecules. Proteins are the second most common type of molecule by number.

Membrane proteins carry out many functions of the cell surface membrane. They can be divided into two groups, based on their association with the phospholipid bilayer:

- **Integral (transmembrane) proteins** are associated with both sides of the phospholipid bilayer. They are hydrophobic on the part of their surface that interacts with the hydrocarbon chains (non-polar tails) in the membrane.
- **Peripheral (intrinsic) proteins** emerge from only one side of the membrane, and are not embedded in it. Hydrophilic sections of the protein associate with the regions of phosphate heads.

The typical positions of integral proteins and peripheral proteins are labelled in Figure 4.2.

The function of membrane proteins is directly related to whether they are integral or peripheral. Table 4.1 summarises the main roles of proteins found in the cell surface membrane of many cells. More details will follow for each later in this chapter.

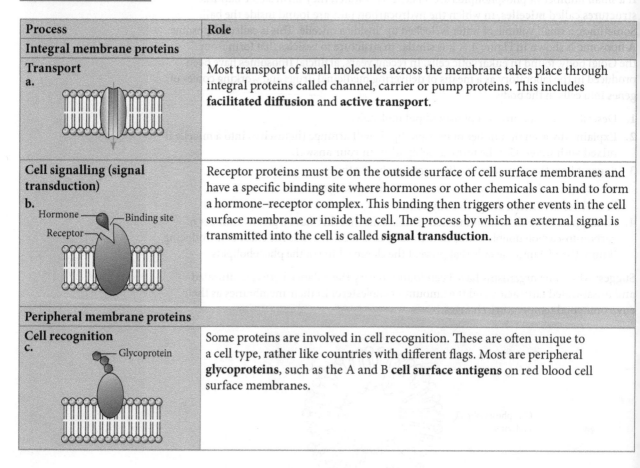

Process	Role
Integral membrane proteins	
Transport a.	Most transport of small molecules across the membrane takes place through integral proteins called channel, carrier or pump proteins. This includes **facilitated diffusion** and **active transport**.
Cell signalling (signal transduction) b. Hormone — Binding site Receptor	Receptor proteins must be on the outside surface of cell surface membranes and have a specific binding site where hormones or other chemicals can bind to form a hormone–receptor complex. This binding then triggers other events in the cell surface membrane or inside the cell. The process by which an external signal is transmitted into the cell is called **signal transduction**.
Peripheral membrane proteins	
Cell recognition c. Glycoprotein	Some proteins are involved in cell recognition. These are often unique to a cell type, rather like countries with different flags. Most are peripheral **glycoproteins**, such as the A and B **cell surface antigens** on red blood cell surface membranes.

Catalysis d.	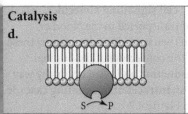	Membrane-bound enzymes catalyse reactions in the cytoplasm or outside the cell, converting substrates (S) into products (P). This includes enzymes such as maltase in the small intestine.
Structure and adhesion e.	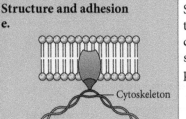	Structural proteins on the inside surface of cell surface membranes are attached to the **cytoskeleton**. They are involved in maintaining the cell's shape, or in changing the cell's shape for cell motility. Structural proteins on the outside surface can be used in cell **adhesion** – sticking cells together temporarily or permanently.

Table 4.1 The structure and function of membrane proteins found in cell surface membranes.

CHOLESTEROL

High cholesterol levels in the diet are associated with the development of coronary heart disease. However, this lipid is vital for health. It is required for the proper functioning of all cell surface membranes.

Cholesterol is a lipid that belongs to the steroid family. Cholesterol molecules are found between phospholipid molecules in membranes. Most of the molecule is hydrophobic, so it associates with the hydrophobic hydrocarbon tails in the middle of the cell surface membrane. However, one end of the cholesterol molecule has a hydroxyl (–OH) group, which is hydrophilic. This associates with the phosphate heads on the outer surfaces of the membrane (see Figure 4.2).

The amphipathic nature of cholesterol means that it closely associates with phospholipids to regulate fluidity. At low temperatures, the cholesterol molecules increase fluidity by preventing the close packing of the hydrocarbon tails of the phospholipids. This enables greater movement of molecules in the membrane. At high temperatures, however, cholesterol molecules are thought to have the opposite effect. They stabilise the membrane by decreasing its fluidity.

Cell surface membranes of animal cells are particularly rich in cholesterol, whereas organelle membranes contain less cholesterol. One further property of cell surface membranes that have a higher percentage of cholesterol is that they are far less permeable to ions or polar substances. Similar molecules play similar roles in plant and prokaryotic cells.

POLYSACCHARIDES

Short-chain polysaccharides, branched polymers of simple sugars, emerge from the outer surface of some membranes like antennae. They are never found on the inner surface. They can be attached to phospholipids, forming **glycolipids**, or much more commonly to proteins, forming glycoproteins ('glyco' means 'related to sugar'). They have roles in interacting with molecules that arrive at the cell surface from its environment. Their composition and branching pattern are often unique to a cell type, which may explain how cells are able to recognise each other. This has important implications: for example, allowing the immune system to tell the difference between body cells and invading bacteria.

CELL SIGNALLING

The structure of the cell surface membrane, particularly the presence of membrane proteins and glycoproteins, is crucial to the interactions of cells with their external environments. As you saw in Figure 4.1(b), a number of membrane proteins, called **cell surface receptors**,

Link

In Chapter 11 you will learn about phagocytosis, a key process in the immune response. Fluidity of the surface membrane of the cells involved in this process is very important.

Link

In Chapter 15 you will see how the myelin sheath, which surrounds neurones, has a high cholesterol content. This prevents the leakage of ions that would slow down nerve impulses.

Link

Glycoproteins help cells to recognise each other. However, some viruses, including HIV, use these 'labels' to attach to cells before infecting them, as you will see in Chapter 10.

exist in the cell surface membrane. These bind to specific **cell signalling molecules**, called **ligands**, which have been secreted by cells elsewhere and transported in the blood or through the tissue fluid. The cell surface receptor protein then transduces (transmits) these signals into the cell cytoplasm and achieves a response. The conversion of the original signal into a message inside the cell is called **signal transduction** and often occurs by way of a series of chemical reactions in the cytoplasm called an **intracellular signaling cascade**. The importance of signal transduction is summarised in Figure 4.7.

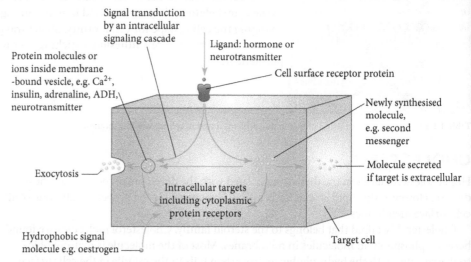

Figure 4.7 A summary of the methods of cell signalling, also known as signal transduction.

Link

A particular mechanism of cell signalling is described in Chapter 14. You will see how the hormones insulin and glucagon, acting as ligands, control the glucose concentration of the blood.

Importantly, only some cells will be able to detect and respond to signals. This is because, like enzymes, cell surface receptors are specific. They are complementary, and bind to, only one type of signalling molecule. Cells that have these receptors are called **target cells** and their response to that particular signalling molecule is therefore specific.

Worked example

Using your knowledge of the effect of cholesterol on cell surface membranes, suggest an explanation for the following findings.

a. Red blood cells (erythrocytes) have a very high percentage of cholesterol in their cell surface membranes.

Answer

Erythrocytes must squeeze through capillaries (which are the same diameter as these cells), so must be flexible. A very high proportion of cholesterol in the cell surface membranes of these cells means that the membrane will be flexible, and the cell will be able to change shape easily in order to fit through these narrow pathways.

b. Vesicles and lysosomes contain a greater percentage of cholesterol in their membranes than mitochondria and chloroplasts.

Answer

Vesicles and lysosomes must fuse with other membrane-bound structures. A higher proportion of cholesterol in the cell surface membranes of these organelles means that they can fuse easily with others. Mitochondria and chloroplasts do not need to interact with other organelles directly in such a way, so their cell surface membranes can be less fluid, and will hence have a lower proportion of cholesterol.

Figure 4.8 shows a diagram of the cell surface membrane.

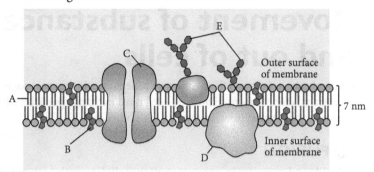

Figure 4.8

5. Cholesterol reduces the permeability of the cell surface membrane to hydrophilic molecules and ions. Suggest why.
6. Identify the molecules labelled A to E.
7. Give **two** possible functions of peripheral proteins such as those labelled D and E.
8. Read the following passage and then answer the questions that follow.

   'Cells can communicate with each other by a process called signal transduction. This involves the transmission of a signal, in the form of a ligand, to an intracellular signalling cascade, which causes a response in the target cell.'

   a. Explain what is meant by the terms:
      i. 'ligand'
      ii. 'intracellular signalling cascade'
      iii. 'response'.

   b. Suggest **one** reason that signal transduction is the subject of significant scientific research.

   c. Muscle cells (myocytes) respond to the hormone adrenaline much more rapidly than most cell types. Suggest why.

Key ideas

→ The cell surface membrane forms an effective barrier that separates the contents of a cell from its external environment.

→ The cell surface membrane contains a variety of molecules, including phospholipids, proteins and cholesterol.

→ The cell surface membrane is described as fluid, in that the molecules can move relative to each other.

→ The cell surface membrane is described as a mosaic in that different types of molecules are interspersed.

→ Phospholipids are amphipathic: they have a hydrophobic head (consisting of a phosphate group) and two hydrophobic hydrocarbon tails.

→ There are two groups of membrane proteins – integral and peripheral – which have a range of functions, including transport, signal transduction, cell recognition, catalysis, and structure and adhesion.

→ Cholesterol associates with the phospholipids to reduce the fluidity of the membrane and provide strength.

→ In signal transduction, receptor proteins in the cell surface membrane bind to signalling molecules called ligands found outside the cell. Ligands are secreted by other cells and transported in the blood or tissue fluid. The receptor protein then transduces (transmits) these signals into the cell cytoplasm.

→ Cells that have cell surface receptors that are complementary to a particular signalling molecule are called target cells.

4.2 Movement of substances into and out of cells

The cell surface membrane acts as an effective barrier to separate the internal contents of a cell from its external environment. However, a cell must exchange particles with its environment to survive and, sometimes, to interact with other cells.

SIMPLE DIFFUSION

Particles of all liquids and gases are in constant motion, colliding with each other and changing direction. Over time, this results in a net movement, or simple **diffusion**, of particles from a region of high concentration to a region of lower concentration. The ultimate effect of diffusion is to spread the particles out evenly in their container or cell (Figure 4.9). This is a **passive** process, requiring no additional input of energy from ATP. The rate of diffusion does, however, depend on the temperature of the material, which affects the kinetic energy of the particles.

Particles can also diffuse across cell surface membranes (Figure 4.10). Small substances, especially those that are hydrophobic (non-polar), can diffuse directly through the phospholipid bilayer because they are lipid-soluble. This applies to many extremely small hydrophobic molecules, including oxygen, carbon dioxide and ethanol. Surprisingly, even water molecules can pass directly through the phospholipid bilayer. It is thought that this molecule is small enough to fit through the small gaps between the phospholipid molecules. However, as you'll see in Chapter 14, special membrane proteins form channels to transport water. These are called aquaporins.

Figure 4.9 Illustrating diffusion with a tea bag; the molecules responsible for the colour and flavour of the tea move randomly in the water, eventually spreading out evenly.

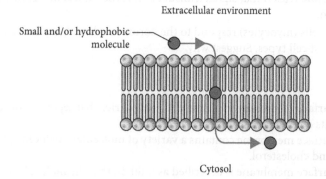

Extracellular environment

Small and/or hydrophobic molecule

Cytosol

Figure 4.10 Some small, hydrophobic molecules can diffuse directly through the phospholipid bilayer.

Larger molecules, and particles that carry significant charges such as ions, cannot diffuse directly through the phospholipid bilayer. Ions are water-soluble and thus unable to diffuse through the hydrophobic phospholipid bilayer. These substances can only diffuse through the membrane if there are special membrane proteins that allow them to do so.

Simple diffusion into and out of cells can be modelled in the laboratory using cubes of agar. Agar is a jelly-like substance that can be set into different shapes, including cubes, and when solid it is fully permeable to small molecules. Model cells can be made of differing sizes, and the surface area to volume ratio can be calculated.

Larger cubes have a smaller surface area to volume ratio than smaller cubes. For example:

- a cube of side length 2 mm has a surface area of 24 mm² ($2^2 \times 6$) and a volume of 8 mm³ (2^3); its surface area to volume ratio is therefore 3 ($24 \div 3$)
- a cube of side length 5 mm has a surface area of 150 mm² ($5^2 \times 6$) and a volume of 125 mm³ (5^3); its surface area to volume ratio is therefore 1.2 ($150 \div 125$).

The relationship between the length of a side of a cube and its surface area to volume ratio is shown in Figure 4.11. As the side length increases, the surface area to volume ratio decreases.

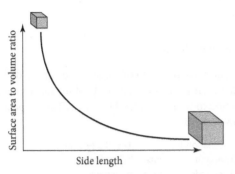

Figure 4.11 The effect of increasing size on surface area to volume ratio.

To measure the rate of diffusion, a pH indicator is added to the agar before it sets into cubes. Indicators change colour depending on the pH of the solution that surrounds them. The agar cubes are placed into separate beakers with enough acid to cover them (Figure 4.12). The time taken for the cube to become completely clear is then measured.

Figure 4.12 Investigating the effect of surface area to volume ratio on the rate of diffusion using agar blocks. In this experiment, an indicator is used that is pink in neutral conditions but becomes colourless in solutions of acidic pH.

The results of investigations such as that in Figure 4.12 show that as the surface area to volume ratio of agar blocks decreases, the rate of diffusion across their surface decreases. This explains why cells have an upper size limit, above which the rate of obtaining required molecules (and releasing waste) would be too slow to support life. You will now explore this in the following Assignment.

Link

Diffusion is not effective over larger distances. In Chapters 7 and 8 you will see how transport systems have evolved in multicellular organisms to bring nutrient and waste molecules closer to the cell surface membrane.

Assignment 4.1: Modelling cells to investigate the importance of a high surface area to volume ratio

A student made up a quantity of agar by dissolving agar powder in slightly alkaline water. A fixed volume of cresol red dye, an indicator solution that is red in neutral and alkaline conditions but becomes orange in solutions of acidic pH, was added to the water. This made the agar red.

QUESTIONS

A1. The student used a ruler to measure the agar, and cut cubes from it using a scalpel. The cut cubes had sides of 5.0 cm, 4.0 cm, 3.0 cm, 2.0 cm and 1.0 cm.

    a. Calculate the surface area of each cube.

    b. Calculate the volume of each cube.

    c. Calculate the surface area to volume ratio of each cube.

A2. The student made three cubes of each dimension, then placed one cube into a glass dish, and poured dilute hydrochloric acid around it. As the acid diffused into the cube, its red colour became orange.

    The student measured the time for the cube to turn completely orange. They repeated this with each of the three cubes. The student's results are shown below:

5 cm cube: 14 minutes, 15 seconds; 12 minutes, 50 seconds; 13 minutes, 10 seconds.
4 cm cube: 8 minutes, 50 seconds; 9 minutes, 25 seconds; 8 minutes, 50 seconds.
3 cm cube: 5 minutes, 40 seconds; 5 minutes, 30 seconds; 6 minutes, 5 seconds.
2 cm cube: 3 minutes, 30 seconds; 3 minutes, 30 seconds; 3 minutes, 20 seconds.
1 cm cube: 1 minute, 15 seconds; 0 minutes, 55 seconds; 1 minute, 10 seconds.

    a. Draw a results table. Convert the time values into seconds, and record the student's results. Remember:

- draw the table with a ruler
- head each column and/or row fully, including units
- make sure that each entry for a particular set of readings is given to the same number of decimal places, which is determined by the method of measuring the value
- when calculating a mean value, the number of significant figures should be the same as (or one more than) the value with the fewest number of significant figures used in your calculation.

    b. Plot a graph of mean time taken for the colour to change (*y*-axis) against the surface area to volume ratio of the cube (*x*-axis). Join the points with a line of best fit.

    c. Use your graph to write a brief conclusion for the student's experiment. Your conclusion should summarise how the dependent variable is affected by changes in the independent variable.

    d. The student stated that the rate of diffusion – the distance travelled by the acid in a certain amount of time – was probably the same for all of the cubes. Do you agree with this statement? Explain your answer.

A3. The student was asked to identify significant sources of error in the experiment, and to state whether each error was random or systematic.

- **Systematic errors** are generally caused by lack of accuracy or precision in measuring instruments, and by the limitations of reading the scale. They tend to be always the same size, and act in the same direction, on all of the readings and results.
- **Random errors** are often caused by difficulties in keeping control variables (standardised variables) constant. They can also be caused by difficulties in measuring the dependent variable. They may be of different sizes, and act in different directions, at different times or stages of the experiment.

Below are the sources of error that the student listed. For each one:

- state whether you think this was likely to be a significant source of error or not
- if you think it was a significant source of error, state whether it was a random or systematic error.

    1. It was difficult to cut the agar blocks to the exact size.

    2. Temperature changes might have affected the rate of diffusion.

    3. One side of the block was resting on the base of the dish, so was not exposed to the sodium hydroxide solution.

9. List organs in which exchange of materials by simple diffusion is important.

10. A student investigated the relationship between the size of agar cubes containing indicator, and the time taken for the indicator to completely change colour.

   a. Describe the relationship between the surface area to volume ratio of a cell and the rate of diffusion of a substance into that cell.

   b. State two variables that the student should standardise during this investigation. For each variable, explain why this is necessary.

   c. The student found it difficult to identify the point at which the indicator changed colour, and consistently timed each experiment for 10 seconds longer after this had happened. This introduced a source of error into the experiment. Which statements about this error are correct?

      1. The effect of the error will be reduced if the student performs three repeats for each agar cube.

      2. The error will prevent the student from identifying the relationship between surface area to volume ratio and the rate of diffusion.

      3. The error is systematic as the student consistently timed each experiment for 10 seconds after the end-point.

      A. 1 and 2

      B. 1 and 3

      C. 2 and 3

      D. 3 only

   d. Apart from their larger size, describe two ways in which the agar blocks used in this experiment did **not** accurately model a cell and how it exchanges substances with its environment.

FACILITATED DIFFUSION

Glucose molecules are relatively large and have small charges on their hydroxyl (–OH) groups. Their size and polar nature makes it impossible for them to pass between the phospholipid tails in a cell surface membrane. So how are larger or polar particles exchanged by diffusion between the inside of a cell and its environment? This is the job of some of the integral membrane proteins found in the cell surface membrane.

Three types of transport protein are involved in this mechanism, and they differ by how much which they interact with the particle they are transporting. These are **channel proteins**, **ion channels** and **carrier proteins**. Each type of protein provides help, to facilitate, the movement of these particles. This process is called facilitated diffusion. It still involves movement down a concentration gradient and requires no input of metabolic energy, derived from ATP, to transport the particles.

Channel proteins and ion channels have a water-filled pore or 'hole' that is lined with amino acids whose variable R groups are hydrophilic. This enables them to interact with polar and water-soluble particles, facilitating their movement across the cell surface membrane. However, despite this, membrane proteins cannot transport every ion or molecule. They are highly specific, having a tertiary structure that is complementary to the shape of the particle that it will transport. For example, ion channels generally transport only one type of ion, such as Ca^{2+}, Na^+, K^+ or Cl^-. Similarly, the channel protein that allows glucose to move into and out of cells is called glucose permease (Figure 4.13). Some channel proteins are gated. This means that part of the protein molecule on the inside surface of the membrane can move to close or open the pore, like a gate.

Tip

To 'facilitate' means to 'help'. In facilitated diffusion, carrier proteins help molecules to cross the membrane.

Link

The difference in concentration of a substance between two regions is called a concentration gradient. In Chapter 7, you will see how keeping this can maintain rapid diffusion of molecules across membranes.

Link

You will encounter two examples of gated channel protein in Chapter 15, one that allows the entry of sodium ions into neurones, and one that enables the release of potassium ions.

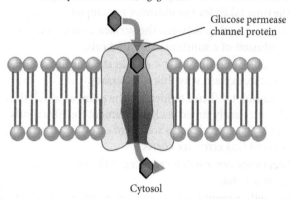

Figure 4.13 Facilitated diffusion using a channel protein. Protein channels and ion channels allow polar molecules and charged ions to move through.

Link

Transmission of nerve impulses, owing to the regulation of ion gradients between their cytoplasm and their external environment by gated ion channels, is discussed in detail in Chapter 15.

Carrier proteins also have a pore that is lined with amino acids whose variable R groups are hydrophilic. The difference between these and channel proteins is that when the diffusing molecule binds to a carrier protein, the protein undergoes a change in shape (a conformational change). This brings the diffusing particle to the other side of the membrane, where it is released (Figure 4.14). Some ion channels can also open and close, owing to small changes in the shape of the protein, to control the passage of charged particles. Channels that can be open or closed are called **gated ion channels**.

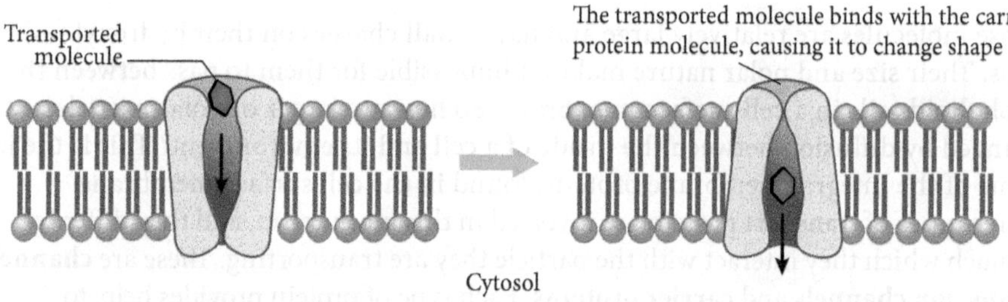

Figure 4.14 Facilitated diffusion using a carrier protein. The diffusing molecule interacts with the carrier protein, causing a change in shape which 'squeezes' the molecule through the channel.

ACTIVE TRANSPORT

Unlike simple diffusion and facilitated diffusion, which are passive processes, there are some situations that require an input of energy from ATP to bring about the movement of particles. This occurs if cells often move substances across membranes that are only present in small concentrations outside them, or vice versa. In this situation, the cell must use some energy, from the hydrolysis of ATP, to move particles against their concentration gradient. The process is therefore called active transport, because the cell is actively making it happen.

The carrier proteins responsible for active transport are sometimes referred to as **pump proteins**, to distinguish them from the carrier proteins involved in facilitated diffusion. An example of active transport using a pump protein is shown in Figure 4.15.

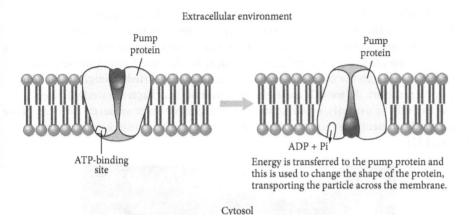

Figure 4.15 Active transport using a pump protein. The molecule to be transported interacts with the pump protein, which hydrolyses ATP to change shape and brings the molecule through the cell surface membrane.

Pump proteins engaged in active transport are often involved in a process called **cotransport**. This involves the transport of two particles at the same time, with the movement of each dependent on the other. Two important examples are given below.

- In the surface membranes of intestinal epithelial cells, sodium ions are pumped out of the epithelial cell cytoplasm and into the surrounding tissue fluid by active transport using ATP. This causes a sodium ion gradient to build up. The **glucose-sodium cotransporter protein** then transports glucose molecules and sodium ions into the cytoplasm of the epithelial cell by facilitated diffusion at the same time (cotransport). This is shown in Figure 4.16(a).

- In the surface membranes of companion cells, hydrogen ions are pumped out of the epithelial cell cytoplasm and into the cell wall by active transport using ATP. This causes a hydrogen ion gradient to build up. The **sucrose-hydrogen cotransporter protein** then transports sucrose molecules and hydrogen ions into the cytoplasm of the companion cell by facilitated diffusion at the same time (cotransport). This is shown in Figure 4.16(b).

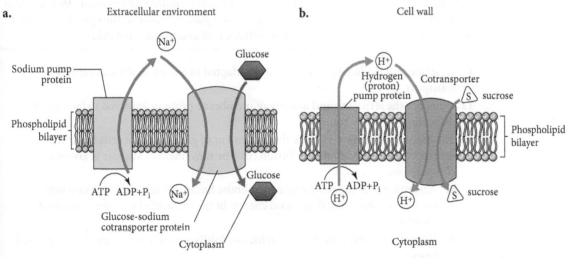

Figure 4.16 Cotransport in an animal cell (a) and a plant cell (b). The glucose-sodium cotransporter protein transports glucose and sodium ion into the cell at the same time (a). The sucrose-hydrogen cotransporter protein transports sucrose and hydrogen ion into the phloem vessel at the same time (b).

The mechanism of glucose-sodium cotransport has led to the development of many applications, ranging from effective sports drinks to important medical developments. Oral rehydration therapy for sufferers of chronic (excessive, ongoing) diarrhoea involves providing drinks to restore the sodium ions and chloride ions. As the glucose molecules are absorbed through the intestinal wall, sodium ions are carried through at the same time by cotransport. As we will see in the next part of this chapter, water will then follow passively by **osmosis**, avoiding the loss of excessive water as faeces. This can prove vital in the treatment of a number of infectious diseases that cause chronic diarrhoea (Figure 4.17).

Figure 4.17 Around five million children each year die from the effects of diarrhoea, through infectious diseases. Poor sanitation, such as obtaining drinking water from water sources contaminated with human excrement, often underlie outbreaks of cholera.

Worked example

Phosphorylated particles (those with a phosphate group added) cannot pass easily through cell surface membranes. Suggest why glucose molecules are phosphorylated as soon as they enter a cell.

Answer

The immediate phosphorylation of glucose after it enters a cell will prevent it from diffusing out of the cytoplasm and back into the extracellular environment. This will also maintain a low concentration of free, unphosphorylated glucose in the cytoplasm, which will maintain a high rate of (facilitated) diffusion of glucose into the cell.

11. Describe and explain how a cell may be adapted to absorb molecules rapidly by active transport.

12. Compare the structure and activity of membrane proteins involved in facilitated diffusion and active transport.

☆ 13. The graphs in Figure 4.18 show the movement of glucose out of cells by active transport (a) and facilitated diffusion (b). The movement of sucrose is also shown for comparison.

   a. Assuming that the cell surface membrane has no proteins able to transport sucrose, explain why there is no change in the intracellular concentration of sucrose.

   b. Describe and explain the similarities and differences in the shapes of the plotted lines.

   c. Predict the difference in the shapes of the two graphs if a respiratory poison such as cyanide was used in these investigations. Explain your answers.

a.

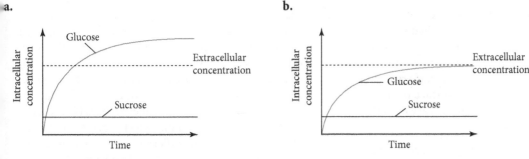

b.

Figure 4.18 How the intracellular concentration of glucose changes over time as it is moved out of the cell by active transport (a) or facilitated diffusion (b).

Experimental skills 4.1: The history of modelling the cell surface membrane

In this exercise, you will be asked to analyse the data of scientists working in the 20th century to draw conclusions about the structure of the cell surface membrane.

With the cell surface membrane being so narrow and difficult to see, scientists needed to use indirect approaches to determine its structure. Early cell biologists worked out its structure by investigating its properties.

In 1925, Dutch scientists Evert Gorter and François Grendel extracted phospholipids from the 'ghosts' of red blood cells (erythrocytes). These are the cell surface membranes that remain when cells are lysed. They calculated that the area that these phospholipids occupied when arranged in a single layer (monolayer) was twice as large as the total area of the cell surface membrane of the cells.

Ten years later, Hugh Davson and James Danielli also purified erythrocyte ghosts in England. They conducted experiments and proposed a model that placed the phospholipid bilayer at the centre of a 'sandwich' between two thin protein outer layers. Electron micrographs produced in the 1950s gave support to this model, showing two dark bands with a lighter and wider band in between.

QUESTIONS

P1. Suggest reasons why red blood cells, rather than cells of other types, were used by early cell biologists to investigate the structure of the cell surface membrane.

P2. Davson and Danielli discovered that proteins exist in the cell surface membrane when they identified the presence of nitrogen atoms in the structure. Describe how, if provided with a sample of cell surface membrane from red blood cell 'ghosts', you could confirm the presence of protein using a simple laboratory test.

P3. Use Figure 4.19 to suggest a feature of the cell surface membrane that cannot be explained by the Davson–Danielli model.

The model proposed by Davson and Danielli persisted for several decades. However, a change in opinion was to come, based on the results of two significant discoveries in the 1960s.

- **Freeze-fracture electron micrographs:** A piece of cell surface membrane is frozen rapidly using liquid nitrogen, then split down the middle between the phospholipid bilayer. An electron microscope is then used to view the inner surface, which shows 'bumps' and 'dents' in a scattered pattern.

a. Davson–Danielli Model (1935)

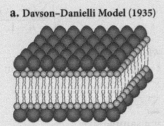

b. Singer–Nicolson Model (1972)

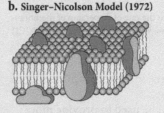

Figure 4.19 The Davson–Danielli and Singer–Nicolson models of cell surface membrane structure.

- **Fluorescent antibody tagging:** Different-coloured fluorescent markers were attached to antibodies that bind to membrane proteins on different cells. If the cells are fused together, then within 1–2 hours the red and green markers were mixed throughout the entire cell surface membrane of the fused cell. This was the case even if ATP synthesis was blocked.

The fluid mosaic model was proposed by 1972 by American cell biologists S. Jonathan Singer and Garth Nicolson.

P4. Explain how freeze-fracture electron micrographs support the idea that the cell surface membrane has a mosaic structure.

P5. Explain how fluorescent antibody tagging supports the idea that the cell surface membrane is a fluid structure.

A Level analysis

P6. A group of scientists carried out an investigation using the fluorescent antibody tagging technique. However, instead of fusing cells, they fused a range of membrane-bound organelles together. They measured the time it took for the differentially-tagged membrane proteins to completely spread out across the fused structures. For each organelle, they carried out the experiment 10 times and calculated a mean value for the time taken (Figure 4.20).

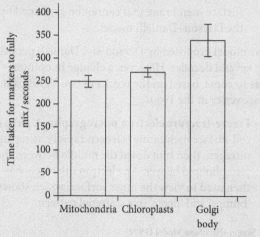

Figure 4.20 An investigation into the rate of dispersion of membrane proteins between fused organelles. 95% confidence interval (95% CI) error bars are shown.

Calculating standard error, 95% confidence intervals and error bars

Descriptive statistics can be used to provide more information on data collected in an investigation. For example, a mean and a sample standard deviation can be calculated, to assess the reliability of a specific dataset. Or, to gain an insight into the reliability of a mean value, the standard error of the mean (SE) can be calculated. This provides an estimate of the range within which the true mean is likely to be found. Next, 95% confidence interval error bars (found by ±2SE) should be drawn on a plotted graph to assess the significance of the differences between the mean values for each solution. If the error bars between any two points overlap, then there is unlikely to be a significant difference between the sample means.

1. How to calculate the mean of a dataset:

   The sample mean is the most commonly used average value in most scientific investigations. It can be calculated by dividing the sum of all the data points by the number of data points:

   $$\bar{x} = \frac{\sum x}{n}$$

   where
   $\bar{x}$ = mean,
   $\sum$ = sum of,
   n = sample size.

2. How to calculate the standard error of a mean of a dataset:

   When it is impractical to record data for all members of a population, the data is collected for just a sample of that population. In these cases, it is important to determine how close the sample mean is to the population mean; this measure is called the standard error of a mean. The standard error provides an indication of the degree of similarity of the mean of the sample population to the mean of the whole population. This is important, because it is usually not possible, or practicable, to measure the value of every individual or cell in a population or tissue sample.

   The standard error of the mean (SE) is found by first calculating the sample **standard deviation** (s) using the following formula:

   $$s = \sqrt{\frac{\sum (x - \bar{x})^2}{n - 1}}$$

   where
   s = sample standard deviation,
   $\bar{x}$ = mean,
   $\sum$ = sum of,
   n = sample size.

The value of the sample standard deviation is then divided by the square root of the sample size. This provides the value for the **standard error**:

$$SE = \frac{s}{\sqrt{n}}$$

where
SE = standard error,
s = sample standard deviation,
n = sample size.

3. How to calculate 95% confidence intervals (CI) for the sample mean of a dataset and plot these as error bars on a graph:
The standard error can be used to calculate the 95% CI for a sample mean:

$$95\% \text{ confidence interval} = \text{mean} \pm 2\,SE$$

The 95% CI represents the range of the sample data in which the true value of the population mean lies, with 95% probability. This can be indicated using an error bar on a graph or chart. For a given sample mean, the error bar extends to the value of 95% CI either side of the sample mean.

In the example bar chart shown in Figure 4.21, the values of three sample means have been plotted, with error bars included for each mean.

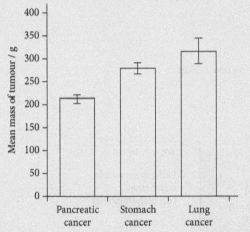

Figure 4.21 A study into the mean mass of tumours found in three organs of people with cancer.

If the error bar is small, then the calculated mean is close to the true mean and the data is reliable. This is the case for the pancreatic cancer tumour mean mass.

Smaller error bars are less likely to overlap. The greater the overlap between any two bars, the greater the probability that there is not a significant difference between the two datasets. This suggests that any perceived difference between the sample means may be due to chance (although a statistical test should be conducted to confirm this).

In the chart above:

- there is likely to be a significant difference between the population (actual) mean of the mass of pancreatic and lung tumours, and pancreatic and stomach tumours, as the sample means of these tumours do not overlap although the mean values for stomach and lung tumours are different, the significance of this difference is questionable, because the error bars for these samples slightly overlap
- the error bar for lung tumours is particularly wide, suggesting that its sample mean may not be a true representation of the population mean for all lung tumours.

a. Describe how the 95% confidence interval error bars (Figure 4.21) provide an indication of the reliability of the mean values for the three different experiments.

b. Identify a statistical test that could be used to assess the significance of difference between the rate of diffusion in mitochondria and the Golgi body. Explain your answer.

☆ c. Suggest reasons to explain the difference in the rates of diffusion seen between the organelles used in this investigation.

OSMOSIS

All cells contain cytoplasm, and most of this cytoplasm consists of a solution of many different dissolved solutes. The cytoplasm is separated from the environment of the cell by the cell surface membrane.

Solute particles and water molecules continuously collide with the cell surface membrane on both sides. The cell surface membrane is permeable to some particles, and not others, meaning that over time, changes in the relative number of all particles on either side of the membrane may occur.

Tip

A **solute** is dissolved in a **solvent** to form a **solution**.

Tip

Never refer to 'water concentration'. The term 'concentration' should be used for referring to solutes only. Water potential is the measure of the purity of a solution in terms of water.

Tip

When comparing two negative values, the more negative value is smaller (or more negative) than the value that is closer to zero. For example, –9 is lower than –3.

Tip

Dialysis tubing is a partially permeable membrane. Cell surface membranes are selectively or differentially permeable. Unlike dialysis tubing, they control the exchange of factors based on more than just their size.

Tip

The polarity of water molecules is vital to many biological processes, including the movement of water through the xylem tissue of plants. You will encounter this topic in Chapter 7.

As we have seen, water molecules can pass directly across the phospholipid bilayer by simple diffusion without the help of membrane proteins. This process is called osmosis. In the case of water, diffusion happens from a region of higher solute concentration to lower solute concentration. The term 'concentration' describes the number of solute particles in relation to the volume of water in a solution. We therefore need a different term to describe the proportion of a solution that is made up by water. This term is **'water potential'**.

Osmosis tends to happen through a **partially permeable membrane**. This is a material that contains pores large enough for water molecules to pass through, but not large enough for the solute molecules to pass through. If two solutions with different water potentials are separated by such a membrane, the solution that has a lower concentration of solute molecules will gradually lose water molecules to the solution with a higher concentration. This is true of all situations involving osmosis: water molecules move from higher to lower water potential, or down a water potential gradient.

Dialysis tubing (also called **Visking tubing**) can be used to demonstrate the effect of osmosis. One set-up is called an osmometer (Figure 4.22). This uses a small piece of dialysis tubing attached to a long piece of glass apparatus with a thistle bulb at one end. The osmometer is filled with a solution of known concentration and placed inside another solution. If the water potential of the solution in which it has been immersed is higher than its contents, then there will be a net flow of water into the osmometer, and the level of the fluid will rise. If the water potential of the external solution immersed is lower, then there will be a net flow of water out of the osmometer, and the level of the fluid will fall. It is possible to calibrate an osmometer using solutions of known water potential.

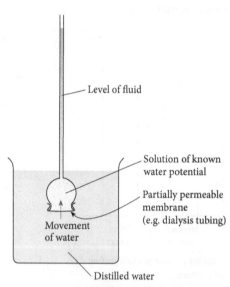

Figure 4.22 The use of an osmometer to demonstrate the process of osmosis.

It is possible to think of water potential as a measure of how 'free' water molecules are to move. Water molecules in a solution with few solute particles (a dilute solution) can move more freely than water molecules in a concentrated solution. This is because water molecules have a small charge and are attracted to soluble molecules and ions, which therefore restrict their ability to move and pass across a partially permeable membrane, through which the solute particles cannot move. Figure 4.23 shows this phenomenon.

a.

Before osmosis

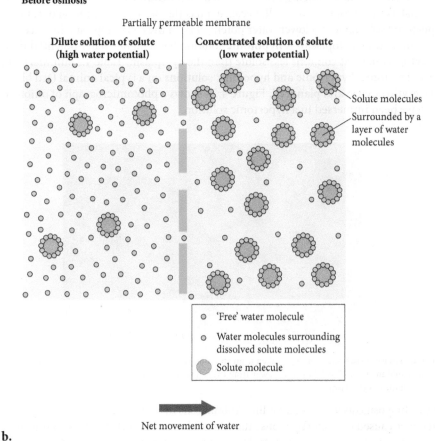

Partially permeable membrane

**Dilute solution of solute
(high water potential)**

**Concentrated solution of solute
(low water potential)**

Solute molecules

Surrounded by a
layer of water
molecules

○ 'Free' water molecule

○ Water molecules surrounding
dissolved solute molecules

● Solute molecule

Net movement of water

b.

After osmosis

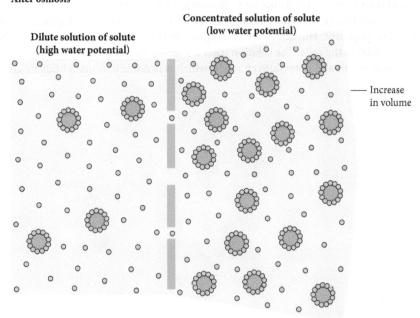

**Concentrated solution of solute
(low water potential)**

**Dilute solution of solute
(high water potential)**

— Increase
in volume

Figure 4.23 The presence of solute molecules reduces the freedom of movement of water molecules and their ability to leave a solution, and so reduces water potential. If separated by a partially permeable membrane, water molecules move, by osmosis, from the solution of a higher water potential to the solution of lower water potential, down a water potential gradient.

Tip

When referring to hypertonic or hypotonic solutions, ensure you refer in a comparative way. For example, 'seawater is hypertonic to plant root hair cells', rather than 'seawater is hypertonic'.

Tip

Prokaryotic cells are affected by osmosis in a similar way to plant cells. As you will see in Chapter 10, osmosis explains how some antibiotics, including penicillin, kill bacteria.

Tip

If a cell is isotonic to its surrounding solution, water moves in and out at an identical rate; if hypertonic, more will enter than exit; if hypotonic, more will go out than in.

Tip

Only use the word 'because' when you are explaining something, never when you are describing.

The direction in which there is a net flow of water between two solutions depends on the difference in water potential between them. An **isotonic** solution has the same water potential as the cytoplasm of a cell, because it has the same concentration of solute. A **hypertonic** solution has a lower water potential than the cell cytoplasm, i.e. it contains more solute than cytoplasm. A **hypotonic** solution has a higher water potential than the cell cytoplasm, i.e. it contains less solute than the cytoplasm. Table 4.2 summarises the effect of isotonic, hypertonic and hypotonic solutions on a typical animal (red blood cells, erythrocytes) and plant cell. Figure 4.24 shows a photomicrograph of onion cells that have been immersed in a hypertonic solution.

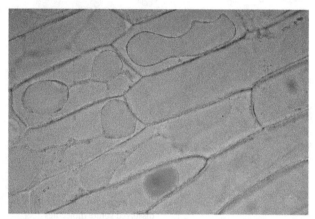

Figure 4.24 Plasmolysed cells in a tissue taken from an onion root exposed to seawater for one hour. The cell surface membrane in every cell has pulled away from the cell wall, so the contents of the cell are contained in a smaller volume of cytoplasm.

Controlling osmosis is essential for life. Animal cells in particular are very susceptible to damage caused by osmosis because they do not have a cell wall. For this reason, all animals must osmoregulate, or control the water potential of, the blood and tissues that surround their cells. You will encounter this concept in Chapter 14. As we saw at the start of this chapter, this also has implications for modern medicine, including the latest organ transplant equipment. Human tissue and organs must be stored in a solution with the same water potential as the cells. The solution used is usually Ringer's solution, a 0.9% solution of sodium, potassium and potassium chloride, which is isotonic to human body fluids.

Cell type	Effect of isotonic solution		Effect of hypotonic solution		Effect of hypertonic solution	
	Description	Explanation	Description	Explanation	Description	Explanation
Animal	A red blood cell has a biconcave shape, roughly circular with a central indentation.	Because the rate of entry and exit of water molecules across the cell surface membrane is identical.	Red blood cell swells and bursts, releasing its contents, including haemoglobin.	Because the water potential of the surrounding solution is higher, the cell experiences a net gain of water by osmosis and grows. As membranes have little mechanical strength, and animal cells lack a cell wall, the blood cell bursts, or **lyses**.	Erythrocytes shrink and become crinkled, or **crenated**. They take on a spiky appearance.	Because the water potential of the surrounding solution is lower, there is a net loss of water from the red blood cells by osmosis.
Plant	A palisade mesophyll cell has a roughly cuboidal shape with a visible vacuole and the cell surface membrane touching the cell wall.	Because the rate of entry and exit of water molecules across the cell surface membrane is identical.	The plant cell begins to swell. The vacuole and cytoplasm enlarge until the cell surface membrane presses firmly against the cell wall. It is described as **turgid**.	Because the cell absorbs water by osmosis and ultimately an equilibrium is reached when the osmotic forces drawing water in are balanced by the wall pressure resisting further expansion.	The vacuole shrinks and the cell becomes **flaccid**. The cell surface membrane and its contents eventually become detached from the cell wall, at which point it is referred to as plasmolysed. Partial **plasmolysis** can lead to a 'star' shaped cytoplasm, but full plasmolysis will cause the cytoplasm to occupy a rounded shape in the middle of the cell (Figure 4.24).	Because the water potential of the surrounding solution is lower than the contents of the cell, there is a net loss of water from the plant cell by osmosis.

Table 4.2 The effect of solutions of different water potentials on animal (red blood cell, or erythrocyte) and plant (palisade mesophyll) cells.

Worked example

An experiment to investigate osmosis was conducted using a bag made from dialysis (Visking) tubing. Dialysis tubing is a sheet of plastic that is partially permeable. It allows water molecules through, but has pores that are too small for larger molecules, such as starch, to pass through. The apparatus was set up as shown in Figure 4.25.

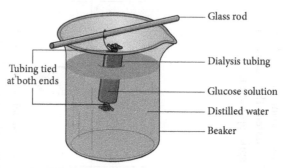

Figure 4.25 An experiment using dialysis tubing.

Using the term 'water potential' in your answer, explain why the volume of glucose solution inside the bag shown in Figure 4.25 will increase over time.

Answer

The water potential of the solution inside the bag is lower than the water potential of distilled water that surrounds the bag. This means that water will move by osmosis, from a region of higher water potential to a solution of lower water potential, causing the volume of glucose solution inside the bag to increase over time.

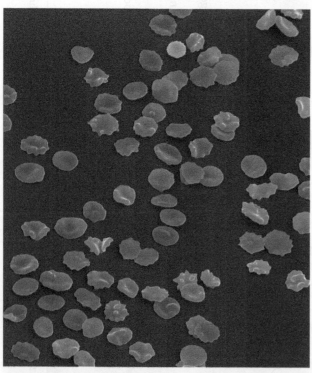

Figure 4.26 The effect of of sodium chloride solution on human erythrocytes.

14. Figure 4.26 shows the effect of a solution of concentrated sodium chloride on human red blood cells (erythrocytes). The cells have taken on a 'spiky' shape and are small in size. Explain these observations.

15. Many species of water-dwelling nematode, including the flatworm, have specialised cells called flame cells. These actively pump out ions from the cytoplasm into the surrounding environment. Using this information, suggest and explain whether the usual habitat of the flatworm is fresh water or salt water.

16. Most foods will spoil when exposed to the environment, because bacteria land on them and divide. Suggest why bacteria are not able to reproduce in and spoil sugary foods such as fruit preserves and syrups.

★17. Suggest why sports scientists recommend that the maximum concentration of ions in sports drinks should be 200 mg dm^{-3}.

Experimental skills 4.2: Determining the water potential of potato tissue

A class of students investigated the water potential of potato tissue. To do this, they produced a series of solutions of different known concentrations of sucrose. They cut pieces of potato of similar mass and immersed these for 3 hours in test-tubes containing the different solutions, as shown in Figure 4.27. They used the final masses of these potato pieces to draw a calibration curve that could be used to estimate the water potential of this plant tissue.

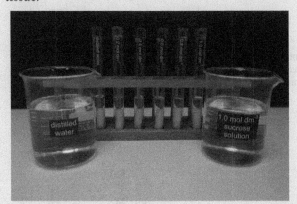

Figure 4.27 An investigation to estimate the water potential of potato tissue.

The students were provided with a solution of sucrose of concentration $1.0 \, mol \, dm^{-3}$ and distilled water.

QUESTIONS

P1. Describe how the students could use the proportional dilution method to produce a series of four more solutions, of $20 \, cm^3$ each, of sucrose concentrations 0.8, 0.6, 0.4 and $0.2 \, mol \, dm^{-3}$.

The students decided to plot a graph of sucrose concentration against the percentage change in mass of the potato tissue. To find this, they calculated the percentage change in mass using this formula:

$$\text{percentage change in mass} = \frac{\text{final mass} - \text{original mass}}{\text{original mass}} \times 100$$

P2. Explain in detail why some pieces of potato were found to have a positive value for the percentage change in mass, whereas others had a negative value.

P3. Using the term 'valid' in your answer, explain why the students plotted the percentage change in mass of potato tissue instead of the final mass of potato tissue.

The students obtained the results shown in Table 4.3.

Concentration of sucrose / mol dm⁻³	Percentage change in mass / %
0.0	19.50
0.2	8.15
0.4	−5.45
0.6	−12.50
0.8	−19.25
1.0	−25.55

Table 4.3 The effect of sucrose concentration on the percentage change in mass of potato tissue.

Tip

To ensure that an investigation is valid, all other variables must be standardised (kept the same). The independent variable should be the only factor that is changed.

P4. Outline how, by plotting a calibration curve, the students could estimate the concentration of sucrose solution that has a water potential equivalent to that of potato tissue.

Tip

To ensure that the data obtained from an investigation is reliable, repeat measurements must be conducted and a mean calculated. This will reveal anomalous results and make conclusions replicable.

P5. Suggest **two** reasons to explain why the students may not be able to generalise their findings to all plant tissues.

Tip

Accurate measurements are close to the actual values of a factor. Using more precise equipment can increase the accuracy of measurements obtained in an investigation.

P6. State the level of risk in this investigation and explain how **two** hazards could be minimised.

ENDOCYTOSIS AND EXOCYTOSIS

Simple diffusion, facilitated diffusion, active transport and osmosis apply to individual molecules or ions across the cell surface membrane. **Endocytosis** and exocytosis (Figure 4.28) enable a cell to transport larger volumes of material (solids and liquids) between the internal and external environments. These are both active processes – they require ATP to occur.

Link

As you will see in Chapter 5, when plant cells divide, exocytosis is used to release the materials required to build the cell wall between the two new (daughter) cells.

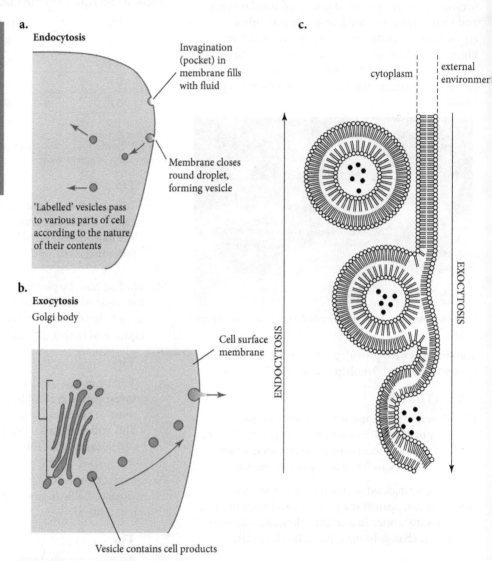

a. Endocytosis

Invagination (pocket) in membrane fills with fluid

Membrane closes round droplet, forming vesicle

'Labelled' vesicles pass to various parts of cell according to the nature of their contents

b. Exocytosis

Golgi body

Cell surface membrane

Vesicle contains cell products

c.

cytoplasm

external environmer

ENDOCYTOSIS

EXOCYTOSIS

Figure 4.28 Endocytosis (a) and exocytosis (b). In endocytosis, a cell's cytoplasm flows round material and completely surrounds it. A segment of cell surface membrane is pinched off and a vesicle is formed within the cytoplasm. In this way, some cells engulf (takes in) particles such as bacteria and fragments of cells in order to destroy them. In exocytosis, materials enclosed in vesicles are expelled from the cell when the vesicle membrane fuses with the cell surface membrane; this is how the cell secretes many of its synthesised products, such as digestive enzymes and hormones. (c) Both methods of transport are dependent on the fluid nature of the cell surface membrane, which means that vesicles can fuse with it (endocytosis) or bud off from it (exocytosis).

Key ideas

→ Simple diffusion is caused by the random motion of particles of gases, liquids and solutes in solution. The net movement of the particles is from high to low concentration; down their concentration gradient. It is a passive process, relying only on the kinetic energy of the moving particles.

→ Small molecules, and larger ones with no charge, are able to diffuse freely through the phospholipid bilayer of cell surface membranes.

→ Other particles, such as large molecules and those with significant charges, can only diffuse through protein channels in the cell surface membrane. This is called facilitated diffusion and is a passive process.

→ Proteins that simply provide a channel through which particles move are called channel proteins. If they interact with the particles and change shape while doing so, they are called carrier proteins. Many carrier proteins transport ions and are called ion channels.

→ Active transport is the movement of particles across cell surface membranes against a diffusion gradient. This process requires energy, usually in the form of ATP. It is achieved by special carrier proteins called pump proteins in cell surface membranes.

→ In cotransport, two substances (for example, sodium ions and glucose) are transported together through the same carrier protein. Depending on the molecules, this can involve either facilitated diffusion, active transport or both.

→ Osmosis is a particular type of diffusion, in which water molecules diffuse through a selectively permeable membrane, such as a cell surface membrane.

→ The tendency of water to leave a solution is known as its water potential. Adding a solute decreases water potential.

→ Osmosis is the diffusion of water from a region of higher water potential to a region of lower water potential across a partially permeable membrane.

→ An isotonic solution has the same water potential as the cytoplasm of a cell, because it has the same concentration of solute. A hypertonic solution has a lower water potential than the cell cytoplasm, i.e. it contains more solute than cytoplasm. A hypotonic solution has a higher water potential than the cell cytoplasm.

→ Endocytosis is the passage of droplets of fluid (pinocytosis) or solid particles (phagocytosis) into the cell by becoming enveloped in cell surface membrane.

→ Exocytosis is the passage of material (usually a cellular secretion) out of the cell.

Assignment 4.2: When do we sleep?

Many people work at night-time, or travel long distances between time zones on aircraft (Figure 4.29). How does this affect our natural sleep patterns?

The hormone melatonin, a hydrophilic hormone released from the pineal gland in the brain, is known to play a very important role in controlling sleep patterns. As concentrations of the hormone rise, we feel sleepier. This happens naturally when light levels drop at the end of a day. Just before we wake, the secretion of melatonin by the pineal gland reduces.

QUESTIONS

A1. Explain what is meant by the term 'hydrophilic'.

A2. Melatonin acts as a ligand and has its effects on target cells via a cytoplasmic enzyme called phospholipase C (PLC). This is an enzyme that

Figure 4.29 Understanding how our sleep patterns are controlled can help those who work unusual shifts, or who suffer from jet lag owing to long-distance travel on aircraft.

phosphorylates (adds a phosphate group to) its substrate, a molecule called diacyl glycerol. Phosphorylated diacyl glycerol stimulates channel proteins to transport calcium ions into the cytoplasm.

a. The term 'ligand' is derived from the Latin word 'ligare', which means 'to bind'. Explain why this is an appropriate term for this component of the cell signalling mechanism.

★ b. Using this information, sketch a diagram to show how melatonin causes an uptake of calcium ions from the environment of the cell by endocytosis.

c. Describe one other change that ligands can bring about in their target cells via cell signalling, in order to bring about a response.

A3. Melatonin can also control blood glucose concentration by affecting target cells in the pancreas. It triggers the release of insulin that has been stored in vesicles in the cytoplasm. Insulin reduces blood glucose concentration.

a. Name the process by which large hydrophilic molecules such as insulin could be released from a cell by storage vesicles, and describe in detail how this occurs.

b. Just before waking, an individual's blood glucose concentration starts to rise. Using the information provided throughout this assignment, suggest how this mechanism is controlled.

Experimental skills 4.3: Investigation into the effect of temperature on the permeability of cell surface membranes

As we saw earlier in this chapter, the identity of the molecules in the cell surface membrane was for a long time studied using indirect methods. Similarly, its permeability has also been investigated indirectly, by measuring the effect of environmental factors on the integrity of the cell surface membrane.

One method used to investigate membrane permeability is by measuring the release of detectable substances from the cell. Some cells contain coloured substances, which are released as the cell surface membrane becomes more permeable. The cells of beetroot plants, for example, contain the water-soluble, dark-red pigment betalain in their vacuoles, as shown in Figure 4.30. The colour intensity of a solution into which a beetroot cylinder has been immersed correlates well with the permeability of its tonoplast (vacuolar membrane) and cell surface membrane.

Temperature is one factor that affects membrane permeability. At cold temperatures, the vacuolar pigment cannot leave through the tonoplast and cell surface membrane. Therefore, little or no pigment will enter the surrounding solution and its colour intensity will be low. However, at warmer temperatures, the kinetic energy of both the betalain molecules and molecules in the tonoplast and cell surface membrane increases. This increases the likelihood that more betalain molecules will be able to move by diffusion from the vacuole and into the cytoplasm, and then through the cell surface membrane to enter the surrounding solution. Therefore, the colour intensity of the surrounding solution will be higher, as shown in Figure 4.31.

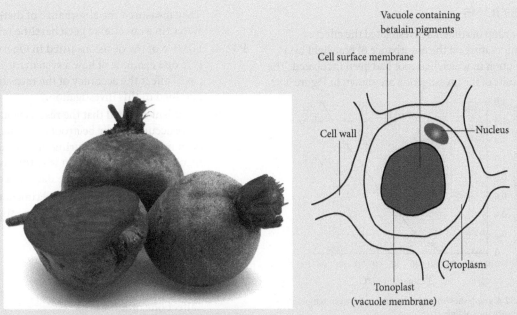

Figure 4.30 Beetroot is commonly used to investigate cell surface membrane permeability. (a) A beetroot and its soft flesh, exposed after cutting. (b) A diagram of a beetroot cell and the location of the red-purple betalain pigments. The tonoplast has a similar molecular structure to that of the cell surface membrane.

At higher temperatures, membrane proteins in both the tonoplast and the cell surface membrane will denature. This will result in the formation of 'holes' in the structure through which greater concentrations of betalain can diffuse out. This means that the colour intensity of the surrounding solution at these temperatures will be the highest, as shown in Figure 4.31.

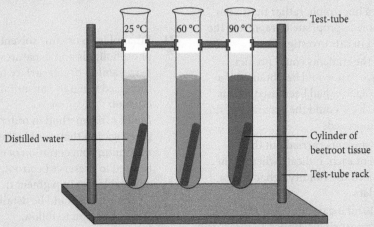

Figure 4.31 The effect of temperature on the amount of betalain pigment leaking from beetroot cylinders. The more damage to the membrane, the darker the surrounding solution will become.

Link

Look back to Topic 3.2 for a description of how a colorimeter works, and how it should be used.

A series of colour standards for comparison can be used to provide qualitative data in this investigation, but a colorimeter is preferable.

The basis of this method will enable you to exercise your skills in planning an investigation. You can carefully consider how many measurements to take and with which instruments, and how and why other variables are controlled. This investigation also provides good opportunities to consider the difference between random and systematic errors in practical science.

QUESTIONS

P1. A group of students investigated the effect of temperature on the absorbance of blue light by a solution in which beetroot had been incubated. The results of the investigation are shown in Figure 4.32.

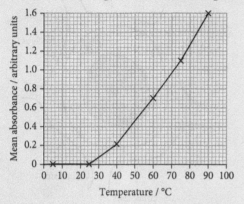

Figure 4.32 A graph showing the relationship between temperature and the absorbance of light.

a. Describe in detail the relationship between temperature and the absorbance of the surrounding solution, as shown in Figure 4.32.

b. Explain in detail how the graph shown in Figure 4.32 supports the fluid mosaic model of cell surface membrane structure.

c. Explain why a line graph, rather than a bar chart, was more appropriate to represent the data collected in this investigation.

d. Outline how the students could predict the absorbance that would be obtained if a temperature of 65 °C had been used in this investigation. How could the accuracy of this estimate be improved?

P2. Explain why the students carried out the following steps. For each, indicate whether the procedure enhanced the validity, reliability or accuracy of the data.

1. They used a 10 ml measuring cylinder, rather than a 50 ml measuring cylinder, to measure 25 ml distilled water into each boiling tube.

2. They cut all beetroot cylinders with a cork borer to ensure that they were the same size, and removed all pieces of skin from the tissue.

3. They washed the beetroot cylinders under cold, running water for at least 2 minutes before they were put into the distilled water.

4. They set the colorimeter to transmit blue–green light.

5. They measured the absorbance of distilled water (in a cuvette) to get a baseline reading.

P3. a. For two of the decisions listed in Question P2, give **one** example of how a systematic error could affect the accuracy of the measurements obtained in this investigation.

b. A student claimed that the results would be more accurate if the beetroot and water had been mixed regularly during the 20-minute incubation period. Explain why this would improve the accuracy of the data collected.

c. Describe a suitable control experiment that the students could have used in this investigation.

Tip

Errors reduce data accuracy: **random errors** affect one or more measurements and generally cannot be avoided; **systematic errors** affect all measurements in a similar way and can usually be avoided.

P4. Ethanol is an organic solvent that dissolves phospholipids and denatures proteins. The permeability of cell surface membranes is increased when the membranes are exposed to ethanol.
Modify this method in order to plan an investigation that would allow you to find the minimum concentration of ethanol that can be used to preserve beetroot, without causing a detectable loss in pigment from cells.
Your plan should be detailed enough for another person to follow.

A Level analysis

P5. The data obtained by the class of students is shown in Table 4.4.

a. Copy and complete the table by calculating the missing values (for the experiment conducted at 5 °C) for standard error and the values required to plot 95% confidence interval error bars on a graph. Refer to the exercises you completed earlier in this chapter for further guidance.

b. If a statistical test were to be carried out between different mean values of absorbance, identify for which **two** temperatures there is unlikely to be a significant difference in the mean absorbance values. Explain your answer.

☆ **c.** Identify a statistical test that can be used to determine whether the difference between two means is significant.

Temperature / °C	Mean absorbance / arbitrary units	Sample standard deviation (s) / arbitrary units	Standard error (SE)/ arbitrary units	Upper limit of 95% confidence interval	Lower limit of 95% confidence interval
5	0.034	0.011			
25	0.041	0.004	0.002	0.045	0.037
40	0.290	0.046	0.026	0.293	0.187
60	0.747	0.257	0.148	1.043	0.450
75	1.113	0.086	0.050	1.213	1.014
90	1.690	0.066	0.038	1.766	1.614

Table 4.4 The data obtained in an investigation into the effect of temperature on membrane permeability in beetroot cells.

CHAPTER OVERVIEW

Review the mini mind map to link new learning with your prior learning, and to highlight which of the key concepts underpin the chapter.

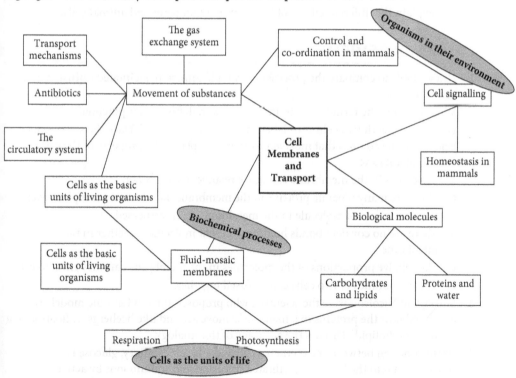

Try copying this mini mind map and expanding upon it. Use your notes from other chapters to help you explore how the essential ideas, theories and principles can be linked further together.

WHAT YOU HAVE LEARNED

Now that you have finished this chapter you should be able to:

- describe the fluid mosaic model of membrane structure with reference to the hydrophobic and hydrophilic interactions that account for the formation of the phospholipid bilayer and the arrangement of proteins
- describe the arrangement of cholesterol, glycolipids and glycoproteins in cell surface membranes
- describe the roles of phospholipids, cholesterol, glycolipids, proteins and glycoproteins in cell surface membranes, with reference to stability, fluidity, permeability, transport (carrier proteins and channel proteins), cell signalling (cell surface receptors) and cell recognition (cell surface antigens)
- outline the main stages in the process of cell signalling leading to specific responses:
  - secretion of specific chemicals (ligands) from cells
  - transport of ligands to target cells
  - binding of ligands to cell surface receptors on target cells
- describe and explain the processes of simple diffusion, facilitated diffusion, osmosis, active transport, endocytosis and exocytosis
- investigate simple diffusion and osmosis using plant tissue and non-living materials, including dialysis (Visking) tubing and agar
- illustrate the principle that surface area to volume ratios decrease with increasing size by calculating surface areas and volumes of simple 3-D shapes
- investigate the effect of changing surface area to volume ratio on diffusion using agar blocks of different sizes
- investigate the effects of immersing plant tissues in solutions of different water potentials, using the results to estimate the water potential of the tissues
- explain the movement of water between cells and solutions in terms of water potential, and explain the different effects of the movement on plant and animal cells.

CHAPTER REVIEW

1. Draw a table to compare the processes of simple diffusion, facilitated diffusion and active transport.
2. Without using the term 'osmosis' in your answer, define 'water potential'.
3. Draw a table with two columns, labelled 'Fluid' and 'Mosaic'. Place the statements below into the column that most appropriately explains the property of the cell surface membrane.
   a. Molecules in the membrane can move relative to each other.
   b. There are many separate proteins in the membrane, rather than a complete layer.
   c. Different types of molecule in the membrane are interspersed
   d. There are no covalent bonds holding different molecules together in the membrane.
   e. The relative proportions of the molecules in the membrane can vary from to cell.
   f. The membranes of two cells can be fused together.
4. Singer and Nicolson were the scientists who proposed the fluid mosaic model. They described how the proteins in a membrane move around like 'icebergs … floating in a sea of phospholipid'. Discuss the accuracy of this analogy.
5. At the junction between the epithelial cell and a blood capillary, glucose is transported into the blood by facilitated diffusion and sodium ions by active transport. Suggest why different processes are required to transport these two substances from the epithelial cell into the blood.

6. Aquaporins are a family of membrane proteins that enable cells to absorb and release water molecules more rapidly. Suggest **two** structural features of aquaporins.

7. Some insects produce a venom that contains an enzyme called phospholipase. This enzyme hydrolyses phospholipids and can destroy red blood cells. Suggest why this occurs.

8. Refer back to Figure 4.25 to help you answer this question.

   A student suggested that increasing the concentration of glucose in the dialysis tubing would increase the rate of osmosis and the volume of solution in the bag.

   a. Suggest a null hypothesis for this experiment.

   b. Construct a table to list the independent, dependent and standardised variables in this investigation. Assume that the same apparatus is used between the different experiments.

   c. For all of your standardised variables, describe the apparatus you would use to ensure they are kept constant.

Chapter 55

The mitotic cell cycle

Henrietta Lacks died of cancer in 1951. Without her permission, doctors took a sample of her cancer cells and grew them in the laboratory. In the decades that followed, these cells, named HeLa after her, were provided to thousands of scientists around the world. These cells are still dividing today and continue to be used in medical research.

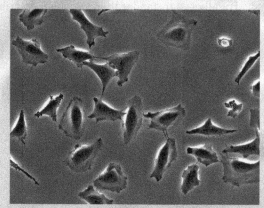

HeLa cells growing in a laboratory culture plate.

Today, we now know much about the role that cell division plays in growth, repair and reproduction, thanks to the work carried out using HeLa cells. Understanding when, how and why cells divide has led to great discoveries related to ageing and problems that arise owing to abnormal cell division – including the disease that led to the death of Henrietta Lacks.

Prior understanding

From your work in Chapter 1, you may remember the structure and function of organelles as seen in electron photomicrographs. You may also recall how to use a light microscope and make temporary preparations of living material. You may have calculated the magnification and actual sizes of specimens from images, and practised using the units (millimetre, micrometre and nanometre) used in cell studies. You may be familiar with calibrating and using a stage micrometer and an eyepiece graticule.

Learning aims

In this chapter, you will learn about the process of cell division, including mitosis. Crucial to mitosis are chromosomes, and you will explore their behaviour during this process. There will be further opportunities to reinforce and develop your skills in using a microscope, a stage micrometer and an eyepiece graticule.

5.1 Replication and division of nuclei and cells (Syllabus 5.1.1–5.1.6)

5.2 Chromosome behaviour in mitosis (Syllabus 5.2.1, 5.2.2)

5.1 Replication and division of nuclei and cells

The invention of the microscope was perhaps the most important development in practical cell biology. It showed for the first time that all organisms are composed of cells. We are now very familiar with the idea of cells as the building blocks of life.

Some scientists, such as Walther Flemming, explored cells further. They proposed that cells originated from pre-existing cells, and that material was passed from one cell to another during the process by which cells make new cells.

Imagine Flemming at work in his laboratory in the late 19th century. Making careful observations using a microscope, he spent hours drawing what he saw. His notebooks provide a record of a now very recognisable set of actions that occur when any eukaryotic cell divides. Whenever one cell became two, Flemming saw threads of material acting out a specific and predictable sequence of movements. He proposed the use of the term **mitosis** for what he saw, from the Greek word 'mitos', or 'thread'.

THE STRUCTURE OF CHROMOSOMES

The threads that Flemming observed were later recognised as being involved in the transfer of genetic material from one eukaryotic cell to another. Soon after Flemming's work, these threads were given their own name, chromosomes, which means 'coloured bodies'. This is because they absorb special stains when they are applied to cells.

In one nucleus of diameter 6 μm, there is DNA of around 1.8 m in length. This is the equivalent of trying to pack a piece of narrow string 18 km in length into a ping-pong ball 6 cm in diameter. Therefore, chromosomes consist of highly compacted DNA. In each chromosome, a single, immensely long DNA molecule is wound around positively charged **histone proteins** like cotton thread around a reel. DNA has a negative charge so it is attracted to the histone protein. These tightly compacted units, are further coiled around themselves to form a material called chromatin. This is said to be **supercoiled**. In most human cells, there are 46 supercoiled chromosomes (Figure 5.1). This number is characteristic of a species and is usually different for different species. For example, chimpanzees have 48, mice have 40 and fruit flies have 8. Figure 5.2 summarises the number and structure of the human chromosomes.

> **Tip**
> Chromosomes are transparent. Therefore, they must be stained to become visible using a microscope. Methylene blue, toluidine blue and acetic orcein have been used for many years to visualise cells.

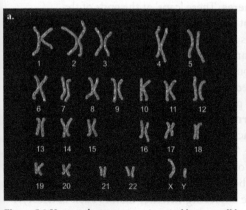
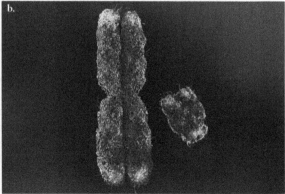

Figure 5.1 Human chromosomes. A typical human cell has 46 chromosomes (a), which are arranged in order of decreasing size. In males, the 23rd pair of chromosomes consists of an X and a Y chromosome (b), which are of greatly different sizes.

Before a cell has replicated its DNA, chromosomes consist of two arms that are pinched inward at a site called the **centromere**. Although it is always in the same position on a particular chromosome, the two arms will not necessarily be equal in length. This is because the centromere can be found anywhere along the length of the chromosome. The centromere plays a very important role during mitosis, as we will see later. After DNA replication, each chromosome consists of two such structures, **sister chromatids**, each held together by the centromere. Both of the sister chromatids have an identical DNA sequence.

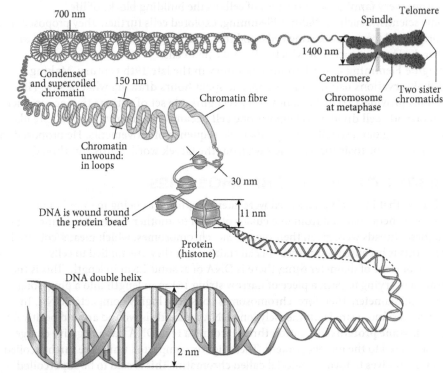

Figure 5.2 The structure of a chromosome.

Telomeres

In the late 1970s, Elizabeth Blackburn, an Australian researcher working at Yale University in the USA, co-discovered **telomeres**. These 'caps' at the ends of the chromosome arms have since attracted great attention. Telomeres gradually reduce in size during the lifetime of an organism. This is because when DNA is copied before cell division, the very end of the chromosome cannot be replicated and so a small section is lost each time a cell divides. Once the telomere is lost, vital genes contained in the chromosome can be lost when DNA is copied.

Telomeres are thought to be involved in protecting the loss of genes found in the DNA further down the chromosome arms. They have been compared with the protective plastic tips on the ends of shoelaces. When the telomeres become too short, the cell usually stops dividing and may die. This explains why most cells can only divide a maximum of around 50 times. Telomere length is an accurate indicator of cellular ageing: there is a good correlation between the length of telomeres and lifespan in many species (Figure 5.3a). Scientists are exploring whether factors such as lifestyle and diet have a direct effect on how quickly telomeres reduce in length (Figure 5.3b). This could mean that slowing down or stopping the shortening of telomeres could slow down the ageing process. However, other factors are also likely to be involved in growing old.

Tip

A correlation does not necessarily imply a causative relationship. Another factor may in fact be responsible for the variation in a dependent (measured) variable.

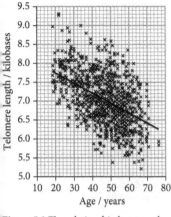

Figure 5.3 The relationship between the age of an individual and the average length of their telomeres (a). The Nicoya Peninsula in Costa Rica. The telomeres of people who live here are longer than average. This could be because of aspects of lifestyle, diet or other factors in the region (b).

1. State the name of the proteins involved in the packaging of DNA in eukaryotic cells.
2. Describe the structure of a chromosome, without using a diagram in your answer.

THE IMPORTANCE OF MITOSIS

Flemming recognised that the structure of chromosomes and their behaviour was crucial for successful cell division. Later work showed that the passage of chromosomes from a **parent cell** to **daughter cells** during mitosis is important in the production of genetically identical cells (**clones**). To achieve this, one cell replicates its chromosomes and then divides to become two cells, distributing one full set to each cell. Mitosis is required for a wide variety of purposes, from embryonic development and **growth** in multicellular organisms, to cell replacement and repair of tissues. Some organisms, such as lizards and starfish, have the ability to regenerate whole body parts by rapid mitosis.

In plants, mitosis occurs in the tissue at the **apical meristem**, found at the end of the root and shoot tips. It is also found in the lateral **meristem**, on the sides of the plant stem. We will revisit the meristem tissue of plants in Topic 5.2 as part of a practical activity.

Mitosis is also the basis of **asexual reproduction** in some organisms. It is the process that occurs when a unicellular organism or simple multicellular organism undergoes cell division to form a new, genetically identical individual of the species. For example, asexual reproduction occurs when unicellular yeast or an *Amoeba* cell divides, or when a bud forms and detaches from a microscopic multicellular organism such as *Hydra*. Asexual reproduction also occurs when new plants develop from vegetative organs such as bulbs.

Tip
The meristem regions in plants consist of **meristematic tissue**.

STEM CELLS AND CANCER

Growth in organisms is the result of repeated cell divisions by mitosis of the zygote, the first cell that forms after fertilisation. As we saw in Chapter 4, cells will need to divide after reaching a certain size, or surface area-to-volume ratio, as in larger cells it will take longer to exchange nutrients and wastes. As we will see later, plants grow by rapid division of meristem cells in the tips of shoots and roots, which remain active during the whole of a plant's life. However, in many animals, including humans, the equivalent cells, **embryonic stem cells**, only exist in the very early life of the organism. They are formed soon after fertilisation, and divide rapidly to form an embryo. Early embryonic stem cells are described as **totipotent**, as they have the ability to **differentiate** into any specialised cell type in the body (Figure 5.4). In a mature human body, most specialised cells have lost their potency and are unable to divide.

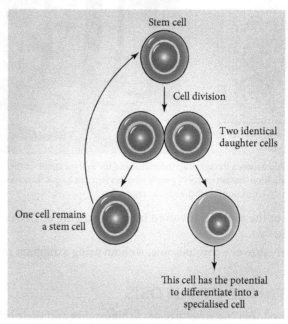

Figure 5.4 Stem cell division. The division of the parent cell forms two daughter cells. One is able to differentiate into a specialised cell, whereas the other remains as a stem cell to replace the parent cell.

Figure 5.5 A computer-generated image of a dividing breast cancer cell.

Small populations of cells in many animals retain the ability to divide in order to allow **tissue repair** (healing). These cells are described as **multipotent**, because they may be able to produce one or very few specialised cell types. Multipotent cells in adult mammals are found in the gut, in the liver, and in the bone marrow, which produces red blood cells that have a lifespan of only a few months and are lost at the rate of more than 250 billion per day. Skin cells are also lost constantly – most of the dust in a house consists of dead human skin cells from those who live there! In the case of all **stem cells**, it is thought that this potency is maintained by keeping certain genes active. When these genes are switched off, the cell begins its journey towards **differentiation**. Research into stem cells has suggested that they may have some exciting medical applications. **Stem cell therapy**, the treatment of damaged organs using stem cells to grow new tissue, promises a cure for diseases ranging from diabetes to stroke, and from blindness to cardiovascular disease.

Continued cell division is not always a good thing, however. Uncontrolled cell division can result in the formation of a **tumour**, an abnormal mass of cells that can develop at any stage of life and in any region of the body (Figure 5.5).

Mutations are random changes to the base sequence of DNA. A number of genes are involved in the control of the **cell cycle**. If these are mutated, the control of the cell cycle can be lost, and cancer is the result. Mutation can be caused by a number of factors, including **ionising radiation** and certain chemicals called **mutagens** or **carcinogens**. A single **cancer cell** will then divide in an uncontrolled way to produce a tumour.

Some tumours are **benign** – they grow slowly and do not affect other parts of the body. However, some are **malignant**. Cancer cells from malignant tumours can enter the bloodstream and the lymphatic system, in a process called **metastasis**. These can form new tumours elsewhere in the body, which can stop vital organs from working properly. This explains the deadliness of cancer: globally, cancers account for about one in six deaths. This sequence of events is summarised in Figure 5.6. Henrietta Lacks, who we met at the beginning of this chapter, developed a malignant cervical cancer in her late twenties. HeLa cells are still used to explore how well new cancer treatments control the rate of cell division and metastasis.

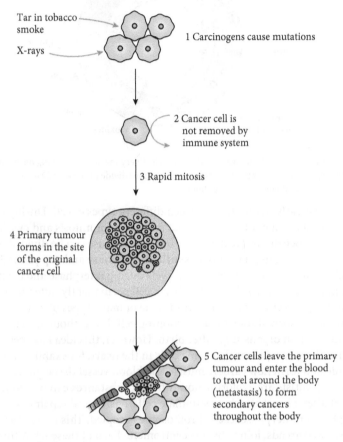

Tar in tobacco smoke

X-rays

1 Carcinogens cause mutations

2 Cancer cell is not removed by immune system

3 Rapid mitosis

4 Primary tumour forms in the site of the original cancer cell

5 Cancer cells leave the primary tumour and enter the blood to travel around the body (metastasis) to form secondary cancers throughout the body

Figure 5.6 How cancer develops.

3. Compare the features of stem cells with cancer cells.

4. Stem cells in the germinal epithelium of the testes, responsible for producing sperm, divide from puberty for the entire lifetime of a human male.

   Scientists have identified an enzyme, called telomerase, which is produced in germinal epithelium cells but not in most body cells.

   a. Suggest the role of the telomerase enzyme in germinal epithelium cells.

   b. State another cell type in humans that is likely to produce the telomerase enzyme. Explain your answer.

THE MITOTIC CELL CYCLE

The process by which one cell becomes two is called the mitotic cell cycle. This begins when two new cells form from a parent cell and ends when the two daughter cells begin the process of dividing again. All cells will be in a given time period, or **phase**, of the cell cycle at any one time. However, not all cells in a tissue are in the same phase. Figure 5.7 shows a summary of the mitotic cell cycle.

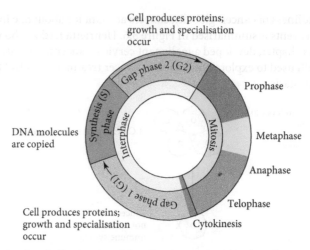

Figure 5.7 The mitotic cell cycle. The duration of the cell cycle is very variable, depending on environmental conditions and the type of cell. For example, root tip cells in onions divide once every 20 hours, whereas epithelial cells in the human intestines divide every 10–12 hours.

Interphase varies greatly in duration between different types of cell. During this time, the cell carries out a number of processes that prepare it for mitosis and **cytokinesis**. We will explore these processes in greater depth later. Interphase can be divided into three periods, with the order: **gap phase 1 (G₁)**, **synthesis (S) phase** and **gap phase 2 (G₂)**.

It is possible for a cell to exit from the cell cycle during interphase and enter another state called **gap phase 0 (G₀)**. This occurs when a cell permanently differentiates, or specialises, and will never divide again. This is true of many types of cell, most notably neurones. This is the normal state for a functioning cell. It was thought that cell division in many mature human organs stops after birth. However, this idea has been challenged in recent years. There is evidence that some cells in the heart, for example, retain the ability to divide and form new cardiac muscle and blood vessel tissue under certain circumstances. Scientists are exploring how different substances can be applied to a damaged heart after a heart attack, to stimulate cell division and repair of cardiac muscle.

During S phase of interphase, each chromosome is copied. This means that they appear to have two threads, joined by one centromere. Each of these DNA threads in a replicated chromosome is called a chromatid. One chromosome has two chromatids, called sister chromatids. These sister chromatids, which consist of copied DNA molecules, will be separated later during mitosis to become chromosomes again.

Link

You will see in Chapter 6 how DNA is replicated during S phase. This ensures that each daughter cell obtains one copy of each chromosome from the parent cell.

Worked example

In an investigation into the effectiveness of a cell cycle inhibitor, a researcher counted how many HeLa cells in a sample were in different phases of the cell cycle. Results were recorded in a tally chart, as shown in Table 5.1.

Stage of cell cycle	Tally count
interphase	ЖЖ ЖЖ ЖЖ ЖЖ ЖЖ ЖЖ ЖЖ ЖЖ ЖЖ ЖЖ |||
prophase	ЖЖ ||
metaphase	|||
anaphase		
telophase	ЖЖ ||
cytokinesis	||||

Table 5.1 The results of an investigation into the effectiveness of a cell cycle inhibitor on mitosis in HeLa cells.

a. Calculate the percentage of cells that were in mitosis in this investigation.

Answer

Prophase, metaphase, anaphase and telophase are phases of mitosis. Therefore, 7 + 3 + 1 + 7 cells were in mitosis (= 18), which as a percentage of the total number of cells counted (75) is 24%.

b. Display the results of this investigation as a bar chart.

Answer

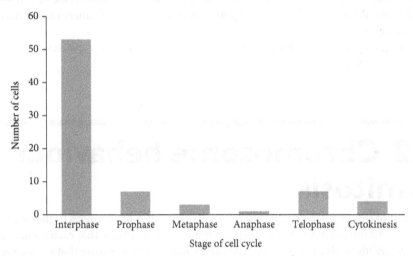

Figure 5.8 Bar chart to show the results of an investigation into the effectiveness of a cell cycle inhibitor on mitosis in HeLa cells.

5. Answer the following three questions by choosing the correct answer from the list of four options.

a. At the start of mitosis, chromosomes consist of two sister chromatids. At which stage of the cell cycle is the second sister chromatid produced?

A. G_1 C. G_2

B. S   D. Prophase

b. Which property of cancer cells is **not** shared by stem cells?

A. They are differentiated.   C. They are immortal.

B. They have a rapid   D. They may be found in bone
   cell cycle.   marrow tissue.

Key ideas

→ Cells contain chromosomes, which contain DNA. Chromosomes have a similar structure, consisting of two sister chromatids (after S phase), which are identical DNA molecules, held together by a centromere.

→ In a eukaryotic cell, DNA is found as very long, linear molecules associated with proteins called histones.

→ A single DNA molecule is wound around histone proteins to enable its packaging into chromosomes

→ New cells are made when one parent cell becomes two genetically identical daughter cells in a process of cell division called mitosis.

→ Mitosis is the division of the nucleus and its chromosome contents.

→ Mitosis is important for embryonic development, growth, repair and replacement, and asexual reproduction.

→ Stem cells are undifferentiated cells that can divide to form specialised cells. They exist in the embryo in animals and in specific locations in the adult body.

→ Meristem tissue is found in the root tips of growing plants and consists of rapidly dividing plant stem cells.

→ The mitotic cell cycle is normally controlled to ensure that cells divide as and when required. If this control goes wrong, uncontrolled division of cancer cells can cause tumours to form.

→ During interphase, the cell prepares for mitosis, including the replication of its DNA in S phase.

5.2 Chromosome behaviour in mitosis

In the rest of this chapter we will look in detail at exactly what happens to the contents of the nucleus during mitosis. The purpose of mitosis is to ensure that each daughter cell is genetically identical to the parent cell. This is achieved by ensuring that each daughter cell has the same number and type of chromosomes as the original parent cell. As you will see, the behaviour of the chromosomes during mitosis is a predictable event; each step has a specific purpose. Flemming, in his notebook, described the process as 'the dance of the chromosomes'.

Figure 5.9 summarises the key events that occur during mitosis.

Although mitosis is a continuous process, biologists refer to four distinct sub-stages, much like screenshots from a movie. In each of these short periods, significant events occur. They are called **prophase**, **metaphase**, **anaphase** and **telophase**.

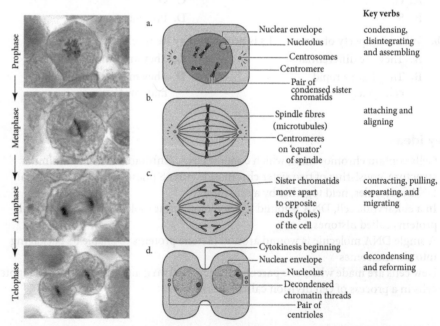

Figure 5.9 Photomicrographs (first column) showing the four stages of mitosis in animal cells, with matching diagrams (second column). Note that the cell shown has just four chromosomes; human cells have 46.

THE STAGES OF MITOSIS

Prophase

At the start of mitosis (Figure 5.9a), the chromosomes in the nucleus become shorter and thicker, or **condense**. If a stain is applied at this point, they become visible as thread-like forms under the microscope, just as Flemming reported. Because the cell has already replicated its DNA during S phase, the chromosomes appear as two sister chromatids, joined by a centromere.

The nuclear envelope disintegrates at the end of prophase, and protein microtubules begin to assemble and radiate from structures called **centrosomes** at the cell **poles**. Centrosomes exist only in animal cells and consist of a pair of cylindrical structures called **centrioles**. The microtubule network consists of **spindle fibres**, which make up the spindle. The **spindle** attaches to the centromeres of all chromosomes and gently pulls them towards the middle of the cell.

Metaphase

The action of the spindle fibres causes the chromosomes to line up (align) along the **equator** (middle) of the cell (Figure 5.9b). The two attachment points on opposite sides of each centromere allow the sister chromatids of each chromosome to attach to spindle fibres that radiate from different poles.

Anaphase

At the start of anaphase (Figure 5.9c), each centromere splits into two. The spindle fibres contract by shortening, which pulls the sister chromatids apart from one another to migrate to the opposite poles of the cell. This is the fastest part of the whole process, and takes no more than 30 minutes in most cells. The chromosome arms are 'pulled' behind the leading centromeres, giving the sister chromatids a characteristic 'V-shape', as seen in Figure 5.10. Once the sister chromatids are separated from each other, we may once again refer to them as chromosomes.

(1) Chromosome (before S phase)

(2) Chromosome consisting of two sister chromatids (after S phase but before anaphase)

(3) Separated sister chromatids / two chromosomes (after anaphase)

— Centromere

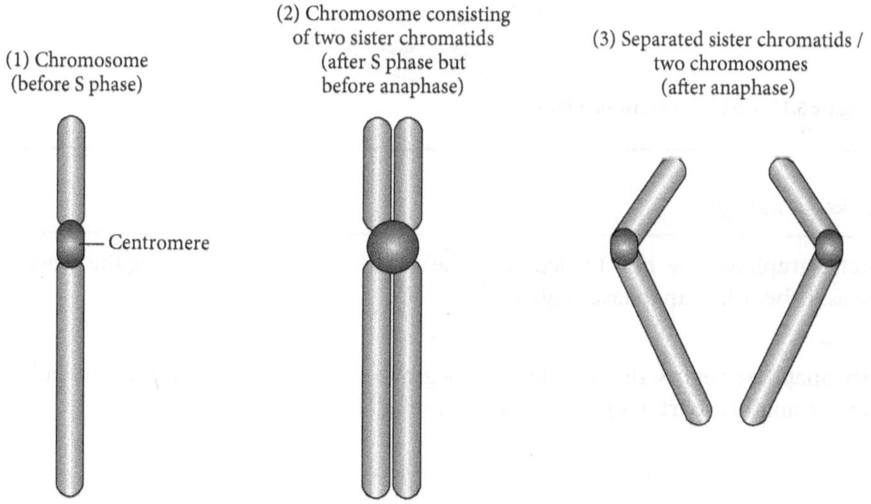

Figure 5.10 The changes in the shape of chromosomes during the cell cycle.

Telophase

After they arrive at the opposite poles of the cell, the chromosomes begin to **decondense** and once again become invisible (Figure 5.9d). At the same time, a new nuclear envelope begins to reform around both of them.

> **Tip**
> Chromosomes of the parent cell replicate before cell division. After S phase, when each chromosome consists of two sister chromatids, the parent cell contains twice its normal DNA content.

> **Tip**
> Use *PMAT* to remember mitosis stages. **P**rophase and **t**elophase are the **p**reliminary and **t**erminal stages. During **m**etaphase, chromosomes align at the cell's **m**iddle (equator). Sister chromatids move **a**part at **a**naphase.

> **Link**
> In Chapter 6 you will see how, during interphase, the DNA in chromosomes is used to produce proteins. Chromosomes decondense (become invisible) so that the enzymes can 'read' the information.

Worked example

Sketch a simple diagram of a cell that contains only two pairs of chromosomes:
a. at metaphase of mitosis
b. at anaphase of mitosis.

Answer

a. A cell should be drawn without a nuclear envelope, in which both pairs of chromosomes consist of two sister chromatids joined at a shared centromere. All four chromosomes should be positioned lined up along the equator of the cell. The presence of microtubules (the spindle fibres), attached to the centromeres, is optional. For example:

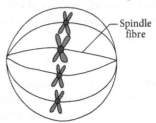

Spindle fibre

Figure 5.11 A cell at metaphase of mitosis.

b. A cell should be drawn in which both pairs of chromosomes are positioned one above the other. Spindle fibres are attached to the centromeres, which have now duplicated and separated, and the two sets of four sister chromatids are being pulled to opposite poles of the cell. For example:

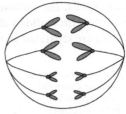

Figure 5.12 A cell at anaphase of mitosis.

Worked example

Sketch a graph to show how the length of the spindle fibres change during the process of mitosis. Label when anaphase begins.

Answer

When anaphase begins, the spindle fibres begin to shorten as they contract and pull the sister chromatids apart to opposite poles of the cell (Figure 5.13).

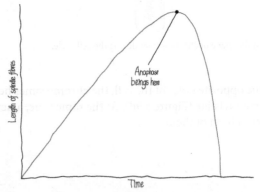

Figure 5.13 A graph to show how the length of the spindle fibres changes during mitosis.

6. Describe and explain how the nuclear envelope changes during the process of mitosis.

7. Explain the difference between each of these pairs of terms:
   a. G_1 phase and G_2 phase
   b. chromosome and sister chromatid
   c. mitosis and cytokinesis
   d. benign and malignant
   e. metaphase and anaphase
   f. telomere and centromere.

8. To test a new chemotherapy drug, a scientist investigated the changes in DNA quantity and chromosome behaviour in a sample of cancer cells growing in the laboratory. The cells had been grown such that their cell cycles were co-ordinated. This means that they started and completed the different phases of the cell cycle at the same time.
   a. The scientist extracted the DNA from samples of cells over a period of 10 hours. The results are shown in Figure 5.14.

   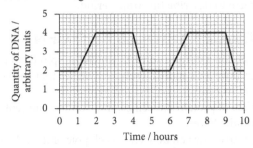

   Figure 5.14 The changes in the quantity of DNA in cultured cells over a period of time.

   i. Describe the results of this study.
   ii. Explain the results of this study.

   b. Just before the cells from Figure 5.14 had been incubated for 4 hours, a sample of the cells was extracted and observed using a light microscope. One of the cells from the sample is shown in Figure 5.15.

   Figure 5.15

   i. Describe the appearance of this cell.
   ii. Explain the appearance of this cell.

9. Arrange the diagrams of the stages of mitosis shown in Figure 5.16 into the correct order.

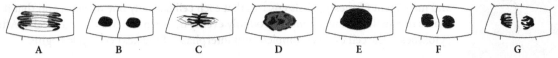

Figure 5.16

Tip

To effectively communicate observations and measurements, it is important to summarise and express key figures, changes and trends accurately. However, describing and explaining data are very different actions.

Tip

Quantitative data relates to numerical measurements, whereas qualitative data usually depends on visual details, such as an image. Descriptions and explanations of qualitative data are also important in scientific communication.

10. Figure 5.17 shows a sketch drawn by a student of a cell in prophase.

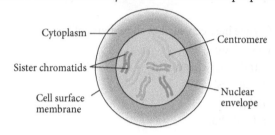

Figure 5.17

a. Suggest reasons to explain why condensation of chromosomes is necessary at the very start of prophase.

b. How does the quantity of DNA in a nucleus differ between prophase and after cytokinesis?

c. Draw a diagram to show how the cell would appear in metaphase.

11. Walther Flemming coined the term 'resting phase' for interphase. Explain why this is an inaccurate term to describe this time period.

CENTROSOMES, CENTRIOLES AND CENTROMERES

The centromere is the site on a chromosome to which the spindle fibres attach in early mitosis. However, if we look more closely, we will see that each chromosome has two **kinetochores** at its centromere, one found on each sister chromatid. It is these kinetochores, made of protein molecules, that connect the centromere to the microtubules in the spindle fibres.

During mitosis, a centrosome can be seen at each pole of the cell. This consists of a pair of centrioles, which are small, cylindrical structures, surrounded by a large number of proteins. These proteins control production of the microtubules.

The microtubules attached to the kinetochore pull the kinetochore towards the centrosomes found at the poles of the spindle. The rest of the sister chromatid drags behind, giving the characteristic V-shaped appearance of the sister chromatids during anaphase. The pulling action is achieved by a shortening of the microtubules from both ends. This is shown in Figure 5.18.

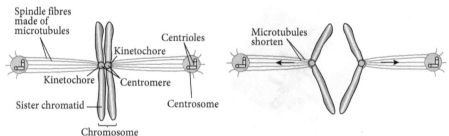

Figure 5.18 The relationship between the centromeres, kinetochores and microtubules during anaphase.

CYTOKINESIS

At the same time that telophase is occurring, the cytoplasm and its contents will begin to divide by cytokinesis. This word derives from the Latin and Greek terms for cell ('cyto') and movement ('kinesis'). This ensures that each cell has only one nucleus, and that it shares organelles of the parent cell roughly evenly between the two daughter cells. After cytokinesis, the cell cycle is complete and interphase of the next cell cycle begins.

In animals, cytokinesis involves the formation of an inward depression (invagination) in the cell surface membrane, called a **cleavage furrow**. This grows deeper and eventually cuts across the equator of the parent cell. The original membrane is separated as the cell is pinched into two, as shown in Figure 5.9 earlier and also in Figure 5.19.

a.

b.

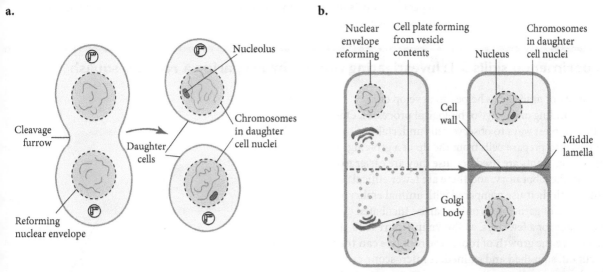

Figure 5.19 Cytokinesis in animal cells (a) and plant cells (b).

In plant cells, which have a cell wall, cytokinesis does not involve the formation of a cleavage furrow. Instead, as mitosis concludes, vesicles containing molecules used to make microtubules and structural proteins are brought to the equator. These coalesce (merge together) to form a temporary structure called the **cell plate**. The cell plate grows until it merges with the cell surface membrane of the parent cell, completing the division of the cytoplasm (Figure 5.19). Next, cellulose is laid down between the two membranes to form the cell wall. The middle lamella, consisting of substances including pectin, then develops between the two walls, which divides the cytoplasm and shares out the organelles between the two daughter cells.

Cytokinesis marks the end of cell division. In both animals and plants, each of the daughter cells produced in this way has an identical set of chromosomes to those in the parent cell prior to DNA replication.

Link

In Chapter 1 you learned how the Golgi body packages materials, usually proteins, into vesicles. Vesicles transport molecules around the cell, and sometimes out of the cell by exocytosis.

Worked example

Describe how the process of cytokinesis differs between animal and plant cells.

Answer

During the process of cytokinesis at the end of the cell cycle, the cytoplasm and its contents are divided roughly equally between daughter cells. The cleavage furrow is an infolding of the cell surface membrane that occurs during telophase in animal cells. It deepens and eventually causes the two cells to 'pinch off' from one another as the cell surface membranes separate. The cell plate is a structure formed by the merging of vesicles from the Golgi body that occurs during telophase in plant cells. This develops into a new cell wall.

Tip

Cytokinesis is not the final stage of mitosis. It occurs at the same time as telophase, so that the cell divides as the nuclear envelopes reform.

🔍**12.** Mitochondria have been shown to cluster around the spindle fibres just before anaphase, around the separated chromosomes during telophase, and at the edges of the cell surface membrane during cytokinesis. Suggest an explanation for these observations.

⭐**13.** Some unusual structures in animals are myocytes (muscle fibres). They are formed by repeated mitosis in the absence of cytokinesis. Suggest and explain some of the features of myocytes.

Experimental skills 5.1: Investigating mitosis by preparing a root tip squash

The study of mitosis has helped to develop our understanding of a range of biological processes. One of the simplest ways to observe cells undertaking mitosis is to prepare cells from the tip of a growing plant root. Plants are used because they are easier to grow in the laboratory, and there are fewer ethical issues with their use compared with animal embryos.

A clove of garlic or an onion can be incubated over water for a few days, as shown in Figure 5.20, to encourage the growth of roots. The root tips can then be cut off, squashed and stained. A microscope can then be used to view cells in all four stages of mitosis, judged on the basis of the characteristic behaviour of chromosomes. This experiment can be used as a basis of an investigation to determine the effect of a named variable on the rate of cell division.

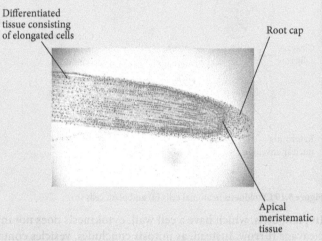

Figure 5.21 The position of apical meristem (meristematic tissue) in a growing plant root tip. Magnification ×40.

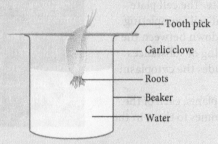

Figure 5.20 Incubating a garlic clove for a few days can lead to the growth of small roots. The tips of these roots contain apical meristematic tissue, which is a small mass of actively dividing cells.

An investigation using a root tip squash will enable you to revisit the microscope skills you developed in Chapter 1. You can make observations and drawings using a light microscope at high power and low power, to show the mitosis stages visible in temporary root tip squash preparations. You will also be able to revisit the use of an eyepiece graticule and stage micrometer to assess the size of a specimen.

To do this, you will need to identify a region of a plant in which mitosis is occurring. Unlike animals, where mitosis can occur in cells throughout the body, cell division and growth in plants is mostly restricted to tissue in the very tips of the shoots and roots, called the apical meristem (Figure 5.21).

Because the plant tissue is no longer living, viewed meristem cells are fixed in the stages of mitosis that they occupied when the tissue was killed (Figure 5.22). The division of cells at any one time is not synchronised – different cells will be at various stages of the cell cycle. It is much like taking a photograph of people at a restaurant. Some will be ordering their food, some eating and some paying for their meal. The phase that is most commonly seen is likely to be the longest phase.

The meristem cells are small and approximately square or rectangular, and during interphase they have a large nucleus relative to the size of the cell. Once you have found some cells using a low magnification, you then use the highest objective lens to further magnify the field of view in this region. You will need a magnification of at least ×400 to see chromosomes in the nucleus and identify the four stages of mitosis.

Link

Plant growth is a topic that has implications for food production and sustainability. In Chapter 15 you will consider the role of plant co-ordination systems in regulating root and shoot development.

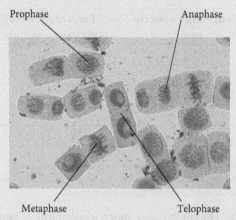

Figure 5.22 Photomicrograph of cells taken from the root tip meristematic tissue of a garlic clove. Magnification ×400.

QUESTIONS

P1. Suggest the purpose of the following steps in the method that could be used to produce temporary preparations of root tip meristematic tissue:

  a. incubating the roots in hot hydrochloric acid

  b. washing the roots to remove hydrochloric acid

  c. incubating the tissue with the stain for a few minutes

  d. lowering the coverslip gently onto the root tissue before squashing

  e. squashing the root tissue to form a very thin layer.

☆**P2.** In addition to the greater ethical concern with using animal tissue, suggest and explain one reason why animal cells would be more difficult to prepare for viewing using this method.

P3. A scientist carried out an investigation into the effect of vincristine on the rate of mitosis in root tip meristematic tissue. Vincristine is a drug used to slow down the division of cancer cells. The scientist decided to incubate one onion in water and another in a solution of vincristine, as shown in Figure 5.23.

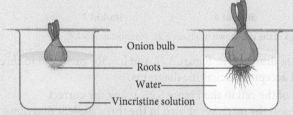

Figure 5.23 An investigation into the effect of the chemotherapy drug vincristine on the mitotic index of root tip meristematic tissue.

The scientist measured the effect of vincristine on the rate of mitosis in root tip tissue by calculating the **mitotic index** for both the untreated and treated garlic cloves. The mitotic index of a tissue sample is the proportion of cells that are undergoing mitosis at a given time, and it is expressed as a percentage. It can be calculated using the following formula:

$$\text{mitotic index} = \left(\frac{\text{number of cells under going mitosis}}{\text{total number of cells counted}} \right) \times 100$$

The mitotic index is much higher in actively dividing cells such as stem cells, and tissues such as tumours.

Figure 5.24 shows two fields of view of root tip meristematic tissue from the onions.

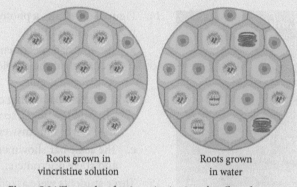

Roots grown in vincristine solution   Roots grown in water

Figure 5.24 The results of an investigation into the effect of vincristine on the mitotic index of root tip meristematic tissue.

  a. Calculate the mitotic index of the root tissue grown in water. Only count cells in which you are able to clearly see the nucleus or chromosomes.

  b. Calculate the mitotic index of the root tissue grown in vincristine solution. Only count cells in which you are able to clearly see the nucleus or chromosomes.

  c. Explain how the scientist could improve the reliability of these calculated mitotic index values.

  d. Explain how the scientist could improve the validity of any comparisons that are made between the calculated mitotic index values.

  e. Give one safety precaution that the scientist should have taken to minimise risk in the investigation.

f. Vincristine stops cell division by affecting spindle fibres. Describe the role of spindle fibres during mitosis.

g. The scientist concluded that vincristine prevents spindle fibre formation during mitosis. Explain how the images in Figure 5.20 provide evidence to support this conclusion.

The researcher carried out further work to investigate the effect of vincristine on cell division. They prepared microscope slides of tissue taken from five measured points from the end of a root tip and counted the number of cells that were undertaking mitosis. The results are shown in Table 5.2.

Distance from root tip / mm	Percentage of cells undergoing mitosis / %
0.2	67
0.4	41
0.6	32
0.8	17
1.0	11
1.2	7

Table 5.2 The relationship between the distance of cells from the end of a root tip and the percentage of cells in mitosis.

h. On graph paper, plot a straight line of best fit using this data and use it to estimate the distance from the end root tip at which all cells will have differentiated.

Link

In Chapter 1, you used an eyepiece graticule and stage micrometer scale to measure the size of cells. You also practised using the units (millimetre, micrometre, nanometre) used in cell studies.

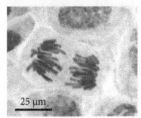

25 µm

Figure 5.25 Photomicrograph of a plant cell undergoing anaphase.

Tip

When drawing low power diagrams of a tissue, you do not need to draw individual cells. Instead, delimit (show the boundaries of) the tissue layers.

14. Figure 5.25 shows a photomicrograph of a meristem cell, undertaking anaphase, from a garlic root tip.

a. The size of a specimen viewed down a light microscope can only be measured accurately if the eyepiece graticule is calibrated. Describe the purpose of an eyepiece graticule.

b. If you are examining cells with a ×10 eyepiece lens combined with a ×10 objective lens and the image spans five small lines on the graticule, what is its approximate size? Give your answer in both mm and µm.

c. The scale bar shown on the image was included after using the eyepiece graticule and stage micrometer scale. Calculate the magnification of the image.

15. Three students were asked to draw a diagram of the cell as seen in the previous field of view (Figure 5.25). Their work is shown in Figure 5.26.

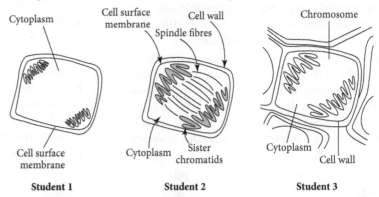

Cytoplasm   Cell surface membrane   Cell wall   Chromosome
Spindle fibres

Cell surface membrane   Cytoplasm   Sister chromatids   Cytoplasm   Cell wall

Student 1   Student 2   Student 3

Figure 5.26 Three students' drawings of the cell shown in Figure 5.25. Magnification ×400.

a. Consider the three student drawings in Figure 5.26 and list, for each, two to three reasons why they are not acceptable scientific diagrams.

b. Draw a labelled diagram of the cell in the Figure 5.25, using the correct techniques. Try not to repeat the mistakes shown in the three student diagrams.

When drawing scientific diagrams using a microscope, you should ensure that:

- lines are continuous, rather than feathery or broken
- shading is not used
- you draw what you have been asked to focus on – just one cell in this example

- all contents of the subject are drawn, that they are in the correct proportion and in the correct positions
- you have included correct labels with lines that are drawn with a ruler, and without arrowheads
- you have used as much of the available space as possible.

Key ideas

→ The separation of chromosomes and nuclear division is called mitosis, which consists of four stages: prophase, metaphase, anaphase and telophase.

→ Chromosomes condense by supercoiling during prophase, before aligning along the equator of the cell at metaphase.

→ The sister chromatids of chromosomes are pulled apart by the spindle at anaphase and are surrounded by two separate nuclei during telophase.

→ Mitosis is followed by cytokinesis, in which the cell surface membrane is pinched inwards (in animals), or a cell plate is laid down between the two daughter cells (in plants).

→ Staining is required to see chromosomes under a microscope, as they are transparent to light.

→ The size of a specimen viewed down a light microscope can only be measured accurately if an eyepiece graticule is calibrated using a stage micrometer.

→ Being able to accurately communicate visual observations as scientific diagrams is very important in the study of cells and cell division.

Assignment 5.1: Investigating a mitotic inhibitor

A researcher planned an experiment using the root tip squash method to investigate the following research question:

Are higher concentrations of paclitaxel more effective at inhibiting mitosis in root tip meristematic tissue than lower concentrations of paclitaxel?

Figure 5.27 The pacific yew tree, source of the chemotherapy drug paclitaxel.

Paclitaxel is a chemotherapy drug that interferes with the action of the spindle, and in so doing reduces the rate of mitosis in dividing plant or animal tissue. It can be extracted from certain plant species (Figure 5.27).

The researcher has access to the equipment and materials used in a typical root tip squash procedure. They also have a solution of paclitaxel of concentration $1.0 \, mg \, cm^{-3}$. The garlic plant *Allium sativum* was chosen as a model organism.

Carefully consider the researcher's plan, and then answer the questions that follow to evaluate its effectiveness and whether it is detailed enough for another person to follow.

First, two solutions of paclitaxel should be prepared, one of a low concentration ($0.1 \, mg \, cm^{-3}$) and one of a high ($1.0 \, mg \, cm^{-3}$) concentration. A control of tap water should also be used. The solution of low concentration can be prepared by mixing the original stock solution of paclitaxel with some tap water.

All three solutions are incubated at 30 °C and one clove of garlic from different bulbs is then placed into each of the solutions so that a section of the garlic clove of the same length is touching the surface of each of the solutions. These are left to develop roots and all of the cloves are removed from the solutions at the same time for analysis.

The root tip squash method is then used to prepare a microscope slide for each of the garlic cloves and they are observed under a light microscope at ×400 magnification.

This is a low risk experiment, but care will be taken to handle sharp instruments carefully, and eye protection will be worn when handling the hot acid and the stain.

The number of cells undergoing mitosis in a field of view is counted for both slides and compared.

QUESTIONS

A1. Identify the dependent variable in this investigation – the factor that the researcher would need to measure.

A2. Serial and proportional dilution methods can be used to produce a series of volumes of a paclitaxel solution of known concentrations, to appropriately vary the independent variable. Consider how this might be used to enhance the accuracy of measurements of the dependent variable.

A3. The bar chart in Figure 5.28 was produced by this researcher, as they are limited to two different concentrations.

    a. Describe what else would be required to produce a line graph in this study.

    b. Explain the advantages of drawing a line graph rather than a bar chart.

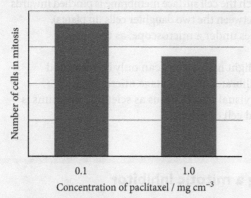

Figure 5.28 The effect of paclitaxel on the number of cells undertaking mitosis.

    c. Are there any other details missing from the researcher's plan that you would like to know in order to carry out the investigation again? For example, the researcher should have standardised (kept constant) variables other than paclitaxel concentration. Comment on the validity of this study.

    d. Describe the reliability of the data obtained by the researcher. Could the researcher identify anomalous results if they were obtained in this investigation? If not, how could the investigation be improved?

    e. The researcher has included a control experiment. What is the purpose of a control, and is the nature of the control they used appropriate? If not, why not?

A Level analysis

A research question is often phrased in terms of a null (H_0) and alternative (H_1) hypotheses. The advantage of this is that the data can be directly used to prove or disprove a clear statement. This reduces the possibility of bias in an investigation: that is, a prior assumption. In the researcher's original investigation, the research question could be phrased as follows:

The **null hypothesis (H_0)** is that there is no significant difference between the rate of mitosis in root tip meristematic tissue grown in low or high concentrations of paclitaxel.

The **alternative hypothesis (H_1)** is that there is a significant difference between the rate of mitosis in root tip meristematic tissue grown in low or high concentrations of paclitaxel.

The researcher extended the investigation to consider four other plant-derived molecules used in chemotherapy treatments. The investigation aimed to find out whether there was a correlation between the number of days of incubating the garlic clove for each of the five molecules and the percentage of cells in mitosis. Solutions of the five chemotherapy drugs of equal concentration were used, and cloves of equal mass were incubated in them for 1, 2, 3, 4 or 5 days.

> **Tip**
>
> The term 'significant' relates to whether the difference between two datasets has a scientific basis. This can be used to prove or disprove a hypothesis and draw a conclusion.

Spearman's rank correlation test was used for this analysis. The results are shown in Table 5.3.

Chemotherapy drug	Spearman's rank correlation coefficient (r_s)
paclitaxel	0.20
vincristine	0.52
vinblastine	0.13
podophyllotoxin	0.81
camptothecin	0.45

Table 5.3

A4. State the null and alternative hypotheses for the Spearman's rank correlation test for this study.

A5. Suggest why the Spearman's rank correlation test was chosen for use in this study.

Table 5.4 shows the critical values for r_s at five levels of significance for the data collected in this study.

Level of significance (p)	0.20	0.10	0.05	0.02	0.01
Critical value of r_s	0.240	0.306	0.362	0.425	0.467

Table 5.4

A6. Using Tables 5.3 and 5.4, identify which chemotherapy treatments showed a statistically significant correlation between the number of days of incubating the garlic clove for each of the five chemotherapy treatments and the percentage of cells in mitosis. Give a reason for your answer.

A7. Explain why, when publishing their work in the scientific community, researchers would prefer to obtain data that is statistically significant at a value of $p = 0.01$ rather than $p = 0.05$. Use the terms 'probability' and 'chance' in your answer.

Pearson's linear correlation coefficient can also be used to investigate the significance of a correlation between two factors.

A8. Refer back to Figure 5.3a. The Pearson's linear correlation coefficient was found to be $r = -0.40$ for the association between the age of an individual and their mean telomere length.
   a. Explain what the value of $r = -0.40$ indicates about the association between the age of an individual and their mean telomere length.
   b. Suggest why the Pearson's linear correlation coefficient (r) was calculated in the study in Figure 5.3a, rather than the Spearman's rank correlation coefficient (r_s).

Link

In Chapter 18 you will explore how to calculate the Spearman's rank correlation coefficient (r_s) and how it can be used in ecological studies to test for association between different species.

Tip

A good way to decide whether data is normally distributed is to look at how the dependent variable was measured. If it was measured, it is normally distributed. If it was counted, it is not.

CHAPTER OVERVIEW

Review the mini mind map to link new learning with your prior learning, and to highlight which of the key concepts underpin the chapter.

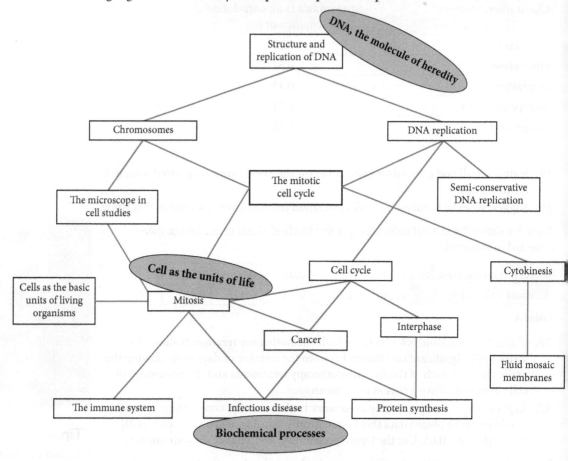

Try copying this mini mind map and expanding upon it. Use your notes from other chapters to help you explore how the essential ideas, theories and principles can be linked further together.

WHAT YOU HAVE LEARNED

Now that you have finished this chapter you should be able to:

- describe the structure of a chromosome, limited to:
  - DNA
  - histone proteins
  - sister chromatids
  - centromere
  - telomeres

- explain the importance of mitosis in the production of genetically identical daughter cells during:
  - growth of multicellular organisms
  - replacement of damaged or dead cells
  - repair of tissues by cell replacement
  - asexual reproduction

- outline the mitotic cell cycle, including:
    - interphase (growth in G_1 and G_2 phases and DNA replication in S phase)
    - mitosis
    - cytokinesis
- outline the role of telomeres in preventing the loss of genes from the ends of chromosomes during DNA replication
- outline the role of stem cells in cell replacement and tissue repair by mitosis explain how uncontrolled cell division can result in the formation of a tumour
- describe the behaviour of chromosomes in plant and animal cells during the mitotic cell cycle and the associated behaviour of the nuclear envelope, the cell surface membrane and the spindle (names of the main stages of mitosis, prophase, metaphase, anaphase and telophase are expected)
- interpret photomicrographs, diagrams and microscope slides of cells in different stages of the mitotic cell cycle and identify the main stages of mitosis.

CHAPTER REVIEW

1. With reference to stem cells in your answer, outline the functions of mitosis in a multicellular organism such as a human.

2. Figure 5.29 shows a clock face showing the series of events that an epithelial cell from the lining of the small intestine undertakes in a 12-hour period, the duration of one cell cycle. Use it to help you answer the following questions.
    a. What phase will the cell occupy at time = 5:20 a.m.?
    b. How many minutes does this cell take to replicate its DNA?
    c. Give a time when the cell is duplicating organelles to share between daughter cells.
    d. What phase will the cell occupy 15 hours after time = 8:30 a.m.?
    e. At approximately what time will cytokinesis occur?

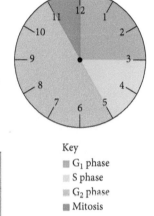

Key
- G_1 phase
- S phase
- G_2 phase
- Mitosis

Figure 5.29

3. Match each word with its correct definition.

(a) Centromere	(i) Type of protein around which DNA is wound during the process of supercoiling of chromosomes.
(b) Sister chromatid	(ii) Ends of the cell, to which position the sister chromatids are pulled during anaphase.
(c) Telomere	(iii) Molecule of heredity; contains genes that encode proteins.
(d) DNA	(iv) Constriction in the middle of a chromosome that holds together the sister chromatids.
(e) Histone	(v) End of a chromosome arm. Its shortening has been implicated in cellular ageing.
(f) Spindle	(vi) Middle of the cell, where the chromosomes align during metaphase.
(g) Equator	(vii) Made of microtubules, this attaches to the centromeres of all chromosomes at prophase.
(h) Poles	(viii) Consists of a structure attached at the centromere to another structure which is genetically identical to it.

4. Which of the following statements are true and which are false?
    a. Centrosomes are replicated before mitosis begins.
    b. Sister chromatids contain DNA that is identical.
    c. The microtubules attached to a kinetochore connect to both poles of the spindle.
    d. Microtubules are assembled and disassembled during S phase of the cell cycle.

    e. Kinetochores are found in the centrosomes.

    f. Telomeres are the sites of attachment of microtubules during mitosis.

    g. Sister chromatids remain paired as they line up on the spindle at metaphase.

5. Cancer cells can arise in plants but are unable to spread around the organism as they do in animals. Suggest reasons to explain this.

6. A student carried out a root tip squash on cells from a garlic plant. One hundred cells were counted and the following numbers of cells were identified in each of the different stages.

Cell cycle phase	Number of cells
interphase	81
prophase	9
metaphase	4
anaphase	2
telophase	4

Assume that a cell cycle in garlic apical meristem takes 24 hours. Using this information, estimate the period of time taken for cells in this tissue to undertake mitosis. Give your answer to the nearest minute.

7. Different species have different numbers of chromosomes. The cells of the giraffe (*Giraffa camelopardalis*) contain 30 chromosomes. Answer the following questions using this information.

    a. In prophase during the mitotic cell cycle, how many of the following structures would be present?

        i. Sister chromatids

        ii. Centromeres

        iii. Telomeres

    b. Identify whether the following statements are true or false. Give a reason for your answer.

        i. During metaphase, microtubules attach to the telomeres of chromosomes.

        ii. The nuclear envelope reforms during metaphase.

        iii. During cytokinesis, chromosomes consist of pairs of sister chromatids.

    c. In organisms with a greater number of chromosomes, prophase often takes a longer period of time. Suggest why.

8. A chromosome is a structure made up of a long molecule of DNA associated with proteins. In the early 1900s, biologists Walter Sutton and Theodor Boveri observed the behaviour of chromosomes in the grasshopper and the sea urchin.

    a. Name the type of proteins that are associated with DNA to form a chromosome.

    b. State two functions of centromeres during nuclear division.

    c. Explain why chromosomes are not visible with an optical microscope during interphase.

    d. Sketch a chromosome found in a cell in prophase, and show the positions of the centromere and telomeres.

    e. In their studies, Sutton and Boveri made two important observations:

      • The number of characteristics in an organism is greater than its number of chromosomes.

      • If the nucleus is removed from an egg, and this egg is fertilised by two sperm cells, the offspring have some growth abnormalities and inherit characteristics from just the male parent.

    Suggest three conclusions regarding chromosomes that Sutton and Boveri made from these observations.

Chapter 6

Nucleic acids and protein synthesis

All cells contain DNA, a molecule inherited during cell division and between generations. An important function of DNA is to store the information required for the production of proteins.

Amazingly, the way in which DNA determines the sequence of amino acids in a protein is universal: it is almost identical in all living organisms. This represents strong evidence that all life on Earth has a common ancestor.

DNA is also a very stable molecule. Some scientists claim that it might be possible to use preserved DNA to regenerate preserved organisms. These include the extinct woolly mammoth, whose 40 000-year-old DNA has been found in frozen remains in the Arctic, and possibly even animals that have been trapped in amber for millions of years.

Prior understanding

You have probably learned that DNA is a genetic material that holds information which can be transferred during cell division and from parents to offspring. From Chapter 2 you may remember that many biological molecules are polymers, made from many monomers joined by condensation reactions. You may also remember that each protein, composed of 20 different types of amino acid arranged in a specific sequence, has an important role in the cell.

Learning aims

In this chapter, you will learn about the structure of the polynucleotides DNA and RNA, their nucleotide monomers, and the nucleotide derivative ATP. You will see how cells are able to replicate DNA with very high accuracy, and why this is important for inheritance. You will also find out how the sequences of bases in DNA molecules guide the cell to make proteins with a specific order of amino acid residues.

6.1 Structure of nucleic acids (Syllabus 6.1.1–6.1.3, 6.1.5)

6.2 Replication of DNA (Syllabus 6.1.4)

6.3 Protein synthesis (Syllabus 6.2.1–6.2.5)

6.4 Mutation (Syllabus 6.2.6, 6.2.7)

6.1 Structure of nucleic acids

DNA, and a similar molecule called RNA, are both nucleic acids. The term 'nucleic acid' relates to the weak acidity of these molecules and the fact that they were first found in the nucleus. All nucleic acids contain carbon, hydrogen, oxygen, nitrogen and phosphorus. Nucleic acids store and carry information, which is used to control the cell's activities, and are the basis for inheritance. We will encounter RNA in greater depth later in this chapter.

DNA, shown in Figure 6.1, is the nucleic acid that stores genetic information in every living cell. As you saw earlier, it has essentially the same structure in eukaryotes and prokaryotes. It has survived throughout evolution as the one substance that can store information that enables inheritance during cell division and from parent to offspring.

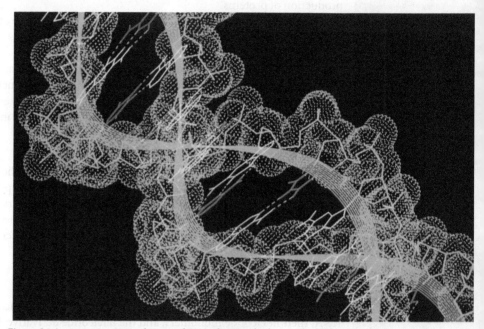

Figure 6.1 A computer-generated image of DNA, showing the famous double helix structure.

Understanding the structure of DNA allows an understanding of its role in the storage of genetic information and how that information is used in the synthesis of proteins. We will learn about the process of **protein synthesis** later in this chapter.

DNA CONTAINS TWO POLYNUCLEOTIDES

The acronym DNA stands for the term 'deoxyribonucleic acid'. DNA is a polymer of smaller molecules called **nucleotides**, which are joined together by **phosphodiester bonds**. Nucleotides react in condensation reactions, in which the phosphate group of one nucleotide joins to the **deoxyribose** sugar of another. DNA is therefore a **polynucleotide**. It is also classified as a macromolecule, because it is a giant molecule that contains many thousands of atoms.

Link

In Chapter 19, you will see that the process of genetic modification relies on the fact that most organisms share DNA as their genetic material.

Russian biochemist Phoebus Levene proposed the structure of nucleotides in 1919. In DNA, these monomers, or building blocks, are specifically called **deoxyribonucleotides**. As shown in Figure 6.2, deoxyribonucleotides have three parts:

- a five-carbon sugar (a pentose), called deoxyribose
- a phosphate group
- a nitrogenous (nitrogen-containing) base, which is either adenine, guanine, cytosine or thymine (in diagrams, these are usually abbreviated to A, G, C or T).

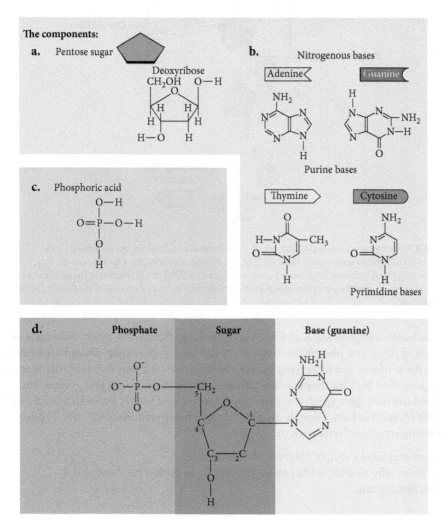

Figure 6.2 Nucleotides consist of three components – a pentose sugar called deoxyribose (a), one of four different nitrogenous bases that fall into two classes, purines and pyrimidines (b) and a phosphate group (c). These can be joined together in condensation reactions to form a nucleotide (d). Note that the structure of nucleotides, and therefore nucleic acids, can be shown in diagrams using simple symbols for the subunits: circles for phosphates, pentagons for the pentose sugar and rectangles with slightly different shapes for the four different bases. Note also that the carbon atoms have been numbered in this diagram – this will be important later.

Nucleotide monomers can combine together by condensation reactions. This forms phosphodiester bonds between the sugar and phosphate molecules. The resulting molecule is a polymer called a polynucleotide (Figure 6.3).

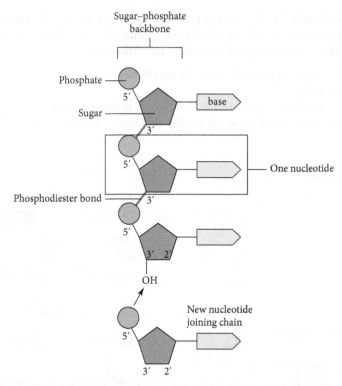

Figure 6.3 Polymers called polynucleotides, which are polymers of nucleotides, can be formed by the condensation of separate nucleotides. These are built by adding nucleotides in a 5′ to 3′ direction. This means that the connecting phosphodiester bond forms between the hydroxyl (OH) group attached to the third carbon (3′ carbon) of the chain, and the phosphate group attached to the fifth carbon (5′ carbon) of the free nucleotide which is to be added.

Tip

To remember which bases are pyrimidines and which are purines, concentrate on the letter Y: thymine and cytosine are pyrimidines.

Polynucleotides such as DNA and RNA are very long, thread-like macromolecules, with alternating sugar and phosphate components that make up a **sugar–phosphate backbone**. This is the uniform and unvarying part of the molecule. The four different **nitrogenous bases**, so-called because they contain nitrogen, can be divided into two groups. You do not need to know their structural formulae; however, you should know that the first pair, **purines** (A and G), have a double-ring structure, while **pyrimidines** (C and T) have a single-ring structure (refer back to Figure 6.2).

1. Draw and label a simple diagram of a nucleotide.
2. Explain why nucleic acids can be referred to as both a polymer and a macromolecule.

THE DISCOVERY OF THE STRUCTURE OF DNA

'We have discovered the secret of life.'

These were the immortal words of Francis Crick on 28 February 1953, after a long day working with his colleague James Watson in the Cavendish Laboratory at the University of Cambridge. It was the conclusion of a story that illustrates much of what makes research science successful. As a result of some imaginative model-building and ongoing team work, one of the great puzzles of the scientific world of the time was solved.

In the 1940s and the early 1950s there were many leaps forward in our understanding of the structure of DNA. Erwin Chargaff, working in the USA, showed that DNA samples taken from different cells of the same species have the same proportions of the four bases. For example, in humans DNA has about 30% each of adenine and thymine, and 20% each of guanine and cytosine. The figure is different for other organisms, but the amounts of A and T are always the same, as are the amounts of C and G. This became known as Chargaff's rule, and this is explored further in Assignment 6.1.

At roughly the same time, Rosalind Franklin, a biophysicist working at King's College in London under the direction of Maurice Wilkins, was working on X-ray diffraction images of DNA. Crick and Watson examined Franklin's X-rays, without her permission. They finally put together the three-dimensional, double-helical model for the structure of DNA in February 1953 (Figure 6.4).

Using cardboard cut-outs representing the individual chemical components of the four bases and the sugar and phosphates, Watson and Crick arranged the model as though putting together a puzzle. When they placed the sugar–phosphate backbone on the outside of the **double-stranded** model, the complementary bases fitted together perfectly (A with T and C with G), with each pair held together by hydrogen bonds. These are called **base pairs** and are located on the inside of the DNA model. There are two H-bonds between A and T and three between C and G. Notably, the proportions of the bases were in agreement with Chargaff's rule.

Figure 6.4 James Watson (left) and Francis Crick (right). They used an X-ray diffraction image produced by Rosalind Franklin to decipher the structure of the DNA molecule.

DNA IS A DOUBLE HELIX

Figure 6.5 shows how the base pairs within DNA fit together to form a double-stranded helix. The sides are formed by alternating sugar–phosphate units to form a sugar–phosphate backbone, while the base pairs form the cross-bridges, like the rungs of a ladder. The molecule is not flat, however. Each base pairing causes a twist in the helix and there is a complete 360° turn every 10 base pairs. This shape is called a **double helix**.

The two strands are held together by hydrogen bonds between the bases, with purines facing pyrimidines. As discovered by Watson and Crick, there are two hydrogen bonds between adenine and thymine, and three hydrogen bonds between cytosine and guanine. Hydrogen bonds can occur only between a hydrogen atom on one base and either an oxygen or nitrogen atom on the other base. This explains why only two hydrogen bonds can form between A and T and three can form between G and C, because a hydrogen bond can form only where a hydrogen atom is close to an oxygen or nitrogen atom of a base on the opposite strand. These are therefore called **complementary** base pairs and this is the **base-pairing** rule:

- adenine always pairs with thymine, forming two hydrogen bonds
- cytosine always pairs with guanine, forming three hydrogen bonds.

The structure of DNA also explains why the two polynucleotide strands run in opposite directions. They are **antiparallel**: one strand of the DNA double helix runs in the 5′ to 3′ direction, and the other runs in the 3′ to 5′ direction (the numbers 5 and 3 refer to the carbon atoms in the deoxyribose that are involved in forming the sugar–phosphate backbone). These two long strands of polynucleotides then twist together to form a double helix.

As we will see later, this complementary base pairing is key to the ability of DNA to store vast amounts of information that can be passed on accurately from one cell to another. The structure of DNA is also crucial to its ability to be accurately replicated and passed from generation to generation during cell division.

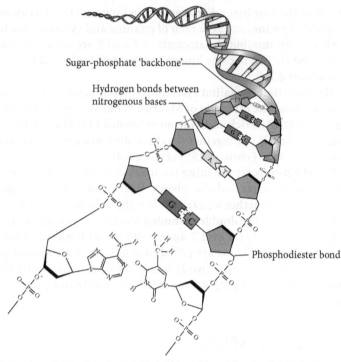

Sugar-phosphate 'backbone'

Hydrogen bonds between nitrogenous bases

Phosphodiester bond

Figure 6.5 DNA consists of two polynucleotide strands wound around each other in the shape of a double helix.

Watson, Crick and Wilkins won the Nobel Prize in Physiology or Medicine in 1962 for their work on the discovery of the structure of DNA. Many people feel that the work of Franklin was not appropriately acknowledged. Sadly, she died in 1958 and the Nobel Prize cannot be awarded posthumously.

Worked example

Two pieces of double-stranded DNA were analysed:

- the first piece of DNA contained 31% adenine
- the second piece of DNA contained 27% cytosine.

a. Identify which piece of DNA contains a greater percentage of thymine.

Answer

The percentage of thymine in a piece of DNA will be equal to the percentage of adenine. Therefore, the percentage of thymine in the first piece of DNA will be 31%. The percentage of thymine in the second piece of DNA can be found by calculating the percentage of DNA that consists of both C and G (27 × 2 = 54%) and then subtracting this from 100 and dividing by 2. The answer is 23%. Therefore, the first piece of DNA contains a greater percentage of thymine.

b. Two hydrogen bonds form between adenine and thymine, but three form between cytosine and guanine. Using this information, identify and explain which of the pieces of DNA would require more energy to break apart into two separate strands.

Answer

There are three hydrogen bonds found between a C–G base pair, and only two between an A–T base pair. Therefore, samples of DNA that have a greater C–G percentage will require a greater amount of energy to break into separate strands (unzip). In this case, the second piece of DNA will require a greater amount of energy to break apart.

3. Identify the 'odd one out' in the following sets of words. The first example has been done for you.
   Example: ribose, deoxyribose, adenine
   Answer: Adenine is the 'odd one out' because it is **not** a pentose sugar.
   a. cytosine, thymine, guanine
   b. phosphate, hydrogen, phosphodiester
   c. nucleic acid, nucleotide, polynucleotide
4. Give a detailed description of the structure of a DNA molecule, using all of the following terms in your answer: nucleotides, hydrogen bonds, phosphodiester bonds, polymer, antiparallel.
5. Look at the diagram of DNA in Figure 6.5. Referring to the structures of the four different bases, give two reasons that explain why base pairing must be complementary in a DNA molecule.

RNA IS A POLYNUCLEOTIDE

Ribonucleic acid (RNA) is another type of nucleic acid. Like DNA, it is also a polynucleotide, but it is made up of monomers with a slightly different structure, called **ribonucleotides**. Ribonucleotides have a very similar structure to deoxyribonucleotides, but with four important differences:

- The pentose sugar in RNA is **ribose**, not deoxyribose. The structure of ribose is shown in Figure 6.6a.
- The pyrimidine base thymine is replaced by a different pyrimidine base, called **uracil**; so the four bases in RNA are abbreviated to A, U, C and G. The structure of uracil is shown in Figure 6.6b.
- Most forms of RNA are made up of a single strand, rather than two antiparallel strands joined together. RNA can, however, fold into three-dimensional structures.
- RNA molecules are usually shorter than DNA molecules: they contain fewer nucleotides joined together. Whereas a DNA molecule can consist of over a billion nucleotides, RNA molecules usually consists of a few hundred, which varies depending on the precise role they have in the cell. RNA molecules are also less stable and are easily broken down, and this is appropriate given their short-term functions in the cell.

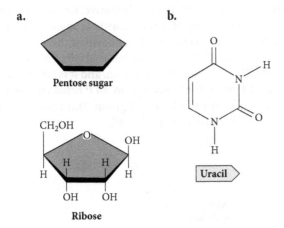

Figure 6.6 The structures of the pentose sugar ribose (a) and the nitrogenous pyrimidine base uracil (b), which are found in RNA. Compare these structures with the pentose sugar deoxyribose and the base thymine, which are their equivalent molecules in DNA (Figure 6.2).

Just as in DNA, ribonucleotides can be joined together through phosphodiester bonds to form long chains. Figure 6.7 shows a short section of an RNA polynucleotide.

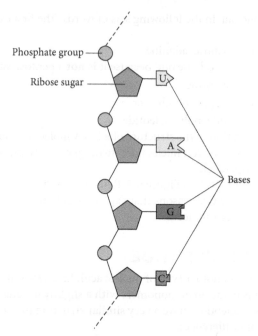

Figure 6.7 An RNA polynucleotide, formed by the condensation of ribonucleotides. As shown in this example, the base uracil replaces the thymine base found in DNA.

RNA comes in several different forms, including **messenger RNA (mRNA)** and **transfer RNA (tRNA)** molecules. You will encounter these later in this chapter when we consider protein synthesis.

ADENOSINE TRIPHOSPHATE (ATP) – A NUCLEOTIDE DERIVATIVE

Nucleotides are not just found in DNA and RNA. All cells contain another vitally important and similar molecule, called **adenosine triphosphate**, or **ATP**.

Like DNA and RNA nucleotides, an ATP molecule is made up of an organic base (adenine), a pentose sugar (ribose) and phosphate. The big difference is that this molecule has not one, but three, phosphate groups arranged into a linear sequence (Figure 6.8). For this reason, it is sometimes called a nucleotide derivative, rather than simply a nucleotide.

ATP is a very important molecule. It is the immediate source of energy for almost every process that takes place in a cell. It is often known as the 'energy currency' of cells.

The energy contained in an ATP molecule is released when one of its phosphate groups is removed (Figure 6.9). This is a hydrolysis reaction, and it is catalysed by an enzyme called **ATP hydrolase (ATPase)**. The **hydrolysis of ATP** breaks it down to form adenosine diphosphate, ADP, plus an inorganic phosphate group. This releases energy, which can be used to drive energy-requiring processes in the cell.

Link

In Chapter 14, you will see how another nucleotide derivative, cyclic AMP (cAMP), acts as an important signalling molecule in cells, called a second messenger.

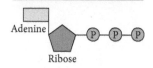

Figure 6.8 The simplified structure of adenosine triphosphate (ATP).

Tip

You will not be expected to draw the structure of ATP, but it is important to know that it consists of a deoxyribonucleotide with three phosphate groups.

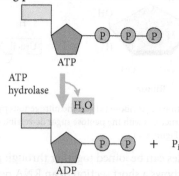

Figure 6.9 ATP can be hydrolysed in a reaction catalysed by ATP hydrolase. This releases energy, and produces an adenosine diphosphate (ADP) and an inorganic phosphate (P_i) molecule.

What happens to the products of this reaction? The phosphate group can be used to add to other substances, which often makes them more reactive. This lowers the activation energy of any (enzyme-catalysed) reactions they are involved in. For example, before glucose can be used in respiration or polymerised, it first has to have phosphate added to it. ATP is referred to as the 'energy currency' of the cell because, like money, it can be used for many different purposes, and is constantly recycled. That is, ADP and phosphate can be joined together again to resynthesise ATP. This is done using an enzyme called **ATP synthase**.

6. Describe how these structures differ:

   a. DNA and RNA

   b. adenosine triphosphate (ATP) and a nucleotide found in DNA.

Link

In Chapters 12 and 13, you will see how ATP can be resynthesised from the products of its hydrolysis. This happens during respiration and photosynthesis in a condensation reaction.

Assignment 6.1: Chargaff's rule

Erwin Chargaff was a biochemist born in what was the Austro-Hungarian empire, in a city that is now in Ukraine. Working in the USA, he discovered that in a sample of DNA from any species, the percentage of adenine is similar to that of thymine, and that the percentage of cytosine is similar to that of guanine. Put another way, for any species, the proportion of purines (A + G) and the proportion of pyrimidines (C + T) are usually equal (Figure 6.10).

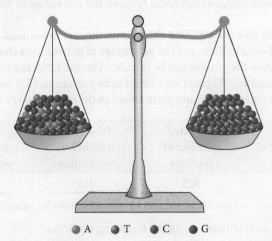

● A ● T ● C ● G

Figure 6.10 Chargaff's rule. In a double-stranded section of DNA, the proportion of purines (A + G) and the proportion of pyrimidines (C + T) are usually equal.

An experiment was conducted using four different samples of DNA to find the percentages of each of the four bases. The percentage of purines and pyrimidines in these samples was also calculated, as shown in Table 6.1.

Organism	A %	T %	G %	C %	Purine %	Pyrimidine %
bacteriophage φX174 (virus)	24.0	31.2	23.3	21.5		
E. coli (bacterium)	24.7	23.6	26.0	25.7	50.7	49.3
sea urchin	32.8	32.1	17.7	17.3	50.5	49.4
human	29.3	30.0	20.7	20.0	50.0	50.0

Table 6.1 The results of an analysis of the base percentages of four different species. The values for the purine and pyrimidine percentages are missing for the virus. A = adenine; T = thymine; G = guanine; C = cytosine.

QUESTIONS

A1. Calculate the percentages of purine and pyrimidine DNA bases in the virus.

A2. Explain how the results for the *E. coli* bacterium, sea urchin and the human support the model of DNA structure proposed by Watson and Crick.

★A3. The minor deviations from a 50:50 ratio between the percentages of purines and pyrimidines in most species are due to measurement error. However, the deviation for the virus has a biological explanation. Suggest a reason for this.

A Level analysis – the chi-squared (χ^2) test

The data in Table 6.1 shows that the results for the virus do not agree perfectly with Chargaff's rule. This is because the percentages of purine and pyrimidine bases are not as close to equal as the other organisms.

Nominal data such as the frequency of bases is not normally distributed. It is counted, not measured. Therefore, the chi-squared (χ^2) test should be used to determine the significance of any difference between the observed frequencies and expected frequencies. The χ^2 test can be used to determine whether or not the results obtained in an investigation are significantly different from those we would expect if a hypothesis is correct. If this is the case, then there is likely to be an underlying scientific reason for the difference. If not, then the difference may have arisen by chance, or could be due to experimental error.

In order to be able to use the χ^2 test, a null hypothesis should be generated. This states that there is no difference in the proportions in the different categories. Chargaff's rule would suggest that the percentage of purines in a sample of DNA is equal to the percentage of pyrimidines. This can be phrased as a null hypothesis:

In a sample of DNA, there is no significant difference between the percentage of purines and the percentage of pyrimidines.

Next, we use this hypothesis to state the expected numbers. If the null hypothesis were true, then the percentage of purines in a sample of DNA should be 50%, and the percentage of pyrimidines should also be 50%.

The following example shows how the χ^2 value can be calculated for the DNA sample obtained from the sea urchin. In a sample of DNA from this organism, 332 bases were found to be purines and 318 were pyrimidines. The expected and observed values for the numbers of purines and pyrimidines in the sea urchin are shown in in Table 6.2.

Observed number of purines	Expected number of purines	Observed number of pyrimidines	Expected number of pyrimidines
332	325	318	325

Table 6.2 The observed and expected purine and pyrimidine bases in a sample of DNA from the sea urchin.

The χ^2 statistical value can now be calculated using the following equation:

$$\chi^2 = \sum \frac{(O - E)^2}{E}$$

where χ^2 = chi-squared, O = observed result, E = expected result.

So, in this example:

$$\chi^2 = \frac{((332 - 325)^2 + (318 - 325)^2)}{325}$$

$$\chi^2 = 0.301$$

To interpret the value of χ^2, the next step is to work out the **degrees of freedom**. These take into account the number of comparisons that are being made. To determine the number of degrees of freedom, the number of classes of data are identified, and 1 is subtracted from this value. The following equation is used to calculate this:

$$v = c - 1$$

where v = degrees of freedom and c = number of classes of data.

In this investigation, because there are two classes of data (purine percentage and pyrimidine percentage), then there are $(2 - 1) = 1$ degree of freedom.

Finally, it is now possible to determine whether our results show a significant difference from that expected. Table 6.3 shows a selection of **critical values** for χ^2, which correspond to two probability levels ($P < 0.05$ and $P < 0.01$). The first row, showing probability values for a single degree of freedom, is relevant here.

Degrees of freedom	Probability (P)	
	0.05	0.01
1	3.841	6.635
2	5.991	9.210
3	7.815	11.345
4	9.488	13.277
5	11.071	15.086

Table 6.3 The probability values for the one to five degrees of freedom at $P < 0.05$ and $P < 0.01$, as required for the analysis of the calculated χ^2 value.

If the calculated χ^2 value is greater than or equal to the critical χ^2 value, then there is a significant difference between our observed results and our expected results. That is, the difference between the actual data and the expected data is probably too great to be attributed to chance. If the calculated χ^2 value is less than the critical χ^2 value, then there is no significant difference between the observed and expected data, and the difference is likely to be due to chance at this probability level. So, we conclude that our sample does not support the hypothesis of a difference, and we accept the null hypothesis.

We usually look at the value of probability of $P < 0.05$. Comparing the value of 0.301 with the critical values on the χ^2 table for 1 degree of freedom, we can see that it is less than 3.841, so the probability of this result being due to chance is less than 0.05. It is in fact less than the critical value for χ^2 at the probability level of $P < 0.01$, which is 6.635. Therefore, we can accept the null hypothesis, and assume that the difference between the observed and expected ratios in the sea urchin DNA sample is due to chance, or possibly measurement error. The χ^2 test provides evidence that the deviation from a 50:50 ratio is not great enough to suggest that Chargaff's rule does not apply to this DNA sample.

A4. **a.** Explain what is meant by 'significant at $P < 0.05$'.
   b. A scientist wanted to find out whether Chargaff's rule is true for a virus. In a sample of DNA from bacteriophage φX174, 451 bases were found to be purines and 525 were pyrimidines. Calculate the χ^2 value for the purine and pyrimidine percentage values of this virus.
   c. With reference to the probability values listed in Table 6.3, explain whether the null hypothesis should be accepted or rejected at the probability level of 0.05.

Tip

A probability (P) of <0.05 means that we would expect this difference to occur less than five times out of every 100 experiments by chance. If $P < 0.01$, then this would be less than once per 100.

Tip

You saw in Chapter 5 that a null hypothesis is a statement that indicates an expectation that there is no significant relationship between two factors.

6.2 Replication of DNA

You will be aware that DNA is passed down from parents to offspring, and this is crucial to the inheritance of characteristics over the generations. With each new life, a zygote divides to form two cells, which then divide over and over again, eventually forming every cell that makes up a multicellular organism. All of these cells contain identical DNA. This can only happen if the DNA in a cell can be perfectly copied before it divides, so that each 'daughter' cell obtains a complete set of identical genetic information.

It was not until the work of James Watson and Francis Crick, using the experimental data of Rosalind Franklin and others, that the method of DNA replication was proposed.

Link

In Chapter 5, you will have learned that the period during interphase of the cell cycle when DNA replication occurs is called S phase.

To these scientists, it was clear that the two polynucleotide strands could split apart, allowing new nucleotides to line up along each strand, opposite their complementary partners as templates. Figure 6.11 shows what happens in this process, called **semi-conservative DNA replication**.

1 DNA helicase unwinds the DNA double helix and breaks the hydrogen bonds holding the two strands together

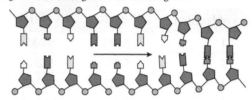

2 Free nucleotides are attracted to their complementary exposed bases, and form hydrogen bonds with them

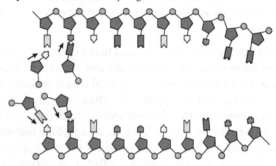

3 DNA polymerase joins the new lines of nucleotides together. Now there are two complete two-stranded DNA molecules, each of which twists to form a double helix

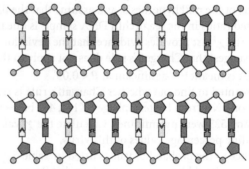

Figure 6.11 The process of DNA replication.

Link

In Chapter 19, you will encounter the polymerase chain reaction (PCR), a procedure that enables scientists to replicate DNA artificially in the laboratory.

During DNA replication, the DNA double helix unwinds as the hydrogen bonds that connect the base pairs are broken. This is carried out by an enzyme called **DNA helicase**, which uses energy from ATP. The two strands separate, exposing the bases on each strand. The phosphodiester bonds between the sugar and phosphate groups in the polynucleotide strands are strong, and they keep the separate strands intact. Exact copies can then be produced, since specific base pairing means that each exposed base will combine with only one of the four types of nucleotide. In other words, both single strands are now used as templates for the production of two new, complementary strands.

Free DNA nucleotides can now attach to the exposed bases on each strand. These nucleotides are said to be 'activated' because they have been phosphorylated by enzymes. Only the complementary bases will fit together. Another enzyme, called **DNA polymerase**, joins the new nucleotides to each other, by catalysing condensation reactions to form phosphodiester bonds between the deoxyribose sugar of one and a phosphate group of the other. A new strand is built on each of the original strands, so that the two new DNA molecules are exactly the same as the original.

As is shown in Figure 6.12, each of the two new molecules of DNA has one of the original polynucleotide strands and one newly-synthesised one made from the supply of free activated nucleotides in the cell. For this reason, the mechanism is called semi-conservative replication, because half (semi) of the original molecule is kept (conserved) in each of the new molecules.

Shortly after Watson and Crick published their theories and suggested semi-conservative replication, US scientists Matthew Meselson and Franklin Stahl set out to investigate whether DNA was replicated using such a mechanism. Figure 6.13 summarises their experiment.

First, Meselson and Stahl grew *E. coli* bacteria in a culture (food) medium containing ammonium chloride with atoms of 'heavy nitrogen'. Heavy nitrogen (^{15}N) contains eight neutrons instead of seven in the more common nitrogen atoms, so it is denser. After time was provided for the bacteria to replicate, the nitrogen in the nucleotides of the bacterial DNA contained ^{15}N.

Next, the scientists took bacteria from the ^{15}N medium and grew them on a medium containing the common 'light' nitrogen (^{14}N) for around an hour, which is just long enough for the cells to replicate once. Other samples of bacteria were left long enough to replicate their DNA twice, three or more times.

The scientists then extracted the DNA from these bacteria and used a technique of density gradient centrifugation, which separates molecules on the basis of their density. They centrifuged the DNA in a solution of caesium chloride. Any substance centrifuged with the caesium chloride solution becomes concentrated at a level corresponding with its density. The DNA molecules came to rest at a position in the centrifuge tube depending on the mass of the molecule. The heavier molecules settled near the bottom of the tube and the lighter molecules near the top.

As shown in Figure 6.13, the results showed that in bacteria that contain ^{15}N DNA, one lighter band was present after growing in a ^{14}N medium for a single generation, and another even lighter band was seen after growing in a ^{14}N medium for another generation or more.

How can this be explained?

- The single, lower band shows that all bacteria in the original sample had DNA containing only ^{15}N.
- The single, higher band in the bacterial sample from the first generation shows that all bacteria now have DNA that has become less dense. Comparing the position of this band with the bands in the other tubes shows that it is not the least dense, suggesting that this DNA contains both ^{15}N and ^{14}N.
- The two bands in the bacterial sample from the second generation shows that some bacteria have DNA that contains both ^{15}N and ^{14}N (lower band), but also that some bacteria have DNA that contains only ^{14}N.

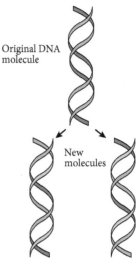

Figure 6.12 DNA replication is semi-conservative. One polynucleotide strand of the original parent DNA molecule (shown in red) is found in each of the two daughter DNA molecules. The blue strands are newly made from nucleotide monomers during the process.

Original DNA molecule

New molecules

Tip

Centrifugation is a laboratory procedure in which a mixture of suspended substances is spun at high speed and the different components separated by density. The denser the substance, the lower it will form a band.

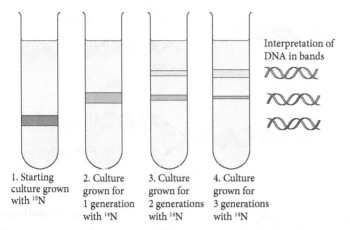

Interpretation of DNA in bands

1. Starting culture grown with ^{15}N

2. Culture grown for 1 generation with ^{14}N

3. Culture grown for 2 generations with ^{14}N

4. Culture grown for 3 generations with ^{14}N

Figure 6.13 The Meselson–Stahl experiment. DNA was isolated from bacteria grown for different numbers of generations. After centrifugation, the bands of DNA formed at different levels, owing to their different densities. Bands tended to form in three specific positions, which are referred to in the text.

- The same two bands in the bacterial sample from the third generation shows that some bacteria have DNA that contains both ^{15}N and ^{14}N, but also that some bacteria have DNA that contains only ^{14}N. However, the thickness of the upper (less dense) band is now larger, suggesting that more bacteria, compared with the sample in the second generation, contain only ^{14}N.

The results of Meselson and Stahl's simple but effective experiment provided strong evidence for the semi-conservative nature of DNA replication.

Worked example

Explain why it is important that weak hydrogen bonds, rather than strong covalent bonds, hold together the two polynucleotide strands found in DNA.

Answer

DNA must be unzipped in order to be replicated. This means that its hydrogen bonds must be broken, to expose the bases on both strands. Base pairs can be separated to allow this to happen very easily, because the hydrogen bonds that hold them together are individually very weak, and the enzyme DNA helicase is able to break them. If these bonds were strong covalent bonds, then the process of DNA replication would require much more energy.

---

7. Figure 6.14 shows three models of DNA replication that were proposed in the 1950s, before the work of Meselson and Stahl was carried out.
   a. Refer to Figure 6.13 and Figure 6.14 to explain whether the results of Meselson and Stahl falsify any of these models (dispersive, conservative or semi-conservative) after the bacteria has grown for:
      i. one generation
      ii. two generations.
   b. Predict what would be seen after 10 generations.
   ★ c. Sketch a line graph to show how the proportions of DNA containing ^{15}N, $\frac{14}{15}$N and ^{14}N vary over four successive generations.

First generation

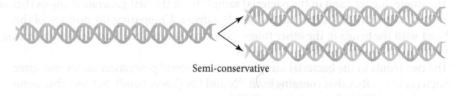

Semi-conservative

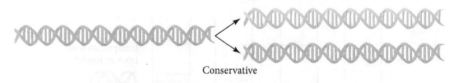

Conservative

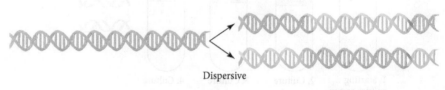

Dispersive

■ Newly synthesised strand
■ Original template strand

Figure 6.14 Three theories of DNA replication – dispersive, conservative and semi-conservative. The mechanism that occurs in living organisms is semi-conservative.

LEADING AND LAGGING DNA STRAND REPLICATION

The semi-conservative replication of DNA explains how the information it contains, in the form of base pairs, is conserved during cell division and from parents to offspring. However, there is a significant problem that must be overcome when a molecule of DNA is replicated. This relates to the structure of the DNA molecule itself.

You will remember that DNA consists of two strands that are arranged in an antiparallel fashion. When a strand of DNA is synthesised, DNA polymerase joins together nucleotides in the 5′ to 3′ direction (Figure 6.3). That is, the phosphate group of a free nucleotide is attached to the 3′ carbon of the deoxyribose of the nucleotide at the end of the existing chain. It is not possible for DNA polymerase to work in the opposite direction and add nucleotides to the 5′ end of an existing molecule. Because of the different organisation of the nucleotides in these two antiparallel strands, this means that they need to be replicated in different ways. The overall process of DNA replication occurs as shown in Figure 6.15.

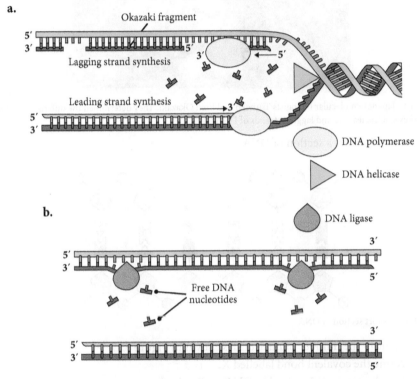

Figure 6.15 The process of leading and lagging DNA strand replication in DNA replication. (a) The leading strand is replicated by DNA polymerase in a continuous process, as free DNA nucleotides are added to the growing strand in a 5′ to 3′ direction. The lagging strand must be replicated in a discontinuous process, in which DNA polymerase adds nucleotides to the growing strand in a 5′ to 3′ direction (away from the site of DNA unwinding), detaching at irregular positions to form Okazaki fragments. (b) To fill in the gaps between the Okazaki fragments on the strand complementary to the lagging strand, the enzyme DNA ligase catalyses the formation of phosphodiester bonds between free nucleotides and the residues at the ends of the fragments.

The two daughter strands are referred to as the **leading strand** and the **lagging strand** and are both made by catalysing the condensation of free nucleotides that have attached by complementary base pairing with the exposed bases on the parental strands.

- The leading strand is made in the 5′ to 3′ direction. By using the parental DNA strand as a template, DNA polymerase joins together nucleotides to make a new daughter strand, in the same direction in which the parental DNA strands are being separated.
- The lagging strand is also made in the 5′ to 3′ direction. However, in order to form the strand in this direction relative to the other parental strand, DNA polymerase must join together nucleotides in the opposite direction compared with the direction in which the parental DNA strands are being separated.

Link

In Chapter 19, you will explore how enzymes such as DNA polymerase and DNA ligase are used in Genetic Technology.

There is one further problem for the cell when it replicates the lagging strand. Because DNA polymerase cannot undertake replication in a **continuous** process, it must stop and restart its activity as the parental molecule is unzipped. This **discontinuous** process produces a series of fragments, called **Okazaki fragments**, named after the scientists who discovered them (Figure 6.16). These are later joined by another enzyme called **DNA ligase**, which catalyses the formation of phosphodiester bonds between them.

Tip

Because the process is discontinuous in nature, the formation of the lagging strand takes a longer time, or 'lags behind,' the leading strand.

Figure 6.16 Japanese molecular biologists Tsuneko and Reiji Okazaki. They determined the differences between the synthesis of the leading and lagging strands of DNA.

Tip

Single-stranded binding proteins attach to the separated DNA strands during replication, to prevent the double helix reforming.

8. Figure 6.17 shows a section of DNA.

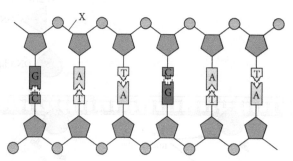

Figure 6.17 A short section of DNA.

    a. Identify the number of nucleotides shown in Figure 6.17.
    b. Name the covalent bond labelled X.
    c. List the enzymes involved in DNA replication.
  ★ **d.** Use Figure 6.17 and your knowledge of enzyme action and DNA replication to suggest why new nucleotides can only be added in a 5′ to 3′ direction.

Key ideas

→ DNA and RNA are nucleic acids, polymers of nucleotides. RNA consists of one polynucleotide strand; DNA consists of two, which are usually much longer than those in RNA.

→ Each nucleotide is formed from a pentose (5-carbon) sugar, a phosphate group and an organic base.

→ The organic bases in DNA are adenine, cytosine, guanine and thymine. In RNA, uracil occurs instead of thymine.

→ DNA bases are either purines (A and G), which have a double-ring structure, or pyrimidines (C and T), which have a single-ring structure.

→ Two nucleotides can join together by a condensation reaction, forming a phosphodiester bond between the pentose group of one of them and the phosphate group of the other.

→ The two polynucleotide strands in DNA run in opposite directions (antiparallel), joined by hydrogen bonds between the bases.

→ There is complementary base pairing in a DNA molecule: A always joins with T, and C always joins with G.

→ DNA is copied perfectly by semi-conservative replication, which ensures genetic continuity between generations of cells.

→ DNA helicase unwinds the DNA helix and breaks the hydrogen bonds between the bases.

→ Free nucleotides bind to exposed bases on both strands of the unwound DNA by complementary base pairing.

→ DNA polymerase joins the new lines of nucleotides by condensation reactions, forming two new polynucleotide strands.

→ The leading and lagging strands of DNA must be replicated in different ways, because DNA polymerase can only add nucleotides in a 5′ to 3′ direction.

6.3 Protein synthesis

The structure of DNA is not suited only to being replicated. It is also an ideal molecule for use in information storage. But how does a single type of molecule such as DNA control the huge range of chemical reactions and processes in a cell?

The answer relates to the molecules that we encountered in Chapter 2 and Chapter 3 – enzymes and other proteins. DNA contains codes to make proteins – the vast family of molecules with different structures and functions, and therefore the identity and function of cells. The two stages required to produce a protein from the DNA base sequence of a gene are called **transcription** and **translation** (Figure 6.18), and these will be considered later in this chapter. As we will see, RNA is crucial to the process of protein synthesis.

Figure 6.18 The production of proteins by a cell is called protein synthesis.

GENES

A polypeptide is coded for by a gene, which is defined as a sequence of nucleotide bases that forms part of a DNA molecule. Polypeptides are variable in size, as we saw in Chapter 2, and so therefore the genes that encode them also vary in size. While there are some genes that are unusually small or large, most are around 1000 to 2000 nucleotide bases in length.

In living cells, a gene for making a particular protein is always found in the same position on the same chromosome. And this is also true for all cells of a multicellular organism. Although it is not quite true in males, owing to the existence of the Y chromosome, humans have 23 different chromosomes, with two copies of each one, making 46 in all. The position of a gene on a chromosome is called its **locus**. For example, the gene that determines the structure and function of the β-chain polypeptide of the haemoglobin molecule is found on chromosome 11. There are many different loci on each chromosome – for example, the genes for both the hormone insulin and the digestive

Tip
The order of transcription and translation in protein synthesis occurs alphabetically: transcription before translation.

Tip
The plural of locus is loci.

enzyme pepsin are also on chromosome 11, at different loci. Genes for keratin (we have several different genes that contribute to making this important structural protein) are found on chromosomes 12 and 17, and so on.

THE GENETIC CODE

Proteins play extremely important roles in living organisms. Many proteins are enzymes, controlling all of the metabolic reactions that take place in a cell. Others may be hormones, antibodies, oxygen-transporters (such as haemoglobin) or structural proteins (such as collagen and keratin). The vast range of proteins that a cell makes, and when it makes them, determines the structure and functions of that cell.

We saw in Chapter 2 that a protein is a polymer built up from units called amino acids. The order of amino acids in a protein (its primary structure) determines its three-dimensional structure (its tertiary structure) and therefore its function. In 1959, Frederick Sanger published the order of amino acids found in a protein for the first time – the 51 subunits that make up the insulin molecule, the hormone that regulates blood glucose concentration.

The sequence of nucleotide bases in a DNA molecule codes for the sequence in which a cell puts together amino acids when it builds protein molecules. There are only four different bases in a DNA molecule (A, C, T and G) but there are 20 naturally occurring amino acids. So how does the code work? As we will see in the Experimental skills section later, scientists discovered in the 1960s that a sequence of three bases, rather than a single base, codes for each different amino acid. A sequence of three bases on a DNA molecule, coding for one amino acid, is called a **triplet**.

By using three bases (which gives 64 possible triplets), there are plenty of sequences to spare. Some DNA triplets code for the same amino acid as others, and this feature of the code is described as being **degenerate**. Some combinations of bases act as 'START' and 'STOP' **codons** when transcribed into RNA. They indicate where the cell should start and stop the protein synthesis. If we think of the triplets that code for amino acids as being like words in a sentence, then the 'START' codon is like the capital letter at the start of a sentence, while 'STOP' codons are like full stop symbols. Table 6.4 on page 171 shows the triplets in RNA, a molecule produced from the DNA template, that correspond to specific amino acids. This is called the **genetic code**, and, as we saw in the introduction, it is **universal** (shared) among all living organisms.

You will remember that a DNA molecule is made of two strands of bases wound together into a helix. Only the sequence on one of the strands is used as a code for making proteins. This strand is called the **template strand** or **transcribed strand**. We sometimes talk about 'base pairs' rather than just bases when describing the code carried by the DNA, meaning the bases on both of the strands rather than just on the template strand – but it is important to remember that only the sequence of bases on one strand is actually used for making proteins. The other strand is thought to serve to stabilise the molecule, and is important for the process of DNA replication. This is called the **non-transcribed strand**.

Figure 6.19 shows a section of the gene that codes for the first four amino acids of insulin. The first triplet of bases, AAA, codes for phenylalanine, the next for valine (GTC), and so on. The triplets in a DNA strand are read sequentially, and the code is said to be **non-overlapping**. This means that three bases make up one triplet code for one amino acid, and the next three codes for the next one, and so on. So, in this example, the length of the first six bases in the DNA tells the cell to join the amino acid valine to the amino acid phenylalanine.

So how does the sequence of bases in a DNA molecule determine which proteins are made in the cell? As you have seen, this sequence of amino acids – known as the primary structure – ultimately determines the three-dimensional shape of the protein, and therefore its function. Hence the sequence of A, T, C and G in the DNA in a cell's nucleus carries information about all the proteins that will be made in that cell, and this in turn determines the structures that are formed in the cell, and the metabolic reactions that it will carry out.

Link

You may remember how the 20 different amino acids differ in structure. The basic skeleton of the molecule is similar, whereas the R group varies (see Chapter 2, Figure 2.30).

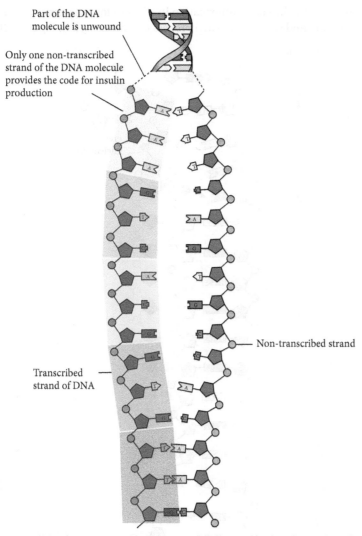

Part of the DNA molecule is unwound

Only one non-transcribed strand of the DNA molecule provides the code for insulin production

Non-transcribed strand

Transcribed strand of DNA

Figure 6.19 How DNA encodes proteins. The template strand (left) is read by the cell in triplets (shaded different colours), with each triplet responsible for encoding one specific amino acid. The strands unwind in the region of the gene that will be read to make a protein.

TRANSCRIPTION AND THE PRIMARY TRANSCRIPT

DNA is immensely valuable – if it gets damaged, then the cell does not have correct instructions for making proteins, and is likely to die. The DNA is therefore kept in the nucleus, often packaged tightly into chromosomes. However, in order to make a protein, the information that this 'master copy' carries must be transferred to the ribosomes, which are found in the cytoplasm. To do this, copies of the base sequence are produced, in the form of a molecule of RNA, called the primary RNA transcript, or **primary transcript** for short. This RNA molecule can be thought of as a mobile copy of a gene, from which it has been copied.

As we saw earlier in this chapter, the structure of RNA is similar to that of a single strand of DNA, except that the sugar ribose replaces deoxyribose and the base uracil replaces thymine. Uracil and thymine molecules are similar in size and shape, and uracil forms a complementary base pair with adenine.

All primary transcripts are produced by joining together ribonucleotides containing specific bases in a particular order – an order complementary to the sequence of the gene in the DNA. These molecules of RNA are small enough to pass through pores in the nuclear envelope and they move from the nucleus into the cytoplasm and are then used as guides to manufacture the protein encoded in their sequence of bases. However, unlike DNA, RNA molecules are short-lived and can be easily broken down; this enables the cell to change protein production to suit its needs.

The process of using the coded information in DNA to form a primary transcript is called transcription, and this step-by-step process is summarised in Figure 6.20.

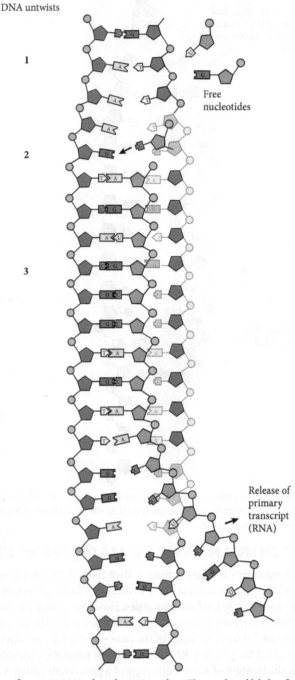

DNA untwists

Free nucleotides

Release of primary transcript (RNA)

Figure 6.20 The process of transcription in the eukaryotic nucleus. The numbered labels refer to the accompanying text.

1. An enzyme, DNA helicase, catalyses the breaking of hydrogen bonds between the two DNA strands. This causes the DNA in the region of a gene to unzip.
2. The bases of free RNA nucleotides then join, by complementary base pairing, with the exposed DNA bases on the template strand.
3. Another enzyme, **RNA polymerase**, then links the RNA nucleotides together by catalysing the formation of phosphodiester bonds between their ribose and phosphate groups. As the RNA polymerase moves along the template strand, it produces a single-stranded molecule of RNA called the primary transcript.

The primary transcript carries coded information in a similar way to DNA. The order of bases on this molecule is therefore complementary to those on the template strand of DNA; and the sequence of bases in the primary transcript is therefore complementary to the sequence of bases in the DNA. Remember, however, that uracil bases are used in place of thymine. A sequence of three bases on the primary transcript, coding for one amino acid, is known as a **codon**.

As an example, a section of the DNA template strand shown in Figure 6.20 has the sequence of bases AAAGTCACG. When transcribed, the sequence of bases on the mRNA is UUUCAGUGC. The three mRNA codons are therefore UUU, CAG and UGC.

MODIFICATION OF THE PRIMARY TRANSCRIPT TO FORM mRNA

When transcription is complete, the primary transcript detaches from the DNA and passes out of the nucleus through the nuclear pores. However, in eukaryotic cells, before it reaches its ultimate destination – the ribosomes – it is usually modified in a process called **splicing**.

To understand what happens during splicing, we must look more closely at the structure of the gene that was transcribed. Figure 6.21 shows that within the sequence of bases on the DNA that makes up a gene, only some of the triplets actually code for amino acids that will become part of the final protein molecule. These parts are known as **exons**, and the parts that do not code for amino acids are called **introns**. This is not to say that introns are unimportant. In the past two decades, more and more functions for introns have been discovered, some of which relate to gene control, or when and how genes are expressed.

All of the sequence of DNA bases in a gene, both exons and introns, are transcribed to form the primary transcript. This therefore contains the complementary RNA base sequence of both exons and introns.

As it leaves the nucleus, the primary transcript is processed to remove the introns and stick the exons together during splicing. The resulting molecule is shorter and contains only base sequences that will encode amino acids. It is now called messenger RNA (mRNA).

> **Link**
>
> In Chapter 16, you will learn more about when and why genes are expressed, or 'switched on and off'. This is called gene control.

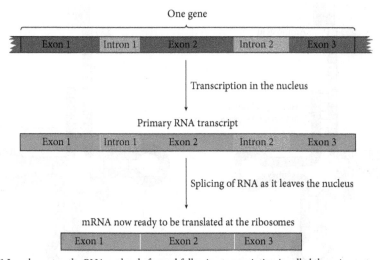

Figure 6.21 In eukaryotes, the RNA molecule formed following transcription is called the primary transcript. This is modified by the removal of non-coding sequences (introns) and the joining together of coding sequences (exons) to form messenger RNA (mRNA).

Splicing of the primary transcript does not occur at all in prokaryotes because there are no introns. After transcription, this molecule travels straight to a free ribosome in the cytoplasm to be translated. In eukaryotes, ribosomes are often attached to the rough endoplasmic reticulum, which is attached to the nuclear envelope (Figure 6.22).

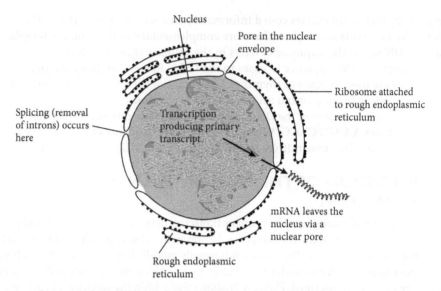

Figure 6.22 The process of transcription and splicing in eukaryotic cells. Structures are not to scale.

TRANSLATION AND TRANSFER RNA (tRNA)

In translation, the code on the mRNA is used to assemble the amino acids of the protein in the correct order. This involves another type of RNA, called transfer RNA (tRNA). Every cell contains thousands of tRNA molecules, each around 80 nucleotides long. These are made of a single strand of RNA nucleotides, folded round to make three lobes, held together by hydrogen bonds between complementary base pairs (**intramolecular hydrogen bonds**). The tRNA molecules are often described as being shaped like a clover leaf. Each tRNA has three unpaired bases, called an **anticodon**, usually shown at the base of the molecule. Figure 6.23 shows two different tRNA molecules, which differ in the sequence of bases found at the anticodon.

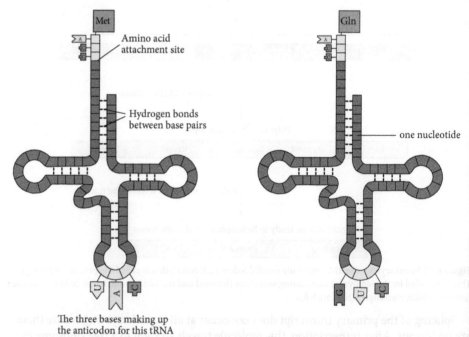

Figure 6.23 The structure of two transfer RNA (tRNA) molecules. Intramolecular hydrogen bonds form between complementary base pairs that are close to each other in the single strand. The three unpaired bases that are closest to the amino acid attachment site are always ACC. The three bases called the anticodon differ between different tRNA molecules, and they correspond to the amino acid that is being carried. Note that only the bases at the amin acid attachment site and the anticodon are shown.

At the other end of the molecule is a site onto which an amino acid can be attached. This is always at the end of three RNA nucleotides with the bases ACC. There is a specific tRNA for each amino acid. Each attachment of a tRNA to an amino acid is carried out by a specific enzyme (called aminoacyl transferase), which requires energy derived from the hydrolysis of ATP. A tRNA with a particular anticodon will only bind with the amino acid that corresponds to that anticodon. For example, a tRNA molecule with the anticodon GAA carries only the amino acid leucine. In this way, the sequence of nucleotide bases in the mRNA molecule determines the sequence of amino acids in the polypeptide chain (Table 6.4).

First base	Second base				Third base
	G	A	C	U	
G	glycine	glutamic acid	alanine	valine	G
	glycine	glutamic acid	alanine	valine	A
	glycine	aspartic acid	alanine	valine	C
	glycine	aspartic acid	alanine	valine	U
A	arginine	lysine	threonine	methionine/START	G
	arginine	lysine	threonine	isoleucine	A
	serine	asparagine	threonine	isoleucine	C
	serine	asparagine	threonine	isoleucine	U
C	arginine	glutamine	proline	leucine	G
	arginine	glutamine	proline	leucine	A
	arginine	histidine	proline	leucine	C
	arginine	histidine	proline	leucine	U
U	tryptophan	STOP	serine	leucine	G
	STOP	STOP	serine	leucine	A
	cysteine	tyrosine	serine	phenylalanine	C
	cysteine	tyrosine	serine	phenylalanine	U

Table 6.4 The genetic code for mRNA codons. To read the table, look up a codon using the first, second and third bases – for example, to identify the amino acid encoded by the RNA codon ACU, find the first base A (look down the first column), second base C (look along the top row) and third base U (look down the right-hand column). This codon, ACU, codes for threonine.

9. During transcription, the following events occur:
   1. Hydrogen bonds are broken between complementary bases.
   2. Hydrogen bonds form between complementary bases.
   3. Phosphodiester bonds form.
   4. Free nucleotides pair with complementary nucleotides.

   Before the primary transcript leaves the nucleus, which events will have occurred twice?
   A. 1 and 2 only
   B. 1, 2, 3 and 4
   C. 1, 3 and 4 only
   D. 2, 3 and 4 only

10. Compare the features of introns and exons.
11. Give two differences between the structure of a molecule of mRNA and the structure of a molecule of tRNA.

Translation is a step-by-step process that occurs as follows. The numbered steps refer to the process shown in Figure 6.24.

1. One end of the mRNA strand attaches to the ribosome. This is possible because the ribosome also contains RNA molecules, which have a sequence complementary to a section of all mRNA molecules.
2. A tRNA molecule, with an anticodon that has a complementary base sequence to the first codon on the mRNA, attaches to the codon by way of hydrogen bonding. This

Tip
You are not expected to know which amino acids correspond to each codon, but you may be asked to use a similar table to determine this.

tRNA molecule carries a specific amino acid, the identity of which is related to its anticodon sequence.

3. A second tRNA molecule, also carrying its amino acid, then binds to the next codon on the mRNA. So, two tRNA molecules bind to the ribosome at once.

4. The amino acids at the far end of the molecules are very close together. The amino acids are then joined together by a peptide bond, a process that requires a condensation reaction. Energy from ATP is needed for this reaction to occur.

5. The ribosome then moves along the mRNA. The first tRNA molecule, minus its amino acid, detaches and the second tRNA molecule takes its place.

6. The next codon becomes available to bind a tRNA molecule with the next amino acid, which is then added to the growing polypeptide chain.

7. Eventually, after this continues for hundreds of amino acids, a STOP codon is reached and the ribosome detaches, releasing the finished polypeptide. This will then take on secondary, tertiary and possibly quaternary structures to form a functional protein.

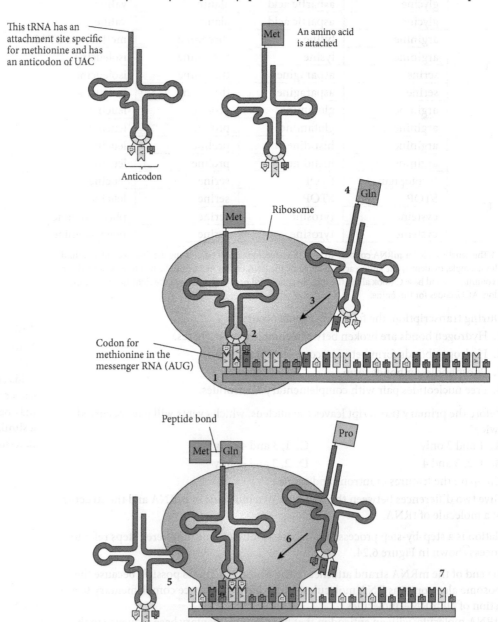

Figure 6.24 The process of translation. The numbered labels refer to the accompanying text.

Through the mechanism of translation, it is possible to imagine how the order of codons on the mRNA molecule determines which tRNA molecules attach. Because the tRNA molecules determine which amino acids are brought to the ribosome, the process ensures that the amino acids are assembled in the correct sequence to make the polypeptide chain encoded by the original gene on the DNA molecule.

Many ribosomes often work together on the same length of mRNA, each translating a particular part of it, so that multiple copies of a protein can be made simultaneously from just one strand of mRNA. A group of ribosomes working like this is called a polyribosome or **polysome** (Figure 6.25). Cells that need or secrete large quantities of a specific protein will use polyribosome assemblies to make many copies of the mRNA for that protein, e.g. insulin in the beta cells of the pancreas.

Figure 6.25 A polyribosome – a group of ribosomes (purple) arranged along an mRNA strand (red) that are producing multiple copies of the same polypeptide (green).

Worked example

Figure 6.26 shows the process of translation.

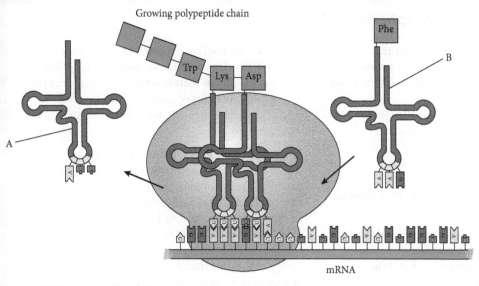

Figure 6.26 The process of translation.

a. Explain the difference in the appearance between the molecules labelled A and B.

Answer

Molecule A is a tRNA that is not carrying an amino acid. It released its amino acid when it was attached to the ribosome, and the ribosome incorporated this into the growing polypeptide chain. Molecule B is a tRNA with its amino acid attached. This will soon attach to the ribosome and will align with the codon on the mRNA that has a complementary base sequence to its anticodon.

b. Explain how the ribosome will release the growing polypeptide chain.

Answer

The ribosome will recognise a STOP codon in the mRNA sequence. There are three STOP codons: UAA, UGA and UAG.

Tip

Remember from your work in Chapter 2 that a polypeptide will only become a protein when it takes on a three-dimensional (tertiary) structure.

12. **a.** One strand of DNA has the following sequence of bases:
CATCATAGATGAGAC
Use the genetic code in Table 6.4 to determine the amino acid sequence that this sequence of bases would code for. Remember that the table shows the mRNA codons, not DNA triplets.

★ **b.** A polypeptide molecule contains the amino acid sequence glycine–leucine–lysine–valine. Use the genetic code in Table 6.4 to determine which anticodons on the transfer RNA molecules (A, B, C or D) would be needed for the synthesis of this polypeptide.
   A. CCC GAA AAA CAA
   B. CCC GAA UUU CAA
   C. GGG CUU AAA GUU
   D. GGG CUU UUU GUU

Experimental skills 6.1: Deciphering the genetic code

The involvement of mRNA in protein synthesis is now very well established. However, who first provided the evidence for its role as a messenger between the DNA and the ribosomes?

In 1961, US biochemists Marshall Nirenberg and Heinrich Matthaei undertook experiments to investigate the role of mRNA in protein synthesis. Their aim was to show that mRNA is vital for protein synthesis.

Nirenberg and Matthaei undertook the following steps in their investigation:

- They purified extracts from bacterial cells that contained ribosomes, amino acids and tRNA molecules. However, they ensured that these extracts did not contain mRNA.
- To these extracts, they added an mRNA that contained hundreds of nucleotides with a single base – uracil. These molecules were called poly-U mRNA, or simply 'poly-U' for short.
- They found that after incubating the mRNA with the extracts, a protein was produced that consisted entirely of phenylalanine amino acid residues joined by peptide bonds.

QUESTIONS

P1. In addition to the experiments described, Nirenberg and Matthaei conducted a trial in which all ingredients were added except mRNA. Explain why no protein synthesis occurred in the investigations without mRNA.

P2. Unlike tRNA, the poly-U mRNA remains linear and does not form intramolecular hydrogen bonds. Explain why.

P3. Describe another experiment that Nirenberg and Matthaei could undertake to strengthen the validity of their investigation, and what the expected results of the experiment would be. You should use Table 6.4 to help you answer this question.

The genetic code explains how mRNA codons specify the 20 amino acids, and how these subunits are assembled into different polypeptides. There are four bases and 20 amino acids found in most organisms, so it is not possible for one base to code for one amino acid. There are 16 combinations of two bases (4^2), which is still insufficient to code for all 20 molecules. Living organisms therefore use a triplet code, with groups of three bases coding for each amino acid. But how was this 3:1 ratio uncovered?

Nirenberg and another colleague, Phillip Leder, then devised a second experiment. Their aim was to identify how many mRNA bases are required to encode one amino acid – the length of a codon. The scientists predicted that the minimum number of nucleotides required to code for 20 amino acids would be three bases.

To undertake this experiment, Nirenberg and Leder purified ribosomes, poly-U mRNA and tRNA molecules carrying radioactive phenylalanine. The scientists also used a type of filter paper, made of nitrocellulose. This had special properties so that it could bind and hold ribosomes, but not RNA molecules, which pass straight through the paper. Figure 6.27 summarises the use of the nitrocellulose membrane.

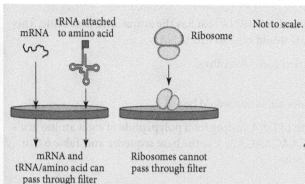

mRNA tRNA attached to amino acid

Ribosome

Not to scale.

mRNA and tRNA/amino acid can pass through filter

Ribosomes cannot pass through filter

Figure 6.27 The equipment used by Nirenberg and Leder to work out the number of mRNA bases in a codon. A special filter made of nitrocellulose is permeable to free RNA and free amino acids, but not to the larger ribosomes, to which it binds.

P4. Identify the **two** independent variables in this experiment.

P5. Identify one factor that must be standardised in this method.

P6. Describe an appropriate control for this experiment and explain why it is important to include this.

🔍 **P7.** Draw a graph of the data provided in Table 6.5.

☆ **P8.** Using your graph, explain how the results of this experiment provide evidence that proves the prediction of Nirenberg and Leder.

The researchers varied the concentration of radioactive poly-U mRNA molecules of different lengths. Table 6.5 shows the results of a similar experiment.

Concentration of poly-U RNA/arbitrary units	Radioactivity / arbitrary units				
	U_2	U_3	U_4	U_5	U_6
0	0	0	0	0	0
5	1	8	8	9	19
10	2	15	17	17	35
15	3	20	22	22	48
20	3	30	31	34	68
25	3	45	46	47	88
30	4	48	49	50	97

Table 6.5 The results of an experiment similar to that undertaken by Nirenberg and Leder. Different concentrations of poly-U RNA molecules, of different lengths (U_2–U_6), were added to a mixture containing ribosomes and tRNA molecules carrying radioactive phenylalanine. After allowing for a period of incubation, the mixtures were poured through a nitrocellulose membrane. This membrane was then washed before being analysed for its radioactivity.

Worked example

Table 6.6 shows some mRNA codons and the amino acids for which they code.

mRNA codon	Amino acid	Code
CUU	valine	Val
GUU	leucine	Leu
GCC	alanine	Ala
ACC	threonine	Thr
AUU	isoleucine	Ile

Table 6.6

a. A tRNA molecule has the anticodon UAA. Which amino acid does the tRNA molecule carry?

Answer

The anticodon UAA would be found on a tRNA that has the amino acid isoleucine. This is because the mRNA codon AUU would bind to this tRNA.

b. List the DNA base sequence that codes for valine.

Answer

The DNA base sequence that codes for valine would be GAA.

c. The base sequence of a section of DNA coding for a polypeptide of eight amino acids is TGGTGGTAAGAACAACAAGAACAA. Use the base sequence and Table 6.6 to deduce the order of amino acids.

Answer

First, you should write the DNA base sequence arranged into triplets:

TGG TGG TAA GAA CAA CAA GAA CAA

Then, work out the base sequence of the RNA molecule that would be transcribed from the DNA base sequence:

ACC ACC AUU CUU GUU GUU CUU GUU

The order of amino acids, as deciphered from the code listed, would be:

Thr–Thr–Ile–Val–Leu–Leu–Val–Leu

13. The following statements show some of the events that happen during protein synthesis:
    A. The primary transcript leaves the nucleus.
    B. tRNA molecules bring specific amino acids to the mRNA molecule.
    C. The two strands of a DNA molecule separate.
    D. Peptide bonds form between the amino acids.
    E. A ribosome attaches to the mRNA molecule.
    F. RNA nucleotides join with the exposed DNA bases and form a molecule of mRNA.
    G. The primary transcript undergoes splicing, and intron sequences are removed.

    a. Deduce the correct order of statements that describes how protein synthesis occurs, starting with the earliest.
    b. State in which of these stages condensation reactions happen.
    c. State which of these stages occurs in the cytoplasm of the cell.
    d. Identify the stage at which translation begins.

14. DNA molecules are much more stable in cells – they last for longer than RNA. One reason that may explain the difference in stability between DNA and RNA is that the enzymes that digest nucleic acids in cells, called nucleases, are more able to recognise RNA compared with DNA.

    a. Using your knowledge of enzyme action from Chapter 3, suggest two features of RNA that are not found in DNA, which may be recognised by nuclease enzymes.
    ★ ◯ **b.** Referring to the process of protein synthesis in your answer, explain why it is important that RNA is shorter lived than DNA in the cell.

Key ideas

→ DNA carries the genetic code. Its structure allows it to store information that can be used to make proteins, and to copy itself, allowing the genetic code to pass to new cells.
→ A gene can be defined as a base sequence of DNA that codes for the amino acid sequence of a protein, or that codes for a functional RNA.
→ Ribonucleic acid (RNA) has two forms, messenger RNA (mRNA) and transfer RNA (tRNA), both of which are involved in protein synthesis.

→ mRNA is a copy of a gene, which allows genes to be used as templates for protein synthesis. tRNA molecules bring amino acids to the ribosome during protein synthesis.

→ A sequence of three bases on an mRNA molecule is called a codon. A sequence of three bases on a tRNA molecule is called an anticodon.

→ Protein synthesis begins in the nucleus with transcription of the DNA code of a complete gene to a complementary primary transcript, made of RNA. This contains exons – sections of the gene that encode amino acids – and introns – which do not.

→ The primary transcript then moves out of the nucleus. Its introns are removed and the exons are spliced together to make mRNA. The mRNA attaches to a ribosome, where translation takes place.

→ The tRNA molecules with complementary anticodons to the first two mRNA codons bond with them in the ribosome. This brings their amino acids close together, and a peptide bond is formed between them. This process continues until the polypeptide is fully made.

6.4 Mutation

As we saw in Chapter 5, all of the DNA in a cell is replicated before cell division. During mitosis, these chromosomes are shared out equally between the daughter cells. However, as in any complex process, errors do occur. These can result in changes in the sequence of base pairs in a DNA molecule that may result in an altered polypeptide. These changes are called mutations.

Mutations occur spontaneously, and at random. **Gene mutations** most often occur during DNA replication. They can happen at any time that a cell is replicating its DNA, which can be just before either mitosis or meiosis takes place.

The risk of gene mutation is also increased with exposure to factors called **mutagenic agents**. These include ionising radiation, such as alpha, beta, gamma and X-rays, which damages DNA. For example, ultraviolet radiation from the sun damages DNA in the skin. Many different chemicals also act as mutagens. For example, cigarette smoke contains numerous mutagenic agents.

There are three types of gene mutation:

- **nucleotide substitution** – a nucleotide is replaced by a nucleotide with a different base
- **nucleotide deletion** – one or more nucleotides are removed, so the sequence has fewer bases
- **nucleotide insertion** – one or more extra nucleotides are inserted, so extra bases are added to the sequence.

The effect that each of these types of mutation have on the 'meaning' of the code carried in the gene varies. Different mutations have different effects on how much of the code is disrupted. We can illustrate this using a sentence of three-letter words in which changes in some of the letters represent changes in bases. Consider the following statement:

THE OLD MEN SAW THE BOY

A nucleotide substitution (M to H) would result in the statement taking on a similar but slightly different meaning:

THE OLD HEN SAW THE BOY

However, because the code is read in groups of three (triplets), nucleotide insertions and deletions result in more muddled statements that have no meaning:

Nucleotide (O) deletion – THE LDM ENS AWT HEB OY
Nucleotide (C) insertion – THE COL DME NSA WTH EBO Y

Link

The special type of cell division that produces gametes, called meiosis, will be explored in Chapter 16.

Link

In Chapter 16, you will see how the symptoms of the blood disorder sickle cell anaemia are caused by a substitution mutation in the β-globin gene.

Unlike substitutions, the deletion or insertion of one nucleotide to the code changes all of the subsequent triplets in the gene. This is called a **frameshift** and causes the entire sentence to become meaningless. All the triplets after the mutation are changed, so a whole string of 'wrong' amino acids are used to build the polypeptide on a ribosome. Often, by chance, the change in base sequence may result in the introduction of a premature STOP codon. This means that the encoded protein will be much shorter (truncated) than expected. Deletion and insertion mutations, therefore, tend to have serious effects on the structure and function of the protein for which that piece of DNA is coding, unless they happen to be very close to the end of the sequence. The resulting protein has a completely different primary structure, and therefore its three-dimensional structure and its function are completely lost or altered.

Substitution mutations, on the other hand, may have no effect at all. You may remember that several different triplets code for the same amino acid – the genetic code is degenerate. A substitution of just one base can therefore mean that there is no change in the amino acid for which that triplet codes. This is sometimes called a **silent mutation**, as it has no effect on the encoded protein. For example, with reference to Table 6.4, a substitution mutation that changes the mRNA codon UCC to UCA will still result in serine being positioned in the polypeptide chain. All of the subsequent triplets remain unchanged, so all of the rest of the protein will be as it should be. There is a significant possibility that, even with one 'wrong' amino acid, the protein may still be able to carry out its normal function, depending on where the amino acid change occurs in the protein.

In summary, most mutations that affect the function of the cell will bring this change about by causing the changes in the sequence shown in Figure 6.28.

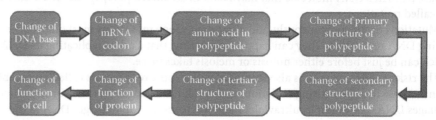

Figure 6.28 How a change in the base of DNA (gene mutation) causes a change in the function of a cell.

Worked example

Explain why a mutation that causes the deletion of a nucleotide may have a greater effect than a mutation in which one nucleotide is substituted for another.

Answer

Deletion of a nucleotide causes a frameshift. This means that the base sequence is changed from the point of the mutation, resulting in a change to the amino acid sequence of the protein from that point onwards. Nucleotide substitution alters only one base, which means that only one amino acid is altered in the protein sequence (assuming that the new codon encodes a different amino acid, which it may not, owing to the degeneracy of the genetic code).

Key ideas

→ Gene mutations involve a change in the sequence of base pairs in a molecule of DNA.
→ Gene mutations involving **nucleotide substitution** often have no effect on the polypeptide made, because of the degenerate nature of the genetic code.
→ **Nucleotide additions** and **nucleotide deletions** are more likely to have an effect, because they change the whole set of triplets in DNA that follow them, and therefore greatly affect the sequence of amino acids in the polypeptide that is built.
→ Gene mutations can occur spontaneously. The risk is increased by mutagenic agents.

CHAPTER OVERVIEW

Review the mini mind map to link new learning with your prior learning, and to highlight which of the key concepts underpin the chapter.

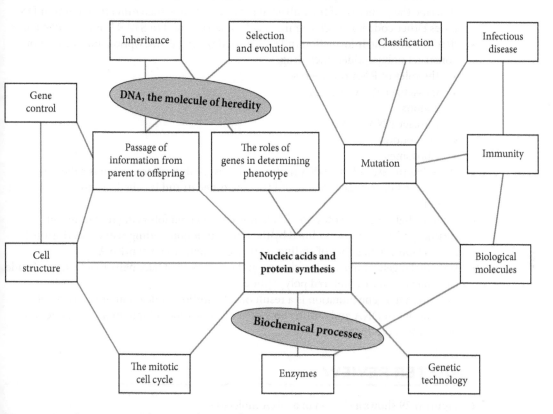

Try copying this mini mind map and expanding upon it. Use your notes from other chapters to help you explore how the essential ideas, theories and principles can be linked further together.

WHAT YOU HAVE LEARNED

Now that you have finished this chapter you should be able to:

- describe the structure of nucleotides, including the phosphorylated nucleotide ATP (structural formulae are not expected)
- state that the bases adenine and guanine are purines with a double-ring structure, and that the bases cytosine, thymine and uracil are pyrimidines with a single-ring structure (structural formulae for bases are not expected)
- describe the structure of a DNA molecule as a double helix, including:
  - the importance of complementary base pairing between the 5′ to 3′ strand and the 3′ to 5′ strand (antiparallel strands)
  - differences in hydrogen bonding between C–G and A–T base pairs
  - linking of nucleotides by phosphodiester bonds
- describe the semi-conservative replication of DNA during the S phase of the cell cycle, including:
  - the roles of DNA polymerase and DNA ligase (knowledge of other enzymes in DNA replication in cells and different types of DNA polymerase is not expected)
  - the differences between leading strand and lagging strand replication as a consequence of DNA polymerase adding nucleotides only in a 5′ to 3′ direction

- describe the structure of an RNA molecule, using the example of messenger RNA (mRNA)
- state that a polypeptide is coded for by a gene and that a gene is a sequence of nucleotides that forms part of a DNA molecule
- describe the principle of the universal genetic code in which different triplets of DNA bases either code for specific amino acids or correspond to START and STOP codons
- describe how the information in DNA is used during transcription and translation to construct polypeptides, including:
  - the roles of RNA polymerase
  - messenger RNA (mRNA)
  - codons
  - transfer RNA (tRNA)
  - anticodons
  - ribosomes
- state that the strand of a DNA molecule that is used in transcription is called the transcribed or template strand and that the other strand is called the non-transcribed strand
- explain that, in eukaryotes, the RNA molecule formed following transcription (primary transcript) is modified by the removal of non-coding sequences (introns) and the joining together of coding sequences (exons) to form mRNA
- state that a gene mutation is a change in the sequence of base pairs in a DNA molecule that may result in an altered polypeptide
- explain that a gene mutation is a result of substitution or deletion or insertion of nucleotides in DNA and outline how each of these types of mutation may affect the polypeptide produced.

CHAPTER REVIEW

1. Figure 6.29 shows a section of a DNA molecule.
   a. Identify the components labelled A, B and C.
   b. Describe two ways in which the structure of the DNA molecule suits its function.
   c. Using the terms 'hydrogen bond' and 'phosphodiester bond' in your answer, describe the process of semi-conservative DNA replication.

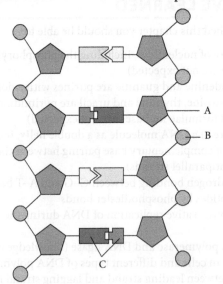

Figure 6.29

2. Copy and complete Table 6.7 to compare the processes of leading and lagging strand synthesis in DNA replication by placing a tick (✔) or cross (✘) into the table.

Feature	Synthesis of leading strand	Synthesis of lagging strand
uses free DNA nucleotides		
is a continuous process		
requires DNA polymerase		
requires DNA ligase		
happens in the 5′ to 3′ direction		

Table 6.7

3. Gemcitabine is a chemotherapy drug used to treat a number of types of cancer. It works by blocking the synthesis of new DNA, which results in the death of cells. Figure 6.30 shows the structures of gemcitabine (a) and a DNA nucleotide containing the base cytosine (b).

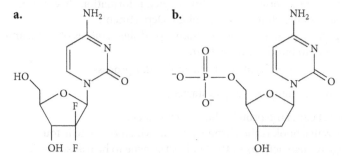

Figure 6.30 The structures of gemcitabine (a) and a DNA nucleotide containing the base cytosine (b).

With reference to the structure of both molecules, suggest how gemcitabine prevents DNA replication.

4. A short sequence of mRNA is shown below:
   AUGGCCUCGAUAACGGCCACCAUG
   a. What is the maximum number of amino acids that could be coded for by this section of mRNA?
   b. How many different types of tRNA molecule would be used to produce a polypeptide from this section of mRNA? Refer to Table 6.4 to help you.

5. Transfer RNA (tRNA) molecules are synthesised inside the nucleus of eukaryotic cells. Outline the process by which tRNA molecules are produced.

6. Draw a table to compare the processes of DNA replication and DNA transcription.

7. Match the terms with their corresponding definitions.

(a) DNA replication	(i)	mRNA is decoded and directs the sequencing of amino acids to make a polypeptide chain.
(b) mRNA	(ii)	A change in base sequence.
(c) anticodon	(iii)	The site of synthesis of ATP, which is required as a source of energy for protein synthesis.
(d) ribosome	(iv)	Material of which genes are made.

(e) DNA	(v)	Triplets of bases on one end of tRNA.
(f) codon	(vi)	The synthesis of mRNA by RNA polymerase, which happens in the nucleus.
(g) mitochondrion	(vii)	Carries DNA code from nucleus to cytoplasm.
(h) transcription	(viii)	mRNA triplet of bases that determines which amino acid will be used during translation.
(i) translation	(ix)	Where translation takes place.
(j) mutation	(x)	DNA makes a copy of itself.

8. Read the following passage, which describes an analogy of protein synthesis.

 A student wants to make a model car in the Technology Department at school. The instructions to make the car are found in a reference book that cannot be removed from the Library.

 The student opens the reference book and finds the correct page. A photocopier is used to take a photocopy, and the photocopied information sheet is taken out of the Library to a workbench in the Technology Department.

 Reading the instructions on the photocopied sheet, pieces of wood and plastic are joined together to build the model car.

 a. Link the items and events with the molecular process.

 Here are three examples to get you started:

 • The model car is the protein that is to be made.
 • The reference book is the chromosome in which the gene is found.
 • The page of instructions in the book is the gene to be transcribed.

   i. The Library
   ii. The Library door
   iii. The photocopier
   iv. The photocopied information sheet
   v. The workbench
   vi. The pieces of wood and plastic that make the model car

 b. The transfer RNA (tRNA) molecules that are involved in translation do not feature in this analogy. Explain the role of tRNA molecules in protein synthesis.

 c. Another process occurs in protein synthesis, but only in eukaryotes. This is not accurately represented by this analogy. Describe this process and explain why it is necessary.

9. Explain how a mutation in a gene can result in a genetic disease.

Chapter 7

Transport in plants

Australian eucalyptus trees can take in large amounts of water through their roots. The water then travels up to their leaves, from where it evaporates. This process is known as transpiration. Although transpiration occurs in most plants, eucalyptus trees do it so efficiently that they are sometimes deliberately planted in areas of swamp or marshland to dry it out, for example, to remove potential breeding grounds for malaria-carrying mosquitoes.

It has also been discovered that their roots, which can be tens of metres long, can absorb gold from deposits deep below the surface. Minute amounts of gold are transported to the leaves but, unlike water, the gold collects there. Testing eucalyptus leaves for the presence of gold might be a more efficient way of finding the sites of new deposits than more traditional methods such as drilling.

Prior understanding

From your previous studies, you may be familiar with the roles of root hair cells, xylem and phloem in transporting substances through a plant. You may have also investigated how the rate of transpiration is affected by external factors. In Chapter 1, you used microscopy to study the structure of plant cells, and in Chapter 4, you looked at the processes by which substances move into and out of cells, by diffusion, osmosis and active transport.

Learning aims

In this chapter, you will learn about the structure and function of the plant transport tissues – xylem and phloem – including using microscopy to study their fine structure. You will also learn about the processes of transpiration and translocation that occur through these tissues.

7.1 Structure of transport tissues (Syllabus 7.1.1–7.1.4)

7.2 Transport mechanisms (Syllabus 7.2.1–7.2.8)

7.1 Structure of transport tissues

Most plants, like most multicellular organisms, have transport systems to move substances from one part of the organism to another. Some simple plants, such as mosses (Figure 7.1a), do not have specialised transport systems and are described as **non-vascular**. (They do not have transport vessels.) However, most plants do have specialised, tube-like transport vessels and are described as **vascular**. In this chapter you will learn about the transport systems found in the most common group of flowering plants, known as **herbaceous dicotyledonous** plants (Figure 7.1b). (Herbaceous means non-woody, so does not include trees, and dicotyledonous means they have two **cotyledons**, or embryonic leaves, in each seed.)

a.  b.

Figure 7.1 (a) Moss, a non-vascular plant, growing on a tree trunk. **(b)** Sunflower plants – an example of a herbaceous dicotyledonous plant.

The two transport tissues found in herbaceous dicotyledonous plants are the **xylem** and **phloem**. Xylem tissue transports water and dissolved mineral ions from the roots, up the stem, to the leaves, where the water evaporates into the atmosphere. This process is called **transpiration**. The flow of water is called the **transpiration stream**.

Phloem tissue transports dissolved **organic compounds**, such as sucrose and amino acids, from regions where the substances are made or stored, to places where they are used or stored. This process is called **translocation**. Transpiration and translocation are both examples of **mass flow** processes. Mass flow is the movement of a liquid, for example, along a pressure gradient. It is not to be confused with the movement of substances within a liquid, for example, by diffusion (see Chapter 4).

Worked example

Explain why diffusion within a liquid is not an example of mass flow.

Answer
When a substance diffuses, its molecules move from an area of higher concentration to an area of lower concentration; however, the liquid as a whole does not move. In mass flow, the liquid itself, and anything it contains, moves.

1. Why are xylem and phloem tissues described as vascular?
   A. They are made up of specialised cells.
   B. They are only found in plants.
   C. They contain tube-like structures.
   D. They transport substances.
2. Suggest why most plants need specialised transport systems, but why some plants, such as mosses, do not.

THE DISTRIBUTION OF XYLEM AND PHLOEM TISSUES IN HERBACEOUS DICOTYLEDONOUS PLANTS

Xylem and phloem tissues are found together, in arrangements called **vascular bundles**. The 'veins' that you can easily see in many leaves are vascular bundles (Figure 7.2).

Vascular bundles contain xylem and phloem tissue, as well as a third tissue, **cambium**. As a plant grows, cambium cells divide and differentiate, forming either xylem or phloem. (Cambium cells are plant stem cells – see Chapter 5.) Vascular bundles are arranged in distinctive, different ways in stems, roots and leaves (Figure 7.3). This is because, in addition to their transport functions, vascular bundles also help support the plant's structure.

Figure 7.2 The vascular bundles ('veins') are easily visible on this leaf.

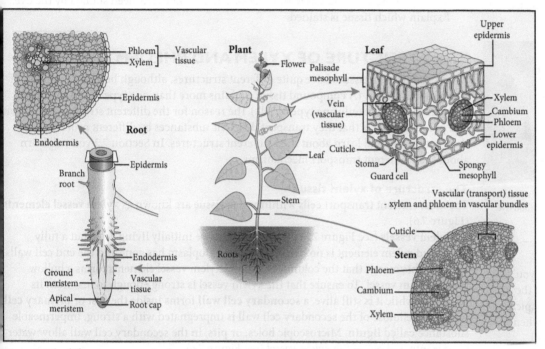

Figure 7.3 The different arrangements of vascular tissues in the parts of a herbaceous dicotyledonous plant.

As well as transporting materials to and from a leaf, the vascular bundles ('veins') also help support the weight of the leaf. They are therefore often more obvious (stick out more) on the lower surface. Within the bundles, the xylem tissue lies closer to the upper surface of the leaf, and the phloem is closer to the lower surface. Cambium tissue is found between the xylem and phloem.

Within a stem, separate vascular bundles are typically found closer to the outer surface than the centre. This 'cylindrical' arrangement provides structural support for a stem, helping keep it upright far more effectively than if the bundles were clustered in the centre. (In the same way as a hollow cylindrical scaffolding pole provides more support than a solid, narrow pole made of the same amount of metal, which is much more likely to bend under strain.)

Unlike the arrangement in a stem, within a root the vascular tissue is typically found at the centre. While the roots do not have to support any weight, and do not have to grow in a more or less straight line like the stem, they do have to have a high tensile strength to anchor the plant in the ground. (In the same way that even a very flexible metal cable can withstand a great deal of tension (pulling) force without breaking.)

3. In vascular bundles, cambium tissue is found between the xylem and phloem tissues. Suggest why it is found here.

Tip

Do not confuse stem cells, meaning cells that can differentiate, with the cells found in a plant stem. Undifferentiated stem cells are found in all parts of a plant, not just the stem.

Tip

Diagrams showing cross sections can either show **transverse sections** (TS) (across the length) or **longitudinal sections** (LS) (along the length).

Figure 7.4 An aphid feeding on the lower surface of a leaf.

Figure 7.5 A celery stalk that has been left in red food dye.

Tip

In Chapter 1, you learned about the structure of typical plant cells – when learning about specialised cells, such as in xylem and phloem tissues, concentrate on how they differ from these.

🔍 **4.** Vascular bundles are usually more obvious on the lower surface of leaves. This arrangement helps support the weight of the leaf tissue above. Suggest one other advantage to the plant of the vascular bundles being closer to the lower surface than the upper one.

5. Aphids are insects that feed from leaves by inserting their piercing mouthparts into the vascular bundles (Figure 7.4). The aphids take in the liquid sap, which contains foods such as the sugar sucrose.

One reason why aphids are more commonly found feeding on the lower surface of a leaf than on the upper surface is that they are less visible to predators such as birds. Suggest two other reasons.

6. Figure 7.5 shows a celery stalk (stem) that has been left standing in a beaker of red food dye. One of the tissues within the vascular bundles has been stained by the dye. Explain which tissue is stained.

THE STRUCTURE OF XYLEM AND PHLOEM TISSUES

Xylem and phloem tissues have quite different structures, although both are examples of a **compound tissue**. (A compound tissue contains more than one type of cell. A **simple tissue** is made up of only one type of cell.) The reason for the different structures of xylem and phloem tissue is that they transport different substances by different processes. In this section you will learn about their different structures. In Section 7.2 you will learn about their different transport mechanisms.

The structure of xylem tissue

The most efficient transport cells within xylem tissue are known as **xylem vessel elements** (Figure 7.6).

Xylem vessels (see Figure 7.7) form from what are initially living cells, but a fully developed xylem element is no longer alive: the cytoplasm has gone and the end cell walls are lost. This means that the column of stacked xylem vessel elements forms a hollow tube: a xylem vessel. To ensure that the xylem vessel is strong enough to maintain its structure, while it is still alive, a **secondary cell wall** forms inside the first or **primary cell wall**. The cellulose of the secondary cell wall is impregnated with a strong, impermeable substance called **lignin**. Microscopic holes, or **pits**, in the secondary cell wall allow water to pass into and out of the xylem vessel (see Figure 7.8).

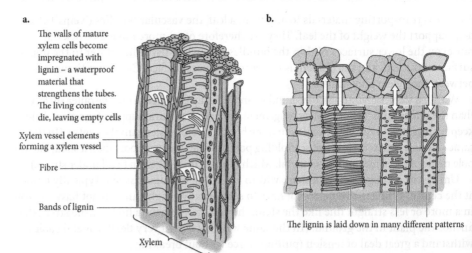

a.

The walls of mature xylem cells become impregnated with lignin – a waterproof material that strengthens the tubes. The living contents die, leaving empty cells

Xylem vessel elements forming a xylem vessel

Fibre

Bands of lignin

Xylem

b.

The lignin is laid down in many different patterns

Figure 7.6 Xylem vessel elements are stacked on top of each other forming long tubes called xylem vessels.

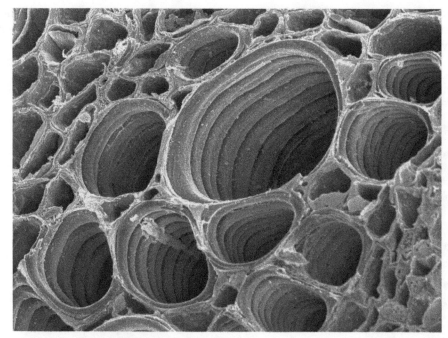

Figure 7.7 Scanning electron micrograph of xylem vessels.

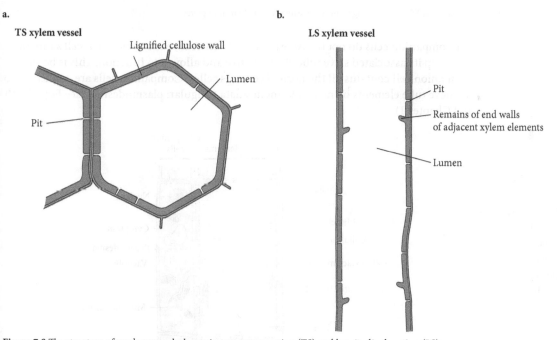

a.

TS xylem vessel

Lignified cellulose wall

Lumen

Pit

b.

LS xylem vessel

Pit

Remains of end walls
of adjacent xylem elements

Lumen

Figure 7.8 The structure of a xylem vessel, shown in transverse section (TS) and longitudinal section (LS).

The structure of phloem tissue

Unlike in xylem tissue, all the cells in phloem tissue are living. Phloem contains two main types of cell: **phloem sieve tube elements** and **companion cells** (Figures 7.9 and 7.10). The sieve tube elements contain cytoplasm but lack most of the usual cell organelles, including a large permanent vacuole and nucleus. This is because their main function is to allow the movement of organic compounds and this would be made difficult by the presence of many cell structures. The end walls of sieve tube elements are called **sieve plates**. This is because they contain many small pores through which the cytoplasm of one sieve tube element connects with the cytoplasm of the next. The sieve pores allow organic compounds such as sugars to move from one sieve tube element to the next.

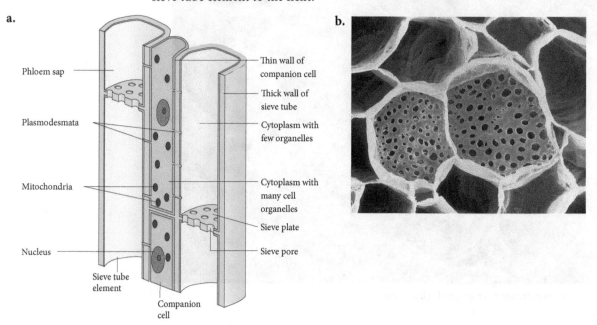

Figure 7.9 (a) The structure of phloem tissue. (b) Section through phloem tissue showing two sieve plates.

The companion cells do not transport materials. Instead, each companion cell's function is to keep its associated sieve tube element alive and allow it to function. This is because a companion cell contains all the normal cell organelles. Companion cells are connected to the sieve tube elements by many plasmodesmata (singular: plasmodesma) (see Figure 7.10 and Chapter 1).

Tip

Do not confuse the plasmodesmata, which connect sieve tube elements and companion cells, with sieve pores, which connect one sieve tube element to the next.

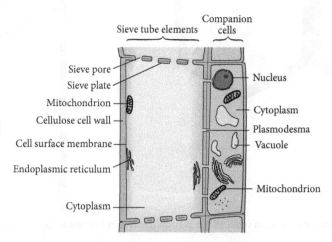

Figure 7.10 A phloem sieve tube element and its associated companion cell.

7. The secondary cell wall of xylem vessel elements contains lignin and pits. Explain the importance of each.

8. Explain why phloem sieve tube elements need companion cells, but xylem vessel elements do not.

9. Suggest whether substances will be transported more quickly through xylem tissue or phloem tissue. Explain your answer.

Experimental skills 7.1: Observing and drawing cells and tissues

In Chapter 1 you made biological drawings of cells from microscope slides and micrographs. In this chapter you will make biological drawings to show the details of individual cells as well as the distribution of different tissues.

Making biological drawings of cells

You should be able to draw and label xylem vessel elements, phloem sieve tube elements and companion cells from microscope slides, photomicrographs and electron micrographs.

Remember, when making biological drawings:

- they should identify clearly and simply the main features of what is being observed;
- only draw what you can actually see – do not draw what you think you should see;
- there should be no shading; drawings and labels should be in pencil;
- use a ruler for label lines;
- drawings should be given a title;
- and show the magnification where possible.

QUESTIONS

P1. Figure 7.11 shows a section through a sieve plate observed using a transmission electron microscope.

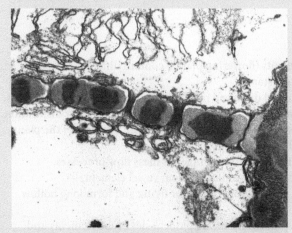

Figure 7.11 A transmission electron micrograph showing a section through a sieve plate between two phloem sieve tube elements.

- **a.** Produce an accurate biological drawing of the sieve plate shown in Figure 7.11.
- **b.** Does Figure 7.11 show a transverse section or a longitudinal section? Explain your answer.
- **c.** Explain why in Figure 7.11 the pores in the sieve plate appear to be of different diameters.

P2. Figure 7.12 shows xylem tissue observed using an electron microscope.

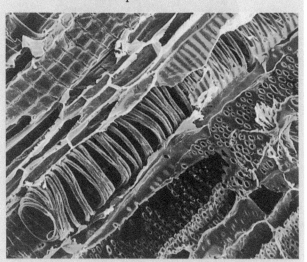

Figure 7.12 An electron micrograph of a longitudinal section through several xylem vessels.

- **a.** Produce an accurate biological drawing to show the wall of one of the xylem vessels shown in Figure 7.12.
- **b.** What type of electron microscope was used to produce the image shown in Figure 7.12? Explain your answer.

Making plan diagrams

When observing biological specimens, you might be more interested in the distribution of different tissues rather than in the individual cells. In this case, a **plan diagram** may be used. A plan diagram shows the outlines of different tissues, but does not show individual cells.

You should be able to use microscope slides and photomicrographs to make plan diagrams of transverse sections of roots, stems and leaves.

When making a plan diagram, the same rules apply as to any biological drawing, except that the individual cells are not shown and the different tissues may be distinguished by different shading or colour. Figure 7.13 shows a photomicrograph of a transverse section through a plant stem, together with a student's plan diagram showing the vascular bundles and the positions of the different tissues. (Note that there are some errors in the student's drawing – see Question P3.)

a.

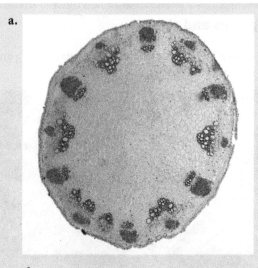

b.

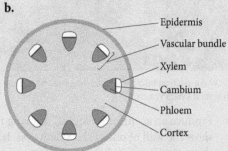

Epidermis

Vascular bundle

Xylem

Cambium

Phloem

Cortex

Figure 7.13 (a) A photomicrograph of a transverse section through the stem of a herbaceous dicotyledonous plant. (b) A student's plan diagram of the photomicrograph. NB. The diagram contains errors.

P3. Look at the plan diagram in Figure 7.13b. Describe any errors you can see in the diagram.

P4. Figure 7.14 shows a transverse section through a root.

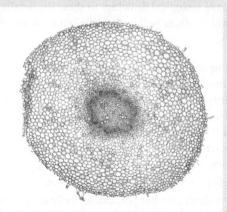

Figure 7.14 A photomicrograph of a transverse section through the root of a herbaceous dicotyledonous plant.

Draw a plan diagram of the plant root shown in Figure 7.14.

Key ideas

→ Herbaceous dicotyledonous plants (flowering plants) contain different transport tissues to transport different substances.

→ Xylem tissue transports water and dissolved mineral ions from the soil to the leaves, where the water evaporates into the atmosphere. This process is called transpiration.

→ Phloem tissue transports dissolved organic compounds from one part of the plant to another. This process is called translocation.

→ Transpiration and translocation are both examples of mass flow processes.

→ The main type of cell in xylem tissue is called a xylem vessel element.

→ Fully developed xylem vessel elements are no longer living and form long hollow tubes.

→ The two main types of cell in phloem tissue are phloem sieve tube elements and companion cells.

→ Phloem sieve tube elements are living cells although they lack some cell organelles. They are stacked into tubes; each sieve tube element is connected to the next by a perforated sieve plate, allowing materials to pass along the tube.

→ Each sieve tube element has an adjacent companion cell that allows it to carry out its functions.

→ Xylem and phloem tissues are distributed differently within roots, stems and leaves.

→ Plan diagrams show the distribution of different tissues.

7.2 Transport mechanisms

In the mammalian circulatory system (see Chapter 8) blood moves through the transport vessels because it is pumped by the heart. Different mechanisms are involved in transport through plants. In this section you will learn how water and mineral ions move through xylem tissue in the process of transpiration, and how organic compounds move through phloem tissue by the process of translocation.

TRANSPORT OF WATER AND MINERAL IONS

Water, together with dissolved mineral ions, enters plant roots through the root hair cells to reach the xylem tissue. The water and ions move up through the xylem vessels to the leaves, where water vapour passes into the atmosphere. Different mechanisms are involved at each stage.

Transport from the soil to the xylem

Root hair cells are cells on the surface of plant roots (see Figure 7.15). The 'hairs' are extensions of the cells, which are adapted to increase the surface area to volume ratio for absorption. Water enters root hair cells from the soil by the process of osmosis. Mineral ions are absorbed by active transport. Water enters by osmosis because the root hair cells have a lower water potential than the soil. As water enters the root hair cells this increases their water potential, which means that water will now move from them to adjacent **cortex** cells (see Figure 7.16) which now have a lower water potential. In this way water moves further into the root. The movement of the water and ions between adjacent cells is aided by the plasmodesmata connecting the cytoplasm of each cell (see Chapter 1).

Figure 7.15 A photomicrograph of plant root hairs.

In addition, water and ions can also travel from the soil into the roots to reach the xylem by moving through the network of spaces between cells, and within cell walls between the cellulose microfibrils that make up the wall (see Chapter 1).

The two routes for the movement of water and ions from soil to xylem are:

- the **symplast** (or **symplastic**) **pathway**, which consists of the cytoplasm of all the cells involved;
- the **apoplast** (or **apoplastic**) **pathway**, which comprises the network of spaces between and within cell walls, but not cell cytoplasm.

The inner and outer parts of the root are separated by a layer of cells called the **endodermis**. Running through the cell walls of the endodermis is an impenetrable layer called the **Casparian strip**, composed mainly of a waxy substance called **suberin**. The impermeable Casparian strip means that water and ions passing through the apoplast

Tip

Remember that water moves by osmosis from an area of higher (less negative) water potential to an area of lower (more negative) water potential.

pathway are blocked and they have to move into the symplast pathway and travel through the cytoplasm of the endodermis and other cells to reach the xylem. The Casparian strip provides a means of controlling which substances will enter the vascular tissue at the centre of the root. The pathways by which water enters and moves through a root to the xylem are shown in Figure 7.16.

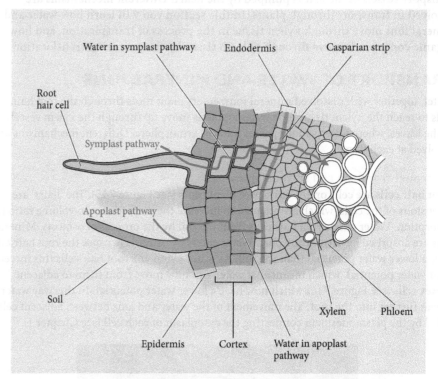

Figure 7.16 The pathways by which water moves from the soil to the xylem at the centre of the root. The apoplast pathway consists of the intercellular spaces and spaces between the microfibrils in the cell walls. The symplast pathway is formed of the cytoplasm of all the living cells interconnected by plasmodesmata.

Transport through the xylem from the roots to the leaves

Once water (and dissolved mineral ions) has reached the xylem tissue, there are several mechanisms involved in moving water up the plant through the xylem vessels:

- capillary action – this is the process by which water rises up a narrow capillary tube or up tissue paper. It is caused by weak electrostatic bonds called hydrogen bonds forming both between different water molecules, and between water molecules and the material the water is moving through,
- by root pressure – an upwards pressure caused by water entering roots;
- by **transpiration pull** – the main process by which water travels up through xylem.

Capillary action and root pressure can only move water a short distance. Transpiration is driven by the loss of water from leaves, which 'pulls' more water up through the xylem.

To understand transpiration, it is easiest to first consider what happens to water in the leaves. When water reaches leaves through the xylem it moves out of the xylem tissue into the surrounding mesophyll cells by osmosis. It will eventually evaporate from the surfaces of these cells into the internal air spaces of the leaf (the **spongy mesophyll** layer). This leads to a high concentration of water vapour in the air spaces, and so the water vapour diffuses through the **stomata** out of the leaves and into the atmosphere (Figure 7.17).

Link

Chapter 4 describes the process of diffusion, in which molecules move from an area of higher concentration to an area of lower concentration, down a concentration gradient.

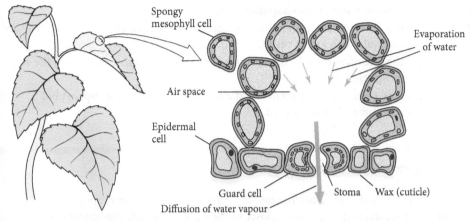

Figure 7.17 Loss of water from leaves by transpiration involves both evaporation and diffusion.

As water evaporates from the mesophyll cells, the water potential of these cells is lowered. This causes more water to move into the mesophyll cells by osmosis from the xylem vessels in the leaf. As water moves out of a xylem vessel, this in turn pulls up the column of water behind it. The reason for this is that the hydrogen bonds between the water molecules produce a **cohesion** force that holds together the separate water molecules. There are other forces also acting on the column of water. There is its own weight caused by gravity, which acts to pull down on the column at the same time as it is being pulled up by transpiration. These opposing forces put a **tension** force on the column of water that might cause it to break. This is partly overcome by cohesion between the water molecules, but there is also another force of attraction between the water molecules and the cellulose in the walls of the xylem vessels, called **adhesion**. Together, cohesion and adhesion overcome gravity and tension and the columns of water are pulled up the xylem vessels. This mechanism is known as **cohesion-tension**.

The tension force can be strong enough to 'stretch' the column of water in each xylem vessel, which causes the water column to become thinner. This in turn can cause the xylem vessels themselves to become narrower as their walls are pulled inwards by the adhesion forces between the water and the xylem walls. It is vital that the xylem vessels do not completely close due to these forces and this is another benefit of the extra strength provided by the secondary cell walls of the xylem vessels. Even so, when, for example, a tree is rapidly transpiring, the trunk can measurably decrease in diameter because of the combined effect of all the xylem vessels narrowing.

Worked example

Describe the apoplast and symplast pathways in a plant root.

Answer

When water and dissolved minerals travel further into a root through the apoplast pathway, the water moves through the spaces in between cells, as well as through the spaces within cell walls, but does not enter the cytoplasm of any cells. By contrast, the symplast pathway is through cytoplasm, passing between cells by the plasmodesmata which connect the cytoplasm of one cell with that of the next.

10. Where is the Casparian strip found?
    A. cortex
    B. endodermis
    C. epidermis
    D. xylem vessel element

11. Describe the function of the Casparian strip.

12. Explain the differences between the cohesion, adhesion and tension forces acting on a column of water in a xylem vessel.

13. Explain the role of osmosis in the movement of water through a plant from the soil to the atmosphere.

★ 14. Explain why xylem vessel elements need a secondary cell wall for support but phloem sieve tube elements do not.

Assignment 7.1: Measuring rate of transpiration

A potometer (or bubble potometer) can be used to measure the rate of transpiration. As water evaporates from the leaves, the bubble of air in the capillary tubing moves closer to the leaves. A simple potometer can be made from capillary tubing, as shown in Figure 7.18a. Figure 7.18b shows a more complex potometer, which incorporates a reservoir: the tap on the reservoir can be opened to help move the air bubble away from the leaves at the start of the investigation. Some potometers may use a ruler to measure the movement of the bubble.

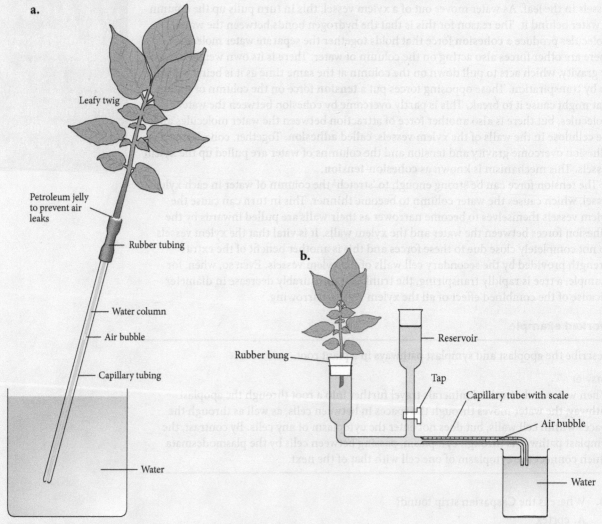

Figure 7.18 (a) A simple potometer. **(b)** A more complicated potometer.

QUESTIONS

A1. Explain why the rate of movement of the air bubble gives a measurement of the rate of transpiration.

There is also a different type of potometer called a mass potometer. A plant growing in a container is placed onto a mass balance and any change in mass is measured over a period of time.

A2. Explain why a change in mass would give a measure of the rate of transpiration.

A student investigated the effect that the number of leaves has on the rate of transpiration. The student began with a shoot that had five leaves all of a similar size. Using a potometer, the distance the air bubble moved in one hour was measured. This was repeated but each time another leaf was removed. Their results are shown in Table 7.1.

Number of leaves remaining on the shoot	Distance air bubble moved in one hour (mm)
5	15
4	11
3	8
2	6
1	3
0	1

Table 7.1

A3. Describe and explain the results shown in Table 7.1.

A4. Suggest how the student could have improved the procedure.

A5. The internal diameter of the capillary tube was 2.0 mm. Use the results when there were five remaining leaves to calculate the mean rate of water uptake per leaf in mm³ per minute. Give your answer to three significant figures.

☆**A6.** A bubble potometer actually measures the rate of water uptake, not the rate of transpiration. Explain why these might not be exactly the same value.

Tip

Volume of a cylinder $= \pi r^2 h$

Reducing transpiration

The loss of water from leaves by transpiration can be both beneficial and detrimental to a plant. It provides transpiration pull, bringing up more water containing mineral ions from the soil. Furthermore, evaporation of the water can have a cooling effect on the plant, which may be important in hot conditions. However, one disadvantage associated with transpiration is that it may cause excessive water loss from the plant if the roots cannot take in enough water to replace that lost from the leaves. In this case, the plant may start to wilt (droop) and could even eventually die.

For this reason, all plants have some adaptations to reduce water loss by transpiration, for example:

- a waxy waterproof cuticle to restrict evaporation from epidermal cells (it may also help to provide protection from pests)
- stomata positioned on the underside of leaves where it is cooler, so reducing evaporation
- closing stomata in hot, dry conditions (which may have the disadvantage of reducing carbon dioxide uptake for photosynthesis).

Link

In Chapter 14, you will learn more about the mechanism and functions of stomatal opening and closing.

Figure 7.19 The leaves of an Australian eucalyptus tree hang vertically. This is so that the leaves are mainly exposed to sunlight during the cooler parts of the day, in the morning and evening. During the middle of the day, very little leaf area is directly exposed to the Sun, so the leaves are cooler than they otherwise would be, so reducing evaporation.

Figure 7.20 Marram grass.

Tip

You should be able to make annotated drawings of xerophyte leaf sections. If it is an unfamiliar plant, just apply what you already know about xerophytes to what you observe.

Transpiration tends to occur most rapidly in the following conditions:

- high temperature – as evaporation occurs more rapidly
- low humidity – this increases the concentration gradient of water between the inside and outside of a leaf so water vapour diffuses more rapidly
- windy – air movement removes transpired water vapour, maintaining the concentration gradient for diffusion
- high light intensity – stomata are more likely to be open to take in carbon dioxide for photosynthesis.

In some plants, known as **xerophytes**, which live in very dry conditions, the adaptations to reduce water loss by transpiration become more extreme (Figure 7.19). These features can include:

- a reduced surface area to volume ratio – for example by reducing leaves to spines
- a thickened waxy cuticle – to reduce evaporation
- sunken stomata in pits or grooves – to create an area of high humidity immediately outside the stomata to reduce further loss of water vapour
- leaf hairs – to reduce air movement to create an area of high humidity next to the stomata.

In addition to reducing transpiration, xerophytes may also be adapted to maximise water uptake by having a very extensive root system, or to store water in succulent tissues. Figure 7.20 shows marram grass, which is a xerophyte that lives on sand dunes. Although a dune may be close to seawater, the plant still lives in very dry conditions as sand drains very rapidly. Also, a low water potential caused by the presence of salt may mean that there is a tendency for plant roots to lose water by osmosis.

The leaves of marram grass contain cells known as 'hinge' cells. When the plant loses water, the 'hinge' cells shrink, causing the leaf to roll up, so all the stomata are on the inner surface.

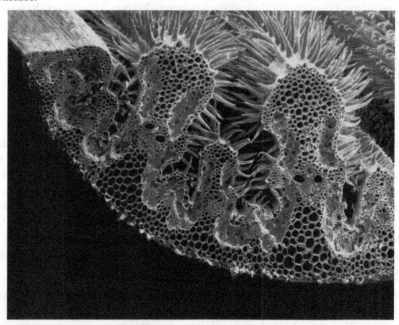

Figure 7.21 A scanning electron micrograph of a section through a marram grass leaf.

15. Use Figure 7.21 to draw a labelled plan diagram of a transverse section through a marram grass leaf. Annotate it to explain how the marram grass leaf is adapted to reduce water loss by transpiration.

TRANSPORT OF ORGANIC COMPOUNDS

Many organic compounds need to be transported around a plant. For example, compounds made in the leaves by photosynthesis need to be moved to growing regions, such as new buds, flowers or fruit, or to storage regions, for example within seeds or roots. Materials that have been stored in one part of a plant, for example in a bulb, may later need to be transported to new growing regions. Organic compounds like these are sometimes called **assimilates** because they have been assimilated by the plant. **Assimilation** means that the substances have been produced by the plant from materials originally taken in (absorbed) from the environment. Examples of assimilates include sucrose, produced from the glucose made in photosynthesis, and amino acids, which have been made from the products of photosynthesis together with mineral ions absorbed from the soil.

Transport from sinks to sources

Assimilates, dissolved in water, move through phloem sieve tubes in the process known as translocation. The places the assimilates move from are known as **sources**; the places they move to are known as **sinks**. The same part of a plant can be both a source and a sink at different times. For example, if assimilates are moved to a root for storage, the root is acting as a sink. If the stored materials are later moved from the root to growing regions, the root is acting as a source. This means that the liquid in phloem sieve tubes, known as sap, can move both upwards and downwards in a plant (although not in the same sieve tube at the same time) – unlike movement in xylem vessels, which is only ever upwards.

Movement from sources to sinks is by mass flow down a **hydrostatic pressure** gradient. At the source, assimilates are moved into the phloem sieve tube elements. This lowers the water potential in the sieve tube elements which causes water to enter by osmosis from surrounding tissue. The increase in water volume increases the hydrostatic pressure in the sieve tubes. At the sink, assimilates leave the sieve tube elements to be used or stored. This causes water to follow by osmosis, so reducing the hydrostatic pressure in the sieve tube. The hydrostatic pressure at the source end is higher than at the sink end. The difference in hydrostatic pressure between the source and the sink (the hydrostatic pressure gradient) is what causes the mass flow of phloem sap through the sieve tubes from sources to sinks.

The processes involved in translocation are not yet fully understood, but the mechanism by which hydrostatic pressure is thought to be increased at the source is shown in Figure 7.22.

> **Tip**
> Remember, when substances are taken in, they are absorbed. When the substances are used to make new organic substances, they have been assimilated.

> **Link**
> In Chapter 13 you will learn how the products of photosynthesis are used to produce amino acids and carbohydrates like sucrose.

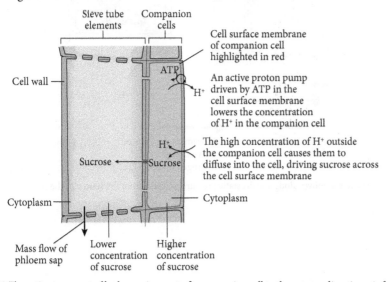

Figure 7.22 The active transport of hydrogen ions out of a companion cell to the surrounding tissue is followed by the cotransport of hydrogen ions and sucrose back into the companion cell. The sucrose then diffuses through plasmodesmata into the phloem sieve tube element from the companion cell.

Proton pumps are carrier proteins that actively transport hydrogen ions (protons). The pumps use ATP to move hydrogen ions out of the companion cells into surrounding tissue. This produces a concentration gradient for the hydrogen ions. The hydrogen ions then diffuse back into the companion cells down this gradient through **cotransporter proteins**. As the hydrogen ions diffuse back into the companion cells sucrose molecules move in with them. This is an example of the cotransport of two different substances, and this cotransporter protein is called a **sucrose–hydrogen cotransporter** (see Figure 4.16b, Chapter 4). The diffusion of hydrogen ions down their concentration gradient provides the energy needed to move sucrose against its concentration gradient. The concentration of sucrose increases in the companion cells, which causes some of the sucrose to diffuse into the phloem sieve tube elements through the connecting plasmodesmata.

Figure 7.23 summarises the process of translocation.

Link

Chapter 4 describes the processes of diffusion, active transport involving carrier proteins, and cotransport involving cotransporter proteins. You will learn how ATP transfers energy from respiration in Chapter 12.

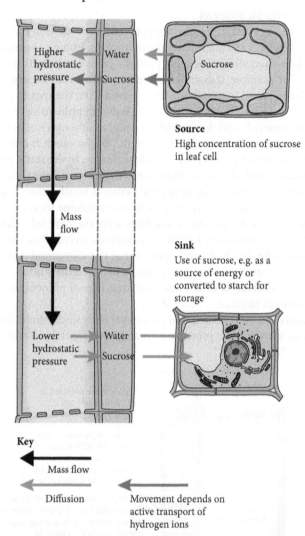

Figure 7.23 Phloem sap moves along a hydrostatic pressure gradient from the source to the sink.

Worked example

Explain why a leaf can be both a source and a sink.

Answer

When a leaf is growing it needs to be provided with the substances it requires to build new cells and tissues. In this case it is acting as a sink, as organic compounds are being transported to the leaf. When a leaf is photosynthesising, the products of photosynthesis will be transported to other parts of the plant for storage or for new growth. In this case the leaf is acting as a source.

16. Plants take in carbon dioxide during photosynthesis.
    a. Is the carbon dioxide an assimilate? Explain your answer.
    b. Carbon dioxide molecules contain carbon, as do all organic compounds. Explain why carbon dioxide is not therefore classed as an organic compound.
17. Explain why transpiration can only move water up a plant, whereas translocation can move organic compounds both up and down a plant.
18. What is the function of the proton pumps in a companion cell?
    A. To move hydrogen ions into the companion cell.
    B. To move hydrogen ions out of the companion cell.
    C. To move sucrose into the companion cell.
    D. To move sucrose out of the companion cell.
19. A plant is treated with a metabolic poison that prevents respiration. Explain the effects this will have on both transpiration and on translocation.
20. Figure 7.24 shows a young photosynthesising tree that has had a ring of bark removed. This has removed the phloem tissue, which lies close to the surface, but has not damaged the xylem tissue, which lies deeper.

Ring of bark and phloem removed from a woody stem

Figure 7.24 A ring of bark has been removed from a young tree.

A few days after the bark has been removed, samples of sap are taken from the phloem tissue above and below the ring and tested for the presence of sucrose. Describe and explain the expected results.

Key ideas

→ Water and dissolved mineral ions enter roots from the soil and are transported to the xylem via both the apoplast and symplast pathways.

→ The apoplast pathway is made up of intercellular spaces and cell walls.

→ The symplast pathway is made up of the cytoplasm of adjacent cells.

→ In transpiration, water and dissolved mineral ions are pulled up the xylem vessels by water evaporating and diffusing out of leaves into the atmosphere.

→ The water columns in xylem vessels are subject to different forces: transpiration pull, cohesion, adhesion, gravity and tension.

→ The mechanism by which water moves through xylem is called cohesion-tension.

→ Xerophytic plants are especially adapted to reduce water loss by transpiration.

→ In translocation, dissolved organic compounds (assimilates) move through phloem sieve tubes from sources to sinks down hydrostatic pressure gradients.

→ Translocation begins when companion cells transfer assimilates to phloem sieve tube elements.

CHAPTER OVERVIEW

Review the mini mind map to link new learning with your prior learning, and to highlight which of the key concepts underpin the chapter.

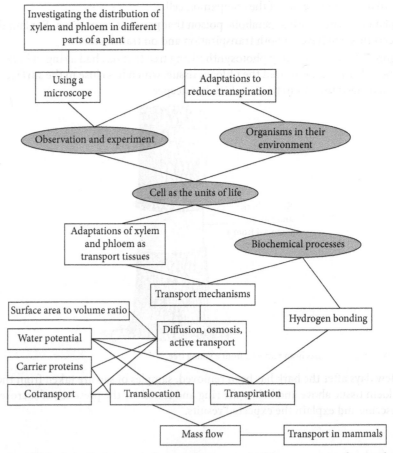

Try copying this mini mind map and expanding upon it. Use your notes from other chapters to help you explore how the essential ideas, theories and principles can be linked further together.

WHAT YOU HAVE LEARNED

Now that you have finished this chapter you should be able to:

- draw plan diagrams of transverse sections of stems, roots and leaves of herbaceous dicotyledonous plants from microscope slides and photomicrographs
- describe the distribution of xylem and phloem in transverse sections of stems, roots and leaves of herbaceous dicotyledonous plants
- draw and label xylem vessel elements, phloem sieve tube elements and companion cells from microscope slides, photomicrographs and electron micrographs
- relate the structure of xylem vessel elements, phloem sieve tube elements and companion cells to their functions
- state that some mineral ions and organic compounds can be transported within plants dissolved in water
- describe the transport of water from the soil to the xylem through the:
    - apoplast pathway, including reference to lignin and cellulose
    - symplast pathway, including reference to the endodermis, Casparian strip and suberin
- explain that transpiration involves the evaporation of water from the internal surfaces of leaves followed by diffusion of water vapour to the atmosphere
- explain how hydrogen bonding of water molecules is involved with movement of water in the xylem by cohesion-tension in transpiration pull and by adhesion to cellulose in cell walls
- make annotated drawings of transverse sections of leaves from xerophytic plants to explain how they are adapted to reduce water loss by transpiration
- state that assimilates dissolved in water, such as sucrose and amino acids, move from sources to sinks in phloem sieve tubes
- explain how companion cells transfer assimilates to phloem sieve tubes, with reference to proton pumps and cotransporter proteins
- explain mass flow in phloem sieve tubes down a hydrostatic pressure gradient from source to sink.

CHAPTER REVIEW

1. What substance are the secondary cell walls of xylem vessel elements impregnated with?
   A. lignin
   B. protein
   C. suberin
   D. sucrose

2. What name is given to the perforated end walls of phloem sieve tube elements?
   A. pits
   B. plasmodesmata
   C. microfibrils
   D. sieve plates

3. What force holds water molecules together?
   A. adhesion
   B. cohesion
   C. tension
   D. transpiration

4. Explain what is meant by mass flow.

5. Describe the distribution of xylem and phloem in:
   a. a root
   b. a stem
   c. a leaf.
6. Describe and explain two adaptations each of:
   a. xylem vessel elements
   b. phloem sieve tube elements.
7. Describe the function of the plasmodesmata in root cortex cells.
8. Describe the processes involved in the movement of water into a leaf through the xylem and out of a leaf through the stomata.

Transport in mammals

The human circulatory system starts working only 22 days after fertilisation and keeps performing its vital job through a whole lifetime. The red box in this picture contains a frozen human heart ready for a transplant operation in Mumbai, India. Why do we need a heart and a circulation system? How does the heart work? How are the vessels that carry blood around the body adapted to do their jobs? How is oxygen carried by the blood and released at the correct time?

Prior understanding

You may recall information about the structure and function of the mammalian circulatory system, comprising the heart, blood vessels and the blood, from previous courses. From Chapter 1, you may recall the structure of typical animal cells. Chapter 7 described the two transport systems in plants: xylem and phloem. The mammalian circulatory system is another example of mass flow; however, unlike in plants, movement through the circulatory system is driven by the heart, and it is a closed system.

Learning aims

In this chapter, you will learn about the structure and function of the double circulatory system used in mammals. You will learn how the structures of arteries, veins and capillaries are related to their functions. You will describe blood as a tissue and be able to recognise the main types of blood cells from photomicrographs. You will discover how oxygen is collected at the lungs and transported to respiring tissues, being released only when required; and how carbon dioxide is transported. You will learn about the structure of the heart and how the beating of the heart is initiated and controlled.

8.1 The circulatory system (Syllabus 8.1.1–8.1.7)

8.2 Transport of oxygen and carbon dioxide (Syllabus 8.2.1–8.2.6)

8.3 The heart (Syllabus 8.3.1–8.3.4)

8.1 The circulatory system

THE CLOSED, DOUBLE CIRCULATION SYSTEM

Mammals have a closed, **double circulation system**. It is described as closed because blood remains within blood vessels: arteries, capillaries and veins. It is described as a double circulation system because blood passes twice through the heart in one complete circulation: once for the **systemic** (body) system and once for the **pulmonary** (lung) system. This is summarised in Figure 8.1.

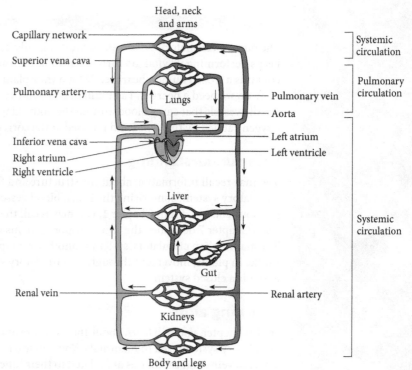

Figure 8.1 The double circulation system.

Figure 8.1 The double circulation system.

The major blood vessels of the double circulation system are:

- the **aorta** – carries oxygenated blood away from the heart to the systemic part of the system
- the **pulmonary artery** – carries deoxygenated blood away from the heart to the pulmonary part of the system
- the **vena cava** – returns deoxygenated blood to the heart from the systemic part of the system
- the **pulmonary vein** – returns oxygenated blood to the heart from the pulmonary part of the system.

THE STRUCTURE AND FUNCTION OF BLOOD VESSELS

Blood vessels are classified into three groups according to their function.

Arteries carry blood away from the heart under high pressure and have a pulse.

Capillaries are the smallest blood vessels, forming networks through tissues, have very thin walls and are the site of exchange of substances between the blood and the surrounding cells.

Veins carry blood back to the heart under low pressure and have valves to prevent blood flowing in the wrong direction.

Figure 8.2 shows how arteries, capillaries and veins are organised in the circulatory system.

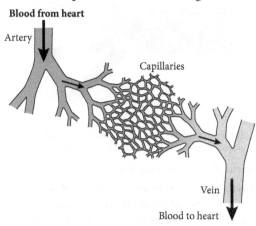

Figure 8.2 The relationship between different types of blood vessels.

1. Mammals have a closed, double circulation system.
   a. Explain what is meant by 'closed'.
   b. Explain what is meant by 'double circulation'.
2. Which of the following carry blood away from the heart?

   aorta   pulmonary vein   pulmonary artery   vena cava

Arteries

Arteries have the thickest, strongest walls of all the blood vessels. This enables arteries to withstand the high pressure of blood inside them. For example, blood leaves the heart in the aorta at a pressure of around 16 kPa. The structure of an artery wall is made up of three layers, shown in Figure 8.3:

- **endothelium**, which is the innermost layer consisting of very smooth endothelial cells to lower resistance for blood flow
- a middle layer containing **elastic fibres**, **smooth muscle** and collagen
- an outer layer containing elastic fibres and collagen.

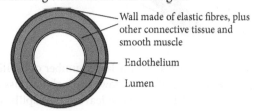

Figure 8.3 Diagram of a transverse section through an artery.

The elastic fibres and collagen in an artery wall give it strength. The elastic fibres allow the artery to expand slightly with each new high-pressure rush of blood from the heart. When the pressure drops slightly between each surge, the elastic fibres cause the artery wall to recoil inward again. This recoil helps to push the blood along and also helps to make the pressure and speed of blood flow more constant. When you feel your pulse, you feel this expanding and recoiling of the artery wall. Arteries such as the aorta are classed as elastic arteries because they can withstand high pressure, stretching slightly and then recoiling again.

As arteries branch into smaller vessels called **arterioles** further from the heart, the blood pressure decreases. This is because the cross-sectional area of the **lumens** of the arterioles are, together, greater than the cross-sectional area of the main artery, as shown in Figure 8.4. The increased resistance to flow caused by the increased surface area in contact with the blood also contributes to reducing the pressure.

Tip

Blood pressure is also measured in units called millimetres of mercury, mmHg, even though these are not SI units. 1 mmHg is equivalent to 0.13 kPa.

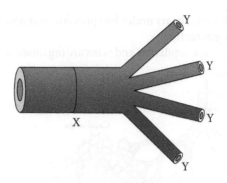

The total cross-sectional area at points Y is much greater than the cross-sectional area at point X

Figure 8.4 Blood pressure in arterioles is less than that in an artery.

Link

Smooth muscle is different from skeletal, or striated, muscle (Chapter 15) in that it can contract uniformly for long periods. Smooth muscle is found in the walls of other vessels, such as in the airways (Chapter 9).

Arterioles, which are further from the heart, have fewer elastic fibres than arteries closer to the heart. The walls of arterioles also contain smooth muscle.

The smooth muscle in arterioles can contract to reduce the diameter of the lumen, so constricting the arteriole. The smooth muscle can also relax to increase the diameter, or dilate, the lumen. This gives the ability to control the quantity of blood flowing to different organs. For example, when lifting weights, the arterioles supplying muscles in the arms are dilated while those supplying the small intestine are constricted. Arteries such as these are classed as muscular arteries as they can undergo vasoconstriction and vasodilation.

Arterioles branch into smaller and smaller vessels inside an organ, eventually forming capillaries.

Capillaries

Capillaries are the smallest of all the blood vessels in the body. The function of the capillaries is to bring blood as close as possible to the cells in the organs and tissues and then to allow exchange of substances.

The walls of capillaries are very thin (Figure 8.5). They are made from only one layer of endothelial cells and are less than 1 μm thick. The diameter of capillaries is also very small; the lumen diameter of about 8 μm is just wide enough for a red blood cell to pass through.

Endothelial cells

Figure 8.5 The structure of a capillary.

Blood flows very slowly in capillaries, but the large number of capillaries in any one tissue means that a large total volume of blood flows through that tissue. The pressure of blood entering capillaries is around 4.6 kPa; when leaving the capillary to return to veins, the pressure has dropped to around 1.3 kPa.

Exchange of substances such as oxygen, carbon dioxide and glucose occurs through capillary endothelial walls. Capillaries have small gaps between the endothelial cells. These gaps, or clefts, allow faster diffusion of substances between blood in the capillaries and the surrounding **tissue fluid**. The clefts also allow some of the liquid part of blood to flow out into surrounding tissue and form tissue fluid. Some capillaries have larger gaps, or **fenestrations**, which allow more rapid exchange. Fenestrated capillaries are found in the kidneys and small intestine. The fenestrations are too small for cells and large proteins to pass through, so they remain in the blood.

As capillaries leave an organ or tissue, they join back together to form larger vessels called **venules**.

Veins

Venules that carry blood away from a tissue join together to form veins. Veins carry blood back to the heart and carry blood at the lowest pressure of all the blood vessels. Pressure in veins is around 0.7 kPa. Blood in veins also flows much more slowly and at a more constant speed than in arteries. Veins therefore have thinner walls than arteries and contain fewer elastic fibres and less smooth muscle. Veins have the same three layers as arteries: endothelium, a middle layer and an outer layer. The middle layer in veins is thinner than that of arteries. The lumen of a vein is larger than that of an artery of the same size (Figure 8.6).

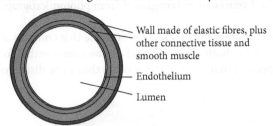

Wall made of elastic fibres, plus other connective tissue and smooth muscle

Endothelium

Lumen

Figure 8.6 Diagram of a transverse section through a vein.

As the pressure in veins is so low, there are potential difficulties in keeping blood moving in the correct direction. For example, when you are standing, the blood in veins coming upwards from your feet would tend to fall back under gravity and flow in the wrong direction. Four factors contribute to maintaining the flow of blood in veins:

- pressure from the blood coming out of capillary networks
- presence of valves in veins
- action of muscles in the legs when moving
- low pressure in the atria where blood enters the heart during the atrial diastole.

The valves in veins are **semilunar valves**. These open when the pressure of blood is greater in the direction from which the blood is supposed to flow. The valves close when the pressure of blood coming from the wrong direction is greater. This is similar to pushing on a door. If you apply pressure in the correct direction, the door will open. If you apply pressure in the wrong direction, you will push the door closed. The action of valves in veins is shown in Figure 8.7.

Link
Atrial diastole and the events in the cardiac cycle will be covered in Section 8.3.

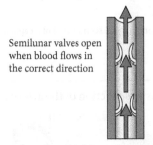

Semilunar valves open when blood flows in the correct direction

The valves close when blood flows the wrong way

Figure 8.7 Valves in veins prevent backflow of blood.

Figure 8.8 shows how the pressure and speed of blood flow is related to the cross-sectional area of blood vessels.

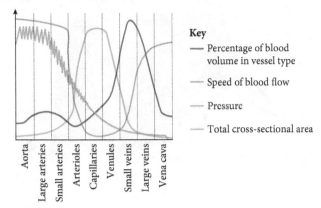

Key
— Percentage of blood volume in vessel type
— Speed of blood flow
— Pressure
— Total cross-sectional area

Figure 8.8 The relationship between the pressure and speed of blood flow and the cross-sectional area of blood vessels.

3. With reference to Figure 8.8, explain:
   a. which type of blood vessel has the largest total cross sectional area
   b. why the pressure in arteries fluctuates, whereas the pressure in veins does not.

Recognising blood vessels

Arteries, capillaries and veins can be recognised from photomicrographs and electron micrographs.

Figure 8.9 is a photomicrograph of a section of tissue that contains an artery and a vein. Transverse sections of both types of vessel can be seen together for comparison. The endothelium is present in both vessels but is too thin to be distinguished from the middle layer.

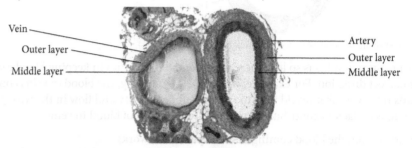

Figure 8.9 Photomicrograph showing transverse sections (TS) through an artery and a vein.

Figure 8.10 is a transmission electron micrograph showing red blood cells in a capillary.

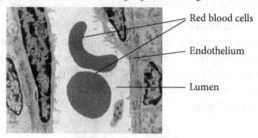

Figure 8.10 Transmission electron micrograph showing a longitudinal section (LS) of a capillary.

Worked example

Make a large, labelled plan drawing of the transverse section of the artery in Figure 8.11.

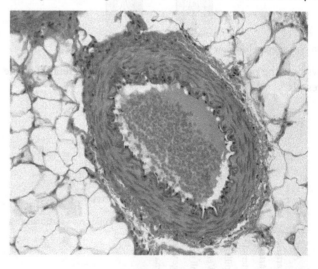

Figure 8.11

Answer

A plan drawing has:

- clear, continuous, unbroken lines
- lines only at tissue boundaries
- no shading anywhere
- no individual cells
- proportions similar to the original image.

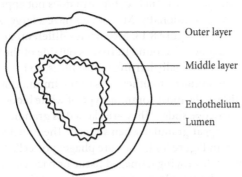

- Outer layer
- Middle layer
- Endothelium
- Lumen

Tip

Draw one layer at a time, probably starting with the outermost layer, then add labels.

4. List these blood vessels in order from those with the highest blood pressure to those with the lowest pressure.

   artery supplying the kidney   vena cava   capillary in the lungs   aorta

5. Gills are the organs of gas exchange in a fish. In the circulation system of a fish, blood flows through the following path:

   heart → gills → body → heart

   Describe how this differs from the path taken by blood in a mammal.

6. Explain why arteries have thicker walls than veins of the same size.

7. Explain how capillaries are adapted for exchange of substances between blood and surrounding tissues.

8. Explain why valves are needed in veins but not in arteries.

BLOOD AND TISSUE FLUID

Blood is a tissue composed of a cellular component and a liquid component. Tissue fluid is formed when some components of the liquid part of blood are forced out through gaps in capillary walls.

The cellular component of blood

The cellular part of blood comprises:

- red blood cells
- white blood cells, which are monocytes, neutrophils and lymphocytes.

Red blood cells are red because of the red globular protein haemoglobin, whose main function is the transport of oxygen from the lungs to respiring tissues. An average adult has around 3×10^{13} red blood cells.

    Red blood cells have a structure that is adapted to their function (Figure 8.12):

- biconcave disc shape – this increases the surface area to volume ratio for faster diffusion of oxygen
- small size – red blood cells have a diameter of approximately 7 μm compared to about 40 μm for an average animal cell; this keeps all haemoglobin molecules close to the surface and allows the red blood cells to fit though narrow capillary lumens
- no nucleus, mitochondria or endoplasmic reticulum – so there is more room for haemoglobin molecules.

Link

The way in which haemoglobin binds and releases oxygen will be described in Section 8.2.

Cell surface membrane   **Cross-section**

Cytoplasm with dissolved haemoglobin, no nucleus, no mitochondria, no endoplasmic reticulum

Surface view

7 µm

Figure 8.12 Red blood cell structure.

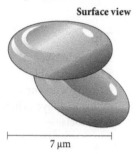

Figure 8.13 Monocytes are characterised by their large kidney-shaped nucleus.

Figure 8.14 Neutrophils are characterised by their large, lobed nucleus.

Figure 8.15 Lymphocytes are characterised by their large, round nucleus.

As red blood cells have no nucleus, they cannot undergo cell division. Red blood cells are formed in the bone marrow and their nucleus is removed before they are released into blood. Red blood cells also have a relatively short lifespan of around 120 days. Old red blood cells are destroyed by the lymph nodes and the spleen.

Monocytes are one of the types of white blood cells. They are the largest of the white blood cells and can be up to 40 µm in diameter. Monocytes are classed as agranulocytes as their cytoplasm does not appear granular when they are stained conventionally. Monocytes have a large, kidney-shaped nucleus, as shown in Figure 8.13. Monocytes differentiate into **macrophages,** which are phagocytic cells found in tissues outside of the blood. Monocytes engulf foreign or potentially harmful substances, so have many lysosomes. They also present antigens to other cells of the immune system.

Neutrophils are the most common type of white blood cells seen in blood smears on microscope slides. They are classed as granulocytes, as their cytoplasm can appear granular when stained. Their nucleus is large and lobed, as shown in Figure 8.14. They are phagocytic cells that can enter tissues. Neutrophils are capable of destroying damaged tissue or bacteria and they die after one such activity. Granules in the cytoplasm of neutrophils contain proteases and bactericidal substances.

Lymphocytes are usually only slightly larger than red blood cells. They look similar to monocytes, except the nucleus of a lymphocyte is round and so large that it almost fills the cytoplasm, as shown in Figure 8.15. Lymphocytes are divided into B- and T-lymphocytes. Often, very little cytoplasm can be seen, but they are classed as agranulocytes as the cytoplasm does not appear granular. T-lymphocytes are involved in cell-mediated immunity. B-lymphocytes are involved with the production of antibodies. Both B- and T-lymphocytes have the ability to recognise non-self antigens, during antigen presentation, which is carried out by monocytes. There are many polyribosomes in the cytoplasm of lymphocytes for the polypeptide synthesis required for antibody formation.

Each type of white blood cell can be recognised using these characteristic features from photomicrographs of blood smears (Figures 8.16 and 8.17).

9.   Which of these processes can occur in a red blood cell?
     A. respiration
     B. diffusion
     C. mitosis
     D. protein synthesis

10. **a.**   State the name given to the shape of a red blood cell.
     b.   Explain how this shape is an adaptation to the function of a red blood cell.

11. There are approximately 3×10^{13} red blood cells in the body. If an average adult has a total blood volume of $5\,dm^3$, calculate the number of red blood cells in $1\,cm^3$ of blood.

12. Use Figures 8.13–8.15 to make drawings of each of the following cells:
     a.   monocyte
     b.   neutrophil
     c.   lymphocyte

The liquid component of blood

The liquid component of blood is called **plasma**. The average adult has about $5\,dm^3$ of blood, more than half of which is plasma. Plasma is 95% water and the remaining 5% is dissolved substances. Water is ideal for this role, as:

- it is an excellent solvent for polar and ionic solutes
- it has a high specific heat capacity, meaning it resists change in temperature.

Plasma is a pale-yellow coloured liquid. The substances dissolved in plasma include:

- nutrients such as glucose and amino acids
- mineral salts, such as Na^+, K^+ and Cl^-, sometimes called electrolytes
- metabolic wastes, such as carbon dioxide and urea
- hormones
- antibodies
- other plasma proteins including blood clotting factors.

The functions of plasma are to provide a liquid medium to enable red and white blood cells to flow through the blood vessels and also to transport substances in solution.

The blood pressure at the arterial end of a capillary network causes fluid from the plasma to be forced out through clefts into the surrounding tissues. Once outside the capillaries, this liquid is called tissue fluid.

Link

The roles of B- and T-lymphocytes will be described in more detail in Chapter 11.

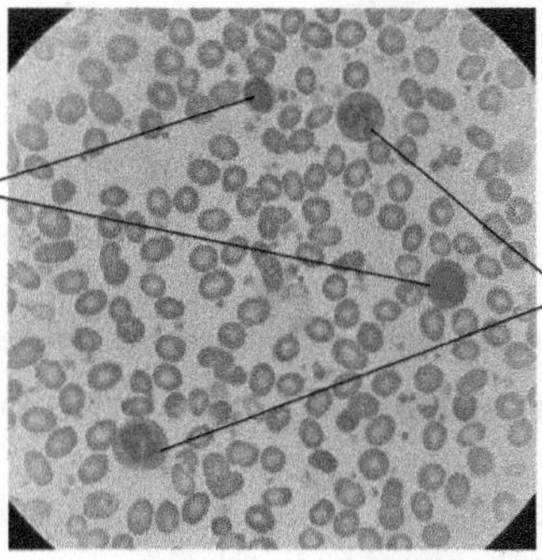

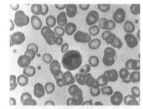

Figure 8.16 Blood smear showing red blood cells, a lymphocyte and a neutrophil.

Tissue fluid

Tissue fluid surrounds almost every cell in the body and provides the medium for exchange of substances in and out of cells. From tissue fluid, these substances are then exchanged to and from the blood. The composition of tissue fluid is mostly the same as plasma, but with fewer large proteins. The larger the substance in plasma, the more difficult it is for that substance to enter tissue fluid. Hence, there are no red blood cells in tissue fluid and very few proteins, such as albumin, which has a relative formula mass of 69 000.

Some white blood cells have the ability to change shape and squeeze through the gaps in capillary walls. This enables them to move around in tissue fluid and protect against infection outside of the blood.

The movement of water from the capillaries into the tissues is greatest at the arterial end because the blood pressure is greater here than at the venous end. The removal of water from the plasma, and the fact that large proteins remain in plasma, increases the concentration of these proteins. The higher concentration of proteins lowers the water potential of plasma relative to that of tissue fluid. The lower water potential of plasma causes some of the water in tissue fluid to move back into capillaries by osmosis. This osmotic effect is greater at the venous end of the capillary network. However, the net movement of water is from the capillaries into the tissues as the effect of the pressure in the blood is greater than that of osmosis. This is summarised in Figure 8.18.

Figure 8.17 Blood smear showing red blood cells and a monocyte.

Link

The solvent properties of water, polarity and specific heat capacity were discussed in Chapter 2.

Link

In Chapter 4, you discovered that water moves by osmosis from a solution of higher water potential (hypotonic solution) to a solution of lower water potential (hypertonic solution).

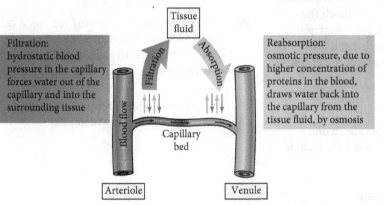

Filtration: hydrostatic blood pressure in the capillary forces water out of the capillary and into the surrounding tissue

Reabsorption: osmotic pressure, due to higher concentration of proteins in the blood, draws water back into the capillary from the tissue fluid, by osmosis

Tissue fluid

Filtration

Absorption

Blood flow

Capillary bed

Arteriole

Venule

Figure 8.18 Movement of water between capillaries and surrounding tissue is controlled by hydrostatic and osmotic pressures. There is a net movement of water and small solutes from the blood into tissues as the hydrostatic pressure of the blood is greater than the osmotic pressure.

If the volume of tissue fluid were allowed to steadily increase, it would result in swelling, called oedema. To avoid this, excess tissue fluid is drained away by open-ended vessels called lymphatics. These lymphatic vessels return the excess fluid back into the blood by connecting with veins in the upper body.

It is vital that the composition of tissue fluid remains relatively constant in terms of water potential, glucose concentration and pH. Maintaining this constant environment for cells is part of homeostasis (see Chapter 14).

13. State two differences between plasma and tissue fluid.
14. **a.** Name two nutrients transported in plasma.
    b. Name two metabolic wastes transported in plasma.
15. Explain how tissue fluid is formed.

Experimental skills 8.1: Red blood cell count

A student obtained a sample of mammalian blood from a local animal hospital.

The student diluted the blood by taking 1 μl of the blood sample and mixed this with 199 μl of a dilution fluid.

The student placed one drop of the diluted blood into a special type of microscope slide called a haemocytometer. A haemocytometer has a square grid printed on the glass.

Figure 8.19 shows how red blood cells appeared in five of the smallest squares of the haemocytometer grid. The squares shown are **not** adjacent to each other.

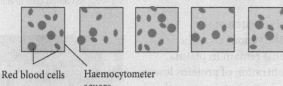

Red blood cells   Haemocytometer square

Figure 8.19

All cells completely within a haemocytometer square are counted. Two sides of the square are chosen in advance. Cells that overlap or touch these two chosen sides are counted. Cells that overlap or touch the other two sides are not counted. Usually, cells overlapping the top side and right side are included and those overlapping the bottom side and left side are excluded from the count.

QUESTIONS

P1. Calculate the mean number of cells in one haemocytometer square in Figure 8.19. Include cells that are overlapping the top and right sides of squares.

P2. The squares shown in Figure 8.19 are 0.05 mm × 0.05 mm in size. The depth between the base of the slide and the coverslip is 0.1 mm.
  a. Calculate the volume of dilution fluid and cells contained in each square in Figure 8.19. Give your answer in mm³.
  b. Blood cell counts often use the volume 1 ml, which is equivalent to 1 cm³. Express your answer to (a) in ml.

P3. Use your answers to questions P1 and P2 (b) to calculate the number of red blood cells per ml of the sample on the haemocytometer.

P4. Calculate the number of red blood cells per ml in the original sample. Remember that the original sample was diluted in dilution fluid.

P5. **a.** Suggest why the original blood sample was diluted before loading onto the haemocytometer.
  b. Explain why the sample cannot be diluted with distilled water.

P6. No stain is used for counting red blood cells. White blood cells can also be seen using a haemocytometer. Explain why the different types of white blood cells are easier to identify using a stain.

P7. Describe one safety precaution that should be taken when using blood samples.

Key ideas

→ Mammals have a closed, double circulation system.
→ The circulatory system consists of the heart, arteries, arterioles, capillaries, venules and veins.

→ The main blood vessels in mammals are the pulmonary artery, pulmonary vein, aorta and vena cava.

→ The structure of arteries, capillaries and veins are related to their functions.

→ Arteries, capillaries and veins can be recognised from photomicrographs and electron micrographs.

→ Blood has a cellular component and a liquid component.

→ The cellular component contains red blood cells and white blood cells.

→ Red blood cells have a structure that is adapted to their function of transporting oxygen.

→ The main types of white blood cells are monocytes, neutrophils and lymphocytes.

→ Each type of cell in the blood can be recognised in photomicrographs by its characteristic features.

→ The liquid component of blood is called plasma.

→ The functions of plasma are to provide a liquid medium within which cells can flow and to transport solutes.

→ Tissue fluid is almost identical in composition to plasma.

→ Tissue fluid forms when components of plasma are forced out of capillaries due to the pressure in the blood.

8.2 Transport of oxygen and carbon dioxide

Transport of oxygen from the lungs to respiring tissues is one major function of the mammalian circulation system. In addition, carbon dioxide must be removed from respiring tissues and transported to the lungs for removal from the body.

HAEMOGLOBIN AND TRANSPORT OF OXYGEN

Haemoglobin is a globular protein with a quaternary structure composed of four polypeptides, each with one haem group.

One haem group can bind one oxygen molecule or two oxygen atoms.

One molecule of haemoglobin can therefore bind four molecules of oxygen, or eight atoms of oxygen to form **oxyhaemoglobin**. This binding is reversible. When oxygen is released from oxyhaemoglobin, this is called **dissociation**.

This can be summarised in the equation:

$$Hb \quad + \quad 4O_2 \quad \leftrightarrows \quad HbO_8$$

haemoglobin + oxygen $\leftrightarrows$ oxyhaemoglobin

Haemoglobin molecules can bind with fewer than four oxygen molecules. Those that have bound the maximum possible number of oxygen molecules are referred to as 100% saturated. Those that have bound fewer oxygen molecules are less than 100% saturated.

The quantity of oxygen present in blood or in air is referred to as **partial pressure**. The partial pressure of oxygen in air at sea level is approximately 21 kPa. In air that has been inhaled into alveoli, the partial pressure of oxygen is around 14 kPa.

When purified haemoglobin is exposed to different partial pressures of oxygen, the percentage saturation can be measured. When these results are plotted on a graph, the graph is called a **dissociation curve**. The dissociation curve for haemoglobin is shown in Figure 8.20.

Link

Globular and fibrous proteins, together with the different levels of protein structure and the structure of haemoglobin, were covered in Chapter 2.

Tip

As oxygen, O_2, is a diatomic molecule, each oxygen molecule contains two atoms of oxygen.

Link

There are different types of haemoglobin. The assignment in Chapter 16 looks at the expression patterns of some of these types during early development.

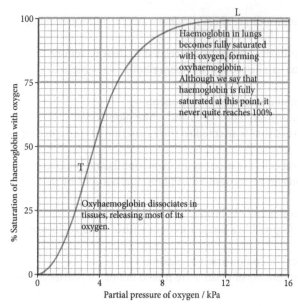

Key
T – conditions in respiring tissues
L – conditions in the lungs

Figure 8.20 The adult haemoglobin dissociation curve.

The dissociation curve in Figure 8.20 shows that the higher the partial pressure of oxygen, the greater the percentage saturation of haemoglobin with oxygen. This is not a linear, or proportional, relationship as shown by the S-shaped, or sigmoid, curve.

When the first oxygen molecule binds with a haemoglobin molecule, the binding causes a slight change in the three-dimensional shape of the protein. This change makes it slightly easier for the second oxygen molecule to bind. The binding of the second and third oxygen molecules also causes similar changes, making it progressively easier for each oxygen molecule to bind. At the lungs, shown by the letter L in Figure 8.20, haemoglobin becomes almost 100% saturated.

The same process occurs during dissociation of oxyhaemoglobin. Dissociation of the first oxygen molecule makes it easier for the next one to dissociate.

The steep part of the curve, labelled T in Figure 8.20, shows the situation in respiring tissues. The steepest part of the curve is where a very small change in oxygen partial pressure produces a large change in percentage saturation of haemoglobin.

16. Which row in the table correctly describes haemoglobin?

	Number of polypeptides	Number of haem groups	Maximum number of oxygen molecules that can bind
A	2	1	8
B	2	4	4
C	4	1	8
D	4	4	4

★ 17. Explain why the haemoglobin dissociation curve is an S-shape and not a straight line.

The Bohr shift

At respiring tissues, carbon dioxide is being released. Carbon dioxide dissolves in tissue fluid and diffuses into blood plasma in solution.

$$\text{in water: } CO_{2\,(g)} \rightleftharpoons CO_{2\,(aq)}$$

However, when carbon dioxide dissolves in water, some of it reacts with the water to form **carbonic acid**. This reaction happens very slowly. Red blood cells contain the enzyme **carbonic anhydrase**, which catalyses this reaction, speeding it up greatly.

$$CO_{2\,(aq)} + H_2O_{(l)} \rightleftharpoons H_2CO_{3\,(aq)}$$

Carbonic acid then dissociates in water to produce hydrogen ions and **hydrogencarbonate ions**.

$$H_2CO_{3\,(aq)} \rightleftharpoons H^+_{(aq)} + HCO_3^-{}_{(aq)}$$

The presence of hydrogen ions dissolved in the water of the red blood cell cytoplasm creates acidic conditions, which could be harmful. However, haemoglobin has a high affinity for binding hydrogen ions. The binding of these ions removes them from solution and so reduces the acidity. In this way haemoglobin acts as a buffer, keeping the pH inside the red blood cell relatively constant. When hydrogen ions are bound, haemoglobin becomes **haemoglobinic acid**, HHb.

$$Hb + H^+ \rightleftharpoons HHb$$

Hydrogencarbonate ions are negatively charged. They leave the red blood cells and move into blood plasma through an ion exchange protein in the cell surface membrane of the red blood cell. About 10–30% of carbon dioxide is transported as hydrogencarbonate ions within red blood cells. A further 60–85% is carried as hydrogencarbonate ions in plasma. Carbonic anhydrase is only found in red blood cells, which means that the hydrogencarbonate ions in plasma were formed in red blood cells. For every hydrogencarbonate ion that leaves the red blood cell, the ion exchange protein allows one chloride ion, Cl^- to enter. This prevents the cytoplasm of the red blood cell from becoming positively charged and is called the **chloride shift**.

Higher carbon dioxide partial pressure at respiring tissues causes the tertiary structure of haemoglobin to change slightly. This reduces its affinity for oxygen, causing dissociation. This change in affinity for oxygen is called the **Bohr shift**. Higher carbon dioxide partial pressure causes the dissociation curve to move to the right, as shown in Figure 8.21.

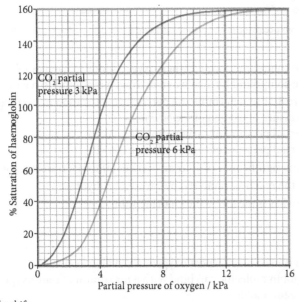

Figure 8.21 The Bohr shift.

The Bohr shift ensures that oxygen remains bound to haemoglobin and is only dissociated when required at respiring tissues.

TRANSPORT OF CARBON DIOXIDE

The majority of carbon dioxide is transported as hydrogencarbonate ions in blood plasma as described above.

About 5–10% of carbon dioxide does not dissociate and travels as a solution of carbon dioxide molecules in plasma.

The remaining 10% of carbon dioxide is transported by haemoglobin in red blood cells. Instead of reacting with water in the presence of carbonic anhydrase to form carbonic acid, this carbon dioxide reacts with the amino terminal groups at the ends of some of the haemoglobin polypeptides. The substance formed is called **carbaminohaemoglobin**.

At the lungs, the low partial pressure of carbon dioxide in alveoli causes carbon dioxide to move from blood plasma into the alveolar air space by diffusion. This causes the reactions described above to reverse: hydrogencarbonate and hydrogen ions react to produce carbon dioxide and carbon dioxide leaves carbaminohaemoglobin.

Figure 8.22 summarises the role of red blood cells in the transport of oxygen and carbon dioxide.

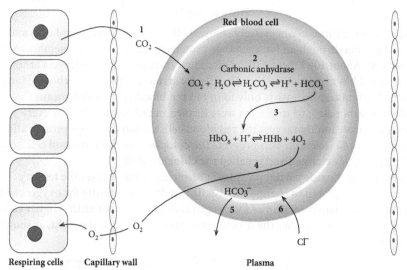

Figure 8.22 The reactions occurring in a red blood cell as oxygen is delivered to respiring tissues and carbon dioxide is removed.

18. **a.** Name the compound formed when oxygen binds to haemoglobin.
    b. Name the compound formed when hydrogen ions bind to haemoglobin.
19. **a.** Use Figure 8.21 to determine the percentage saturation of haemoglobin at a partial pressure of oxygen of 4 kPa when:
    i. the partial pressure of carbon dioxide is 3 kPa
    ii. the partial pressure of carbon dioxide is 6 kPa.
    b. Explain the significance of your answers to (a) to the transport of oxygen.
20. **a.** Describe the reaction catalysed by carbonic anhydrase.
    b. Describe what is meant by the chloride shift.

Key ideas

→ Haemoglobin is used to transport oxygen in red blood cells.
→ Each haemoglobin molecule can bind with a maximum of four oxygen molecules, forming oxyhaemoglobin.
→ A dissociation curve can be drawn for haemoglobin.
→ Carbonic anhydrase catalyses the reaction between carbon dioxide and water to form carbonic acid.
→ Hydrogen ions from dissociated carbonic acid react with haemoglobin to form haemoglobinic acid.

→ Chloride ions are moved into red blood cells to replace hydrogencarbonate ions in the chloride shift.

→ The Bohr shift explains the difference in affinity of haemoglobin for oxygen at different partial pressures of carbon dioxide.

→ The Bohr shift ensures that oxygen dissociates at respiring tissues.

→ Carbon dioxide is transported in solution, as hydrogencarbonate ions and as carbaminohaemoglobin.

8.3 The heart

The heart is the organ that puts the blood under pressure to create the flow of blood through the rest of the circulatory system. The heart is made mostly of muscle tissue and is approximately the same size as your fist (Figure 8.23).

The type of muscle in the heart is called **cardiac muscle**, which is different from smooth muscle found in arteries and veins, and also different from skeletal muscle. Cardiac muscle is much better at conducting electrical impulses and also contains a structure that initiates the contraction of the muscle. Cardiac muscle is also adapted to contract and relax over and over again. Skeletal muscle and smooth muscle require stimuli from the nervous system in order to contract, and they would not be able to perform life-long repeated contractions in the same way. Table 8.1 summarises some of the differences between the different types of muscle.

Figure 8.23 A human heart.

Type of muscle	Location	Appearance under light microscope	Mode of control	Function
skeletal muscle	attached to the skeleton	striated	voluntary	locomotion and body movement
smooth muscle	inner walls of hollow organs	non-striated	involuntary	control of lumen diameter and transport of digested products
cardiac muscle	heart	striated	involuntary	contractions of the heart in the cardiac cycle

Table 8.1 A comparison of skeletal muscle, smooth muscle and cardiac muscle.

STRUCTURE OF THE HEART

The mammalian heart has four chambers. The two upper chambers are called **atria** and receive blood from the veins. The two lower chambers, which are called **ventricles**, receive blood from the atria and then pump this blood out through arteries. A wall called a **septum**, which prevents the mixing of oxygenated and deoxygenated blood, separates the left and right sides of the heart. Oxygenated blood travels through the left side of the heart and deoxygenated blood travels through the right side. Valves separate the atria from the ventricles and also separate the ventricles from the arteries, which are connected to the ventricles. Figure 8.24 shows the internal structure of the heart. The labels 'left' and 'right' refer to the sides of the heart as they are in the animal rather than as we look at the diagram on the page.

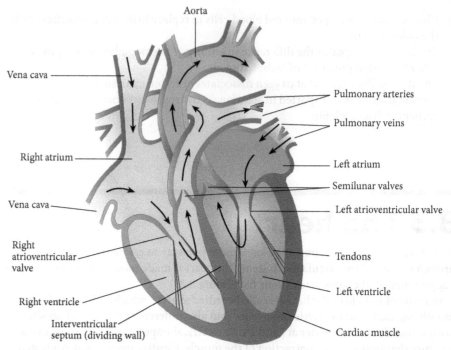

Figure 8.24 The internal structure of the heart as seen in longitudinal section.

THE CARDIAC CYCLE

The **cardiac cycle** is the name given to the events that occur in the heart during one complete heartbeat. When you are at rest, there are around 70 cardiac cycles, or beats, each minute.

> **Tip**
>
> time taken for one beat in seconds $= \dfrac{60}{\text{(number of beats per minute)}}$
>
> number of beats per minute $= \dfrac{60}{\text{(time for one beat in seconds)}}$

The cardiac cycle is initiated by a structure called the **sinoatrial node** (SAN). The SAN is located in the wall of the right atrium. The SAN acts as the pacemaker for the heart, as it generates a wave of electrical activity that spreads downwards across the walls of the atria.

As this wave of excitation spreads across the cardiac muscle tissue, it causes contraction. The result is the contraction of the atria from the top down. This forces blood from the atria down into the ventricles. This event is called **atrial systole** because the word 'systole' means 'to contract'. The contraction of the atria increases the blood pressure within the atria. At the same time, the ventricles are relaxed, which is called **ventricular diastole**. The pressure of blood in the atria becomes greater than that in the ventricles, so the **atrioventricular valves** are forced open and blood moves down to fill the ventricles.

The wave of excitation that has spread from the SAN down across the atria is stopped from passing down across the ventricles by a band of insulating tissue. This is important in preventing both the atria and the ventricles from contracting at the same time.

In the middle of this band of insulating tissue, located on the outer part of the septum and between the atria and the ventricles, is a structure called the **atrioventricular node** (AVN). Whereas the SAN acts as a pacemaker, the AVN acts as a relay station, picking up the wave of electrical activity and passing it on to the ventricles while introducing a short delay.

The AVN conducts the electrical impulse and passes it to conducting fibres called **Purkyne tissue**. These fibres terminate at the apex of the heart. The apex is the bottom part of the heart, between the two ventricles. Here, the Purkyne fibres release the electrical impulse, which spreads upward across the ventricle walls. This makes the ventricles contract from the bottom upwards, known as **ventricular systole**. The delay introduced by the AVN and Purkyne tissue is about 0.1 s, which is enough to separate the contractions of the upper and lower parts of the heart.

The contraction of the ventricles is more powerful than the contraction of the atria because the ventricles have much thicker muscular walls. Ventricular systole raises the blood pressure in the ventricles suddenly. The atria, ventricles and the arteries are each separated by valves that only open in one direction. When the pressure in the ventricles becomes greater than the pressure in the atria, which are now in diastole, the atrioventricular valves are forced closed and blood is pumped out into the aorta and pulmonary artery. After the ventricles have contracted, they relax again. Blood is prevented from flowing back into the ventricles from the arteries by semilunar valves at the point where the arteries join with the ventricles.

Figure 8.25 shows the locations of the SAN, the AVN and the Purkyne tissue.

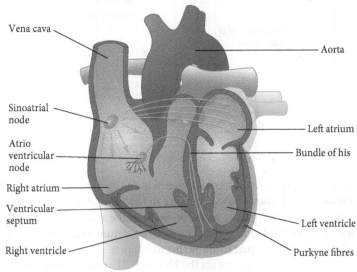

Figure 8.25 Structures involved in control of the cardiac cycle.

Figure 8.26 shows the changes that occur during the cardiac cycle. Only the left side of the heart is shown. The events that occur on the right hand side of the heart are the same in each of the stages.

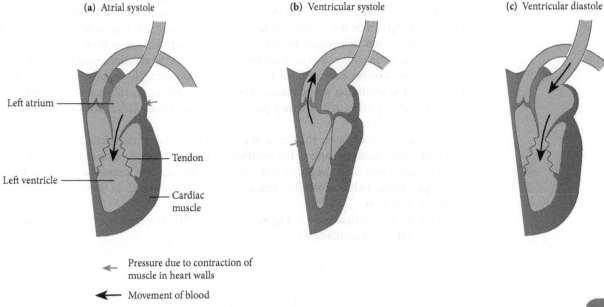

Figure 8.26 The cardiac cycle.

PRESSURE CHANGES IN THE CARDIAC CYCLE

The changes in pressure of blood in the various chambers of the heart during the cardiac cycle are important in making blood flow in the correct direction and also in the opening and closing of valves. These pressure changes are best represented on the same graph, as shown in Figure 8.27.

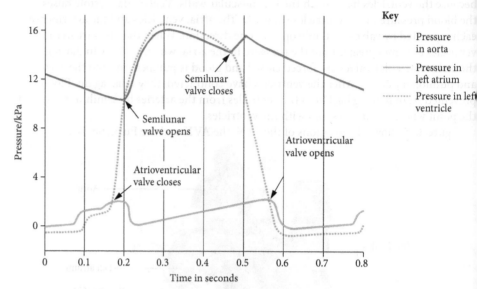

Figure 8.27 Pressure changes during one cardiac cycle in the left side of the heart.

Tip

Any liquid will flow from a region of higher pressure to a region of lower pressure, unless prevented from flowing by a closed valve.

Look at the blue line, which shows blood pressure in the atrium, and compare this line with the green line, which shows blood pressure in the ventricles. At the start of the cardiac cycle as shown in this graph, blood pressure in the atrium is higher than that in the ventricle. That means blood is pushed from the atrium to the ventricle and the atrioventricular valve will be open to allow this to happen.

Just before 0.2 s, the green line crosses the blue line. That means blood pressure in the ventricle becomes higher than in the atrium. This forces the atrioventricular valve closed. The tendons shown in Figure 8.26 prevent the valve from opening the wrong way under the surge of pressure from the blood in the contracting ventricles.

Now compare the green line with the red line, which shows pressure in the aorta. At 0.2 s, the blood pressure in the ventricle is rising rapidly, and becomes higher than the blood pressure in the aorta, where the two lines cross. This forces the semilunar valve open, so blood can be forced out into the aorta. As the ventricle begins to relax, the blood pressure in the ventricle begins to fall. Just before 0.5 s, it falls lower than the pressure in the aorta, again where the two lines cross. This forces the semilunar valve closed, to prevent blood flowing back into the ventricle from the aorta. These valves do not need tendons as the semilunar shape means they contain pockets. As the two pockets either side of the artery fill with blood they press against each other with more force to ensure the valve stays closed.

Now, after 0.5 s, compare the blue and green lines again. As ventricular diastole continues, the blood pressure in the ventricle continues to fall. Just before 0.6 s, this pressure falls lower than the blood pressure in the atrium. This forces the atrioventricular valve open again, so blood can flow from the atrium, now in systole, down into the ventricle once more.

One complete cardiac cycle in Figure 8.27 is 0.7 s. The graph extends to 0.8 s to show the start of the next cardiac cycle.

Pressure in the left and right sides of the heart

The pressure changes shown in Figure 8.27 occur in a similar pattern in both the left and the right sides of the heart. However, you can see by looking at the internal structure of the heart in Figure 8.24 that the thickness of the ventricle walls is different.

The wall of the left ventricle is much thicker than that of the right ventricle. This is because of the destinations for the blood from each side. The left ventricle pumps blood out through the aorta and this blood must continue throughout the whole body. The right ventricle pumps blood out through the pulmonary artery, which goes to the lungs.

The thicker wall of the left ventricle means that it has more cardiac muscle. More muscle can exert more force, and more force means higher blood pressure from the left ventricle. The difference in pressure is needed because:

- the left ventricle must pump the blood a greater distance all around the body compared with the right ventricle, which pumps only to the lungs, only a few centimetres from the heart
- the left ventricle must overcome greater resistance to blood flow than the right ventricle as there are more blood vessels all around the body than in the lungs
- blood leaving the left ventricle must have sufficient pressure to return against gravity from organs below the level of the heart, whereas the lungs are either side of the heart on the same level
- blood at the same high pressure from the left ventricle going to the lungs would cause damage to lung capillaries and result in excessive quantities of tissue fluid forming in the lungs.

21. Name:
    a. the chambers of the heart that receive blood from veins
    b. the internal wall that separates the left and right sides of the heart
    c. the valves that stop blood flowing from the ventricles into the atria
    d. the artery where blood leaves the heart from the right ventricle.

22. Which of these contain oxygenated blood?
    A. the left atrium and left ventricle
    B. the left atrium and right ventricle
    C. the right atrium and left ventricle
    D. the right atrium and right ventricle

23. Describe the role of the sinoatrial node in the cardiac cycle.

24. Use Figure 8.27 to calculate the number of beats per minute made by this heart.

25. The volume of blood pumped out by a particular human heart in one beat is $70\,cm^3$. The heart rate of this heart is 95 beats per minute.
    a. Calculate the volume of blood pumped out by this heart in one minute. Give your answer in dm^3.
    b. The total volume of blood in this person is $5.1\,dm^3$. Calculate the number of heartbeats required to pump this total volume once. Give your answer as a whole number of beats.

Assignment 8.1: Stroke volume

A student obtained a sheep heart and prepared a vertical section of this heart. The drawing in Figure 8.28 shows the appearance of the heart after the section was prepared.

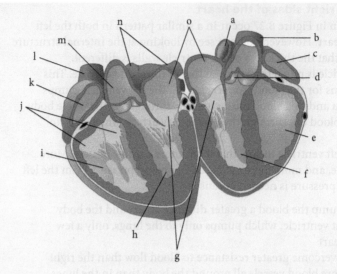

Figure 8.28

QUESTIONS

A1. Use the drawing in Figure 8.28 to identify the parts labelled with the letters a–o.

As the drawing includes both parts of the section, some parts may be named twice.

Stroke volume is the volume of blood pumped by the left ventricle in one cardiac cycle.

A2. In one particular person, during ventricular systole, 70% of the blood that filled the ventricle is pumped out. The person's stroke volume is $120 \, cm^3$. Calculate the volume of blood that filled the ventricle.

A3. Figure 8.29 shows how stroke volume varies with heart rate for a person who has trained as an athlete and an untrained person.

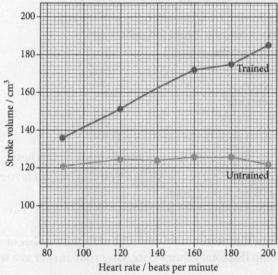

Figure 8.29

a. Describe the trends shown in Figure 8.29.

b. Cardiac output (CO) is related to stroke volume (SV) and heart rate (HR) by the equation:

$$CO = SV \times HR$$

i. Calculate the cardiac output for the trained person when their heart rate is 200 beats per minute. Give your answer in litres per minute.

ii. Calculate the cardiac output for the untrained person at a heart rate of 200 beats per minute. Give your answer in litres per minute.

iii. Explain the advantages to an athlete of having a higher cardiac output than an untrained person.

Key ideas

→ The mammalian heart has four chambers: two atria and two ventricles.

→ The left and right sides of the heart are separated by the septum.

→ The atria and ventricles are separated by atrioventricular valves.

→ Semilunar valves are found between the ventricles and their connecting arteries.

→ The sinoatrial node, located in the wall of the right atrium, initiates the cardiac cycle.

→ The atrioventricular node acts as a relay station and delays the contraction of the ventricles until after the atria have contracted.

→ Blood pressure changes within the heart ensure the blood flows in the correct direction and the valves open and close at the appropriate times.

→ The muscle walls of the ventricles are thicker than those of the atria to pump blood further.

→ The muscle wall of the left ventricle is thicker than the right to pump blood all around the body as opposed to just the lungs.

CHAPTER OVERVIEW

Review the mini mind map to link new learning with your prior learning, and to highlight which of the key concepts underpin the chapter.

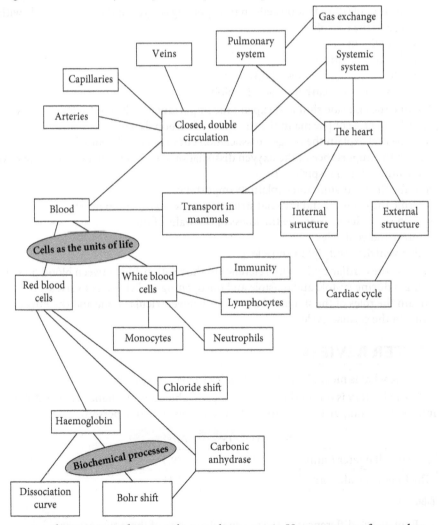

Try copying this mini mind map and expanding upon it. Use your notes from other chapters to help you explore how the essential ideas, theories and principles can be linked further together.

WHAT YOU HAVE LEARNED

Now that you have finished this chapter you should be able to:

- state that the mammalian circulatory system is a closed double circulation consisting of a heart, blood and blood vessels including arteries, arterioles, capillaries, venules and veins
- describe the functions of the main blood vessels of the pulmonary and systemic circulations, limited to pulmonary artery, pulmonary vein, aorta and vena cava
- recognise arteries, veins and capillaries from microscope slides, photomicrographs and electron micrographs and make plan diagrams showing the structure of arteries and veins in transverse section (TS) and longitudinal section (LS)
- explain how the structure of muscular arteries, elastic arteries, veins and capillaries are each related to their functions
- recognise and draw red blood cells, monocytes, neutrophils and lymphocytes from microscope slides, photomicrographs and electron micrographs
- state that water is the main component of blood and tissue fluid and relate the properties of water to its role in transport in mammals, limited to solvent action and high specific heat capacity
- state the functions of tissue fluid and describe the formation of tissue fluid in a capillary network
- describe the role of red blood cells in transporting oxygen and carbon dioxide with reference to the role of:
  - haemoglobin
  - carbonic anhydrase
  - the formation of haemoglobinic acid
  - the formation of carbaminohaemoglobin
- describe the chloride shift and explain the importance of the chloride shift
- describe the role of plasma in the transport of carbon dioxide
- describe and explain the oxygen dissociation curve of adult haemoglobin
- explain the importance of the oxygen dissociation curve at partial pressures of oxygen in the lungs and in respiring tissues
- describe the Bohr shift and explain its importance
- describe the external and internal structure of the mammalian heart
- explain the differences in the thickness of the walls of the:
  - atria and ventricles
  - left ventricle and right ventricle
- describe the cardiac cycle, with reference to the relationship between blood pressure changes during systole and diastole and the opening and closing of valves
- explain the roles of the sinoatrial node, the atrioventricular node and the Purkyne tissue in the cardiac cycle.

CHAPTER REVIEW

1. Describe what is meant by a closed double circulation system.
2. The radial artery is one of the arteries supplying blood to the hand. Table 8.2 shows information comparing the radial artery to the aorta in one person.

	Radial artery	Aorta
Internal diameter / mm	2.35	29.4
Thickness of wall / mm	0.37	2.11

Table 8.2

   a. Explain the differences in the internal diameters of these two arteries.
   b. Explain the difference in the thickness of the walls of these two arteries.

3. Which of these cells has a large nucleus occupying most of the cytoplasm?
   A. red blood cell
   B. monocyte
   C. neutrophil
   D. lymphocyte

4. There are 3×10^{13} red blood cells in a person's body. Each red blood cell has a lifespan of 120 days before it is removed from the blood. Calculate the number of new red blood cells that must be made every second to keep the total number constant.

5. Explain how the difference in composition of plasma and tissue fluid causes the movement of water from tissue fluid into plasma.

6. Name:
   a. the structure in the heart that initiates the electrical impulses
   b. the tissue in the heart that conducts the electrical impulses to the base of the ventricles.

7. Describe what happens in:
   a. atrial systole
   b. ventricular systole.

8. Ventricular septal defect (VSD) is a heart condition that some people are born with. People with VSD have a hole in the septum between the left and right ventricles. Predict the effects that VSD will have on the circulatory system.

Chapter

9

Gas exchange

On average, we breathe around 11 000 litres of air in and out every day. Air contains only about 21% oxygen, so how does this oxygen get from air into our blood? The gas exchange system is adapted to do this. With about 480 000 000 specialised structures called alveoli, our two lungs have an area of about 70 m² for gas exchange to occur.

Prior understanding

You may recall the process of diffusion from Chapter 4. From Chapter 8 you may recall the structure of capillaries and how they are adapted for exchange of substances. Also from Chapter 8 you may recall that red blood cells transport oxygen and the role of blood plasma in the transport of carbon dioxide.

Learning aims

In this chapter, you will learn about the structure of the gas exchange system, together with the distribution of substances and functions of structures within the gas exchange system. You will understand how the structures in the gas exchange system contribute to the maintenance of good health. You will learn to recognise parts of the gas exchange system from photomicrographs and electron micrographs. You will also understand how the structures in the gas exchange system contribute to efficient exchange of oxygen and carbon dioxide between air and blood.

9.1 Structure of the gas exchange system (Syllabus 9.1–9.4)

9.2 Function of the gas exchange system (Syllabus 9.5–9.7)

9.1 Structure of the gas exchange system

Large organisms, such as humans, require a **gas exchange** system because their surface area to volume ratio is small. Large organisms cannot rely on diffusion directly to and from the environment for efficient exchange of gases in all cells. Many cells are located deep in the body, far from the external environment, so diffusion would be too slow and the concentration gradient for diffusion would be too low.

The gas exchange system is the system of organs and associated structures that allows oxygen from the air into blood and carbon dioxide from blood into air.

Figure 9.1 shows the structure of the human gas exchange system.

Tip

Gas exchange and breathing are different processes. Breathing is getting air in and out of the lungs. Gas exchange is the movement of oxygen and carbon dioxide across membranes.

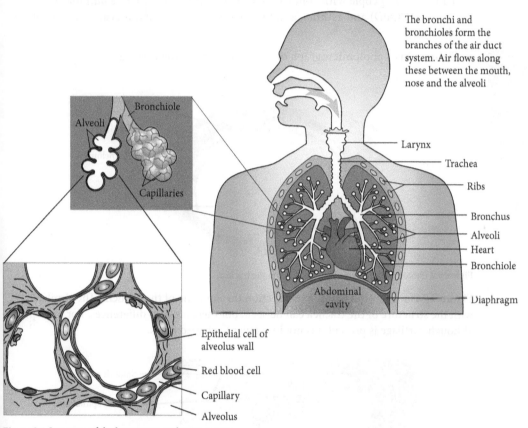

The bronchi and bronchioles form the branches of the air duct system. Air flows along these between the mouth, nose and the alveoli

Bronchiole
Alveoli
Capillaries

Larynx
Trachea
Ribs
Bronchus
Alveoli
Heart
Bronchiole
Diaphragm

Abdominal cavity

Epithelial cell of alveolus wall

Red blood cell

Capillary

Alveolus

Figure 9.1 Structure of the human gas exchange system.

When we breathe in air, the air enters through the nose or mouth and passes into the trachea, bronchi and then bronchioles.

THE TRACHEA AND BRONCHI

The **trachea**, commonly called the windpipe, is a vessel that leads from the larynx in the neck to the two bronchi just above the heart. A typical adult trachea is about 11 cm long. The lower end of the trachea divides into two **bronchi** (singular is bronchus). The bronchi have an internal diameter of around 1.2 cm and have a similar structure to the trachea.

The trachea and bronchi are adapted for carrying air to and from the lungs:

- The trachea has a wide **lumen** with an internal diameter of 1.5–2.0 cm. The lumen has no obstructions to hinder the flow of air.
- The lumens of the trachea and bronchi are held open by **cartilage** in the walls of these structures. Cartilage is a strong type of connective tissue that is more flexible than bone. The presence of cartilage also prevents collapse or over-expansion of the trachea and bronchi during sudden pressure changes that may occur during rapid breathing. Cartilage is present as C-shaped rings in the trachea and as irregular patches in the bronchi.
- The walls of the trachea and bronchi also contain smooth muscle. This can contract or relax to change the diameter of the lumen. These muscles will relax during exercise to allow a greater volume of airflow. Contraction of smooth muscles can occur in people with conditions such as chronic obstructive pulmonary disease (COPD) and asthma, and the narrowing of the airways can cause breathing difficulty.

Figure 9.2 is a photomicrograph of a section through the trachea.

Link

You may find it helpful to look back at Chapter 1 to remind yourself about photomicrographs.

Smooth muscle

Cartilage

Epithelium

Lumen

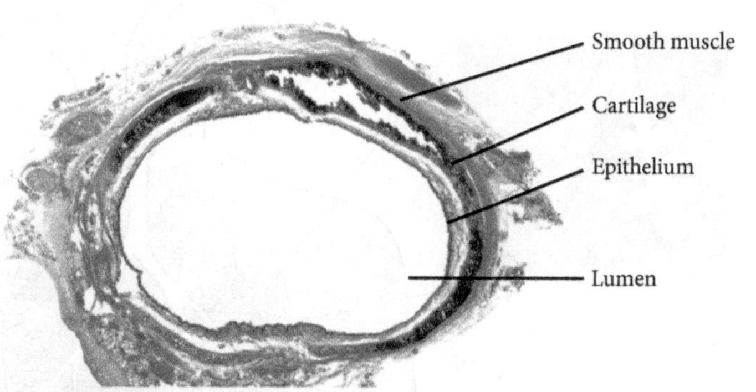

Figure 9.2 Photomicrograph of a section through the trachea.

Figure 9.3 is a photomicrograph of a section through one of the bronchi. The similarity with the structure of the trachea can be seen, but there are also differences. For example, although cartilage is present it is not in regular C-shaped rings.

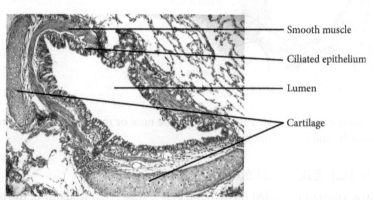

Smooth muscle

Ciliated epithelium

Lumen

Cartilage

Figure 9.3 Photomicrograph of a transverse section through one of the bronchi.

Worked example

Make a large plan diagram of the section of bronchus shown in Figure 9.3.
Label the tissues visible in this section.

Answer

A plan diagram should have:

- clear, continuous lines
- no shading anywhere
- no individual cells
- all lines drawn freehand, so no use of rulers or compasses
- a similar appearance and proportions to the original image.

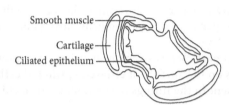

Link

Further guidance
on making plan
drawings is given in
Chapters 1 and 7.

Tip

When adding the
labels, remember that
the lumen is not a
tissue.

Start by drawing the outer boundary of the structure. This is smooth in parts and rough
in other parts.

Then add the two patches of cartilage, which have smooth boundaries.

Add the smooth muscle, which has a slightly wavy edge.

Finish the drawing with the inner layer of epithelium, which is quite convoluted
in places.

BRONCHIOLES

The bronchi divide further into smaller vessels called **bronchioles**. Bronchioles have a
much smaller diameter than bronchi and do not contain cartilage.

There are two types of bronchioles, classified according to diameter and structure:
terminal bronchioles and respiratory bronchioles.

Terminal bronchioles have smooth muscle in their walls, just like the trachea
and bronchi. They have a diameter of about 1 mm. Respiratory bronchioles do
not have smooth muscle and their diameter is about half that of the terminal
bronchioles.

Figure 9.4 is a photomicrograph of a section through a bronchiole. Some smooth
muscle can be seen.

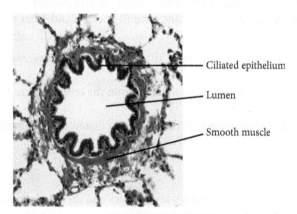

Figure 9.4 Photomicrograph of a transverse section through a bronchiole.

PROTECTING THE LUNGS

Air that we breathe contains many potentially harmful solid particles. These include dust, particulate pollutants, bacteria, viruses and spores from fungi.

Mucus is used to trap solid particles and microorganisms that enter the lungs. The layer of mucus that covers the airways makes it more difficult for pathogens to adhere (stick) to the surface of the airways. The layer of mucus forms a barrier for microorganisms and increases the distance that pathogens must travel to reach the epithelial cells.

Mucus is present in the cavity of the nose and on the linings of the trachea, bronchi and bronchioles. Mucus is a sticky liquid containing mucin, which is a mixture of glycoproteins. Mucus is produced by **goblet cells**, which are located in the epithelium of the trachea and bronchi. Mucus is also produced in **mucous glands**. Mucous glands occur more frequently in the upper airways such as the lining of the nose and trachea. Goblet cells and mucous glands secrete more mucus during infection or irritation of the lungs.

Mucus, with trapped particles, is swept away using structures called cilia. Cilia are hair-like structures on the surface of **ciliated epithelial cells**. These ciliated epithelial cells are present in the linings of the trachea, bronchi and both types of bronchioles.

Figure 9.5 is a scanning electron micrograph showing the epithelial lining of the trachea.

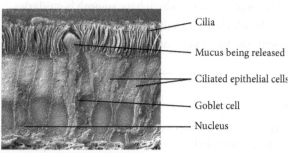

Cilia

Mucus being released

Ciliated epithelial cells

Goblet cell

Nucleus

Figure 9.5 Scanning electron micrograph of the epithelium in the trachea.

Figure 9.6 False-colour scanning electromicrograph showing goblet cells (orange) and cilia (green).

Link

The role of macrophages will be described in more detail in Chapter 11.

Cilia move continually to carry the mucus with the trapped particles up to the back of the throat. Cilia wave backwards and forwards at a rate of 10–20 times per second. They wave in a co-ordinated way to move mucus. Mucus travels at an average speed of about 5 mm per minute. From the throat, the mucus is swallowed. Many of the pathogens in the mucus will then be killed in stomach acid. Figure 9.6 shows cilia and goblet cells in the wall of the trachea.

Cells of the immune system called macrophages are also present on the linings of the airways. Macrophages are phagocytic. They engulf bacteria and other particles such as dust and pollen.

1. A typical adult trachea is 11 cm long. Cilia move mucus at an average speed of 5 mm per minute.
   Calculate the time taken for mucus to move from the bottom of the trachea to the top of the trachea.

2. Make a large labelled drawing of three complete ciliated epithelial cells from Figure 9.5.

ALVEOLI

Terminal bronchioles lead to respiratory bronchioles and these in turn lead to **alveoli**. Alveoli are the structures where gas exchange occurs.

Alveoli have no cartilage or smooth muscle in the walls. Their walls are very thin to make the distance between the air and blood as short as possible. The walls of the alveoli are composed of only one layer of epithelial cells.

A network of capillaries surrounds alveoli. Capillaries also have a wall that is one layer of epithelial cells in thickness.

The walls of alveoli also contain elastic fibres made from a stretchy fibrous protein called **elastin**. These elastic fibres allow the alveoli to expand on inhalation and to recoil again on exhalation. The expansion allows a greater volume of inhaled air into the alveoli and the recoil helps to expel air on exhalation.

Figure 9.7 is a photomicrograph of alveoli and capillaries.

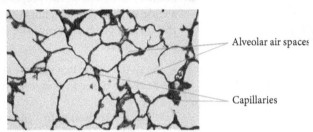

Alveolar air spaces

Capillaries

Figure 9.7 Photomicrograph of lung tissue showing alveoli and surrounding capillaries.

> **Tip**
>
> The words bronchi and alveoli are both plurals. The singular forms of these words are bronchus and alveolus.

Table 9.1 summarises the features of the structures within the gas exchange system.

Structure	Cartilage	Smooth muscle	Goblet cells	Cilia
trachea	✓			
bronchi				
terminal bronchioles				
respiratory bronchioles				
alveoli				

Table 9.1 Summary of the structures in the gas exchange system.

3. Inhaled air enters through the nose. List the structures of the gas exchange system, in the correct order, that air will pass through.
4. List the structures of the gas exchange system that contain:
   a. cartilage
   b. smooth muscle.
5. Gas exchange between air and blood occurs in alveoli. Suggest why gas exchange does not occur in the trachea.
6. Describe how cilia and mucus work together to protect the gas exchange system.
7. Some diseases result in too much mucus being produced. This mucus is also thicker and stickier than mucus in a healthy person. Suggest some problems this could cause.
8. Asthma causes smooth muscles in the airways to contract. One of the drugs used to treat asthma works by causing smooth muscles to relax. Explain how this drug helps people with asthma to breathe more easily.

Experimental skills 9.1: Working with photographs

QUESTIONS

P1. Figure 9.8 is a photomicrograph of part of the wall of the trachea.

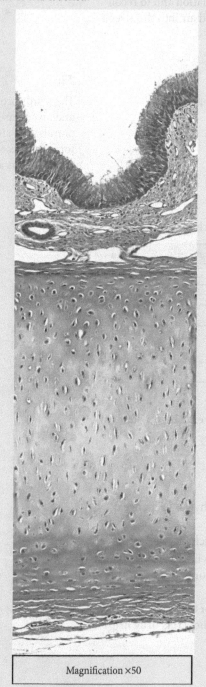

Magnification ×50

Figure 9.8

a. Make a large plan diagram of the section shown in Figure 9.8.

b. Label the epithelium, cartilage and smooth muscle on your diagram.

c. Calculate the thickness of the cartilage in this section using Figure 9.8.

d. Some of the steps in the preparation of a slide are listed below, but they are **not** in the correct order. Arrange the steps in the correct order.

- Soak the tissue in a suitable stain.
- Soak the tissue in preservative to prevent decay.
- Place the section onto a slide and add a coverslip.
- Soak the tissue in paraffin wax to make the tissue easier to cut into sections.
- Cut a thin section from the tissue.

e. Give **two** reasons why a stain is used for the photomicrograph in Figure 9.8.

Key ideas

→ The gas exchange system consists of the lungs, the trachea, bronchi, bronchioles and alveoli.

→ Cartilage is found in the trachea and bronchi.

→ Smooth muscle is found in the trachea, bronchi and bronchioles.

→ Goblet cells are found in the trachea and bronchi.

→ Ciliated epithelium is found in the trachea, bronchi and bronchioles.

→ Mucus, cilia and macrophages are present in lungs to remove pathogens and dust.

9.2 Function of the gas exchange system

The function of the gas exchange system is to allow oxygen into blood and carbon dioxide out of blood.

Gas exchange occurs by simple diffusion. Both oxygen and carbon dioxide are small, non-polar molecules that can pass through the lipid bilayer in cell surface membranes.

If diffusion is to be efficient, four conditions must be met:

- the surface area where diffusion occurs must be large
- the distance that substances have to travel by diffusion must be short
- the difference in concentration on either side of the surface must be large
- the temperature of the substances that are diffusing must be as high as possible.

These conditions are met in the alveoli.

A large surface area is achieved by the huge numbers of alveoli. There are around 5×10^8 alveoli in human lungs that together give a surface area for diffusion of about $70\,m^2$. Elastic fibres in the walls of alveoli allow the alveoli to expand when air is inhaled. When alveoli expand, the area for gas exchange increases further. The shape of alveoli also contributes to the large surface area, as they have rounded surfaces as shown in Figure 9.9.

9. Both human lungs have a total gas exchange surface area of $70\,m^2$. There are 5×10^8 alveoli in total.

   Calculate the surface area of one alveolus. Give your answer in standard form and in m^2. Assume all of the gas exchange surface area is alveoli.

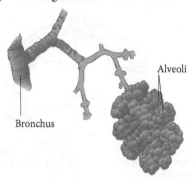

Alveoli

Bronchus

Figure 9.9 Alveoli increase the surface area for gas exchange in the lungs.

The short diffusion distance for gases is achieved in three ways:

1. The walls of the alveoli are very thin because they are made from a single layer of **squamous epithelial cells**. These cells are thin and flat.

2. The capillaries surrounding the alveoli are in very close contact with the outer surface of the alveoli.

3. The walls of the capillaries are also very thin as they too are made of squamous epithelium.

Link

Look back at diffusion in Chapter 4 to recall how some substances can cross cell surface membranes down a concentration gradient. You may also recall the variables that affect the rate of diffusion.

Focus

→ Epithelium is the general name for thin sheets of tissue that cover all body surfaces. The inside of hollow organs, such as the airways and blood vessels, are lined with endothelium, which is a type of epithelium. Squamous epithelium is made from thin, flattened cells, giving the tissue the appearance of scales on a fish. Ciliated epithelium is a type of columnar epithelium, made from tall, thin cells whose nucleus is at the base.

The concentration gradients of oxygen and carbon dioxide are kept as high as possible in two ways:

- The gas exchange surface is well ventilated. That means air is continually being breathed in and out. Inhaled air will contain a higher concentration of oxygen and a lower concentration of carbon dioxide than exhaled air. Table 9.2 shows the composition of inhaled and exhaled air.

Gas	Composition in inhaled air / %	Composition in exhaled air / %
oxygen	21	16
carbon dioxide	0.04	4
nitrogen	79	79

Table 9.2 Composition of inhaled and exhaled air.

- The concentration gradients of the gases are also maintained by having a good circulation system of blood in the capillaries. Deoxygenated blood is continually being brought to the lungs and oxygenated blood is continually being removed.

10. **a.** Present the information in Table 9.2 in one suitable graph.
    b. Explain your choice of type of graph.

The temperature of the air in the lungs is kept as high as possible. During inhalation, air from outside the body is warmed as it passes through the airways. When oxygen and carbon dioxide are at higher temperatures, their molecules have more energy and so will move faster. This means that diffusion will occur faster.

Figure 9.10 shows the direction of diffusion of oxygen and carbon dioxide during gas exchange.

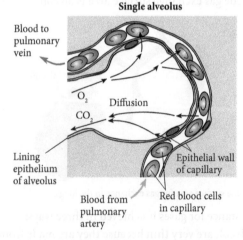

Each alveolus is a tiny air sac that has thin, flat walls. Oxygen from the air dissolves in the liquid that lines the alveolus and then diffuses across into the blood capillary. Carbon dioxide leaves the blood by the reverse route

Figure 9.10 Gas exchange in alveoli occurs by simple diffusion.

Worked example

The mean diameter of one alveolus is 300 μm. Assuming that the alveolus is a sphere, calculate

a. the surface area in m^2
b. the volume in m^3
c. the surface area to volume ratio of the alveolus.

Use the equations:

Surface area of a sphere $= 4\pi r^2$

Volume of a sphere $= \frac{4}{3}\pi r^3$

where r = radius of the sphere.

Answer

$r = \dfrac{300 \times 10^{-6}}{2}$

$= 150 \times 10^{-6}$

$= 1.5 \times 10^{-4} m$

a. Surface area of a sphere $= 4\pi r^2$

$= 4 \times \pi \times (1.5 \times 10^{-4} m)^2$

$= 2.83 \times 10^{-7} m^2$

b. Volume of a sphere $= \frac{4}{3}\pi r^3$

$= \frac{4}{3} \times \pi \times (1.5 \times 10^{-4} m)^3$

$= 1.41 \times 10^{-11} m^3$

c. Surface area to volume ratio $= \dfrac{\text{surface area}}{\text{volume}}$

$= \dfrac{2.83 \times 10^{-7}\, m^2}{1.41 \times 10^{-11}\, m^3}$

$= 2 \times 10^4$ or 20 000

Tip

It is easier to convert units by applying the conversion factor rather than multiplying or dividing. The conversion factor here is 1 μm = 1×10^{-6} m, so 300 μm = 300×10^{-6} m.

Tip

Surface area to volume ratios are usually expressed as numbers, rather than ratios. Here the ratio is 20 000:1 so the answer is given as 20 000.

11. Look at the information in Table 9.2.
    Which statement is true?
    A. All of the oxygen from inhaled air passes into blood.
    B. Nitrogen molecules do not move from air into blood.
    C. There is no net movement of nitrogen from air into blood.
    D. Exhaled air is mostly carbon dioxide.

12. How many cell surface membranes will an oxygen molecule cross as it travels from air into a red blood cell?
    A. Two
    B. Three
    C. Four
    D. Five

13. Describe how gas exchange happens in the lungs.

14. Explain how the concentration gradients of oxygen and carbon dioxide are maintained across the walls of the alveoli.

15. Inhaled air contains 21% oxygen and exhaled air contains 16% oxygen. Calculate the percentage decrease in oxygen content of air between inhalation and exhalation.

Assignment 9.1: The effects of smoking on the lungs

Inhaling tobacco smoke can cause serious damage to the lungs.

One of the components of tobacco smoke is tar. Tar can paralyse cilia (make them unable to move). Smokers whose cilia have been paralysed will cough more frequently than healthy people. The coughing occurs even when there is no infection in the lungs.

QUESTIONS

A1. Explain the effect that paralysing cilia will have on the lungs.

A2. Suggest why many smokers cough more frequently than healthy people.

A3. Smokers are more likely to have infectious diseases of the lungs than non-smokers. Explain why.

A4. Tar in tobacco smoke can also cause a disease called emphysema. Emphysema causes the elastic fibres in the alveolar walls to break down.
   a. Suggest the effect that breaking down the elastic fibres will have on ventilation of the alveoli.
   b. The walls of the alveoli are weakened by emphysema and the walls can break down. This creates fewer, larger air spaces than in healthy lungs.
   Explain the effect this will have on gas exchange.
   c. People who have severe emphysema can be given air that contains around 80% oxygen to breathe. Explain the reason for this.

A5. Tar in tobacco smoke can also cause lung cancer. A study was carried out using a large number of male volunteers between 1951 and 2001. All the volunteers were medical doctors in the UK. The study recorded the survival rates of the volunteers who included:

   - non-smokers, who had never smoked
   - smokers
   - those who had given up smoking between the ages of 25 and 34 (ex-smokers).

   The graph in Figure 9.11 shows the results of the study.

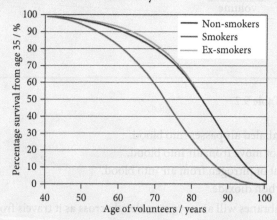

Figure 9.11

   a. Describe the trends shown in these results.
   b. Discuss the limitations of this study.

Key ideas

→ The lungs are adapted for efficient gas exchange of oxygen and carbon dioxide.
→ Gas exchange occurs at the alveoli by simple diffusion.
→ Alveoli provide a large surface area and thin walls for diffusion.
→ The concentration gradients of the exchange gases are maintained by good ventilation and a good blood supply.

CHAPTER OVERVIEW

Review the mini mind map to link new learning with your prior learning, and to highlight which of the key concepts underpin the chapter.

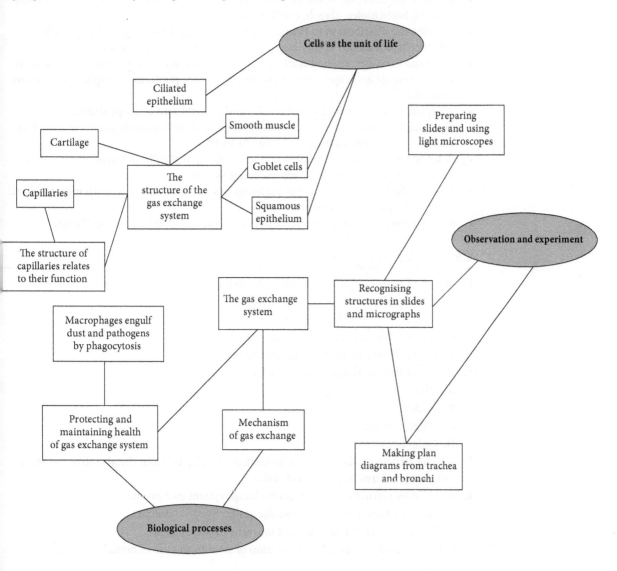

Try copying this mini mind map and expanding upon it. Use your notes from other chapters to help you explore how the essential ideas, theories and principles can be linked further together.

WHAT YOU HAVE LEARNED

Now that you have finished this chapter you should be able to:

- describe the structure of the human gas exchange system, limited to: lungs, trachea, bronchi, bronchioles, alveoli, capillary network
- describe the distribution in the gas exchange system of cartilage, ciliated epithelium, goblet cells, squamous epithelium of alveoli, smooth muscle and capillaries
- recognise cartilage, ciliated epithelium, goblet cells, squamous epithelium of alveoli, smooth muscle and capillaries in microscope slides, photomicrographs and electron micrographs
- recognise trachea, bronchi, bronchioles and alveoli in microscope slides, photomicrographs and electron micrographs, and make plan diagrams of transverse sections of the walls of the trachea and bronchus
- describe the functions of ciliated epithelia cells, goblet cells and mucous glands in maintaining the health of the gas exchange system
- describe the functions in the gas exchange system of cartilage, smooth muscle, elastic fibres and squamous epithelium
- describe gas exchange between air in the alveoli and blood in the capillaries.

CHAPTER REVIEW

1. Which of these contain cartilage in the walls?
    A. trachea only
    B. trachea and bronchi only
    C. trachea, bronchi and bronchioles only
    D. trachea, bronchi, bronchioles and alveoli
2. Which of these is **not** present in goblet cells?
    A. cilia
    B. nucleus
    C. mitochondria
    D. Golgi body
3. List the structures through which exhaled air will pass from the alveoli to the throat. List the structures in the correct order.
4. a. State two structures in the gas exchange system that produce mucus.
    b. State the function of mucus in the gas exchange system.
    c. Describe how mucus is moved upwards in the trachea.
5. Explain the advantage of the relaxation of smooth muscle in bronchioles during exercise.
6. Make a large plan drawing of the section through one of the bronchi in Figure 9.4. Add the labels that are on Figure 9.4.
7. Make a labelled drawing of a ciliated epithelial cell. Include:
    - the outline of the cell
    - the nucleus.
8. a. Describe the role of elastic fibres in the walls of alveoli.
    b. Describe how alveoli are adapted for efficient gas exchange.

Infectious disease

Infectious diseases have had devastating effects on humans throughout recorded history, and even changed the course of world history on more than one occasion. Even today, an estimated 219 million people are affected by malaria alone. Yet many infectious diseases are either preventable or treatable or both. So what are some of the most serious infectious diseases today and how can they be prevented? The picture shows one way of trying to prevent some types of disease. Insecticide is being sprayed to kill flying insects, some of which can transmit malaria.

Prior understanding

You may recall some facts about bacteria and viruses from previous courses. You may also recall some of the differences between prokaryotic and eukaryotic cells from Chapter 1, including the key structural features of bacterial cells.

Learning aims

In this chapter, you will learn what defines an infectious disease. You will then learn about the pathogens and modes of transmission of cholera, malaria, tuberculosis and HIV/AIDS. You will also discover some of the biological, social and economic factors affecting the prevention and control of these diseases. You will also learn how antibiotics such as penicillin work on bacteria and why they are ineffective against viruses. You will find out about the consequences of antibiotic resistance in some bacteria and the steps that can be taken to reduce the impact of antibiotic resistance.

10.1 The causes and control of infectious disease (Syllabus 10.1.1–10.1.4)

10.2 Antibiotics (Syllabus 10.2.1, 10.2.2)

10.1 The causes and control of infectious disease

An **infectious disease** is a disease caused by a pathogen that can be transmitted from an infected person to an uninfected person. The pathogens that cause infectious diseases include bacteria, viruses, fungi and protozoa. A pathogen is a microorganism capable of causing disease. Not all microorganisms are pathogenic.

This chapter covers four of the most important infectious diseases affecting humans in the world today: **cholera**, **malaria**, **tuberculosis** and **HIV/AIDS**.

CHOLERA

There are an estimated 1.3–4.0 million cases of cholera in the world every year, most being in less economically developed countries. The symptoms of cholera include:

- large volumes of watery diarrhoea
- 'rice-water' stools
- vomiting
- increased heart rate
- reduced blood pressure.

Cholera can be fatal within hours if left untreated, mainly due to the dehydration effects of the diarrhoea.

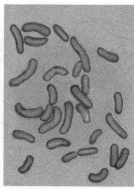

Figure 10.1 Photomicrograph of *Vibrio cholerae*, the pathogen that causes cholera.

Cause and treatment of cholera

Cholera is caused by a bacterial pathogen, *Vibrio cholerae*. These are comma-shaped bacteria that each have one flagellum, as shown in Figure 10.1.

Cholera is transmitted from infected people to uninfected people by the faecal–oral route. Infected people pass *Vibrio cholerae* in their faeces and the bacteria enter uninfected people with contaminated food or drink.

About 75% of the people who are infected with cholera do not show symptoms and will probably not even know that they are infected. These people are called **symptomless carriers** and pose great risk for disease transmission.

Cholera is most frequently transmitted in untreated sewage. Many towns and cities in less-industrialised countries do not have the financial resources to treat the rapidly growing volume of sewage that is produced. If this untreated sewage is discharged into rivers or into the sea, then the water will be contaminated.

Drinking contaminated water, or eating uncooked fish or shellfish from the contaminated water, can transmit cholera.

Cholera is also transmitted when infected people with poor personal hygiene handle food or cooking utensils. Washing hands thoroughly after passing faeces will remove many of the *Vibrio cholerae* from the skin and prevent these bacteria entering food.

In many countries, human faeces are used as fertiliser for crops. Eating unwashed or uncooked vegetables from fields that have been fertilised in this way can also transmit cholera.

Cholera **epidemics** are also common in the aftermath of natural disasters such as earthquakes and floods. Treated drinking water can become contaminated with sewage when pipes are damaged. In addition, people may be housed in temporary accommodation, which is overcrowded and often has poor sanitation.

An epidemic is a widespread outbreak of a disease within a community or geographical region. A **pandemic** is the global spread of a disease.

An example of a cholera pandemic, called El Tor, started in 1961 in Indonesia. The pandemic had spread to Bangladesh by 1963, to Italy by 1973 and to Peru by 1991. In

Peru, the outbreak started by spreading along the coast and then further inland where it caused more than 2000 reported new cases every day. Contaminated seafood is thought to have been the cause of the rapid spread of this cholera outbreak in Peru.

Cholera can be treated with **antibiotics** and the symptoms can be relieved by oral rehydration therapy (ORT). *Vibrio cholerae* releases a toxin called choleragen. Choleragen affects the epithelium of the small intestine by causing water and mineral ions to pass from blood into the lumen of the small intestine. This is the cause of the watery diarrhoea, which can cause dehydration, and may be fatal within hours.

ORT involves drinking water with salt and glucose, called oral rehydration solution (ORS), and can reduce deaths from dehydration by over 90%. The transport of glucose and sodium ions is coupled by use of a cotransporter protein, so glucose accelerates the uptake of sodium ions and consequently water by osmosis. Commercially-available packs of ORS are available in many countries.

Prevention and control of cholera

Cholera is a preventable disease. Evidence for this is that cholera cases in more economically developed countries are extremely rare.

As cholera is transmitted by the faecal–oral route, breaking this **transmission cycle** is key to preventing the spread of the disease.

Proper sewage treatment involves stages where microorganisms are either killed by chemical disinfectants or prevented from multiplying by exposure to ultraviolet radiation. Properly treated sewage should contain very few or no pathogens and so can be safely discharged into the sea or into rivers without contaminating the water with *Vibrio cholerae*.

As well as sewage treatment, treatment of the drinking water supply also plays a key part in breaking the cholera transmission cycle. Water for drinking is chemically treated, usually with chlorine, to kill pathogens. The treated water is then distributed in closed pipes so that it cannot be contaminated before reaching people's homes.

Building sewage treatment and water treatment facilities is expensive, and some countries cannot afford this expenditure.

Good personal hygiene is also essential for breaking the cholera transmission cycle. Disposing of faeces into purpose-built toilets where they cannot contaminate food or water is part of this process. Thoroughly washing hands after passing faeces and before handling food or cooking utensils will help prevent *Vibrio cholerae* from being passed on from infected people.

Personal hygiene is not expensive, and can be promoted by educating people about the dangers from cholera and the mode of transmission of the disease.

A vaccine for cholera is available. The vaccine does not offer 100% protection against cholera because *Vibrio cholerae* multiplies in the small intestine where it is antigenically concealed. Antigenic concealment refers to the location of a pathogen in parts of the body where antibodies and cells of the immune system are not present in high numbers or cannot gain access to the pathogen at all.

1. Name the pathogen that causes cholera.
2. Explain why cholera can be fatal within hours of developing symptoms.
3. Explain why cholera is most prevalent in less-industrialised countries.

MALARIA

At any time, there are an estimated 500 million people infected with malaria, mostly in the tropics. The symptoms of malaria include:

- fever
- headaches
- anaemia
- muscle pain
- shivering.

Link
You will discover more about antibiotics and how they work in Section 10.2.

Link
The mechanisms by which ions cross cell surface membranes and the concept of osmosis were described in Chapter 4.

Link
The topic of cotransporter proteins was covered in Chapter 4.

Link
You will learn about the immune system and how vaccines work in Chapter 11.

In sub-Saharan Africa alone, malaria is responsible for the deaths of 1 million children under the age of five every year.

Cause and treatment of malaria

Malaria is caused by four different species of a eukaryotic *Plasmodium* **parasite**: *P. vivax*, *P. falciparum*, *P. ovale* and *P. malariae*.

The parasite is transmitted from uninfected people to infected people by the female *Anopheles* mosquito. The mosquito is called a **vector** because it transmits the pathogen but the mosquito does not show any disease symptoms when carrying the parasite.

Female mosquitoes feed on human and animal blood to obtain the necessary nutrients for egg development. When a female *Anopheles* mosquito (Figure 10.2) takes a blood meal from a person infected with the malarial parasite, the gametes of the pathogen are taken into the mosquito's digestive system. Here, the gametes fuse. The first infective stage of the *Plasmodium* then develops in the mosquito's salivary glands. The parasite reproduces in the mosquito, increasing greatly in numbers. When the mosquito takes another blood meal, the parasites are introduced to the person's blood through the mosquito's saliva. Figure 10.3 shows the thread-like sporozoites of *Plasmodium* in human blood. The sporozoite is the stage in the parasite life cycle when it is injected into blood from mosquito saliva.

Figure 10.2 A female *Anopheles* mosquito feeding on human blood.

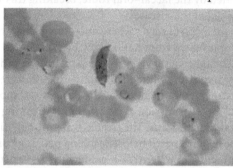

Figure 10.3 *Plasmodium* is visible in the centre of this photomicrograph of human blood.

The *Plasmodium* parasite reproduces in the liver and red blood cells of humans. Rapid reproduction inside red blood cells causes the red blood cells to burst, releasing many more parasites into the blood. This process causes fever and anaemia.

Malaria is endemic to tropical countries due to the distribution of the mosquito vector that cannot survive in milder climates. Climate change and environmental change are together causing an increase in the habitable range for *Anopheles*. This is increasing the global impact of malaria.

People can become immune to malaria if they are repeatedly infected and if they survive beyond the age of about 5 years. People who live in areas where malaria is not endemic are rarely immune to the disease.

Drugs such as chloroquine can be used to treat and also to prevent malaria. However, many *Plasmodium* are now resistant to chloroquine, so new drugs such as mefloquine have been developed.

Prevention and control of malaria

The best way to prevent infection with *Plasmodium* is to avoid being bitten by mosquitoes. Sleeping under mosquito nets, using insecticides and insect repellents, and covering skin after sunset are all effective ways to prevent mosquito bites. Mosquito bed nets can be soaked in insecticide for added protection (Figure 10.4).

The development of a vaccine for malaria has been complicated by the fact that the *Plasmodium* has many different stages in its life cycle, so many different antigens are involved. Antigenic concealment of the parasite within human cells also makes vaccines less effective. The entire genome of *Plasmodium* has now been sequenced and the information gained from this has led to the development of some new vaccines that are being trialled in malaria-endemic areas.

Link

Sickle cell anaemia will be described in Chapter 16. People with this genetic disease have increased resistance to malaria, and carriers of the allele for the disease have some resistance.

Figure 10.4 Sleeping under mosquito nets that are soaked in insecticide is a proven method of reducing the rate of malaria.

Prophylactic use of anti-malarial drugs is another method of preventing the disease. People who are not immune to malaria take the drugs before entering a malaria-endemic area and continue taking the drugs for some days or weeks after leaving. Some anti-malarial drugs can be expensive and have unpleasant side-effects.

Many countries and regions have taken steps to reduce the number of mosquitoes. Mosquitoes lay eggs in water, so filling in areas that fill with water during rainy seasons has been an effective way to reduce numbers. The eggs hatch into larvae in water, so coating water with oil can prevent oxygen from reaching larvae. Water can be stocked with fish such as *Gambusia affinis* that feed on mosquito larvae. The bacteria *Bacillus thuringiensis* can also be introduced to water. *B. thuringiensis* kills mosquito larvae but is harmless to other forms of aquatic life.

Spraying insecticide on a large scale during or after rain can also effectively reduce mosquito larvae.

4. Figure 10.5 shows the global distribution of malaria transmission.

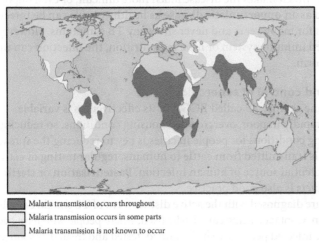

■ Malaria transmission occurs throughout
□ Malaria transmission occurs in some parts
▨ Malaria transmission is not known to occur

Figure 10.5

Explain the trends shown in Figure 10.5.

5. List some precautions that a person can take to avoid being infected with the malarial parasite when in a malaria endemic area.

6. Outline some of the biological issues that have made the development of a malaria vaccine difficult.

TUBERCULOSIS

An estimated 25% of the world population are infected with **tuberculosis**, which is shortened to TB, but not all of these people will become ill. The symptoms of TB include:

- a cough lasting 3 weeks or more
- coughing up blood
- chest pain when breathing
- fever
- weight loss.

In some countries, it is becoming more difficult to treat cases of TB due to an increase in drug resistance in the TB pathogen.

Cause and treatment of tuberculosis

Tuberculosis is caused by the bacteria *Mycobacterium tuberculosis* (Figure 10.6) and *Mycobacterium bovis*.

    M. tuberculosis is transmitted between infected people and uninfected people by the aerosol route. When an infected person coughs, sneezes or even talks, small liquid droplets

Figure 10.6 Scanning electron micrograph of the rod-shaped *Mycobacterium tuberculosis* bacteria.

containing the pathogen are expelled into the air. When the droplets containing the pathogen are inhaled by an uninfected person, the bacteria can be taken into their lungs.

M. bovis is transmitted in contaminated or unpasteurised dairy products, such as unpasteurised milk from infected cattle.

TB primarily affects the lungs and so can be diagnosed with a chest X-ray or by microscopic examination of sputum samples. Many people who are infected with *Mycobacterium* do not show symptoms. People with this inactive form of infection do not transmit the disease.

Although mainly affecting the lungs, TB can also spread to the kidneys, brain and spine.

TB is treated using antibiotics such as rifampicin and isoniazid. Most treatments involve the use of more than one antibiotic taken for up to 12 months. The treatment regimen should be monitored by the DOTS (directly observed treatment short-course) system where someone other than the infected person monitors that the full course of antibiotics is followed. Careless use of antibiotics has resulted in **antibiotic resistance** developing in *Mycobacterium,* which makes the disease much more difficult, or even impossible, to treat.

TB is classed as an **opportunistic disease** as healthy people can be infected with *Mycobacterium* for many years and never show any TB symptoms. However, if people have a weakened immune system or have malnutrition, the infection can turn into the active disease form.

Prevention and control of tuberculosis

There is a vaccine against TB called BCG, but its effectiveness is variable.

TB spreads rapidly in poor, overcrowded housing conditions, so reducing poverty and improving living conditions for people in cities is key to reducing the spread of TB.

As *M. bovis* is transmitted from cattle to humans, regular testing of cattle for TB can help detect a potential source of human infection. **Pasteurisation** or **sterilisation** of milk and dairy products is also good practice.

People who are diagnosed with the active disease form of TB should be isolated to prevent further infection. Contact tracing can also be used, where people who have been in close contact with the infected person can be identified, tested and treated as soon as possible.

7. Describe what is meant by the term infectious disease.

8. Outline how tuberculosis is transmitted.

★9. Suggest why the numbers of cases of tuberculosis in some countries may be much higher than the published numbers.

10. Figure 10.7 shows the change in death rates from TB during the 20th century in the UK.

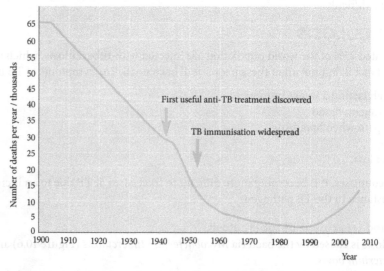

Figure 10.7

Explain the trend shown in Figure 10.7.

HIV/AIDS

HIV/AIDS originated in the early 20th century but the first cases were not identified until 1981. The symptoms of HIV/AIDS include:

- a brief period of influenza-like symptoms 2–3 weeks after infection
- weight loss
- fatigue
- swelling of the lymph nodes (armpits, neck and groin)
- increased susceptibility to other infections.

In 2017 there were an estimated 40 million cases of HIV/AIDS worldwide and in the same year an estimated 1.8 million people became newly infected.

Tip

Numbers of cases of any disease are usually estimates. This is because not all cases are diagnosed and not all diagnosed cases are officially recorded.

Cause and treatment of HIV/AIDS

HIV/AIDS is caused by a virus called the human immunodeficiency virus (HIV). HIV is a small particle of approximately 120 nm diameter, or around one-sixtieth the diameter of a red blood cell. The particle is composed of RNA genetic material and an enzyme called reverse transcriptase surrounded by a protein coat and this, in turn, is surrounded by an outer lipid envelope (Figure 10.8).

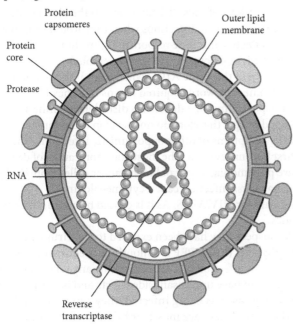

Figure 10.8 Structure of the human immunodeficiency virus particle.

The HIV pathogen is transmitted in three ways: by sexual contact, by blood-to-blood contact and from an infected mother to a child.

The most common ways that people become infected with HIV are:

- unprotected sexual intercourse with an infected person
- sharing contaminated needles for injection and tattooing equipment
- unsterilised surgical equipment
- blood transfusions or tissue transplants that have not been screened for HIV
- from an infected mother to baby during pregnancy (**vertical transmission**) or during childbirth or during breastfeeding.

HIV infects white blood cells called T-helper cells, which are the helper cells that control the immune response to infection (Figure 10.9).

Tip

HIV is the name of the pathogen and HIV/AIDS is the name of the disease caused by the pathogen.

Link

You will learn more about the immune system, immune responses and the role of lymphocytes in Chapter 11.

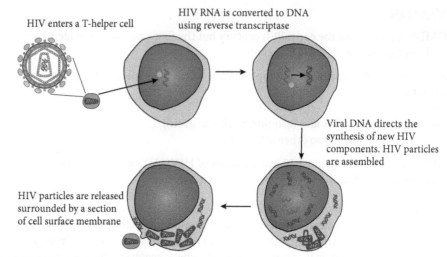

HIV enters a T-helper cell

HIV RNA is converted to DNA using reverse transcriptase

Viral DNA directs the synthesis of new HIV components. HIV particles are assembled

HIV particles are released surrounded by a section of cell surface membrane

Figure 10.9 The replication cycle of the human immunodeficiency virus.

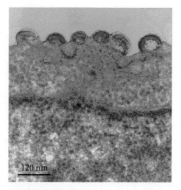

120 nm

Figure 10.10 False-colour scanning electron micrograph of HIV particles being released from an infected lymphocyte.

HIV causes the numbers of these T-helper cells in the blood to decrease. This leaves infected people susceptible to infections when they would otherwise have been able to create an effective immune response. Figure 10.10 shows new virus particles emerging from the lymphocyte where they assembled.

This is the origin of the name AIDS, which stands for acquired immune deficiency syndrome. The word acquired means that they have become infected, whereas some types of immune deficiency disorders are inherited.

Many infected people therefore do not develop the symptoms of HIV/AIDS directly, but the symptoms of other opportunistic infections. These can take up to 10 years to first become evident. The most common causes of death in people with HIV/AIDS are pneumonia, a type of blood cancer called non-Hodgkin lymphoma, and a usually harmless infection called cytomegalovirus.

There is no cure for HIV/AIDS. As it is caused by a virus, it cannot be treated with antibiotics.

People who suspect they have been exposed to the virus can be given **post-exposure prophylaxis (PEP)**. PEP is a month-long course of a combination of drugs that aims to prevent infection of T-helper cells by any virus particles that may have entered the blood. PEP can have unpleasant side effects and is not 100% effective.

Once someone is diagnosed as being infected, they can be given drugs that slow the replication of the virus. These are mostly inhibitors of the enzyme reverse transcriptase. The virus uses reverse transcriptase to make a complementary double-stranded DNA copy of its RNA genome. Reverse transcriptase is not found in any uninfected eukaryotic cells. Inhibitors of a protease that the virus uses to assemble new particles can also be used. Usually, the drugs are given in combinations of different types.

Prevention and control of HIV/AIDS

Some countries have much higher incidences of HIV/AIDS than others. Eight per cent of deaths from HIV/AIDS occur in African countries and in some of these states up to 40% of the population is infected.

In the late 20th century, HIV/AIDS was first identified in the human population, and in most countries where the disease is endemic, roughly equal proportions of males and females are infected.

Link

You may wish to look back at Chapter 3 for details of how enzyme inhibitors work. Reverse transcriptase and its uses in biotechnology will be described in Chapter 19.

Education about the modes of transmission and dangers of HIV/AIDS is key to prevention and control.

People at greatest risk of exposure to the pathogen include those who have unprotected sex and drug users who inject. People can be advised to change their behaviour, such as having only one sexual partner and not injecting drugs. Encouraging people to use barrier methods of contraception, such as condoms and femidoms, will reduce the risk of becoming infected. If a person is known to be infected, then contact tracing is recommended. Contacts who may have been exposed to the virus can be tested for the presence of anti-HIV antibodies and offered early treatment. Use of sterile needles for any form of injection is also essential, as is proper sterilisation of medical and tattooing equipment.

There have been many cases of HIV/AIDS resulting from blood transfusions. These cases mostly resulted from blood that was donated and given to recipients in the 1970s and 1980s before health authorities became aware that HIV/AIDS was transmitted this way. Blood for transfusion is now routinely screened for the HIV virus.

Regular HIV tests for those most at risk can enable early treatment to be started. Also, if people are aware that they are infected, they can take precautions to avoid infecting others.

A summary of the infectious diseases covered in Section 10.1 is given in Table 10.1.

Infectious disease	Pathogen	Mode of transmission	Prevention/control
cholera	the bacterium *Vibrio cholerae*	faecal–oral route	good public sanitation and personal hygiene, careful food preparation
malaria	the parasites *Plasmodium vivax*, *P. falciparum*, *P. ovale* and *P. malariae*	through the bite of the female *Anopheles* mosquito that is carrying the pathogen	insecticides, insect repellents, mosquito nets, prophylactic drugs, draining breeding grounds, biological control of mosquito larvae
tuberculosis (TB)	the bacteria *Mycobacterium tuberculosis* and *Mycobacterium bovis*	aerosol route (*M. tuberculosis*) through meat and unpasteurised dairy products (*M. bovis*)	reduction of overcrowding in housing, contact tracing, screening of cattle
HIV/AIDS	human immunodeficiency virus	sexual contact, blood-to-blood contact and vertical transmission	use of barrier contraception, use of sterile needles, screening of blood for transfusion

Table 10.1 Summary of the infectious diseases described in this chapter.

Worked example

The graph in Figure 10.11 shows the number of people infected with HIV, the number of deaths from HIV/AIDS and the number of new cases of HIV/AIDS between 1990 and 2015.

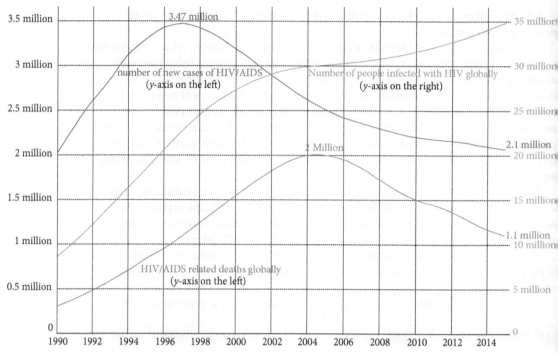

Figure 10.11 Global number of AIDS-related deaths, new HIV infections, and people living with HIV (1990–2015).

Describe the trends shown in Figure 10.11.

Answer

The number of new cases of HIV/AIDS increased from 2 million in 1990 to 3.47 million in 1997, then decreased to 2.1 million in 2015.

The number of people infected with HIV has increased continuously from 8 million in 1997 to 36.7 million in 2015. The increase was fastest from 1990 to 2002.

The number of HIV/AIDS related deaths increased from 0.3 million in 1990 to 2 million in 2005, then decreased to 1.1 million in 2015.

11. Name the pathogen that causes HIV/AIDS.
12. Look at Figure 10.10, which shows HIV particles being released from an infected lymphocyte. Calculate the magnification of this electron micrograph.
13. List some suggestions that could be made when educating people at high risk of HIV/AIDS in order to reduce their chance of being infected or spreading the disease.
14. Explain why deaths of people infected with HIV/AIDS are caused by other diseases.

Key ideas

→ Infectious diseases are those caused by pathogens and can be transmitted from an infected person to an uninfected person.
→ Cholera is caused by a bacterial pathogen, *Vibrio cholerae*.
→ Cholera is transmitted by the faecal–oral route.
→ Malaria is caused by a eukaryotic parasite of the genus *Plasmodium*.
→ Malaria is transmitted by the female *Anopheles* mosquito acting as a vector.
→ Tuberculosis is caused by the bacterial pathogens, *Mycobacterium tuberculosis* and *Mycobacterium bovis*.
→ Tuberculosis is transmitted by the aerosol route.

→ HIV/AIDS is caused by a viral pathogen, human immunodeficiency virus (HIV).
→ HIV/AIDS is transmitted sexually, by blood-to-blood contact and from mother to baby.

10.2 Antibiotics

Antibiotics are substances produced by organisms that can kill or inhibit the growth of microorganisms.

Many fungi that live in soil face competition for nutrients from soil bacteria. Some of these fungi produce substances to kill or inhibit the growth of bacteria to reduce the competition. The first of these substances to be used commercially, **penicillin,** was discovered in the early 20th century. Since that time, many more naturally occurring antibiotics have been discovered. Many more antibiotics have been artificially synthesised based upon the knowledge gained from study of the natural ones.

Antibiotics work by inhibiting processes in bacteria. It is possible to use many of them as drugs because the processes that are inhibited are either unique to prokaryotes or target molecules that are unique to prokaryotes.

Those antibiotics that kill bacteria are called **bactericidal** and those that inhibit the growth of bacteria are called **bacteriostatic**.

The modes of action of some antibiotics are shown in Table 10.2 and Figure 10.12.

Mode of action	Examples
inhibitors of cell wall synthesis	penicillin, vancomycin (both bactericidal)
inhibitors of transcription	rifampicin (bactericidal)
inhibitors of protein synthesis	streptomycin (bactericidal) tetracycline (bacteriostatic)
inhibitors of cell surface membrane function	polymixin B, colistin (both bactericidal)
inhibitors of other metabolic processes	sulfonamides (bacteriostatic).

Table 10.2 Modes of action of some antibiotics.

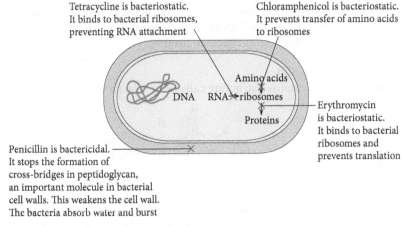

Tetracycline is bacteriostatic. It binds to bacterial ribosomes, preventing RNA attachment

Chloramphenicol is bacteriostatic. It prevents transfer of amino acids to ribosomes

Amino acids

DNA   RNA→ribosomes

Proteins

Erythromycin is bacteriostatic. It binds to bacterial ribosomes and prevents translation

Penicillin is bactericidal. It stops the formation of cross-bridges in peptidoglycan, an important molecule in bacterial cell walls. This weakens the cell wall. The bacteria absorb water and burst

Figure 10.12 Mechanisms of action of some antibiotics.

Some antibiotics are described as **broad spectrum** because they are effective against a wide range of different types of bacteria. Others are **narrow spectrum** as they are only effective against certain types.

Antibiotics are not effective against viruses. Viruses are intracellular parasites and are not independent living organisms. That means viruses require the metabolic processes of the host cell to replicate their genetic material, synthesise protein and assemble new virus particles. Viruses have no metabolic processes of their own, so there is nothing to be inhibited by antibiotics.

PENICILLIN

Historical records show that people in ancient Egypt, China and Europe would sometimes treat infected wounds by applying scrapings from food that had gone mouldy. The mould that forms on food is a type of fungus.

The first antibiotic substance to be discovered was penicillin. Alexander Fleming was professor of bacteriology at a London hospital. In 1928, while studying bacteria called *Staphylococcus*, Fleming noticed that one of his bacterial cultures had been accidentally contaminated with a fungus. Around this fungus was a clear zone where no bacteria were growing. The fungus was identified as *Penicillium notatum* (Figure 10.13). The substance produced by this fungus was found to kill a wide range of harmful bacteria. Extraction and purification of this substance proved difficult, but was eventually achieved in the 1930s.

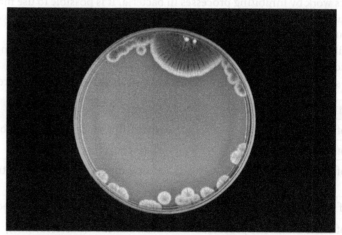

Figure 10.13 *Penicillium notatum* fungus can be seen growing on this plate of solid growth medium.

The discovery of penicillin was one of the greatest advances in medical technology. Until its discovery, people with diseases such as pneumonia, gonorrhoea and tuberculosis could not be cured. The middle part of the 20th century, however, became known as the age of antibiotics, as these and other diseases became curable.

Mode of action of penicillin
The penicillin molecule contains a structure called a beta lactam ring (Figure 10.14).

Beta lactam ring

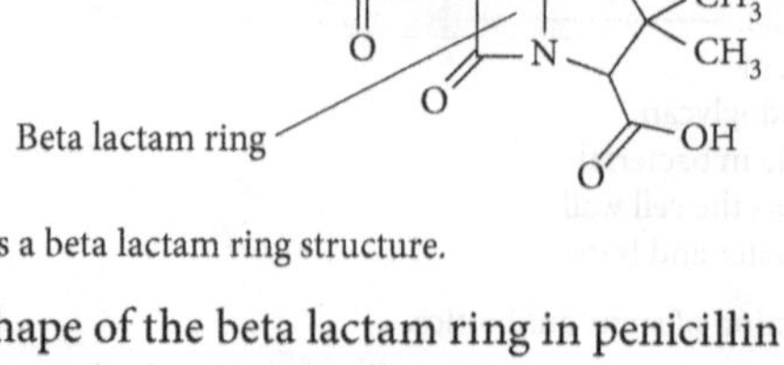

Figure 10.14 Penicillin contains a beta lactam ring structure.

Link

It may be helpful to look back at Chapter 3 to recall how enzymes and competitive inhibitors work.

The three-dimensional shape of the beta lactam ring in penicillin is the same as that of part of a molecule used to make bacterial cell walls.

Penicillin acts as a competitive inhibitor of a bacterial enzyme used to synthesise cell walls. Bacterial cell walls are made from a substance called peptidoglycan (murein). One of the enzymes involved in peptidoglycan synthesis is called DD transpeptidase.

This enzyme forms the cross-links in peptidoglycan that give the cell wall its strength. Penicillin occupies the active site of DD transpeptidase, and therefore blocks the formation of the enzyme–substrate complexes that are required for cell wall synthesis.

Bacteria continually modify their cell walls by making and breaking the cross-links in peptidoglycan. Penicillin has no effect on the enzymes that break the cross-links, so in the presence of penicillin the rate of breaking peptidoglycan cross-links begins to exceed the rate of making cross-links. The cell wall then rapidly weakens and the bacteria die as a result.

Peptidoglycan and the enzyme DD transpeptidase are not found in any eukaryotic cells, so penicillin has little or no effect on human cells.

15. **a.** State what is meant by the term antibiotic.
    b. Explain why antibiotics are not effective against viruses.
16. Which of these is the mode of action of penicillin?
    A. inhibitor of DNA replication
    B. inhibitor of cell wall synthesis
    C. inhibitor of transcription
    D. inhibitor of cell surface membrane function

Experimental skills 10.1: Testing antibiotics

The effectiveness of antibiotics against a particular type of bacteria can be tested using the disc diffusion method.

Bacteria are evenly distributed throughout nutrient agar in a Petri dish. Nutrient agar is a solid, jelly-like, growth medium for bacteria. Before the bacteria are allowed to grow in the agar, discs containing the antibiotics are placed on the surface of the agar.

The discs are made from an absorbent material similar to filter paper and are soaked in solutions of antibiotics.

When placed on the agar, the antibiotics will diffuse outward from the disc through the agar. The bacteria are then allowed to grow, usually by placing the Petri dish in an incubator overnight. As the bacteria grow they make the clear agar look cloudy. The diameter of the clear zone around the discs can be used to compare the effectiveness of the antibiotics.

QUESTIONS

P1. Figure 10.15 shows the results of a disc diffusion test with five antibiotics, A–E. The discs are made from filter paper and soaked in solutions of each antibiotic. The solutions were made using equal concentrations of each antibiotic in distilled water. Disc W has been soaked in distilled water.

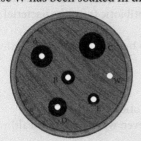

Figure 10.15 Testing the effectiveness of five different antibiotics using the disc diffusion method.

a. State the purpose of disc W in Figure 10.15.
b. The Petri dish in Figure 10.15 has a diameter of 9 cm. Calculate the actual diameter of the clear zone around the disc with antibiotic B.
c. Explain what can be concluded about the result for antibiotic C compared with the other antibiotics.
d. Explain why the disc diffusion method cannot distinguish between bacteriostatic and bactericidal antibiotics.
e. The same disc diffusion method using the same five antibiotics is to be used on a different type of bacteria.

List three variables that should be kept the same in order to make the results comparable.

Preparing serial dilutions

Preparing serial and simple (proportional) dilutions was introduced in Chapter 2.

A group of students is given a 0.1% (w/v) solution of an antibiotic in ethanol. They have been asked to prepare a two-fold serial dilution of the antibiotic in ethanol.

A two-fold serial dilution means diluting the solution by half each time. This is the method that the students follow:

1. Remove 5 cm³ of the 0.1% (w/v) solution and add to 5 cm³ of ethanol to make a 0.05% (w/v) solution.
2. Use separate pipettes for the antibiotic solution and for the ethanol. Mix the new dilution thoroughly.
3. Remove 5 cm³ of the 0.05% (w/v) solution and add to 5 cm³ of ethanol to make a 0.025% solution, again using separate pipettes.

P2. a. Explain why a clean pipette is used for the antibiotic solution in step 3 of the students' method.

b. Extend the students' method to describe how to make two further two-fold serial dilutions. Start with the 0.025% solution and state the concentration of each dilution in standard form.

P3. You are given 10 cm³ of a 1% weight per volume (w/v) solution of antibiotic. Weight per volume in this case means 1 g of antibiotic dissolved in 100 cm³ water. You are required to prepare a 10-fold dilution series of this antibiotic in water, each dilution to be 10 cm³. Describe how to make the first three dilutions in this series and state the percentage concentration of each.

ANTIBIOTIC RESISTANCE

Where many diseases were previously treated and cured quite easily with antibiotics, some of these same diseases are now very difficult, or even impossible to treat. This is due to antibiotic resistance.

Antibiotic resistance in bacteria is caused by the bacteria evolving through genetic variation and natural selection. Bacterial DNA undergoes random mutations, which cause genetic variation. Some of this variation may result in the bacteria becoming resistant to an antibiotic. If this happens, then only the resistant bacteria will survive and continue to multiply, eventually leading to the entire population of bacteria being resistant (Figure 10.16).

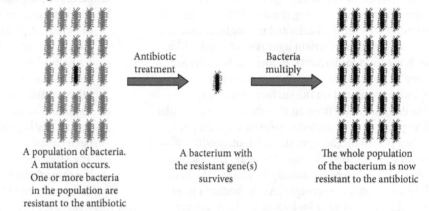

A population of bacteria. A mutation occurs. One or more bacteria in the population are resistant to the antibiotic

A bacterium with the resistant gene(s) survives

The whole population of the bacterium is now resistant to the antibiotic

Figure 10.16 Summary of how antibiotic resistance occurs in a population of bacteria.

The genes required for resistance in bacteria are often carried on **plasmids**. Plasmids are small pieces of DNA in the bacteria cytoplasm that can replicate independently and be passed between bacteria. That means when one bacteria of a particular type becomes resistant to an antibiotic, the gene responsible for the resistance can be passed to others of the same type.

The mechanisms of antibiotic resistance are:

- the production of a bacterial enzyme to break down the antibiotic
- decreased permeability of the cell wall or membrane to the antibiotic
- selective molecular pumps to remove the antibiotic from the bacterial cell
- a change to the target site for the antibiotic
- development of a new metabolic pathway to bypass the inhibited pathway.

Antibiotic resistance is caused by:

- over-use of antibiotics
- patients not completing the course of antibiotics
- routine use of antibiotics in farm animals, even when they are healthy.

An example of antibiotic resistance occurs in the case of tuberculosis. All strains of *Mycobacterium tuberculosis* are now resistant to penicillin, and several other strains are resistant to many antibiotics. These strains are called MDR TB (multiple drug-resistant TB) and XDR TB (extensively drug-resistant TB). The required course of antibiotics for treatment of TB is 9–12 months, but many people stop taking the antibiotics when they feel better. This leaves some bacteria that may be antibiotic resistant and could be passed to other, uninfected, people.

Antibiotic resistance is also common in bacteria found in hospitals where antibiotics are used and bacteria are exposed to them. So-called hospital super-bugs and nightmare-bacteria are bacteria that are multiple drug resistant. Some of these are opportunistic pathogens, such as MRSA (multiple resistant *Staphylococcus aureus*) that are found on the skin of healthy people.

The impact of antibiotic resistance means:

- longer hospital stays
- increased medical costs
- higher mortality from bacterial diseases
- use of last-resort antibiotics that often have serious side effects.

Even if new antibiotics keep being developed, bacteria will adapt and eventually become resistant to these new drugs. To combat antibiotic resistance, our use of antibiotics needs to change. If we do not change our use of antibiotics, we could be back to the pre-antibiotic age where bacterial diseases are once again impossible to cure.

Antibiotics need to be used more selectively and, in the case of some diseases, only as a last resort. People should not be given access to antibiotics when they just feel unwell. Antibiotics should be used under the supervision of healthcare professionals.

Improve personal hygiene, sanitation, food preparation, keep vaccinations up to date and have protected sex in order to reduce the spread of bacterial diseases. If there are fewer cases of bacterial diseases then antibiotics will need to be used less frequently.

Antibiotics are used extensively in the farming of livestock in many countries. Animals are given commercially prepared food that routinely contains antibiotics for prevention of disease. In 2013 a total of 131 000 tonnes of antibiotic, which is approximately 40% of global antibiotic production, was fed to farm animals in this way. Bacteria that are exposed to these antibiotics, either in the animals or in the environment, are becoming resistant. Antibiotics should only be used to cure diseases with farm animals and not be part of daily feeding to increase growth or prevent infection.

Antibiotics for certain bacterial diseases should be used in combination. The probability of bacteria becoming resistant to one antibiotic at random is about 1 in 1000. The probability of becoming resistant to two antibiotics in combination is therefore (1 in 1000) × (1 in 1000) = 1 in a million. If three antibiotics are used together then the probability would be 1 in 1000^3 = 1 in a thousand million, and so on.

Use of programmes such as DOTS to ensure that people complete the course of antibiotics will also reduce the rate at which bacteria become resistant. People should also be educated to not share left-over antibiotics.

17. Describe how bacteria can become resistant to antibiotics.

18. Outline the impact of antibiotic resistance on healthcare.

Assignment 10.1: *Clostridium difficile*

Clostridium difficile is a type of rod-shaped bacteria that lives in the intestines of healthy humans. *C. difficile* is an opportunistic pathogen that causes diarrhoea.

Treatment of *C. difficile* is carried out using an antibiotic.

Figure 10.17 shows the use of an antibiotic A to treat *C. difficile* and the death rates from *C. difficile* in a large country between the years 2001 and 2008.

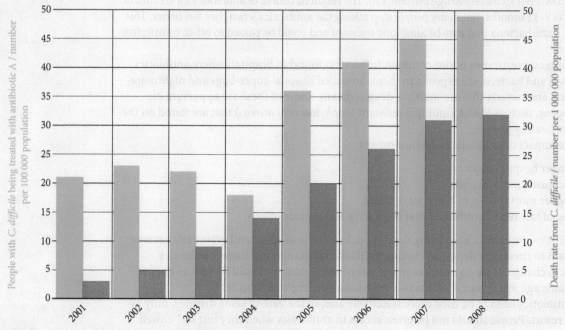

Figure 10.17 The use of antibiotic A to treat *C. difficile* and the death rates from *C. difficile* in a large country between 2001 and 2008.

QUESTIONS

A1. Explain what is meant by the term opportunistic pathogen.

A2. Explain why the values on the graph are numbers per 100 000 or numbers per 1 000 000 and not actual numbers of people.

A3. Describe the trend in the number of people with *C. difficile* being treated with antibiotic A shown in the graph.

A4. Explain what can be concluded from the trend in death rate from *C. difficile* shown in the graph compared with antibiotic A use.

A5. *C. difficile* infections often occur in patients who are receiving hospital treatment for other bacterial infections.

Table 10.3 shows the number of cases per 1000 patients in a hospital in Italy between 2007 and 2011.

a. Calculate the percentage change in number of cases of *C. difficile* in the hospital between 2010 and 2011.

b. Plot a graph of the results in Table 10.3. Put the year on the horizontal axis. Complete your graph by drawing the most appropriate line.

c. Describe the trend shown in these results.

Year	Number of cases per 1000 patients
2007	0.75
2008	2.35
2009	1.61
2010	3.38
2011	9.80

Table 10.3 Cases of *C. difficile* infections per 1000 patients in a hospital in Italy between 2007 and 2011.

Key ideas

→ Antibiotics are substances produced by organisms that kill or inhibit the growth of microorganisms.

→ Antibiotics are active against bacteria as antibiotics inhibit metabolic processes or target structures in bacterial cells.

➡️ Antibiotics are not effective against viruses as viruses do not have any metabolic processes of their own.

➡️ Penicillin works as a competitive inhibitor of an enzyme involved in bacterial cell wall synthesis.

➡️ Antibiotic resistance has led to some bacterial diseases being very difficult or impossible to treat.

➡️ Steps can be taken to reduce the rate at which bacteria are becoming resistant to antibiotics.

CHAPTER OVERVIEW

Review the mini mind map to link new learning with your prior learning, and to highlight which of the key concepts underpin the chapter.

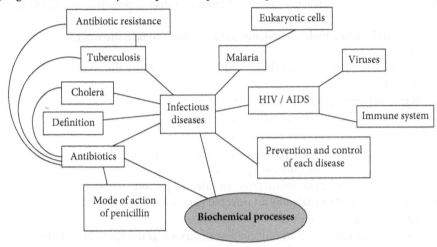

Try copying this mini mind map and expanding upon it. Use your notes from other chapters to help you explore how the essential ideas, theories and principles can be linked further together.

WHAT YOU HAVE LEARNED

Now that you have finished this chapter you should be able to:

- state that infectious diseases are caused by pathogens and are transmissible
- state the name and type of pathogen that causes each of the following diseases:
  - cholera caused by the bacterium *Vibrio cholerae*
  - malaria – caused by the protoctists *Plasmodium falciparum*, *Plasmodium malariae*, *Plasmodium ovale* and *Plasmodium vivax*
  - tuberculosis (TB) – caused by the bacteria *Mycobacterium tuberculosis* and *Mycobacterium bovis*
  - HIV/AIDS – caused by the human immunodeficiency virus (HIV)
- explain how cholera, malaria, TB and HIV/AIDS are transmitted
- discuss the biological, social and economic factors that need to be considered in the prevention and control of cholera, malaria, TB and HIV/AIDS
- outline how penicillin acts on bacteria and why antibiotics do not affect viruses
- discuss the consequences of antibiotic resistance and the steps that can be taken to reduce its impact.

CHAPTER REVIEW

1. Which of these is a way of transmitting cholera?
   A. sharing needles contaminated with the pathogen
   B. being bitten by mosquitoes carrying the pathogen
   C. drinking water contaminated with the pathogen
   D. inhaling droplets containing the pathogen

2. a. Explain why cholera can spread rapidly in refugee camps.
   b. List some facts that could be used to advise someone about reducing their risk of getting cholera.

3. a. Name the two pathogens that cause tuberculosis.
   b. Describe the precautions that can be taken to avoid the spread of tuberculosis from cattle to humans.

4. a. Name two of the pathogens that cause malaria.
   b. Describe how malaria is transmitted from an infected person to an uninfected person.
   c. A small lake in a malaria-endemic area is known to be a site where mosquitoes lay eggs.
      Describe some procedures that could be taken to reduce mosquito numbers at this lake.

5. a. Name the pathogen that causes HIV/AIDS.
   b. Explain why antibiotics are not effective against HIV/AIDS.

6. Which of these activities can transmit HIV/AIDS?
   A. sharing cooking or eating utensils with an infected person
   B. shaking hands with an infected person
   C. receiving an organ transplant from an infected person
   D. inhaling droplets exhaled by an infected person

7. a. Explain how penicillin works.
   b. Explain why the mid-20th century was described as the age of antibiotics.
   c. Explain why some scientists think we are at risk of going back to the pre-antibiotic era.

8. a. A small percentage of the population of a bacterial pathogen is found to be antibiotic resistant. Describe three ways in which the whole population of that pathogen could become antibiotic resistant.
   b. Describe some of the strategies that can be used to reduce the increase of antibiotic resistance in bacterial diseases.

Immunity

Globally, around 80 million people suffer from a skin disease called psoriasis. This condition is characterised by reddened, scaly skin. In some cases, is it associated with arthritis, a painful swelling of the joints.

Psoriasis is a member of a group of conditions called autoimmune disorders. These occur when the body's immune system attacks some of its own tissues as though they were invading cells. In the case of psoriasis, this leads to tissue damage and a continuous replacement of skin cells.

Autoimmune diseases also include multiple sclerosis, insulin-dependent diabetes and myasthenia gravis. It is hoped that understanding how the immune system works may offer more advanced treatments for those who suffer from autoimmune diseases.

Prior understanding

You will have seen that we are surrounded by pathogens that can invade our bodies and make us ill. You may remember that every cell is surrounded by a cell surface membrane that contains protein or glycoprotein molecules. From Chapter 3 you may also remember that some molecules have complementary shapes (for example, enzymes and their substrates).

Learning aims

In this chapter, you will learn how the body is protected against pathogens, cancer cells and toxins. You will see how the process of phagocytosis is a faster, non-specific process, whereas the actions of lymphocytes provide effective defence against specific pathogens. You will also see how the body is protected from further infection by the same pathogen.

11.1 The immune system (Syllabus 11.1.1–11.1.4)
11.2 Antibodies and vaccination (Syllabus 11.2.1–11.2.6)

11.1 The immune system

Our external environment contains bacteria, viruses, fungi and other **foreign** organisms. Some of these, called pathogens, are capable of invading our bodies and causing disease. We are able to overcome infections by pathogens because we have an **immune system**. This is a complex system involving many different cells and tissues that allows us to start an **immune response** and develop **immunity**. An immune response is a series of responses of the body to the entry of a pathogen or other foreign antigen. Immunity is a long-term resistance to infections by a specific pathogen.

Crucial to starting an immune response and the development of immunity are cells that recognise and help to destroy antigens or the foreign cells that carry them. These are the **white blood cells**, or **leucocytes**. Figure 11.1 is a blood smear showing some leucocytes along with many red blood cells. Leucocytes are made in the **bone marrow** and are found throughout the body. They can move and are able to squeeze between cells, passing freely in and out of the capillaries. In addition to the blood, they are found in the lymph, a fluid carried by **lymphatic vessels**, and also in tissue fluid and body cavities (such as alveoli). They gather in a number of lymphoid organs of the **immune system**, including the **lymph nodes**, **spleen** and **thymus** (Figure 11.2).

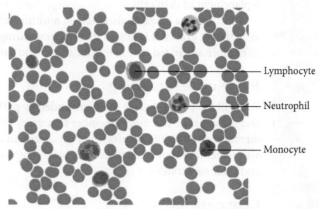

Figure 11.1 A photomicrograph of a very thin film of blood. The red shapes are red blood cells, and you can see that these are by far the most common type of cell in blood. The nuclei of the white blood cells (of which there are three different types in this image) have taken up a purple stain.

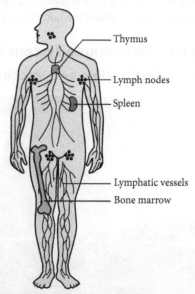

Figure 11.2 The lymphoid organs and lymphatic vessels associated with the immune system.

A key property of leucocytes is that they can recognise foreign organisms and particles that do not normally occur in the body. This is because of the **antigens** they possess. Antigens are macromolecules, usually proteins, glycoproteins, lipoproteins or polysaccharides, that are found in the surface membrane or wall of cells; however, small bodies such as viruses and even individual protein molecules (such as toxins) can also be classed as antigens. Cells from different individuals have different antigens, whereas all the cells of the same individual have the same antigens, i.e. they are **specific**. **Non-self antigens** are substances or cells that are recognised by the immune system as being foreign and will therefore stimulate an immune response. **Self antigens** are substances produced by the body that the immune system does not recognise as foreign, and therefore do not cause an immune response.

1. State the origin of all leucocytes (white blood cells).
2. Psoriasis is a disease in which the immune system treats some self antigens as non-self. Define the term 'non-self antigen'.

Antigens are recognised by leucocytes. Although leucocytes are diverse in their structures and functions, we can group most of them into two main families. These take part in two different but linked processes, the **non-specific immune response** and the **specific immune response**:

- Phagocytes engulf and digest pathogens. They are involved in the non-specific immune response.
- **Lymphocytes** recognise specific pathogens and respond in a number of different ways, some direct and some indirect, to kill the pathogen. They are involved in the specific immune response.

The importance of leucocytes in the immune response is underlined by the fact that there is a surge in their number after infection. Normally, a person's leucocyte count is around 5000 to 10 000 cells per millilitre (ml) of blood. However, during and immediately after an infection, this number can be considerably higher.

PHAGOCYTOSIS AND THE NON-SPECIFIC IMMUNE RESPONSE

The process of **phagocytosis** is an example of a non-specific immune response. It is carried out by two types of cell, called **neutrophils** and **monocytes**, which are both produced in the bone marrow and are commonly referred to as phagocytes. They have a complex cytoskeleton that enables them to move and change shape more rapidly than most cells, and many mitochondria, which provide ATP required for the process.

- Neutrophils have a lobed nucleus, as shown in Figure 11.1. They are found mainly in the blood, but they can leave the capillaries and enter tissue fluid.
- Monocytes have a large, kidney-shaped nucleus, as shown in Figure 11.1. Once they leave the blood they develop into **macrophages**. These are the largest phagocytes and are found in lymph, tissue fluid and the lungs, where they kill microbes before they enter the blood.

Phagocytosis is a modified type of endocytosis. It is a fast process that is carried out by these cells when they meet a pathogen or foreign antigen. All phagocytes surround the particle, take it into their cytoplasm, and digest it with enzymes.

Phagocytosis is summarised in a series of steps in Figure 11.3.

Link

In Chapter 4 you learned about the wide range of functions carried out by proteins in the cell surface membrane. Many of these proteins can be recognised by cells of the immune system.

Link

Remind yourself about the process of endocytosis, in Chapter 4. This involves the invagination of the cell surface membrane around particles outside the cell, to take them into the cytoplasm.

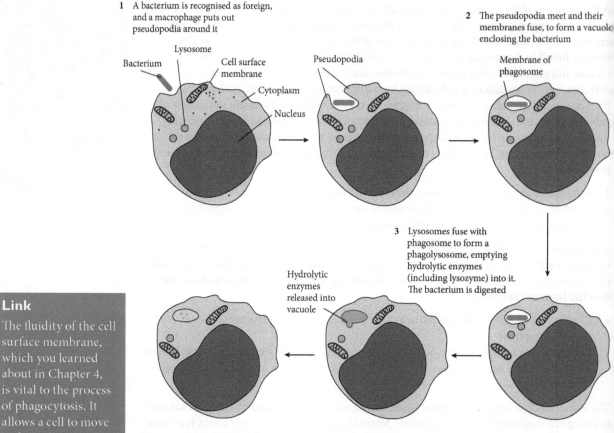

1 A bacterium is recognised as foreign, and a macrophage puts out pseudopodia around it

Bacterium · Lysosome · Cell surface membrane · Cytoplasm · Nucleus

2 The pseudopodia meet and their membranes fuse, to form a vacuole enclosing the bacterium

Pseudopodia · Membrane of phagosome

3 Lysosomes fuse with phagosome to form a phagolysosome, emptying hydrolytic enzymes (including lysozyme) into it. The bacterium is digested

Hydrolytic enzymes released into vacuole

Figure 11.3 The process of phagocytosis. Note that in macrophages, fragments of the digested pathogen will enter the cell surface membrane, as shown later in Figure 11.8. The structures shown in this diagram are not to scale.

Link

The fluidity of the cell surface membrane, which you learned about in Chapter 4, is vital to the process of phagocytosis. It allows a cell to move and rapidly change its shape.

Link

In Chapter 10, you saw how lysozyme destroys bacterial cell walls by hydrolysing murein (peptidoglycan), allowing other hydrolytic enzymes to digest the contents of the pathogen.

1. The membrane of the phagocyte extends around and surrounds the bacterium. This occurs as structures called **pseudopodia** (cytoplasmic 'feet') flow around each side of the pathogen. This is shown in the electron micrograph later in this chapter, in Figure 11.6.

2. The cell surface membranes of the pseudopodia arms then fuse together, so that the pathogen is completely enclosed by a membrane within the cell's cytoplasm. This produces a small, membrane-bound vesicle containing the pathogen, called a **phagosome**.

3. The phagosome moves deeper into the cytoplasm, and fuses with one or more lysosomes, forming a **phagolysosome**. Lysosomes contain enzymes, mainly **lysozyme** and other hydrolytic enzymes, which digest the pathogen into smaller fragments. The pathogen is now said to be phagocytosed.

3. Distinguish between the following terms:
   a. neutrophil and macrophage
   b. phagosome and phagolysosome
   c. lysosome and lysozyme.

In neutrophils and monocytes, the process of phagocytosis ends with the digestion of the pathogen. The cell often dies after this process. However, in macrophages, part of the phagolysosome can bud off, taking some of the fragments of pathogen with it. When this structure fuses with the cell surface membrane, the partially digested bacterial fragments become part of the membrane and stick out from the surface of the phagocyte. This is called **antigen presentation**, and the macrophage is now said to be an **antigen-presenting cell**. This often means that the antigens are more easily recognised by other cells of the immune system, enhancing the subsequent immune response against the pathogen. This process is shown later in this chapter in Figure 11.8.

4. Phagocytosis involves the uptake of a large molecule or organism by a macrophage or neutrophil.

    a. Use your knowledge of cell function to identify which of the following statements about phagocytosis are correct.

        1 It is a process requiring energy in the form of ATP.

        2 It is a form of endocytosis.

        3 Substances brought into a cell using this method are enclosed in a small vacuole.

        A 1, 2 and 3

        B 1 and 2 only

        C 1 and 3 only

        D 2 and 3 only

    ☆ b. Using the information provided in this chapter, draw and label a series of diagrams to show how antigen presentation occurs in a macrophage.

5. Unlike most bacteria, those that cause the disease typhoid can survive and multiply inside phagocytes. Suggest how they are able to achieve this.

> **Tip**
>
> Be careful when describing the action of phagocytes. They engulf (or ingest) pathogens. Do not describe this process as 'eating'.

> **Tip**
>
> Do not confuse these similar terms: phagocyte, phagocytosis, phagocytosed, phagosome, phagolysosome.

Experimental skills 11.1: The importance of phagocytosis in healthy lungs

Chronic obstructive pulmonary disease (COPD) of the lungs is globally one of the most common causes of death. Smoking is the most common risk factor for COPD and contributes to 85% of all cases.

In many patients with COPD, the lungs can be readily infected by bacteria. For example, infections associated with pneumonia are much more common in patients with COPD than those without it.

A group of scientists investigated the activity of macrophages taken from the lungs of patients with COPD compared with non-smokers and smokers (who had not developed COPD). They did this by measuring the mass of a bacterium, *Escherichia coli*, which was phagocytosed in a laboratory culture plate over a 4-hour incubation period. Their results are shown as a bar chart in Figure 11.4.

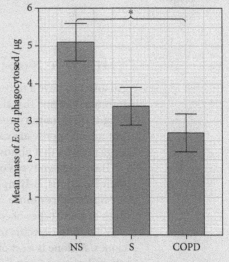

Figure 11.4 Phagocytosis of *Escherichia coli* by macrophages from non-smokers (NS), smokers (S) and patients with chronic obstructive pulmonary disease (COPD); 95% confidence interval error bars are shown.

QUESTIONS

P1. Explain why a reduction in the rate of phagocytosis by macrophages increases the risk of death due to bacterial infection of the lungs.

P2. Apart from the 4-hour incubation period, state two other factors that the scientists would have standardised to enable them to make a valid comparison of the data obtained from the three experiments.

P3. Refer to Figure 11.4 to help you answer the following questions.

    a. Calculate the difference in the mass (in µg) of *E. coli* that were phagocytosed in non-smokers and in patients with COPD.

    ★**b.** It has been estimated that one bacterial cell has a mass of one picogram (pg). This is equal to a thousand nanograms. Using your answer to part (a), estimate the difference in the number of *E. coli* cells that were phagocytosed in non-smokers and in patients with COPD. Express your answer in standard form.

    c. Identify one improvement that could be made to the presentation of the scientist's results.

        A. The bars should be drawn so that they are touching.

        B. A line graph should be drawn.

        C. A value of zero should be added at the origin.

        D. The orientation of the axes should be changed.

★**P4.** There are a number of chemicals in cigarette smoke that attach to proteins on the surface of cells. Using this information, suggest why phagocytes are less able to carry out their usual role in the lungs of smokers and people with COPD.

A Level analysis

To obtain values for the mass of *E. coli* that had been phagocytosed, the scientists took five samples from the culture plate and calculated a mean value for each.

P5. From which group of people did the scientists obtain the most reliable data? Explain your answer.

P6. The star shown on the graph in Figure 11.4 indicates that the difference in mass of *E. coli* that were phagocytosed in non-smokers and in patients with COPD was significant at a level of $P < 0.05$. Explain what this statement indicates.

> **Tip**
>
> The error bars, which extend either side of the sample mean, indicate the values within which the true value of the whole population mean will lie, with 95% confidence.

THE SPECIFIC IMMUNE RESPONSE

All animals have a non-specific immune system, whereas only vertebrates have a specific immune system. This suggests that it arose later in the evolutionary process. It enables long-term immunity against specific pathogens. Leucocytes involved in the specific immune response are called lymphocytes. All lymphocytes are white blood cells with a large nucleus that almost fills the cell, as shown in Figure 11.1.

There are two types of lymphocyte, which look identical and can only be told apart by their functions. **B-lymphocytes**, or **B-cells**, are involved in a defence mechanism called the **humoral response**. **T-lymphocytes**, or **T-cells**, are involved in the **cellular response**. Both of these types of cell exist in a huge number of varieties.

Vital to the specific immune system is the ability of both types of lymphocyte to form clones. A clone is a set of genetically identical individuals. In this context, a clone refers to a population of B- or T-cells, all of which react to a pathogen in a similar way.

The role of lymphocytes in the specific immune system is complex; it is summarised in Figure 11.5. In the following pages, you will learn the individual parts of this mechanism and how it is so effective in defending the body against pathogens.

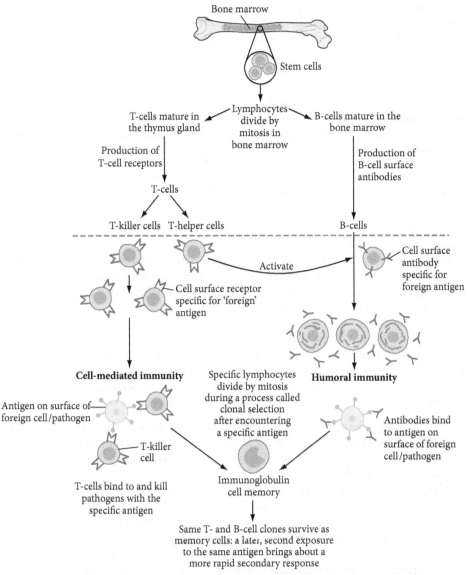

Tip

Note that B-lymphocytes have antibodies in their cell surface membrane, whereas T-lymphocytes have receptor proteins.

Figure 11.5 The formation and development of B- and T-lymphocytes (cells). B-cells are formed and differentiate in the bone marrow, where they develop a cell surface membrane containing antibodies of one specific type. T-cells are formed in the bone marrow, but then they move in the blood to the thymus gland, where they differentiate. During this process, they produce receptor proteins of one specific structure, which are found in the cell surface membrane. After differentiation, B- and T-cells spread throughout the body but are found concentrated in the lymph nodes and spleen. The structures shown in this diagram are not to scale.

B-lymphocytes and humoral immunity

Like all leucocytes, B-lymphocytes (B-cells) are produced in the bone marrow. They remain there for some time to differentiate (mature). During this process, specific antibodies build up in their cell surface membrane.

B-cells respond to the foreign antigens of a pathogen by differentiating into **plasma cells**. Plasma cells are filled with many ribosomes attached to extensive rough endoplasmic reticulum, and mitochondria that provide ATP for protein synthesis (Figure 11.6). This structure is suited to the rapid production of **antibodies**, proteins

Tip

B-lymphocytes are so called because they differentiate (mature) in the bone marrow.

that are released into the blood plasma and carried to the site of infection. The rate of antibody secretion can be as high as 2000 molecules per second per cell. An antibody, or **immunoglobulin**, is a Y-shaped protein molecule that is made by a plasma cell in response to a particular antigen. Its structure is summarised in Figure 11.12. For reasons you will see later, antibodies attach to the antigens on pathogens, to make them harmless.

A single B-cell can divide to form 1×10^6 plasma cells, each of which can release 1×10^3 antibodies every second for 4 days. This is called the humoral response because the foreign cells are killed by soluble antibodies dissolved in the blood plasma (body fluids were called 'humours' in ancient terminology).

6. You will need to refer to Figure 11.6 to answer these questions.
   a. Using the magnification provided in the figure, calculate the actual size of the plasma cell. Show your working.
   b. Suggest why plasma cells contain a large number of mitochondria and rough endoplasmic reticulum.

7. Cancer of the bone marrow is called leukaemia. To treat this condition, chemicals and radiation can be provided that damage and destroy bone marrow cells.
   Explain why a patient who is being treated for leukaemia would be at risk of developing an infectious disease.

T-lymphocytes and cellular immunity

T-lymphocytes (T-cells) are also produced in the bone marrow. During their differentiation in the thymus, protein receptors build up in their cell surface membrane. This process is summarised in Figure 11.5. Like B-cells, T-cells respond to foreign antigens. However, they can form two different types of T-cell:

- **T-helper cells** enhance the specific immune response. They produce chemicals called **cytokines**. Cytokines circulate in the blood and stimulate phagocytes to undertake phagocytosis. They also stimulate B-lymphocytes to divide and to differentiate into plasma cells to produce antibodies.
- **T-killer cells** (also called **cytotoxic T-cells**) attack infected body cells and the cells of some larger pathogens directly. The T-killer cells are specific for the presented antigens and so recognise and kill only body cells that express these antigens, ensuring that only infected cells are destroyed. The two cells face each other, surface membrane-to-membrane, and the T-killer cell secretes hydrogen peroxide or a protein called perforin by exocytosis. These form holes in the target cell surface membrane. This is shown later in Figure 11.8. Cytotoxic T-cells are also stimulated by the cytokines released by T-helper cells.

The specific immune response achieved by T-cells is called the cellular response because the foreign cells are killed directly by the immune cells.

Clonal selection of lymphocytes

It is estimated that, at birth, there are around 1×10^8 (one hundred million) different types of B-cell found in the lymph nodes and spleen, each with its own unique set of antibody molecules in its cell surface membrane. This is also true of T-cells, which differ by the unique set of protein and glycoprotein molecules in their cell surface membrane, which act as receptors. Between them, these antibodies and receptor proteins can therefore bind specifically to 1×10^8 different antigens, so there will be cell surface membrane proteins to match almost every possible antigen that might enter the body.

Tip

Remember the conventions of **standard form** notation. For example, 1×10^3 or $10^3 = 1000$ and 1×10^6 or $10^6 = 1\,000\,000$.

×3000

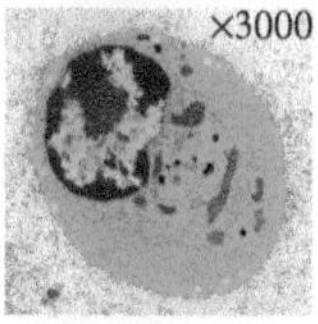

Figure 11.6 An electron micrograph of a plasma cell. The cytoplasm is full of rough endoplasmic reticulum, studded with ribosomes.

Tip

T-lymphocytes are so called because they differentiate (mature) in the <u>t</u>hymus, a gland found behind the ribcage (see Figure 11.2).

If a B- or T-cell encounters an antigen or antigen-presenting cell displaying the antigen that has a complementary shape to the antibody or receptor on its surface, it is stimulated into immediate action. First, the lymphocyte responds by dividing by mitosis, over and over again, to form a clone of genetically identical cells. This process is called **clonal selection**, because only these specific cells divide. The result is a vast number of B- and T-cells with identical binding sites on their cell surface membrane that specifically complement the foreign antigen. This is called **clonal expansion**. After further differentiation, all of the daughter cells carry the same antibody or receptor in their cell surface membrane, so all of them can detect and respond to the specific antigen that stimulated the response. The process of clonal selection, clonal expansion and differentiation in B- and T-cells is summarised in Figures 11.7 and 11.8.

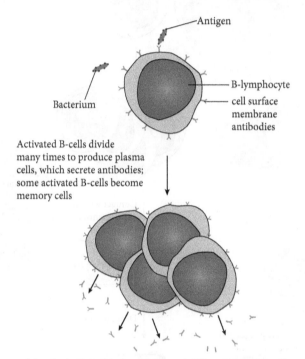

Figure 11.7 The process of clonal selection, clonal expansion and differentiation in B-cells and the humoral response. The structures shown in this diagram are not to scale.

Worked example

Draw a table to summarise the differences between the cellular response and the humoral response.

Answer

Cellular response	Humoral response
involves T-cells	involves B-cells
involves the release of chemicals called cytokines	involves the release of proteins called antibodies
causes death of foreign cells directly by immune cells	causes death of foreign cells indirectly by the use of antibodies

Table 11.1

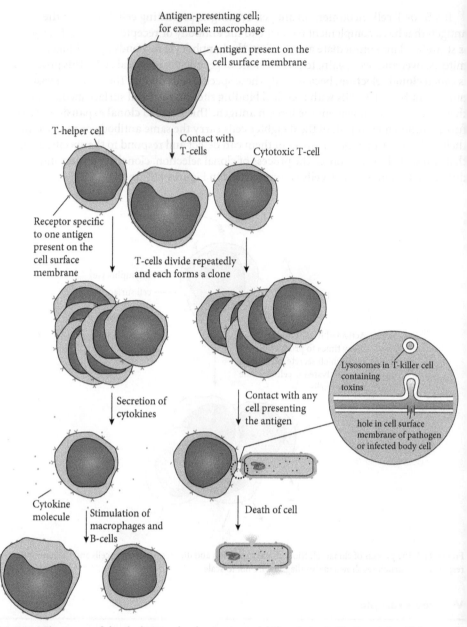

Figure 11.8 The process of clonal selection, clonal expansion and differentiation in T-cells and the cellular response. The structures shown in this diagram are not to scale.

8. Explain why all of the T-cells produced by the division of an activated T-cell carry the same receptors in their cell surface membrane.

9. T-killer cells interact directly with some pathogens and can destroy them by forming holes in the cell surface membrane.
   a. Explain why T-killer cells cannot destroy all pathogens by this method.
   b. Suggest why making holes in the cell surface membrane of a cell causes it to die.
   c. T-killer cells are able to destroy body cells that carry antigens that are not 'self'. Suggest why these cells do not always kill all types of cancer cell.

10. Outline the roles of macrophages and lymphocytes in:
    a. the humoral response
    b. the cellular response.

11. Figure 11.9 shows an interaction between an antigen-presenting cell (APC) and a T-helper cell.

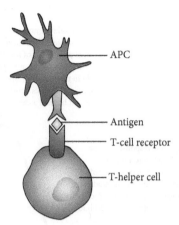

APC

Antigen

T-cell receptor

T-helper cell

Figure 11.9 The interaction between an antigen-presenting cell, such as a macrophage, and a T-helper cell. The structures shown in this diagram are not to scale.

a. Identify an example of a cell that can become an APC.
   A. T-killer cell
   B. T-helper cell
   C. Plasma cell
   D. Neutrophil

b. Describe in detail the steps that will occur as a result of this interaction.

Memory cells and long-term immunity

Plasma cells and activated T-cells do not live for very long and, provided the infection is successfully cleared, disappear within a few weeks. However, the body is able to retain a 'memory' of the antigens of pathogens years into the future. So that the body can respond more quickly next time, a small number of the activated B-cells and T-cells persist in the body for several years. These **memory cells** accumulate in the lymphoid organs. If the pathogen enters the body again, these cells provide an immunological memory. They divide rapidly to produce an even greater number of active cells, which are released into the circulation and are all capable of secreting a specific antibody (in the case of B-cells) or responding in their characteristic ways to pathogens (in the case of T-cells).

Consider Figure 11.10. The first time that the body encounters a new antigen there are only a few lymphocytes able to recognise each antigen, so it can take several days for clonal selection to take place and for a sizeable number of lymphocyte clones to be produced. This slow and weak response to a first infection is called the **primary immune response**. It is during this period that the symptoms of the disease may develop. This is partly owing to toxins and cell death because of the pathogen, and partly due to the immune response itself (e.g. fever, inflammation).

However, imagine that the same pathogen enters the body again, possibly many years later. As there are already many memory cells carrying the antibody or receptor specific to this pathogen, the chances that one of the memory cells will meet quickly with the pathogen and be activated are increased. In the case of plasma cells, their rapidly increasing numbers mean that the specific antibody is usually secreted fast enough to destroy the pathogen before it has time to reproduce – or replicate, if a virus – and cause illness.

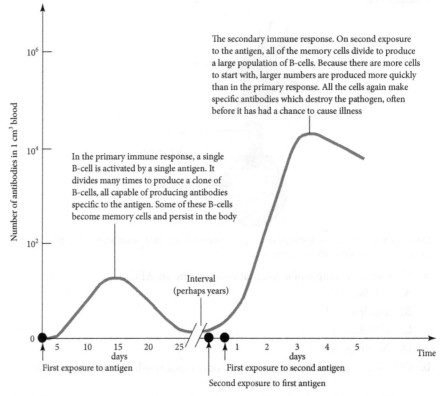

The secondary immune response. On second exposure to the antigen, all of the memory cells divide to produce a large population of B-cells. Because there are more cells to start with, larger numbers are produced more quickly than in the primary response. All the cells again make specific antibodies which destroy the pathogen, often before it has had a chance to cause illness

In the primary immune response, a single B-cell is activated by a single antigen. It divides many times to produce a clone of B-cells, all capable of producing antibodies specific to the antigen. Some of these B-cells become memory cells and persist in the body

Figure 11.10 The immune system has an immunological memory of previously-encountered antigens.

The ability of the body to trigger this **secondary immune response** explains why it is very unusual for some diseases to affect us twice in a lifetime. The response is much greater and much faster than the primary immune response, and often produces antibodies in such large quantities, and so quickly, that the invaders are destroyed before they have time to divide and make us ill. We are said to be **immune** to the disease caused by that pathogen. For example, many children get chickenpox, and become immune to it once they have recovered from the illness.

As we will see later in this chapter, the immunological memory of a pathogen is of vital importance to **vaccination**. A vaccine stimulates the body to produce a primary immune response to a particular pathogen, without becoming infected by it. Later, if the pathogen invades the body again, the immune system can start a very fast secondary response and the person does not become ill.

Worked example

Compare the primary and secondary immune responses, explaining the reasons for the similarities and differences in the processes.

Answer

The primary immune response occurs upon the first exposure to a non-self antigen. Clonal selection of B- and T-cells results in production of antibodies that are complementary to the antigen, which gradually falls in concentration in the coming weeks. If the antigen is met again in the future, then a secondary immune response occurs. During this response, the number of antibodies produced, and the rate at which they are released, is greater than during the primary immune response. This is because after the first exposure, some B- and T-cells differentiated to form memory cells, which are stored in the immune system. They serve to quickly recognise the antigen on the second exposure, resulting in a faster response that may prevent symptoms from appearing.

Key ideas

→ Neutrophils and macrophages, collectively called phagocytes, can carry out phagocytosis.
→ The phagocytic cell extends pseudopodia, which surround the structure to be ingested. The membranes of the pseudopodia fuse together, enclosing the structure in a vacuole called a phagosome.
→ Lysosomes fuse with the phagosome to form a phagolysosome and empty hydrolytic enzymes into it. These include lysozyme, which hydrolyses bacterial cell walls.
→ Macrophages place antigens from structures that they have ingested in their cell surface membranes; this is known as antigen presentation.
→ T-lymphocytes (T-cells) are responsible for the cellular response to an antigen.
→ If a T-cell encounters a presented antigen that has a complementary shape to the specific receptor on its surface, it is activated. The activated T-cell divides repeatedly, forming a clone. Some of the cloned cells remain as memory T-cells.
→ Some T-cells are cytotoxic. They fuse with cells carrying their antigen and kill them.
→ Some T-cells are helper cells. They secrete cytokines that stimulate macrophages and B-cells to act.
→ B-lymphocytes (B-cells) are responsible for the humoral response to an antigen.
→ If a B-cell encounters an antigen that has a complementary shape to the specific receptor on its surface, it is activated. The activated B-cell divides repeatedly, forming a clone.
→ Some of the clones of B-cells become plasma cells and secrete specific antibodies. Some of them remain as memory B-cells.
→ The first time an antigen is encountered, a primary immune response occurs. On the second encounter with the antigen, a secondary immune response occurs.
→ The secondary response is faster than the primary response and produces greater quantities of antibodies. This is because memory cells specific to that antigen are already present in the body.

11.2 Antibodies and vaccination

Plasma cells produce antibodies, or immunoglobulins. These are globular glycoproteins with a quaternary structure. All antibodies have a similar structure composed of four polypeptide chains (two heavy chains and two light chains) joined together by strong disulfide bonds to form a 'Y-shaped' molecule (Figure 11.11).

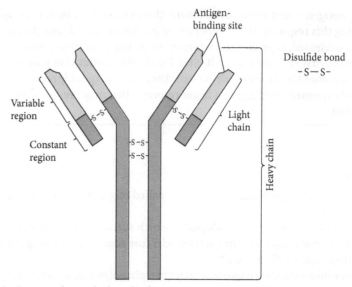

Figure 11.11 The structure of an antibody molecule.

Link

In Chapter 2, you learned how the primary structure of a polypeptide plays a key role in determining its function, by controlling the way it takes on a three-dimensional shape.

Tip

Remember that only enzymes have active sites, complementary to specific substrates. Antibodies have antigen-binding sites, which are complementary to specific antigens.

The stem of the Y (coloured orange in Figure 11.11) is called the **constant region** because in all immunoglobulins it has the same amino acid sequence, and therefore the same structure. This part of the antibody can be recognised by other leucocytes, as we will see later. The ends of the arms of the Y (coloured blue) are called the **variable regions** of the molecule because different immunoglobulin molecules have a different amino acid sequence in those regions and therefore different structures. The variable regions contain a highly specific **antigen-binding site**, which is complementary to only one antigen. Once attached, an **antigen–antibody complex** is formed, in a similar way to an enzyme–substrate complex.

Antibodies play a very important role in the body's defence against pathogens. For example, the attachment of many antibodies to the antigens on the surface of bacteria can cause the bacteria to clump together, effectively stuck to one another and unable to move or reproduce. This is called **agglutination**. Phagocytes are more able to identify agglutinated pathogens, and then destroy them by phagocytosis. There are many other ways in which antibodies are thought to limit a pathogen from causing disease. A number of these methods are shown in Figure 11.12.

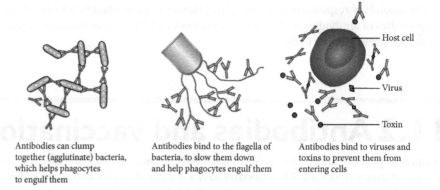

Antibodies can clump together (agglutinate) bacteria, which helps phagocytes to engulf them

Antibodies bind to the flagella of bacteria, to slow them down and help phagocytes engulf them

Antibodies bind to viruses and toxins to prevent them from entering cells

Figure 11.12 Antibodies disable pathogens and neutralise toxins in a variety of ways. Note that the diagrams are not to scale and the antibodies, as well as other cellular structures, have been simplified.

MONOCLONAL ANTIBODIES

Earlier in this chapter, you saw that once a particular B-cell has met a specific antigen, it is activated and divides rapidly by mitosis to form clones of genetically identical cells. Some of these differentiate into plasma cells, which secrete many copies of the antibody that can bind specifically with the antigen. Since the clone is produced from a single plasma cell, all the cells of the clone secrete the same, specific antibody. The antibody that they produce is therefore called a **monoclonal antibody**.

Scientists have developed methods to produce monoclonal antibodies in the laboratory. Monoclonal antibodies have many uses, and these are all based on the same principle. If they are mixed with a sample containing a mixture of antigens, the antibodies will bind specifically and tightly to only one complimentary antigen in the sample, as illustrated in Figure 11.13.

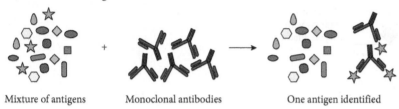

Mixture of antigens   Monoclonal antibodies   One antigen identified

Figure 11.13 The effectiveness of monoclonal antibodies in recognising one specific complimentary antigen is very useful.

This field of research has opened up many new techniques for treatment and diagnosis of medical conditions. Examples include pregnancy tests (see Experimental skills box 11.2), and many others, including:

- **Blood and tissue typing**. This may be necessary before an operation, for example, to ensure that the blood or organ to be provided to a patient will not cause an immune response and rejection.
- **Diagnosis of infectious disease**. Monoclonal antibodies can be produced that are complementary to the pathogen that causes a disease and can be used to test whether this pathogen is present in a sample of blood plasma or another biological fluid. This procedure is shown in Figure 11.14 and can be described in the following steps.

1. The biological fluid that is thought to contain the suspected antigen is placed into one of the wells in a plastic plate.
2. The samples are left for a short while, to give the antigen time to adhere to the plastic (it is immobilised).
3. An enzyme is attached to the monoclonal antibody specific to the pathogen to make a monoclonal antibody-enzyme complex.
4. The monoclonal antibody–enzyme complex is added to the wells, and then the plate is washed so that any unbound antibody–enzyme molecules are washed away and only those bound to the suspected antigen remain in place. If the suspected antigen is present in any of the wells, the monoclonal antibodies will bind to it.
5. The substrate of the enzyme is added. This is often a substance that changes colour when the enzyme acts on it. If the colour of the liquid in any of the wells changes, then this shows that the antibody–enzyme complex is present, meaning that it has bound to the specific antigen in the well.

a.

Virus particles are attached to a solid surface

The substrate for the enzyme is added and is changed to a product that has a different colour

Monoclonal antibody with enzyme attached

b.

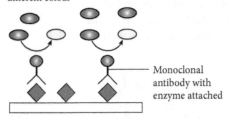

Key

◆ Virus (antigen)

⬭ Substrate for enzyme

⬭ Product

Figure 11.14 The process of diagnosis using monoclonal antibodies (a). This procedure is often carried out for a large number of people at once in a single plate with many wells. A colour change indicates infection; the wells that are colourless did not contain the virus, but those that are yellow did (b).

- **Location and treatment of cancer**
  Examples include:
  1. Monoclonal antibodies can be produced that have complementary shapes to specific abnormal proteins on the surface membrane of cancer cells. The monoclonal antibody molecules can be attached to a medicinal drug or particle that emits radiation. These will bind to the cancer cells and specifically deliver the drug to them – avoiding potential harm to healthy body cells that do not have the same antigen. Alternatively, an inactive drug can be injected into the bloodstream but this is only activated at the site of action by an enzyme attached to the monoclonal antibody.
  2. Monoclonal antibodies can be produced that attach to and block molecules on the surface of cancer cells that signal the cancer cell to divide.
  3. Radioactive or detectable chemicals can be attached to monoclonal antibodies and can be used to find the locations of cancer cells in the body.

Clearly, monoclonal antibodies have very important applications. However, B-lymphocytes are short-lived and can only be used to produce small quantities of the antibodies required. The work of German scientist Georges J. F. Köhler and Argentinian scientist César Milstein led to a method that is still used today to produce large quantities of monoclonal antibodies. Working in Cambridge, UK, the scientists were awarded a share of the 1984 Nobel prize in Physiology or Medicine for their work. Their procedure involved fusing the B-lymphocyte or plasma cell that produces the required antibody with a cancer cell (specifically, a malignant **myeloma** of a B-cell). It is summarised in Figure 11.15.

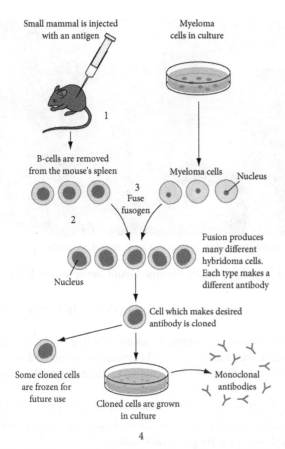

Figure 11.15 The procedure of Köhler and Milstein that is used to make monoclonal antibodies in the laboratory. Tumour cells, which divide uncontrollably, are fused with lymphocytes that produce antibodies. The hybrid cell produced shares the properties of the two cells that formed it. It divides continuously (it is immortal) and it produces a specific antibody. The structures shown in this diagram are not to scale.

1. Inject a mouse or other small mammal with the antigen for which complementary antibodies are required. The mouse will show a primary immune response and clonal selection will occur to produce B-cells with antibodies specific for that antigen.
2. After a few days, extract B-cells or plasma cells from the mammal's spleen or lymph nodes. These tissue samples will contain a mixture of thousands of different B-cells, each making their own specific antibodies, so we need to isolate the B-cell required.
3. Fuse the activated B-cells or plasma cells with myeloma cells using an electric current or a chemical substance called a **fusogen**, such as polyethylene glycol (PEG). This will produce **hybridoma cells**, which are separated and cultured individually to produce large numbers of clones.
4. The hybridoma cell clones are then screened to test for production of the specific antibody required and then these cells are grown in a culture flask or fermenter to 'scale up' production (Figure 11.16). They multiply by mitosis, making millions of identical cloned cells, each secreting identical antibodies, called monoclonal antibodies.

As a result of this method, the immortal hybridoma cells continue to divide and secrete antibodies over long periods of time and in high quantities.

Tip

Be careful when describing the injection of antigens into the small mammal. This is not vaccination, as the main aim of the procedure is not the production of memory cells.

Figure 11.16 A fermenter used to produce large quantities of monoclonal antibodies.

🔍 **12.** Suggest why B-lymphocytes are obtained from the spleen or lymph nodes, rather than the blood.

13. Explain the reason for the following steps in the production of monoclonal antibodies.

  a. A few days are allowed to elapse between injecting the small mammal with the antigens and removing cells from the spleen or lymph nodes.

  b. Cells producing antibodies are screened to identify whether the antibody they produce is able to bind to the antigen.

  c. Selected cells are taken from the well on a plate and grown in a fermenter.

⭐ **14.** Monoclonal antibodies attached to medicinal drugs are sometimes called 'magic bullets'. Suggest why they are given this name.

15. Some antibodies are described as polyclonal.

  a. Suggest the difference between monoclonal and polyclonal antibodies.

  b. State one advantage of using monoclonal antibodies in diagnostic tests for infectious diseases.

  c. State one advantage of using polyclonal antibodies in diagnostic tests for infectious diseases.

Experimental skills 11.2: Optimising the pregnancy testing kit

One very common use of monoclonal antibodies is in pregnancy test kits. During pregnancy, the developing embryo or fetus produces the hormone **human chorionic gonadotrophin (hCG)**. This enters the blood and urine of the woman and can be detected using the test kit (Figure 11.17).

Monoclonal antibodies complementary to hCG have been produced that also have coloured granules attached. In a simple version of the test kit, the appearance of a coloured strip in one window provides immediate visual confirmation of a pregnancy, after a sample of urine is applied. Figure 11.18 shows how a typical pregnancy test works.

Figure 11.17 A pregnancy testing kit. In the simplest version, a woman applies a sample of urine to a dipstick and two lines appear in a window if she is pregnant, as shown here. The presence of only one line (in the second window) would indicate that she is not pregnant.

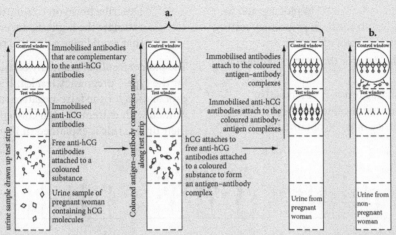

Figure 11.18 How a pregnancy testing kit works for a pregnant woman (**a**) and a woman who is not pregnant (**b**). The structures shown in this diagram are not to scale.

QUESTIONS

P1. Use Figure 11.18 to help you explain in detail:

   a. why the pregnancy test gives a positive result (two bands) if a woman is pregnant

   b. why the pregnancy test gives a negative result (one band) if a woman is not pregnant

   c. why immobilised monoclonal antibodies are used in the second window, despite the fact that they do not indicate whether a woman is pregnant.

Attempts have been made to improve the reliability of the pregnancy testing kit. One longstanding problem has been the fact that at very high concentrations of hCG, the test can give a negative result, even though the woman is pregnant. This is known as a false-negative result.

In a quality control investigation, a pharmaceutical company reviewed two types of pregnancy test that had been reported to give a false-negative result. They obtained 10 testing kits of each type and applied 1 ml (1 cm³) of five solutions of known hCG concentration to five kits. They used a method to determine the intensity of the second line in the window, which indicated pregnancy, and recorded a value. They called this value the positive signal intensity.

P2. Use the information provided to help you answer the following questions.

   a. State one reason why the results of the investigation were considered to be reliable.

   b. Suggest two ways in which the pharmaceutical company could increase the validity of their comparison of the test kits.

   c. Explain why a solution of concentration 0 mIU ml⁻¹ hCG was applied in the investigation into both test kits.

P3. In the investigation, a micropipette was used to measure and apply the solutions of 1 ml to the different pregnancy test kits. This piece of apparatus can precisely measure to the nearest 10 µl.

   a. Calculate the actual error (uncertainty) when using this micropipette.

   b. Calculate the percentage error when using this micropipette.

P4. The results of the investigation are shown in Table 11.2.

Concentration of hCG / mIU ml⁻¹	Positive signal intensity / arbitrary units	
	Type 1 pregnancy test kit	Type 2 pregnancy test kit
0	0.0	0.0
50	75.0	55.0
500	85.0	80.5
5.0×10^3	98.5	90.0
5.0×10^4	97.5	96.0
5.0×10^5	75.0	98.5
5.0×10^6	35.0	99.0

mIU ml⁻¹ = milli-international units per millilitre.

Table 11.2

a. Use the data in Table 11.2 to plot a graph to display the results for both pregnancy testing kits. Draw two curves of best fit.

b. Describe the data for the type 1 pregnancy test kit.

c. For a pregnancy test to give a positive result, a minimum value of 50% positive signal intensity must be obtained by the kit.

Describe how you could use your graph to estimate the highest concentration of hCG that would give a positive result with test kit 1.

The pharmaceutical company considered the results for the type 1 pregnancy test kit. It suggested an explanation for the false-negative results seen by some women with extremely high concentrations of hCG in their urine. This is explained in Figure 11.19.

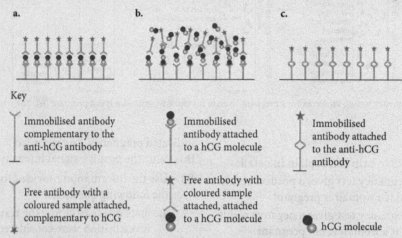

Figure 11.19 A normal positive pregnancy test in which molecules of hCG are recognised by immobilised antibodies and by soluble free antibodies with a coloured substance attached (a). However, if hCG is present at such a high concentration that it binds to both the fixed and the free antibodies, this could prevent the attachment of the antibodies carrying the coloured substance. The effect would be a false-negative result (b). In a true negative pregnancy test, the anti-hCG antibodies (which do not have hCG attached) are bound to the immobilised antibodies in the second window, as shown in (c).

★**P5.** Use Figure 11.19 to suggest why:

a. a false-negative result was obtained when a solution of concentration 5.0×10^6 mIU ml^{-1} was used with test kit 1.

b. the band in the second (control) window may give a higher positive signal intensity (appear darker) if a false-negative result occurs.

A Level analysis

In response to the findings of the investigation, the pharmaceutical company developed a third type of pregnancy testing kit that they claimed would provide a positive result (corresponding to a positive signal intensity of 75 arbitrary units or more) at very high concentrations of hCG. Ten replicates were conducted for the new test using a concentration of 5.0×10^6 mIU ml^{-1} hCG. Table 11.3 shows the results of the investigation.

Concentration of hCG / mIU ml^{-1}	Positive signal intensity of type 3 pregnancy test kit / arbitrary units				
5.0×10^6	Replicate 1	Replicate 2	Replicate 3	Replicate 4	Replicate 5
	98.5	96.0	88.0	96.0	96.0
	Replicate 6	Replicate 7	Replicate 8	Replicate 9	Replicate 10
	95.5	95.5	95.0	99.0	95.0

Table 11.3

P6. Identify one result in Table 11.3 that may be anomalous and state how the pharmaceutical company could deal with this anomaly.

P7. The pharmaceutical company decided to calculate the standard deviations of their results using the formula:

$$s = \sqrt{\frac{\sum(x - \bar{x})^2}{n - 1}}$$

where s = standard deviation, x = a result, $\bar{x}$ = mean, Σ = sum of, n = sample size.

Table 11.4 and use the formula above to calculate the standard deviation for the data in Table 11.3. Give your answer to two decimal places.

Replicate	x	$\bar{x}$	$(x - \bar{x})^2$
1	98.5		
2	96.0		
3	88.0		
4	96.0		
5	96.0		
6	95.5		
7	95.5		
8	95.0		
9	99.0		
10	95.5		
$\bar{x}$			
Σ			

Table 11.4 Calculating the standard deviation of a dataset. The unit of all values is mIU ml⁻¹.

P8. The pharmaceutical company decided to undertake a t-test to compare the significance of the difference between the results obtained with the type 3 pregnancy test kit and the result

obtained with the type 1 pregnancy test kit when an hCG concentration of 5.0×10^6 mIU ml⁻¹ was used.

a. State one feature of this data that allows the use of the t-test.

b. State one feature of this data that may not allow the use of the t-test.

☆c. The standard deviation for the positive signal intensity the type 1 pregnancy test kit when a hCG concentration of 5.0×10^6 mIU ml⁻¹ was 1.89 (arbitrary units). Using the formula below, calculate the t-value for this study. Give your answer to one decimal place.

$$t = \frac{|\bar{x}_1 - \bar{x}_2|}{\sqrt{\left(\frac{s_1^2}{n_1} + \frac{s_2^2}{n_2}\right)}}$$

where t = t-value, x_1 = mean of dataset 1, x_2 = mean of dataset 2, s_1 = standard deviation of dataset 1, s_2 = standard deviation of dataset 2, n_1 = number of samples in dataset 1, n_2 = number of samples in dataset 2.

☆d. Explain why, when referring to a data table to determine the significance of the difference between kit 1 and kit 3, the critical value found for 18 degrees of freedom should be chosen.

e. Describe how, using a probability table, the pharmaceutical company could conclude whether there is a significant difference between the performance of the two test kits.

Tip

You will have encountered probability tables previously for different statistical tests. An example is in the Experimental skills section of Chapter 2.

TYPES OF IMMUNITY

Earlier in this chapter you saw how the human body reacts to a pathogen when it infects the body. Provided the infection is removed, and the host survives, the immunological memory of a pathogen remains and the immune system will have some immunity for the future. Because the person is exposed to the antigen, and makes their own antibodies against it and retains memory cells for a long time, this is called active, natural immunity: **active** because an immune response has occurred to produce memory cells and permanent immunity; **natural**, because the immunity is gained by direct interaction between one organism and another.

In addition to active, natural immunity, there are a number of other mechanisms by which a body can gain immunity to a given pathogen. Table 11.5 summarises these types of immunity.

	Active immunity (antigens received)	Passive immunity (antibodies received)
Natural	achieved through the primary immune response following an infection	achieved through the passing of antibodies from mother to child through the placenta and milk
Artificial	achieved through injection of modified antigens (vaccination)	achieved through injection of antibodies

Table 11.5 Types of immunity.

Vaccination: active artificial immunity

When exposed to a pathogen, the body has a range of defence mechanisms that it uses to stay healthy. However, some bacteria and viruses cause significant harm while the primary response is happening, before antibody concentration increases, and the disease they cause may even kill or permanently disable the person. Examples of such diseases are cholera, smallpox and diphtheria. For these and other infectious diseases, vaccinations have been developed. They are usually administered via an injection into the bloodstream using a syringe (Figure 11.20). They are said to provide active, artificial immunity. As before, this is an active process because an immune response has occurred to produce memory cells and permanent immunity. It is **artificial**, because the immunity is not gained by direct interaction between one organism and another: the antigen responsible for the disease has not entered the body.

The basis of a vaccine is that it contains some form of the pathogen, or its antigens, so that it stimulates specific memory cells to develop. These will remain in the body for a long time, ready to destroy the real pathogen should it be met later in life. Obviously, the vaccine cannot simply be the pathogen itself, or the toxins it makes. Instead, a vaccine contains antigens that come from pathogenic organisms. Some vaccines contain just the antigen found on the surface of a pathogen, whereas some contain weakened, or **attenuated**, forms of the virus or bacterium that are no longer able to cause the disease.

When injected into an individual, the vaccination stimulates a primary immune response that results in the formation of memory cells. Later, possibly tens of years in the future, a secondary immune response is triggered if the individual is then infected by the same pathogen. In effect, the immune system is 'fooled' into making memory cells so that if the actual pathogen infects later, it is killed before it can cause the disease.

Vaccination programmes have been put in place for many years around the world and in one case – smallpox – what was once a widespread and debilitating disease has been completely eliminated. However, as long as most people are vaccinated, it does not usually matter if a few are not. This is because it would be difficult for a virus or bacterium to move from one non-vaccinated person to another, if there are not many of these non-vaccinated individuals in the population. This is sometimes called **herd immunity**. For example, it is estimated that as long as 94–95% of people are vaccinated against measles, it is very unlikely that anyone – even the non-vaccinated people – will get the disease. Herd immunity is vital in a population as it protects very young children, before they have reached an age at which they can be protected, and the elderly, who have a weakened immune system. However, in recent decades, some people have chosen not to provide their children with vaccines, on the basis of some claims that they have associated health risks. However, almost all of the studies that claim a health risk have been found to be false, or the risks are significantly outweighed by the benefits of not contracting the infectious disease.

Worked example

Spores of the bacteria that cause tetanus, *Clostridium tetani*, are found in soil, dust and animal faeces. When they enter a deep flesh wound, spores grow into bacteria that can produce a powerful nerve toxin.

Although many children are vaccinated against tetanus, booster vaccinations should be given every few years.

Figure 11.20 Vaccinations are an effective way of stimulating the body's own defences so that we need not suffer the infectious diseases, such as measles, mumps and whooping cough, that used to be a common feature of childhood.

a. Explain how vaccination can prevent the onset of symptoms in later life if a person is infected with spores from *Clostridium tetani*.

Answer

A vaccination introduces an antigen similar or identical to those of the actual pathogen into the bloodstream. In this case, the antigens are similar to the spores of *Clostridium tetani*. These are recognised as non-self by the body's immune system, which starts a primary immune response and produces memory cells that remain in the body well into the future. If *Clostridium tetani* spores later enter the body, the memory cells initiate a secondary immune response, which involves the production of antibodies at a faster rate and in higher numbers, which prevents the pathogen from causing damage to the body tissues and therefore avoids symptoms.

b. Sketch a graph, from age 0 to 25, to show the relative number of memory cells in the body of a 25-year-old adult who received their original vaccination for tetanus at age 2 but received a booster vaccine at age 20. There is no need to show values on the y-axis.

Answer

An example graph is shown in Figure 11.21.

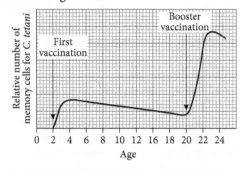

Figure 11.21

---

16. Explain why vaccinations are usually administered into the bloodstream, rather than orally (by mouth).

17. Many childhood vaccinations are given in two doses, separated by several months or years. Suggest why this is done.

18. The graph in Figure 11.22 shows the level of antibodies in a person's blood during a vaccination programme.

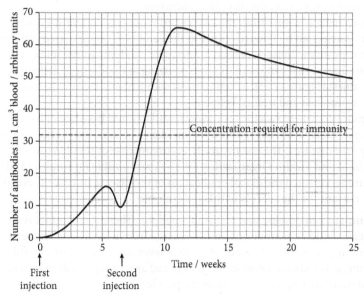

Figure 11.22

a. How long after the second injection did it take for the level of antibodies to reach the concentration required for immunity?
b. By how many arbitrary units did the antibody level rise after the second injection?
c. After several weeks, the concentration of antibodies may drop below the concentration required for immunity. However, the person may still be immune to the disease. Explain why.

Mother-to-infant: passive natural immunity

Table 11.5 showed that in addition to active immunity, it is possible to gain immunity passively. An example is the transfer of antibodies from mother to infant, which is known as passive natural immunity: **passive**, because an immune response has not occurred and no memory cells have been produced, resulting in temporary immunity; natural, because the immunity is gained by direct interaction between one organism and another.

When it is born, a baby emerges from the protective environment of its mother's uterus and is suddenly exposed to many potential pathogens. During the first few days of breastfeeding, the mother's breasts produce a high-protein, low-fat liquid called **colostrum**. This, and the breast milk that she produces later, contains many 'readymade' antibodies, produced by the mother as a result of infections and vaccinations that she has had during her lifetime. Antibodies pass across the placenta before birth and the supply continues through breast milk after the birth. Babies have very porous intestines that can absorb these large proteins directly into the bloodstream without digesting them.

Antiserum: passive artificial immunity

Passive immunity is also used to treat some types of poisoning, such as snake bites. This treatment is in the form of **antiserum**, which can be derived from blood that contains antibodies specific to a particular snake venom. The antiserum is produced in a similar way to monoclonal antibodies (Figure 11.15) or is prepared from the blood of a person who has previously been infected with the pathogen. This passive, artificial immunity can also be used to treat people who have been exposed to pathogens that cause diseases which rapidly affect the host's tissues. For example, antibodies can be injected that immediately neutralise the tetanus toxin, or prevent the rabies or hepatitis viruses from entering cells. This can prevent the person from falling ill and the disease spreading to others. Trials are also currently in progress for antiserum vaccines specific to the virus that causes Ebola, a disease that has had outbreaks in parts of West Africa in the past decade. Like the natural immunity transferred from mother to infant, artificial passive immunity is always temporary. This is because no memory cells are produced and the supply of injected antibodies will slowly fall in number.

The effectiveness of vaccination programmes

The term 'vaccination' comes from the Latin word *vacca* ('cow'). This is because the first vaccination, carried out by Edward Jenner in the early 19th century, used material taken from cowpox pustules. These were used to vaccinate children against smallpox, a much more serious disease that can result in severe scarring of the skin and lifelong disabilities such as blindness (Figure 11.23). In comparison with Jenner's, today's vaccines are thoroughly researched and trialled before use.

In communities where vaccines are widely available and have been given to more than 90% of people, vaccination has reduced the incidence of some previously very dangerous and widespread diseases. The last naturally occurring case of smallpox was diagnosed in 1977 and the World Health Organization (WHO), responsible for funding and organising the programme, certified the global eradication of the disease in 1980. There were a number of reasons for the success of this programme:

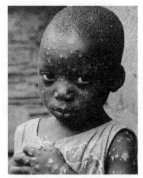

Figure 11.23 A child with smallpox from Central Africa. This photograph was taken in the 1960s, before smallpox had been eradicated.

- Only humans can develop and transmit smallpox. There is no animal reservoir where the disease could be maintained and re-emerge from later. This is one reason that vaccination programmes to eradicate yellow fever have been unsuccessful.
- Symptoms of infection emerge quite quickly and are readily visible, which allows medical teams to 'ring vaccinate' all of the people who may have come into contact with

the affected person. This is one reason that vaccination programmes to eradicate polio have been unsuccessful: the virus does not always cause readily identifiable symptoms.

- The smallpox virus did not persist in the body after an infection and become active later to form another reservoir of infection. This is one reason that vaccination programmes to eradicate HIV have been unsuccessful.

Unfortunately, not all vaccination programmes are likely to be as successful as that for smallpox. This is often owing to limited resources, and an inability to reach every person that needs vaccination. In many countries, the measles vaccine has been provided to more than 95% of children, resulting in herd immunity, but in others it is less than 50%. However, other reasons for the failure of vaccination programmes depend on the pathogen for which the vaccination has been developed:

- **Antigenic variation**. Some pathogens, such as the influenza virus, regularly change their antigens as a result of a high rate of genetic mutation. Others, such as the *Plasmodium* protoctist responsible for malaria, are eukaryotes and so have many genes that encode many different antigens, often expressed at different stages of its life cycle. Therefore, any memory cells produced from a previous encounter with a pathogen may no longer recognise the pathogens responsible for a later infection.
- **Antigenic concealment**. Some pathogens exist for a very short time in the blood plasma before they invade body cells, or exist entirely inside the digestive tract. This means that the leucocytes that need to recognise the antigens do not encounter them and therefore an immune response is not initiated. For example, *Plasmodium* spends part of its life cycle within host red blood cells and liver cells. The same is also true of pathogens that surround themselves with a portion of the cell surface membrane of their host, such as HIV.

Link

In Chapter 6, you saw how a mutation of the DNA base sequence can cause a change in the amino acid sequence of the encoded polypeptide, with effects on its function.

Worked example

In late 2015 and early 2016, a virus called Zika was linked to a significant increase in the number of babies born in Brazil with a serious brain defect called microcephaly. Figure 11.24 shows the number of pregnant women with Zika virus disease during a period in 2015–2016 and the number of babies born with microcephaly.

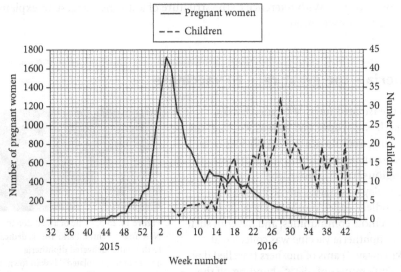

Figure 11.24 A graph showing how the numbers of pregnant women with Zika virus disease, and the number of children with microcephaly, varied between 2015 and 2016 in Brazil.

a. Calculate the number of weeks between the peak number of pregnant women with Zika virus disease and the peak number of children with microcephaly.

Answer

The peak number of children with microcephaly was seen at 27 weeks in 2016. The peak number of pregnant women with Zika virus disease was seen in Week 4 of 2016. The difference between these two time points is 23 weeks.

b. In response to the outbreak of Zika virus disease, urgent scientific research was undertaken to develop a vaccine that contained antigens from the Zika virus. Assuming that the Zika virus is responsible for causing microcephaly, explain how vaccinating a woman prior to pregnancy could prevent her future children being born with this defect.

Answer

Assuming the Zika virus causes microcephaly, the vaccination of a woman prior to pregnancy would result in a primary immune response specific to this antigen, and therefore the production of memory cells. If she is exposed to the Zika virus during pregnancy, then the antibodies complementary to this pathogen would be produced rapidly and in great numbers, which would prevent the virus from causing damage to the developing fetus.

---

19. The Zika virus is carried by mosquitoes. The increase in the number of children with microcephaly received widespread media attention in the first 2 months of 2016.
    a. You will need to refer to Figure 11.24 to help you answer this question.
       i. Calculate the percentage decrease in the number of cases of pregnant women with Zika virus disease between Week 5 and Week 12 of 2016.
       ii. Suggest reasons for this decrease.
    ★**b.** Some commentators stated that a correlation between the number of cases of Zika virus disease in pregnant mothers and the number of children with microcephaly does not necessarily imply a causative relationship. Evaluate this claim, using the data provided in Figure 11.24.
    c. An outbreak of Zika virus in French Polynesia in 2013–2014 resulted in a large number of adults developing an autoimmune disease called Guillain–Barré syndrome (GBS). The symptoms of GBS, which include muscle weakness, are due to antibodies mistakenly binding to self antigens on the cell surface membranes of neurones. With reference to the structure of antigens, suggest an explanation for this observation.

Assignment 11.1: Diphtheria – the great race to vaccinate

In the winter of 1925, a doctor in the isolated town of Nome, Alaska, detected an outbreak of diphtheria. It was winter and he was worried: the hospital had a limited stock of vaccine, and the nearest supply was over 2 weeks' journey away. The town's population of 10 000 residents was at serious risk from this highly contagious bacterial infection.

In response to his calls for help, a relay was started involving the best sled dogs (Figure 11.25) and their handlers, called mushers. These dogs were normally used to deliver mail to the remote communities, and never in such a hurry. On 27 January, a package of diphtheria vaccine was collected at the nearest train station, over 250 km away. Teams of mushers travelled day and night, enduring blizzards and temperatures of –50 °C, handing off the package to fresh teams. Eventually, just under 6 days later, the lifesaving race was completed – and the residents of Nome were saved.

Figure 11.25 A statue of a sled dog in New York's Central Park. This is dedicated to those that delivered diphtheria antitoxin to an isolated Alaskan town in the winter of 1925.

QUESTIONS

A1. Diphtheria causes a very swollen neck in infected individuals. Suggest why.

In the 1920s, an estimated 100 000 to 200 000 diphtheria cases occurred in the USA annually. Even with treatment, the disease killed as many as 13 000 to 15 000 people annually. The publicity from this outbreak prompted efforts to develop vaccines with a longer shelf life, and a national campaign in the 1930s to use diphtheria vaccine.

A2. Modified diphtheria toxin (called a toxoid) is used to vaccinate against diphtheria.

    a. Identify the type of immunity that is provided by a diphtheria toxoid vaccine.

        A. Artificial active

        B. Artificial passive

        C. Natural active

        D. Natural passive

    b. Suggest one way in which antibodies may prevent damage to body tissues caused by the diphtheria toxin.

    ⭐**c.** Suggest the advantages of using modified toxins in vaccines, rather than weakened pathogens.

    ⭐**d.** Chemicals called adjuvants are often mixed with toxoid vaccines to increase the size of the immune response caused by the antigen. This is especially the case if these vaccines are administered orally. Suggest one function of an adjuvant used in vaccines.

A3. The graph in Figure 11.26 shows the global number of cases of diphtheria and the global percentage of children immunised against diphtheria over a more recent 25-year period.

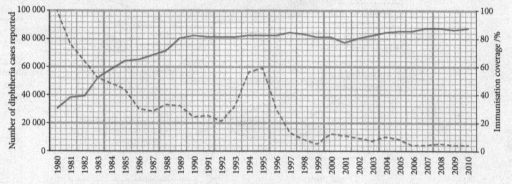

---- Number of cases   —— Global percentage of children immunised

Figure 11.26 Global cases of diphtheria and the global percentage of children immunised.

    a. Describe the relationship between the number of cases of diphtheria and the percentage of children immunised.

    b. Explain why it is not necessary to vaccinate every individual in a population for diphtheria to be eliminated from that population.

Key ideas

→ Monoclonal antibodies are identical antibodies made by a clone of plasma cells.

→ Monoclonal antibodies can be used to target medication to specific cell types.

→ Monoclonal antibodies can also be used to diagnose pregnancy or diseases, by detecting the presence of particular antigens in a sample.

→ Vaccination introduces harmless antigens into the body, causing a primary response and the production of memory cells, without causing illness.

→ Active immunity is the result of the response of the immune system to antigens, and involves the production of antibody and memory cells.

→ Passive immunity is the result of acquiring antibodies from another organism (for example, a baby's mother), and does not involve production of antibody or memory cells.

→ Herd immunity results if enough members of a population have immunity to a particular pathogen (for example, because they have been vaccinated against it) to make it difficult for the pathogen to pass from one non-immune person to another.

CHAPTER OVERVIEW

Review the mini mind map to link new learning with your prior learning, and to highlight which of the key concepts underpin the chapter.

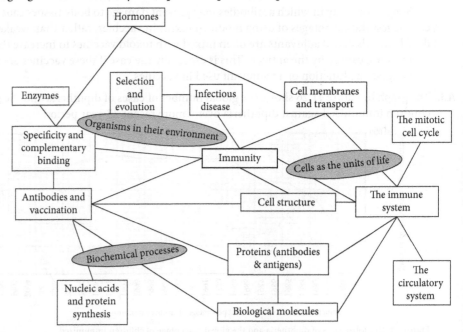

Try copying this mini mind map and expanding upon it. Use your notes from other chapters to help you explore how the essential ideas, theories and principles can be linked further together.

WHAT YOU HAVE LEARNED

Now that you have finished this chapter you should be able to:

- describe the mode of action of phagocytes (macrophages and neutrophils)
- explain what is meant by an antigen and state the difference between self antigens and non-self antigens
- describe the sequence of events that occurs during a primary immune response with reference to the roles of:
  - macrophages
  - B-lymphocytes, including plasma cells
  - T-lymphocytes, limited to T-helper cells and T-killer cells
- explain the role of memory cells in the secondary immune response and in long-term immunity

relate the molecular structure of antibodies to their functions
outline the hybridoma method for the production of monoclonal antibodies
outline the principles of using monoclonal antibodies in the diagnosis of disease and in the treatment of disease
describe the differences between active and passive immunity and between natural and artificial immunity
explain that vaccines contain antigens that stimulate immune responses to provide long-term immunity
explain how vaccination programmes can help to control the spread of infectious diseases.

CHAPTER REVIEW

1. Figure 11.27 shows how the immune system responds to a bacterium when it invades the body.
   a. Identify the process labelled **X**.
   b. The cell labelled **Y** is known as an antigen-presenting cell. Explain why antigen presentation is important in the immune response.
   c. Explain how the secretion of cytokines by T-helper cells is vital for an effective immune response.
   d. Identify the cells labelled **Z**.

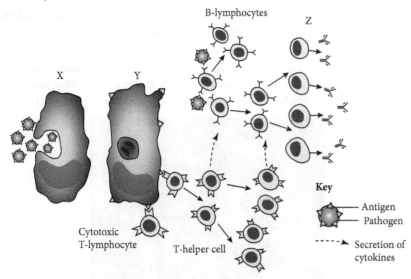

Figure 11.27 The structures shown in this diagram are not to scale.

2. Figure 11.28 shows the structure of an antibody molecule.

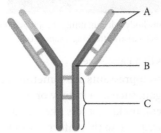

Figure 11.28

   a. With reference to sections **A**, **B** and **C** of the molecule, describe how the structure of an antibody is related to its function.

  b. Describe how an antigen–antibody complex is formed.

  c. Which statement describes how passive natural immunity is obtained?

    A. A vaccination containing dead microorganisms is given.

    B. An immunisation containing specific antigens is given.

    C. Antibodies are passed from mother to developing baby.

    D. Antibodies from another individual are injected.

  d. Name the type of bonds that hold the polypeptide chains together in the antibody structure.

3. In many countries, there appears to be a rising number of children who are born with an allergy to tree nuts such as cashews. After eating them, an allergic person can start an inappropriate immune response, which can lead to complications with breathing.

  a. Some evidence suggests that antigen-presenting cells may be responsible for triggering an inappropriate immune response. Explain how.

  b. One way in which people can be tested to see if they have a nut allergy is to screen a sample of their blood for antibodies complementary to antigens found in nuts. Explain why this method may not identify every person who is allergic to nuts.

4. Draw tables to compare the structure and functions of:

  a. B- and T-lymphocytes

  b. a primary immune response and a secondary immune response

  c. active and passive immunity.

5. Cancer is a disease in which body cells suffer mutation and become immortal (divide continuously) and disrupt normal body functioning.

  a. Suggest why most cancer cells are not eliminated by the immune system.

   New treatments for cancer have been trialled that consist of monoclonal antibodies that are able to recognise these cells. In some cases, after binding to the cancer cells, these antibodies are recognised by T-killer cells (Figure 11.33).

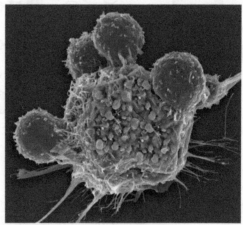

Figure 11.29 T-killer cells will destroy any body cell containing the antigen that is complementary to its specific receptors. Here, four T-killer cells (red) are attacking a cancer cell.

  b. Explain how a T-killer cell can kill a cancer cell.

  c. Medicinal drugs called immunosuppressants are sometimes given to patients who have received a transplanted organ, such as a kidney or heart. These drugs reduce the activity of the immune system, making it less likely that it will reject the organ by recognising its antigens as foreign. Use this information to explain why these drugs cause a slight increase in the risk of developing cancer.

6. Not all vaccines will give lifelong immunity to an infectious disease. Explain in detail why it is not possible to achieve lifelong immunity for some infectious diseases, such as influenza, by vaccination.

7. Match the processes with their corresponding definitions.

(a) Phagocytosis	(i) The display of antigens in the cell surface membrane by macrophages, which allows T-lymphocytes to more easily recognise the antigen
(b) Humoral response	(ii) The stimulation of the immune system that occurs when the body is first exposed to an antigen
(c) Cellular response	(iii) A process in which lymphocytes divide by mitosis to form a clone of identical lymphocytes
(d) Primary immune response	(iv) An immune response mediated by an antibody
(e) Secondary immune response	(v) An immune response involving T-cells
(f) Antigen presentation	(vi) The stimulation of the immune system that occurs when the body is exposed to an antigen for a second time
(g) Clonal selection	(vii) The process by which a cell engulfs large particles or whole cells, as a defence mechanism

Chapter
12

Energy and respiration

If your body mass is 65 kg, the energy contained in about 50 g of chocolate is sufficient for you to run a distance of 1 km! We eat foods containing carbohydrates for energy, but how does the chemical energy in the food get transferred within cells for activities such as running? During the 1 km run, you will use around 17 litres of oxygen. How is oxygen involved in releasing the energy from food?

Prior understanding

Aerobic and anaerobic respiration may have been described in one of your previous courses, along with the summary equations for these reactions. You may recall information about the structure of glucose from Chapter 2. The structure and function of mitochondria were described in Chapter 1, and ATP was covered in Chapter 6. You may also recall some information about oxidation and reduction reactions from a previous chemistry course.

Learning aims

In this chapter, you will learn about the requirement for energy in organisms and the role of ATP in transferring this energy. You will compare the energy contents of carbohydrates, lipids and proteins by calculating the respiratory quotient (RQ) for each of these. You will also describe how a simple respirometer can be used to measure RQ and rates of respiration. You will discover the four stages of aerobic respiration and learn where each of these occurs within the cell. The synthesis of ATP will be described as well as the roles of coenzymes such as NAD and FAD in respiration. Anaerobic respiration will also be described.

12.1 Energy (Syllabus 12.1.1–12.1.7)

12.2 Aerobic respiration (Syllabus 12.2.1–12.2.9, 12.2.11)

12.3 Anaerobic respiration (Syllabus 12.2.10, 12.2.12–12.2.14

12.1 Energy

Energy is a quantity that must be changed or transferred in order to make something happen. In cells, energy is needed for processes such as:

- active transport, where substances are moved against a concentration gradient
- **anabolic** reactions, such as those occurring in protein synthesis and DNA replication.

In whole organisms, energy is needed for:

- movement, where muscles convert chemical energy into kinetic energy
- generating heat, where chemical energy is converted to thermal energy through **exothermic** reactions.

ATP

ATP stands for adenosine triphosphate. ATP is often described as the universal energy currency in cells. This is because ATP is a source of energy for almost all cellular processes that require energy.

An ATP molecule is formed from adenine, ribose and three phosphate groups. The structure of ATP is shown in Figure 12.1.

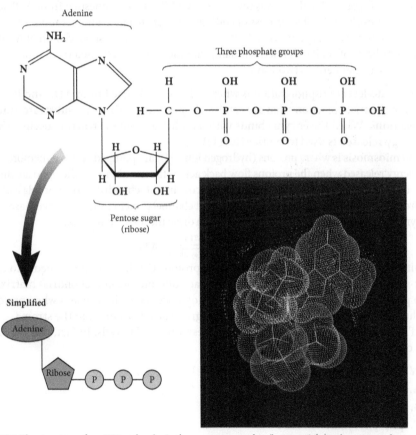

Figure 12.1 The structure of an ATP molecule. In the computer graphic (bottom right), adenosine is shown in blue, pentose in white and the phosphate groups in red.

Chemical energy is released from ATP when the covalent bond to the third phosphate group is hydrolysed. This reaction is catalysed by the enzyme ATP hydrolase, sometimes referred to as ATPase for short.

ATP hydrolysis forms **adenosine diphosphate**, **ADP**, and an inorganic phosphate ion, P_i. This reaction is shown in Figure 12.2.

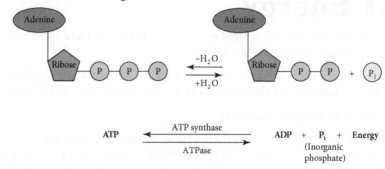

Figure 12.2 Hydrolysis of ATP releases chemical energy for the cell.

This reaction can be written

$$ATP \xrightleftharpoons[\text{ATPase}]{\text{ATP synthase}} ADP + P_i$$

Many other enzymes and proteins have ATPase domains. That means part of the enzyme or protein hydrolyses ATP to release the energy needed for the enzyme to catalyse the main reaction. An example of this is the active transport carrier protein that pumps sodium and potassium ions across cell surface membranes. The carrier protein hydrolyses ATP to obtain the energy to move the ions.

The reaction of ATP hydrolysis is reversible, so ATP can be synthesised from ADP and an inorganic phosphate ion. The process of adding a phosphate ion to another molecule is called phosphorylation, so the synthesis of ATP can be referred to as phosphorylation of ADP.

The synthesis of ATP is carried out using chemical energy released from other reactions. This happens in two ways:

- Substrate-level phosphorylation is where ATP is synthesised from ADP and P_i using chemical reactions that are catalysed by enzymes. These are **substrate-linked reactions**. When the enzyme binds with its substrate and the reaction occurs, the energy released is used to synthesise ATP.
- **Chemiosmosis** is where protons (hydrogen ions, H^+) are pumped across a membrane. The energy released when the protons flow back across the membrane in the opposite direction is used to synthesise ATP. Chemiosmosis occurs in mitochondria and chloroplasts. The protons flow across the membrane through a channel protein called ATP synthase. ATP synthase uses the energy from the protons to drive the forward reaction.

$$ADP + P_i \xrightleftharpoons[\text{ATP synthase}]{\text{ATPase}} ATP$$

In mitochondria, chemiosmosis occurs when protons that have been pumped into the inter-membrane space from the matrix flow back out into the mitochondrial matrix.

In photosynthesising plant cells, chemiosmosis occurs in chloroplasts when protons that have been released from reactions in the grana flow back out into the stroma.

ATP is continually being hydrolysed and resynthesised in cells. In fact, you synthesise your own body mass in ATP every day!

1. Which of the following describes ATP synthesis?
   A. ATP hydrolysis
   B. ADP hydrolysis
   C. phosphorylation of ADP
   D. phosphorylation of ATP
2. Write the equation for:
   a. ATP hydrolysis
   b. ATP synthesis.

ENERGY CONTENT OF FOODS

We will see in Section 12.2 that the hydrogen atoms in biological molecules are responsible for the energy available from aerobic respiration. Therefore, the greater proportion of hydrogen atoms in a molecule, the more energy can be released from it.

We can derive energy from carbohydrates, lipids and proteins (Figure 12.3). Lipids and fatty acids have the highest content of hydrogen atoms of these three types of molecule, so their energy content is highest.

Table 12.1 shows a comparison of the hydrogen atom content of three molecules.

Class of molecule	Formula of example	Proportion of hydrogen atoms in molecule
carbohydrate	$(C_6H_{10}O_5)_n$	0.48
protein	$C_{72}H_{112}N_{18}O_{22}$	0.50
lipid	$C_{56}H_{108}O_6$	0.64

Table 12.1 The higher the proportion of hydrogen atoms in a molecule the greater the energy content of that molecule.

Figure 12.3 Different food groups contain different quantities of energy.

3. A student has three food samples, each with a mass of 1.0 g.

   - olive oil
   - pasta
   - uncooked lean meat.

   List these foods in order from highest to lowest energy content.

RESPIRATORY QUOTIENT

Glucose is classed as a **respiratory substrate** because chemical reactions that release energy for ATP synthesis can start with glucose.

The overall, balanced, equation for the use of glucose in aerobic respiration is:

$$C_6H_{12}O_6 + 6O_2 \rightarrow 6CO_2 + 6H_2O$$

The equation shows that 6 moles of oxygen are needed and 6 moles of carbon dioxide are produced when 1 mole of glucose is used as a respiratory substrate in aerobic respiration.

One mole of any gas occupies the same volume under the same conditions of temperature and pressure. That means the volume of oxygen used and the volume of carbon dioxide produced in this reaction are equal.

We use the term **respiratory quotient** (RQ) to describe this. RQ is calculated using the equation:

$$RQ = \frac{\text{volume of } CO_2 \text{ released}}{\text{volume of } O_2 \text{ taken in}}$$

For glucose in aerobic respiration, the RQ value is 1.0 because the volume of CO_2 released is equal to the volume of O_2 taken in.

All other carbohydrates, such as starch and sucrose, are converted to glucose for respiration, so that means the RQ value for any carbohydrate in aerobic respiration is 1.0.

In general, respiratory substrates that contain fewer oxygen atoms have lower RQ values.

When lipids are used as an energy source, the fatty acids are used as respiratory substrates. As with glucose, the fatty acids will be completely oxidised to carbon dioxide and water.

Worked example

Olive oil contains a fatty acid with the formula $C_{18}H_{34}O_2$ whose oxidation is shown in the equation:

$$2C_{18}H_{34}O_2 + ...O_2 \rightarrow 36CO_2 + 34H_2O$$

a. Balance the equation by writing the number of molecules of oxygen required.

Answer

The number of oxygen atoms on the right-hand side is $(36 \times 2) + 34 = 106$.

There are already $2 \times 2 = 4$ oxygen atoms on the left-hand side in the fatty acid, so $106 - 4 = 102$ atoms required for balancing.

Each oxygen molecule contains two oxygen atoms, so $\frac{102}{2} = 51$ oxygen molecules required.

$$2C_{18}H_{34}O_2 + \underline{51}O_2 \rightarrow 36CO_2 + 34H_2O$$

b. Calculate the respiratory quotient of this fatty acid.

Answer

The RQ value of this fatty acid is therefore:

$$RQ = \frac{\text{volume of CO}_2 \text{ released}}{\text{volume of O}_2 \text{ taken in}}$$

$$RQ = \frac{36}{51} = 0.71$$

The value of 0.7 from the worked example is typical for the use of lipids as energy sources.

When proteins are being used as an energy source, the amino acids are used as respiratory substrates. Complete oxidation of the amino acid glycine is shown in the equation:

$$4C_2H_5NO_2 + 9O_2 \rightarrow 8CO_2 + 10H_2O + 2N_2$$

The RQ of this amino acid is therefore:

$$RQ = \frac{8}{9} = 0.89$$

Values in the range 0.8–0.9 are typical for the use of proteins as energy sources.

4. Palm oil is used in many foods. Palm oil contains the fatty acid palmitic acid. The complete oxidation of palmitic acid is shown in the equation:

$$C_{16}H_{32}O_2 + ... O_2 \rightarrow 16CO_2 + 16H_2O$$

   a. Balance the equation by calculating the number of moles of oxygen needed for the complete oxidation of palmitic acid.

   b. Calculate the RQ value of palmitic acid.

5. Protein can be used as an energy source. One of the amino acids in a protein is threonine. The complete oxidation of threonine is shown in the equation:

$$4C_4H_9NO_3 + 19O_2 \rightarrow ...CO_2 + 18H_2O + 4N_2$$

   a. Calculate the number of moles of carbon dioxide produced in this reaction.

   b. Calculate the RQ value of threonine.

Measuring the respiratory quotient

The respiratory quotient can be measured using a **respirometer**. A respirometer is shown in Figure 12.4.

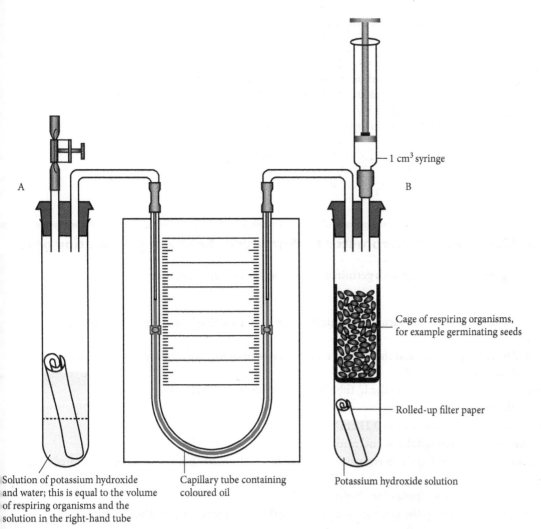

A

1 cm³ syringe

B

Cage of respiring organisms, for example germinating seeds

Rolled-up filter paper

Solution of potassium hydroxide and water; this is equal to the volume of respiring organisms and the solution in the right-hand tube

Capillary tube containing coloured oil

Potassium hydroxide solution

Figure 12.4 One type of respirometer.

The respirometer in Figure 12.4 measures the oxygen uptake of the respiring organisms that are placed in tube B. The potassium hydroxide solution at the bottom of tube B absorbs any carbon dioxide produced by respiration. Potassium hydroxide solution is caustic, so the organisms are supported in a cage above the solution. Rolled-up filter paper placed in the solution increases the surface area of the solution in contact with the air. This improves the efficiency of absorbing carbon dioxide.

The respiring organisms will use oxygen and produce carbon dioxide. The carbon dioxide that is produced will be absorbed by the potassium hydroxide solution. Therefore, the volume of air in tube B decreases. The coloured oil in the capillary U-tube will move upwards on the right, showing the volume change.

To measure the RQ of the organisms, first the rate of oxygen uptake must be found, usually in cm³ min⁻¹, using the method described above. We call this V_1.

Then, under the same conditions, the respirometer is set up again but without the potassium hydroxide solution. The same volume of distilled water can be used in its place.

If the level of the coloured oil in the U-tube stays the same, then the volume of carbon dioxide being produced is equal to the volume of oxygen being used, so the RQ is 1.0.

If the volume of carbon dioxide produced is greater than the volume of oxygen used, then the level of the liquid in the U-tube will go down on the right as the volume of tube B increases. This volume change in cm³ min⁻¹ is called V_2.

The apparatus in Figure 12.4

Tip

V_1 = volume of O_2 taken in, from part 1 of the investigation.

V_2 = volume of CO_2 produced minus the volume of O_2 taken in, from part 2.

So, $V_1 + V_2$ = total volume of CO_2 produced.

The RQ can then be calculated using the equation:

$$RQ = \frac{V_1 + V_2}{V_1}$$

If the volume of carbon dioxide produced is less than the volume of oxygen used, then the level of the liquid in the U-tube will go upwards on the right. This volume change in $cm^3\ min^{-1}$ is called V_3.

The RQ can then be calculated using the equation:

$$RQ = \frac{V_1 - V_3}{V_1}$$

Assignment 12.1: Using a respirometer to determine the RQ of germinating peas

The apparatus in Figure 12.4 was set up with germinating peas as the respiring organism.

QUESTIONS

A1. Temperature is one of the control variables in this investigation. Describe how temperature can be controlled.

Tube A was sealed by closing the tap and the syringe on tube B was used to adjust the level of the liquid in the U-tube to be equal on both sides.

After 15 minutes, the level of the liquid in the U-tube had gone up on the side closer to tube B.

A2. Explain why this happened.

The change in volume after 15 minutes was 0.21 cm^3.

A3. Calculate the rate of oxygen uptake in $cm^3\ min^{-1}$.

The investigation was repeated using the germinating peas in the respirometer but without potassium hydroxide solution.

A4. Suggest what could replace the potassium hydroxide solution.

During this part of the experiment, the germinating peas caused no change in volume after 15 minutes.

A5. Calculate the RQ for the germinating peas.

A6. Explain the conclusion that can be made from this result.

Sunflower seeds contain lipid as an energy store.

★**A7.** Explain how the observations would compare if the investigation with the germinating peas was repeated with germinating sunflower seeds. Include information about the investigation with and without potassium hydroxide solution.

Key ideas

→ The ATP molecule is described as the universal energy currency in cells.

→ ATP contains adenine, ribose and three phosphate groups.

→ ATP is synthesised from ADP and inorganic phosphate.

→ Hydrolysis of ATP yields ADP and inorganic phosphate.

→ The energy content of lipids is greater than that of protein, which in turn is greater than that of carbohydrate.

→ The respiratory quotient (RQ) is the ratio of the volume of carbon dioxide given out to the volume of oxygen taken in during respiration.

→ The RQ value of carbohydrate is 1.0, that of lipid is 0.7 and that of protein is in the range 0.8–0.9.

→ The RQ value for a respiring organism can be determined experimentally using a respirometer.

12.2 Aerobic respiration

Respiration is the name given to enzyme-catalysed reactions in cells that release energy from organic molecules.

Aerobic respiration in eukaryotic cells is often represented by the equation:

$$C_6H_{12}O_6 + 6O_2 \rightarrow 6CO_2 + 6H_2O$$

However, this equation is only a summary of many different reactions that occur in different parts of the cell.

In order to describe the reactions that occur in aerobic respiration, we divide the process into four stages:

- glycolysis
- link reaction
- Krebs cycle
- oxidative phosphorylation.

GLYCOLYSIS

The term **glycolysis** means splitting of a sweet-tasting substance. Glycolysis is a series of enzyme-catalysed reactions that occur in the cytoplasm of cells. Glycolysis has no requirement for oxygen, so it is common to both aerobic and anaerobic respiration.

Glucose contains a large quantity of chemical energy but is not very reactive. In order to be able to release this energy, glucose must be made more reactive. This is done by phosphorylation. Two ATP molecules are used to supply the two phosphate groups to the 6-carbon sugar. Although ATP is used at the start of glycolysis, more ATP is produced later, so there is a net gain of ATP in the process.

The phosphorylated glucose is changed to its isomer, fructose, in the process. Isomers are compounds with the same formula but with different chemical structures. It is the phosphorylated 6-carbon fructose that is split. This molecule is called fructose 1,6 bisphosphate. This results in two 3-carbon (triose) sugars, each of which is phosphorylated.

In the name fructose 1,6 bisphosphate, 'bis' means two. The numbers refer to the positions of the two phosphate groups on the molecule, so one phosphate is on the first carbon and one is on the sixth carbon, as shown in Figure 12.5.

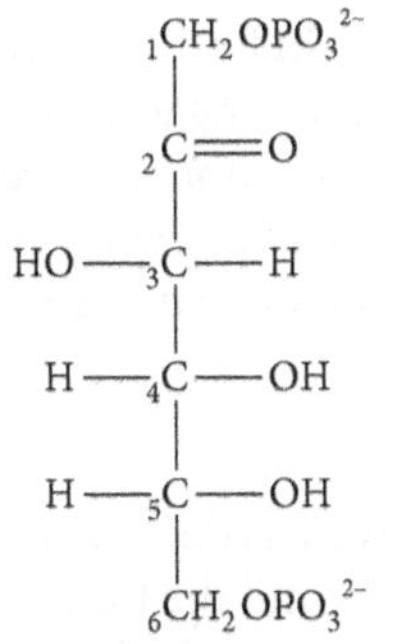

Tip

Respiration is often confused with breathing and gas exchange. These are three completely different processes.

Figure 12.5 The structure of fructose 1,6 bisphosphate showing the two phosphate groups on carbon numbers 1 and 6. You do not need to memorise this structure.

Link

The numbering of carbon atoms in molecules was introduced in Chapter 2.

Tip

In biological molecules, reduction can be recognised as the gain of hydrogen or an electron or the loss of oxygen.

Tip

In biological molecules, oxidation can be recognised as the loss of hydrogen or an electron or the gain of oxygen.

These two phosphorylated triose sugars each undergoes the same series of reactions, in which phosphate is lost, energy is released and hydrogen atoms are removed. When sufficient energy is released, ATP can be synthesised. There is sufficient energy released in the later stages of glycolysis for four ATP molecules to be made. Therefore, the net production of ATP in glycolysis is two molecules.

When the hydrogen atoms are removed from the two triose phosphates, two molecules of NAD are reduced. In these reactions, each of the triose phosphates is converted to a 3-carbon substance called **pyruvate**.

Key definition

→ **Coenzymes** are hydrogen carriers that act as **oxidising agents**. The coenzymes NAD and FAD are used in respiration, and the coenzyme NADP is used in photosynthesis. When these molecules accept hydrogen, they become reduced. When they release the hydrogen again, they become oxidised. They are called hydrogen carriers because they carry hydrogen atoms between the first three stages of respiration to the final stage. These hydrogen atoms, which have come from the respiratory substrate, are used to release most of the energy in aerobic respiration.

Figure 12.6 shows a summary of what happens in glycolysis.

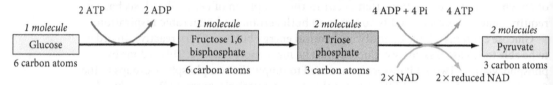

Figure 12.6 A summary of the reactions in glycolysis, which occur in the cytoplasm.

THE LINK REACTION

Pyruvate produced by glycolysis is toxic and must not be allowed to build up in the cytoplasm. In the presence of sufficient oxygen, pyruvate is taken by active transport into the matrix of mitochondria.

Link

When there is insufficient oxygen pyruvate is used in a different way, which will be explained in Section 12.3.

Key definition

→ **Decarboxylation** reactions occur in both the link reaction and in the Krebs cycle. Decarboxylation means removing a carboxyl group, which has the formula COOH. The carbon and two oxygen atoms are released as CO_2 and the hydrogen atom is used to reduce a coenzyme. The removal of the hydrogen is called **dehydrogenation** When a compound is decarboxylated, the number of carbon atoms drop by one.

In the matrix of mitochondria, pyruvate is decarboxylated, leading to the release of carbon dioxide, the reduction of NAD and the formation of a 2-carbon group called acetate. The acetate reacts with a molecule called coenzyme A. The product of this reaction is the 2-carbon compound called **acetyl coenzyme A**. This is called the **link reaction** because it links the process of glycolysis, which occurs in the cytoplasm, with the Krebs cycle, which occurs in the mitochondrial matrix.

The link reaction is shown in Figure 12.7.

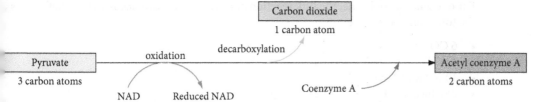

Figure 12.7 The link reaction occurs in the mitochondrial matrix.

THE KREBS CYCLE

When a biochemist called Hans Krebs and his co-workers established what happened in this stage of respiration, they named it the citric acid cycle because the first substance formed in this stage is citrate. As citric acid is a tri-carboxylic acid, the stage is also known as the tri-carboxylic acid (TCA) cycle. The stage was re-named the **Krebs cycle** in his honour and today is mostly known by this name.

The purpose of the Krebs cycle is to release hydrogen atoms from the respiratory substrate. These hydrogen atoms are taken by the coenzymes NAD and FAD to the final stage of respiration. The Krebs cycle occurs in the mitochondrial matrix.

The 2-carbon acetyl coenzyme A produced by the link reaction is attached to a 4-carbon compound called oxaloacetate. This forms a 6-carbon compound called citrate.

The cycle is formed when the 6-carbon citrate is converted back to the 4-carbon oxaloacetate and the acetyl coenzyme A group is released again. As this requires the loss of two carbons, two decarboxylation reactions are needed. This means two molecules of carbon dioxide are released. There are two hydrogen atoms released from these reactions and two additional hydrogen atoms lost from other steps in the cycle. The four hydrogen atoms that are released are accepted by three NAD molecules and one FAD molecule. This forms three reduced NAD and one reduced FAD. Also, in the cycle, one of the reactions yields sufficient energy for one ATP to be formed in substrate-level phosphorylation.

The reactions in the Krebs cycle are summarised in Figure 12.8.

> **Tip**
> The organic acids pyruvic acid and citric acid are usually referred to as pyruvate and citrate – the names of their ionised forms. This is because organic acids react with other substances when in their ionised forms.

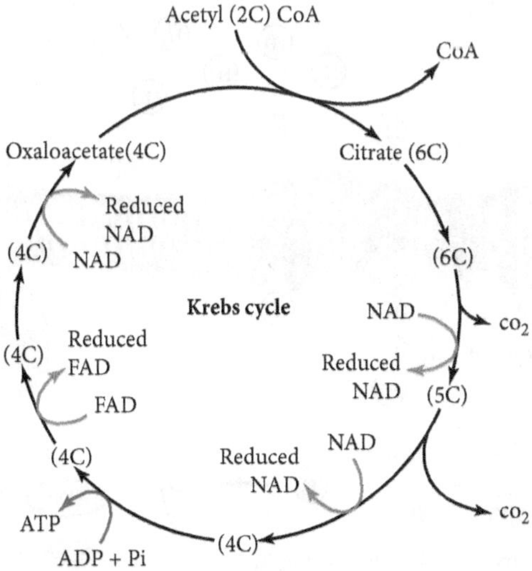

Figure 12.8 The Krebs cycle releases hydrogen from respiratory substrates and occurs in the mitochondrial matrix.

So far through the stages of glycolysis, the link reaction and the Krebs cycle, all of the carbon and oxygen atoms of the respiratory substrate, glucose, have been removed. Only the hydrogen remains, and that has been taken by NAD and FAD.

For one glucose molecule, the link reaction and Krebs cycle can occur twice. That means the total yields so far are:

- 6 CO_2
- 4 ATP
- 10 reduced NAD
- 2 reduced FAD.

OXIDATIVE PHOSPHORYLATION

The hydrogen from the reduced coenzymes NAD and FAD are used to synthesise ATP in the final stage of aerobic respiration. This happens through a series of reduction-oxidation (redox) reactions involving electron carrier proteins in the mitochondrial inner membrane.

The hydrogen atoms are separated into protons, H^+, and electrons, e^-. The electrons are used to reduce the first of the carrier proteins and the protons are released into the solution in the mitochondrial matrix. The electrons are passed between the carrier proteins, causing these proteins to be alternately reduced and then oxidised. This series of reactions is called the **electron transport chain**. Energy released in the electron transport chain as the electrons pass through is used to pump the protons into the intermembrane space of the mitochondria.

As the concentration of protons in the intermembrane space increases, a concentration gradient and an **electrochemical gradient** are set up. The protons are allowed to flow back into the matrix though proton channels that have ATP synthase domains. This process is called chemiosmosis and uses facilitated diffusion. The channels with ATP synthase are sometimes called stalked particles due to their shape. The electrical potential energy from the moving protons is used by the ATP synthase to phosphorylate ADP to form ATP.

The final acceptor of the electron is oxygen, which is present in the mitochondrial matrix. A proton is then used to reduce the oxygen to water. As hydrogen has become oxidised and ADP has been phosphorylated, the process is called **oxidative phosphorylation**.

Oxidative phosphorylation is summarised in Figure 12.9.

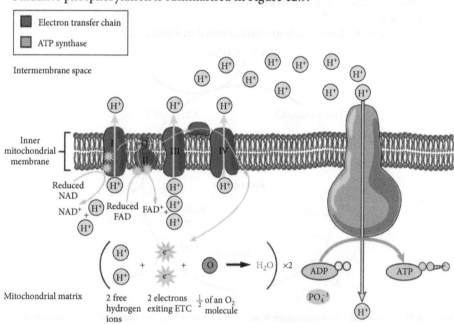

Figure 12.9 Oxidative phosphorylation occurs on the inner membrane of mitochondria.

A maximum of three ATP molecules can be made from one reduced NAD and a maximum of two ATP molecules can be made from one reduced FAD. However, some of

---

Tip

When NAD accepts hydrogen it forms reduced NAD, sometimes referred to as NADH. Reduced NAD, or NADH, can give up its hydrogen again, becoming oxidised back to NAD.

Tip

Mitochondria have two membranes: an outer membrane and an inner membrane. The inner membrane is folded to form cristae. The intermembrane space is the space between the inner and outer membranes.

Link

It may be helpful to look back at Chapter 4 for details of facilitated diffusion.

this ATP is used for the movement of ADP into mitochondria and for the movement of ATP out into the cytoplasm. This lowers the net yield to an average of:

- $2\frac{1}{2}$ ATP per reduced NAD

- $1\frac{1}{2}$ ATP per reduced FAD.

As we have 10 reduced NAD and 2 reduced FAD from the previous stages, this gives a net yield from oxidative phosphorylation of 28 ATP molecules per glucose molecule.

These are theoretical figures for ATP yield, and factors such as availability of ADP in the mitochondrial matrix can lower this yield.

The yield of ATP from one molecule of glucose entering glycolysis is summarised in Table 12.2.

Stage of respiration	Number of ATP molecules used	Number of ATP molecules synthesised	Net yield of ATP molecules
glycolysis	2	4	2
link reaction	0	0	0
Krebs cycle	0	2	2
oxidative phosphorylation	0	28	28
overall	2	34	32

Table 12.2 Summary of ATP use and synthesis in respiration from one molecule of glucose used as respiratory substrate.

Tip
You will not be required to recall the total numbers of ATP molecules produced in respiration.

6. State, precisely, where each of the following processes occur in the cell.
   a. glycolysis
   b. the link reaction
   c. the Krebs cycle
   d. oxidative phosphorylation
7. State the number of carbon atoms on one molecule of:
   a. glucose
   b. fructose 1,6 bisphosphate
   c. pyruvate
   d. acetyl coenzyme A
   e. oxaloacetate
   f. citrate.
8. Copy and complete the following equation:
   NAD + H →
9. When reduced NAD releases hydrogen to form NAD, what is this process called?
   A. oxidation
   B. reduction
   C. decarboxylation
   D. phosphorylation
10. What is the main purpose of the Krebs cycle?
   A. to produce ATP using substrate-level phosphorylation
   B. to make carbon dioxide for use in the cytoplasm
   C. to reduce NAD and FAD for use in oxidative phosphorylation
   D. to synthesise oxaloacetate and citrate for use in the cytoplasm

MITOCHONDRIA

Tip

The word 'mitochondria' is the plural form. The singular form is mitochondrion.

Mitochondria are double-membrane-bound organelles found in most eukaryotic cells. The structure of mitochondria is adapted to their function.

Mitochondria have their own DNA, which is circular and lies freely in the matrix in a similar manner to the way bacterial DNA lies in the prokaryotic cytoplasm. Mitochondrial DNA shares many similarities with bacterial DNA. This leads some biologists to think that part of the evolution of eukaryotic cells has involved taking in prokaryotic cells to assist with ATP synthesis – and this, they think, was the origin of mitochondria.

Mitochondrial DNA contains some of the genes necessary for aerobic respiration, but it also allows mitochondria to replicate independently of the cell. That means more mitochondria can be made within the cell if energy demands of the cell increase.

Figure 12.10 shows the structure of a mitochondrion and shows the positions where the reactions of aerobic respiration take place.

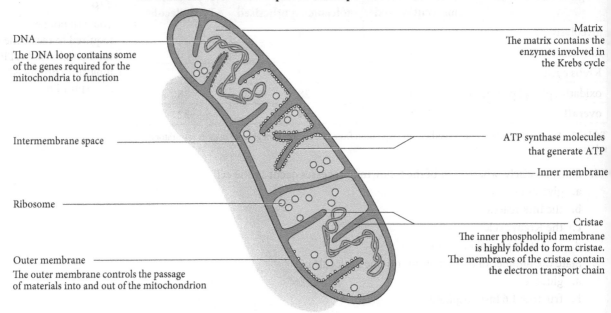

DNA
The DNA loop contains some of the genes required for the mitochondria to function

Intermembrane space

Ribosome

Outer membrane
The outer membrane controls the passage of materials into and out of the mitochondrion

Matrix
The matrix contains the enzymes involved in the Krebs cycle

ATP synthase molecules that generate ATP

Inner membrane

Cristae
The inner phospholipid membrane is highly folded to form cristae. The membranes of the cristae contain the electron transport chain

Figure 12.10 Mitochondrial structure and function.

Figure 12.11 shows a transmission electron micrograph of a mitochondrion.

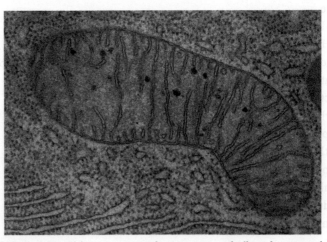

Figure 12.11 The high resolution of this transmission electron micrograph allows the inner and outer membranes of the mitochondrion to be seen.

11. Suggest why the inner membrane of mitochondria is folded.
12. **a.** Name two substances used in oxidative phosphorylation that will cross the mitochondrial outer membrane from the cytoplasm by active transport.
    b. Name one substance used in oxidative phosphorylation that will cross the mitochondrial outer membrane from the cytoplasm by diffusion through the lipid bilayer.
13. Most mitochondria have a shape similar to that in Figure 12.10. Explain why some mitochondria that have this shape appear circular in electron micrographs.
14. More ATP can be synthesised from one molecule of fatty acid than from one molecule of glucose in aerobic respiration. Explain why.

Key ideas

⇒ Respiration is a series of chemical reactions by which cells release energy from organic compounds.

⇒ Aerobic respiration is divided into four stages: glycolysis, the link reaction, the Krebs cycle and oxidative phosphorylation.

⇒ Glycolysis occurs in the cytoplasm, requires no oxygen and has a net production of two ATP, two reduced NAD and two pyruvate molecules per glucose.

⇒ The link reaction occurs in the mitochondrial matrix and converts pyruvate to acetyl coenzyme A while releasing CO_2 and reducing NAD.

⇒ The Krebs cycle occurs in the mitochondrial matrix and produces three reduced NAD, one reduced FAD and one ATP while releasing two molecules of CO_2.

⇒ Oxidative phosphorylation uses the hydrogens from reduced NAD and reduced FAD to synthesise ATP from the energy released from a series of redox reactions; the final electron acceptor is oxygen and water is formed as a result.

⇒ Mitochondria are the organelles for aerobic respiration, whose structure is adapted for this function.

12.3 Anaerobic respiration

Glycolysis has no requirement for oxygen, so can occur in aerobic or anaerobic conditions.

In anaerobic respiration, oxygen is limited so the respiratory substrate cannot be completely oxidised. That means the respiratory substrate cannot be completely broken down to release all the hydrogen atoms. As a consequence, the ATP yield from anaerobic respiration is only two molecules of ATP per glucose.

This yield may seem to be very low compared with the 32 ATP from aerobic respiration. However, the advantages of carrying out anaerobic respiration are:

* the cell can be kept alive in conditions of limited oxygen
* the process of ATP synthesis is faster than aerobic respiration
* the ability to regenerate NAD from reduced NAD.

The processes of anaerobic respiration are slightly different in animal cells and in yeast cells.

ANAEROBIC RESPIRATION IN ANIMAL CELLS

Glycolysis produces two molecules of pyruvate per molecule of glucose. Under anaerobic conditions, pyruvate remains in the cytoplasm. Pyruvate is converted to another 3-carbon compound called **lactate**. As this conversion is a reduction reaction, reduced

NAD is used as a **reducing agent** and NAD is regenerated as shown in Figure 12.12. The regeneration of NAD is important as it allows glycolysis to continue.

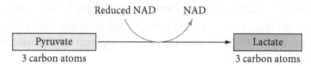

Figure 12.12 Conversion of pyruvate to lactate in animal cells regenerates NAD under anaerobic conditions.

The conversion of pyruvate to lactate in this way is called **lactate fermentation**.

Lactate is toxic to cells. Its build-up in muscles causes muscle fatigue and cramp. Lactate is removed from cells and taken to the liver, where the lactate can be broken down when sufficient oxygen is available.

Build-up of lactate is also a problem in organ transplants. Once an organ is removed from its blood supply, the cells no longer receive oxygen. Respiration in these cells will be anaerobic and the lactate produced cannot be removed. This can cause death of the cells in the organ if it is not used in a recipient's body quickly.

ANAEROBIC RESPIRATION IN YEAST CELLS

Yeast, *Saccharomyces cerevisiae*, is a eukaryotic unicellular fungus. Anaerobic respiration in yeast is one of the oldest applications of biotechnology as it has been used for over 5000 years in bread and wine production.

Yeast cells carry out glycolysis in the same way as animal cells and produce two molecules of pyruvate for every glucose molecule. Anaerobic respiration in yeast converts pyruvate to ethanal, which is a 2-carbon compound, and carbon dioxide. Ethanal is converted, by reduction, to ethanol; this regenerates NAD from reduced NAD. This process is called **ethanol fermentation** and is shown in Figure 12.13.

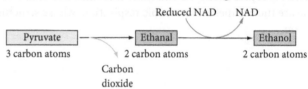

Figure 12.13 Conversion of pyruvate to ethanol and carbon dioxide in yeast cells regenerates NAD under anaerobic conditions.

The products ethanol and carbon dioxide are commercially important, not just in bread, wine and beer production, but also for industrial alcohol and alcohol for use as fuel in road vehicles. Fuels with between 5% and 85% added ethanol are available in most countries (Figure 12.14). The ethanol is produced using yeast cells under anaerobic conditions to ferment cane sugar.

Figure 12.14 The fuel from this pump, called E85, is 85% ethanol and 15% petrol (gasoline).

15. Which of these is **not** a product of anaerobic respiration in animal cells?
    A. NAD
    B. ATP
    C. CO_2
    D. lactate

16. Explain why the yield of ATP per glucose molecule is much lower in anaerobic respiration than in aerobic respiration.

17. Write a word equation for anaerobic respiration in yeast using glucose as a respiratory substrate.

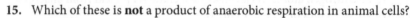 18. When an organ is removed from an organ donor, the organ is kept on ice and used in transplant surgery as soon as possible, otherwise the tissues in the organ will be damaged. Explain what causes the tissues to be damaged.

ADAPTATIONS OF RICE

Cells in plant roots obtain energy from respiration, which is mostly aerobic. In order to gain sufficient oxygen, soil needs to be well aerated. That means soil has many air spaces for gas exchange to occur between cells in the root and the atmosphere. During flood conditions, soil becomes waterlogged. This means that plant root cells cannot gain sufficient oxygen and so anaerobic respiration occurs. Anaerobic respiration in plants is similar to that in yeast, where ethanol and carbon dioxide are produced. Ethanol is toxic to plant cells, so its build-up during flood conditions can be fatal to plants.

Rice, *Oryza sativa*, is a crop plant that is adapted to grow in waterlogged soil.

One of the adaptations of *O. sativa* is the development of a tissue called **aerenchyma** in roots (Figure 12.15). Aerenchyma is spongy tissue formed from a combination of cell death and lysis to create many air spaces in the root. The presence of aerenchyma means that gas exchange can occur at a sufficient rate to support aerobic respiration in the roots of *O. sativa*.

Oryza sativa has a hollow stem. The air space in the middle of the stem provides a channel for gas exchange between the atmosphere and parts of the plant that are submerged. Figure 12.15 is a photomicrograph of a transverse section of a stem of *O. sativa*. Some varieties of rice plant have the ability to elongate the stem rapidly when the water level rises. This keeps the top of the stem above water so that gas exchange through the hollow part of the stem can continue.

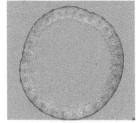

Figure 12.15 The rice plant has a hollow stem to facilitate gas exchange in parts of the plant that are under water.

In addition, cells in the roots of *O. sativa* are more tolerant to ethanol than those of plants growing in aerated soils. This tolerance means that ethanol is not toxic to the cells until it reaches a much higher concentration. Being tolerant to ethanol in this way enables the cells in *O. sativa* roots to perform anaerobic respiration for longer than plants that are adapted to grow in aerated soils. Tolerance to ethanol is achieved by the cells producing an enzyme called **alcohol dehydrogenase**, which converts ethanol back to ethanal using NAD as an oxidising agent.

19. List three adaptations of the rice plant, *Oryza sativa*, to grow in waterlogged soil.

Experimental skills 12.1: Investigating respiration in yeast

Measuring the rate of respiration using an indicator

The indicators DCPIP and methylene blue are called **redox indicators**. When the indicators are oxidised they are blue, and as they gain electrons and become reduced they turn colourless. Either of these indicators can be used to measure the rate of respiration in yeast, as the indicator will accept the electrons from glucose as it is oxidised. The indicator will become reduced in a similar way to NAD becoming reduced during respiration.

Effect of temperature on rate of respiration

A student set up three test-tubes as follows:

Test-tube 1 – 1 g of yeast suspended in distilled water with two drops of methylene blue

Test-tube 2 – 1 g of yeast suspended in 20% glucose solution with two drops of methylene blue

Test-tube 3 – 1 g of boiled yeast suspended in 20% glucose solution with two drops of methylene blue.

QUESTIONS

P1. Explain the purpose of each of the test-tubes.

The three test-tubes were placed in a water bath at 25 °C, as shown in Figure 12.16.

The student measured the time taken for the blue colour to disappear in tube 2.

The investigation was repeated at 30, 35, 40, 45 and 50 °C.

The results are shown in Table 12.3.

Temperature / °C	Time to go colourless / s
25	540
30	425
35	235
40	275
45	325
50	375

Table 12.3

The time taken to go colourless in Table 12.3 can be converted to a rate of respiration using the equation:

$$\text{rate of respiration} = \frac{1}{\text{time taken to go colourless}}$$

when the unit of time is s, the unit of rate of respiration is s^{-1}.

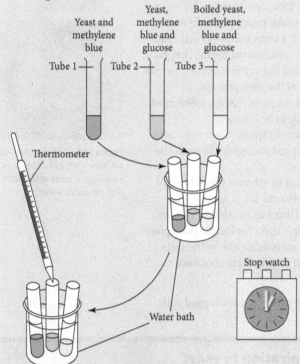

Figure 12.16

P2. Draw a new table showing the rate of respiration at each of the temperatures shown in Table 12.3. Express the rates in standard form.

P3. Plot a graph of rate of respiration against temperature. Complete your graph by drawing a smooth curve of best fit.

P4. The student concluded that the optimum temperature for respiration in these yeast cells is in the range 30–40 °C. Describe how the optimum temperature could be determined more accurately.

P5. One problem in this investigation is judging exactly when the blue colour disappears. An alternative method is to remove samples from tube 2 at regular intervals and place the samples into a colorimeter.

A colorimeter gives a numerical value of the intensity of a colour in a solution compared to another solution called a blank. The blank solution is made by adding all the components of the solution except the component that causes the colour. The colorimeter is set to zero using the blank.

a. Suggest what would be added to make the blank for this investigation.

b. Explain why the use of a colorimeter could give better results in this investigation.

P6. Another investigation that can be carried out using this method is to determine how the rate of respiration depends on the concentration of glucose.
Outline a plan for this investigation.

Key ideas

→ Anaerobic respiration occurs in cells when insufficient oxygen is available.
→ Anaerobic respiration has a much lower yield of ATP than aerobic respiration but enables the regeneration of NAD.
→ Anaerobic respiration in animal cells converts pyruvate to lactate.
→ Anaerobic respiration in yeast cells converts pyruvate to ethanol and carbon dioxide.
→ The rice plant, *Oryza sativa*, shows adaptations to enable aerobic respiration in parts of the plant that are under water, including ethanol tolerance and presence of aerenchyma in stem and roots.

CHAPTER OVERVIEW

Review the mini mind map to link new learning with your prior learning, and to highlight which of the key concepts underpin the chapter.

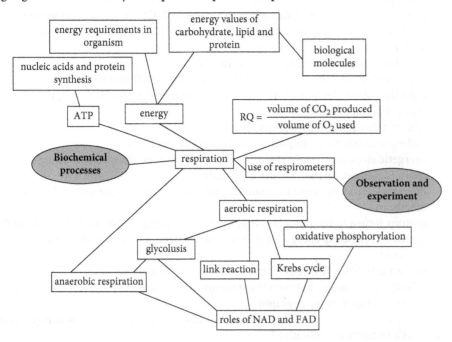

Try copying this mini mind map and expanding upon it. Use your notes from other chapters to help you explore how the essential ideas, theories and principles can be linked further together.

WHAT YOU HAVE LEARNED

Now that you have finished this chapter you should be able to:

- outline the need for energy in living organisms, as illustrated by active transport, movement and anabolic reactions, such as those occuring in DNA replication and protein synthesis
- describe the features of ATP that make it suitable as the universal energy currency
- state that ATP is synthesised by:
    - transfer of phosphate in substrate-linked reactions
    - chemiosmosis in membranes of mitochondria and chloroplasts
- explain the relative energy values of carbohydrate, lipid and protein as respiratory substrates
- state that the respiratory quotient (RQ) is the ratio of the number of molecules of carbon dioxide produced to the number of molecules of oxygen taken in, as a result of respiration
- calculate RQ values of different respiratory substrates from equations for respiration
- describe and carry out investigations, using simple respirometers, to determine the RQ of germinating seeds or small invertebrates
- state where each of the four stages in aerobic respiration occurs in eukaryotic cells:
    - glycolysis in the cytoplasm
    - link reaction in the mitochondrial matrix
    - Krebs cycle in the mitochondrial matrix
    - oxidative phosphorylation on the inner membrane of mitochondria
- outline glycolysis as phosphorylation of glucose and the subsequent splitting of fructose 1,6-bisphosphate (6C) into two triose phosphate molecules (3C), which are then further oxidised to pyruvate (3C) with the production of ATP and reduced NAD

- explain that, when oxygen is available, pyruvate enters mitochondria to take part in the link reaction
- describe the link reaction, including the role of coenzyme A in the transfer of acetyl (2C) groups
- outline the Krebs cycle, explaining that oxaloacetate (4C) acts as an acceptor of the 2C fragment from acetyl coenzyme A to form citrate (6C), which is converted to oxaloacetate in a series of small steps
- explain that reactions in the Krebs cycle involve decarboxylation and dehydrogenation and the reduction of the coenzymes NAD and FAD
- describe the role of NAD and FAD in transferring hydrogen to carriers in the inner mitochondrial membrane
- explain that during oxidative phosphorylation:
  - hydrogen atoms spit into protons and energetic electrons
  - energetic electrons release energy as they pass through the electron transport chain (details of carriers are not expected)
  - the released energy is used to transfer protons across the inner mitochondrial membrane
  - protons return to the mitochondrial matrix by facilitated diffusion through ATP synthase, providing energy for ATP synthesis (details of ATP synthase are not expected)
  - oxygen acts as the final electron acceptor to form water
- describe the relationship between the structure and function of mitochondria using diagrams and electron micrographs
- outline respiration in anaerobic conditions in mammals (lactate fermentation) and in yeast cells (ethanol fermentation)
- explain why the energy yield from respiration in aerobic conditions is much greater than the energy yield from respiration in anaerobic conditions (a detailed account of the total yield of ATP from the aerobic respiration of glucose is not expected)
- explain how rice is adapted to grow with its roots submerged in water, limited to the development of aerenchyma in roots, ethanol fermentation in roots and faster growth of stems
- describe and carry out investigations, using redox indicators, including DCPIP and methylene blue, to determine the effects of temperature and substrate concentration on the rate of respiration of yeast
- describe and carry out investigations, using simple respirometers, to determine the effect of temperature on the rate of respiration.

CHAPTER REVIEW

1. List three processes in cells that require energy from respiration.
2. a. Explain what is meant by the term 'respiratory quotient'.
   b. Explain why the respiratory quotient for lipid differs from that of carbohydrate.
3. a. Name the 3-carbon substance that is the end-product of glycolysis.
   b. Complete these sentences using the correct numbers.
      When one glucose molecule enters glycolysis, ... ATP molecules are used. Later in glycolysis .... ATP molecules are produced. In addition, ... molecules of NAD are reduced in glycolysis.
4. a. Explain the role of FAD in aerobic respiration.
   b. NAD is synthesised from a compound called niacin (vitamin B3), which is present in the diet. NAD is in continuous use in aerobic respiration, yet the dietary requirement for niacin is less than 0.02 g per day. Suggest an explanation for this.
5. a. Name the 6-carbon compound formed in the Krebs cycle by adding an acetyl group onto oxaloacetate.

   b. This 6-carbon compound is used to regenerate oxaloacetate. Explain how this is done in terms of carboxyl groups.

   c. ATP is formed in the Krebs cycle. Give the name for this type of ATP synthesis.

6. a. Explain how chemiosmosis is used to synthesise ATP in mitochondria.

   b. Name the final electron acceptor in the process of oxidative phosphorylation.

   c. Name the product formed when the final electron acceptor is reduced.

7. a. Summarise the process of anaerobic respiration in yeast using glucose as a respiratory substrate.

   b. Explain how the roots of the rice plant, *Oryza sativa*, can survive under water while the roots of many other plants would die.

8. a. Methylene blue is an indicator that can be used in respiration experiments. Name one other indicator that can be used in respiration experiments and state the colours of this indicator under the conditions in a respiration experiment.

   b. Name the equipment used to measure the rate of respiration.

   c. Potassium hydroxide solution or sodium hydroxide solution is often used in respiration experiments. State the purpose of using one of these solutions in respiration experiments.

Chapter 3

Photosynthesis

The largest trees on Earth include the giant sequoia, shown here. The dry mass of these trees can exceed 500 tonnes, but where does this mass come from? Many people think that plants obtain their mass from the soil, but only a tiny proportion of the mass of a plant comes from minerals in the soil. Plants obtain their mass for growth from carbon dioxide gas in the air. As plants are producers in the food chain, this growth process is essential for all eukaryotic life on Earth. But how does it work?

Prior understanding

You may recall some basic facts about photosynthesis, including the equation for the process from previous courses. You may also recall information about the structure of plant cells and chloroplasts from Chapter 1. The synthesis and hydrolysis of ATP was covered in Chapter 12, as was the role of hydrogen-carrying coenzymes and reduction-oxidation reactions.

Learning aims

In this chapter, you will learn how the structure of a chloroplast is related to its function. You will also discover the two stages of photosynthesis – the light-dependent and light-independent stages – and find out where each stage occurs in the chloroplast and what happens in each stage. You will learn about the pigments that are responsible for absorbing light. You will also learn about the processes of ATP and carbohydrate synthesis and the factors that limit the rate of photosynthesis.

13.1 Photosynthesis and energy transfer
(Syllabus 13.1.1–13.1.12)

13.2 Factors limiting the rate of photosynthesis
(Syllabus 13.2.1–13.2.4)

13.1 Photosynthesis and energy transfer

In eukaryotic cells, photosynthesis occurs in chloroplasts. The process of photosynthesis occurs in two stages, which happen in separate parts of the chloroplast. Figure 13.1 shows an overview of the complete process of photosynthesis.

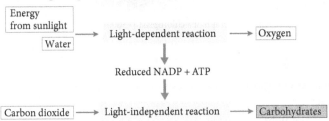

Figure 13.1 Overview of photosynthesis.

> **Tip**
>
> A granum is one stack of thylakoid membranes, and grana is the term for more than one stack of thylakoid membranes.

CHLOROPLASTS

Chloroplasts are double-membrane-bound organelles with an oval shape measuring approximately 7–10 µm in length.

The structure of a chloroplast is shown in Figure 13.2. The stroma is the site of enzyme-catalysed anabolic reactions that use carbon dioxide for the synthesis of carbohydrate. The grana (singular granum) are made from thylakoid membranes. The thylakoid membranes are the sites where light energy is harnessed to make ATP.

> **Tip**
>
> The word stroma is very similar to stoma, which is the singular form of stomata, so take care with spelling these words.

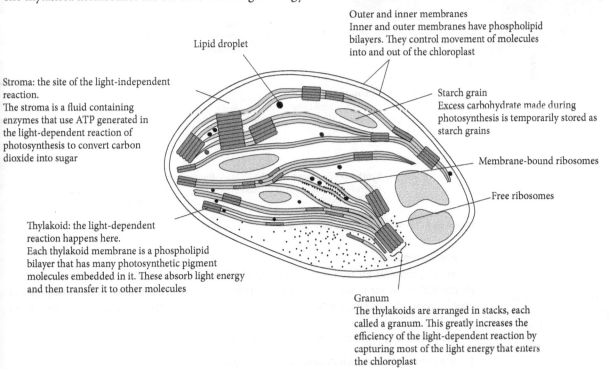

Figure 13.2 Chloroplast structure.

Figure 13.3 is a transmission electron micrograph of a chloroplast from maize, *Zea mays*.

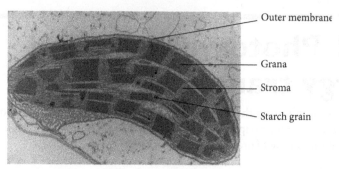

Figure 13.3 Transmission electron micrograph of a chloroplast [×5000].

THE LIGHT-DEPENDENT STAGE

The **light-dependent stage** of photosynthesis occurs in the thylakoids. The light-dependent stage:

- uses energy from light
- synthesises ATP
- reduces a coenzyme called NADP
- uses water to supply hydrogen and electrons
- produces oxygen as a waste product.

Figure 13.4 shows an overview of the light-dependent stage.

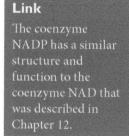

Link

The coenzyme NADP has a similar structure and function to the coenzyme NAD that was described in Chapter 12.

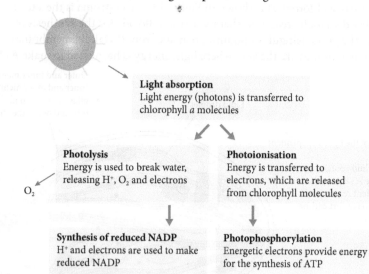

Figure 13.4 Overview of the light-dependent stage of photosynthesis.

Photoactivation

There are pigments located within the thylakoid membranes that absorb light. Each pigment absorbs some wavelengths better than others. The use of more than one pigment enables more wavelengths of light to be absorbed, so more efficient use can be made from sunlight.

The important pigments are divided into two groups:

1. the chlorophylls – **chlorophyll a** and **chlorophyll b**
2. the carotenoids – **carotene** and **xanthophyll**.

The pigments are organised into groups called **photosystems** within the membrane. There are two photosystems: **photosystem I** and **photosystem II**.

Tip

Photosystems I and II are named in the order in which they were discovered, and not in the order in which they are involved in the light-dependent stage.

The graph in Figure 13.5 is called an **absorption spectrum** as it shows the relative absorption of each pigment across the wavelengths present in visible light. The absorption spectrum is made using separate, purified, pigments.

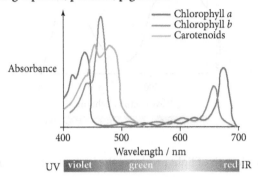

Figure 13.5 The absorption spectrum of photosynthetic pigments.

The graph in Figure 13.6 has a line called an **action spectrum** because it shows the rate of photosynthesis at different wavelengths. An action spectrum is often shown for the whole leaf or for whole chloroplasts, as photosynthesis will not occur only in the presence of a purified pigment. This is because a variety of other molecules are required to harness the energy captured by the pigment. Figure 13.6 is a comparison of an absorption spectrum and an action spectrum for the same leaf.

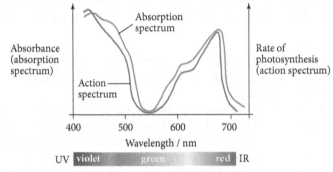

Figure 13.6 Absorption spectrum and action spectrum for a whole leaf.

Worked example

The graph in Figure 13.5 is drawn from results obtained from three photosynthetic pigments that have been extracted from chloroplasts and purified. Explain why the results used for the graph in Figure 13.6 cannot be obtained from purified pigments.

Answer

The graph in Figure 13.5 only shows the absorption of each separate pigment between 400 and 700 nm. The graph in Figure 13.6 includes an action spectrum, which shows how the rate of photosynthesis varies between 400 and 700 nm. The rate of photosynthesis cannot be determined by using purified pigments, as whole chloroplasts or whole leaves are needed. This is because other parts, such as the thylakoid membranes and the stroma are required for photosynthesis. The other line in Figure 13.6 is the absorption spectrum for the whole leaf. This spectrum includes absorption from other leaf components, such as cells walls, and not just photosynthetic pigments.

1. **a.** Explain why an action spectrum can be determined for isolated chloroplasts but not for isolated photosynthetic pigments.
   b. The action spectrum for a leaf is shown in Figure 13.7.

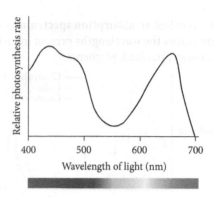

Figure 13.7

i. Which wavelength of light produces the highest rate of photosynthesis in this leaf?

ii. Suggest why the leaf used to produce this action spectrum appears green.

Light of the correct wavelength is absorbed by a complex of very many photosynthetic pigment molecules. The energy transferred to these pigment molecules is, in turn, passed to a chlorophyll molecule. An electron from chlorophyll uses this energy from the light to leave its orbital. This is called **photoexcitation**. The excited electron is taken by an electron acceptor. As the electron has a negative charge, it leaves behind a positively charged chlorophyll ion. The event where the electron is taken by an electron acceptor is called **photoionisation**. It is the energy of these electrons that is used for ATP synthesis.

Photophosphorylation

The use of light energy to phosphorylate ADP forming ATP is called **photophosphorylation**. This process can be cyclic or non-cyclic, depending on the fate of the electron that was released in the photoionisation step.

Cyclic photophosphorylation uses only photosystem I. Light causes electrons to be excited and removed from chlorophyll *a*. These electrons are passed through a chain of electron carriers before returning to the chlorophyll *a*. During the electron transfer process in the carriers, enough energy is released by the electrons to later enable the formation of ATP from ADP and inorganic phosphate. (The section below on chemiosmosis will explain how.) The process is called cyclic because electrons are removed from chlorophyll *a* and the electrons return to chlorophyll *a*. Cyclic photophosphorylation is summarised in Figure 13.8.

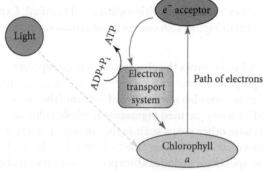

Figure 13.8 Cyclic photophosphorylation.

Non-cyclic photophosphorylation uses both photosystems. It is called non-cyclic because the electrons are not returned to the pigment from which they were removed.

Photolysis occurs first in photosystem II. This is where a water molecule is split, as shown in the equation

$$H_2O \rightarrow 2H^+ + 2e^- + \tfrac{1}{2}O_2$$

Photolysis is catalysed by the oxygen-evolving complex. This is an enzyme that is surrounded by the proteins of photosystem II. When light energy is absorbed by the pigments in photosystem II, this energy is passed to the oxygen-evolving complex to enable the splitting of water. Photolysis is the source of oxygen, which is a waste product of photosynthesis. The two electrons are transferred to photosystem II. When it absorbs light energy, the electrons become 'excited'. Instead of returning to photosystem II, they pass through an electron transport chain to photosystem I. As they do so they yield energy which will later enable the formation of ATP. (The section below on chemiosmosis will explain how.) When light is absorbed by photosystem I, the electrons are excited again and this time used, with the hydrogen ions from water, to reduce NADP.

Non-cyclic photophosphorylation is summarised in Figure 13.9.

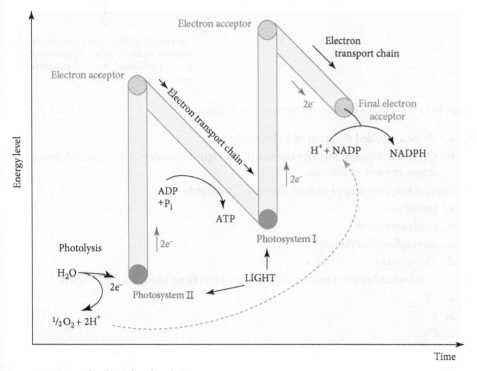

Figure 13.9 Non-cyclic photophosphorylation.

In the case of both cyclic and non-cyclic photophosphorylation, chemiosmosis is used to synthesise ATP. Energy from the electron transport chains is used to pump protons across the thylakoid membranes and into the spaces between these membranes. This builds up a high concentration of hydrogen ions in the thylakoid space. Protons are then allowed to flow out into the stroma of the chloroplast by facilitated diffusion through ATP synthase, which turns ADP into ATP. This process is summarised in Figure 13.10.

Link

The use of an electron transport chain to move protons across a membrane and the process of chemiosmosis was described in Chapter 12 in the context of mitochondria.

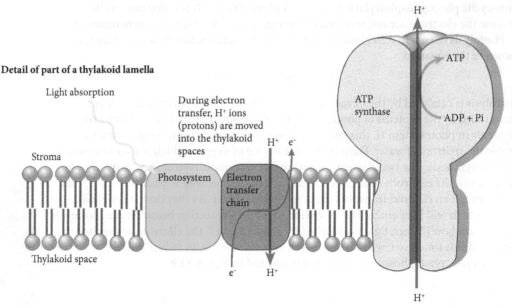

Detail of part of a thylakoid lamella

Light absorption

During electron transfer, H⁺ ions (protons) are moved into the thylakoid spaces

Stroma

Photosystem

Electron transfer chain

Thylakoid space

ATP

ATP synthase

ADP + Pi

The passage of the protons through the ATP synthase molecule causes a conformational (tertiary structure) change that leads to the synthesis of ATP from ADP and Pi

Figure 13.10 Use of chemiosmosis to synthesise ATP in chloroplasts.

2. a. Draw a labelled diagram of a chloroplast.
   b. On your diagram, identify the site of the light-dependent and light-independent stages of photosynthesis.

3. Explain the meaning of each of the following words.
   a. photolysis
   b. photoactivation
   c. photophosphorylation
   d. photosystem

4. Describe what happens to each of these products from the splitting of water.
   a. H^+
   b. e^-
   c. O_2

Experimental skills 13.1: Chromatography of photosynthetic pigments

Health and safety: propanone liquid and vapour is highly flammable, causes drowsiness and dizziness if vapour is inhaled, and causes dryness and cracking of skin on repeated contact with liquid. Use as small a volume as possible. Use in a well-ventilated location. Do not use near naked flames. Wear eye protection and avoid contact with skin.

A student obtained a green leaf from spinach and a green seaweed.

The student extracted the photosynthetic pigments from each of these samples using the following procedure.

- Some green tissue, some sand and a small volume of propanone (acetone) were crushed together using a pestle and mortar.
- Chromatography paper was prepared by drawing a pencil line close to the bottom of the paper. The

letter X was placed, using a pencil, at the positions on the line where the samples from the spinach and the seaweed would be placed.

- A small volume of propanone was placed into a beaker to a depth less than the height of the pencil line from the bottom of the chromatography paper.
- The green extract from the crushed sample was spotted onto the pencil X using the head of a large pin.
- Each spot was allowed to dry before applying more of the extract to the same position. This was repeated many times, to make a small very dark green spot of pigment.
- The chromatography paper was placed into the beaker as shown in Figure 13.11.
- The chromatography paper was removed when the propanone from the bottom of the beaker had travelled almost to the top of the paper.

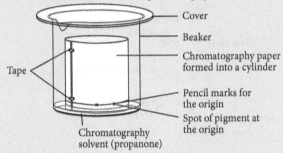

Figure 13.11

The position of the X is called the **origin** and the position on the chromatography paper reached by the propanone is called the **solvent front**.

The pigments are carried different distances from the origin. This is due to differences in solubility in the solvent, differences in molecular sizes and differences in attraction to the paper.

The results, called **chromatograms**, are shown in Figure 13.12.

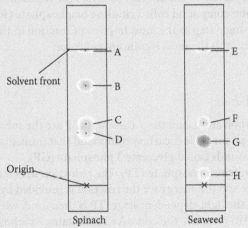

Figure 13.12

QUESTIONS

P1. Suggest why:
   a. the chromatography paper should only be handled with forceps or when wearing gloves
   b. the origin line and X are drawn with pencil and not with pen
   c. the position of the origin must be above the level of the solvent in the beaker.

P2. Explain why it is **not** important to use the same mass of samples from the spinach and the seaweed.

P3. Use the letters in Figure 13.12 to identify:
   a. pigments that are common to both the spinach and the seaweed
   b. pigments that are present in the spinach but not in the seaweed
   c. pigments that are present in the seaweed but not in the spinach.

P4. In chromatography, separated pigments are identified by their R_f **values**. The R_f value of a pigment is calculated using the equation:

$$R_f = \frac{\text{distance travelled by pigment}}{\text{distance travelled by solvent}}$$

Both distances are measured from the origin line.
   a. Calculate the R_f values of each of the pigments in Figure 13.12. For the height of the pigment, measure to the pencil dot in the middle of the pigment spot.

The R_f values of various photosynthetic pigments in propanone are shown in Table 13.1.

Pigment	R_f value
beta carotene	almost 1
chlorophyll *a*	0.45
chlorophyll *b*	0.38
chlorophyll *c*	0.01
fucoxanthin	0.32
xanthophyll	0.74

Table 13.1

   b. Use your R_f values to identify each of the pigments in Figure 13.12.

P5. Describe one difficulty in obtaining an accurate measurement of the distance travelled by the pigments.

THE LIGHT-INDEPENDENT STAGE

The **light-independent stage** of photosynthesis occurs in the stroma. It is also called the **Calvin cycle**. The light-independent stage:

- is a series of enzyme-catalysed reactions
- uses ATP and reduced NADP from the light-dependent stage
- uses carbon dioxide from the atmosphere
- synthesises carbohydrate.

The light-independent stage is so named because light is not directly required for any of the reactions in this stage. However, the reactions in this stage require ATP and reduced NADP from the light-dependent stage. That means the light-independent reactions can occur in the dark, but only for a limited time as the store of ATP and reduced NADP will eventually be used up.

The light-independent stage is summarised in Figure 13.13.

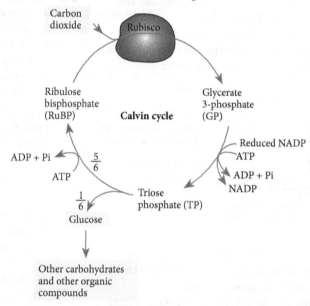

Figure 13.13 The light-independent reactions.

Carbon fixation is the first part of the light-independent stage. As the reactions occur in a cycle, there is no true starting point, but it is logical to begin with carbon fixation.

Carbon dioxide through the atmosphere diffuses into chloroplasts from the cytoplasm of the plant cell. The enzyme **rubisco**, present in the stroma, catalyses the reaction between carbon dioxide and a 5-carbon compound called **ribulose bisphosphate (RuBP)**. Some biologists have referred to this single step as the most important reaction in the biosphere. This is because carbon fixation is the most significant route for:

- taking CO_2 out of the atmosphere
- getting carbon into food chains
- getting chemical energy into food chains.

Carbon dioxide, which contains 1 carbon atom, and the 5-carbon RuBP are the substrates for rubisco. The product of rubisco is an unstable 6-carbon compound that immediately breaks down into two 3-carbon compounds called **glycerate 3 phosphate (GP)**.

GP is reduced to form the 3-carbon **triose phosphate (TP)**. The reduction uses reduced NADP from the light-dependent stage and the energy for the reaction is provided by ATP hydrolysis. The ATP also comes from the light-dependent stage. TP is then used, with energy from ATP, to regenerate RuBP to complete the cycle. As TP contains 3 carbon atoms and RuBP contains 5, then not all of the TP produced is required to regenerate

Tip

The word 'fixation' means a reaction where a gas, in this case carbon dioxide, is attached (or fixed) onto another molecule that is not in the gaseous state.

RuBP. Five out of every six TP molecules are used to regenerate RuBP, and the remainder of the TP molecules are available to synthesise glucose.

Glucose produced in the light-independent stage can be used as a respiratory substrate in the cell or as the starting point for the synthesis of other carbon-containing compounds such as starch, cellulose, and lipids. TP can be used as a starting point for the synthesis of amino acids. Amino acids contain nitrogen that comes from nitrate ions taken in through plant roots.

5. Describe the reaction catalysed by rubisco.
6. Describe how RuBP is regenerated from GP in the Calvin cycle.
7. The enzyme rubisco is not very efficient, and the rate of the catalysed reaction of rubisco is slow. Attempts have been made to artificially modify rubisco to increase its efficiency. Explain the importance of this to agriculture.
8. The typical rate of reaction of rubisco is 3 reactions per second. Assuming $\frac{1}{6}$ of TP molecules can be used to synthesise glucose, calculate the number of glucose molecules that can be made in 20 seconds per rubisco enzyme. Show your working.

Key ideas

→ Chloroplasts are the organelle of photosynthesis and their structure is adapted to this function.
→ The thylakoid membranes, organised into grana, are the site of the light-dependent reactions.
→ The pigments chlorophyll a, chlorophyll b, carotene and xanthophyll are present in thylakoid membranes and absorb light.
→ The pigments are characterised by their different absorption spectra, whereas the rate of photosynthesis at different wavelengths is characterised by an action spectrum.
→ The light-dependent reactions use both cyclic and non-cyclic photophosphorylation to synthesise ATP.
→ Non-cyclic photophosphorylation uses electrons and protons from water to reduce NADP.
→ The stroma is the site of the light-independent reactions.
→ The enzyme rubisco catalyses carbon fixation, in which carbon dioxide reacts with RuBP to form two molecules of GP.
→ GP is reduced to TP using reduced NADP and ATP.
→ Energy from ATP is used to regenerate RuBP from TP.
→ $\frac{5}{6}$ of TP is used to regenerate RuBP.
→ $\frac{1}{6}$ of TP is used to synthesise glucose.
→ Glucose can be used in respiration or as the starting point for synthesis of other organic compounds.
→ TP can be used as a starting point for the synthesis of amino acids.

13.2 Factors limiting the rate of photosynthesis

A **limiting factor** in any process is something that restricts the speed at which the process can occur. An increase in availability of a limiting factor will result in an increase in the overall speed, or rate, of the process. When one factor is limiting, increases in the availability of any of the other factors will have no effect on the overall speed of the process.

Three of the limiting factors for the rate of photosynthesis are:

• light intensity

- carbon dioxide concentration
- temperature.

Control of these limiting factors is important in agriculture, and often plant crops are grown indoors so that these factors can be monitored and changed. If any one factor is limiting, plants will not grow at the maximum possible rate. Increased plant growth rates mean higher profits for growers and a higher rate of food production.

LIGHT INTENSITY

Light intensity is seen as the brightness of light, but it is actually a measure of the light energy per unit area arriving at the plant.

Changes in light intensity can be due to:

- time of day
- weather conditions, e.g. whether there is cloud or fog
- time of year in temperate zones
- presence of geographical features or competing plants.

When there is zero light intensity, or complete darkness, there can be no photosynthesis, so the rate of photosynthesis is zero. As light intensity increases, the rate of the light-dependent reactions can increase, as more energy is available. Hence there is an increase in the rate of photosynthesis. At this point, light intensity is limiting because increasing it will increase the rate of reaction.

As light intensity increases further, the reactions will reach their maximum possible rates under those conditions. This means any further increases in light intensity will not produce any further increase in the rate of photosynthesis. At this point, other factors such as carbon dioxide concentration or temperature become limiting.

Figure 13.14 shows how the rate of photosynthesis depends on light intensity when all other factors are kept constant.

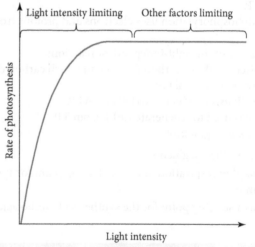

Figure 13.14 Light intensity as a rate-limiting factor in photosynthesis.

Figure 13.15 The bright artificial lights in this greenhouse mean that light is not a rate-limiting factor for photosynthesis.

Light intensity is controlled in greenhouses by installing bright, artificial lights (Figure 13.15).

CARBON DIOXIDE CONCENTRATION

The carbon dioxide concentration in air is about 0.04%. However, when plants first evolved on Earth, the carbon dioxide concentration in air was much higher, around 0.5%.

In many parts of the world today, such as the tropics, temperature and light intensity may not be limiting factors for the rate of photosynthesis, but the concentration of carbon dioxide may be limiting.

When plants are grown indoors in a structure such as a greenhouse, the carbon dioxide concentration can fall significantly below 0.04%. This is because the growing plants are removing carbon dioxide from the air and it is not being replaced as the building is closed to the outside.

As with light intensity, there can be no carbon fixation if there is zero carbon dioxide concentration because rubisco lacks one of its substrates. As carbon dioxide concentration increases, so the rate of photosynthesis can increase. If changing the concentration of carbon dioxide affects the rate of photosynthesis, then carbon dioxide concentration is said to be limiting.

As carbon dioxide concentration increases further, rubisco will reach its maximum rate of reaction under those conditions. Any further increase in carbon dioxide concentration will produce no further increase in the rate of photosynthesis. At this point, other factors are limiting.

Figure 13.16 shows how the rate of photosynthesis depends on carbon dioxide concentration when all other factors are kept constant.

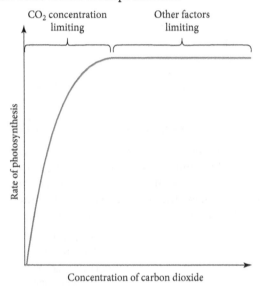

Figure 13.16 Carbon dioxide concentration as a rate-limiting factor in photosynthesis.

Greenhouse heaters can be of the oil-burning type. Oil-burning heaters produce carbon dioxide as a waste product, which raises the carbon dioxide concentration in the greenhouse.

TEMPERATURE

An increase in temperature will increase the rate of chemical processes such as enzyme-catalysed reactions. Physical processes, such as photoactivation, are not affected by temperature changes within the range found in plant habitats. Therefore, changes in temperature have a greater effect on the light-independent reactions than on the light-dependent stage.

An increase in temperature increases the kinetic energy of molecules, so increasing the frequency of collisions between the active sites of enzymes and their substrates. Enzyme–substrate complexes will form more frequently, so the rate of product formation will be greater.

Beyond the optimum temperature for enzymes, a further increase in temperature will cause denaturation of the enzyme. This causes a change in the shape of the active site so it is no longer complementary to the shape of the substrate and the rate of reaction decreases.

Link

The effect of temperature on the rate of enzyme-catalysed reactions was described in Chapter 3.

The optimum temperature for enzymes in plants varies according to the plant and its habitat, but the optimum temperature for photosynthesis in many plants is in the range 20–30 °C.

Figure 13.17 shows the effect of temperature on the rate of photosynthesis.

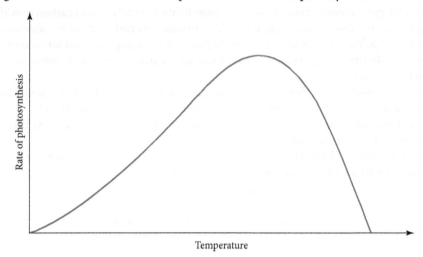

Figure 13.17 Temperature as a rate-limiting factor on the rate of photosynthesis.

CHANGES TO MORE THAN ONE LIMITING FACTOR

When experiments are carried out to measure the rate of photosynthesis with changes to one limiting factor, all other limiting factors are kept constant. However, these experiments can be repeated with a change to one of the other limiting factors.

For example, Figure 13.18 shows the effect of changing light intensity on the rate of photosynthesis at three different carbon dioxide concentrations.

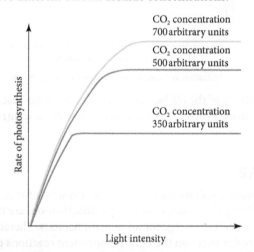

Figure 13.18 Effects of changing light intensity and carbon dioxide concentration on the rate of photosynthesis.

A graph of the effect on photosynthetic rate of changing light intensity at different temperatures would have a similar form to that in Figure 13.18 provided the temperatures did not result in enzyme denaturation.

Figure 13.19 shows the effect of changing temperature on photosynthetic rate at different light intensities.

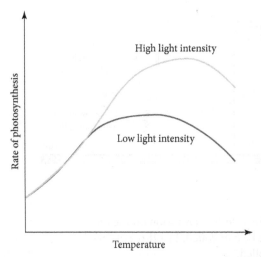

Figure 13.19 Effects of changing temperature and light intensity on the rate of photosynthesis.

A graph of changing temperature at different carbon dioxide concentrations would have a similar form to that in Figure 13.19.

9. On a hot sunny day just after rain, what is the most likely limiting factor for the rate of photosynthesis in a plant?

A	temperature
B	light intensity
C	availability of soil water
D	carbon dioxide concentration

10. An electric heater and an oil-burning heater can each heat the same greenhouse to the same temperature. Explain why plants grow faster when the oil-burning heater is used.

11. **a.** Sketch a graph of the effect of changing temperature on the rate of photosynthesis when other factors remain constant.
    b. Add another line on your graph to show the effect of temperature at a higher carbon dioxide concentration.

Key ideas

→ Light intensity, carbon dioxide concentration and temperature are rate-limiting factors for photosynthesis.
→ Increasing light intensity increases the rate of photosynthesis up to a maximum, after which the rate no longer increases.
→ Increasing carbon dioxide concentration increases the rate of photosynthesis up to a maximum, after which the rate no longer increases.
→ Increasing temperature increases the rate of photosynthesis up to the optimum, after which the rate decreases due to enzyme denaturation.

Assignment 13.1: Investigating photosynthesis

1. Use of a whole plant

A student used a 5 cm length of *Elodea* pondweed to investigate the effect of carbon dioxide concentration on the rate of photosynthesis. *Elodea* lives under water. Adding different masses of potassium hydrogen carbonate to the water will vary the carbon dioxide concentration in the water.

The apparatus was set up as shown in Figure 13.20.

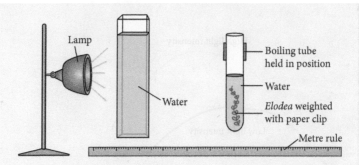

Figure 13.20

QUESTIONS

A1. When investigating the effect of carbon dioxide concentration on the rate of photosynthesis, other variables need to be controlled. List three of these variables and describe how each is controlled.

A2. The student used water that had been boiled and then cooled in the boiling tube with the *Elodea*. Suggest why boiled water was used.

A3. The student measured the rate of photosynthesis in *Elodea*. The student repeated the investigation by adding 0.05 g of potassium hydrogen carbonate each time to increase the carbon dioxide concentration in the water.

Sketch a graph to show your prediction of the student's results. Assume that potassium hydrogen carbonate has no other effect on the plant besides affecting the rate of photosynthesis.

A4. The student measured the rate of photosynthesis by counting the number of bubbles coming from the *Elodea* in one minute.
  🔍 **a.** Describe two limitations of this method of counting bubbles.
  b. Suggest two improvements to the method of counting bubbles.

A5. This apparatus can also be used to investigate the effect of other variables on the rate of photosynthesis.

For each of these variables, describe how the variable would be changed and how other variables would be controlled to determine the effect on the rate of photosynthesis. Make reference to the apparatus in Figure 13.20.
  a. Light intensity.
  b. Temperature.

2. Use of extracted chloroplasts

Chloroplasts can be extracted from leaves and the colour change of an indicator such as DCPIP or methylene blue can be used to measure the rate of photosynthesis. DCPIP and methylene blue are both blue when oxidised. Both these indicators will turn colourless when the light-dependent reactions occur because the indicators will be reduced by the electrons released from photoexcitation.

> **Tip**
> Sketch a graph means that you only need to draw and label axes and draw the trend line. You do not need to add values to the axes or plot individual points.

> **Link**
> The redox indicators DCPIP and methylene blue were introduced in Chapter 12.

QUESTIONS

A6. Suggest three advantages of using extracted chloroplasts and an indicator to measure the rate of photosynthesis rather than a whole plant and counting bubbles.

The method for this investigation is to crush a leaf using a pestle and mortar. A solution called isolation medium is added to the crushed leaf. The isolation medium is added at a temperature of 0 °C and contains:

- distilled water
- 0.4 mol dm^{-3} sucrose
- 0.05 mol dm^{-3} pH 7.0 buffer
- 0.01 mol dm^{-3} potassium chloride.

A7. Suggest the functions of:
    a. the 0.4 mol dm^{-3} sucrose and potassium chloride in the isolation medium
    b. the pH 7.0 buffer.

The extract is then filtered through fine cotton to remove leaf debris.
    Blue DCPIP can be added to the filtered extract, the extract illuminated and the time taken to go colourless is recorded.
    This method is summarised in Figure 13.21.

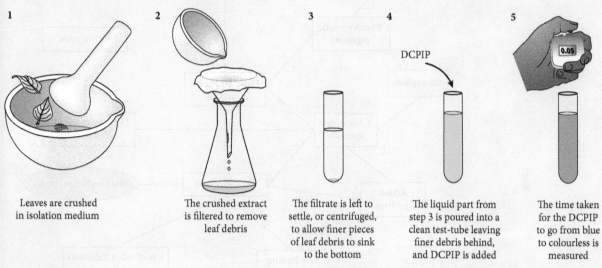

1	2	3	4	5
Leaves are crushed in isolation medium	The crushed extract is filtered to remove leaf debris	The filtrate is left to settle, or centrifuged, to allow finer pieces of leaf debris to sink to the bottom	The liquid part from step 3 is poured into a clean test-tube leaving finer debris behind, and DCPIP is added	The time taken for the DCPIP to go from blue to colourless is measured

DCPIP

Figure 13.21

A8. Describe how to convert the time taken to go colourless in seconds into a rate and state the unit of this rate.
A light filter is a transparent piece of glass or plastic through which light can be made to pass. Coloured light filters can be used with a white light to produce light of any colour.

A9. Describe a method of producing an action spectrum for chloroplasts that have been isolated from a certain species of plant using the procedure described here.
    You may assume that a coloured filter only changes the colour of light and not the intensity of the light.

CHAPTER OVERVIEW

Review the mini mind map to link new learning with your prior learning, and to highlight which of the key concepts underpin the chapter.

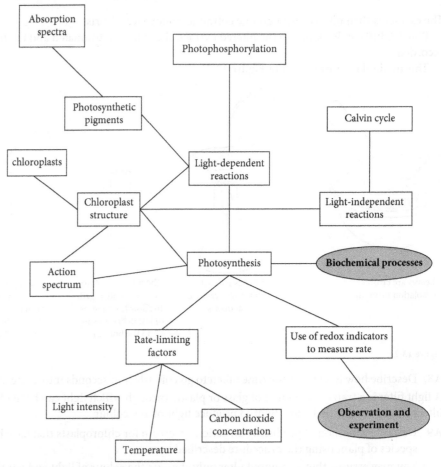

Try copying this mini mind map and expanding upon it. Use your notes from other chapters to help you explore how the essential ideas, theories and principles can be linked further together.

WHAT YOU HAVE LEARNED

Now that you have finished this chapter you should be able to:

- describe the relationship between the structure of chloroplasts, as shown in diagrams and electron micrographs, and their function
- explain that energy transferred as ATP and reduced NADP from the light-dependent stage is used during the light-independent stage (Calvin cycle) of photosynthesis to produce complex organic molecules
- state that within a chloroplast, the thylakoids (thylakoid membranes and thylakoid spaces), which occur in stacks called grana, are the site of the light-dependent stage and the stroma is the site of the light-independent stage
- describe the role of chloroplast pigments (chlorophyll *a*, chlorophyll *b*, carotene and xanthophyll) in light absorption in thylakoids
- interpret absorption spectra of chloroplast pigments and action spectra for photosynthesis

describe and use chromatography to separate and identify chloroplast pigments (reference should be made to R_f values in identification of chloroplast pigments)
state that cyclic photophosphorylation and non-cyclic photophosphorylation occur during the light-dependent stage of photosynthesis
explain that in cyclic photophosphorylation:
- only photosystem I (PSI) is involved
- photoactivation of chlorophyll occurs
- ATP is synthesised

explain that in non-cyclic photophosphorylation:
- photosystem I (PSI) and photosystem II (PSII) are both involved
- photoactivation of chlorophyll occurs
- the oxygen-evolving complex catalyses the photolysis of water
- ATP and reduced NADP are synthesised

explain that during photophosphorylation:
- energetic electrons release energy as they pass through the electron transport chain (details of carriers are not expected)
- the released energy is used to transfer protons across the thylakoid membrane
- protons return to the stroma from the thylakoid space by facilitated diffusion through ATP synthase, providing energy for ATP synthesis (details of ATP synthase are not expected)

outline the three main stages of the Calvin cycle:
- rubisco catalyses the fixation of carbon dioxide by combination with ribulose bisphosphate (RuBP), a 5C compound, to yield two molecules of glycerate 3-phosphate (GP), a 3C compound
- GP is reduced to triose phosphate (TP) in reactions involving reduced NADP and ATP
- RuBP is regenerated from TP in reactions that use ATP

- state that Calvin cycle intermediates are used to produce other molecules, limited to GP to produce some amino acids and TP to produce carbohydrates, lipids and amino acids
- state that light intensity, carbon dioxide concentration and temperature are examples of limiting factors of photosynthesis
- explain the effects of changes in light intensity, carbon dioxide concentration and temperature on the rate of photosynthesis
- describe and carry out investigations using redox indicators, including DCPIP and methylene blue, and a suspension of chloroplasts, to determine the effects of light intensity and light wavelength on the rate of photosynthesis
- describe and carry out investigations, using whole plants, including aquatic plants, to determine the effects of light intensity, carbon dioxide concentration and temperature on the rate of photosynthesis.

CHAPTER REVIEW

1. **a.** Which row of the table shows structures present in in chloroplasts?

A	matrix	double membrane	thylakoids
B	double membrane	thylakoids	DNA
C	cristae	DNA	thylakoids
D	double membrane	thylakoids	cristae

b. State the stage of photosynthesis that occurs in:
- **i.** grana
- **ii.** stroma.

2. In terms of photosynthesis, explain the difference between an absorption spectrum and an action spectrum.

3. Outline the events in cyclic photophosphorylation.

4. Describe the role of water in the light-dependent stage of photosynthesis.

5. **a.** Explain how the products from the light-dependent stage are used in the light-independent stage of photosynthesis.

   b. Describe the carbon-fixation step in photosynthesis.

   c. Draw a diagram to summarise what happens in the Calvin cycle.

6. Suggest why not all of the oxygen produced in a plant cell by photosynthesis is released from the cell.

7. **a.** Explain how the rate of photosynthesis of a plant depends on temperature.

   b. Explain how the rate of photosynthesis of the same plant varies with temperature at a higher carbon dioxide concentration.

8. When DCPIP is added to a chloroplast extract that is exposed to light, the DCPIP changes from blue to colourless. Explain what causes the colour change.

Homeostasis

One morning in 1920, Frederick Banting arose from a restless sleep and hastily scribbled some thoughts in his notebook.

Banting was a Canadian doctor who was trying to find a cure for diabetes. He knew that insulin was involved in regulating blood glucose concentration, but he had been unable to extract a concentrated sample of insulin from animals kept in the laboratory.

Banting's idea was to isolate the insulin-producing cells from the digestive enzyme-producing cells in the pancreas. This proved successful: injecting this extract into a person with diabetes reduced their life-threatening symptoms almost immediately.

In 1951, insulin became the first protein to have its primary structure decoded. It was also the first human protein to be produced by transgenic bacteria, in 1978. Today, people with diabetes are able to self-administer insulin and live normal lives.

Prior understanding

You may remember that hormones are transported in the blood. Hormones, found in animals and plants, modify the activity of target cells by binding to complementary receptors in the cell surface membrane or inside the cytoplasm. You may recall that toxic waste products are excreted from the body by the kidneys.

Learning aims

In this chapter, you will learn how cells function most efficiently if they are kept in near-constant conditions. In mammals, you will explore how blood glucose concentration and blood water potential are maintained within narrow limits. In plants, you will learn how guard cells respond to limited water supply by closing the stomata.

14.1 Homeostasis in mammals (Syllabus 14.1.1, 14.1.2, 14.1.9–14.1.11)

14.2 The urinary system (Syllabus 14.1.3–14.1.8)

14.3 Homeostasis in plants (Syllabus 14.2.1–14.2.4)

Tip

The nervous system and the endocrine system are both referred to as **co-ordination systems**, because they co-ordinate the body's responses to stimuli and changes in the environment

Tip

Negative feedback maintains the temperature of a thermostatically controlled water bath. When the sensor detects a temperature fall, the heater switches on. It switches off when the temperature rises to the set point.

Link

Some metabolic pathways involve negative feedback. In Chapter 3 you saw how end-product inhibition is used to restrict enzyme activity to make products because they co-ordinate the body's responses to stimuli and changes in the environment.

Link

Positive feedback occurs when oestrogen, released by the ovaries, stimulates the secretion of another hormone – FSH, from the pituitary gland.

14.1 Homeostasis in mammals

To function properly, all cells in the body need to be maintained in their optimal (best) physical and chemical conditions. Keeping every cell at the right temperature, with enough water and nutrients, involves many systems in the body. Biologists call these processes **homeostasis**.

The word homeostasis means 'steady state'. However, it is more accurate to say that conditions inside the body need to be maintained within certain limits. Some factors, including core body temperature and blood pH, are kept within very narrow limits. Others, such as **blood glucose concentration**, can vary within a slightly wider range without negative effects.

To understand why homeostasis is so important, think about enzymes. Metabolism – the sum total of the chemical reactions in the body – relies on relationships between many different enzymes, all working together. As you saw in Chapter 3, enzymes are very sensitive to temperature and pH. Enzymes need to work at the correct rate at all times, so that products are made as quickly as they are needed.

When you study how the body controls a body factor such as temperature, concentration of blood glucose or blood pressure, it is important to organise your thoughts. Ask yourself the following questions:

- What conditions bring about a change in the factor being considered? These are **stimuli**, and can be internal (for example, a reduction in blood glucose concentration in mammals) or external (for example, an increase in the humidity of the air in the environment of a plant).
- What detects the change? These are **receptors**, and are usually receptor cells found in organs such as the pancreas or leaf.
- How is the change reversed? In animals, this is co-ordinated by either the **nervous system** or **endocrine system** brought about by **effectors**. These are muscles and glands in mammals, and specialised cells and tissues in plants.

Whenever a physiological (bodily) factor changes and moves away from its **set point**, this change is detected. Then, by using nervous impulse or **hormone** secretion (or sometimes both), the body reverses the change. The correction is achieved by a system called **negative feedback**. This is self-correcting: as the level of the factor returns to normal, the corrective mechanisms are reduced.

The opposite of negative feedback is **positive feedback**. Here, a change is amplified and the change from the normal increases. There are many situations in the body that depend on positive feedback, including childbirth and blood clotting.

REGULATION OF BLOOD GLUCOSE CONCENTRATION

Most tissues can cope with some variation in blood glucose concentration. They can store excess glucose as glycogen and can turn to other energy-providing molecules, such as lipids, when blood glucose concentration falls too low. The brain, however, can do neither.

The blood must contain between 80 mg and 90 mg of glucose per 100 cm^3 plasma to keep brain cells fully functional. If the concentration of blood glucose varies outside these limits, even for a short time, brain tissue can be damaged.

The control of blood glucose concentration is therefore an example of homeostasis. A negative feedback mechanism detects and corrects the concentration of glucose in the blood, keeping it within 'safe' limits.

The **pancreas** is the co-ordination system for the regulation of blood glucose concentration. This leaf-shaped organ is found in the abdomen, between the stomach and the liver. Most of the tissue in the pancreas is responsible for making pancreatic juice. It delivers this digestive juice to the small intestine by means of the pancreatic duct. However, small regions of the pancreas consist of endocrine (hormone-producing) tissue. They are called **islets of Langerhans**, which contain two types of cells (Figure 14.1). These cells can detect slight changes in the concentration of blood glucose, and will respond by secreting a different hormone:

- **alpha (α) cells** produce **glucagon** when the blood glucose concentration decreases
- **beta (β) cells** produce **insulin** when the blood glucose concentration increases.

a.

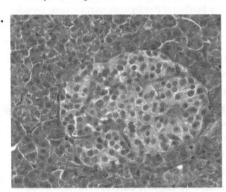

b.

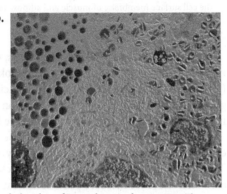

Figure 14.1 The photomicrograph (**a**) shows a section through the islets of Langerhans in the pancreas. The electron micrograph (**b**) shows insulin-secreting beta cells (the green, yellow and brown cells on the right) and glucagon-secreting alpha cells (the red cells on the left).

Insulin and glucagon are hydrophilic (water-soluble) globular proteins. As you saw in Chapter 2, they therefore cannot pass directly through the phospholipid bilayer of their target cells and enter the cytoplasm. Instead, they have specific shapes, and bind to complementary receptors on the surface of their target cells. Remember that it will only be the target cells that respond to these hormones – because only these cells have complementary receptors.

What happens when blood glucose concentration increases?

Blood glucose concentration may briefly rise up to 150 mg per 100 cm³ shortly after a meal. This is because the digestion of carbohydrates produces glucose, which is absorbed from the small intestine. Prolonged exposure of cells to plasma of this concentration can cause dehydration, which is dangerous. Negative feedback is able to bring the concentration back to normal within about 2 hours.

The β cells detect increases in blood glucose concentration and respond by secreting insulin. This hormone travels to all parts of the body in the blood, but it mainly affects target cells in the muscles, liver and adipose tissue. Insulin increases the rate at which adipose and muscle cells absorb glucose, and therefore the rate at which it is removed from the surrounding tissue fluid and the blood. In liver cells, insulin activates metabolic reactions that encourage continued glucose uptake.

How does insulin increase the permeability of cell surface membranes to glucose? The key is facilitated diffusion. Glucose can only pass from the tissue fluid into cells through specific glucose transport (carrier) proteins in cell surface membranes. Therefore, the more proteins there are, the faster glucose leaves the blood and enters cells. Target cells can boost their permeability to glucose. When insulin binds to its receptors in the cell surface membrane, vesicles that contain extra glucose transport proteins (called Glucose transporter type 4, GLUT4) are stimulated to fuse with the cell surface membrane (Figure 14.2). It is like the cell has thrown open more 'doors' to let glucose in at a faster rate.

Tip

If blood glucose concentration rises too high, the water potential of the blood reduces (becomes more negative) and surrounding cells will lose water by osmosis and become dehydrated.

Tip

Refer to blood glucose concentration, rather than levels. Concentration has units, usually presented as mg of glucose per 100 cm³ blood.

Tip

The regulation of blood glucose concentration involves cells that are both the receptor of a stimulus (change in glucose concentration) and the effector of a response (secretion of a hormone).

Link

Where does glucose in our diet come from? In Chapter 2, you saw that plant material contains starch, and meat contains glycogen. Both are polymers, containing thousands of glucose residues.

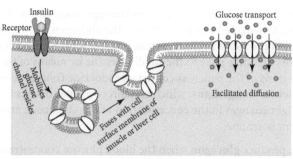

Figure 14.2 Insulin has a specific shape that is complementary to the binding site of the receptor proteins found on the cell surface membrane of muscle and adipose cells. The binding of insulin to its receptor causes vesicles containing glucose channel proteins (GLUT4) to fuse with the cell surface membrane and deliver extra channel proteins specific to glucose.

On the surface membrane of liver cells, insulin also binds to its complementary receptors. This activates the enzyme glucokinase, which adds a phosphate group to (phosphorylates) glucose that enters by facilitated diffusion. Conversion of glucose to glucose phosphate keeps the concentration of glucose in these cells low, which encourages further passive uptake of glucose from the bloodstream.

In addition to the mechanisms described, insulin also activates enzymes inside the cells. One of these enzymes is **glycogen synthase**, which converts glucose to glycogen. This is a process called **glycogenesis**, which means 'formation of glycogen'. Other activated enzymes increase protein and fat synthesis to rapidly convert the absorbed glucose into other macromolecules for storage, as well as respiration of glucose in an attempt to use it up at a faster rate.

The effects of insulin cause a fall in blood glucose concentration. In keeping with the mechanism of negative feedback, the pancreatic β cells detect this and reduce their secretion of insulin accordingly.

1. Why do you think homeostatic mechanisms are sometimes described as detection–correction systems?
☆ 2. The hepatic portal vein transports blood from the small intestine to the liver and pancreas before returning to the heart. With reference to the regulation of blood glucose concentration, explain the advantage of blood following this route.

What happens when blood glucose concentration decreases?

After prolonged, intense physical activity, the blood glucose concentration may fall to as low as 70 mg 100 cm^{-3}. How does the body react to this change? The answer is α cells in the pancreas and their secretion of **glucagon**.

The target cells for glucagon are **hepatocytes** (liver cells). These have vast stores of glycogen that can be hydrolysed, releasing glucose, in a process called **glycogenolysis**. 'Lysis' means 'to split', as in hydrolysis. The glucose then diffuses into the blood, increasing the blood glucose concentration towards its set point. When blood glucose concentration starts to rise again, this is also detected by the α cells. Due to negative feedback, their secretion of glucagon falls. Note that there are very few glucagon receptors on muscle cells. It is important that muscle cells do not convert their glycogen stores to glucose, except during vigorous exercise.

Glycogenolysis is not the only way in which the body can increase the glucose concentration of the blood. Glucagon can also stimulate cells to produce glucose by **gluconeogenesis**. The word means 'generating new glucose' and involves making glucose from non-carbohydrate sources. For example, this new glucose is sometimes made from pyruvate, and so is essentially the reverse of glycolysis, but the process involves different enzymes. Another source of pyruvate is the metabolism of excess amino acids, which we will encounter later in this chapter.

Worked example

Insulin and glucagon are said to be 'antagonistic' because they have opposite effects on blood sugar concentrations. Explain why.

Answer

Both hormones are released by the pancreas in response to changes in blood glucose concentration. Insulin is released when this rises, and stimulates liver and muscle cells to increase their absorption of glucose and conversion into glycogen (glycogenesis). This reduces blood glucose concentrations back to the set point. Glucagon is released when blood glucose concentration falls, and stimulates liver cells to break down their stores of glycogen (glycogenolysis). This releases more glucose into the blood.

3. The **glucose tolerance test** is used to help diagnose **diabetes**. After drinking a concentrated solution of glucose, the blood glucose concentration is measured at regular intervals over a number of hours. Table 14.1 shows the results of this investigation.

Time / hours	Blood glucose concentration / mg 100 cm^{-3}	
	Person with diabetes	Person without diabetes
0	140	90
1	240	135
2	290	100
3	250	95
4	220	90
5	175	90

Table 14.1

a. Plot a suitable graph of the data in Table 14.1.
b. Compare the blood glucose concentrations of the person with diabetes and the unaffected person for the first 2 hours of the study.
c. Suggest an explanation for the differences in blood glucose concentration between the two people.
d. In this investigation, state three factors that should be standardised between the two individuals in order to make a valid comparison of these results.
e. Explain why a person with diabetes must be very careful to avoid injecting themselves with too much insulin.

4. 'The mechanism by which glucose carrier proteins are inserted into the cell surface membrane is an example of exocytosis.' Discuss this statement.

5. Identify the incorrect statement about the reduction of blood glucose concentration by insulin.
   A. The pancreas contains receptor cells
   B. The α cells are the effectors
   C. The response is consistent with negative feedback
   D. The stimulus is an increase in blood glucose concentration

The mechanism by which glucagon stimulates its target cells to undertake glycogenolysis occurs in a series of steps, each one of which is dependent on the last. It is a good example of signal transduction and cell signalling, concepts you met in Chapter 4. This mechanism is summarised below and in Figure 14.3.

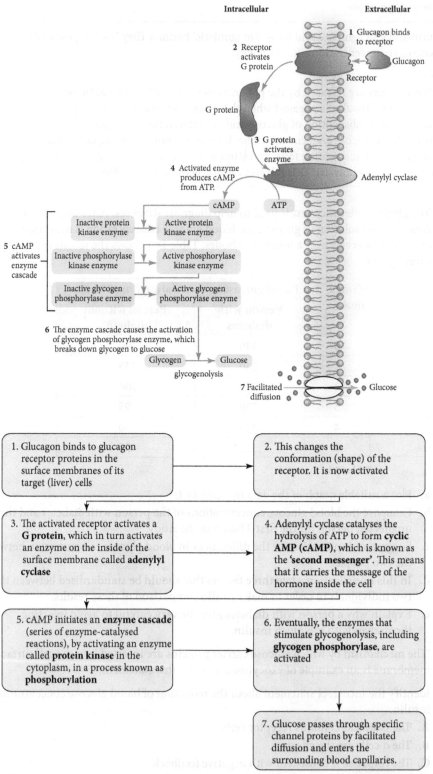

Figure 14.3 The water-soluble hormone glucagon binds to a complementary receptor protein in the surface membrane of a target cell, e.g. a liver cell. This signal is then transduced and produces a second messenger molecule (cAMP) inside the cell to bring about a response.

Cell signalling cascades have unexpected advantages. As you can see in Figure 14.3, glucagon does not directly activate the enzymes responsible for glycogenolysis. Instead, a series of steps occurs before these enzymes are activated. This is called a signalling cascade and serves to achieve a process called **amplification**, because it boosts the size of the signal. As shown in Figure 14.4, this explains how some hormones, such as glucagon, adrenaline and abscisic acid (see later in this chapter) have such rapid effects on their target cells.

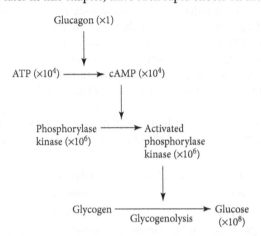

Figure 14.4 Amplification during glycogenolysis. One hormone molecule is able to have an enormous effect on a cell by activating a large number of enzymes inside the cell.

How can blood glucose concentration be monitored?

As we saw at the start of this chapter, the discovery of insulin by Banting and Best in the 1920s brought a life-saving treatment to the lives of millions. Since then, the source of the insulin used to treat people with diabetes has changed. So, too, has the medical technology used to help them monitor their blood glucose levels.

Glucose **test strips** have been used to monitor blood glucose concentration for a long time. The absorbent pad of the test strip can be dipped in blood, or more commonly, urine. Figure 14.5 shows how a glucose test strip works.

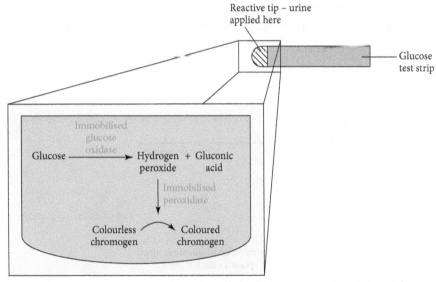

Figure 14.5 How a glucose test strip works. First, the absorbent pad of the stick is lowered into the urine. **Glucose oxidase** is an enzyme that has been immobilised on the stick. This oxidises glucose if present in the urine, forming two products, called gluconic acid and hydrogen peroxide. Hydrogen peroxide is a strong oxidising agent, and reacts with a coloured chemical called a chromogen on the pad in the presence of another immobilised enzyme called **peroxidase**. This brings about a colour change, which is then recorded and matched against a chart to indicate the concentration of glucose in the urine.

Tip

The amplification of hormone signals is sometimes referred to as the 'avalanche effect'. This reflects the increase in the size of an avalanche from a small stimulus as it runs down a mountainside.

Tip

In addition to this mechanism, remember that glucagon will also increase blood glucose concentration by activating the enzymes involved in gluconeogenesis.

Link

Insulin-dependent diabetes is often caused by an individual's immune cells attacking self antigens on the surface of β cells. In Chapter 11, you encountered an autoimmune disorder, psoriasis.

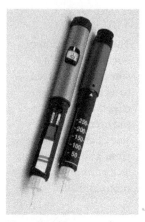

Figure 14.6 Two types of insulin pen, used by people with insulin-dependent diabetes to help control blood glucose concentration.

More recently, a glucose **biosensor** has been developed for the same purpose. This is an electronic device that works in a similar way to the glucose test strip, but it generates a small electric current instead of a colour change. Interestingly, the electric current is brought about by the same two enzymes inside the device – glucose oxidase and peroxidase. The current is then detected by an electrode and is converted into a digital reading. An image of a glucose biosensor is shown at the very start of this chapter.

Although the glucose test strip is non-invasive and painless, the electronic glucose biosensor has significant advantages. Subjectivity in taking a reading is reduced. The re-useable electronic biosensor provides an immediate numerical reading of the current blood glucose concentration and does not involve colour matching. The value is therefore more accurate.

If a person with diabetes detects that their blood glucose concentration is high, they may administer a small volume of insulin solution by injection from a pen. Two examples are shown in Figure 14.6.

6. List the similarities and differences between the following terms:
   a. glucose and glycogen
   b. gluconeogenesis and glycogenolysis
   c. hormone and receptor
   d. signal transduction and amplification
   e. glycogen phosphorylase and phosphorylase kinase.

7. Diabetes is a disorder in which a person cannot reduce blood glucose concentration after a meal.
   a. Explain why insulin must be injected into the blood using an insulin pen, rather than ingested in the form of a tablet.
   b. Explain why it is recommended that people with diabetes eat starchy foods, rather than sugary foods.
   c. Suggest explanations for the following symptoms of diabetes:
      i. excessive thirst
      ii. weight loss
      iii. fatigue
      iv. breath smelling fruity (due to ketones, which are a by-product of lipid metabolism)
      v. excessive urination.
   d. Figure 14.7 shows a graph comparing the effect of insulin on the rate of glucose intake.

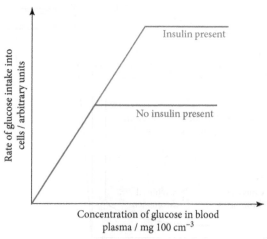

Figure 14.7 The effect of insulin on glucose intake into cells.

      i. State the method by which glucose enters cells.
      ii. Explain the difference between the rate of glucose intake into cells in the presence and in the absence of insulin.
      iii. The glucose biosensor used to measure the concentration of glucose in blood plasma had a percentage error of 0.5% of glucose per 100 cm³ blood plasma. Explain what this means.

8. Copy and complete the following passage describing the mechanism that occurs in a biosensor.

   People with diabetes can use a biosensor that is sensitive to glucose. The biosensor contains an immobilised enzyme called _____ _____. The substrate for this enzyme is _____ and this is converted into two products: _____ _____ and _____ _____. The synthesis of these products generates a very small electrical _____, which is proportional in size to the concentration of glucose in the blood. The current is then read by a meter, which displays a quantitative (_____) reading for blood glucose _____.

Key ideas

→ Homeostasis describes the processes that maintain conditions in the body between narrow limits (a set point).

→ The body detects a change in a particular internal factor, such as core body temperature, and then activates a corrective mechanism to reinstate the normal level. This is called negative feedback.

→ The pancreas monitors and controls blood glucose concentration; the brain is not involved.

→ Alpha (α) cells in the pancreas detect a fall in blood glucose concentration and respond by secreting the hormone glucagon.

→ Glucagon activates enzymes in the liver that convert glycogen to glucose.

→ The action of glucagon occurs via a second messenger molecule called cyclic AMP (cAMP). This activates protein kinase, which initiates an enzyme cascade that stimulates the action of the enzymes that break down glycogen.

→ Beta (β) cells in the pancreas detect a rise in blood glucose concentration and respond by secreting the hormone insulin.

→ Insulin increases the rate of uptake of glucose by body cells, by stimulating the movement of carrier protein molecules from vesicles in the cytoplasm to the cell surface membrane. These carrier protein molecules move glucose into the cells by facilitated diffusion.

→ Insulin also activates enzymes in the liver, which catalyse condensation reactions that convert glucose to glycogen. This is called glycogenesis.

→ The actions of both insulin and glucagon are part of a negative feedback mechanism, bringing the blood glucose concentration back to within normal limits.

→ Glucose test strips and biosensors, which both use immobilised enzymes, can be used to measure the blood glucose concentration of the blood or urine.

14.2 The urinary system

Excretion is the removal of **metabolic waste**. Metabolic waste is a by-product of the metabolic reactions that occur in cells. Organs of excretion in the human body are the lungs (which excrete CO_2 and H_2O), the skin (which excretes ions and H_2O) and the **kidneys**, which we will explore in this section.

The kidneys are paired organs found in the abdominal cavity. Put your hands on your hips, wiggle your thumbs – that is roughly where your kidneys are. Because the kidneys are not fully protected by the skeleton, they are surrounded by a **fibrous capsule**. This is a layer of tough connective and adipose tissue that provides some protection from trauma and damage. The kidneys produce **urine**, and in so doing achieve two important functions:

1. Excretion of **nitrogenous waste**.
2. **Osmoregulation**: the control of blood volume and blood water potential.

Nitrogenous waste refers to nitrogen-containing compounds. These must be excreted, as they are often toxic. In mammals, the main nitrogen-containing compound is **urea**, which has the formula $CO(NH_2)_2$.

Why do nitrogen-containing compounds need to be excreted? The answer relates to amino acids. Mammals cannot store these molecules. Therefore, on average, an adult human needs a minimum amount of protein each day (40 g to 60 g, about the weight of one egg) to provide the amino acids the body needs to repair and grow new cells. However, most people consume more than this in their diet.

The nitrogen content of urea comes from the breakdown of excess amino acids. This is brought about by enzymes in the **liver** that remove the amine (NH_2) group from amino acids to form ammonia. This is a very toxic compound, so is converted immediately into urea, which is relatively safe. This process, called **deamination**, is summarised in Figure 14.8. The urea is carried by the blood from the liver to the kidneys.

Tip

Remember that deamination takes place in the liver, not in the kidneys.

Tip

Although they contain some waste products of metabolism, faeces mainly consist of undigested material that is removed in a process called egestion.

Link

Refresh your knowledge of water potential and the process of osmosis in Chapter 4. Remember that cells surrounded by a very hypertonic or a hypotonic solution will experience damage.

Tip

To carry out their function, the kidneys receive the largest blood supply of any organ, per gram of tissue. About 1200 cm^3 of blood flows to each of them every minute.

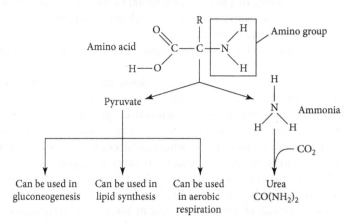

Figure 14.8 Deamination of excess amino acids produces urea, which is excreted, and pyruvate, which can be used in a number of metabolic reactions.

You probably notice the effects of your kidneys' functions each day. On a hot day, or after an extended period of physical activity, your urine is darker than normal. On the other hand, when you drink a large volume of fluid, your urine looks just like water. What you see is due to your kidneys trying to keep the water potential of your blood and tissue fluid as constant as possible. This need to balance the solute concentration of body fluids is called osmotic regulation or osmoregulation. Like the control of blood glucose concentration, it is an important example of homeostasis.

The human body consists mainly of water. A person weighing 65 kg is made up of about 40 litres of water, of which 28 litres is intracellular (inside cells). The rest, the extracellular fluid, is made up of nine to 10 litres of tissue fluid and two to three litres of blood plasma. The kidneys play a major role in regulating the volume and composition of these body fluids. They excrete or conserve water and salt so that their volume and composition remain relatively constant.

The kidneys (Figure 14.9a) receive blood from the two **renal arteries** that branch off the aorta. Blood leaves the kidneys in the two **renal veins**. The urine made by the kidneys drains into a structure called the renal pelvis, and from there into **ureters**. These vessels push the fluid by **peristalsis** (which involves slow rhythmic muscular contractions) to the **bladder**. As the bladder fills, stretch receptors in the bladder walls inform the brain of the situation. When convenient, rings of muscle at the top of the bladder relax, and urine is released to the outside via a single urethra.

Figure 14.9b shows a diagram of the gross structure of a kidney cross-section. The gross structure of an organ shows what can be seen with the naked eye. Figure 14.9c shows how the gross structure looks during a dissection. Each kidney consists of a collection of about a million microscopic **nephrons** that are tightly packed together and surrounded by a dense network of blood vessels (Figure 14.9d). The different regions of the kidney relate to the different regions of the nephrons. The cortex of the kidney consists of the **Bowman's capsules**, **glomerulus** (plural: glomeruli) and the tubules (PCT, **proximal convoluted tubule**; and DCT, **distal convoluted tubule**), while the **medulla** consists of the **loops of Henlé** and **collecting ducts**.

a.

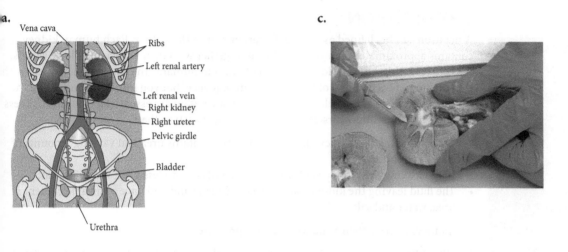

c.

b.

d.

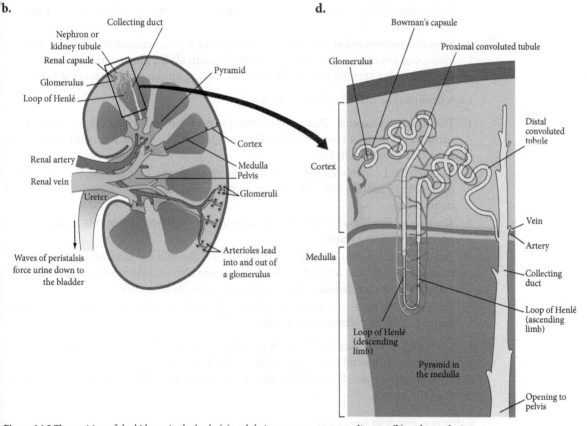

Figure 14.9 The position of the kidneys in the body (**a**) and their gross structure as a diagram (**b**) and seen during a dissection (**c**). The fine structure of the kidney consists of filtration units called nephrons (**d**). Note how the position of the loop of Henlé and the collecting ducts are the only parts of the nephron that pass through the medulla of the kidney.

9. State the two parts of the nephron that are found in the medulla of the kidney.
10. A person has 5 dm³ of blood, and the kidneys filter 1.2 dm³ of blood per minute.
    a. How many times on average does the total volume of blood in the body pass though the kidneys every hour?
    ☆ b. This blood consists of 0.7 dm³ of plasma and 125 cm³ of this plasma passes into the Bowman's capsules. Calculate the percentage of plasma that enters the Bowman's capsules. State your answer to one decimal place.

THE NEPHRON

Link

Urine mostly consists of water, salts and urea. However, other substances can be present, including hormones such as human chorionic gonadotrophin (hCG), the indicator of pregnancy described in Chapter 11.

A nephron can be defined as a single filtration unit in the kidney. Each human kidney contains approximately a million nephrons, together with their associated blood vessels. A nephron is thin but very long, up to 30–40 mm in humans. In this section, we deal with the function of each region of the nephron in sequence, but you must remember that the nephron functions as a whole: the activities of one region are essential to the effectiveness of others. Also remind yourself as you read the next section that:

- Blood entering the kidney via the renal artery contains urea and varying amounts of water and salt.
- Blood leaving the kidney via the renal vein will contain less urea, water and salt.
- The fluid leaving the kidney via the ureter is urine and will have a variable content of urea, water and salt.

The kidneys achieve their functions via two processes:

1. **Ultrafiltration**. This produces a fluid called **glomerular filtrate** (usually abbreviated to filtrate), which has nearly the same composition as the tissue fluid that surrounds all cells in the body.
2. **Selective reabsorption**. The nephron will reabsorb some substances back into the blood, but not others. For example, it will reabsorb all the glucose and amino acids. It will reabsorb a varying amount of water depending on the body's state of dehydration.

Each nephron consists of five parts (Figure 14.9d):

1. The Bowman's capsule. This is the start of the nephron and is cup-shaped. The 'cup' surrounds a mass of blood vessels called the glomerulus. This is where the blood is filtered to produce filtrate.
2. The proximal convoluted tubule (PCT). The name means 'first or near, twisted tube'. Most of the filtrate is immediately reabsorbed from here into the blood capillaries that surround it.
3. The loop of Henlé – a hairpin-shaped loop that plays an important role in water retention (keeping water).
4. The distal (or second) convoluted tubule (DCT). The name means 'second or far, twisted tube'. Here, a lot of the fine-tuning of the filtrate takes place. Substances are reabsorbed or excreted according to the precise needs of the body.
5. The collecting duct – the final part of the nephron that reabsorbs water. The fluid that passes out of the collecting duct is urine.

Look once again at Figure 14.9d. See how the blood vessels that come out of the glomerulus wrap around the rest of nephron. It is this close association between the nephron and these blood vessels that allows the exchange of substances according to the needs of the body.

ULTRAFILTRATION

Ultrafiltration, which means 'filtration under pressure', occurs between the glomerulus and the Bowman's capsule. Figure 14.10 shows this structure in detail. It is cup-shaped and surrounds a central, densely packed mass of blood vessels called the glomerulus (Figure 14.11). In simple terms, the high hydrostatic (fluid) pressure of the blood in the glomerulus 'squeezes' some of its contents through microscopic pores and into the Bowman's capsule. Figure 14.12a shows a transverse section through the cortex of the kidney, showing a Bowman's capsule surrounding a glomerulus.

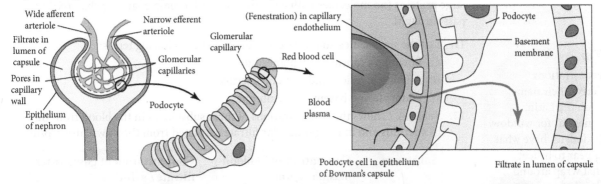

Figure 14.10 The fine structure of the Bowman's capsule and glomerulus, a region of the kidney adapted for ultrafiltration of the blood. Note the difference between the width of the lumen of the afferent and efferent arteriole.

The kidneys receive blood from the renal arteries, which are the first branches from the aorta. Therefore, the blood they receive is already under pressure. However, in the nephrons, this pressure is further increased because the **afferent arteriole**, the blood vessel that takes blood into the glomerulus, has a wider lumen than the **efferent arteriole** that takes blood away. This bottleneck generates a high hydrostatic pressure that acts to force blood against a filter that consists of three layers (see Figure 14.10):

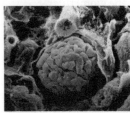

Figure 14.11 A scanning electron micrograph reveals a glomerulus with part of the torn Bowman's capsule (white-coloured) around it.

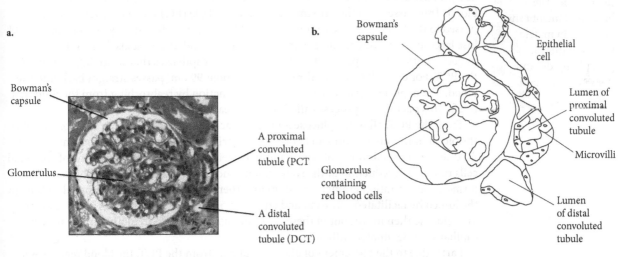

Figure 14.12 In this photomicrograph, three renal capsules (containing glomeruli) can be seen, surrounded by sections of tubules (**a**). A diagram of a small section of the photomicrograph shows the structures typically seen in the cortex of the kidney (**b**).

1. The **endothelium** of the glomerular capillaries. Unlike the walls of most capillaries, there are gaps, or slits, between the endothelial cells called fenestrations. The fenestrations, which are about 100 nm in diameter, enable small molecules in the plasma to pass through.
2. The epithelial cells lining the tubule. These are called **podocytes** that wrap around the glomerular capillaries to increase the surface area over which selective reabsorption occurs, and to provide a tight junction.
3. A **basement membrane**, which acts as a filter. This is a continuous sheet of glycoprotein between the other two layers of cells. It forms a continuous mesh and as such is the finest part of the selective filter.

Tip

You will be expected to identify the parts of a nephron and its associated blood vessels in diagrams, photomicrographs and electron micrographs – see Assignment 14.1.

The basement membrane acts as a fine filter and is therefore mostly responsible for the chemical composition of the filtrate. At this stage, the filtrate in the Bowman's capsule is identical to tissue fluid. It consists of a fluid similar to blood plasma, without cells and large proteins (no molecules with a relative molecular mass of greater than 68 000 can pass through).

11. List the similarities and differences between the following terms:
    a. Bowman's capsule and glomerulus.
    b. Afferent arteriole and efferent arteriole.
    c. Podocyte and basement membrane.

12. Table 14.2 shows the concentrations of three substances in the blood from the glomerulus and in the glomerular filtrate extracted from the Bowman's capsule.

Substance	Concentration in blood plasma / g dm^{-3}	Concentration in glomerular filtrate / g dm^{-3}
urea	0.4	0.4
protein	82.0	0.1
glucose	0.9	0.9

Table 14.2

a. Compare the content of the blood with that of the glomerular filtrate.
b. Explain the differences you described in part (a).
c. Suggest why it is important that proteins remain in the efferent arteriole.

Selective reabsorption

The product of ultrafiltration is the filtrate, fluid that has been forced out of the blood. This passes into the proximal convoluted tubule (PCT). If this filtrate simply passed to the bladder for excretion, we would become dehydrated very quickly and would lose large amounts of vital nutrients such as glucose and amino acids. In fact, the vast majority of the filtrate passes straight back into the capillaries that wrap tightly around the PCT. For every 100 cm^3 of filtrate formed, about 99 cm^3 passes straight back into the blood. The nephron performs its homeostatic function by reabsorbing from the fluid that is left. This occurs in a process called selective reabsorption.

From the PCT, all of the glucose and amino acids, and some ions, are reabsorbed into the blood in the surrounding network of capillaries. Active transport of sodium ions by ATP-dependent sodium / potassium pump proteins in the surface membrane of epithelial cells results in a reduction of the sodium ion concentration inside their cytoplasm. This forms a concentration gradient. Sodium ions then re-enter the tubule epithelial cells from the lumen by facilitated diffusion and co-transport glucose into the cells at the same time. This glucose then moves out of these cells and into the blood circulating in the nearby capillaries by facilitated diffusion.

Partly due to the movement of glucose and ions from the PCT, the blood reabsorbs a large percentage of the water by osmosis. This is also achieved by the presence of plasma proteins that remain in the capillaries after ultrafiltration.

Figure 14.13 shows the epithelial cells that line the proximal convoluted tubule and their relationship with the surrounding capillaries. The epithelial cells lining the PCT have adaptations that enable them to carry out active transport:

- microvilli, which provide a large surface area for increased absorption
- many mitochondria to provide ATP for active transport using sodium / potassium pumps
- no gaps between them. These tight junctions mean that fluid cannot pass between the cells, which means that substances must pass through the cytoplasm.

What remains in the filtrate then passes from the PCT to the loop of Henlé. This was named after the German doctor Friedrich Henlé, who identified it for the first time in

Tip

Fenestrations derive their name from the Latin, 'fenestra', for window. Podocytes have what appear to be feet that wrap around the glomerular capillaries. 'Podos' in ancient Greek means 'foot'.

Tip

Remember that the basement membrane is different from the cell surface membrane. The basement membrane is a protein matrix that contains no phospholipids, and does not surround a cell.

the 1870s. The loop of Henlé is a long U-shaped region of the nephron. It consists of a descending limb, which takes fluid down into the medulla, and an ascending limb, which brings it back to the cortex (see Figure 14.9d).

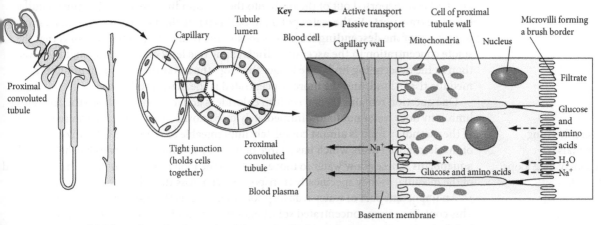

Figure 14.13 Most of the filtrate formed in the renal capsule is reabsorbed from the proximal convoluted tubule (PCT). Amino acids, glucose and sodium move from the tubule to the capillaries by active transport, and water follows passively by osmosis. The methods by which these substances move across the cell surface membranes of the PCT epithelium and the capillary endothelium are shown by the two arrows.

The purpose of the loop of Henlé is to create a region of very high solute concentration (low water potential) in the medulla. Because the final part of the nephron, the collecting duct, passes through this region, the water potential gradient between the inside of the collecting duct and the outside acts to draw water out of the duct by osmosis. As we will see later, the urine therefore becomes more and more concentrated (compared with body fluids) as it passes down the duct. This loop is therefore a vital adaptation that benefits humans and other land-living organisms: it allows us to get rid of waste without losing too much water.

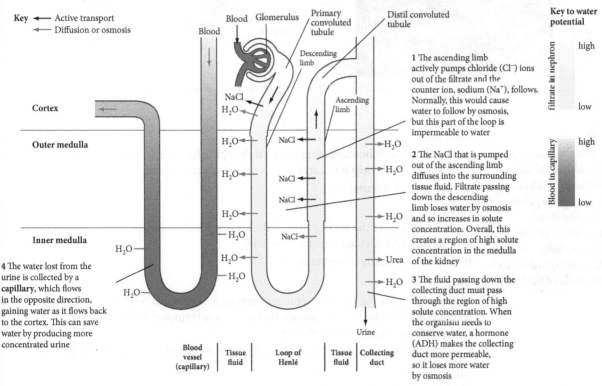

Figure 14.14 The loop of Henlé. This structure, which extends into the medulla of the kidney, is called a counter-current multiplier because it consists of two tubes that carry fluid that travel in opposite directions. The result of this arrangement is that tissue fluid with a very low water potential (dark blue shading) is produced in the medulla (outside of the blood and nephron). This is vital for the regulation of blood water potential by the collecting duct.

Figure 14.14 outlines how the loop of Henlé works. Fluid in the two 'limbs' of the loop flow in opposite directions. We describe this sort of arrangement as a **counter-current multiplier**. As fluid travels up the ascending limb, sodium chloride (NaCl, as ions Na⁺ and Cl⁻) is transported out of the limb into the surrounding tissue fluid of the medulla by active transport. As the water potential of this tissue fluid decreases, this causes water to pass out of the **descending limb** by osmosis. The result of this mechanism is that the solute concentration in the **ascending limb** at any one level of the loop is slightly lower than in the descending limb. The longer the loop, the lower the water potential in the medulla tissue fluid, and the more concentrated is the urine produced.

One question remains. When sodium chloride is actively removed from the ascending limb, why does water not follow immediately by osmosis? The answer is because the wall of the ascending limb is almost completely impermeable to water. This is unlike the wall of the descending limb, which has a large number of specific membrane channels called **aquaporins**. These allow water to move out from the loop of Henlé and into the tissue fluid of the medulla, as they specifically transport water across the cell surface membrane.

An important part of water reabsorption, therefore, is the tissue fluid of the medulla. This contains a very concentrated solution of sodium and chloride ions, and therefore a low (more negative) water potential. As the filtrate passes down the collecting duct, water can pass out of it via its selectively permeable membrane by osmosis. The reabsorbed water is then carried away by the blood in the capillaries. Figure 14.15 shows a transverse section through the medulla of the kidney, showing the ascending and descending loops of Henlé and the collecting ducts.

Tip

Desert-dwelling mammals have extremely long loops of Henlé.

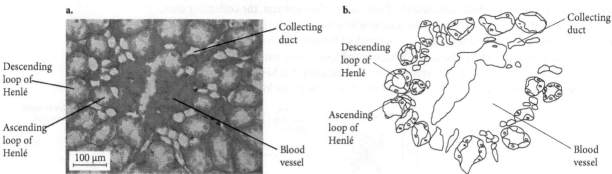

Figure 14.15 A photomicrograph (**a**). and diagram (**b**). of a transverse section from the medulla of a kidney.

Tip

While the primary convoluted tubule reabsorbs most of the filtrate, absorption from the distal convoluted tubule is far more selective and changes according to the needs of the body.

Worked example

Explain how the following adaptations of the epithelial cells lining the proximal convoluted tubule help it to carry out its role in selective absorption.

a. Presence of microvilli.

Answer

Microvilli increase the surface area of these cells, which increases their capacity to absorb sodium ions (Na⁺), glucose and amino acids.

b. Presence of many mitochondria.

Answer

Mitochondria are the site of aerobic respiration, and produce ATP. This is required for the sodium–potassium (Na⁺/ K⁺) pumps that use active transport to move these ions across the epithelial cell surface membrane.

c. Tight junctions between epithelial cells of the PCT and those in the capillary wall.

Answer

The tight junctions between cells mean that adjacent cells are held closely together, so that fluid cannot pass between the cells and hence must pass through them.

13. The loop of Henlé is sometimes described as a counter-current multiplier. Explain why.
14. Sketch a graph to show the rate of ATP hydrolysis (*y*-axis) changes with distance along a nephron (*x*-axis).
15. Identify the 'odd one out' from these series of terms. Explain your choice.
    a. Glomerulus, Bowman's capsule, proximal convoluted tubule.
    b. Protein, urea, glucose.
    c. Ultrafiltration, selective reabsorption, osmosis.
    d. Cortex, medulla, renal pelvis.
16. Figure 14.15 shows a transverse section through the medulla of the kidney. Explain why there are no distal convoluted tubules seen in this figure, despite these structures joining together the loops of Henlé and collecting ducts.

Assignment 14.1: Dirty air, damaged kidneys?

Air pollution is bad for health. It is linked with asthma and lung disorders. But more and more evidence suggests that it may also damage the kidneys.

Air pollution contains fine particulate matter (PM), called $PM_{2.5}$. This has a diameter of 2.5 µm or less. Because these particles have such a low mass, they can stay in the air longer.

QUESTIONS

A1. Suggest how $PM_{2.5}$ may enter the kidneys.

A large study in the USA evaluated the effects of air pollution on kidney function.

In this study, the scientists:

- analysed the residential address and health of 2.5 million people
- included a negative control, the levels of sodium ions in the atmosphere, and found no association between these and incidence of CKD
- concluded that 45 000 new cases of CKD and 2500 new cases of kidney failure may be caused by high levels of air pollution.

A2. With reference to this information:
    a. calculate the ratio of new cases of CKD and new cases of kidney failure that were observed in this study
    b. state one reason why the results of the investigation were considered to be reliable
    c. state the purpose of the negative control used in this study.

Figure 14.16 shows a photomicrograph of tissue from a kidney of a person with CKD. This has the following abnormalities:

1. shrunken glomerulus
2. enlarged Bowman's capsule
3. enlarged proximal convoluted tubule (PCT) lumen.

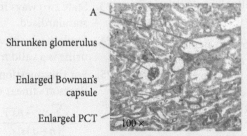

Figure 14.16 A photomicrograph of kidney tissue taken from a patient with chronic kidney disease (CKD).

A3. Refer to Figure 14.16 to answer this question.

    a. Calculate the actual width of the structure labelled A. Show your working.

    b. Has the tissue sample shown in Figure 14.16 been taken from the medulla or the cortex of the kidney? Explain your answer.

    c. Draw diagrams to show two of the three defects that are common in patients with CKD. Use the information in Figure 14.15 to help you.

  ⭐**d.** An enlarged PCT is also associated with death of epithelial cells lining this structure. Suggest why this would reduce the ability of the kidney to carry out its function.

    e. Suggest how the PCT will look different in a longitudinal section of the kidney.

Experimental skills 14.1: Comparing the urine of rodents

About 40% of all mammal species are rodents. They are found in a diverse range of habitats, from semi-aquatic to dry desert.

Rodents such as the kangaroo rat can survive without water for months. They can achieve this partly because they can produce urine with a water potential over 14 times more negative than their blood plasma. How? Their nephrons have loops of Henlé that are extremely long, and extend very deep into the medulla of the kidney.

QUESTIONS

P1. Explain why it is important that the kangaroo rat produces urine with a very low water potential.

disrupt the formation of hydrogen bonds between the water molecules. Freezing requires additional hydrogen bonding, so adding solute lowers the freezing point.

- Animals were kept in a cage placed at a temperature of 25 °C for 6 hours with a plentiful water supply.
- After 6 hours, the urine collected in a dish placed beneath the cage was collected.
- 1 cm³ of this urine was placed into a freezer compartment and a probe was placed into the urine to detect at which temperature the solution begins to turn into a solid.
- This process was repeated on five successive days for each species of rodent to obtain five repeat values.

Rodent species	Common name	Habitat	Mean body mass/ g	Mean RMT
Arvicola amphibius	water vole	freshwater	140	1.5
Ondatra zibethicus	muskrat	freshwater	700	2.1
Rattus rattus	black rat	land	250	5.9
Mus musculus	house mouse	land	19	8.1
Jaculus jaculus	jerboa	desert	57	9.0
Dipodomys deserti	kangaroo rat	desert	110	12.4

Table 14.3 The characteristics of a range of rodents from diverse habitats.

Table 14.3 shows the mean body mass and relative medullar thickness (RMT) for six species of rodent. RMT is a measure of the proportion of the kidney that consists of the medulla. The larger the value, the greater the thickness of the medulla tissue in relation to the whole kidney.

P2. Sketch three diagrams to show the nephrons from the water vole, house mouse and kangaroo rat to compare their structure and position relative to the cortex and medulla of the kidney.

A researcher investigated the relationship between the relative medullar thickness and the freezing point of urine in rodents. When solutes are added to water, they

- A scatter graph was plotted, which suggested a strong negative correlation between the two variables. The Pearson's correlation coefficient was calculated.

P3. State two ways in which the method has been standardised.

P4. Explain why measuring the freezing point of urine is a valid measurement of its water potential.

P5. For this question, you will need the equation for Pearson's linear correlation (r):

$$r = \frac{\sum xy - n\overline{x}\overline{y}}{(n-1)s_x s_y}$$

a. Explain what is meant by a strong negative correlation.

b. Copy and complete Table 14.4 and then substitute the values into the formula to show that the Pearson's correlation coefficient (r) is −0.93.

c. Calculate the degrees of freedom for this investigation and use the probability table (see Table 2.10, Chapter 2) to find out if r = −0.93 is significant.

d. The researchers attempted to use their graph to estimate the freezing point of urine from *Notomys alexis*, the spinifex hopping mouse. Using the terms interpolation and extrapolation, describe how this could be done.

e. Spearman's rank can also be used to assess the significance of a linear relationship. Explain why the Pearson's linear correlation test was used, rather than the Spearman's rank.

☆ **P6.** The researchers concluded that the larger the RMT, the lower the freezing point of the urine. Explain these results.

P7. Which of the following further investigations would provide the best evidence for this conclusion?

A. Measuring the volume of urine produced by each animal over a period of time.

B. Measuring the body temperature of each animal.

C. Measuring the time taken for urine to freeze when incubated in a freezer at a set temperature.

D. Measuring the hours that each animal sleeps in a 24-hour period.

🔎 **P8.** Using all of the information provided, describe two reasons why the conclusion drawn in this study may not be valid.

Rodent species	Mean relative medullar thickness/no units, x	Mean freezing point of urine/ °C, y	xy
Arvicola amphibius	1.5	−1.4	
Myocastor coypus	2.8	−2.9	
Rattus rattus	6.0	−3.8	
Mus musculus	6.3	−4.1	
Jaculus jaculus	9.8	−4.2	
Dipodomys deserti	12.2	−5.6	
mean	$\bar{x} =$	$\bar{y} =$	
$n\bar{x}\bar{y}$	=		$\Sigma xy =$
standard deviation	$s_x =$	$s_y =$	$r =$

Table 14.4 Calculating the Pearson's correlation coefficient (r).

HORMONAL CONTROL OF BLOOD WATER POTENTIAL

Table 14.5 shows a typical water 'balance sheet' for an average person, assuming normal activity and a comfortable external temperature. Each day, round 2.3 litres of water is exchanged with the environment. On average, we get almost two-thirds of our water from drinks and a third from food. We obtain a small volume of water, around 200 cm³ per day, from our own metabolic reactions.

Water gain	Volume / cm3	Water loss	Volume / cm3
food and drink	2100	through skin	350
metabolic water	200	sweat	100
		in breath	350
		urine	1400
		faeces	100
Total	2300	Total	2300

Table 14.5 The water balance sheet for a 24-hour period.

Some of the water loss shown in Table 14.5 is unavoidable. Metabolic waste such as urea must be removed in solution, and so some water loss in urine is inevitable. Similarly, water is always lost from the lungs as we breathe out. A significant amount of water is also lost by diffusion through our skin (this is not the same as sweating). However, at certain times during a typical 24-hour period, water consumption will drop or will increase. How does the body control the kidneys to modify how much water is lost in the urine?

The answer is a hormone, called **antidiuretic hormone (ADH)**. This is part of a negative feedback system, similar to the regulation of blood glucose concentration. In this mechanism, **osmoreceptor cells** in a part of the brain called the **hypothalamus** are important because they continuously monitor the water potential of the blood. When they detect that the water potential decreases, indicating that more water has been lost than taken in, the hypothalamus responds in two ways:

- it triggers feelings of thirst in the brain
- it stimulates specialised nerve cells in the **posterior pituitary gland** to release ADH.

Antidiuretic hormone (ADH) acts on the epithelial cells in the wall of the collecting duct (and to a lesser extent the distal convoluted tubule) to reduce the volume of urine produced. It does this by increasing the permeability of the wall to water in the following way (Figure 14.17):

1. ADH leaves the cells lining the capillaries by facilitated diffusion and enters the tissue fluid surrounding the collecting duct. It binds to specific complementary receptors on the cell surface membranes of the epithelial cells lining the collecting duct.
2. This activates a series of enzyme-catalysed reactions in the cytoplasm of the epithelial cells.
3. A phosphorylase enzyme is activated, which causes vesicles containing aquaporin water protein channels to move to the cell surface membrane and fuse with it.
4. This increases the permeability of the wall of the tubule to water.
5. Water therefore moves out of the lumen of the collecting duct, down the water potential gradient and into surrounding capillaries by osmosis.

The mechanism of antidiuretic hormone action is summarised in Figure 14.18. When more fluid is lost than taken in, the hypothalamus detects this and it increases ADH secretion. ADH increases the permeability of the distal convoluted tubule and the collecting duct to water. Remember that the loop of Henlé has reduced the water potential of the tissue fluid in the medulla surrounding the collecting duct. Therefore, the increased permeability of the collecting duct results in more water leaving the nephron by osmosis, and moving back into the blood. Much more concentrated urine is produced (with a lower water potential) and vital water is conserved. Conversely, when more fluid is taken in than lost, the hypothalamus detects this and it reduces ADH secretion. The action of ADH on the kidneys lessens, because fewer aquaporins are placed into the cell surface membrane. This results in less water reabsorption and the production of larger volumes of dilute urine (with a higher water potential).

Tip

'Diuresis' is the medical term for excessive urination. So, antidiuretic hormone acts to reduce the volume of urine produced.

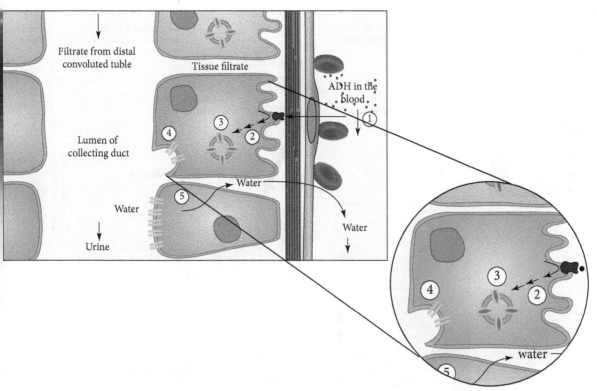

Figure 14.17 The action of antidiuretic hormone on the epithelial cells of the collecting duct wall.

Tip
Avoid referring to the 'epithelial cell walls of the collecting duct' (incorrect). Instead, refer to the 'surface membranes in the epithelial cells of the wall of the collecting duct' (correct).

Link

In negative feedback systems, receptors and effectors are involved. In osmoregulation, the receptors are cells in the hypothalamus, and the effectors are the epithelial cells of the collecting duct.

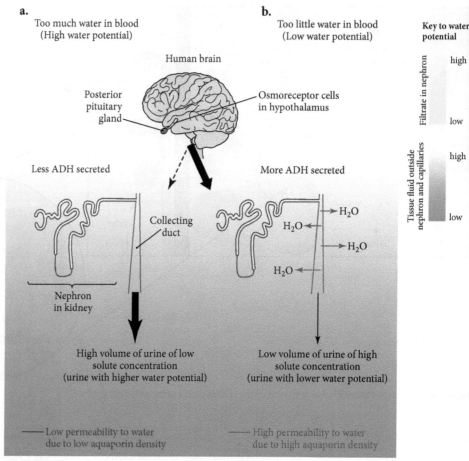

Figure 14.18 (a) When the solute concentration of the blood rises, osmoreceptor cells in the hypothalamus stimulate secretion of ADH from the posterior pituitary gland. ADH makes the distal convoluted tubule and collecting duct more permeable to water, so more water passes from the filtrate into the blood. (b) When there is too much water in the blood (high water potential), less ADH is released, meaning the collecting duct has a low permeability to water, so less water passes from the filtrate into the blood.

Finally, we reach the conclusion of the story. The collecting ducts from many nephrons drain into the renal pelvis of the kidney. The remains of the filtrate leave the kidneys via a ureter and enter the bladder. This waste, a solution containing urea, salts and various other substances, is the urine.

Worked example

Tip

Note that the kidney removes some metabolic waste at all times, in solution. This means that some water must be lost no matter how dehydrated you are.

Some drugs affect the secretion of ADH from the pituitary gland. Predict and explain the effect on blood water potential of:

a. Caffeine, which reduces ADH production.

Answer

Reduced ADH production will cause a reduction in the permeability of the collecting ducts. Less water will be reabsorbed from the filtrate in the nephron into the blood, so the water potential of the blood will decrease (become more negative).

b. Nicotine, which increases ADH production.

Answer

Increased ADH production will cause an increase in the permeability of the collecting ducts. More water will be reabsorbed from the filtrate in the nephron into the blood, so the water potential of the blood will increase (become less negative).

17. Suggest why the following scenarios could prevent the kidneys from properly carrying out their function:

   a. losing a large volume of blood

   b. drinking seawater

   c. ingesting a diet extremely rich in protein.

18. People with the disorder diabetes insipidus cannot produce antidiuretic hormone (ADH) because they have a faulty pituitary gland.

   a. Suggest two symptoms of diabetes insipidus.

   b. Suggest how diabetes insipidus could be treated.

19. Figure 14.19 shows how the relative water potential of the filtrate changes as it passes through a nephron.

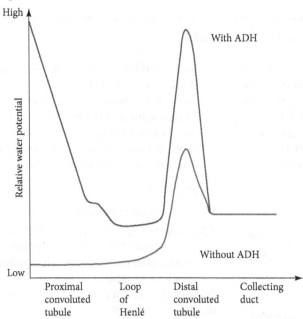

Figure 14.19 How the relative water potential of the filtrate changes as it moves through a nephron.

   a. Outline why there is a difference between the water potential in the collecting duct in the presence and in the absence of ADH.

   b. Copy the graph and sketch lines to show how the concentration of the following substances would change in the filtrate as it passes from the PCT to the collecting duct:

     i. urea   **ii.** sodium chloride   **iii.** glucose.

Key ideas

→ The kidneys remove metabolic waste and control water and solute levels in the body. As a result of these functions, the kidney also plays a vital role in the control of blood volume and pressure.

→ Each kidney is made from around one million nephrons, narrow tubules closely entwined with blood vessels.

→ At one end of the nephron, the renal capsule, the kidney filtrate is formed by ultrafiltration (pressure filtration) of the blood.

→ The first filtrate has the same composition as tissue fluid. As it passes along the nephron, the composition of the filtrate is altered by various active transport mechanisms that reabsorb some substances while allowing others to pass into the urine.

→ A large amount of filtrate is formed, but over 99% of it is reabsorbed in the first convoluted tubule, mainly by active transport mechanisms and osmosis. Usually, all glucose and amino acids are reabsorbed into the blood.

→ The movement of solutes from the loop of Henlé creates a region of high solute concentration in the medulla, through which the collecting ducts must pass. As filtrate (now urine) flows along the collecting ducts, water leaves it by osmosis. The resulting urine is hypertonic (more concentrated than body fluids).

→ The second convoluted tubule is involved in several homeostatic mechanisms including the regulation of salt, water and pH levels.

→ When the water potential of the blood drops (e.g. when we dehydrate) it is detected by the hypothalamus, which stimulates ADH release from the posterior pituitary. ADH makes the second convoluted tubule and the collecting duct more permeable to water, so water leaves the filtrate and enters the blood.

Experimental skills 14.2: A thirst for the truth

Some foods, such as cheese, anchovies and miso, have a distinctly savoury taste called 'umami'. In 1908, the Japanese chemist Kikunae Ikeda patented a process for extracting the substance responsible, glutamate, from seaweed. He stabilised it with sodium ions to create monosodium glutamate, or MSG. Today, this is a widely used food additive, commonly found in dishes from east and southeast Asia.

A student wanted to investigate the following research question:

Does ingestion of monosodium glutamate cause greater feelings of thirst and lower hydration levels compared with ingestion of sodium chloride (table salt)?

In this investigation:

- 200 g plain (unflavoured) noodles were provided to 20 volunteers. Ten of these volunteers were provided with noodles with 1 g MSG, and 10 with noodles with 1 g sodium chloride. The meals should look identical.

- Test subjects should not know whether MSG or sodium chloride has been added to their meal.

- To estimate the feeling of thirst, the test subject is asked to score their feeling on a scale of 1 to 10 (with 1 indicating no feelings of thirst, and 10 indicating an extreme feeling of thirst).

- To estimate hydration levels, the colour of urine – taken at 1-hour intervals after eating the meal – was compared with a colour chart.

QUESTIONS

P1. State a null hypothesis for this investigation.

P2. Suggest why it is important that the test subjects should not know whether they are eating noodles containing monosodium glutamate or with sodium chloride.

Some of the student's results for the investigation are shown in Table 14.6.

P3. Refer to Table 14.6 to help you answer this question.

a. With reference to the terms range and interval, explain whether this investigation is likely to accurately identify a trend in the data.

Time after ingestion of meal/ hours	Mean feeling of thirst after ingesting MSG/ arbitrary units	Standard error/ arbitrary units	95% confidence interval/ arbitrary units	Upper limit of error bar	Lower limit of error bar	Mean feeling of thirst after eating NaCl/ arbitrary units	Standard error/ arbitrary units	95% confidence interval/ arbitrary units	Upper limit of error bar	Lower limit of error bar
1	3.3	0.10				2.8	0.21			
2	5.7	0.21				4.9	0.26			
3	7.5	0.15				7.1	0.14			
4	8.9	0.12				7.6	0.07			
5	9.1	0.13				9.3	0.16			

Table 14.6 The results of an investigation into the effects of monosodium glutamate or sodium chloride on thirst and hydration levels.

☆ **b.** Calculate the 95% confidence interval values for the data. The confidence interval = mean ± 2SE. Next, calculate the upper and lower limits of the error bars for each mean value at each time point. Copy and complete Table 14.6.

c. The confidence limit of each of the mean values is 95%. Explain what this tells you about each mean value.

d. Plot a graph to compare the effect of MSG and sodium chloride on feeling of thirst, including standard error bars.

e. Identify the mean value that has the highest reliability. Explain your answer.

f. Identify the time point at which there may be a significant difference between the feelings of thirst of volunteers that ingested MSG and those who ingested sodium chloride. Explain your answer.

🔍 **g.** The student decided to conduct a *t*-test on the data for this time point, to compare the two mean values of continuous (not discrete) data samples that have a normal distribution. They found that the test led to an acceptance of the null hypothesis. Copy and complete the following paragraph to explain how the *t*-value was used to assess the significance of a difference between these two mean values:

'The calculated *t*-value is [greater than / less than] the critical value for [10 / 20 / 18 / 19 / eight / nine] degrees of freedom. This means there is a probability of [<0.05 / >0.05] that the difference in reaction time is due to chance. This means there is [no significant / a significant] difference and the null hypothesis is [accepted / rejected].'

P4. Explain why the results for both dependent variables in this investigation are described as semi-quantitative.

P5. Suggest one way in which a more accurate estimation for hydration levels could be obtained.

14.3 Homeostasis in plants

Plants also need to respond to changes in their environment in order to ensure optimum conditions for their cells. This often relates to providing the best conditions for photosynthesis, while minimising the loss of water due to transpiration.

Cells in the leaves require a constant supply of carbon dioxide if they are to make best use of light energy for photosynthesis. They obtain this gas via stomata, consisting of two **guard cells** either side of a pore. These are usually found on the lower surface of a leaf (Figure 14.20).

Tip

Stomata (plural) are groups of stoma (singular).

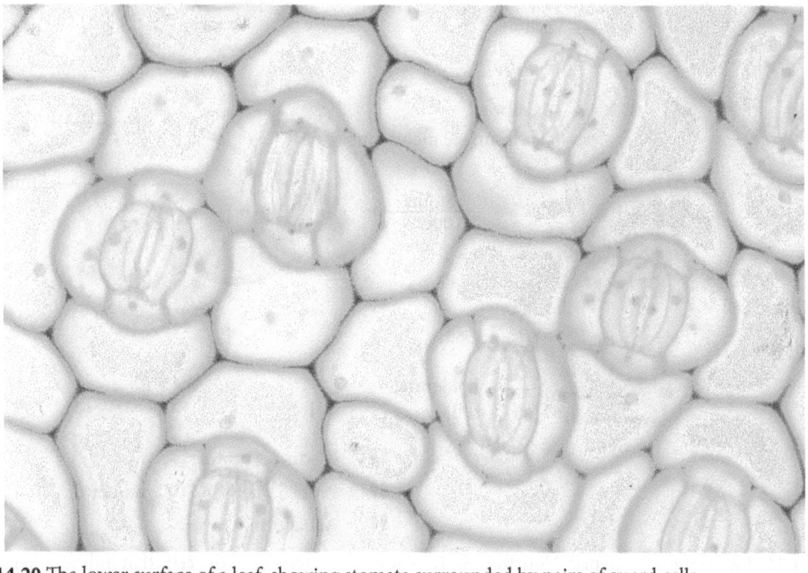

Figure 14.20 The lower surface of a leaf, showing stomata surrounded by pairs of guard cells.

As you saw in Chapter 7, stomata control the movement of gases between the atmosphere and the air spaces inside a leaf. They allow:

- the inward diffusion of carbon dioxide to be used in photosynthesis
- the outward diffusion of water vapour during transpiration.

Table 14.7 summarises the environmental factors that cause the opening and closing of stomata.

Stomata open in response to...	Stomata close in response to...
increasing light intensity	darkness
low carbon dioxide concentrations in the air spaces within the leaf	high carbon dioxide concentrations in the air spaces in the leaf
	low humidity
	high temperature
	limited supply of water from the roots and/or there are high rates of transpiration

Table 14.7 The response of stomata to various environmental conditions.

Guard cells are very sensitive to changes in light intensity and carbon dioxide concentrations inside the leaf, and respond by regulating the opening and closing of the stoma. The disadvantage of reducing the stomatal aperture (closing the stomata) is that the supply of carbon dioxide decreases so the rate of photosynthesis decreases. The advantage is that more water is retained inside the leaf, which is important in times of **water stress**. A plant will need to balance these two exchanges at all times and in response to external stimuli, although in general most stomata follow a daily rhythm in which guard cells open the stomata during the day and close them at night.

During the day, a reduction in the water potential of the guard cells causes water to move in by osmosis. This causes the cells to become turgid, and the stomatal aperture increases (the stomata open). During the night, the opposite occurs: their water potential increases and water leaves the guard cells by osmosis. They become flaccid and the stoma closes. The full mechanism of stoma opening depends on protein channels and the movement of ions in the cell surface membrane of guard cells (Figure 14.21).

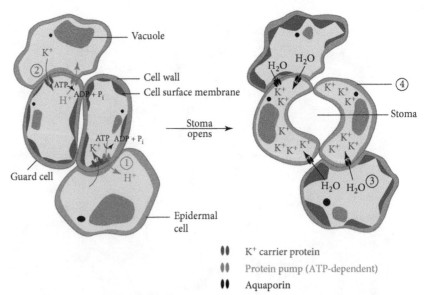

Figure 14.21 How a stoma is opened. Note the relative sizes of the vacuoles in the cells that indicate their degree of turgidity. The numbers relate to the text below.

1. Proton pump proteins in the guard cell surface membrane actively transport hydrogen ions, H⁺ (protons) into neighbouring epidermal cells as channel proteins close. This requires ATP.
2. The decrease in the hydrogen ion concentration inside the cells causes channel proteins specific for potassium (K^+) ions in the cell surface membrane to open. These ions move into the cell down an **electrochemical gradient**.
3. The extra potassium ions inside the guard cells lower the water potential in the cytoplasm. Water therefore moves in by osmosis through aquaporins in the cell surface membrane, down the water potential gradient.
4. The increase in volume (turgor) of the guard cells causes the stoma to open.

To close an open stoma, the reverse occurs. The proton pump proteins stop actively transporting hydrogen ions out of the guard cell and potassium ions begin to leave the guard cell and enter neighbouring epidermal cells as channel proteins close. As the water potential in the guard cells increases, water molecules leave them by osmosis. This causes the cells to become flaccid, and they return to their original shape and close the stoma.

As you saw in Table 14.7, there are some times when stomata need to be closed. Although this reduces the rate of photosynthesis, as carbon dioxide concentration in the leaf will fall, this process conserves vital water during times of water stress.

The onset of stomatal closure is due to a hormone called **abscisic acid** (ABA). As it is thought to be released by chloroplasts when a plant experiences water stress, it is called a **stress hormone**. The presence of high concentrations of ABA in leaves causes stomata to close, reduces the rate of transpiration and conserves water inside the plant. It is a rapidly acting hormone; if ABA is artificially applied to a leaf, its stomata close within just a few minutes. This suggests that the response of guard cells to ABA involves a second messenger and amplification of the signal occurs (Figure 14.22). Exactly what triggers ABA secretion is not known, but its levels rise during times of water stress.

> **Link**
>
> Guard cells do not have plasmodesmata, which you encountered in Chapter 7. Therefore, the exchange of all substances with their surroundings occurs across the cell surface membranes.

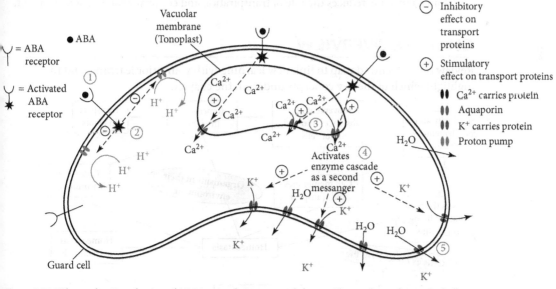

Figure 14.22 The mechanism of action of ABA in stimulating stomatal closure. The numbers relate to the bullet points below.

1. ABA is released from surrounding cells and binds to complementary receptors on the cell surface membrane of guard cells. This has two effects.
2. Firstly, this inhibits the proton pumps. Hydrogen ions are no longer pumped out. This allows the concentration of hydrogen ions to increase, causing a high positive charge inside the cell. To balance this, potassium ions leave the cell down the electrochemical gradient by facilitated diffusion.
3. Secondly, this activates calcium ion channel proteins in the tonoplast, the vacuolar membrane. The release of calcium ions from the vacuole and influx into the cytoplasm occurs by facilitated diffusion.

353

4. Calcium ions are a second messenger. Their rising concentration in the cytoplasm triggers a signalling cascade that results in potassium ion efflux (potassium ions leaving the cell) by facilitated diffusion via specific protein channels.

5. The movement of potassium ions out of the cytoplasm rapidly increases the water potential of the contents of the guard cells, causing water to move out by osmosis. This reduces the volume of the guard cells, which become flaccid, causing the stoma to close.

20. Stomata are opened and closed due to the establishment of two gradients – one is electrochemical (a difference in the concentration of ions) and the other relates to water potential.

    a. Explain how an electrochemical gradient is produced across the guard cell surface membrane during the opening of a stoma.

    b. Describe how the electrochemical gradient leads to the formation of a water potential gradient, and explain how this causes a stoma to open.

    c. During times of water stress, a plant will close its stomata. Outline the role of abscisic acid (ABA) in this process.

Key ideas

→ A stoma, which consists of two guard cells either side of a pore, allows the inward diffusion of carbon dioxide to be used in photosynthesis and the outward diffusion of water vapour in transpiration.

→ To increase the stomatal aperture (open the stomata), efflux of hydrogen ions and influx of potassium ions across the cell surface membrane of guard cells decrease their water potential so water enters by osmosis; the cells become turgid. To close the stoma, the opposite events occur to make the guard cells become flaccid.

→ Abscisic acid (ABA) is a plant hormone that is secreted by plant cells in stress conditions. The presence of high concentrations of abscisic acid in leaves stimulates stomata to close, reduces the rate of transpiration and conserves water inside the plant.

CHAPTER OVERVIEW

Review the mini mind map to link new learning with your prior learning, and to highlight which of the key concepts underpin the chapter.

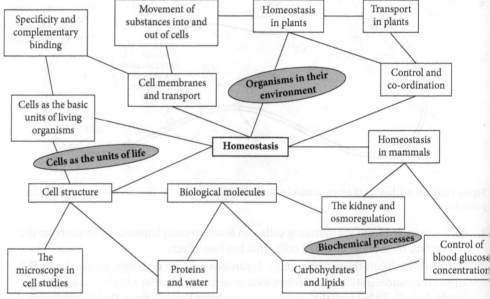

Try copying this mini mind map and expanding upon it. Use your notes from other chapters to help you explore how the essential ideas, theories and principles can be linked further together.

WHAT YOU HAVE LEARNED

Now that you have finished this chapter you should be able to:

- explain what is meant by homeostasis and explain the importance of homeostasis in mammals
- explain the principles of homeostasis in terms of internal and external stimuli, receptors, **co-ordination systems** (nervous system and endocrine system), effectors (muscles and glands) and negative feedback
- state that urea is produced in the liver from the deamination of excess amino acids describe the structure of the human kidney
- Identify, in diagrams, photomicrographs and electron micrographs, the parts of a nephron and its associated blood vessels and structures
- describe and explain the formation of urine in the nephron
- relate the detailed structure of the Bowman's capsule and proximal convoluted tubule to their functions in the formation of urine
- describe the roles of the hypothalamus, posterior pituitary gland, antidiuretic hormone (ADH), aquaporins and collecting ducts in osmoregulation
- describe the principles of cell signalling using the example of the control of blood glucose concentration by glucagon, limited to:
  - formation of the second messenger, cyclic AMP (cAMP)
  - activation of protein kinase A by cAMP leading to initiation of an enzyme cascade
  - amplification of the signal through the enzyme cascade as a result of activation of more and more enzymes by phosphorylation
- explain how negative feedback control mechanisms regulate blood glucose concentration, with reference to the effects of insulin on muscle cells and liver cells and the effect of glucagon on liver cells
- explain the principles of operation of test strips and biosensors for measuring the concentration of glucose in blood and urine, with reference to glucose oxidase and peroxidase enzymes
- explain that stomata respond to changes in environmental conditions by opening and closing and that regulation of stomatal aperture balances the need for carbon dioxide uptake by diffusion with the need to minimise water loss by transpiration
- explain that stomata have daily rhythms of opening and closing
- describe the structure and function of guard cells and explain the mechanism by which they open and close stomata
- describe the role of abscisic acid in the closure of stomata during times of water stress, including the role of calcium ions as a second messenger.

CHAPTER REVIEW

1. Explain what is meant by the term homeostasis.
2. Using the key terms 'stimulus', 'receptor' and 'effector' in your answers, outline the control of:
   a. blood glucose concentration by insulin and glucagon
   b. water potential by antidiuretic hormone
   c. stomatal closure by abscisic acid (ABA).
3. Draw a table to compare insulin and glucagon. You should include their origins, mode of action, target cells and functions.
4. Explain how a glucose biosensor works.
5. The kidney is an organ that carries out osmoregulation.
   a. Explain what is meant by osmoregulation.
   b. Explain why osmoregulation is an example of negative feedback.
   c. Both the nervous and endocrine systems co-ordinate the control of osmoregulation. Explain how.
   d. Explain how ADH increases the permeability of the collecting duct to water.

6. Identify whether the following statements regarding the regulation of stomata opening and closing are true, sometimes true, or false. If your answer is 'sometimes true', provide an explanation.
   a. Abscisic acid binds to specific receptors on the surface membrane of guard cells.
   b. Potassium ions move out of the guard cell by facilitated diffusion.
   c. Regulation of stomata opening and closure is controlled by negative feedback.
   d. Water moves into guard cells by osmosis.
   e. Calcium ions move out of the vacuole by facilitated diffusion.
   f. Stomata are found in fewer numbers on the lower surface of a leaf than the upper surface.

Chapter 15

Control and co-ordination

In high-level athletics events, any runner who leaves the starting blocks less than 100 ms after the starting gun is classed as making a false start. Why? Analysis of reaction times shows that it takes a minimum of 110–120 ms to react to the sound of the starting gun. What happens during that time? Why is it not possible, even with practice, to react faster than this? This chapter will examine the workings of the nervous system to enable you to answer these questions.

Prior understanding

You may recall information about the hormones insulin, glucagon and antidiuretic hormone (ADH) from Chapter 14. You may also recall some basic facts about the nervous system from previous courses, such as the relationship between the brain and spinal cord, sensory and motor neurones and possibly the reflex arc. It may be helpful to look back at Chapter 4 to remind yourself about simple diffusion, facilitated diffusion, exocytosis and active transport. In previous courses you may have looked at the responses of plants to light and gravity.

Learning aims

In this chapter, you will learn how the nervous and the endocrine (hormonal) systems work together to control and co-ordinate mammalian systems. You will learn about the structure and function of the nervous system. You will also discover how muscles are able to contract and relax at the molecular level. You will find out how the Venus fly trap plant responds to touch and how plants control the germination of seeds and the growth of stems.

15.1 Control and co-ordination in mammals (Syllabus 15.1.1–15.1.4)

15.2 Transmission of a nerve impulse (Syllabus 15.1.5 – 15.1.9)

15.3 Muscle action (Syllabus 15.1.10–15.1.12)

15.4 Control and co-ordination in plants (Syllabus 15.2.1–15.2.3)

15.1 Control and co-ordination in mammals

Mammals are large, complex organisms that need to respond to changes in their external environment and co-ordinate their internal environment. Mammals do these things using a system of hormones and nerves that work together in different ways to achieve control and co-ordination.

THE ENDOCRINE SYSTEM

The endocrine system is a system of glands, called **endocrine glands**, which produce hormones. These hormones have effects on other organs, called **target organs**.

A hormone is a chemical messenger that travels in the blood.

The word 'endocrine' means 'secreting on the inside'. This refers to the fact that the endocrine glands produce their hormones and release them directly into the blood in the capillaries passing through that gland. Figure 15.1 shows some of the endocrine glands in humans.

> **Link**
>
> You have already met the hormones insulin, glucagon and ADH in Chapter 14.

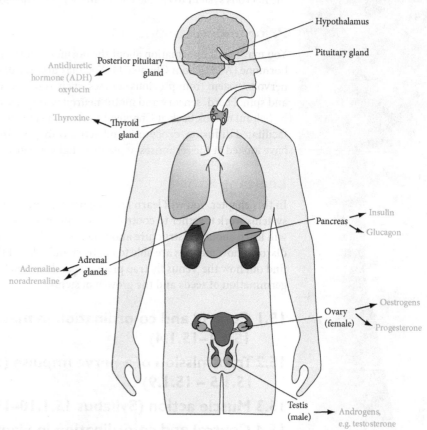

Labels: Hypothalamus · Pituitary gland · Posterior pituitary gland · Antidiuretic hormone (ADH) oxytocin · Thyroxine · Thyroid gland · Pancreas · Insulin · Glucagon · Adrenaline noradrenaline · Adrenal glands · Ovary (female) · Oestrogens · Progesterone · Testis (male) · Androgens, e.g. testosterone

Figure 15.1 Some human endocrine glands.

Many hormones are either polypeptide chains or **steroids**. Steroids are derived from cholesterol. Examples of polypeptide hormones are insulin, glucagon and ADH. Examples of steroid hormones are the sex hormones **oestrogen** and **testosterone**.

Table 15.1 shows some endocrine glands, their hormones and the roles of these hormones.

Endocrine gland	Hormone	Effect of hormone
adrenal glands (adrenal medulla)	adrenaline and noradrenaline	'fight or flight', increased heart rate, increase in blood glucose
posterior pituitary	antidiuretic hormone (ADH)	regulation of body water content
thyroid	thyroid hormone	regulation of metabolism
α-cells of pancreas	glucagon	increase in blood glucose through breakdown of glycogen in liver
β-cells of pancreas	insulin	decrease in blood glucose through uptake by liver and muscles
ovaries (female)	oestrogen	development of female secondary sexual characteristics, regulation of menstrual cycle
testes (male)	testosterone	development of male secondary sexual characteristics

Table 15.1 Some glands and hormones of the endocrine system.

The endocrine system and the nervous system work in different ways and control different types of responses. Table 15.2 compares the functions of both systems.

Endocrine system	Nervous system
no structural connection between endocrine gland and target organ(s)	specific structural connection between receptors, neurones and target cells
response is slower and of longer duration	response is faster and of short-duration
controls longer-term and systemic effects, such as growth, blood glucose and water content	controls short-term and precise effects, such as individual muscle movements
hormones are secreted into blood and travel through whole body	neurotransmitters released only between neurones and diffuse very short distances
hormones bind to specific receptors on target cells	neurotransmitters bind to specific receptors at synapses

Table 15.2 Comparison of the endocrine and nervous systems.

THE NERVOUS SYSTEM

The mammalian nervous system is divided into two main structural parts. The brain and spinal cord comprise the **central nervous system** and the receptors, sensory neurones and motor neurones comprise the **peripheral nervous system**.

Sensory neurones

Sensory neurones are nerve cells that carry electrical impulses, called **action potentials**, from receptor cells to the central nervous system. The receptor cells can be located in the periphery (meaning outer parts) of the body, so some sensory neurones in humans are almost 1 metre long. The structure of a sensory neurone is shown in Figure 15.2.

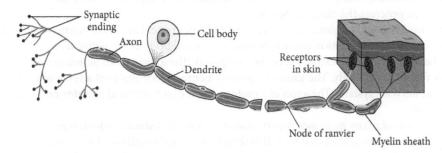

Figure 15.2 Structure of a sensory neurone.

Motor neurones

Motor neurones are nerve cells that carry action potentials from the central nervous system to muscles, or effectors. Figure 15.3 shows the structure of a motor neurone.

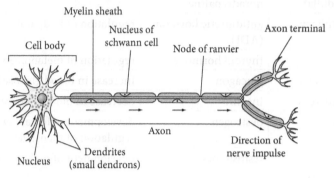

Figure 15.3 Structure of a motor neurone.

Intermediate neurones

Intermediate neurones, sometimes called relay neurones or interneurones, are found within the spinal cord. They care called intermediate because they form connections, or **synapses**, between sensory neurones and motor neurones. Intermediate neurones are much shorter than most sensory or motor neurones. Figure 15.4 shows the structure of an intermediate neurone.

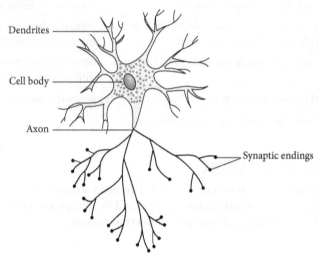

Figure 15.4 Structure of an intermediate neurone.

Structural features of neurones

All neurones have a **cell body**. This part contains the nucleus and other organelles, usually including many mitochondria. Mitochondria are used to produce the ATP needed to generate the electrical impulses.

Dendrites are extensions of the cytoplasm of a neurone. They form connections with other cells and carry action potentials towards the cell body of the neurone.

The **axon** is another extension of the cytoplasm. The axon carries action potentials away from the cell body. The axon of a sensory neurone terminates at a synapse in the central nervous system. The axon of a motor neurone terminates at a **motor end plate** in a muscle.

The axon of many neurones is surrounded by a **myelin sheath**, which is made from **Schwann cells**. Neurones that have this sheath are called **myelinated neurones**. The Schwann cell grows around the axon when the neurone is first formed. The Schwann cell wraps around the axon many times, surrounding the axon with multiple layers of cell

Link

Synapses will be described in Section 15.2.

Tip

Dendrites carry action potentials towards the cell body and **a**xons carry action potentials **a**way from the cell body.

surface membranes of the Schwann cell, as shown in Figures 15.5 and 15.6. Myelin, which is a fatty mixture of lipids and proteins, is produced between the layers formed by the Schwann cells.

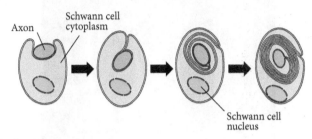

Figure 15.5 Growth of a Schwann cell around an axon seen in transverse section.

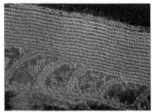

Figure 15.6 False-colour transmission electron micrograph showing the layers of Schwann cell surface membranes around an axon of a human neurone.

The Schwann cell's surface membranes provide a layer of electrical insulation around the axon, enabling the action potential to travel faster than in non-myelinated neurones.

The Schwann cells do not completely cover the axon. There are small gaps between the Schwann cells where the axon is exposed. These gaps are called **nodes of Ranvier** (see Figures 15.2 and 15.3).

Sensory receptor cells

Sensory receptor cells are located in sense organs and are adapted to detect stimuli. The receptor cells are classified according to the type of stimulus they detect. Some of these are summarised in Table 15.3.

Class of receptor	Stimulus	Example	Used for
mechanoreceptor	• pressure • sound	• Pacinian corpuscle • hair cell	• touch • hearing
photoreceptor	light	rods and cones	vision
osmoreceptor	changes in water potential	osmoreceptors in hypothalamus	regulation of body water content
chemoreceptors	chemical	• taste buds • peripheral chemoreceptors	• taste • CO_2 concentration in blood
thermoreceptors	temperature change	warm and cold receptors in skin	regulation of body temperature
nociceptors	• excessive pressure • temperature change	temperature nociceptors in skin	detecting pain or potential tissue damage

Table 15.3 Some mammalian sensory receptor cells.

Sensory receptor cells work by receiving a **stimulus**. If this stimulus is greater than a certain minimum, or threshold, value then an action potential will be generated.

Take the chemoreceptor cells in taste buds (Figure 15.7) as an example. The chemoreceptor cells inside the taste buds are called gustatory cells. Each taste bud contains up to 500 gustatory cells, which detect different tastes, such as sweet, bitter and salty.

Link

The action potential and the way in which the myelin sheath speeds up transmission of the action potential are explained later in section 15.2.

Link

The role of osmoreceptors in the hypothalamus was described In Chapter 14.

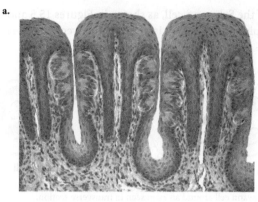

a.

b.

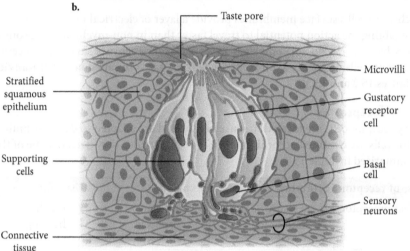

Taste pore

Microvilli

Stratified
squamous
epithelium

Gustatory
receptor
cell

Supporting
cells

Basal
cell

Sensory
neurons

Connective
tissue

Figure 15.7 Taste buds. A photomicrograph of a section from part of the tongue (a) and a drawing showing the structure of a taste bud (b).

There are chemoreceptors for five tastes: sour, salty, sweet, bitter and umami. The chemoreceptors detect either ionic or chemically complex stimuli: sour and salty are ionic stimuli, whereas sweet, bitter and umami are chemically complex stimuli. Numerous microvilli are at the end each gustatory cell. Their membranes contain specialised protein molecules that help to detect the different taste stimuli.

The sour or salty taste mechanism involves detection of H^+ ions, or acidity, for sour tastes and the detection of Na^+, from sodium chloride, for salty tastes. Each ion has its own specific ion channel in the gustatory cells. The movement of these ions through the channels causes a change in the charge difference across the cell surface membrane, called **depolarisation**. The depolarisation causes other ion channels, called **voltage-gated channels**, to open. These voltage-gated channels are specific for calcium ions, Ca^{2+}. Calcium ions move into the cell from the outside.

The sweet or bitter taste mechanism involves specific receptors on the cell surface membranes of the gustatory cells. These receptors bind common sweet or bitter substances. Binding of a substance to the receptor activates a G-protein, which in turn triggers the signal amplification pathway via a second messenger. This pathway causes increased concentration of Ca^{2+} inside the cell.

The umami, or savoury, taste involves receptors that bind amino acids, especially glutamate. These receptors are G-protein coupled receptors that, when activated, cause the release of Ca^{2+} within the cell.

Link

Refer back to Chapter 14 to remind yourself about G-protein coupled receptors and how signals are amplified in a series of steps in a cell. Ion channels are covered in Chapter 4.

Link

Ion channels are covered in Chapter 4.

Increased concentration of Ca^{2+} inside the cells triggers the release of compounds called **neurotransmitters**. Neurotransmitters diffuse across narrow gaps to the nearby sensory neurones where an action potential can be generated. Separate sensory neurones make connections with each type of gustatory cell. These sensory neurones carry the action potentials to the gustatory cortex in the brain where the sensation of taste is identified.

1. Which type of receptors are contained in taste buds?
   A. Chemoreceptors
   B. Nociceptors
   C. Osmoreceptors
   D. Photoreceptors
2. Describe the meaning of each of the following terms:
   a. endocrine gland
   b. hormone
   c. target organ for a hormone.
3. Give the site of release and the target for each of the following hormones:
   a. insulin
   b. glucagon
   c. antidiuretic hormone (ADH).
4. Suggest whether each of the following processes is controlled mainly by the nervous system or by the endocrine system. Give reasons for each of your answers.
   a. Control of growth in young mammals.
   b. Blinking when an object approaches the eye of a mammal.
   c. Response to changing day length with the seasons in mammals.
   d. Carnivorous mammals finding and catching prey.
5. a. Describe the structure of the myelin sheath found in some neurones.
   ☆ b. Suggest why a sensory neurone connecting the fingertip to the spinal cord is myelinated whereas the intermediate neurone in the spinal cord with which it connects is non-myelinated.
6. People with the disease multiple sclerosis (MS) have antibodies specific for myelin. Suggest the effect of these antibodies on the nervous system.

Key ideas

→ The endocrine system uses endocrine glands to secrete hormones, such as insulin, glucagon and ADH, into the blood.
→ Hormones are chemical messengers that act on specific target organs or target cells.
→ The endocrine system and nervous system work together, but in different ways, to achieve control and co-ordination in a mammal.
→ The nervous system contains specialised cells including sensory, motor and intermediate neurones.
→ Receptor cells in the nervous system receive stimuli and generate action potentials.

15.2 Transmission of a nerve impulse

The electrical impulses that are transmitted by the nervous system are called action potentials. The word potential is another word for a voltage or a difference in charge between two sides of a membrane.

When neurones are not transmitting any impulses, they are maintained at **resting potential**.

THE RESTING POTENTIAL

In an axon at resting potential, the difference in potential between the inside of the axon and the outside is about –70 mV. This potential means that the inside of the axon is 70 mV lower potential than the outside.

As this potential difference occurs across the cell surface membrane, we say that the membrane is polarised. Polarised in this case means the charge is different on each side.

The membrane becomes polarised through a combination of active transport and facilitated diffusion of two ions: sodium, Na^+, and potassium, K^+.

A carrier protein called a **sodium-potassium-ATPase pump** actively moves Na^+ out of the axon and K^+ into the axon. The rates of movement of the two ions, however, are not equal. Each pump moves about 200 Na^+ out and about 130 K^+ in every second.

As well as this pump, there are two specific channel proteins in the axon membrane. One of the channels is specific for Na^+ and the other for K^+. The K^+ channel is partly open, so most of the K^+ ions that have been actively pumped in are free to move out again by facilitated diffusion. However, there is a slightly higher concentration of K^+ inside the axon than outside. The Na^+ channels are more fully closed, so Na^+ is less able to move back into the axon down its concentration gradient. This causes a build-up of Na^+ outside the axon, leaving the inside of the axon more negative. The potential of the inside of the axon at resting potential is about –70 mV. The process of maintaining the resting potential is summarised in Figure 15.8.

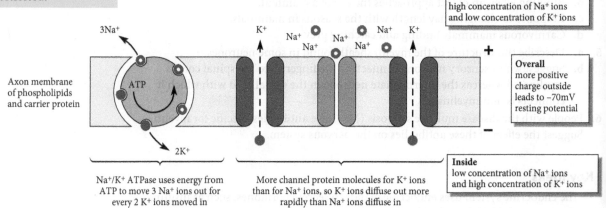

Outside
high concentration of Na^+ ions and low concentration of K^+ ions

Overall
more positive charge outside leads to –70 mV resting potential

Inside
low concentration of Na^+ ions and high concentration of K^+ ions

Axon membrane of phospholipids and carrier protein

Na^+/K^+ ATPase uses energy from ATP to move 3 Na^+ ions out for every 2 K^+ ions moved in

More channel protein molecules for K^+ ions than for Na^+ ions, so K^+ ions diffuse out more rapidly than Na^+ ions diffuse in

Figure 15.8 Maintenance of the resting potential. Na^+ channels are shown in green, K^+ channels in brown.'

THE ACTION POTENTIAL

When a stimulus exceeds a minimum value, called the **threshold**, an action potential will be generated.

Sodium channels in the axon membrane, which have been almost fully closed in the maintenance of resting potential, open fully. This causes a rapid rush, called an influx, of Na^+ into the axon through these open channels. The movement is rapid because of both the concentration gradient and the charge difference. Na^+ ions on the outside are attracted to the negative charge inside the axon.

This sudden influx removes the charge difference across the membrane and we say the membrane has been depolarised. In fact, so many Na^+ rush in that the charge on the inside of the membrane goes past zero and into a positive state. The action potential is typically measured at about +40 mV. The potential inside the axon goes from –70 to +40 mV in about 1 ms.

MEMBRANE REPOLARISATION

As soon as the Na$^+$ channels open for the action potential, the K$^+$ channels also open fully. This allows K$^+$ ions to move out of the cell, down their concentration gradient. This begins to restore the resting potential. In addition, once the action potential has occurred, the Na$^+$ channels close again. The pump continues to operate, moving Na$^+$ back outside to help in restoring the resting potential. The K$^+$ channels remain open until the resting potential has been established again. This causes a brief period of **hyperpolarisation** where the potential inside the axon briefly dips below –70 mV. The time during which hyperpolarisation occurs is called the **recovery period**. This process of restoring the resting potential is called **membrane repolarisation** as the charge difference across the membrane is established once more.

The events that occur in the action potential only happen at a very localised part of the axon membrane. In myelinated neurones, this is at a node of Ranvier, which is the boundary between neighbouring Schwann cells. The effects of these events on membrane potential are summarised in Figure 15.9a. The variation of relative permeability of the membrane to Na$^+$ and K$^+$ with time during an action potential is shown in Figure 15.9b.

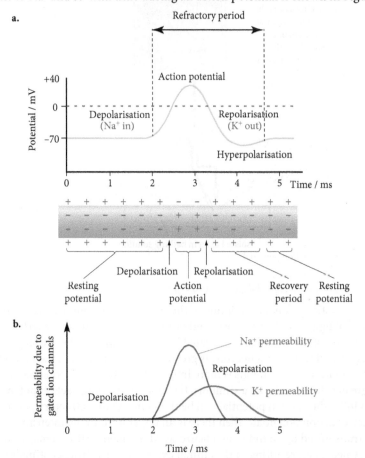

Figure 15.9 Changes in membrane potential before, during and after an action potential (a). Changes in membrane permeability through an action potential (b).

THE REFRACTORY PERIOD

Once the events for an action potential have been initiated, it is not possible to initiate another action potential. Also, after repolarisation, both Na$^+$ and K$^+$ channels are closed for a short time. That means there is a period of about 2 ms following the start of an action potential when another stimulus will not result in a second action potential. This time is called the **refractory period**. The refractory period means that every action potential is separated and action potentials cannot overlap. This separation of action potentials allows them to be distinguished from one another. The number of action potentials arriving in a certain time period can then be used to determine the strength of the stimulus. The refractory period also ensures that action potentials can only travel in one direction along the axon. This is because the region behind an action potential will be in refractory period while the region in front will be ready for stimulation.

EXPERIMENTAL OBSERVATION OF THE ACTION POTENTIAL

Some of the early work on action potentials was carried out on one of the axons of the longfin inshore squid, *Doryteuthis pealeii*. This axon is the largest known, and can be up to 1.5 mm in diameter. It is large enough to insert an electrode into the lumen of the axon. Another electrode is placed outside the axon. Both electrodes are connected to an oscilloscope so the change in potential with time can be measured, as shown in Figure 15.10.

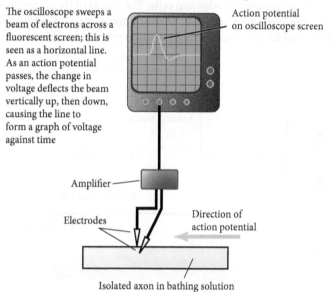

The oscilloscope sweeps a beam of electrons across a fluorescent screen; this is seen as a horizontal line. As an action potential passes, the change in voltage deflects the beam vertically up, then down, causing the line to form a graph of voltage against time

Action potential on oscilloscope screen

Amplifier

Electrodes

Direction of action potential

Isolated axon in bathing solution

Figure 15.10 Measuring an action potential.

THE ALL OR NOTHING PRINCIPLE

The intensity of a stimulus is a description of the strength of a stimulus. For example, the brightness of light, the loudness of sound or the amount of pressure exerted on skin. The intensity of the stimulus is indicated by the frequency of the action potentials, not their size. If a stimulus for a neurone is greater than the threshold, then an action potential will occur. If that stimulus is more intense, then the action potential will be the same. The greater the intensity of a stimulus, the more frequent the action potentials will be, but each individual action potential will be the same. As the refractory period limits the frequency of action potentials, then there is an upper limit to the intensity of stimulus that can be transmitted by an individual neurone. More intense stimuli may also affect more sensory neurones, resulting in the animal experiencing a stronger stimulus. If the intensity of the stimulus drops below the threshold, then there will be no action potential. This is known as the **all or nothing principle**.

THE ADVANTAGE OF MYELINATION

In a myelinated neurone, the insulation provided by the Schwann cells prevents the movement of ions into or out of the axon. However, there is no insulation at the tiny gaps between Schwann cells – the nodes of Ranvier. That means the action potential jumps between the nodes of Ranvier rather than working its way slowly along the axon. This mode of conduction of the impulse is called **saltatory** conduction, from the Latin word *saltare*, meaning 'jumping'. This greatly speeds up the transmission of the impulse compared with that in non-myelinated neurones, as shown in Figure 15.11.

Non-myelinated axon

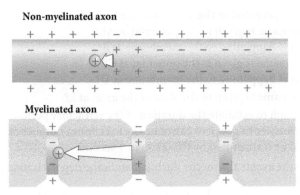

Myelinated axon

Figure 15.11 Myelinated neurones can transmit action potentials many times faster than non-myelinated ones.

Worked example

Describe how the resting potential is maintained in an axon.

Answer

The sodium-potassium-ATPase pump in the membrane of the axon pumps Na^+ out and K^+ into the axon using energy from hydrolysis of ATP. Three Na^+ are pumped out for every two K^+ that are pumped in. This occurs by active transport in the nodes of Ranvier. Facilitated diffusion channels for K^+ are open, so these ions can move down their concentration gradient, and back into the axon. Facilitated diffusion channels for Na^+ are closed, so these ions accumulate outside the axon. This leads to a charge difference across the membrane, and the membrane is said to be polarised. The inside of the axon, at resting potential, is around $-70\,mV$ compared with the outside.

7. Which ions have a net movement into an axon during membrane depolarisation?
   A. K^+
   B. Na^+
   C. Ca^{2+}
   D. Mg^{2+}
8. Name the mechanism by which:
   a. sodium ions move outside an axon to maintain the resting potential
   b. potassium ions move outside an axon to restore the resting potential.
9. Figure 15.12 shows how the potential inside an axon varies with time during an action potential.

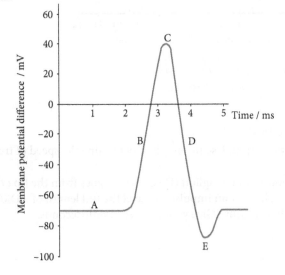

Figure 15.12

     a. What is the potential in the axon measured relative to?

     b.   **i.**   What name is given to the potential at A?

          ii. What is happening to the membrane at B?

          iii. What name is given to the potential at C?

          iv. What is happening to the membrane at D?

          v.   What name is given to the state of the axon at E?

     c. Use the graph to estimate the duration of the refractory period.

     d. Use your answer to part c to calculate the maximum number of action potentials that could be transmitted by this axon in 1 second.

10. **a.** Describe the events that occur during membrane depolarisation in an action potential.

     b. Calculate a typical value for the change in axon potential between resting and action potentials.

11. Describe what is meant by the term saltatory conduction.

Assignment 15.1: Myelinated and non-myelinated neurones

Mammals have myelinated sensory and motor neurones, whereas invertebrates have non-myelinated neurones.

QUESTIONS

Figure 15.13 shows the variation in speed of action potential transmission with both the presence and absence of myelination and with axon diameter.

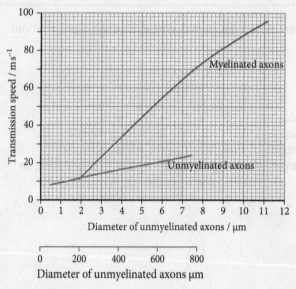

Figure 15.13

A1. Explain the information shown in Figure 15.13.

A2 **a.** Many human axons have a diameter of 10 μm. Use the graph to determine the speed of transmission in these axons.

     b. The longest axon in humans is the dorsal root ganglion (DRG), which goes from the tip of the toe to the base of the brain. Calculate the time taken for an impulse to travel the full length of a DRG that is 1.76 m long. Assume the diameter of the DRG is 10 μm. Give your answer in milliseconds.

A3. Squid are invertebrates. Some larger squid have axons over 1.0 mm in diameter. Use Figure 15.13 to estimate the speed of transmission in these axons.

A4. The earthworm is an invertebrate. The axon that is used to transmit the reflex response that withdraws the head of the worm from touch has a diameter of 90 µm. The axon that is used to transmit the reflex response that withdraws the tail has a diameter of 50 µm. Suggest a reason for the difference and explain your suggestion.

> **Tip**
>
> A reflex response uses the reflex arc. This transmits the impulse from sensory to motor neurone via an intermediate neurone in the spinal cord and bypasses the brain.

SYNAPSES

A synapse is a junction between neurones. The neurones do not touch but have a very narrow gap, called the **synaptic cleft**, between them. Four synapses between one dendrite and four axons are shown in Figure 15.14.

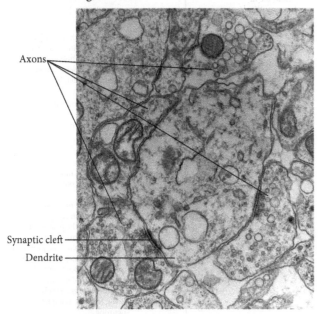

Axons

Synaptic cleft

Dendrite

Figure 15.14 Transmission electron micrograph of synapses.

Inside the axons at the top and at the right of Figure 15.14, vesicles are clearly visible. These vesicles contain a chemical called a neurotransmitter. It is the neurotransmitter that relays the information across the synapse.

One of the most extensively studied neurotransmitters is called acetylcholine. Synapses that use acetylcholine are called **cholinergic synapses**.

Structure and function of a cholinergic synapse

The end of the axon that forms part of the synapse is called the **synaptic knob** or **synaptic button**. The side of the axon where the action potential arrives is called the pre-synaptic side and the other side of the synaptic cleft is called the post-synaptic side.

The pre-synaptic side has many vesicles containing neurotransmitter. The post-synaptic side has receptor proteins in the membrane that faces the synaptic cleft. These receptor proteins are specific for the neurotransmitter.

The function of a synapse is summarised in Figure 15.15.

- An action potential arrives at the pre-synaptic knob.
- This action potential causes channels specific for Ca^{2+} to open in the membrane of the pre-synaptic knob.

- Influx of Ca^{2+} occurs across the membrane of the pre-synaptic knob.
- Presence of Ca^{2+} in the pre-synaptic knob causes the vesicles containing neurotransmitter to move to the pre-synaptic membrane facing the synaptic cleft.
- The vesicles fuse with the membrane, releasing neurotransmitter into the synaptic cleft by exocytosis. The neurotransmitter is later retrieved by endocytosis.
- Neurotransmitter diffuses across the synaptic cleft and binds with the specific receptor proteins on the post-synaptic side.
- This causes opening of channels for Na^+ on the post-synaptic side. If the quantity of Na^+ reaches a threshold, a new action potential is set up.

Link

The merging of the membranes of the vesicles with the pre-synaptic membrane results in exocytosis of the neurotransmitter. Exocytosis was described in Chapter 4.

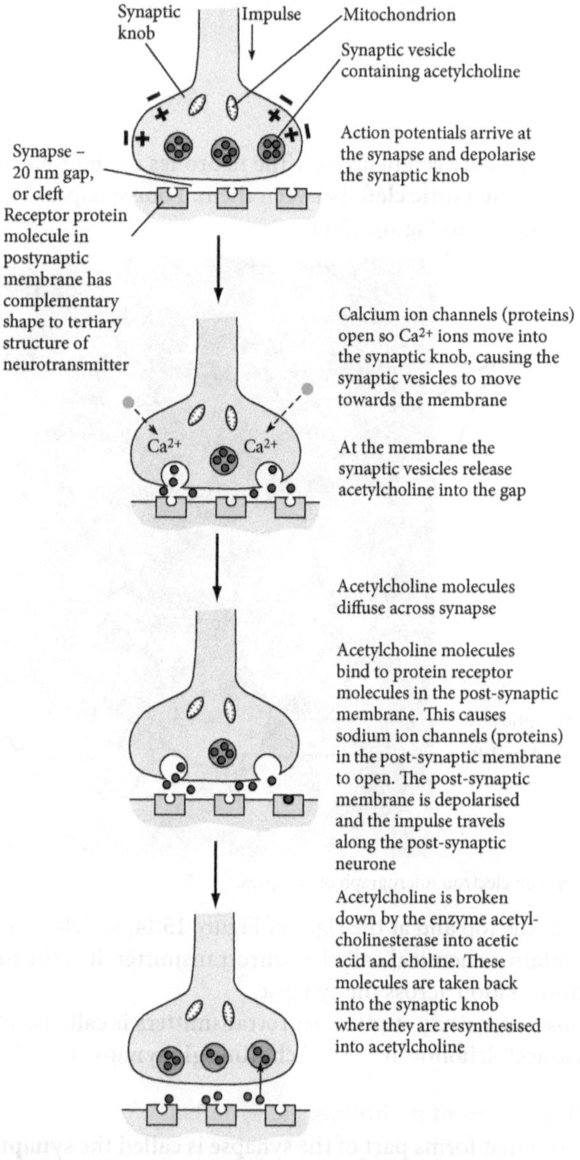

Figure 15.15 Function of a synapse.

Action potentials would continue to be produced on the post-synaptic side as long as the neurotransmitter remained bound to the receptors. This may not be required, as possibly only one action potential may need to be relayed. Hence, an enzyme is present to break down the neurotransmitter in a time period less than the refractory period of the neurone.

In the case of acetylcholine, the enzyme is called **acetylcholinesterase**. The breakdown products of the neurotransmitter are taken into the pre-synaptic side and used to synthesise more neurotransmitter. As these products are taken in by specific carrier proteins and energy from ATP hydrolysis is required for neurotransmitter synthesis, a large number of mitochondria are found in the pre-synaptic knob.

12. **a.** Name the substance that moves into the synaptic knob to stimulate neurotransmitter release.
  b. Name the process by which the neurotransmitter:
  i. is released from the pre-synaptic knob   **ii.** crosses the synaptic cleft.
  c. Explain the role played by acetylcholinesterase in a cholinergic synapse.
  d. Alzheimer's disease is associated with reduced levels of acetylcholine in the brain. Alzheimer's disease can be treated with acetylcholinesterase inhibitors. Suggest how these inhibitors help to restore normal synapse function.

13. One of the neurotransmitters in the brain is glutamate, which acts at 90% of the brain's synapses. Glutamate is found in the food additive monosodium glutamate (MSG). Suggest how consuming food containing MSG can alter brain function.

Key ideas

→ The resting potential in axons is maintained by the active transport and facilitated diffusion of sodium and potassium ions through specific channels.
→ The action potential is caused by sodium channels opening.
→ The resting potential is restored following an action potential after the refractory period.
→ The refractory period limits the frequency of action potentials and ensures that action potentials move in the correct direction.
→ Action potentials move by saltatory conduction in myelinated neurones, which is faster than conduction in non-myelinated neurones.
→ Synapses are junctions between neurones where there are narrow gaps, called synaptic clefts, between neurones.
→ Action potentials arriving at synapses stimulate the influx of calcium ions, which in turn causes release of neurotransmitter.
→ Neurotransmitter diffuses across the synaptic cleft, binds to receptors on the post-synaptic side and stimulates a new action potential.

15.3 Muscle action

Motor neurones transmit action potentials from the central nervous system to effectors, which are the **striated muscles**.

The synapses between motor neurones and muscles are called **neuromuscular junctions**. The structure of a neuromuscular junction is shown in Figure 15.16.

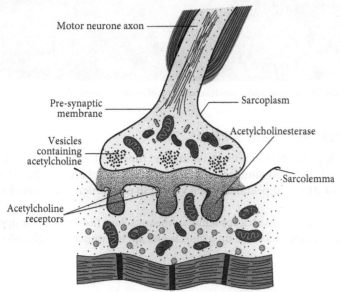

Figure 15.16 A neuromuscular junction.

The events that occur when an action potential arrives at a neuromuscular junction are the same as that in a synapse between two neurones. However, the post-synaptic side of a neuromuscular junction is the membrane of a muscle fibre, called the **sarcolemma**. Muscle fibres are large muscle cells, called myocytes, with many nuclei. Muscle fibres contain **myofibrils**, which are the basic, cylinder-shaped, unit of muscle cells.

The arrival of an action potential at a neuromuscular junction triggers the release of Ca^{2+} within the muscle fibre. The presence of Ca^{2+} within the fibre causes the muscle to contract.

MECHANISM OF TRIGGERING CA²⁺ RELEASE IN A MUSCLE FIBRE

The binding of neurotransmitter, released from the motor neurone, causes depolarisation of the sarcolemma. A wave of depolarisation spreads across the membrane and then into the fibre via the **T-tubule system**. The T-tubules are extensions of the sarcolemma running into the muscle fibre to bring the action potential events closer to the **sarcoplasmic reticulum**. Sarcoplasmic reticulum is similar to smooth endoplasmic reticulum, but acts as a reservoir for Ca^{2+} within the muscle fibre. When relaxed, Ca^{2+} is held within the sarcoplasmic reticulum. In order to start the events needed for contraction, the Ca^{2+} must be released from the sarcoplasmic reticulum into the muscle fibre. The presence of the action potential close to the sarcoplasmic reticulum causes Ca^{2+} channels to open in the membranes of the sarcoplasmic reticulum. The rapid movement of Ca^{2+} down its concentration and electrical gradients causes a sudden increase in Ca^{2+} concentration in the muscle fibre. This process is summarised in Figure 15.17.

Link

The structure of smooth endoplasmic reticulum was described in Chapter 1.

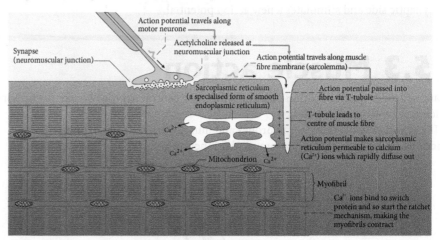

Figure 15.17 Events at a neuromuscular junction on arrival of an action potential.

STRIATED MUSCLE STRUCTURE

Striated muscle is different in both structure and function to either smooth muscle or cardiac muscle. Striated muscle is under voluntary control and is called striated because it has a striped appearance when viewed in photomicrographs or electron micrographs.

Figure 15.18 uses the quadriceps femoris, the muscle at the front of the upper leg, as an example of how myofibrils make up a muscle and appear under the light microscope.

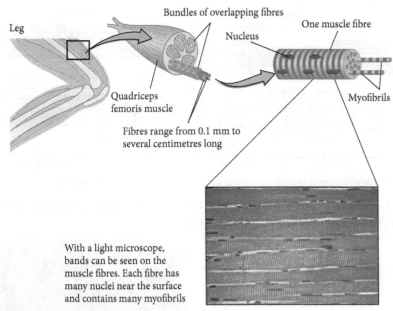

Figure 15.18 The structure of striated muscle.

The striated, or banded, pattern visible in photomicrographs of striated muscle is due to the organisation of actin and myosin filaments within a **sarcomere**. A sarcomere is the repeating unit of the myofibril that causes the contraction. This is shown in Figure 15.19.

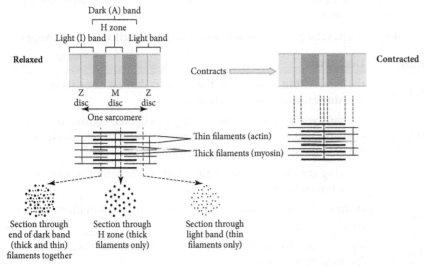

Figure 15.19 The striated appearance of a myofibril is due to actin and myosin. The Z disc is the membrane to which the actin filaments are attached. Z discs form the boundaries of sarcomeres. The M disc is in the middle of the sarcomere. The M disc is where the myosin filaments are attached to each other.

Figure 15.20 shows how the sliding filament model accounts for the changes in appearance of the banding between a relaxed and a contracted myofibril.

The diagram shows how the appearance of a sarcomere changes during contraction. During contraction:

- a sarcomere becomes shorter
- the light bands become shorter
- the dark bands remain the same length
- the distance between the Z discs becomes shorter.

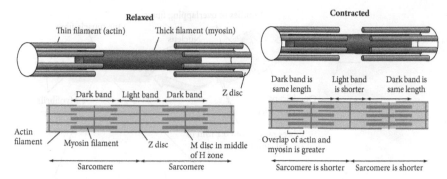

Figure 15.20 The arrangement of actin and myosin causes the banding pattern in striated muscle.

Figure 15.21 shows the relationship between the bands in striated muscle and how these bands appear in a transmission electron micrograph.

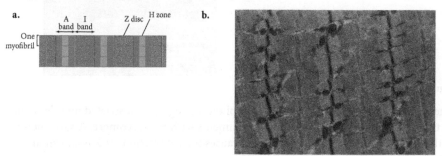

Figure 15.21 The relationship between (a) the bands in striated muscle and (b) the appearance of striated muscle in a false-colour electron micrograph.

Table 15.4 summarises the parts of the sarcomere that are labelled in these images.

Feature	Description
dark band	the band within a sarcomere that appears darkest, consists of myosin filaments, also called the A band
light band	the band within a sarcomere that appears lighter, consists of actin filaments that are not overlapping with myosin, so gets shorter as the muscle contracts, also called the I band
Z disc	marking ends of a sarcomere, this is where the actin filaments are cross-linked to each other
M disc	in the middle of a sarcomere, this is where the myosin filaments are cross-linked to each other
H zone	region either side of the M disc where myosin filaments are not overlapping with actin

Table 15.4

Link

The ability to contract is a property of the spindle fibre described in Chapter 5.

Link

Many cells joining together is part of the formation of xylem vessels that were described in Chapter 7.

THE SLIDING FILAMENT MODEL OF MUSCLE CONTRACTION

The contraction of striated muscle occurs through the contraction of individual sarcomeres within myofibrils of the muscle. Sarcomeres contain four proteins involved in regulating and generating the contraction.

Muscle cells are adapted to their function by having their cell contents organised in such a way to enable contraction. The muscle cells have also joined together to create long fibres.

Myosin is a thick fibrous protein with filamentous tails and globular heads at one end of each of the polypeptide chains. Between the head and the tail is a flexible neck region.

Actin is a thinner fibrous protein that contains binding sites for the myosin heads.

Tropomyosin is a thin fibrous protein with shorter polypeptide chains than actin. In relaxed muscle, tropomyosin covers the binding sites for the myosin heads on actin.

Troponin is a small protein that has a binding site for Ca^{2+} and that is attached to tropomyosin. When troponin binds Ca^{2+} this causes the troponin to displace tropomyosin and expose the binding sites for the myosin heads on actin.

Figure 15.22 summarises the roles of Ca^{2+}, troponin and tropomyosin in regulation of muscle contraction, and shows the arrangement of actin and myosin in transverse section through a sarcomere.

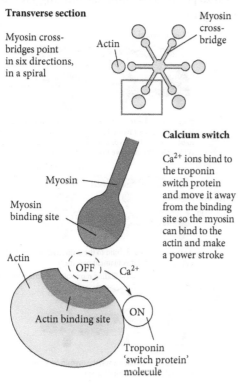

Figure 15.22 Regulation of muscle contraction by Ca^{2+}, troponin and tropomyosin.

The events that enable muscle contraction are together called the sliding filament model. These events, shown in Figure 15.23, are:

- Myosin's binding site on the actin filament is blocked by tropomyosin, which is held in place by troponin. The actin filaments are held in place by the Z discs.
- An action potential stimulates the release of Ca^{2+} into the myofibril's sarcoplasm from the sarcoplasmic reticulum by the mechanism described above.
- Ca^{2+} binds to troponin, causing it to change shape. This pulls the attached tropomyosin away from the actin-myosin head binding site on the actin filament.
- The binding sites for the myosin heads on actin are now exposed. This enables the myosin heads to attach to the binding sites, forming **actin-myosin cross bridges**. At the end of the previous power stroke a molecule of ATP attached to a myosin head.
- On binding with actin, ATPase in the myosin heads hydrolyses ATP, causing the myosin heads to 're-cock' by bending backwards about 5 nm towards the next set of binding sites, further along the actin filament.
- The ADP and P_i remain attached to the myosin head, keeping it in the 'ready' ('cocked') position. If the actin binding sites are still available they reattach further along the actin molecule.

- When they reattach, the ADP and P$_i$ become displaced from the myosin head, resulting in it flipping back to its former position. It rotates though an angle of about 45°. This pulls the actin filaments, shortening the sarcomere by about 5 nm each time. This is known as the **power stroke**.
- Another ATP molecule becomes attached to the myosin head and hydrolysis of this ATP provides the energy to make the myosin head release the actin and once again 're-cock' to its 'ready' position.

This cycle of events repeats until either the muscle becomes fully contracted or action potentials no longer arrive. Once action potentials stop, Ca^{2+} is pumped back into the sarcoplasmic reticulum by active transport. Troponin will have no Ca^{2+} bound to it, so tropomyosin moves back into position blocking the binding sites for the myosin heads. The muscle is now relaxed again.

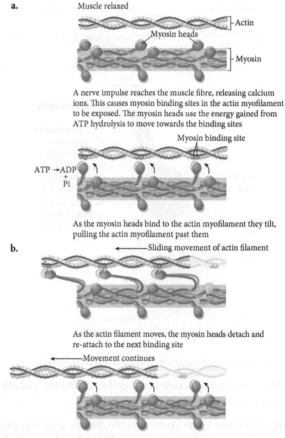

Figure 15.23 (a) The mechanism of the sliding filament model. (b) The location of actin and myosin in a sarcomere.

Even though each movement of a myosin head is only through a distance of 5 nm and creates a pulling force of between 1 and 5 pN, the repeated action of many myosin heads can create considerable contraction distance and considerable force.

Worked example

In Figure 15.23a the myofibril is relaxed. The distance from a Z disc to the ends of the myosin filaments is 0.3 μm. When fully contracted, the ends of the myosin filaments move to be in contact with the Z discs. There are 4000 identical sarcomeres in the length of the myofibril. Calculate the total distance of the myofibril contraction.

Answer

Each sarcomere contracts $0.3\,\mu m \times 2 = 0.6\,\mu m$

$4000 \times 0.6\,\mu m = 2400\,\mu m = 2.4\,mm$

Tip

Remember that a sarcomere contracts equally at both ends.

14. Explain the function of each of the following in the sliding filament model of muscle contraction.
    a. Calcium ions
    b. Troponin
    c. Tropomyosin
    d. ATP

15. In striated muscle, as shown in Figure 15.21a, state what is found at:
    a. the A band
    b. the I band
    c. the H zone.

16. Explain why, during contraction of striated muscle:
    a. the dark band stays the same length
    b. the light band becomes shorter.

17. Duchenne muscular dystrophy is a genetic disease affecting the function of the protein dystrophin. Dystrophin forms part of the structure of the sarcoplasmic reticulum and is involved in connecting actin filaments to other support proteins in the muscle fibre.
    Suggest the effects on muscle action of a lack of functional dystrophin.

18. When a certain mammalian muscle contracts, it shortens by a total distance of 7.8 cm. In one power stroke, myosin heads move the actin filaments by 5 nm.
    a. Calculate the number of power strokes needed for this contraction.
    b. The time taken for this muscle to contract through 7.8 cm is 250 ms. Calculate the number of power strokes needed per second for this contraction.

Key ideas

→ Neuromuscular junctions are synapses between motor neurones and muscles.
→ Action potentials arriving at muscles stimulate the release of calcium ions from the sarcoplasmic reticulum.
→ Calcium ions in a myofibril bind to troponin, causing tropomyosin to move, exposing the binding sites for myosin heads.
→ The sliding filament model explains how muscle contraction occurs using the proteins actin and myosin.

15.4 Control and co-ordination in plants

In order to increase their chances of survival, plants must also respond to changes in their internal and external environments. Plants to not have nervous or endocrine systems, but they do use movement of ions and chemicals that stimulate changes in cells for control and co-ordination.

THE VENUS FLY TRAP

The Venus fly trap, *Dionaea muscipula*, is a plant adapted to grow in waterlogged soil that has very low concentrations of nitrate ions. Nitrate ions are used by plants as a source of nitrogen to synthesise amino acids, nucleotides and other biological molecules containing nitrogen. *D. muscipula* is adapted by its ability to trap insects using the lobes of modified leaves. The plant then digests the insect protein, releasing ammonium ions that act as a nitrogen source.

D. muscipula shows sensitivity to touch and a response of rapid leaf lobe movement.

Figure 15.24 shows *D. muscipula* with leaf lobes in the open and closed positions. The closure of the leaf lobes is triggered when hair-like structures called **trichomes** on the inside of the trap are touched. Two trichomes must be touched in rapid succession, or the one trichome touched twice to trigger the closing action. Once triggered, the lobes close within 0.1 s.

Figure 15.24 The Venus fly trap, *D. muscipula*, in the open (left) and closed positions.

Bending of the trichomes causes distortion of some specialised cells at the base of the trichome. These cells generate action potentials when distorted due to the opening of channels for Ca^{2+} in their cell surface membranes to allow Ca^{2+} to enter. There are two possible mechanisms for the opening of these channels. One of these is the movement causing mechanical deformation of the cell surface membrane. This deformation could cause the calcium channels to become squashed or pulled and therefore allow Ca^{2+} to enter. Another possibility is that the movement in the cell surface membrane physically moves something that otherwise blocks the calcium channel. These proposed mechanisms are very different from the voltage-gated calcium channels found, for example, in axon terminals at synapses.

The specialised cells in the Venus fly trap are more complex than axons because the action potentials are generated at two membranes: the cell surface membrane and the vacuolar tonoplast. When sufficient Ca^{2+} ions are present in the cytoplasm, voltage-gated channels open in the cell surface membrane, allowing Cl^- to flow out and further depolarise this membrane. One theory is that the Ca^{2+} ions diffuse through the cytoplasm towards the vacuole and voltage-gated channels for Cl^- in the tonoplast also open. Cl^- ions flow from the vacuole to the cytoplasm, causing hyperpolarisation of the tonoplast. The Ca^{2+} ions are also thought to open K^+ channels in the cell surface membrane and cause another action potential there. The action potential at the cell surface membrane is complete when K^+ leaves the cytoplasm, and the tonoplast action potential is complete when K^+ moves from the vacuole to the cytoplasm. The duration time of the action potential is 1.5 ms and it has a velocity of 10 ms^{-1}. The mechanism for generating the action potential due to a touch stimulus is summarised in Figure 15.25.

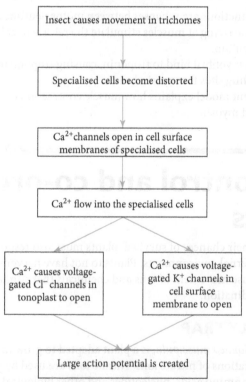

Figure 15.25 The proposed mechanism for generation of an action potential in the Venus fly trap.

There is some evidence that the action potential moves between leaf cells along the plasmodesmata to the base of the leaf trap. Once a threshold value of the charge has been reached, ATP hydrolysis and fast proton transport begin. These enable the opening of aquaporin (water) channels, and water flows down an osmotic gradient from the cells on the inside of the trap to those on the outside. This causes the balance of hydrostatic pressure to change as the cells on the outside of the trap leaf become more turgid.

One action potential is not sufficient to produce a response, and so a second trichome needs to be touched or the same trichome touched again within about 20–30 seconds to elicit a response.

The closure response in *D. muscipula* is caused by the change in curvature of cells close to the bases of the leaf lobes. The mechanism for this change is similar to that by which guard cells open and close stomata. These hinge cells are several millimetres away from the specialised cells that produce the action potentials, so it is likely that a wave of depolarisation spreads from the specialised cells to these hinge cells.

The mechanisms by which the Venus fly trap responds to touch are not yet fully understood. The events described here have been discovered by using techniques such as adding inhibitors that prevent the opening of ion channels.

Two substances, jasmonic acid (JA) and abscisic acid (ABA), are involved in the closure mechanism. JA and ABA cause some cells to change curvature. These cells are called motor cells, as they cause the movement of the leaf lobes.

AUXIN AND ELONGATION GROWTH IN PLANTS

Auxin is the name given to a plant growth regulator compound called indole acetic acid (IAA). IAA was the first such compound to be isolated from plants. Since its isolation, more compounds with roles in plant growth regulation have been discovered, so the term auxins is used to describe a group of compounds.

Auxin is produced in the growing tips of plant roots and stems and is involved in stimulating or inhibiting plant cell elongation. Auxin is present in all parts of the plant, but at different concentrations. It is the concentration of auxin that determines its effect.

Cell elongation in growing roots and stems can be used to produce two different growth effects:

- elongation, where the root or stem gets longer by cells all around the circumference of the root or stem elongating equally
- bending, where the root or stem changes direction of growth. This happens when cells on one side of the root or stem elongate more than cells on the other side.

The bending effect is used by plants to respond to the direction of light and gravity, known as **phototropism** and **gravitropism**, respectively. The gravitropic response of the root and stem in a seedling is shown in Figure 15.26.

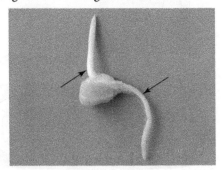

Figure 15.26 This seedling is mounted on vertical card. The positions of greatest cell elongation that cause bending are shown with arrows.

The effect of auxin on roots and stems is different. In roots, auxin inhibits elongation of cells, whereas in stems, auxin stimulates cell elongation.

Link
The mechanism of opening and closing stomata was described in Section 14.2.

Tip
JA and ABA are not plant hormones. Although they are produced in one part of the plant and affect other parts, there are no endocrine glands or circulatory systems in plants.

Mechanism of action of auxin in stems

In order for a plant cell to elongate, the cell wall must be extended. The cell wall cannot be completely broken down to achieve this, as the cell would be in danger of undergoing lysis. Instead auxin makes the cell wall more flexible.

The presence of auxin increases the activity of a proton pump in the cell surface membrane. This pump moves protons, or hydrogen ions, from the cytoplasm into the cell wall. The presence of protons in the cell wall causes acidity and the breakdown of the cross-links between cellulose molecules.

Auxin also increases the expression of the genes coding for proteins called expansins. The expansins are non-enzymatic proteins that help to further increase the flexibility of the cell wall.

The presence of auxin causes channels for potassium ions, K^+, to open in the cell surface membrane. K^+ moves into the cytoplasm from outside, and this lowers the water potential of the cytoplasm. Increased water uptake by osmosis increases the cell's turgor. This, together with the flexibility of the cell wall, causes the cell to elongate. This process is summarised in Figure 15.27.

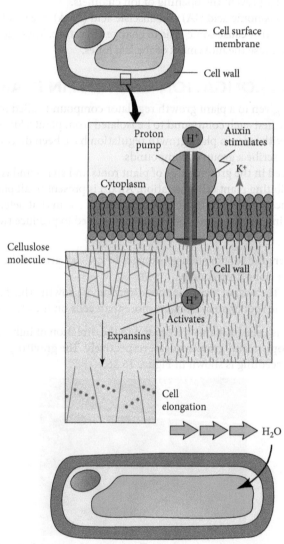

Figure 15.27 Mechanism of auxin action on cells in the plant stem.

Mechanism of action of auxin in roots

Compared with stems, auxin has the opposite effect on root cell elongation, in that the presence of auxin inhibits elongation in roots. Where auxin is present in higher concentration in root cells, the cell walls become alkaline due to increased presence of hydroxide, OH⁻ ions in the cell wall. When the cell wall becomes alkaline, it becomes more rigid. Where auxin is present in lower concentrations, the cell wall becomes acidic and more flexible. In a similar way to stems, the cell wall can be extended when it is flexible but not when it is rigid.

Figure 15.28 shows how varying the concentration of auxin produces different effects in stems and roots.

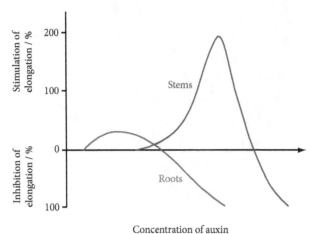

Figure 15.28 Auxin stimulates cell elongation in stems and inhibits cell elongation in roots.

GIBBERELLIN AND GERMINATION

Gibberellins are another class of plant growth regulators. In some plants such as the food crop barley, *Hordeum vulgare*, gibberellins play an important role in breaking dormancy in seeds.

When seeds of a plant such as *H. vulgare* are dropped from the parent plant, the seed enters a state of dormancy in which the water content of the seed is very low and metabolic reactions are almost stopped. This dormant state allows the seed to survive conditions such as cold winters in temperate climates. **Germination** is the process that begins when dormancy comes to an end. Water content increases, metabolic reactions increase and root and shoot growth begins. The role of gibberellins in germination of barley seeds has been extensively studied because of the commercial importance of barley as a food crop.

When soil becomes moist, the dormant seed begins to absorb water. The embryo within the seed synthesises gibberellin in response to increased water content. The embryo is the part of the seed that will grow to form the new plant. The embryo is surrounded by a food store called the **endosperm** that is high in starch content. The endosperm, in turn, is surrounded by a layer high in protein, called the **aleurone layer**. The gibberellin causes the cells in the aleurone layer to start synthesising amylase. Gibberellin acts by increasing transcription of the genes coding for amylase, which catalyses the hydrolysis of starch to maltose. Maltase is also produced, which catalyses the hydrolysis of maltose to glucose. Glucose acts as a respiratory substrate for the germinating seed. This process is summarised in Figure 15.29.

Link

Control of gene expression by regulation of transcription will be described in more detail in Section 16.3.

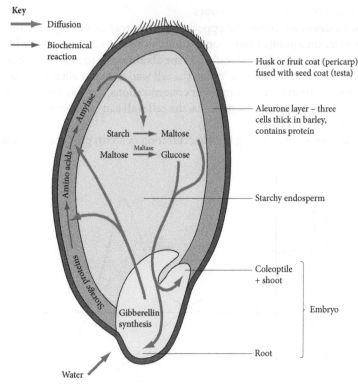

Key
→ Diffusion
→ Biochemical reaction

Husk or fruit coat (pericarp) fused with seed coat (testa)

Aleurone layer – three cells thick in barley, contains protein

Amylase

Amino acids

Starch → Maltose

Maltose —Maltase→ Glucose

Starchy endosperm

Storage proteins

Coleoptile + shoot

Embryo

Gibberellin synthesis

Root

Water

Figure 15.29 The role of gibberellin in the germination of a barley seed.

19. **a.** Name the parts of the Venus fly trap plant that can show rapid movement.
    b. State the stimulus for the rapid movement.
    c. Receptor cells in the Venus fly trap undergo ion movements. What name is given to ion movements that decrease any charge difference across a membrane?
    d. Name two of the ions involved in generation of an action potential in the Venus fly trap.

20. Explain why the term 'plant growth regulator' is a better description for auxin than 'plant hormone'.

21. State the effect of auxin on:
    a. plant stems
    b. plant roots.

22. Name the two substances that when present in a plant cell wall increase the flexibility of the cell wall.

23. **a.** Name the plant growth regulator that breaks dormancy in barley seeds.
    b. Describe what stimulates the production of this growth regulator.
    c. Explain how this growth regulator increases metabolic rate in barley seeds.

24. The function of the tonoplast in isolated cells that produce action potentials in the Venus fly trap can be carried out by making the cell surface membranes of these cells completely permeable to ions. This allows scientists to control the concentrations of ions in the cytoplasm outside of the tonoplast. Explain why the cell surface membranes must be made completely permeable in order to do this.

Experimental skills 15.1: Investigating the effect of gibberellin on germination

A student investigated the effect of gibberellin, GA, on the germination of seeds of the plant *Lepidium virginicum*.

It was suggested that seeds of *L. virginicum* required a short exposure to light to germinate and that germination was temperature-dependent. The student prepared two solutions:

- $0 \, mol \, dm^{-3}$ GA (distilled water)
- $2.0 \times 10^{-3} \, mol \, dm^{-3}$ GA.

The student soaked seeds of *L. virginicum* in either solution for 48 hours.

Seeds that were exposed to light were exposed to red light for 60 seconds.

Seeds were then incubated at 15, 20, 25 or 35 °C on filter paper that was kept moist.

Seeds were treated to the various conditions in batches of 100. After 14 days, the number of seeds germinated in each batch was recorded as a percentage.

Table 15.5 shows the results.

Concentration of GA / $mol \, dm^{-3}$	Light or dark	% Germination at temperature			
		15 °C	20 °C	25 °C	35 °C
0	dark	0	0	0	0
0	light	1	7	1	0
2×10^{-3}	dark	93	99	30	0
2×10^{-3}	light	98	100	56	0

Table 15.5

QUESTIONS

P1. In this investigation, state the:
  a. independent variables
  b. dependent variable.

P2. Draw a suitable graph to display these results.

P3. Describe the patterns shown in the results.

P4. One of the improvements to this experiment would be a statistical analysis of the results.
  a. Describe the statistical test that would be used to test for significance in the results.
  b. Suggest three other improvements to this experiment.

Key ideas

→ The Venus fly trap is a plant that uses touch-sensitive receptors to trigger rapid movement of its leaf lobes.
→ The Venus fly trap has specially adapted hair-like trichomes that cause action potentials when bent.
→ Auxin regulates elongation in plant cells.
→ Auxin stimulates elongation of cells in stems and inhibits elongation of cells in roots.
→ Auxin causes elongation in cells of the stem by causing events that make the cell wall more flexible.
→ Auxin inhibits elongation in cells of the root by causing events that make the cell wall more rigid.
→ Gibberellin is involved in germination of barley seeds by stimulating transcription of genes coding for amylase.
→ The amylase produced in the presence of gibberellin causes starch to be broken down to maltose and then glucose for use in respiration for the germinating seed.

CHAPTER OVERVIEW

Review the mini mind map to link new learning with your prior learning, and to highlight which of the key concepts underpin the chapter.

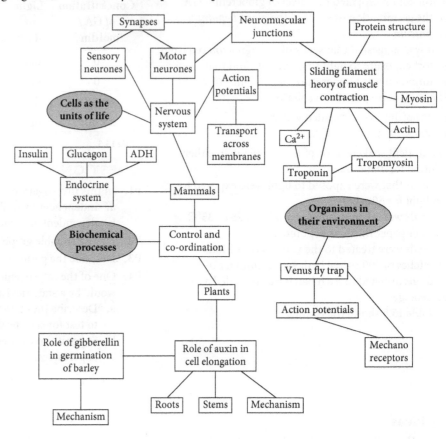

Try copying this mini mind map and expanding upon it. Use your notes from other chapters to help you explore how the essential ideas, theories and principles can be linked further together.

WHAT YOU HAVE LEARNED

Now that you have finished this chapter you should be able to:

- describe the features of the endocrine system with reference to the hormones insulin, glucagon and ADH
- compare the features of the nervous system and the endocrine system
- describe the structure and function of a sensory neurone and a motor neurone and state that intermediate neurones connect sensory neurones and motor neurones
- outline the role of sensory receptor cells in detecting stimuli and stimulating the transmission of impulses in sensory neurones
- describe the sequence of events that results in an action potential in a sensory neurone, using a chemoreceptor cell in a human taste bud as an example
- describe and explain changes to the membrane potential of neurones, including:
  - how the resting potential is maintained
  - the events that occur during an action potential
  - how the resting potential is restored during the refractory period
- describe and explain the rapid transmission of an impulse in a myelinated neurone with reference to saltatory conduction
- explain the importance of the refractory period in determining the frequency of impulses

- describe the structure of a cholinergic synapse and explain how it functions, including the role of calcium ions
- describe the roles of neuromuscular junctions, the T-tubule system and sarcoplasmic reticulum in stimulating contraction in striated muscle
- describe the ultrastructure of striated muscle with reference to sarcomere structure using electron micrographs and diagrams
- explain the sliding filament model of muscular contraction including the roles of troponin, tropomyosin, calcium ions and ATP
- describe the rapid response of the Venus fly trap to stimulation of hairs on the lobes of modified leaves and explain how the closure of the trap is achieved
- explain the role of auxin in elongation growth by stimulating proton pumping to acidify cell walls
- describe the role of gibberellin in the germination of barley.

CHAPTER REVIEW

1. **a.** Explain why insulin is described as a hormone.
    b. Explain how some hormones, such as adrenaline, can have effects on many different types of cells in different parts of the body.
2. Describe two ways in which the endocrine system and nervous system achieve control and co-ordination differently.
3. Explain the purpose of each of the following in a sensory neurone:
    a. mitochondria
    b. dendrites
    c. voltage-gated sodium channels.
4. Explain what is meant by refractory period in a neurone.
5. Describe how an impulse is transmitted across a cholinergic synapse.
6. **a.** Describe how the sliding filament model explains the shortening of a myofibril during muscle contraction.
    b. Following the death of a mammal, the muscles stiffen and become rigid. This is called rigor mortis. Use the sliding filament model to suggest how rigor mortis occurs.
7. Explain how auxin causes a plant cell to elongate.
8. Malt is used in the food industry and is produced from barley seeds. When making malt, barley seeds are harvested, placed on a damp surface and allowed to germinate. Suggest how adding gibberellin could make this process more efficient.

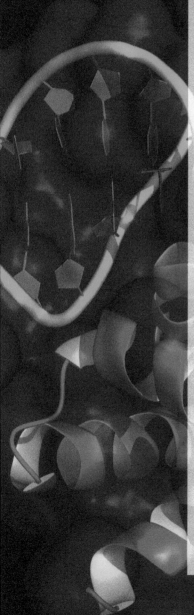

Chapter

16 Inheritance

'Restarting cells'

In 2012, the scientists John Gurdon and Shinya Yamanaka were awarded a Nobel Prize for work they conducted nearly 50 years apart. Gurdon's work in the 1960s demonstrated that an intestinal cell from an adult frog still has all the information needed to form a new tadpole. In 2006, Yamanaka showed that skin cells from a mouse could be reprogrammed to become cells able to form different tissues.

Many degenerative diseases could be cured by replacing damaged or lost cells with new ones. But our ability to 'press restart' and reprogramme a specialised cell is yet to be achieved. The background image shows two proteins, called transcription factors, binding to DNA. They are able to activate genes that are expressed in embryonic stem cells.

Prior understanding

You may remember that mitosis is a type of nuclear division that produces genetically identical daughter cells, and that chromosomes are structures that contain genes, which are passed on during mitosis to daughter cells. You may recall that protein synthesis requires transcription of DNA into RNA, which determines the order of amino acids in polypeptides during translation.

Learning aims

In this chapter, you will learn how genetic information is transmitted between generations to maintain the continuity of life. In sexual reproduction, you will see how meiosis introduces genetic variation so that offspring resemble their parents but are not identical to them. You will explore how genetic crosses reveal how some features are inherited, and how the phenotype of organisms is determined partly by the genes that they have inherited and partly by the effect of the environment.

16.1 Passage of information from parent to offspring (Syllabus 16.1.1–16.1.7)

16.2 The roles of genes in determining the phenotype (Syllabus 16.2.1–16.2.6)

16.3 Gene control (Syllabus 16.2.7, 16.3.1–16.3.4)

16.1 Passage of information from parent to offspring

Humans, like other **sexually reproducing** organisms, inherit characteristics from their parents via the sex cells (**gametes**). As you saw in Chapter 5, each of the cells in our body has two complete sets of chromosomes – 23 inherited from our mother and 23 from our father, making 46 in all. Cells with two sets of chromosomes are **diploid** cells, whereas cells with just one member of each pair, called gametes, are **haploid** cells.

Each chromosome has a matching partner of a near-identical size, so chromosomes belong to **homologous** pairs. Homologous pairs of chromosomes share the same **genes** at the same positions. These positions are called **loci** (singular: **locus**). There are two chromosomes assigned the number 1, two chromosomes with the number 2, and so on, all the way up to 22. These are the **autosomes**. The final pair, the **sex chromosomes**, are not homologous as they are different sizes and have different genes. When they are placed together, the homologous pairs of chromosomes are called a **bivalent**. Figure 16.1 shows the chromosomes found in a diploid human cell and summarises the difference between chromosomes, sister chromatids and bivalents.

Tip

<u>D</u>iploid cells have <u>d</u>ouble the number of chromosomes compared to <u>ha</u>ploid cells, which have just <u>ha</u>lf the number.

Link

You will recall from Chapter 6 that a gene is sometimes defined as a length of DNA that codes for the production of a particular protein.

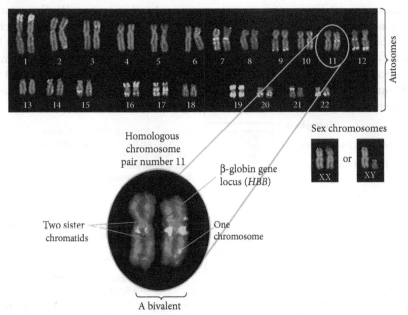

Figure 16.1 The full complement of human chromosomes arranged in numbered homologous pairs. These are metaphase chromosomes. These pictures are prepared by stimulating cells to divide and then adding a chemical that stops spindle movement, thus 'freezing' the cells in mid-mitosis. The cell is then photographed and the chromosomes are arranged by cutting and pasting, before they are assigned artificial colours.

Therefore, when the nuclei of two haploid gametes fuse together at **fertilisation**, they form a cell with pairs of homologous chromosomes – a diploid **zygote**. It is important that one copy of each chromosome, and hence of each gene, is passed on from each parent; otherwise there would be an incomplete set of instructions in the zygote.

THE PROCESS OF MEIOSIS

Meiosis is a type of nuclear division that produces haploid cells from diploid cells. Although it occurs in all organisms that reproduce sexually, you will most often consider the mechanism in humans in this chapter. Here, it only takes place in the **ovaries** and **testes**, to produce eggs and sperm. It involves two cell divisions, one after the other, and in so doing it halves the genetic material in diploid cells.

Tip

Meiosis **m**akes **e**ggs **i**n the **o**varies, and **s**perm **i**n the testes which are held in the **s**crotum.

Figure 16.2 shows how meiosis fits into the life cycle of a sexually reproducing animal. We can use the letter n to show the number of chromosomes in a set. A haploid cell therefore has n chromosomes, and a diploid cell has 2n chromosomes. In humans, n is 23. Different species have different chromosome numbers; in fruit flies, for example, n is 4. In humans, a cell about to begin meiosis has 46 chromosomes. When two gametes fuse during fertilisation, a diploid zygote is formed.

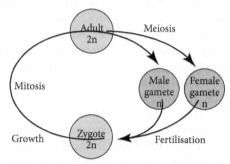

Figure 16.2 The fusion of a haploid male and a haploid female gamete during fertilisation produces a diploid zygote.

Link

Remember that between S-phase and cytokinesis, each chromosome consists of two sister chromatids; revisit Chapter 5 to refresh your memory of the cell cycle.

Meiosis has two separate divisions. Both have the same names as the stages that occur in mitosis, but the events are different. They have a number to denote whether they refer to the first or second division (for example, prophase I and metaphase II). The events in prophase, metaphase, anaphase and telophase mostly mirror those in mitosis, which you encountered in Chapter 5. In the same way as it happens in advance of mitosis, the DNA in each chromosome duplicates during the S-phase of the preceding cell cycle, forming two identical sister chromatids that remain attached at the centromere. Figure 16.3 shows a simple summary of how meiosis happens, using an imaginary diploid cell with just two chromosomes as an example.

Tip

During the first division of meiosis a process called crossing over occurs. You will explore the importance of this later in this section.

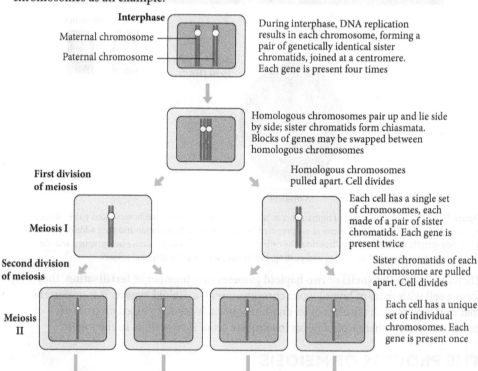

These cells develop and become gametes

Figure 16.3 The stages of meiosis in an imaginary animal cell that has just one pair of homologous chromosomes (2n = 2). In the first meiotic division (prophase I to telophase I), one member of each homologous pair of chromosomes passes into each new cell. In the second meiotic division, the individual sister chromatids are pulled apart so that one chromatid (now called a chromosome) goes into each daughter cell. The end result of meiosis is four genetically unique haploid cells, each containing a single set of chromosomes.

During prophase I, the nuclear envelope disintegrates, and the chromosomes condense and thicken. They begin to move towards the equator of the cell, pulled by spindle fibres that have attached to their shared centromere (see Figure 5.9, Chapter 5). However, then something different happens, compared to mitosis. Each chromosome finds and pairs up with its homologous partner, so that homologous pairs of chromosomes (bivalents) are formed and arrange themselves either side of the equator during metaphase I. Figure 16.4 shows the differences between metaphase of mitosis and metaphase I of meiosis in a cell of an organism with a diploid number of 8.

Tip

The centre of the cell should be referred to as the equator, and the ends should be referred to as the poles.

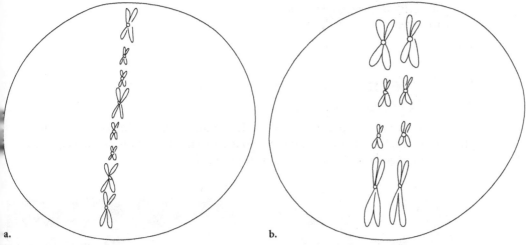

a.   b.

Figure 16.4 A student's sketch that shows the location of chromosomes during metaphase of mitosis (a) compared with metaphase I of meiosis (b) in an organism with a diploid number of 8. Spindle fibres are not shown. In humans, 23 homologous pairs of chromosomes line up along the equator in both processes, but they are unpaired in mitosis, unlike in meiosis.

During anaphase I, the members of each bivalent are then pulled apart from each other and towards opposite poles of the cell. This means that one chromosome from each homologous pair goes to each pole. The spindle fibres then disassemble as the cell undertakes telophase and the nuclear envelope reforms, followed by cytokinesis to produce two new cells. Each of these cells has a nucleus containing 23 chromosomes, which consist of one member of each original homologous pair. The chromosomes decondense as the cell enters interphase. In humans, the length of time this cell spends in interphase between the cell divisions called meiosis I and meiosis II ranges from a few minutes to many decades.

Meiosis II begins when the nuclear envelope disintegrates and the chromosomes again condense in prophase II. During metaphase II, the chromosomes line up along the equator of the cell but at 90 degrees to the equator in meiosis I. Again, this is brought about by the formation of the spindle fibres that connect to the centromeres of the chromosomes. This process then resembles mitosis, but with half the number of chromosomes of a diploid cell. Centromeres divide, spindle fibres contract, and sister chromatids are pulled to opposite poles of the cell. Telophase and cytokinesis then occur, during which the spindle disappears, the nuclear envelope reforms, chromosomes decondense and cytokinesis occurs.

1. Why is meiosis sometimes described as a reduction division?
2. Copy and complete Table 16.1 to summarise the differences between mitosis and meiosis.

Tip

Meiosis II (the second meiotic division) is certainly not a repeat of meiosis I (the first meiotic division). Look at Figure 16.4 carefully and make a note of the differences.

Mitosis	Meiosis
one division	two divisions
two daughter cells produced	
daughter cells have the same number of chromosomes as the parent cell	
homologous chromosomes do not pair	
no crossing over takes place	
daughter cells are genetically identical to the parent cell	

Table 16.1

INTERPRETING PHOTOMICROGRAPHS AND DIAGRAMS OF MEIOSIS

In Chapter 5, you developed your skills in identifying the stage of mitosis of cells in photomicrographs, with justifications for your answers. These skills should help you to identify the stages of meiosis during gamete production.

a.

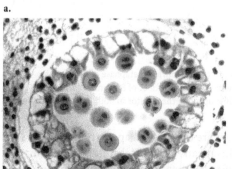

b.

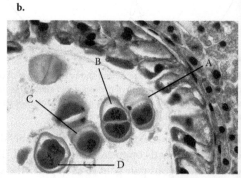

Figure 16.5 The process of meiosis in the anther of a lily (a) and the seminiferous tubules of the testes (b). Pollen grains and sperm are forming from a layer of cells on the inner surface of the pollen sac or tubule and undertake meiosis as they reach the centre. It is possible to distinguish the chromosomes in each cell and therefore determine the stage of meiosis (see text below).

Look carefully at the cells labelled A to D in Figure 16.5. Cell A contains chromosomes that still consist of two sister chromatids joined by a centromere. Therefore, this cell must still be undertaking events in meiosis I. The position of the chromosomes – which are a characteristic V-shape and appear to be separating from each other – suggests that this cell is undertaking anaphase. Cells B, C and D contain chromosomes that consist of only one sister chromatid. Cell B is undertaking telophase I, and will shortly undergo meiosis II. Cell C is about to undertake metaphase II, and the chromosomes are being pulled towards the equator of the cell. Cell D is undertaking anaphase II as the sister chromatids that formed each chromosome are being pulled apart to opposite poles of the cell.

Correctly identifying the stages of meiosis from photomicrographs and diagrams can be difficult, because you must remember whether or not homologous chromosome pairs have been separated. Figure 16.6 shows a flow diagram that will help you to identify the stage of meiosis that a cell is undertaking.

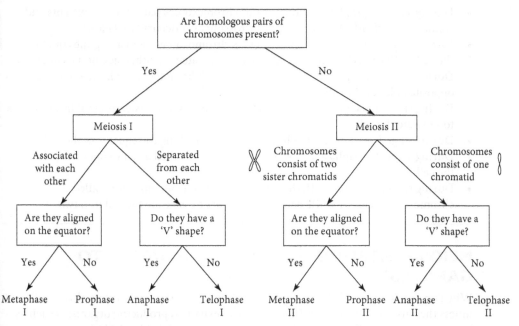

Figure 16.6 How to identify the stages of meiosis.

Worked example

The images in Figure 16.7 show photomicrographs of cells from the anther of a lily plant at various stages of meiosis.

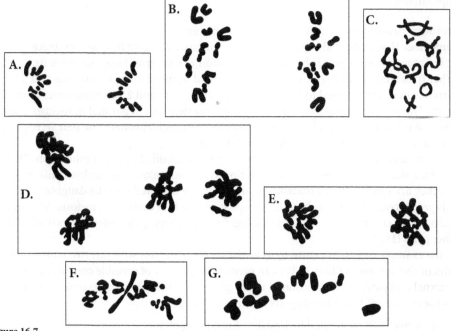

Figure 16.7

Starting with image C, arrange the images into the correct order to describe the behaviour of chromosomes in meiosis.

Answer

The order of photomicrographs should be: C → G → B → E → F → A → D.

- During stage C, prophase I, homologous chromosomes pair to form bivalents and condense (thicken). During this stage, crossing over occurs (see below).
- During stage G, metaphase I, homologous chromosomes align along the equator of the cell, with each chromosome randomly orientated on either side of the equator.
- During stage B, anaphase I, the members of each homologous pair are pulled apart to opposite poles of the cell.
- During stage E, telophase I, the separated chromosomes cluster as cytokinesis occurs to separate the cell into two.
- During stage F, metaphase II, chromosomes align along the equator of the cell.
- During stage A, anaphase II, sister chromatids are pulled apart to opposite poles of the cell.
- During stage D, telophase II, the separated sister chromatids (now called chromosomes) cluster while cytokinesis occurs to separate the cell into two.

MEIOSIS INTRODUCES GENETIC VARIATION TO GAMETES

Why don't we look exactly like our parents? Why are brothers and sisters always different, unless they are identical twins? This is because sexually reproducing organisms such as humans are **genetically unique**. You have never been exactly identical to anyone before you, and there will never be someone exactly the same as you.

As we will see in the following section of this chapter, meiosis is mainly responsible for the 'shuffling' of many of the traits between one generation and the next. Hence, no two gametes will ever be the same. However, the fact that fertilisation of two gametes is also a random process means that which traits are passed to the next generation can be unpredictable.

Random orientation (independent assortment)

Remember that in diploid organisms such as humans, a zygote has two complete sets of chromosomes, one inherited from its mother (maternal) and one from its father (paternal). In humans, which have a diploid number of 46, this means that 23 chromosomes were passed to the zygote from the sperm, and 23 chromosomes from the ovum. During the formation of these gametes, bivalents are separated in meiosis I into different cells. The process by which the members of each chromosome pair were passed into these cells is random. How does this happen?

The maternal and paternal chromosomes can be reshuffled in any combination. That is, on which side of the equator they are positioned during meiosis I is random. This means that they are **randomly orientated**, or **independently assorted** into the daughter cells. This process occurs at metaphase and anaphase of the first division of meiosis. Maternal and paternal chromosomes of one homologous pair behave quite independently of all the other pairs (Figure 16.8).

The example shown in Figure 16.8a is of a simple organism with just two homologous pairs of chromosomes (2n = 2). We can predict the number of possible combinations of maternal and paternal chromosomes in the gametes by using the expression 2^n, where n = the number of homologous pairs of chromosomes:

- in Figure 16.8a, n is equal to 2, so $2^2 = 4$ unique gametes
- in Figure 16.8b, n = 3, so $2^3 = 8$ unique gametes
- in human cells, which have 23 pairs of chromosomes, n = 23. Therefore, the number of unique gametes that could arise due to random orientation is $2^{23} = 8\ 388\ 608$!

So, more than 8 million genetically unique gametes could be produced by a human by simply distributing different parental chromosomes into different gametes. But this figure is, in fact, much greater. We will explore this in the next section.

Tip

You can use many different words to describe the characteristics of an organism. These include 'traits', 'qualities' and 'features'.

Link

The variation due to meiosis is mainly responsible for maintaining variation between individuals of a species. This reduces the likelihood of extinction, as you will see in Chapter 17.

Tip

Random orientation (independent assortment) of homologous chromosomes occurs during meiosis I and of sister chromatids during meiosis II.

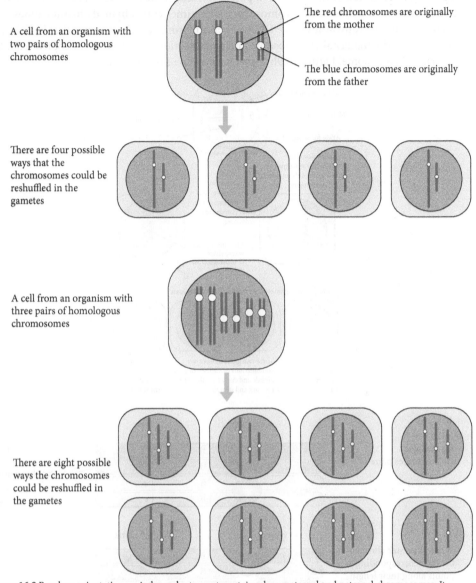

a.

A cell from an organism with two pairs of homologous chromosomes

The red chromosomes are originally from the mother

The blue chromosomes are originally from the father

There are four possible ways that the chromosomes could be reshuffled in the gametes

b.

A cell from an organism with three pairs of homologous chromosomes

There are eight possible ways the chromosomes could be reshuffled in the gametes

Figure 16.8 Random orientation, or independent assortment, involves maternal and paternal chromosomes. It happens during the first division of meiosis. In this example, imaginary cells from organisms with just two pairs (**a**) or three pairs (**b**) of homologous chromosomes undertake meiosis I.

Crossing over

So, can we assume that a human is able to produce 8 388 608 unique gametes? This is actually a vast underestimate. Although the number seems large, it is actually less than the number of sperm released by a man during one ejaculation. Therefore, if random orientation was the only method by which variation was assigned to gametes, it would be perfectly possible for two sperm to be identical.

Even more variation is introduced into gametes during prophase I of meiosis by a process called **crossing over**. This is shown in Figure 16.9a. Once a homologous pair of chromosomes have paired to form a bivalent, the sister chromatids from the two non-sister chromosomes wrap around each other. At the points where the sister chromatids touch each other (or 'cross over'), they may break and re-join. The positions at which non-sister chromatids appear joined to each other are called **chiasmata** (singular – **chiasma**, which means 'crosspiece'). When the chromosomes separate again during anaphase, parts of the sister chromatids of homologous chromosomes are swapped from one chromatid to another. The position of chiasma formation varies and is random, so this can produce a great variety of new combinations of parental traits in the gametes. As a result, when pulled apart at anaphase, the chromosomes, sometimes called **hybrid chromosomes**, or **recombinant chromosomes**, have DNA containing some blocks of genes that have been copied from the maternal chromosome and some sections that have been copied from the paternal chromosome. Figure 16.9b shows a close-up of this process.

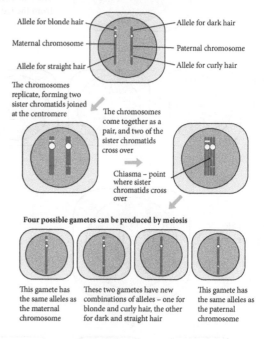

a.

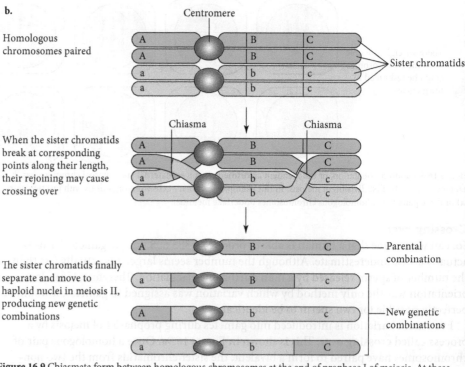

b.

Tip

Be very careful when describing crossing over. It is defined as the 'exchange of genetic material between non-sister chromatids of homologous pairs'.

Tip

Crossing over does not happen between chromosomes during fertilisation. The genetic material in chromosomes inherited from *your* parents is exchanged in *your* testes or ovaries during gamete formation.

Figure 16.9 Chiasmata form between homologous chromosomes at the end of prophase I of meiosis. At these points, parts of the maternal chromosome separate and join with the paternal chromosome, and vice versa. This produces new 'hybrid' chromosomes, which are genetically different from each other and from the chromosomes that formed them. The example chosen, for hair colour and texture, is for example purposes only (**a**). A close-up of crossing over shows that chiasmata form between non-sister chromatids of homologous chromosomes (**b**).

Random fusion of gametes at fertilisation

Random orientation leading to independent assortment of chromosomes, and crossing over, are two methods by which gametes are produced that are genetically unique. However, beyond meiosis, there is another means by which it is possible for the combination of alleles to be further mixed. This is due to the process of fertilisation itself. In humans, fertilisation is random – any female gamete can join with any male gamete.

Worked example

Drosophila melanogaster is the binomial name of the fruit fly. Figure 16.10 shows two individuals with different genetic traits.

Figure 16.10 The fruit fly, *Drosophila melanogaster*. Shown here are a red-eyed female and a white-eyed male.

Fruit flies have a haploid number (n) of 2. Sketch diagrams to show the following. In your diagrams:

- draw the maternal and paternal chromosomes in different colours
- assume that the chromosomes are condensed (short and thick)
- try to illustrate how crossing over changes the appearance of the chromosomes
- ignore the existence of sex chromosomes.

a. The chromosomes found in a cell just before prophase I of meiosis.

Answer

A total of four chromosomes (two homologous pairs) are present, each consisting of two sister chromatids. The members of each pair are inherited from the mother (white) or father (black).

b. The chromosomes found in one cell just before prophase II of meiosis.

Answer

A total of four chromosomes are present, each consisting of two sister chromatids. At various regions, exchange of genetic material has occurred between non-sister chromatids of homologous pairs.

c. The chromosomes found in a gamete after meiosis.

Answer

A total of two chromosomes are present (two pairs), which now consist of structures with only two arms. They are recombinant chromosomes as they contain both maternal and paternal DNA. Note that the other three cells (gametes) that form are not shown.

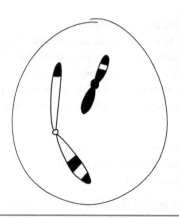

3. Match the statements to make five correct sentences that describe some of the events of meiosis.

1. At the end of meiosis I ...	**A.** ... is due to random alignment of bivalents on either side of the equator in metaphase I.
2. At the end of meiosis II ...	**B.** ... there are four cells, each containing one chromosome, derived from a sister chromatid of the bivalent found in the original cell.
3. Random orientation ...	**C.** ... means that any paternal gamete can fertilise any maternal gamete.
4. Crossing over ...	**D.** ... is dependent on the close proximity of bivalents on either side of the equator in metaphase I.
5. Random fertilisation ...	**E.** ... there are two cells, each containing one copy of each chromosome (consisting of two sister chromatids) derived from each homologous pair.

4. An analysis of cells from a species of bear showed that most have 74 chromosomes.
   a. State the haploid number (n) of bears.
   b. Meiosis introduces genetic variation into gametes through a number of processes.
      i. Calculate the possible number of unique gametes in a bear that could be produced by random orientation of chromosomes during metaphase I of meiosis. Express your answer in standard form.
      ★ ii. Apart from mutation, explain how further genetic variation between gametes can be introduced during meiosis.
   🔍 c. Suggest why prophase I in bears is likely to take slightly longer than prophase I in humans.

Key ideas

→ Gametes are haploid cells (n) that fuse during fertilisation to form a diploid zygote (2n).

→ Homologous pairs of chromosomes have the same size, and contain the same genes at the same loci (positions).

→ A reduction division during meiosis is required to halve the number of chromosomes in the production of gametes, so that when fertilisation occurs the diploid number is restored.

→ During meiosis, chromosomes in plant and animal cells behave in characteristic ways and cells undertaking the sub-stages of this process can be identified in photomicrographs and diagrams of meiosis. Homologous pairs of chromosomes (bivalents) are separated into two different cells during meiosis I, while sister chromatids are separated into different cells during meiosis II. The disintegration and reforming of the nuclear envelope and the assembly and disassembly of spindle fibres are important in this process.

→ Crossing over and random orientation (independent assortment) of homologous chromosomes and sister chromatids during meiosis produces genetically different gametes.

→ The random fusion of gametes at fertilisation produces genetically different individuals.

16.2 The roles of genes in determining the phenotype

Link

You will encounter selective breeding, also called artificial selection, in Chapter 17.

It may surprise you to hear that humans have been applying the science of **inheritance**, also called **genetics**, for over 10 000 years. Most of the domestic animals and crop plants that exist today are very different from their wild ancestors, and have been gradually developed by the process of selective breeding (Figure 16.11).

As we have already seen in this chapter, all sexually reproducing organisms undertake meiosis to produce gametes. This enables them to pass on information to their offspring in the next generation. Can we predict how offspring will differ from each other, and from their parents?

KEY TERMS IN INHERITANCE

An important message from our work on meiosis earlier in this chapter is that diploid organisms such as humans have two copies of each gene. One copy is inherited from the mother (via an ovum) and one from the father (via a sperm or pollen grains). Remember that most organisms have several thousand genes. The two sets of genes combine to form the genes of the zygote, and were copied over and over again as the zygote divided by mitosis to form all the cells in the new individual.

Many genes occur in more than one version. Different versions of the same gene are called **alleles**. A simple example would be a gene for flower colour with two alleles, one coding for red flowers and one coding for white flowers. Both are versions (alleles) of the same gene – for flower colour – but give different colours. An organism's **genotype** is a description of which two alleles, or 'gene types', it has inherited.

Because diploid organisms such as humans have two copies of each gene, one on each member of a homologous pair, the alleles they possess for one gene may be different or the same. If they are the same, the individual is said to be **homozygous** for these alleles, whereas if they are different, the individual is **heterozygous** for these alleles. It makes

Figure 16.11 The three colours of Labrador dog – black, gold and brown. The different traits have been produced by many centuries of selective breeding. But is it possible to accurately predict the colour of offspring born to parents of different coat colours?

things easier if we use shorthand symbols to represent different alleles of a gene. The convention is to use one letter for the two alleles. The upper-case letter is used for the **dominant** allele, and the lower-case letter for the **recessive** allele. As there are two copies of each gene in each cell, there are three possible combinations of alleles.

Let's assume that in Labrador dogs, coat colour is controlled by one gene with two alleles A and a (you will see later that this is not actually the case). If this were the case, the genotypes for coat colour in the three dogs in Figure 16.11 can be described as:

- homozygous dominant (AA)
- heterozygous (Aa)
- homozygous recessive (aa).

The **phenotype** is defined as the observable or measurable features that an organism possesses. In the case of Labradors, 'black coat' would be an example, whereas in humans, 'dark hair' and 'left handed' would be examples. Most phenotypic features can also be affected by the environment, for example, skin colour becomes darker when exposed to the ultraviolet radiation in sunlight, although they usually have a significant genetic basis.

Many genetic diseases are causes by mutant alleles that do not code for the required protein. **Albinism** is a good example. The individual will not have albinism unless they inherit two faulty alleles, one from each parent. People who have homozygous recessive alleles will have pale skin and eyes (Figure 16.12). People who have heterozygous alleles for the albinism gene do not show the characteristic features of the albinism phenotype because one functioning allele is all you need to produce pigment. This means that the albinism allele is recessive: it must be present twice in the genotype in order to affect the phenotype.

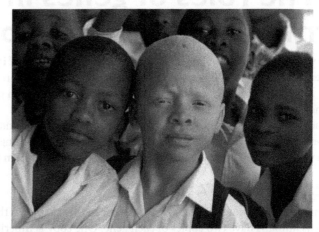

Figure 16.12 A South African schoolboy with albinism, surrounded by his classmates.

Huntington's disease, a severe neurological disorder, is different. Here, the mutant allele codes for a protein that actively causes damage and is dominant, and so is not masked by a normal allele. It can be present just once in the genotype in order to affect the phenotype. In the case of Huntington's, individuals only need to have one copy of the allele to have the disease. So, in the case of Huntington's disease, people with a homozygous recessive genotype (hh) do not have the condition. A heterozygous genotype of Hh or dominant homozygous HH means that the individual will develop the disease at some stage in their life.

When you are considering the inheritance patterns of specific genes you may find it helpful to start by jotting down all the different possible genotypes and their corresponding phenotypes, as shown in Table 16.2.

a.

Genotype	Phenotype
HH	Huntington's disease
Hh	Huntington's disease
hh	unaffected

b.

Genotype	Phenotype
AA	unaffected
Aa	unaffected
aa	albinism

Table 16.2 The relationship between genotype and phenotype with a dominant trait, Huntington's disease (**a**) and a recessive trait, albinism (**b**). We do not need to write down hH or aA as well, because these genotypes are the same as Hh or Aa.

However, not all alleles behave in a straightforward dominant or recessive pattern when they combine in the genotype of an individual. Some are **codominant**, which means that if two different alleles are present (in the genotype), they will both have an effect on the phenotype. The snapdragon plant (*Antirrhinum sp.*) is a simple example. Petal colour is controlled by two alleles, one for red and one for white. If a plant is heterozygous it will have pink flowers because both alleles are expressed, and therefore shown in the phenotype (Table 16.3). Note that the letters for codominant alleles are often shown as superscript to another letter, which in this case is C.

Genotype	Phenotype
$C^R C^R$	red
$C^R C^W$	pink
$C^W C^W$	white

Table 16.3 The relationship between genotype and phenotype for a codominant trait, petal colour in snapdragon plants.

The human ABO blood group system is another example of codominance. Every individual can be placed into a group: A, B, AB or O, on the basis of the proteins they carry on the surface membranes of their red blood cells. Group A individuals have proteins called A antigens on their red cells, Group B people have B antigens, AB have both, O have neither. This is controlled by one gene with three alleles:

- allele I^A codes for the A antigen
- allele I^B codes for the B antigen
- allele I^o codes for no antigen.

I^A and I^B are codominant to I^o. When the two alleles I^A and I^B are present together, both of them affect the phenotype. The person has blood group AB. This explains why they are written using the same letter with a different superscript. If we used just the letters A and B, this would imply that they were different genes, not just different alleles of the same gene. If we used A to represent the allele giving blood group A, and a to represent the allele giving blood group B, this would imply that allele A is dominant and allele a is recessive, which is not true. So, this is why choosing symbols for codominant alleles, use one symbol to represent the gene, and then different superscripts to code for its different alleles.

All the possible genotypes and phenotypes for blood group are shown in Table 16.4.

Genotype	Phenotype
$I^A I^B$	blood group AB
$I^A I^A$	blood group A
$I^A I^o$	blood group A
$I^B I^B$	blood group B
$I^B I^o$	blood group B
$I^o I^o$	blood group O

Table 16.4 The relationship between genotype and phenotype for a codominant trait with multiple alleles, the human ABO blood group system.

Tip

The superscript in I^o in Table 16.4 is actually a lowercase 'o', not an uppercase 'O'. This is because this allele is recessive to both I^A and I^B.

Dominant   Recessive

Plant height

Tall   Dwarf

Flower position

Axial   Terminal

Flower colour

Purple   White

Pea shape

Round   Wrinkled

Pea colour

Yellow   Green

Pod shape

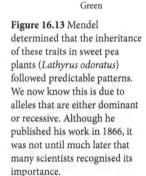

Smooth   Ridged

Pod colour

Green   Yellow

Figure 16.13 Mendel determined that the inheritance of these traits in sweet pea plants (*Lathyrus odoratus*) followed predictable patterns. We now know this is due to alleles that are either dominant or recessive. Although he published his work in 1866, it was not until much later that many scientists recognised its importance.

If a gene has three or more alleles, it is said to have **multiple alleles**. This is the case with the human ABO blood group system.

5. Define the following terms:
   a. allele
   b. dominant
   c. codominant.

6. Pandas in the Qinling Mountains in China usually have fur with black patches on a white background. Some pandas, however, have brown patches of fur, rather than black. Coat colour in these pandas is thought to be controlled by a gene with two alleles, B and b, where the pandas with brown fur have the genotype bb.
   a. Which allele is dominant? Explain your answer.
   ⭑ b. In the past decade, brown-and-white pandas have become more common in the Qinling Mountains. Explain why scientists have concluded on this basis that breeding between closely related pandas is increasing in this region.

7. A faulty allele of the gene that codes for haemoglobin produces a form of this protein that cannot transport oxygen when oxygen concentrations are low. A person with two of these faulty alleles has a disease called sickle cell anaemia. A person with one copy of the faulty allele and one normal allele has a mild form of the disease called sickle cell trait.
   a. Do these alleles show dominance or codominance? Explain your answer.
   b. Choose suitable symbols for the sickle cell allele and the normal allele. Write down the possible genotypes and the phenotypes that these alleles produce.

> **Tip**
>
> When describing what is meant by dominant, codominant and recessive, always include the terms 'genotype' and 'phenotype' in your description.

THE MECHANISMS OF INHERITANCE

Surprisingly, the researcher who first worked out the underlying rules of genetics did so without any knowledge of genes, chromosomes, DNA or cell division. Known as the Father of Genetics, Gregor Mendel was a Czech monk. In the 1860s, he studied a number of characteristics in pea plants (Figure 16.13).

For years, Mendel bred and studied varieties of the edible pea plant and took detailed records of what he saw. He chose these plants because they had easily observable features, they were easy to cultivate, they had rapid life cycles, and their pollination could be easily controlled. His breeding experiments centred on the inheritance of pairs of contrasting characteristics such as tall and dwarf plants, and round and wrinkled peas. His suggestion was that 'factors' exist in the plants that influence characteristics. Importantly, these do not blend – they can appear in the shape or colour of the plant, or remain hidden between generations.

In one of Mendel's crosses, he considered petal colour. The first cross was between parents with two different phenotypes (purple flowers and white flowers) that are both homozygous for different alleles. This produces a new generation, sometimes called **first filial**, **F1**, which refers to the offspring of the parental generation. This consists of plants with only the purple phenotype. Crossing two plants from this generation produces a second generation (called **second filial**, **F2**). In the F2 generation, the white phenotype,

last seen in one of the grandparents, reappears in one in four individuals, on average.
Table 16.5 shows some of Mendel's data.

Trait	Parent generation	First filial generation (F1)	Second filial generation (F2)	Ratio of traits in the F2 generation
seed shape	round × wrinkled	all round	5474 round 1850 wrinkled	2.96:1
seed colour	yellow × green	all yellow	6022 yellow 2001 green	3.01:1
flower colour	red × white	all red	705 red 224 white	3.15:1
plant height	tall × short	all tall	787 tall 277 short	2.84:1

Table 16.5 Some examples of breeding experiments conducted by Mendel in the 1860s. In all cases, there is a ratio that is close to 3:1 in the phenotypes of the offspring.

Look at the number of individuals Mendel counted in the F2 generation. Why did he have to conduct so many crosses and count so many seeds before he became convinced of the underlying ratios? To answer this question, we need to consider the probabilities of events occurring.

If you flip a coin, which side is showing? If you toss the coin 10 times, you might expect it to land on each side five times each. But it might not, as we are dealing with chance and probability. You could also get three of one and seven of the other, or nine and one. Or, the result of all 10 throws could be all of the same side! It is entirely possible. However, the more times you toss the coin, the more likely you are to get to a 1:1 ratio of each side.

This coin-flipping analogy applies to the study of inheritance. When there are large numbers of offspring, we tend to see the ratios that we would predict (the expected ratio). When there are small numbers, we should not be surprised if these ratios are not achieved in the observed ratio. This is due to chance. In the example of flower colour, for example, the offspring of the F1 generation have a fixed probability of 0.25 that every further plant they produce will have white flowers. The chance of that individual having white flowers is not affected by whether or not their other offspring have that trait. This is because male and female gametes fuse randomly at fertilisation. The greater the number of offspring, the higher the chance that the numbers reflect the expected ratio.

Monohybrid inheritance – studying the transmission of one gene

Clearly, the science of inheritance has certain rules: fixed probabilities apply to the numbers of offspring with given phenotypes in different generations. But what determines these probabilities, and how can they be used to help people at risk of having children with genetic diseases? The key is to know whether the allele for a particular characteristic is dominant, recessive, codominant or sex-linked. In this section, we will consider dominant and recessive alleles.

Earlier in the chapter we encountered two genetic disorders that affect humans (Table 16.2). These were albinism, a disorder that affects the pigmentation of people who are homozygous recessive only, and Huntington's disease, a neurological disorder that affects those who are heterozygous and homozygous dominant. Table 16.6 compares the two disorders in greater detail.

	Albinism	Huntington's disease
Worldwide prevalence	1 in 17 000	1 in 10 000
Age of onset of symptoms	Before birth	Usually from middle age onwards
Symptoms	White hair, light-coloured eyes, and very pale skin	Involuntary movements (chorea), memory loss and changes in personality
Gene name	*TYR*	*HTT*
Gene locus	Long arm of chromosome 11	Short arm of chromosome 4
Gene product (protein)	Tyrosinase, an enzyme located in melanocytes, which are specialised cells that produce a pigment called melanin in the skin, hair follicles, the coloured part of the eye (the iris), and the retina	Huntingtin, a protein whose function is not completely understood but which is required for normal brain function
Mutation	Recessive, so present only in individuals who are homozygous recessive	Dominant, so it will affect individuals who are heterozygous and those who are homozygous dominant
Mutation effect on protein	Most *TYR* mutations eliminate the activity of tyrosinase.	One region of the *HTT* gene contains repeated CAG triplets that appear multiple times. Normally, there are 10 to 35 repeats. The mutation increases the size of this segment to 36 repeats or more that causes the production of an abnormally long version of the huntingtin protein containing a long stretch of glutamine amino acid residues.
Mechanism of disease	Tyrosinase catalyses the first step of melanin production. It converts tyrosine, an amino acid, to another compound called dopaquinone. Other chemical reactions convert dopaquinone to melanin. This prevents melanocytes from producing any melanin throughout life.	The expanded CAG segment leads to the production of an abnormally long version of the huntingtin protein containing a long stretch of glutamine amino acid residues. The elongated protein accumulates in neurones, disrupting the normal functions of these cells.

Table 16.6 A summary of the features of the genetic disorders albinism (recessive) and Huntington's disease (dominant).

Geneticists use three different types of diagram to analyse inheritance patterns (Figure 16.14). Remember that each gamete has only 23 chromosomes, rather than 46. Because of this, each gamete has only one copy of each gene, rather than two.

1. We can show all this in a particular format known as a **genetic diagram**.
2. The **Punnett square**. This is a 'shorthand' version of a genetic diagram that illustrates how two gametes, each containing a specific allele, combine in the genotype of the individual.
3. The **pedigree diagram** (also called 'family tree'). By convention, the oldest individuals are at the top, males are squares and females are circles. The rows below show their children/offspring.

In practice, pedigree diagrams are usually used to show how a characteristic has changed over many generations. In the following example, the allele that causes albinism is recessive. It codes for a protein (tyrosinase) that does not work. We can use the symbol A for the normal (dominant) allele, and a for the faulty one.

In the case of Huntington's disease, however, an affected individual can have the genotype HH or Hh. If a parent has the genotype HH, each of the gametes they produce will contain one copy of the H allele. Remember from Table 16.6 that this allele encodes a protein with 36 or more glutamine repeats, which is responsible for the symptoms of the disease. However, if the parent has the genotype Hh, approximately half of the gametes will contain an H allele, and the other half will have an h allele. Understanding

Tip

When choosing a letter to represent alleles, pick one that looks different in upper and lower-case. A / a and H / h are good examples, unlike C / c or S / s.

this enables us to predict the genotypes – and therefore the phenotypes – of the offspring of a couple. Let's take an example of a woman with the genotype hh and a man with the genotype Hh (Figure 16.15).

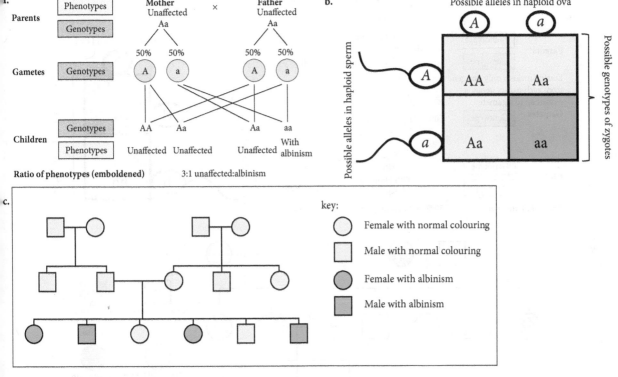

Figure 16.14 In a genetic diagram, a series of statements are made in the order (from top to bottom) parent phenotypes, parent genotypes, gamete genotypes, offspring genotypes, and offspring phenotypes (**a**). In a Punnett square, the two gametes are shown on different sides of a box, and the combinations within the box are the genotypes of the individuals that they could produce (**b**); the pedigree (**c**) can be used to illustrate how the inheritance of a trait or disorder is inherited between different generations.

As shown in Figure 16.15, each of the woman's ova (eggs) will contain one copy of an h allele. However, half of the man's sperm will be genotype H, and half will have the h allele. If a sperm fertilises an egg in the woman's oviduct, there is an equal chance that it will be an H sperm or an h sperm. This means that there is an equal chance that the child will have the genotype Hh (and suffer from Huntington's disease) or hh (and be unaffected).

> **Tip**
>
> When drawing genetic diagrams, be sure to link offspring genotypes with their correct phenotypes.

8. In pea plants, the allele for purple flowers (F) is dominant to the allele for white flowers (f) (refer to Figure 16.13).
   a. Draw a table to show the possible genotypes and phenotypes of pea plants for flower colour.
   b. If a heterozygous plant with purple flowers is crossed with a white-flowered plant, what ratio of phenotypes would you expect in the first generation? Draw a genetic diagram to show how you worked out your answer.
   c. A researcher carried out the cross described in (b) and found that 952 offspring were purple and 1020 offspring were white.
      i. Calculate the ratio of purple to white flowers in the offspring in this investigation.
      ii. Discuss whether the observed ratio is close to the expected ratio, and suggest reasons to explain any differences.
9. Explain which two genotypes of the parents could result in offspring in a family with all four blood groups. Use a Punnett square to help you with your answer.

> **Tip**
>
> In this example, probability can be written as 0.25, 25% or 1 out of 4; the ratio of unaffected to affected is written 3:1.

10. The ability to taste some bitter substances is controlled by a dominant allele for the *TAS2R38* taste receptor gene. 'Supertasters' are people able to taste these bitter substances. A non-supertaster child is born to parents who are supertasters. What is the probability of the parents having a child who is a supertaster? Use a Punnett square to help you work out your answer.

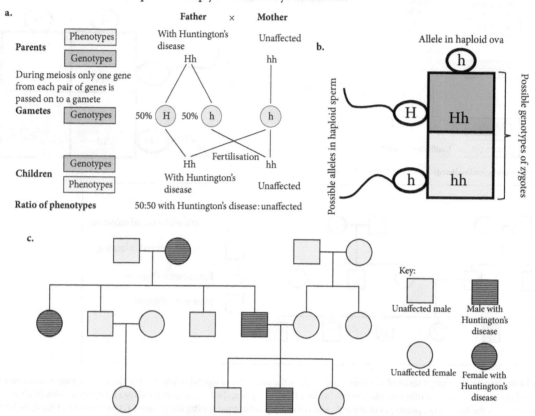

Figure 16.15 A genetic diagram (a), Punnett square (b), and pedigree diagram (c) showing the inheritance of Huntington's disease.

It is sometimes important to know if an individual displaying a dominant trait is homozygous or heterozygous for a particular allele. For example, how can we tell if a tall pea plant is homozygous dominant, or if it is heterozygous and carrying a recessive dwarf (short) allele? To find out, we can do a **test cross**. This involves crossing the tall plant with a homozygous dwarf plant whose genotype is known (tt). If the tall plant is homozygous dominant (TT), then all of the offspring will be tall (genotype Tt). However, if the genotype of the tall plant is heterozygous (Tt), approximately half of the offspring will be tall (Tt) and half will be dwarf (tt). Another example of a test cross is shown in Figure 16.16

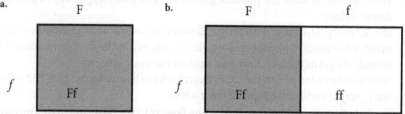

Figure 16.16 A test cross is a method used to identify whether an individual with a trait caused by a dominant allele is homozygous dominant or heterozygous. It is carried out by crossing the individual with an individual that is homozygous recessive. In this Punnett square example, a pea plant with purple flowers would give rise to no offspring with white flowers if it were a homozygous dominant (FF; **a**), whereas it will give rise to some offspring with white flowers if it were heterozygous (Ff; **b**). If enough offspring are counted, a ratio of 1:1 should be observed.

As we saw earlier with snapdragon petal colour and human ABO blood group system, some alleles are codominant: they are both able to affect the phenotype in a heterozygous individual. **Sickle cell anaemia** is another example of a genetic disorder that is codominant in its inheritance pattern. When oxygen concentrations are low, for example in actively respiring tissues, the red blood cells of affected individuals take on a sickle shape instead of a round, biconcave shape (Figure 16.17). They are no longer able to flow easily through capillaries and may get stuck, starving tissues of oxygen. This can lead to extreme pain and even death. The anaemia that results, due to reduced oxygen-carrying capacity, occurs because the spleen breaks down these cells at a much faster rate than normal.

Link

In Chapter 2 you saw that red blood cells contain haemoglobin, a soluble, globular protein. This consists of two α-chains and two β-chains, each combined with an oxygen-binding haem group.

Figure 16.17 Red blood cells are usually biconcave in shape and able to squeeze through narrow capillaries. However, as shown in this diagram, people with sickle cell anaemia have many red blood cells that become distorted and take on a sickle shape. These cells are far less able to move smoothly through capillaries, which results in many of the symptoms of this disorder.

The majority of individuals with sickle cell anaemia have a substitution mutation in the β-globin gene (which has the abbreviation *HBB*). Figure 16.18 explains how this simple substitution mutation in this gene has such a significant effect on the health of an individual. The two different alleles of the β-globin gene are Hb^A – the normal allele – and Hb^S, the sickle cell anaemia allele. You may want to remind yourself of the stages of protein synthesis in Chapter 6 before you consider this diagram.

So how does sickle cell anaemia illustrate a codominance pattern of inheritance? Heterozygous individuals are not **asymptomatic** (without symptoms). They may suffer from anaemia when oxygen concentrations in their surroundings are low, such as at high altitude. The fact that it is also remarkably common in many populations, especially in sub-Saharan Africa and in tropical Asia, also gives a clue to the fact that heterozygous individuals have phenotypic differences. Indeed, it was established some time ago that a person with one copy of the allele has a survival advantage in these parts of the world, where malaria is a common and frequently fatal disease. So, having one copy of the sickle cell allele gives a survival advantage when malaria is present.

The genetic diagram in Figure 16.19 shows how a man and a woman, who are both carriers of the Hb^S allele, could have a child with sickle cell anaemia.

Link

In Chapter 6 you learned about the different types of mutation, including base substitution, base addition and base deletion. Protein synthesis was also described.

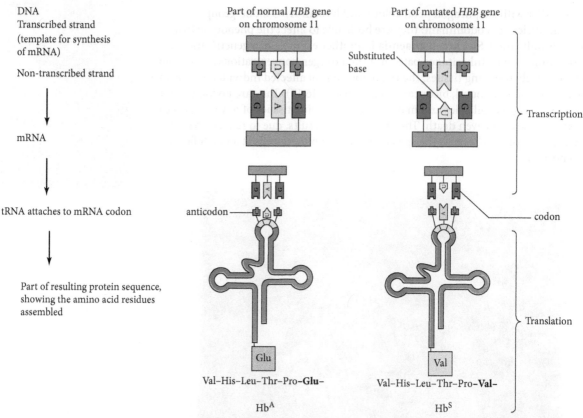

DNA
Transcribed strand
(template for synthesis
of mRNA)

Non-transcribed strand

mRNA

tRNA attaches to mRNA codon

Part of resulting protein sequence,
showing the amino acid residues
assembled

Part of normal *HBB* gene
on chromosome 11

Part of mutated *HBB* gene
on chromosome 11

Substituted base

Transcription

anticodon

codon

Translation

Glu

Val

Val–His–Leu–Thr–Pro–**Glu**–

Val–His–Leu–Thr–Pro–**Val**–

Hb^A

Hb^S

Figure 16.18 How a mutation in the DNA base sequence causes a change in the primary structure (amino acid sequence) of a protein. The sixth DNA base triplet in the *HBB* gene on chromosome 11 that codes for the β-globin polypeptide reads GAG in the Hb^A (normal) allele, but a substitution mutation of A to T has changed this to GTG in the Hb^S (mutant) allele. This means that one DNA triplet is different. The sixth codon on the mRNA of *HBB* Hb^S will be translated into the hydrophobic amino acid valine in place of the hydrophilic amino acid glutamic acid. The protein takes on a different tertiary and quaternary structure, in which the sixth residue is found on the outside surface of the structure. Because valine has a much more hydrophobic R group than glutamic acid, the solubility of the protein is lowered. If oxygen concentrations fall, the haemoglobin molecules stick together, forming a big chain of molecules that are not soluble and therefore forms long fibres. This pulls the red blood cells, in which haemoglobin is found, out of shape, making them sickle-shaped. The haemoglobin in sickle cells is also much less efficient at binding and transporting oxygen.

Link

In Chapter 17, you will see how mutation changes allele frequencies in a population, and how this is crucial for the process of natural selection.

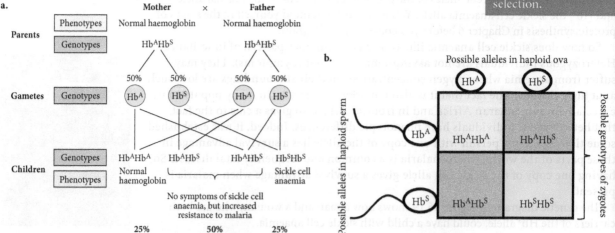

a.

		Mother	×	Father
Parents	Phenotypes	Normal haemoglobin		Normal haemoglobin
	Genotypes	Hb^AHb^S		Hb^AHb^S

Gametes — Genotypes: Hb^A (50%) Hb^S (50%) Hb^A (50%) Hb^S (50%)

Children	Genotypes	Hb^AHb^A	Hb^AHb^S	Hb^AHb^S	Hb^SHb^S
	Phenotypes	Normal haemoglobin			Sickle cell anaemia

No symptoms of sickle cell anaemia, but increased resistance to malaria

25%   50%   25%

b.

Possible alleles in haploid ova

	Hb^A	Hb^S
Hb^A	Hb^AHb^A	Hb^AHb^S
Hb^S	Hb^AHb^S	Hb^SHb^S

Possible alleles in haploid sperm

Possible genotypes of zygotes

Figure 16.19 The inheritance of sickle cell anaemia illustrates codominant inheritance and can be shown in a genetic diagram (**a**) and a Punnett square (**b**). When two parents heterozygous for the Hb^S allele have a child, there is a probability of 0.25 that their child will suffer from sickle cell anaemia. On average, half of their children will have the genotype Hb^AHb^S and will therefore, like them, have an increased resistance to malaria.

So far, we have considered the inheritance of dominant, recessive and codominant alleles. There is a final type of monohybrid inheritance pattern that considers the inheritance of one gene. This occurs in humans and other animals in which different chromosomes determine sex.

Remember that in humans, each cell has 46 chromosomes: 22 pairs of autosomes and one pair of sex chromosomes. The autosomes are found in homologous pairs. Females have two X chromosomes, which are also homologous – they have the same genes at the same loci. Males have one X and one Y, which are not homologous. Hence in humans, sex is determined by the **non-homologous** X and Y chromosomes.

Genes that are found on the sex chromosomes are said to be sex-linked. It is estimated that there are nearly 1000 protein-encoding genes on the human X chromosome, compared with fewer than 100 on the Y chromosome.

Given the number of genes on the X chromosome, there are many genetic disorders that are far more common in boys than girls. This is because females have two copies of each gene on the X chromosomes, but males only have one. So, if a man has a faulty allele on his X chromosome, he will have whatever trait or disease that allele codes for. This is because there is no other allele to 'mask' its effect, as is usually the case in females who have a recessive allele. It is important to remember that as far as X-linked genes are concerned, males will always display the phenotype associated with the single allele they carry.

An example of an X-linked gene is *F8*. Mutations in this gene cause a disease called **haemophilia** (Figure 16.20). Haemophilia is more common in boys than in girls. Sufferers of this disorder have blood that fails to clot quickly, because the product of the *F8* gene is a vital blood clotting protein called Factor VIII. This means that they are at risk of internal bleeding, especially in the joints, even during light exercise. Fortunately, today, most people with haemophilia can live a full and active life by having regular injections of Factor VIII protein.

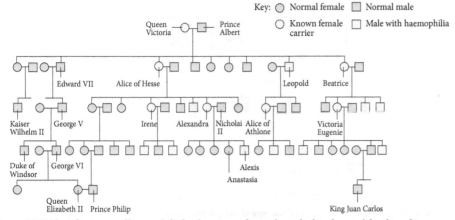

Figure 16.20 The inheritance of haemophilia has been most famously studied in the royal families of Europe. Queen Victoria had nine children and passed the haemophilia allele on to two daughters and one son, Leopold. Since Leopold, like all males, had only one X chromosome, he had only one copy of this gene – the defective allele. Her carrier female daughters, Alice of Hesse and Beatrice, passed on the mutant allele to the next generations. No living member of the present reigning royals of Europe is known to have symptoms of haemophilia nor is believed to carry the mutant allele for it.

The inheritance pattern of haemophilia, like other X-linked genes, is quite unusual. Look at Figure 16.21, which shows how a man and a woman without haemophilia could have a child with the disorder. Because the *F8* gene is found on the X chromosome, a different way is used to illustrate it in genetic diagrams and Punnett squares. Like Huntington's disease, we can use the symbol H for the normal (dominant) allele, and h for the faulty one. However, because the alleles are found on the X chromosome, they are shown as superscript to the letter X. The normal dominant allele that produces factor VIII is written as X^H, and the recessive allele that results in failure to produce factor VIII is X^h. The Y chromosome is written as Y or Y–, to show that it does not carry that gene.a.b.

Tip

The number of genes on the X chromosome is much higher than those on the Y chromosome; therefore, most 'sex-linked' genes are 'X-linked' genes. However, Y-linked inheritance patterns do exist.

Tip

Sex-linked genes are found on the sex chromosomes. However, on the autosomes, genes are linked with other autosomal genes.

Tip

Our definition for recessive alleles must be adapted for those that are sex-linked. In males, a recessive allele on the X chromosome can be present only once in his genotype to affect his phenotype.

Link

In Chapter 19, you will see how haemophilia can be treated using recombinant human Factor VIII synthesised by genetically modified bacteria.

a.

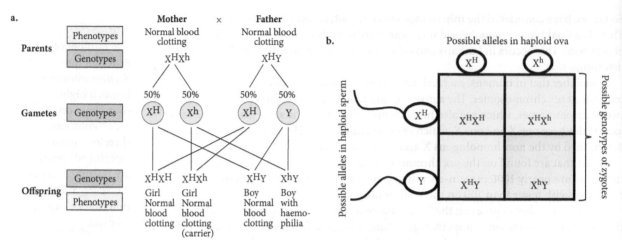

b.

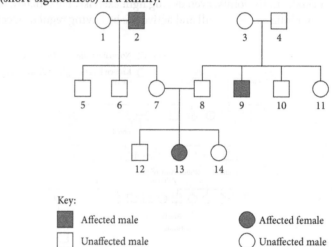

Figures 16.21 a. and b. The inheritance of haemophilia. In this example, two parents who do not have the disorder have a son with the disorder. Notice how it will always be a male child that has the disorder, and that males have a one-in-two chance of inheriting the condition if their mother carries the mutant allele.

> **Tip**
>
> Although they are both disorders of the blood, do not confuse the cause or symptoms of sickle cell anaemia (recessive, autosomal) and haemophilia (X-linked, autosomal).

> **Tip**
>
> The Y chromosome does not carry any of the genes on the X chromosome. Do not make the mistake of writing a superscript letter next to the Y in genetic diagrams.

Figure 16.22 A person with red-green colourblindness cannot distinguish between red and green. On these test cards a person with normal vision would see numbers, but a person with red-green colour-blindness only sees a random pattern of dots.

Other sex-linked traits in humans, which affect more boys than girls, include Duchenne muscular dystrophy (DMD) and red–green colourblindness (Figure 16.22).

Worked example

The pedigree diagram in Figure 16.23 shows the pattern of inheritance of one type of myopia (short-sightedness) in a family.

Key:
- ▦ Affected male
- ⬜ Unaffected male
- ⬤ Affected female
- ⭕ Unaffected male

Figure 16.23 The pattern of inheritance of one type of myopia in a family.

Give one piece of evidence from the diagram that suggests:

a. that the allele for myopia is recessive.

b. that the allele is not X-linked.

Answer

In this case, affected individuals 9 and 13 are both born to unaffected parents. In both cases, the parents must be carrying both alleles. Therefore, it is recessive as otherwise it would affect their phenotype. Males inherit their Y chromosome from their father (otherwise they would not be male) and so must get their X chromosome from their mother. Using the notation N for the dominant allele and n for the recessive allele, individual 13 must be X^nX^n. However, her father (8) only has one copy of the allele and he is not affected. He must have the genotype X^NY in order to be unaffected, so he has not got an n allele to pass to his daughter. Therefore, it cannot be X-linked.

11. The inheritance of coat colours in some cats is caused by a sex-linked gene. This has two codominant alleles, X^B, which gives a black coat, and X^O, which gives orange. When black cats ($X^B X^B$ or $X^B Y$) are mated with orange cats ($X^O X^O$ or $X^O Y$), the female offspring are always tortoiseshells (showing black and orange patches), whereas the male cats have the same phenotype as their mothers. Construct two genetic diagrams to show how these phenotypes in the offspring could arise.

12. Look back to Table 16.3, which shows the genotypes controlling petal colour in snapdragon plants.
    a. Draw a genetic diagram to show the phenotypes of offspring formed from a cross between a red and a white snapdragon plant (the F1 generation).
    b. Draw another genetic diagram to show what ratio of phenotypes would be expected in the F2 generation if the F1 generation are self-pollinated.
    c. Draw a Punnett square to predict the ratio of genotypes and phenotypes that would be expected in the F2 generation.

13. Females with red-green colourblindness, caused by a recessive allele of a gene on the X chromosome, are rare. Suggest the genotypes of parents of a girl born with red-green colourblindness.

14. Can a man with haemophilia pass the allele that causes the disorder to his:
    a. son?   b. daughter?   c. grandson?
    Explain your answers.

The chi-squared test and inheritance

Like all statistical tests, scientists can use the chi-squared test to tell if their results are likely to be significant, which means that there is an underlying scientific basis. Otherwise, these could be due to chance or experimental error. Specifically, this test can be used to see whether there is a significant difference between observed results (what you saw in an investigation) and the expected results (what you expected from theory alone). In genetics, it can be used to compare the **goodness of fit** of observed phenotypic ratios with expected ratios. We will now explore this using a simple worked example.

A plant with red flowers was crossed with a plant with yellow flowers. In this experiment, the resulting offspring (F1 generation) were all orange. This would suggest that the alleles that control flower colour (R and Y) are codominant and that a genotype of RY results in orange flowers. If this conclusion were true, we would expect a ratio of 1 red:2 orange:1 yellow in the F2 generation (see Figure 16.19 for a similar example).

A scientist crossed orange flowers which produced 320 offspring. However, when these plants were bred together and the phenotypes of the plants in the next generation were compared, the results in Table 16.7 were obtained.

Phenotype	Expected ratio	Expected numbers	Observed numbers
red	1	80	77
orange	2	160	172
yellow	1	80	71
total		320	320

Table 16.7 The observed and expected ratios of flower colour in the offspring of a cross between two plants with orange flowers. The expected numbers were calculated by applying the ratio of 1:2:1 to the total number of offspring counted (320).

As you can see, the ratios of offspring are very similar, but not exactly the same. Are the differences just due to chance or does an unusual type of inheritance pattern apply to flower colour? This is where the chi-squared test comes in.

As we saw in Chapter 6, the formula for chi-squared, χ^2, is:

$$\chi^2 = \sum \frac{(O - E)^2}{E}$$

where χ^2 = chi-squared; O = observed result; E = expected result; and $\sum$ represents 'the sum of'. Using a table to lay out our working can help to calculate the χ^2 value (Table 16.9).

Tip

Chi-squared is always calculated from actual numbers and not percentages or mean values. Sample size is therefore very important.

Phenotype	Observed (O)	Expected (E)	$(O - E)^2$	$\dfrac{(O - E)^2}{E}$
red	77	80	9	0.11
orange	172	160	144	0.90
yellow	71	80	81	1.01

$$\sum \frac{(O - E)^2}{E} = 2.02 \ (\chi^2)$$

Table 16.8 The calculation of χ^2 for the observed and expected ratios of phenotypes in the offspring of a cross between two orange-flowered plants.

Therefore, $\chi^2 = 2.02$. We now need to look up the values on a table of values of χ^2. We must first work out the number of **degrees of freedom** in order to look in the right row of the table. The number of degrees of freedom is calculated using this equation:

$$v = c - 1$$

where v = degrees of freedom and c = number of classes of data.

In this investigation, because there are three different phenotypes, or classes of data (red, orange and yellow flowers), then there are $(3 - 1) = 2$ degrees of freedom.

Finally, it is now possible to determine whether our results show a significant difference from what is expected. However, to do this, a null hypothesis should be generated. The null hypothesis in a χ^2-test is that the observed results are not significantly different from the expected results, and any difference is due to chance. Table 16.9 shows a selection of critical values for χ^2, which correspond to two probability levels ($P < 0.05$ and $P < 0.01$). The second row, showing probability values for two degrees of freedom, is relevant here.

	Probability (P)	
Degrees of freedom	**0.05**	**0.01**
1	3.84	6.64
2	5.99	9.21
3	7.82	11.35
4	9.49	13.28
5	11.07	15.09

Table 16.9 The probability values for the one to five degrees of freedom at $P < 0.05$ and $P < 0.01$, as required for the analysis of the calculated χ^2 value.

If the calculated χ^2 value is greater than or equal to the critical χ^2 value, then there is a significant difference between our observed results and our expected results. That is, the difference between the actual data and the expected data is probably too great to be attributed to chance. If the calculated χ^2 value is less than the critical χ^2 value, then there is no significant difference between the observed and expected data, and the difference is likely to be due to chance at this probability level. So, we conclude that our sample does not support the hypothesis of a difference, and we accept the null hypothesis.

We usually look at the value of probability value of $P < 0.05$ or $P < 0.01$. Comparing the calculated value of 2.02 with the critical values on the χ^2 table for two degrees of freedom we can see that it is less than 5.99, and less than 9.21. Therefore, the probability of this result being due to chance is less than 0.01. Therefore, we can accept the null hypothesis, and assume that the difference between the observed and expected ratios in the colour of the offspring is due to chance. The χ^2 test provides evidence that the deviation from a 1:2:1 ratio is not great enough to reject the idea that these two alleles are codominant.

Experimental skills 16.1: *Drosophila*: the perfect model organism in genetics

Have you ever seen flies buzzing around ripening fruit? Chances are that these are *Drosophila melanogaster*. These organisms were the subject of work that gave rise to the first Nobel prize for genetics in 1933, for linking the inheritance of a specific trait with a particular chromosome.

Researchers have used a wide range of model organisms to study inheritance patterns. These included the pea plants of Mendel, but also Sutton's grasshoppers and Boveri's sea urchins. Thomas Hunt Morgan preferred fruit flies, which he bred in his 'fly room' at Columbia University in the United States. These flies have a number of useful characteristics:

- they are approximately 2.5 mm in length and males are slightly smaller than females
- like humans, male fruit flies have the sex chromosomes XY and the females have XX
- males have a distinct black patch on their bodies that is absent in females
- females lay up to 400 eggs, which develop into adults in 7–14 days
- in the laboratory, fruit flies will survive and breed in small flasks containing a simple nutrient medium.

One day in 1910, Morgan looked down a hand lens at a male fruit fly and saw something surprising. Instead of having the normally bright red eyes, this fly had white eyes (look back to Figure 16.10).

QUESTIONS

P1. To find out more about the inheritance of this unusual trait, Morgan crossed these white-eyed males with homozygous, red-eyed females and observed the phenotype ratios of the F1 offspring.
  a. Explain what is meant by:
      i. homozygous   ii. F1.
  b. This cross is similar to a test cross. Explain why it is not a test cross.

P2. Morgan determined that a gene for eye colour is carried on the X chromosome, and that the allele for red eyes, R, is dominant to the allele for white eyes, r. When he crossed heterozygous red-eyed females with red-eyed males, white-eyed males appeared in the offspring.
  a. Explain what is meant by a dominant allele.

b. Suggest why the recessive allele gives a phenotype with no pigment.
c. Copy and complete the genetic diagram for the following cross.

Phenotypes of parents:	red-eyed female	× red-eyed male
Genotypes of parents:	....	....
Gametes:	..........and..........	............and............
Genotypes of offspring:	..........and..........and............and..........	
Phenotypes of offspring:	..........and..........and............and..........	

d. Using the information in the genetic diagram, explain why it is more likely that male fruit flies will show a phenotype produced by a recessive allele carried on the X chromosome than female flies.

P3. A scientist carried out a cross in an attempt to recreate the results of Morgan's famous experiment.
  a. State the null hypothesis for this investigation.
  b. Using the terms probability and chance in your answer, explain what is meant by $P < 0.05$.
  c. Phenotypes of flies born to red-eyed males and heterozygous red-eyed females were counted and the following were observed:
      - red-eyed males: 61
      - red-eyed females: 101
      - white-eyed males: 58
      i. Calculate the expected number of flies of each phenotype.
      ii. Draw a table similar to Table 16.8 and calculate the chi-squared value.
      iii. Use your calculated chi-squared value and Table 16.9 to determine if these phenotype ratios significantly differ from the ratios expected.

P4. Using all of the information given, explain two reasons why the fruit fly is suited as a model organism for studying genetic crosses.

Dihybrid inheritance – studying the inheritance of two unlinked genes simultaneously

Earlier in this chapter, we considered the inheritance of one gene with two alleles. Things get a little more challenging when we look at how two genes, each with two alleles, are inherited at once.

We will start by looking at genes on different chromosomes, called **unlinked genes**. As a result of meiosis, the alleles of unlinked genes can pass into the gametes in any combination. However, if the genes are on the same chromosome they are said to be **linked**. That is a different situation, which is covered later in this chapter.

The steps for working out the outcomes of a dihybrid cross are below:

1. Identify the genotypes of both parents, using their phenotypes to help you.
2. Work out the possible gametes and put a circle around them.
3. Identify the outcomes of all possible fertilisations.
4. Translate the genotypes into phenotypes and work out the ratios/probabilities.

Let's now use these steps to consider the inheritance of two traits in mice:

- Hair length is controlled by one gene with two alleles: allele A for short hair is dominant to allele a, for long hair.
- Coat colour is controlled by a different gene on another chromosome. The allele B for black fur is dominant to b for brown fur.

What happens when a homozygous black, short-haired mouse mates with a homozygous brown, long-haired mouse? Will the specific combination of coat colour and hair length always be inherited together, or can they be separated? Is it possible to produce mice with long black fur, or short brown fur, for example? The answer is yes, and these individuals are called **recombinants**. These are formed because of the process of crossing over in meiosis, which produces new combinations of alleles. Furthermore, because fertilisation is always random (any male gamete can combine with any female gamete), all combinations of these traits may occur in the offspring. Figure 16.24 shows the process. In the F1 generation all the gametes are the same (AB and ab) and so all fertilisations produce genotypes of AaBb. They all look the same – black with short fur. However, the phenotypes become more varied when mice with genotype AaBb reproduce with each other.

If you start with a genotype of AaBb, you have four different alleles, each of which is on a separate chromosome. Due to the random orientation and independent assortment of homologous chromosomes during meiosis, either of the first two alleles can go into the gamete with either of the second two alleles. This gives gametes of AB, Ab, aB and ab from both male and female. In other words, one allele from each pair passes in the gamete, and all four combinations are possible. Fertilisation gives 16 (4×4) different possibilities. A Punnett square (Figure 16.24) gives you the precise ratio of phenotypes that you would expect in a cross like this.

15. In sweet pea plants (*Lathyrus odoratus*), height and petal colour are controlled by two genes found on different chromosomes. The allele for tall stem is dominant to the allele for dwarf (short) stems, and the allele for purple flowers is dominant to the allele for white flowers.
    a. Draw a genetic diagram to show the expected genotypes and phenotypes that would occur if two plants that are both heterozygous for both genes are crossed. Select appropriate letters to represent the alleles.
    b. Use a Punnett square to predict the ratio of phenotypes that would be expected if these two plants were crossed.

Tip

When you write down a dihybrid genotype, write the two alleles of one gene, followed by the two alleles of the second gene. For example, AaBb.

Tip

A cross between two individuals heterozygous for two unlinked genes is expected to give a ratio of 9:3:3:1 in the phenotypes of the offspring.

1. 'Deduce the parents' genotypes from their phenotypes'

2. 'Work out the possible gametes and put a circle around them'

3. 'Deduce the outcomes of all possible fertilisations'

4. 'Translate the genotypes into phenotypes and work out the ratios/probabilities'

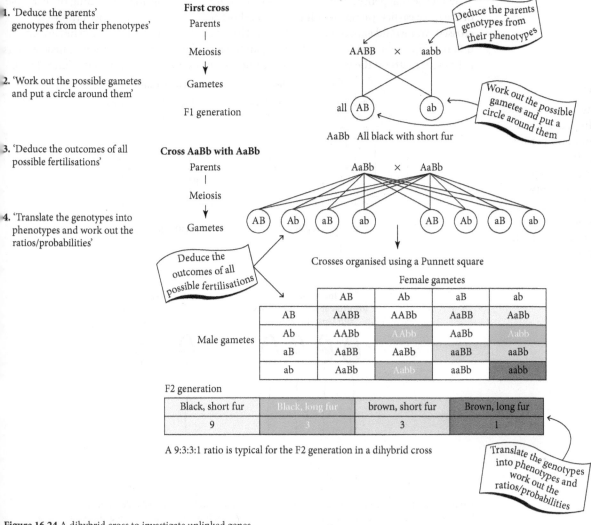

First cross

Parents

Meiosis

Gametes

F1 generation

AABB × aabb

Deduce the parents genotypes from their phenotypes

all AB   ab

Work out the possible gametes and put a circle around them

AaBb   All black with short fur

Cross AaBb with AaBb

Parents

Meiosis

Gametes

AaBb × AaBb

AB Ab aB ab   AB Ab aB ab

Deduce the outcomes of all possible fertilisations

Crosses organised using a Punnett square

	Female gametes			
	AB	Ab	aB	ab
AB	AABB	AABb	AaBB	AaBb
Ab	AABb	AAbb	AaBb	Aabb
aB	AaBB	AaBb	aaBB	aaBb
ab	AaBb	Aabb	aaBb	aabb

Male gametes

F2 generation

Black, short fur	Black, long fur	brown, short fur	Brown, long fur
9	3	3	1

A 9:3:3:1 ratio is typical for the F2 generation in a dihybrid cross

Translate the genotypes into phenotypes and work out the ratios/probabilities

Figure 16.24 A dihybrid cross to investigate unlinked genes.

Dihybrid inheritance – studying the inheritance of two autosomal linked genes simultaneously

So far, we have considered the inheritance of two genes on different chromosomes that could be separated at meiosis. However, the ratios expected from theory in genetic crosses are not always seen in practice. As we learned earlier, this was first noticed by Thomas Hunt Morgan, who was surprised to note that eye colour and sex in fruit flies were not inherited independently. This suggests that some genes cannot undergo independent assortment: they are usually inherited together.

Remember that humans have around 20 000 genes, which are present on just 23 pairs of homologous chromosomes. Therefore, any gene is found on the same chromosome as a few thousand others. As we saw earlier, these genes are said to be linked. This means that they will usually be inherited together. For example, if a gene for hair colour is on the same chromosome as a gene for height, then it would be likely that all tall individuals have the same colour of hair, and all short individuals have a different colour of hair. The only way that this might not be true is if the alleles for these two genes are separated by crossing over between (non-sister) chromatids on homologous chromosomes during prophase I of meiosis. The formation of chiasmata at random points along the homologous chromosomes may swap 'blocks' of genes, and the result is chromosomes with new allele combinations, called hybrid chromosomes or recombinants.

Evidence that genes are linked comes by looking at the offspring. Extending the example of the mice earlier, we saw that if two black, short-haired mice of genotype AaBb were crossed, you would expect a ratio of 9:3:3:1. However, this assumes that the genes are on different chromosomes, and so can be separated by independent segregation in meiosis. It is a different matter if the genes are on the same chromosome. This is shown in Figure 16.25.

Tip

A clue that two genes are linked is when greater numbers of offspring have the same phenotypes as their parents, and fewer are phenotypically different from their parents.

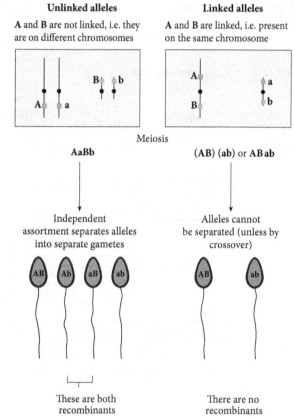

Figure 16.25 If two genes are on different autosomes, then they will be inherited independently as they can be passed on to different gametes during meiosis (**a**). However, if two genes are close together on the same chromosome, they are almost always inherited together: they are linked (**b**). The probability of them being separated by crossing over depends on the distance between them. This diagram shows the process of meiosis in the male to produce sperm.

To determine whether it is likely that offspring ratios are different from those expected by chance, or whether there is another factor at play, we can use the chi-squared test to analyse data (see Experimental skills box).

It is important to note that if two genes are located at opposite ends of the chromosome, they will almost always be separated by crossing over. This means that they will behave as if they are unlinked and will be inherited independently. The closer they are together, the less likely it will be that they will be separated by crossing over. This fact can be used to work out the relative positions (loci) of genes on chromosomes.

Tip

Your final answer for chi-squared should have the same number of decimal places as, or one more than, the figures you have added together.

Worked example

In the sweet pea plant (*Lathyrus odoratus*), the allele for purple flowers is dominant to the allele for red flowers, and the allele for elongated pollen grains is dominant to rounded pollen grains. When two plants heterozygous for both genes are bred together, offspring with the following phenotypes were obtained.

Phenotype	Observed numbers	Expected numbers (assuming no linkage)
purple, elongated	376	405
purple, rounded	99	95
red, elongated	80	109
red, rounded	145	91
total	720	720

Table 16.10

a. Calculate the percentage of recombinant plants in Table 16.10.

Answer

The total number of offspring is 720, of which 59 (31 + 28) are recombinants. Therefore, the percentage of recombinant plants is $(59 \div 720) \times 100 = 8\%$.

b. Use the chi-squared test to identify whether the genes that control flower colour and pollen grain shape are linked. Use a table to help you calculate the value of chi-squared.

Answer

A table lays out the calculation in a stepwise fashion, and allows us to calculate separate values for the expression $(O - E)^2 \div E$, which are then added together:

Phenotype	Observed number (O)	Expected number (E)	($O - E$)	($O - E$)^2	$\dfrac{(O-E)^2}{E}$
purple, elongated	376	405	−29	841	2.08
purple, rounded	99	95	4	16	0.17
red, elongated	80	109	−29	841	7.72
red, rounded	145	91	54	2916	32.04

$$\sum \frac{(O-E)^2}{E} = \underline{\quad\quad} \chi^2$$

Therefore, adding together the values for $(O - E)^2 \div E$ gives a value of $\chi^2 = 42.01$. This is much greater than the critical value of 9.210 (see Table 16.9) which applies to 3 degrees of freedom (because there are four classes of data) at $P < 0.01$. Therefore, we can assume that the genes are linked, because the ratio is significantly different from that which we would expect.

---

16. Figure 16.26a shows some of the genes found on chromosome 2 in *Drosophila melanogaster*. The alleles on the chromosome on the left list the dominant traits; the alleles on the chromosome on the right list the recessive traits. Figure 16.26b shows two flies with the corresponding phenotypes.

   a. Choose appropriate letters to represent the genotype of flies with the following phenotypes:
   i. a fly with long legs and a grey body
   ii. a fly with long aristae and brown eyes
   iii. a fly with curly wings and a black body.

   b. The genes shown in Figure 16.26a are linked.
   i. Explain what is meant by linked genes.
   ii. Outline why it is important to know whether genes are linked when predicting the ratio of offspring with phenotypes dependent on these genes.

Tip
Recombinant offspring are noticeable because they have a phenotype that is different from their parents.

Tip
You will always be provided with the table of critical values and probabilities, but you will need to be able to calculate degrees of freedom.

iii. In a cross between two flies heterozygous for all of these genes, the percentage of recombinant flies for eye colour and aristae length would be larger than the percentage of any other recombinants. Explain why.

⭐**iv.** A fly was found that had long legs and a grey body. What are the possible genotypes for this fly, and how could you find out which genotype this fly had? Use Punnett squares to illustrate your answer.

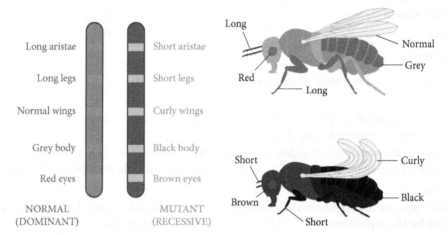

Figure 16.26 Linked genes on chromosome 2 of *Drosophila melanogaster*. When two genes are close together on the same chromosome, they are almost always inherited together. They can only be separated by crossover. The probability of being separated by crossover depends on the distance between them.

Tip

Do not confuse the interactions between different alleles of a gene at one locus with epistasis. This term is reserved for the interactions between alleles of genes at different loci.

Tip

You will not be expected to recall specific ratios of crosses involving epistasis. However, you should be able to show how these arise in offspring when provided with appropriate information.

Interactions between genes: epistasis

So far, we have discussed the inheritance and effect of genes that encode proteins that act on their own. In practice, however, the proteins encoded by different genes interact with each other. Therefore, the presence of one allele for one gene can affect whether an allele for another gene can affect the phenotype. This type of interaction is called **epistasis**, which derives from the Greek word for 'to stop', or 'stand upon'.

An example of epistasis is coat colour in Labrador dogs; as we saw earlier, there are three colour varieties: gold, brown and black (Figure 16.11). The analysis of offspring phenotypes using the chi-squared test (see Experimental skills box) suggested that the trait is not completely controlled by codominant alleles. Instead, epistasis operates between two different genes, found on different pairs of autosomes.

- One gene controls the colour of the pigment in the fur. Allele B codes for black pigment, and allele b codes for brown pigment.
- However, without the presence of an allele from a second, controlling gene, the pigment cannot be deposited in the fur. When there is no pigment, the coat appears gold.
  - Allele E causes the pigment to accumulate in the hairs.
  - Allele e prevents the pigment from accumulating in the hairs.

The following worked example illustrates how the phenotypes of the various genotypes can be determined. Note that all dogs that do not inherit the E allele will be gold.

Worked example

Given the information provided, identify the phenotype of the Labrador dogs with the following genotypes. Explain your answers.

a. BbEe

Answer

The presence of allele B gives black hair, which can accumulate in the hairs as one copy of allele E is present.

b. BBee

Answer

Although the presence of allele B could give black hair, the absence of allele E means that the pigment is prevented from accumulating in the hairs. The hair is therefore gold.

c. bbEE

Answer

The presence of two copies of the recessive b allele gives brown hair, which can accumulate in the hairs as allele E is present.

To explore epistasis in greater detail, let's consider the example of Labrador dogs again. In a cross between two Labradors that are heterozygous for both loci (BbEe), we would expect the phenotypes of the offspring to be in a ratio of 9 black:4 gold:3 brown puppies in the F2 generation (Table 16.11).

	BE	Be	bE	be
BE	BBEE	BBEe	BbEE	BbEe
Be	BBEe	BBee	BbEe	Bbee
bE	BbEE	BbEe	bbEE	bbEe
be	BbEe	Bbee	bbEe	bbee

Table 16.11 A cross between two Labradors of genotype BbEe.

17. In white clover (*Trifolium repens*), two genes control the production of cyanide. This tastes bitter and stops animals from eating the plant. Whether the plant is able to produce cyanide depends on two different genes, found on different pairs of autosomes.
    - One gene, coding for an enzyme called linamarase, has two alleles:
      ○ the dominant allele, L, codes for a functional enzyme
      ○ the recessive allele, l, codes for a non-functional enzyme.
    - A second gene, coding for the presence or absence of the enzyme's substrate, called cyanogenic glucoside, also has two alleles:
      ○ presence of the dominant allele, G, results in the production of the substrate
      ○ presence of the recessive allele, g, results in no substrate being formed.
      These unlinked genes cause three phenotypes:
    - clover that has both L and G alleles gives off cyanide as soon as its leaves are crushed
    - clover with allele L but not G releases cyanide slowly when its leaves are crushed
    - clover that does not have allele L cannot release cyanide.
    a. How do different alleles of a gene arise?
    b. Explain how the absence of allele L prevents cyanide production.
    ☆**c.** Using the symbols provided, draw a genetic diagram to show what proportion of the three phenotypes would be expected if a plant heterozygous for both genes was self-pollinated. State the expected ratio of the different offspring.
    d. In an experiment where individuals that are double heterozygous were self-pollinated the following numbers of offspring were obtained.

Phenotype	Observed number (O)	Expected number (E)	($O - E$)	($O - E$)^2	$\dfrac{(O - E)^2}{E}$
rapid cyanide release	145				
slow cyanide release	43				
no cyanide release	52				
total number					

$$\sum \frac{(O - E)^2}{E} = \_\_\_ \, \chi^2$$

Table 16.12

Copy and complete Table 16.12 to help you answer this question.

i. Using the ratio from part (c) calculate the expected number of each phenotype of the offspring, and enter the values in Table 16.12.

ii. Use the other columns in the table to carry out a chi-squared test. Using the probability level of $P < 0.05$ and the correct number of degrees of freedom, state whether you would accept or reject the null hypothesis for this cross and explain why. You will need to use Table 16.9 to help you answer this question.

18. A dog breeder investigated the inheritance of coat colour in Labradors (Table 16.11). Labradors have three coat colours – black, brown and gold. A prediction was made that there are two codominant alleles, and heterozygous individuals are brown. Phenotypes of puppies born to different pairs of black and gold dogs over several years were counted and the following phenotypes were observed:
   ○ black puppies: 53
   ○ gold puppies: 10
   ○ brown puppies: 17

   The breeder wondered whether these numbers are in the ratio of 9 black:4 gold:3 brown in the offspring of this cross, which would indicate epistasis.

   a. State a null hypothesis for this investigation.

   ★b. Using a similar table to Table 16.12, calculate the chi-squared value and use Table 16.9 to decide whether to accept or reject your null hypothesis.

   c. Explain why the data may not be sufficient to draw a conclusion from this calculation.

Key ideas

➜ The observable characteristics of an organism are known as its phenotype.
➜ Phenotype = genotype + environment.
➜ A diploid cell contains two sets of chromosomes. There are therefore two copies of each chromosome in a set, and two copies of the genes that they carry.
➜ Homologous chromosomes carry the same genes at the same loci.
➜ Genes often come in different forms, called alleles.
➜ The alleles of a gene that an organism has are known as its genotype. If the two alleles are the same, it is homozygous. If they are different, it is heterozygous.
➜ An allele is said to be dominant if it exerts its full effect on the phenotype even when a different allele of the same gene is also present. An allele that only has an

effect when no other allele is present is said to be recessive. If both alleles have an effect in an individual with a heterozygous genotype for that gene, the alleles are codominant.

→ Although a diploid individual has a maximum of two different alleles for a given gene, some genes have three or more different alleles, known as multiple alleles.

→ There is an equal chance of any gamete from one parent fusing with any gamete from the other. We can show this in a genetic diagram. The genetic diagram shows the different genotypes that can arise in the offspring, and the relative chances of each genotype occurring.

→ If an individual is homozygous for a particular gene, both alleles are the same. If they are heterozygous, the alleles are different.

→ The offspring ratios expected by genetic diagrams are only probabilities, and the observed (actual) results may not be exactly the same, especially if small numbers of offspring are involved.

→ In a situation involving dominance, if two heterozygous organisms are crossed we would expect a ratio of three offspring showing the dominant characteristic to one showing the recessive characteristic in their phenotype (a 3:1 ratio). If a heterozygous organism were crossed with a homozygous recessive organism, then we would expect a 1:1 ratio of dominant:recessive characteristic in the offspring.

→ In dihybrid inheritance, two genes are inherited. If both parents are heterozygous for both alleles, giving a genotype of, say, AaBb, there will be gametes of AB, Ab, aB, and ab. This gives 16 different combinations and, typically, a phenotypic ratio of 9:3:3:1.

→ Numerous genes that are found on the X chromosome are not found on the Y chromosome. A male has only one copy of each X-linked gene. A recessive X-linked allele will therefore always show up in a male whereas it will often be masked by a dominant allele in a female. Examples of sex-linked conditions are haemophilia and red-green colour blindness.

→ Genes that are located on the same chromosome are said to be linked. When this is the case, a dihybrid cross will produce far more offspring with the same genotypes as their parents, and fewer recombinants (offspring with new combinations of parental alleles).

→ Any recombinants produced are the result of crossing over in meiosis.

→ The chi-squared test can be used to assess the 'goodness of fit' of the results of genetic crosses.

16.3 Gene control

You may remember from Chapter 6 that mutations will usually have no effect at all on an organism. Imagine, for example, that a mutation occurs in the gene for hair colour or keratin production in a cell in your heart. Neither of those genes is expressed (switched on and used to make proteins) in that cell, so mutations in them would have no effect. Switching genes on and off in this way is called gene regulation, or **gene control**.

Our current estimate is that humans have approximately 20 000 different genes. These are spread out over the 23 human chromosomes. A typical human cell normally expresses just 3% to 5% of its genes at any given time. By switching off genes when they are not needed, cells can prevent resources from being wasted. There is no use in producing insulin in the retina, or myosin in neurones, for example. This **spatial** (location-dependent) regulation of gene expression is very important. In addition, changes in gene

Link

To switch a gene 'on' is a very informal way of referring to initiating (starting) its transcription or expression (see Chapter 6).

expression are known to control the differentiation of stem cells. As we saw at the start of this chapter, this **temporal** (time-dependent) regulation of gene expression earned Professors John Gurdon and Shinya Yamanaka a Nobel Prize in 2012. Note that the functioning of gene control mechanisms is a good example of interactions between genes and the environment on the phenotype.

An example of gene control is the control of coat colour in Himalayan rabbits. An allele codes for an enzyme that synthesises black pigment, but this enzyme is very temperature sensitive and only works at below body temperature. The mutant allele codes for an enzyme with a tertiary structure that is slightly different from that of the normal enzyme, but molecules of this enzyme happen to unfold (denature) at about 37 °C. The Himalayan rabbit is therefore only black in the cooler parts of the body, such as the tail, ears and lower legs. The rest of the coat is pale cream coloured. This shows how both the genotype and the environment can affect the phenotype.

Gene expression is controlled by a variety of mechanisms that range from those that prevent transcription from happening, to those that act after the protein has been produced. In this chapter, we will only consider **transcriptional mechanisms**: those that prevent or promote transcription, and thereby turn off or turn on the synthesis of mRNA.

In both prokaryotes and eukaryotes, whether or not a gene is transcribed (expressed) by a cell is controlled by the presence of **transcription factors**. Transcription factors are proteins that bind to DNA at specific regions called **promoters**. It is thought that this enables RNA polymerase, the enzyme involved in transcription, to bind to the DNA and transcribe a gene nearby, or 'downstream'. Transcription factors are clearly important: it is estimated that 10% of all human genes encode transcription factors, and some can bind to the promoters of more than 100 different genes. As we saw at the start of this chapter, Shinya Yamanaka showed that adding transcription factors Oct1 and Sox2 to mouse skin cells in the laboratory caused the cells to reprogram into stem cells.

When transcription factors were first discovered, it was thought that they all stimulated transcription. Further research has shown that there are some factors that inhibit transcription, and these are just as important in gene control. These **repressors** can bind to a section of DNA near the promoter and interfere with the ability of RNA polymerase to transcribe the gene.

To understand how gene expression is controlled, a distinction must be made between **structural genes** and **regulatory genes**:

- Structural genes are those that code for proteins required by a cell for its normal function. They include fibrous proteins that build structures, including keratin and collagen, but also non-structural globular proteins including enzymes and haemoglobin.
- Regulatory genes code for proteins that regulate the expression of other genes (e.g. transcription factors and repressors).

There must also be a distinction made between **repressible enzymes** and **inducible enzymes**, and how they are produced by the cell:

- Repressible enzymes are synthesised only in the absence of a repressor.
- Inducible enzymes are synthesised only in the presence of their substrate, which is called the **inducer**.

We will consider two examples of gene control in the next part of this section – one that applies to prokaryotic cells and one that applies to eukaryotic cells. Throughout, remind yourself of these four key terms.

Link

Cancer, which you encountered in Chapter 5, can be due to the incorrect control of genes involved in cell division.

Link

In Chapter 4, you saw how lipid-soluble hormones pass across the cell surface membrane. They attach to a cytoplasmic receptor to form a transcription factor, which binds to promoters in DNA.

THE *LAC* OPERON IN PROKARYOTES

In prokaryotes, an **operon** is a length of DNA that:
- contains a number of structural genes that are under the control of a single promoter
- has structural genes whose expression is controlled by the presence or absence of the products of regulatory genes.

In the 1960s, the French scientists François Jacob and Jacques Monod deciphered the first example of gene control in *Escherichia coli*. Like most bacteria, this cell can use glucose in respiration to make ATP. However, when they are unable to obtain glucose from their environment, *E. coli* are able to make use of other sources. One of these sources is lactose, and the events that occur when a bacterium encounters lactose is dependent on the *lac* operon, found in its DNA. This consists of a combination of three structural and two regulatory genes working together (Figure 16.27).

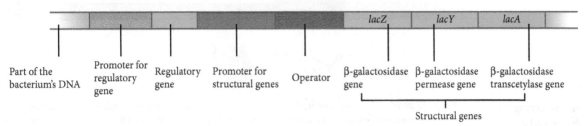

Figure 16.27 The *lac* operon in prokaryotes such as the *E. coli* bacterium. This consists of two regulatory genes, and three structural genes called *lacZ*, *lacY* and *lacA*. In this diagram, the sense strand of DNA (the strand that is read by RNA polymerase in transcription) is shown as a blue ribbon.

Lactose is a disaccharide consisting of glucose joined to galactose by a glycosidic bond. The molecule can be hydrolysed into galactose – which can be converted into glucose, and glucose – which can be used in respiration. The *lac* operon has three structural genes that enable the bacterium to absorb and hydrolyse lactose:

- The *lacZ* gene encodes an enzyme called β-**galactosidase**. This catalyses a hydrolysis reaction to break the α(1,4) glycosidic bond between glucose and galactose to produce these monomers from lactose.
- The *lacY* gene encodes a cell surface membrane protein called **lactose permease**. This is a channel protein that increases the permeability of the cell to lactose by allowing lactose to move into the bacterium by facilitated diffusion.
- The *lacA* gene encodes an enzyme called **lactose transacetylase**. This is required for the metabolism of lactose.

Protein synthesis places a demand on a bacterium's resources. It would be a waste of ATP, nucleotides and amino acids to make β-galactosidase and the two other structural gene products if no lactose is present in the surroundings. The bacterium would be able to use this energy and resources for other purposes instead. Therefore, a mechanism has evolved that allows a bacterium to prevent transcription of the gene encoding the enzyme unless lactose is present (Figure 16.28).

Tip

β-galactosidase in bacteria plays a very similar role to lactase in humans. They both catalyse the hydrolysis of lactose.

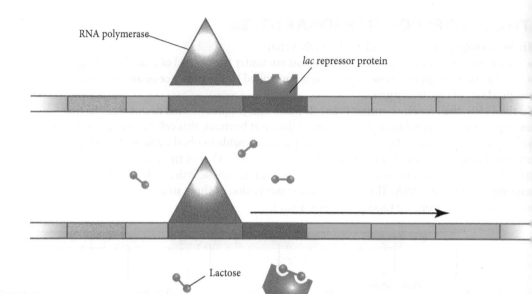

Figure 16.28 When lactose is absent, the regulatory gene is expressed to form a protein called the repressor. The repressor binds to the operator, a region of DNA next to the promoter. Although RNA polymerase can attach to the promoter, it cannot now move past the operator, and therefore is unable to transcribe the three structural genes (**a**). However, when lactose is present, some will move by facilitated diffusion into the bacterial cell. In addition to its DNA-binding site, the repressor has a second binding site, which is complementary to lactose. Lactose binds to the repressor, and causes a conformational change, which means that it is no longer complementary to the operator. It detaches from the operator, and RNA polymerase can now move along the DNA from the promoter. The structural genes are then transcribed to form mRNA, which will then be translated (**b**).

In short, this mechanism allows the bacterium to produce β-galactosidase and the two other structural proteins required for lactose absorption only when lactose is available in the surrounding medium.

GENE CONTROL IN EUKARYOTES: A CASE STUDY OF GIBBERELLIN

The control of gene expression is complex in eukaryotes, but researchers have determined how some mechanisms work. As we saw in the case of the *lac* operon in prokaryotes, some genes are switched on only at certain times. In multicellular eukaryotes, some genes are only expressed in specific cell types. Transcription factors ensure that appropriate genes are transcribed in the appropriate cells at the appropriate time and at the appropriate rate.

One example of the role transcription factors play in eukaryotic gene expression is the action of the plant growth regulator, **gibberellin**, on seed germination. This mechanism involves the following sections of DNA and proteins:

- The amylase gene. This encodes the amylase enzyme, which catalyses the breakdown of starch in the endosperm energy store. This releases maltose molecules that are further hydrolysed to form glucose to use in respiration.
- **Phytochrome-interacting protein (PIF)**. This is a transcription factor that is specific to promoters upstream of the gene encoding the enzyme amylase.
- **DELLA proteins**. These attach to PIF and prevent it from attaching to the promoter upstream of the amylase gene. This prevents germination. They have this name because of the first five amino acids in the polypeptide chain.
- **Gibberellin receptor**. When attached to gibberellin, this protein acts as an enzyme able to catalyse the breakdown of DELLA proteins.
- RNA polymerase. As in prokaryotes, this enzyme is required for the transcription of genes to form mRNA.

With these facts in mind, the list below summarises the process by which gibberellin stimulates seed germination:

1. The seed absorbs water from its environment.
2. The embryo secretes gibberellin.
3. Gibberellin diffuses into the aleurone layer.
4. Gibberellin attaches to its receptor in cells, which forms an active enzyme.
5. The enzyme breaks down DELLA proteins.
6. PIF is able to bind to its promoter, upstream of the amylase gene.
7. The amylase gene is transcribed to form mRNA, which is then translated into amylase enzyme.
8. Amylase hydrolyses starch into maltose.
9. Maltose is hydrolysed into glucose.
10. Glucose is respired by the embryo during germination.

Link

In Chapter 15, you saw how gibberellin stimulates seed germination. Refer to Figure 15.29 to remind yourself of the processes involved.

As well as starting seed germination, gibberellin has other roles in plants during later stages of their life cycle; including the elongation of the stem. The height of many plants, including the pea plant, is controlled by a gene with two alleles – **Le and le** (Table 16.13).

Genotype	Phenotype
Le/Le	tall
Le/le	tall
le/le	short (dwarf)

Table 16.13 The genotype and phenotype of plants controlled by the gene encoding an enzyme required for gibberellin synthesis.

How is gibberellin involved in this process? The protein product of the gene is an enzyme that catalyses the final reaction in a metabolic pathway that produces gibberellin. If present, the hormone stimulates cell division and cell elongation in the stem, which causes a plant to grow taller. An amino acid substitution close to the active site of this enzyme is responsible for most of the *le* alleles that have been identified, which produce enzymes that cannot catalyse this process (Figure 16.29). Homozygous recessive plants are therefore short.

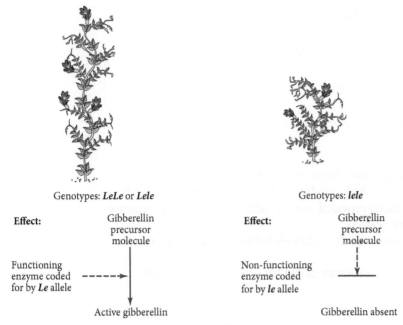

Figure 16.29 The genetic basis of height controlled by gibberellin in plants.

19. Copy and complete the table to summarise the identity of three enzymes involved in the regulation of lactose hydrolysis in prokaryotes. Add a tick (✓) or cross (✗) to each box.

Enzyme name	Encoded by a gene which is		Acts as an enzyme which is	
	regulatory?	structural?	repressible?	inducible?
β-galactosidase				
lactose transacetylase				
RNA polymerase				

★ 20. Make simple sketches of the region of DNA that contains the amylase gene to illustrate how gibberellin release causes amylase secretion in the aleurone layer.

21. In pea plants, gibberellin is required for the elongation of stems. A dominant allele, *Le*, is associated with tall plants, while a recessive allele causes a short stem phenotype.
    a. Identify the phenotypes of plants with the following genotypes:
        i. *LeLe*   ii. *Lele*   iii. *lele*
    b. Applying gibberellin to dwarf plants can stimulate them to grow taller. Suggest what this observation indicates about the function of the protein encoded by the *le* allele.

Assignment 16.1: Reawakening sleeping genes

Haemoglobin is a protein with a quaternary structure, composed of two pairs of globin polypeptides. During the process of development from fertilisation to adulthood, human haemoglobin changes in composition. This is due to temporal (time-dependent) gene control.

As you know, adult haemoglobin consists of two alpha- and two beta-globin molecules. However, in addition to the alpha- and beta-globin genes, a number of other globin genes exist, including zeta-, beta-, delta-, epsilon- and gamma-globin. Table 16.14 summarises the polypeptides that are predominantly found in haemoglobin and Figure 16.30 illustrates the changes in expression of some of the globin genes over time.

Life stage	Haemoglobin	Polypeptides in human haemoglobin
embryo	embryonic haemoglobin (HbE)	2 epsilon- and 2 zeta-globin molecules
fetus	fetal haemoglobin (HbF)	2 alpha- and 2 gamma-globin molecules
adult	adult haemoglobin (HbA)	2 alpha- and 2 beta-globin molecules

Table 16.14 The polypeptides found in human haemoglobin at various stages of life.

QUESTIONS

A1. Using your knowledge of oxygen dissociation curves from Chapter 8, suggest reasons for the changes in haemoglobin expression before and after birth.

A2. With reference to the information provided, state whether the following statements are true or false. Explain your answers.
    a. The percentage increase in beta-globin is 128.6% between birth and six months of age.
    b. By the time birth occurs, all fetal haemoglobin has been replaced by adult haemoglobin.

c. By 10 weeks after fertilisation, all embryonic haemoglobin has ceased to exist.

d. The first globin genes to be expressed after fertilisation are alpha- and gamma-globin.

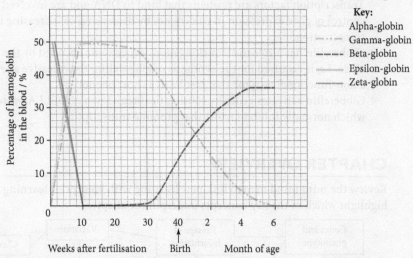

Figure 16.30 The percentage of haemoglobin found in human blood from fertilisation to 6 months of age.

★A3. Globally, more than 100 million people have sickle cell anaemia. In the United States, common treatment is with injections of a medical drug called hydroxyurea (HU). For reasons not yet fully understood, HU has an effect on cells in the bone marrow, which give rise to red blood cells. HU stimulates them to change their expression patterns of genes. Daily use of HU correlates with increased fetal haemoglobin production. Using this information, explain in detail how hydroxyurea treatment reduces the symptoms of sickle cell anaemia.

A4. Figure 16.31 shows a simplified diagram of the α-globin gene cluster on human chromosome 16. The genes in the α-globin gene cluster and the β-globin cluster found on chromosome 11 are both expressed in order, from left to right, during development. This is because a transcription factor called GATA1 binds to a promoter upstream of each gene during different periods of time during the lifetime of an individual.

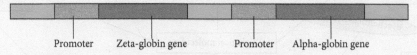

Promoter   Zeta-globin gene   Promoter   Alpha-globin gene

Figure 16.31 A simplified diagram of the α-globin gene cluster on human chromosome 16.

a. Identify a time shown in Figure 16.30 when GATA is attached to the promoter upstream of the zeta gene.

b. Explain why this cluster of genes is not described as an operon.

★ c. Using a similar style to Figure 16.31, draw the β-globin cluster found on chromosome 11. Note that the β-globin cluster contains genes for epsilon-, gamma- and beta-globin (as you saw earlier, this last gene of the three has the code *HBB*).

A5. A type of hereditary anaemia can arise due to mutations in the gene that encodes GATA1, which is found on the X chromosome. This is characterised by the production of abnormal, and fewer, red blood cells.

a. Explain why this hereditary anaemia is more common in newborn boys than girls.

★ b. Suggest how a mutation in a gene coding for a transcription factor could stop it working properly.

Key ideas

→ Structural genes code for proteins required by a cell for its normal function, whereas regulatory genes code for proteins that regulate the expression of other genes.

→ Repressible enzymes are synthesised only in the absence of a repressor, whereas inducible enzymes are synthesised only in the presence of their substrate.

→ Protein production is controlled in prokaryotes by operons including the *lac* operon.

→ Transcription factors are proteins that bind to DNA and are involved in the control of gene expression in eukaryotes by decreasing or increasing the rate of transcription.

→ Gibberellin has a role in stem elongation. An enzyme encoded by the dominant allele, *Le*, is required for the synthesis of gibberellin; the recessive allele, *le*, codes for a non-functional enzyme that results in a short phenotype.

→ Gibberellin activates genes by causing the breakdown of DELLA protein repressors, which normally inhibit factors that promote transcription.

CHAPTER OVERVIEW

Review the mini mind map to link new learning with your prior learning, and to highlight which of the key concepts underpin the chapter.

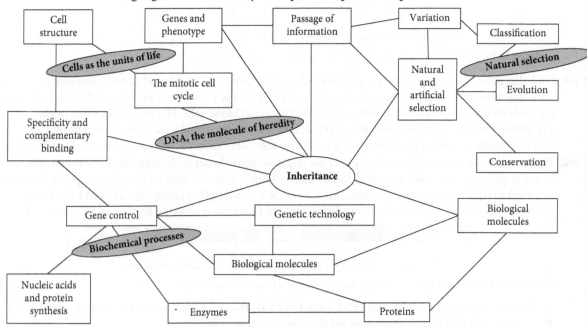

Try copying this mini mind map and expanding upon it. Use your notes from other chapters to help you explore how the essential ideas, theories, and principles can be linked further together.

WHAT YOU HAVE LEARNED

Now that you have finished this chapter you should be able to:

- explain the meanings of the terms haploid (n) and diploid (2n)
- explain what is meant by homologous pairs of chromosomes
- explain the need for a reduction division during meiosis in the production of gametes
- describe the behaviour of chromosomes in plant and animal cells during meiosis, and the associated behaviour of the nuclear envelope, the cell surface membrane and the spindle (names of the main stages of meiosis, but not the sub-divisions of prophase II, are expected: prophase I, metaphase I, anaphase I, telophase I, prophase II, metaphase II, anaphase II and telophase II)
- interpret photomicrographs and diagrams of meiosis and identify the main stages of meiosis

- explain that crossing over and random orientation (independent assortment) of homologous chromosomes and sister chromatids during meiosis produces genetically different gametes
- explain that the random fusion of gametes at fertilisation produces genetically different individuals
- explain the terms gene, locus, allele, dominant, recessive, codominant, linkage, test cross, F1 and F2, phenotype, genotype, homozygous and heterozygous
- interpret and construct genetic diagrams, including Punnett squares, to explain and predict the results of monohybrid crosses and dihybrid crosses that involve dominance, codominance, multiple alleles and sex linkage
- interpret and construct genetic diagrams, including Punnett squares, to explain and predict the results of dihybrid crosses that involve autosomal linkage and epistasis
- interpret and construct genetic diagrams, including Punnett squares, to explain and predict the results of test crosses
- use the chi-squared test to test the significance of differences between observed and expected results
- explain the relationship between genes, proteins and phenotype with respect to the:
  o *TYR* gene, tyrosinase and albinism
  o *HBB* gene, haemoglobin and sickle cell anaemia
  o *F8* gene, factor VIII and haemophilia
  o *HTT* gene, huntingtin and Huntington's disease
- explain the role of gibberellin in stem elongation including the role of the dominant allele, *Le*, that codes for a functioning enzyme in the gibberellin synthesis pathway, and the recessive allele, *le*, that codes for a non-functional enzyme
- describe the difference between structural genes and regulatory genes and the difference between repressible enzymes and inducible enzymes
- explain genetic control of protein production in a prokaryote using the *lac* operon
- state that transcription factors are proteins that bind to DNA and are involved in the control of gene expression in eukaryotes by decreasing or increasing the rate of transcription
- explain how gibberellin activates genes by causing the breakdown of DELLA protein repressors, which normally inhibit factors that promote transcription.

CHAPTER REVIEW

1. Using the key terms 'reduction' and 'variation', in your answer, explain the importance of meiosis in sexually reproducing organisms.
2. Identify how many of the following would be associated with a cell from a chicken (2n = 78).
   a. Bivalents formed during meiosis I.
   b. Centrosomes that form during metaphase I.
   c. Centrosomes that form during metaphase II.
   d. Genetically unique gametes formed due only to random orientation and independent assortment. Write your answer in standard form.
3. Trisomy disorders are caused by the inheritance of two homologous chromosomes from one parent. An example is Down's syndrome, which is due to the inheritance of three copies of the autosome, chromosome 21. The zygote therefore contains three copies of a homologous chromosome, which can result in effects on the phenotype.
   a. Explain what is meant by:
      i. homologous chromosomes
      ii. autosome
      iii. phenotype.

    b. Suggest how both homologous chromosomes could be inherited from one parent. Use the terms centromere, spindle and cell poles in your answer.

    c. There are a number of genetic disorders caused by recessive mutant alleles of genes found on chromosome 21. Explain why it would be less likely for an individual with Down's syndrome to suffer from such a disorder than an individual without Down's syndrome.

4. Identify the stage of meiosis during which the following events occur. The first one has been done for you.

    Chromosomes condense. **Answer**: prophase I

    a. Centromeres divide.

    b. Bivalents line up on equator.

    c. Chiasmata form.

    d. Sister chromatids separate.

    e. Nuclear envelope reforms in a haploid cell.

    f. Spindle contracts to pull apart homologous pairs.

5. Draw a Venn diagram to compare meiosis I and meiosis II. You should include the following in your answer:

    o the sub-stages of meiosis

    o the structures that are separated.

6. Identify whether the following statements regarding inheritance are always true, sometimes true or never true. If your answer is 'sometimes true', provide an explanation.

    a. Human males have 23 pairs of homologous chromosomes

    b. Crossing over occurs during prophase II of meiosis

    c. All children born to a woman with haemophilia will have haemophilia

    d. To perform a test cross in plants, the individual with unknown genotype is mated with a homozygous recessive individual

    e. Disorders caused by a dominant X-linked allele will be more common in males than females

7. Identify the 'odd one out' from these series of terms. Explain your choice. The first one has been done for you.

    sperm, egg, zygote. **Answer**: the odd one out is the zygote, because it is diploid (unlike haploid gametes).

    a. chromosome 4, chromosome 11, chromosome X

    b. ABO blood group in humans, albinism in humans, petal colour in snapdragons

    c. dominant, heterozygous, homozygous

    d. gene, genotype, allele

    e. prophase I, metaphase I, anaphase II

8. Describe the genetic crosses in which the following ratios of phenotypes would be expected:

    a. 3:1   **b.** 1:1   **c.** 1:2:1   **d.** 9:3:3:1   **e.** 9:4:3

9. Rearrange the following four statements to accurately describe how a bacterium hydrolyses lactose.

    1. RNA polymerase transcribes structural genes.

    2. Repressor undergoes a conformational change.

    3. Lactose binds to repressor.

    4. Repressor detaches from the operator.

10. Copy and complete the table to compare the methods of gene regulation in a bacterium and a plant. Write a tick (✓) or a cross (✗) in each box.

	Regulation of:	
	lactase synthesis in bacteria	amylase synthesis in plants
Involves the expression of structural genes		
Involves the expression of regulatory genes		
Requires repressible enzymes		
Requires inducible enzymes		
Involves negative feedback		

Selection and evolution

With its colourful, pulsating eye stalks, you might mistake this snail for an unusual species. But this is in fact a common snail – infected by an unusual parasite.

Leucochloridium paradoxum is a parasitic worm that causes extraordinary changes in its host. After hatching from eggs ingested by the snail, tubular structures squeeze into the eyestalks of its victim. These begin to move back and forth, resembling a crawling caterpillar. The snail loses its dislike of bright light and moves into the open. Here, it becomes an irresistible meal for a bird. The worms can then begin the final stage of their life cycle in another host.

Host-parasite co-evolution is remarkably common in nature. Scientists believe that these relationships may have given rise to sexual reproduction. This maximises genetic variation between individuals and may have evolved to better prepare species to resist their biological enemies.

Prior understanding

You may remember from Chapter 16 that the inheritance of some characteristics in organisms is due to genes, which you first encountered in Chapter 6. You may recall from prior studies that the environment of an organism also has an influence on its features.

Learning aims

In this chapter, you will learn how all species are composed of populations that show variation. This results in a difference between the ability of individuals to survive and reproduce, and hence pass on their alleles to the next generation. You will see how this has been exploited by humans, through selective breeding, and how it can result in the formation of new species.

17.1 Variation (Syllabus 17.1.1–17.1.3; 17.2.3, 17.2.5)

17.2 Natural selection and evolution (Syllabus 17.2.1, 17.2.2, 17.2.4)

17.3 Selective breeding (artificial selection) (Syllabus 17.1.4, 17.2.6, 17.2.7)

17.4 Species and speciation (Syllabus 17.3.1–17.3.3)

17.5 Molecular comparisons between species (Syllabus 17.3.2)

17.1 Variation

The differences between organisms of the same species are known as **variation**. Significant variation between members of a population is caused by different alleles of particular genes. However, variation is also caused by the effects of the **environment**. For example, a plant that has alleles that allow it to grow tall will only be able to do this if it can absorb enough sunlight. If this is lacking, then it will not grow to the potential size determined by its genotype. The same is somewhat true with humans: a child born to tall parents will never reach their potential height if they are malnourished in childhood. Therefore, the phenotype of an organism depends on the interaction between genetic and **environmental factors**.

In this chapter, we will concentrate on variation in a **population** of a species. This is defined as a group of organisms of the same species occupying a particular space at a particular time that can potentially interbreed. Examples of populations include all of the cheetahs in a game reserve, all of the fleas living in the fur of a cat, or all of the durian trees in a rainforest. As you will already be aware, individuals within a species can vary quite considerably from one another. We are most aware of these differences in our own species, but careful observation also shows that every species has some degree of variation in its characteristics.

TYPES OF VARIATION

Variation in a characteristic may be distinct and distinguishable categories, or **categorical**, such as having different blood groups. This is **discontinuous variation**. An example is the ABO blood groups in humans: A, B, AB or O. Each of us must be one of these four possible phenotypes. You cannot be 'halfway' between groups A and O, or 'just slightly' blood group B. Discontinuous variation is caused by different alleles of genes. As you encountered in Chapter 16, the ABO blood group is determined by a single gene that has three alleles: I^A, I^B and I^o. Different combinations of these alleles give the four possible blood groups. Another example of discontinuous variation relates to spots in cheetahs (Figure 17.1a). A rare recessive allele codes for stripes that run down the animal's back, while the dominant allele codes for a full coat of spots (Figure 17.1b).

Link

In Chapters 6 and 16, you saw how a phenotype is influenced by genotype. An example is albinism in humans, which prevents the formation of the melanin pigment.

Link

In Chapter 16, you saw how meiosis and random fusion of gametes can increase genetic variation in a species. This is also true of mutation, provided it occurs in gamete-producing cells.

a.

b.

Figure 17.1 The cheetah, *Acinonyx jubatus*, is native to south and central Africa (a). Although most cheetahs look very similar, the presence of spots or stripes on the back is a characteristic that shows discontinuous variation (b).

Discontinuous variation is common in a population. More often, however, there is a range of variation, in which there are no definite categories. Instead there is a range of heights between two extremes and this is called **continuous variation**. For example, in humans, a person can be any height between the two extremes of the height range

Link

In Chapter 16, you encountered the phenomenon called epistasis. Just two different genes, at different loci, can interact to produce several phenotypes.

Link

In Chapter 18 you will see how it is very important to take a random sample of a population when investigating the variation of a factor in a population.

(Figure 17.2). It is not only physical characteristics, such as mass, length of limbs, and fur colour that have a range of continuous variation, but also metabolic characteristics. These include heart rate, speed of reaction, muscle efficiency and the ability of the brain to process information. Characteristics that vary continuously are usually the result of several different genes acting in such a complex way that it is not possible to distinguish separate phenotypes. For example, in humans, height depends on many different factors, including the growth rate of several different bones, hormone production and the rate of various metabolic reactions. Each factor may be controlled directly, or indirectly, by the combinative effect of a large number of genes, known as **polygenes**. However, these characteristics are often also strongly influenced by an organism's **environment**. As we saw earlier, an animal may have a collection of alleles that would allow it to grow large, but if it has a poor diet during its growth period, it may never achieve its potential mass.

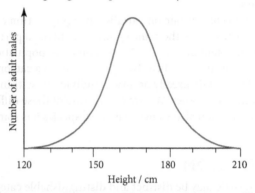

Figure 17.2 A normal distribution curve for height in adult human males. A curve of this shape is very commonly produced when we plot data for a characteristic that shows continuous variation. This symmetrical curve is called a **normal distribution** curve. The greater the standard deviation, the greater the spread of the data on either side of the mean value.

Table 17.1 summarises the two types of variation.

	Continuous variation	**Discontinuous variation**
Description	a range of values falling between two extremes	discrete values that fall into categories
Cause	genetic and environmental	genetic only
Genetic basis	caused by large numbers of genes (polygenes) that have an additive effect in which the gene products interact	caused by one gene
Examples of characteristics in humans	height, skin colour, heart rate	sex, ABO blood group, earwax type
Examples of disorders in humans	psoriasis, diabetes, schizophrenia, asthma	Huntington's disease, sickle cell anaemia, red-green colour-blindness, albinism

Table 17.1 Comparing continuous and discontinuous variation in populations.

Where we see variation in the same feature in different populations, we can use the *t*-test to compare the means of the two populations – for example, the mean height of boys and girls in a class of students, or the mean percentage of individuals with blood type O in the populations of two countries. The results of the test tell you whether the difference between the means is significant, or could just be due to chance. You may recall the *t*-test from Chapter 11 and you will encounter another example in Experimental skills 17.1.

🔍 **1.** Look at Figure 17.1. List all of the characteristics that show continuous variation in this group of cheetahs.

2. 'All continuous variation is quantitative data, which can be shown using histograms, and all discontinuous variation can be represented by qualitative data on bar charts.' Discuss the accuracy of this statement.

3. Explain why most variation in a population caused by the environment cannot be inherited by offspring.

POPULATION GENETICS

In Chapter 16, you explored how the genotype of an individual, defined as the combination of alleles for a specific gene, has an impact on its phenotype. It is also possible for scientists to consider the alleles and genotypes of all individuals in a population.

To study the genetics of a population, **gene pools** and **allele frequencies** are two useful terms. The total of all of the alleles of all of the genes of all of the individuals in a population at a particular time is known as the gene pool of that population. For example, a species of butterfly may be divided into two or more populations that do not meet and interbreed. Each population has its own gene pool. There may be particular alleles of genes that are present in one population's gene pool, but not in the other. Some alleles may be common to both populations, but the proportion may differ. It can be helpful to know something about how common a particular allele is in a population. We call this the frequency of that allele. You will explore the uses of these two terms in the next section of this chapter.

Genetic bottlenecks and founder effects

Sometimes, a population experiences a severe decline in numbers, and then recovers from just a few individuals. This is called a **genetic bottleneck**. Because it was likely due to chance which individuals survived, the new population will usually have a very different gene pool to the original. A good example of this is the cheetah (Figure 17.1). There is very little variation in this species – i.e. the cheetah species has a very small gene pool. It is thought that all the cheetahs in the world are descended from one small family that survived in Africa about 10 000 years ago. As a consequence of passing through a bottleneck, possibly a period of climate change at the end of the last Ice Age, all cheetahs are almost identical. All of them have characteristic health issues, which are associated with inbreeding. Figure 17.3 shows a model that can be used to imagine this scenario.

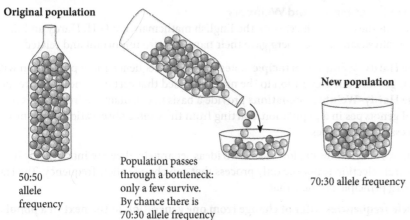

Original population

50:50 allele frequency

Population passes through a bottleneck: only a few survive. By chance there is 70:30 allele frequency

New population

70:30 allele frequency

Figure 17.3 A model of a genetic bottleneck, in which individuals are represented by beads. In the original population, there are 1000 beads; half red, half blue. If 10 beads are extracted at random, there is a high probability that there will not be five of each. If the organisms carrying these alleles then breed and produce (found) a new population, the proportion of colour (the allele frequency) will be significantly changed; in this example, to 7:3.

A related phenomenon is called the **founder effect**. In this scenario, which is common when new islands form, a small number of individuals from a large population leave the population to start (or 'found') a new population elsewhere. It is likely that the founding population was a chance sample of the original population. Therefore, the allele frequencies in the new population will often not reflect those in the original, larger population from which they came. A good example of this are the *Anolis* lizards of the Caribbean (Assignment 17.1). Lizards can be carried between islands on floating plant matter, especially after a hurricane. It is just chance which individual lizards land on an island, and which combination of alleles that this population carries. Figure 17.4 shows a model to represent the founder effect.

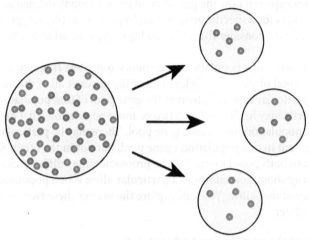

Figure 17.4 A simple illustration of the founder effect. The original population is on the left with three possible founder populations, perhaps representing separate islands, on the right. If individuals are represented by coloured dots, the new populations are all very different from each other, and from the parent population.

Chance, or 'pure luck,' is another phenomenon that can change the gene pool of a population over a number of generations. This is called **genetic drift**. It is most likely to have a significant effect if the population is small. Imagine a very small population of plants in which two plants have white flowers and two plants have purple flowers. Perhaps just by chance the seeds from the white flowers are eaten by a bird, while a few seeds from the purple flowers successfully germinate and grow into adult plants. Over several generations, the alleles for white flowers could be completely lost, just by chance.

The studies of Hardy and Weinberg

Working in the early 20th century, the English mathematician G.H. Hardy and the German physician W. Weinberg gave their names to two important and related ideas:

1. The **Hardy–Weinberg principle** states that allele frequencies in a population will not change from one generation to the next, provided that certain conditions are met.
2. The **Hardy–Weinberg equations** provide a basis to calculate the frequencies of alleles and genotypes in a population, starting from the simple observation of the number of recessive phenotypes.

Although you will now explore these two ideas separately, they are intrinsically linked.

Natural selection is not the only process that can change allele frequency. The Hardy–Weinberg principle predicts that:

'Allele frequencies will not change from one generation to the next in a population.'

However, this statement makes five essential assumptions:

a. The individuals in the population demonstrate **random mating**. The probability of one individual mating with any other individual is equal.
b. There is no **migration**. Individuals with the same or different genotypes cannot move into (immigration) or out of (emigration) the population.
c. There is no mutation, so no new alleles are created.

d. The population is large, so that fluctuations in allele frequency due to chance, called genetic drift, have no substantial effect.

e. There is no **selection**, so no alleles are favoured or eliminated due to the effects that they have on the phenotype, survival and reproduction of the organism that carries them.

However, despite the prediction of the Hardy–Weinberg principle, it is very unusual for it to apply to a population. This is because one, two or even all five of the assumptions are not met in the wild. This means that the gene pool of a population is not stable and allele frequency changes over time (Figure 17.5). This is crucial for a population to evolve and the principle provides a means of quantifying this evolutionary change.

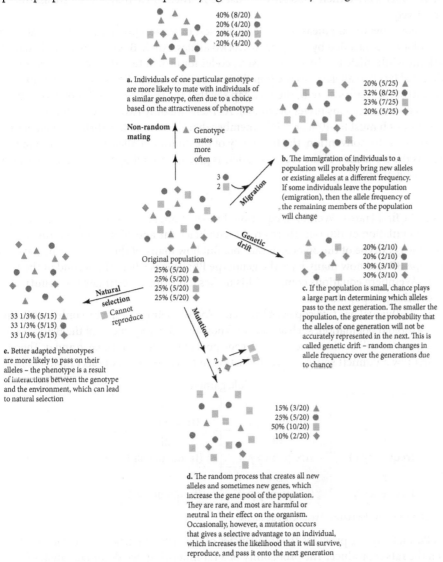

Figure 17.5 The actions that can result in a change in allele frequency. In this hypothetical example of a gene with four alleles, represented by the different coloured shapes, the original (ancestral) population is shown in the middle with equal proportions of each. Although the Hardy–Weinberg principle predicts that allele frequencies will not change over many generations, five mechanisms, (**a**) to (**e**), result in its assumptions not holding true. In addition to these, genetic bottlenecks and the founder effect also have the ability to affect allele frequencies and therefore the gene pool of a population.

As you encountered earlier, it can be helpful to know how common a particular allele is in a population. We call this the **allele frequency**. Any particular allele will have a frequency somewhere between 1 (meaning 100%, that there are no other alleles for this gene) and 0 (0%, that no alleles for this gene are found in a population). For any specific gene, the frequencies of all of its different alleles add up to 1. If there is only one allele, then the frequency is 1. If there are two different alleles, A and a, for example, and they are equally common, then the frequency of each allele is 0.5. If the frequency of one allele is 0.32, you know that the frequency of the other is 0.68, assuming that there are just two alleles. Almost all the organisms you will study are diploid, meaning that their body cells contain two sets of chromosomes and therefore two copies of each gene. So, in a population of 1000 individuals, there will be 2000 copies of any particular gene – assuming that the gene is not sex-linked. If there are two alleles, R and r, and you know that 400 alleles are r, then 1600 alleles must be R. Their allele frequencies are 0.2 and 0.8 respectively.

To show how we can measure allele frequency, imagine a population of 500 rats. Their coat colour is controlled by one gene with two alleles. Allele B is dominant and codes for black fur while allele b, which is recessive, codes for brown fur. The rats are diploid – each individual has two alleles – so the population contains 1000 alleles for coat colour.

If 600 of these alleles are B, then we can say that the frequency of B is 60% or (600 ÷ 1000 =) 0.6 (statisticians prefer decimals). As there are only two alleles, it follows that the frequency of b must be 40% or 0.4. Remember that the sum of the allele frequencies must be 100% or 1. By convention, the frequency of the dominant allele is given the symbol p, whereas the frequency of the recessive allele is given the symbol q. Therefore, we can state the first equation:

$$p + q = 1$$

This is the first Hardy–Weinberg equation. But what does it allow us to do? Let's imagine in one population of 100 rats, there are 64 with black fur and 36 with brown fur. We know the genotype of the 36 brown rats: bb. But how many of the 64 black rats have the genotype BB and how many have the genotype Bb? This is where the second of the two Hardy–Weinberg equations come in. This makes it possible to estimate the **genotype frequencies** in a population.

Each individual has two copies of this gene, and so their genotype can only be BB, Bb or bb. In a diploid population, such as rats, we know that the frequencies of the B and b alleles in the gametes will be the same as in the gene pool of the parent rats, i.e. p and q. Therefore, we can draw a Punnett square for reproduction in this population of rats (Table 17.2).

Tip

The Hardy–Weinberg equations cannot be used when the alleles of a gene are codominant, or if a gene has more than two alleles (multiple alleles) (e.g. the ABO blood group in humans).

		Male gametes	
		B (frequency p)	b (frequency q)
Female gametes	B (frequency p)	BB (frequency p^2)	Bb (frequency pq)
	b (frequency q)	Bb (frequency pq)	bb (frequency q^2)

Table 17.2 Deriving the Hardy–Weinberg equations.

This Punnett square provides the frequencies of the different genotypes in the population when the rats reproduce. The probability of a rat inheriting two dominant alleles from its parents and having the genotype BB is ($p \times p$), which is p^2. The probability of a rat having the genotype bb is ($q \times q$), or q^2. The probability of a rat having the genotype Bb is ($2 \times p \times q$), expressed as $2pq$. This last expression is multiplied by 2 because there are two ways of generating the heterozygous genotype (Bb or bB). This gives rise to the second equation:

$$p^2 + 2pq + q^2 = 1$$

where:
- p^2 is the frequency of the dominant homozygous genotype (here, BB)
- $2pq$ is the frequency of the heterozygous genotype (here, Bb)
- q^2 is the frequency of the recessive homozygous genotype (here, bb).

So, this equation is simply saying that when you add up all the genotypes, you must get the whole population. More importantly, it is possible to use this equation to predict how many individuals have the dominant homozygous genotype and how many have the heterozygous genotype.

Let's now look at how it can be applied to the example we considered earlier regarding the population of black and brown rats.

In this example, if 36 individuals were brown, which is the recessive phenotype. 36 out of 100 is 36%, or 0.36 as a decimal.

So, we know that q^2 is 0.36.

And, therefore, q is $\sqrt{0.36}$, which is equal to 0.6.

Applying the equation that determines allele frequency, we know that $p + q = 1$, so $p = 1 - 0.6 = 0.4$.

The frequency of the Bb genotype is $2pq$, so that is $2 \times 0.6 \times 0.4$, which is 0.48 (or 48%).

So, 48 of the population of 100 rats will have genotype Bb and, therefore, $(64 - 48) = 16$ will have the genotype BB.

There are many applications of the Hardy–Weinberg equations. They can be used to determine whether allele frequencies have changed over generations, providing evidence for natural selection. Furthermore, as you will see in the following worked example, they can be very useful in determining the number of carriers of recessive alleles associated with genetic diseases. This is important in the field of genetic counselling, which you will encounter in Chapter 19.

Worked example

Sickle cell anaemia is a genetic disease caused by a recessive allele. Around 2% of newborn children in Nigeria are affected by sickle cell anaemia. Assuming Nigeria has a population of 2×10^8, estimate the number of individuals who carry the allele that causes sickle cell anaemia. The two Hardy–Weinberg equations are $p + q = 1$ and $p^2 + 2pq + q^2 = 1$.

Answer

The process of working out the allele frequencies using the Hardy–Weinberg equations is just the same as already described, but the numbers are smaller.

In this example, 2% (1 in 50) children are homozygous recessive, so we know that q^2 is $(1 \div 50) = 0.02$.

And, therefore, q is $\sqrt{0.02}$, which is 0.14.

We know that $p + q = 1$, so $p = 1 - 0.14 = 0.86$.

The frequency of the Bb genotype is $2pq$, so that is $2 \times 0.86 \times 0.14$, which is 0.241 (or 24.1%).

Therefore, we would expect $(2 \times 10^8 \times 0.241 =)$ 48.2 million people in this country to be carriers of the allele that causes sickle cell anaemia.

4. Prior to Hardy and Weinberg's work, it was assumed that dominant alleles would increase in frequency over time, and that recessive alleles would decrease in frequency. With reference to Huntington's disease and albinism, explain why this is not always the case.

5. In one study, the frequency of the recessive allele that causes red–green colour-blindness was found to be 0.09. What is the frequency of the dominant allele in this population? The first Hardy–Weinberg equation is $p + q = 1$.

Tip

To determine phenotype frequencies, you must know which allele is dominant, and which is recessive. The frequency of individuals with the recessive phenotype is q^2. The frequency of individuals with the dominant phenotype is $(p^2 + 2pq)$.

Tip

In most Hardy–Weinberg problems, the first step is to find the frequency of individuals with the recessive allele (q). Do this by finding the square root of the frequency of the homozygous recessive phenotype (q^2).

Tip

Note that you do not need to remember the two Hardy–Weinberg equations. They will be provided if you are asked a question on this topic in your examination.

★ **6.** Phenylketonuria (PKU) is a genetic disease caused by a recessive allele. It used to be fatal, but individuals who have the disease can live normal lives if their diet is carefully managed. Heterozygote carriers of the allele do not show symptoms of the disease.

    a. One in every 10 000 babies born in one European population has PKU. Calculate the frequency of heterozygote carriers that would be expected in this population. The two Hardy–Weinberg equations are $p + q = 1$ and $p^2 + 2pq + q^2 = 1$.

    b. In a population in sub-Saharan Africa, the frequency of the recessive allele that causes PKU is 3.0×10^{-3}. Use the Hardy–Weinberg equations to calculate the frequency of newborn babies positive for PKU that would be expected in this population.

    c. It is thought that heterozygote carriers of PKU have some resistance to a toxin, ochratoxin A, which increases the risk of developing cancer. The fungus that produces this toxin can grow on wheat grain that has been stored for food over the winter months. With reference to the Hardy–Weinberg principle, explain how PKU may have become more common in the European population compared with that of sub-Saharan Africa, which historically did not farm or store wheat grain.

Key ideas

→ A population is a group of organisms of the same species occupying a particular space at a particular time that can potentially interbreed. A species may contain many different populations, which interbreed within their population but not often with other populations.

→ In all species, individuals show variation in phenotypes, which is due to a combination of the alleles they have inherited from their parents, and the environment.

→ Phenotypic variation may be continuous, for example the height or mass of an organism, or discontinuous, for example the human ABO blood groups. The extent to which a characteristic controlled by an organism's genotype is shown is also influenced by the environment.

→ Meiosis, random mating and the random fusion of gametes produce genetic variation within populations of sexually reproducing organisms. Variation is also caused by the interaction of the environment with genetic factors, but such environmentally induced variation is not passed on to an organism's offspring.

→ The gene pool is all the alleles of all the genes that are present in a population.

→ The Hardy–Weinberg equations allow us to calculate the frequency of a particular allele in the gene pool of a population, assuming that allele frequencies do not change from generation to generation. They are: $p + q = 1$, where $p =$ the frequency of the dominant allele and $q =$ the frequency of the recessive allele; and $p^2 + 2pq + q^2 = 1$, where p^2 is the frequency of the homozygous dominant genotype (for example, FF), $2pq$ is the frequency of the heterozygous genotype (Ff) and q^2 is the frequency of the homozygous recessive genotype (ff).

→ The allele frequencies and the proportions of genotypes of a particular gene in a population can be calculated using the Hardy–Weinberg equations. Phenomena including genetic bottlenecks, founder effects and genetic drift all change allele frequencies over generations in a species.

17.2 Natural selection and evolution

The work of Hardy and Weinberg suggested that in a species, allele frequencies and gene pools will not change over time. However, all evidence suggests that this is not the case. Therefore, it is a valid prediction to suggest that the characteristics of species change over time. It can lead to the formation of new species from pre-existing ones, as a result of changes to the gene pools of populations over many generations. Indeed, most people today accept this fact. However, for a long time, those who supported the idea were scorned, and Darwin himself was widely ridiculed. However, evidence for the theory is compelling. The discovery of fossils of animals and plants that no longer exist (Figure 17.6) and the development of similar adaptations in animals and plants to cope with similar habitats (Figure 17.7). You will also read about selective breeding later, which provides further tangible evidence for the theory.

Figure 17.6 The discovery of fossils in the 19th century gave rise to the idea that the living world is not static: organisms can change with time. This is the fossilised remains of *Basilosaurus*, an ancestor of modern whales, in Wadi El-Hitan, Egypt.

a.

b.

Figure 17.7 A process called convergent evolution occurs when unrelated species solve the same biological problem by evolving in the same way. Cacti are native to the Americas and have evolved to cope with the high temperature and low rainfall of deserts by reducing surface area and possessing a fleshy stem for storage (**a**). Although it looks like a cactus, this euphorbia from Africa (**b**) is not. However, it has adapted to a similar habitat in a remarkably similar way.

> **Tip**
> 'Struggle for existence' suggests that some individuals physically dominate others. While this is sometimes true, the 'struggle' may also be different – such as the ability of bacteria to withstand an antibiotic.

In 1858, the British naturalists Charles Darwin (Figure 17.8) and Alfred Russel Wallace proposed a theory of **natural selection** to account for the changes in species over time. A year later, Darwin published *On the Origin of Species*. In this book, he described evidence for the way in which an environmental factor called a **selection force** (sometimes called selection pressure or selective agent) could determine which individuals survive and which do not, in a '**struggle for existence**'. This can lead to significant changes in the features of species of organisms over time, which is called **evolution**. As you will read later, this can also lead to the formation of new species from pre-existing species over time.

Darwin put forward the idea of evolution by natural selection. In summary, he observed that:

1. Reproduction generally creates more individuals than can possibly survive.
2. As a consequence, there is competition for food, nutrients, mates, nesting sites, light, and other resources. These are the selection forces.
3. All individuals are different; there is variation within every species.

1881

Figure 17.8 Charles Darwin, the father of evolution. Darwin published his ideas in his book, *On the Origin of Species*.

439

Tip

Natural selection
is a mechanism,
usually too slow
to be observed.
However, the effect
of natural selection,
evolution, consists of
measurable changes
in a species over time.

Link

There are two types
of selection force
(Chapter 18): biotic
factors caused by
other organisms,
such as competition,
predation and disease,
and non-living
abiotic factors, such
as the availability of
sunlight or water.

Link

Darwin and Russel
Wallace developed
their theories with no
knowledge of genes.
Mendel's work with
peas was ongoing, but
did not receive wider
recognition until later
in the 19th century
(Chapter 16).

Tip

Biologists define
fitness with regards
to reproductive
success, not physical
ability. The fittest
organisms survive to
transmit their alleles
to its offspring.

Tip

Advantageous alleles,
not genes, are passed
on by fitter organisms
to the next generation.

In such conditions, individuals that are fortunate enough to possess a favourable phenotype will have a **selective advantage**. They will have more of a chance of surviving, reproducing and passing on the alleles responsible for their selective advantage to the next generation. Therefore, many people simplify the concept of natural section to '**survival of the fittest**'. In the long term, natural selection increases the frequency of alleles conferring an advantage, and reduces the frequency of alleles conferring a disadvantage.

Let's consider an example. Imagine a species of bird living in a forest. In this population, there is variation between individuals. Some have better eyesight than others, some have a more advanced immune system, and so on. This has a direct influence on their ability to compete with other members of the species. The fitter birds may gather more food and so rear larger families than the less fit individuals. Therefore, the fittest individuals pass on their alleles to more of the next generation. Another example is provided in Figure 17.9.

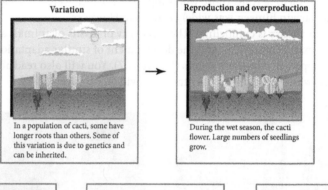

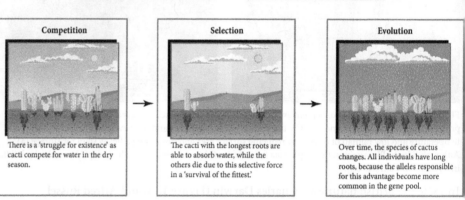

Figure 17.9 Natural selection, as proposed by Darwin and Russel Wallace, leads to evolution. This series of images shows how a species of cactus or euphorbia could evolve to develop deeper roots over a long period of time.

Famously, Darwin's claims were based on specimens he collected from the Galapagos Islands. These are an isolated group (archipelago) of volcanic islands in the Pacific Ocean, about 1000 km from South America. Darwin visited these islands in 1835 and collected many of the small birds called finches and brought them back to London for study. He wondered whether one single ancestral species of finch may have found its way to the archipelago from the mainland a long time ago, and then had somehow become **adapted** to take advantage of the different foods available. He wondered whether these became the 14 species he collected. As he reflected some time later, 'one might really fancy that, from an original paucity (lack) of birds in this archipelago, one species had been taken and modified for different ends'.

Darwin's finches have been closely studied for many years. On the mainland closest to the Galapagos Islands, finches have short, straight beaks to crush seeds. Many of them feed on seeds. Birds with large beaks tend to feed on hard seeds that need to be cracked before they are eaten. Birds with small beaks tend to feed on smaller, softer seeds. Beak size and shape is in fact very diverse, allowing finches to feed on a wide variety of food sources (Figure 17.10).

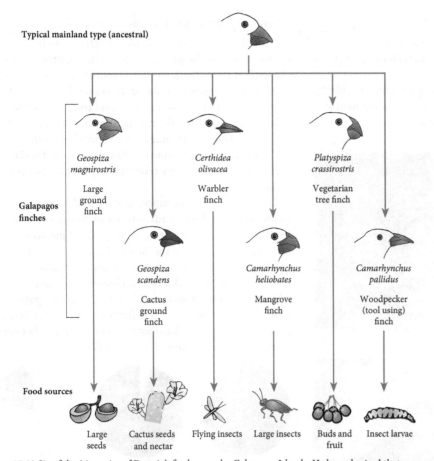

Tip

Natural selection does not always lead to change. Sometimes it can keep things the same: crocodiles and sharks, for example, have not changed significantly since dinosaurs roamed the Earth.

Figure 17.10 Six of the 14 species of Darwin's finches on the Galapagos Islands. He hypothesised that one common ancestor evolved into many different species of finch with characteristic beak shapes, each adapted to consuming a predominant food source.

THE EVIDENCE FOR EVOLUTION

Darwin's ideas were met with ridicule at the time, and even today many people challenge the theories of natural selection and evolution. Precisely because the mechanism takes time, Darwin could not provide direct evidence that natural selection had produced change in a particular species. However, long after Darwin's work, Peter and Rosemary Grant investigated changes on Daphne Major, one of the Galapagos Islands, for over 40 years. They found further evidence that the theory applies in nature.

In 1983, a significant climate change occurred when a warm ocean current brought prolonged rainfall to the normally dry Galapagos Islands. Many of the cacti died in the wet conditions. This reduced the supply of large, hard seeds. On the other hand, plants that produced small, soft seeds flourished. This changed the range of food available to the seed-eating finches considerably. Over the subsequent four years, there was a remarkably rapid response in the population. Whereas birds with large beaks had been particularly successful in dealing with cactus seeds, they could not pick up the smaller seeds of other plants very easily. The population in 1987 had longer and narrower beaks than those in 1983, on average, correlating with the reduction in supply of small seeds. Moreover, as the climate became drier again into the 1990s, the trend was reversed in precisely the expected way. This adaptation to changing conditions was only possible because of the rapid reproductive rate and because there was continuous variation in beak size in the population, and this variation could be inherited.

Other examples of species that have been observed to change in response to specific environmental factors over a short period of time are shown in Table 17.3. Unlike the observations made by the Grants on the Galapagos Islands, in these further cases the selection force has an origin in human activity.

Tip

Changes in environmental factors and selection forces affect the chance of an allele surviving in a population. They do not affect the chance of such an allele arising by mutation.

Species and habitat	Observed changes in response to selection force
The peppered moth, *Biston betularia*, has two phenotypes: normal speckled and melanic (dark) (Figure 17.11). The normal habitat of the peppered moth is birch woodland in temperate climates such as the British Isles. They are predated by birds. **a.**  **b.**  **Figure 17.11** The two variants of the peppered moth: normal speckled (**a**) and melanic (**b**).	During the day, normal speckled moths are well camouflaged on the grey and brown lichen growing on the bark of trees. Dark (melanic) moths are more easily seen by birds and are predated. However, in areas with air pollution, lichen cannot grow and tree trunks take on a dark grey or black colour, which means that melanic moths are more camouflaged than speckled moths against the dark background. This is called industrial melanism because heavy coal burning by factories is the major cause. In areas with a high amount of air pollution, melanic moths are at a selective advantage (are fitter), while in areas with cleaner air the normal speckled moths are fitter. In many cities of the British Isles, the melanic moths predominated after the development of coal-powered factories in the 1800s. However, due to concerns related to human health, the Clean Air Act was passed in 1956. This prohibited the burning of coal in cities. Since then, the normal speckled form of the moth has become more and more common (Figure 17.12). 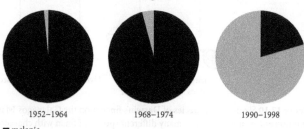 1952–1964 1968–1974 1990–1998 ■ melanic ■ normal speckled **Figure 17.12** Surveys of the numbers of melanic and normal speckled moths in the city of Manchester, U.K. After the Clean Air Act was passed in 1956, air pollution has fallen, and the normal speckled moth has become more common.
The rabbit, *Oryctolagus cuniculus*, was introduced to Australia in the mid-1800s as a source of food and for hunting. Their numbers soon increased rapidly, devastating crops and harming the native Australian wildlife. In the 1950s, farmers in Australia introduced a small number of rabbits infected with a deadly viral disease called myxomatosis in an attempt to kill the rabbits (Figure 17.13).  **Figure 17.13** A rabbit with the disease myxomatosis.	After the significant reduction in population immediately following the introduction of myxomatosis (Figure 17.14), only rabbits with alleles that gave them resistance to myxomatosis were able to survive and pass on the responsible alleles to the next generation. 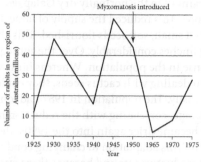 **Figure 17.14** Myxomatosis was introduced to Australian rabbit populations in 1950. But by the late 1960s, rabbits resistant to the myxoma virus had repopulated and today most rabbits in Australia are myxomatosis-resistant.

Pathogenic bacteria, including strains of *Mycobacterium tuberculosis*, the bacterium responsible for tuberculosis (TB; Figure 17.15).

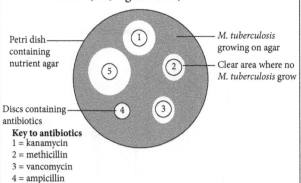

Petri dish containing nutrient agar

M. tuberculosis growing on agar

Clear area where no *M. tuberculosis* grow

Discs containing antibiotics

Key to antibiotics
1 = kanamycin
2 = methicillin
3 = vancomycin
4 = ampicillin
5 = ciprofloxacin

Figure 17.15 Several methods can be used to determine whether a particular population of bacteria is resistant to specific antibiotics. This population of *M. tuberculosis* is completely resistant to ampicillin, because bacteria can grow completely around a small paper disc soaked in this antibiotic.

The widespread use of antibiotics such as penicillin has rapidly led to the development of resistant strains of bacteria (Figure 17.16), which often produce enzymes that break down the antibiotics. In a person that is being treated with an antibiotic, these surviving mutants are at a selective advantage because non-mutated bacteria are inhibited or killed by the antibiotic. The mutants can then grow and reproduce with much less competition for nutrients from other bacteria.

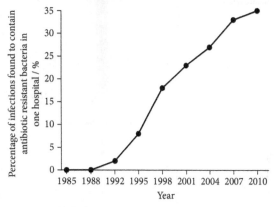

Figure 17.16 The development of antibiotic resistant strains, such as vancomycin-resistant *Staphylococcus aureus*, is a common feature in hospitals.

Table 17.3 Since Darwin's time, a number of species have been observed to change in response to selection forces. This has led to evolution – a change in a species over time. The organisms in these examples have a relatively short life cycle and high reproductive capacity. In the case of bacteria, one resistant bacterium could produce 1×10^{10} descendants within 24 hours. They can therefore respond quickly to the selection forces placed on them. The effects of selection forces on organisms that live longer and have a lower reproductive capacity take much longer to become apparent.

Now you have encountered a range of situations in which natural selection has brought about evolution, think back to some of the most remarkable species you have encountered so far in this book. In Chapter 3, for example, you read how the bombardier beetle produces a jet of boiling-hot liquid to deter a predator. Natural selection provides a means to explain how ingenious adaptations and features could have come about. It might seem improbable, but do not forget that the process takes place over millions of years, through an incremental process of gradual changes and improvements.

Link

As you read in Chapter 10, antibiotics are substances produced by living organisms that kill bacteria but do not normally harm human cells.

TYPES OF SELECTION

There are three basic types of natural selection. These are **stabilising selection**, **directional selection** and **disruptive selection**. The environmental factors that drive these mechanisms are called **stabilising forces**, **directional forces** and **disruptive forces** respectively. The differences between them are summarised in Figure 17.17.

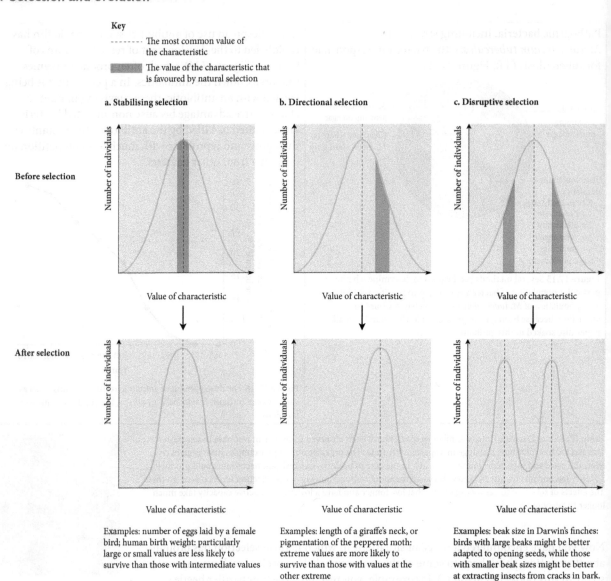

Key

------- The most common value of
the characteristic

The value of the characteristic that
is favoured by natural selection

a. Stabilising selection   **b. Directional selection**   **c. Disruptive selection**

Before selection

Number of individuals

Value of characteristic

After selection

Number of individuals

Value of characteristic

Examples: number of eggs laid by a female
bird; human birth weight: particularly
large or small values are less likely to
survive than those with intermediate values

Examples: length of a giraffe's neck, or
pigmentation of the peppered moth:
extreme values are more likely to
survive than those with values at the
other extreme

Examples: beak size in Darwin's finches:
birds with large beaks might be better
adapted to opening seeds, while those
with smaller beak sizes might be better
at extracting insects from cracks in bark.
Birds with medium-sized beaks are
out-competed by the two extremes:
extreme values are more likely to survive
than those with intermediate values.

Figure 17.17 Types of selection. In an original population with a normal distribution, natural selection can
change the number of individuals with different values of a characteristic. In stabilising selection (**a**), phenotypes
with middle values have an advantage, and individuals with extreme values are less fit. In directional selection (**b**),
the selection force shifts the range of phenotypes towards one end of a variation range by favouring more extreme
phenotypes over others. In disruptive selection (**c**), individuals with both extremes of a phenotypic characteristic
have an advantage over those in the middle.

Worked example

Draw a flow diagram to produce a summary of the process by which a bacterial species undergoes directional selection by exposure to an antibiotic.

Answer

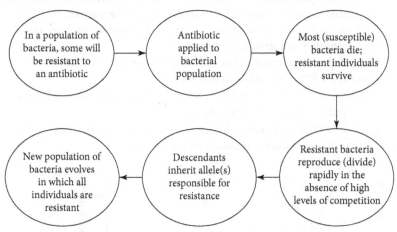

7. The development of antibiotic resistance in bacteria is an example of directional selection. Figure 17.16 shows the numbers of vancomycin-resistant *Staphylococcus aureus* samples collected from hospitals in one country over 20 years.
   a. Describe the trends shown in Figure 17.16.
   b. Suggest reasons to explain why:
      i. natural selection only applies to variation in a population that is controlled by differences in genotype between individuals, rather than their environment
      ii. natural selection happens faster in organisms with short life cycles
      iii. disruptive selection is less common in nature than either stabilising selection or directional selection.

8. The peppered moth, *Biston betularia*, has been extensively studied as a good example of natural selection.
   a. Use Figure 17.11 to produce a summary of the process by which directional selection applied to the peppered moth after it encountered a reduction in air quality. You may draw sketches or list a series of bullet points.
   b. Figure 17.18 shows the mean numbers of moths of these two phenotypic forms caught in 20 surveys conducted in England in the 1960s.

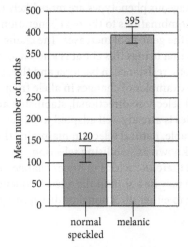

Figure 17.18 Mean number of moths of two phenotypes (The 95% confidence interval error bars are shown).

    i. Calculate the mean percentage of melanic moths captured per survey. Give your answer to the nearest whole number.

    ii. The standard error of the mean (SE) is calculated using the formula $SE = \dfrac{s}{\sqrt{n}}$, where s = sample standard deviation and n = the number of surveys. Calculate the standard deviation for both mean values in Figure 17.18, which were both calculated from the results of 20 surveys. Give your answers to 2 d.p.

    iii. Explain how the 95% confidence interval error bars for each mean value could be determined and drawn on this graph.

    iv. Discuss the significance of the difference between the numbers of the two phenotypic forms of moths in this survey.

  c. Outline experiments that could be conducted to provide evidence for the evolution of the peppered moth by natural selection.

9. Figure 17.19 shows the relationship between the mass of a human baby at birth and the percentage of infant deaths within the first year of life.

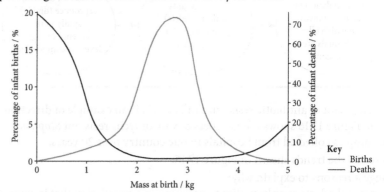

Figure 17.19

Select the statements that explain these observations.

1. Babies born with intermediate birth masses are more likely to reach adulthood.

2. Alleles associated with very low birth mass have become rare.

3. Birth mass in humans has been subject to stabilising selection.

4. Babies that survive infancy only have alleles associated with intermediate birth mass.

A. 1 and 2 only   **B.** 3 and 4 only   **C.** 1, 2 and 3 only   **D.** 2, 3 and 4 only

---

Key ideas

→ Individuals with advantageous phenotypes are more likely to survive and reproduce – passing on their allele combinations to the next generation. The frequency of the advantageous alleles in the gene pool increases. The change in frequency of alleles is evolution – the process that brings this about is known as natural selection.

→ The evolution of antibiotic resistance in bacteria and the spread of industrial melanism in moths are examples of changes in allele frequencies.

→ There are three types of selection: directional, stabilising and disruptive.

→ Normally, when a species is already well adapted to its environment, and that environment is fairly stable, natural selection maintains the frequencies of alleles in the population. This is known as stabilising selection.

→ If the environment and therefore selection forces change, or if a new allele arises by mutation, then there may be a shift in the allele frequencies in subsequent generations. This is due to either directional or disruptive selection.

17.3 Selective breeding (artificial selection)

Familiar animals can have unusual beginnings. The distinctive features of the dogs in Figure 17.20 are the result of **selective breeding**, also called **artificial selection**. This involves processes similar to those seen in natural selection. However, this process is carried out artificially by humans. Over long periods of time, dogs with specific features have been selected, or chosen, as mating partners so that those features are retained and even exaggerated in the resulting litter of puppies. Both of these dogs were bred for hunting. However, the Great Dane was bred for hunting large animals such as wild boar and deer stags for food. The terrier was selected for its ability to follow animal pests, such as rabbits, into underground burrows.

Selective breeding is a technique that originated independently in many areas of the world during the past 10 000 years (Figure 17.21). Importantly, our ancestors did not need a scientific understanding of the mechanism of inheritance to do this. Animals and plants with the most desirable features were simply crossed or mated over many generations, and those that did not have the desired features were not used. In the case of plants, pollen from one plant was dusted onto the stigma of another, and the resulting seeds are collected and sown. In the case of animals, males and females with the desired characteristics were isolated from other individuals to ensure that they had the greatest opportunity to mate.

Figure 17.20 A Great Dane (left) and a Yorkshire terrier (right). The selection of desired features over many generations has produced a wide variety of different breeds of dog.

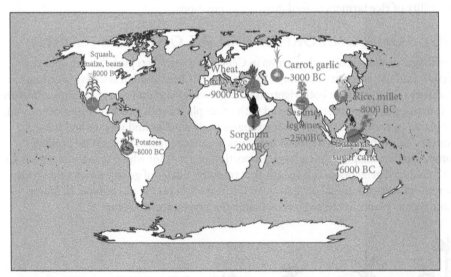

Figure 17.21 The cultivation of crops probably began over 10 000 years ago in independent farming communities all over the world. Today, some cities and regions have been named to recognise their contribution to our modern diet. For example, the city of Almaty in Kazakhstan means 'father of apples', because apple trees were probably domesticated in the nearby Tien Shen mountains.

Selective breeding rather than natural selection can be used to bring about great changes in the phenotype of a species in a short time. Not only has it been used to change the size, shape or features of an organism, it can also bring about changes in metabolic activity, such as a greater resistance to pests and diseases, and good growth in nutrient-poor soils or where water is in short supply. You will now consider three different examples of selective breeding. In all cases the outcomes are different, but the principles used to develop them are the same. Let's consider how a variety of disease-resistant rice could be developed by a farmer.

1. Some individuals in a population of rice plants may have resistance to a disease, while others are killed by it.

> **Tip**
> Selective breeding and genetic engineering are used to change the gene pool of a species, but use different methods. You will encounter genetic engineering, a laboratory technique, in Chapter 19.

Tip

The observable effects of selective breeding show that the characteristics of species are not fixed and can change. This presents further evidence for evolution by natural selection.

2. The farmer selects an individual plant that has better resistance to the disease than others in the population.
3. The farmer then selects another parent plant. This could also have good resistance to disease. However, the farmer might select a parent from a different variety that shows another characteristic, such as high yield, if this is also desirable.
4. The two chosen parents are then bred together.
5. The offspring are then grown and their disease resistance is assessed, perhaps by exposing the offspring to the disease. The individuals that show the greatest resistance are again chosen for breeding.
6. This process continues for many generations, each time selecting the individuals that have the chosen characteristics for breeding, until all individuals show them.

DISEASE RESISTANCE IN WHEAT AND RICE

Link

You will encounter the importance of gene banks in Chapter 18.

The introduction of disease resistance to varieties of wheat and rice has been achieved by selective breeding. Most modern varieties of wheat belong to the species *Triticum aestivum*. Breeding for resistance to various fungal diseases, such as rust and head blight, caused by a fungus, is important, because of the loss of yield resulting from such infections. Rice, *Oryza sativa*, is also the subject of ongoing selective breeding programmes. The yield of rice can be reduced by bacterial diseases such as bacterial blight. The most significant fungal diseases of rice are sheath rot and rice blast. Researchers are using selective breeding to try to produce varieties of rice that show some resistance to all these diseases. The International Rice Research Institute (IRRI), based in the Philippines, holds the rice gene bank and co-ordinates research aimed at improving the ability of rice farmers to feed growing populations.

VIGOROUS, UNIFORM VARIETIES OF MAIZE (CORN)

Tip

Inbreeding depression can occur in nature due to founder effects and population bottlenecks. For example, members of the cheetah species, which you encountered earlier, have numerous health problems.

A wild grass called teosinte that grows in Central America was probably the ancestor of cultivated maize (corn), *Zea mays*. **Inbreeding** has been used in selective breeding to produce varieties that are both vigorous (fast-growing) and uniform (have similar features).

If maize plants are inbred (crossed with other plants with genotypes like their own), the plants in each generation become progressively smaller and weaker and the yield reduces. This **inbreeding depression** occurs because repeated inbreeding produces **homozygosity**: the plant has a large proportion of genes for which it is homozygous due to the reduced gene pool. This means that recessive alleles of genes, which often code for non- or less-functional proteins, can have an effect on the phenotype. In maize, homozygous plants are less vigorous than heterozygous ones (Figure 17.22).

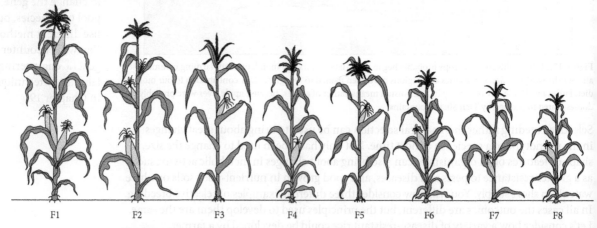

F1   F2   F3   F4   F5   F6   F7   F8

Figure 17.22 Inbreeding depression in *Zea mays*. When two different parent plants are crossed in a process called outbreeding, the first generation (F1) shows hybrid vigour. However, if F1 plants are self-pollinated to produce a second generation (F2), the size of the plant and the yield of maize is reduced. If self-pollination continues from generation to generation, inbreeding depression occurs because repeated inbreeding produces homozygosity.

Outbreeding – crossing with other, less closely related plants – produces heterozygous **hybrid** plants that are healthier, grow taller and produce higher yields. They show **hybrid vigour** because they have a very high proportion of genes in which a dominant allele is expressed. If two unrelated plants are crossed then there is a high chance that one is homozygous recessive and the other could be homozygous dominant. The offspring are therefore more likely to be heterozygous at many loci than either parent.

However, it is important to realise that if outbreeding is done at random, the farmer would have a field full of maize in which there was a lot of variation between the individual plants. Crossing two specific populations of plants therefore produces hybrid plants that are genetically uniform – they all grow at the same time and to the same height, in the same conditions, when sown together. This speeds up the farmer's work at harvest time. Today, farmers are able to buy maize seed from companies that specialise in using inbreeding to produce homozygous maize varieties. Crossing various inbred varieties can produce a large number of different hybrids, suited for different purposes. These F1 hybrids all have the same genotype as they have the same two inbred parents.

MILK YIELD AND QUALITY IN DAIRY CATTLE

In some countries, dairy cattle are kept to produce milk. For people who can digest lactose, milk is a nutritious food. Selective breeding has been carried out using cows that have high milk yields, and bulls whose close female relatives have high milk yields. This has produced a number of cattle breeds, such as the Holstein Friesian, that produce remarkable quantities of milk, rich in protein and other nutrients, over their lifetime (Figure 17.23).

Unlike natural selection, selective breeding often concentrates on just one or two characteristics. So, whereas natural selection tends to result in a species that is well-adapted to its environment, selective breeding risks producing varieties that show one characteristic to an extreme. Meanwhile, other characteristics that would be disadvantageous in a natural situation, perhaps linked to the desired traits, are exaggerated. There is concern, for example, for the welfare of many breeds of dairy cattle. Because they carry such large quantities of milk in their udders (formed from mammary glands), problems with leg joints are more common. In the two dogs you saw in Figure 17.20, common diseases include heart disease (Great Danes) and eye problems (Yorkshire terriers).

Figure 17.23 Holstein cattle grazing near Mount Fuji in Japan. These cows have been selectively bred to have large udders, capable of producing vast volumes of milk.

Experimental skills 17.1: Investigating the durian – the king of fruits

Native to Southeast Asia, species of the *Durio* genus are trees that produce fruits called durians (Figure 17.24). While delicious to many, the flesh of the fruit has a pungent smell. In the wild, the fruit is highly appetising to a diverse range of animals, including herbivores such as squirrels and elephants, but also carnivores such as bears and tigers. These animals can carry the fruit and disperse the seeds far away from the parent plant.

Figure 17.24 The fruits of the durian tree (*Durio sp.*).

QUESTIONS

P1. Most fruits do not appeal to carnivores. Suggest how *Durio* species may have evolved by natural selection to produce fruit that is attractive to carnivores.

P2. A scientist carried out an investigation into two varieties of durian tree. In this investigation, the following characteristics of 25 durian fruits from each tree were measured:

- the number of seeds in the fruits
- the mass of seeds inside the durian fruits
- the mass of the fruits.

  a. Discuss whether each of these three characteristics are examples of continuous or discontinuous variation.

  b. For each characteristic, each of the populations of durians showed a range of different values. Explain one cause of this variation.

  c. Comment on the reliability of the data obtained in this study.

P3. The following values were recorded for the number of seeds in each fruit.

Tree A: 10, 11, 15, 14, 12, 15, 18, 9, 10, 14, 17, 17, 12, 10, 13, 12, 10, 17, 17, 13, 18, 16, 19, 17, 17 (total: 353)

Tree B: 8, 10, 13, 11, 11, 14, 17, 11, 13, 14, 16, 13, 11, 10, 11, 9, 8, 16, 14, 13, 13, 11, 10, 7, 9 (total: 293)

Organise the two sets of measurements for number of seeds into a table. For each of the populations:

  a. Calculate the mean, median and mode values.

  b. Calculate the standard deviation.

  c. Calculate the standard error of the mean. Use the following formulae:

$$s = \sqrt{\frac{\sum\left(x - \overline{X}\right)^2}{n-1}} \quad \text{and} \quad \text{SE} = \frac{s}{\sqrt{n}}$$

where:

s = standard deviation, $\sum$ = 'sum of',
n = sample size (number of observations),
$\overline{x}$ = mean, SE = standard error, and
x, y = observations.

  d. What do the values for the standard error of the mean indicate about the difference between the two varieties of durian tree?

P4. The histograms in Figure 17.25 shows the two sets of results for the mass of seeds found in

the 25 fruits sampled from each variety of durian tree.

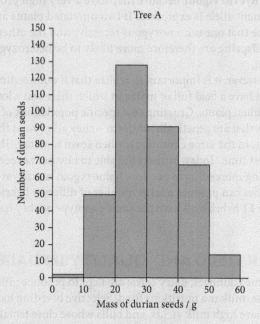

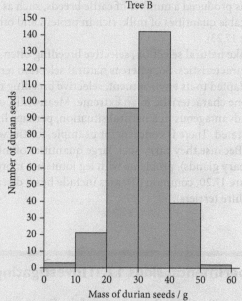

Figure 17.25 The variation in the mass of seeds found in durian fruit samples from two varieties of durian tree.

  a. Give reasons to explain why this data is more appropriately represented by a histogram, rather than a bar chart.

  b. Describe two disadvantages of displaying this data as a histogram.

P5. Some Southeast Asian countries export huge numbers of durians for consumption elsewhere in the world. Figure 17.26 shows the mass of durian fruits exported from one farm during the 20th century.

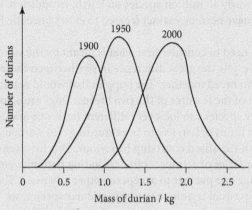

Figure 17.26

a. Describe and explain the changes in the mean mass of durian fruits between 1900 and 2000.

b. Suggest reasons to explain the change in the standard deviation of the three samples.

☆**P6.** A local belief is that consumption of durian with alcohol can be deadly. To investigate this claim, a group of scientists investigated the effect of durian extract (DE) on an enzyme, aldehyde dehydrogenase. This enzyme is required to break down the toxic substance acetylaldehyde, which is a by-product of ethanol (alcohol) metabolism. Ten replicates were conducted with the enzyme with and without durian extract (DE). Table 17.4 shows the mean values of the investigation.

Aldehyde dehydrogenase activity / arbitrary units $\pm s$			
Experimental (active enzyme)		Control (denatured enzyme)	
with DE	without DE	with DE	without DE
0.625 ± 0.066	0.870 ± 0.042	0.135 ± 0.055	0.160 ± 0.040

Table 17.4 The effect of durian extract (DE) on the activity of the enzyme aldehyde dehydrogenase in arbitrary units with standard deviation (s). This enzyme is involved in the metabolism of the products of alcohol breakdown in the liver.

The t-test can be used to determine whether the difference in enzyme activity with or without durian extract is significant. The formulae for the t-test is:

$$t = \frac{|\bar{x}_1 - \bar{x}_2|}{\sqrt{\frac{s_1^2}{n} + \frac{s_2^2}{n_2}}} \qquad v = n_1 + n_2 - 2$$

where:

v = degrees of freedom, s = standard deviation, $\bar{x}$ = mean n = sample size (number of observations).

a. Calculate the value of the t-statistic. Remember that 10 replicates were conducted for both the experimental and control investigations.

b. Calculate the number of degrees of freedom that should be used to find the relevant critical value with which to compare the value.

c. State whether there is a significant difference in enzyme activity in the presence or absence of durian extract. Refer to Table 17.5 for critical values.

d. Evaluate the evidence that durian extract has an effect on aldehyde dehydrogenase activity.

Degrees of freedom	8	10	12	14	16	18	20	22	24	26	28	30	40	50	60
Probability 0.05	2.31	2.23	2.18	2.14	2.12	2.10	2.09	2.07	2.06	2.06	2.05	2.04	2.02	2.01	2.00
Probability 0.01	3.36	3.17	3.06	2.98	2.92	2.88	2.85	2.82	2.80	2.78	2.76	2.75	2.70	2.68	2.66

Table 17.5

Key idea

→ Selective breeding involves the choice by humans of which organisms to allow to breed together, in order to bring about a desirable change in characteristics. Thus, selective breeding, like natural selection, can affect allele frequencies in a population.

17.4 Species and speciation

It has been estimated that there are nearly 10 million species on Earth. In addition, many hundreds of millions are thought to have become extinct (ceased to exist) since life began. Where have they all come from?

To help answer this question, you need to consider first what is meant by the word species. Look back to Figure 17.20. Despite the great difference in size between these two dogs, they would potentially be able to breed together. The puppies that would result from such a mating would show a mixture of the features of the two breeds. Dogs are a good example of how members of the same species can look very different from one another.

However, there are many different kinds of barriers to interbreeding. For animals, successful mating is often the result of ritualised courtship behaviour. This involves the two potential partners carrying out a series of actions – often including both movement, sound, and even light – that encourage one partner to accept the other as a mate. The 'correct' performance of the courtship ritual is unique to one particular species, so even if another species looks similar, mating between two organisms with different courtship behaviours will not take place. Figure 17.27 shows a pair of great crested grebes performing an elaborate 'dance' together, in which the male presents the female with pieces of weed that he has dredged up from the water, and the two of them follow a series of movements. Without this behaviour, the pair would not bond and mating would not occur.

Figure 17.27 Courtship rituals play a part in mating. Here, great crested grebes raise their neck ruffs and move their heads from side to side.

There are examples, however, where members of different species can and do mate and produce offspring. A mule is the result of a mating between a male donkey and a female horse (Figure 17.28). Donkeys and horses would not normally mate in the wild, but will often do so when they are kept together. Genetically, donkeys and horses are quite distinct. Wild horses have 66 chromosomes but donkeys have only 62. While the mule offspring are generally healthy, they have one important deficiency: they are infertile, and so are unable to breed.

a.

b.

c.

Figure 17.28 Horses (**a**) and donkeys (**b**) belong to two different species. However, they are able to breed together to form a mule (**c**).

We can therefore define a species as:

> **'A population or group of populations of similar organisms that are able to interbreed and produce fertile offspring.'**

HOW DO SPECIES ARISE?

Earlier in this chapter you explored natural selection. Together with the founder effect, genetic bottlenecks and genetic drift, this can bring about changes in allele frequencies in a population within a species. In this section, we will consider this idea further, to see how these changes could become so great that a new species evolves, in a process called **speciation**.

In order to create a new species from a pre-existing species, separate groups of individuals in the same population must somehow become unable to breed with each other. For this to happen, there must be some kind of barrier, or restriction, which prevents **gene flow** between two populations. They are said to experience **genetic isolation**. If they live close together and can still interbreed regularly, the populations will continue to exchange alleles. Then they will not separate into populations with distinct sets of alleles. Their gene pools will remain as one, rather than each population having its own gene pool.

Once the individuals of each population are genetically isolated, then familiar mechanisms begin to play a part in their divergence and possible separation into two species as their gene pools change. In these two populations:

- There are different environmental conditions, such as temperature, humidity, predation, disease, and so on. This causes natural selection to act on the two populations in different ways, so those with beneficial alleles are more likely to survive, reproduce and pass on these alleles.
- Genetic drift may occur in which the allele frequencies of the two populations differ, especially if they are small.
- Mutations occur in the two populations that are unique to each of the two populations and are not shared between them.

Eventually, after a long period of time, genetic differences accumulate to a point where individuals from separate populations that come together again, if the barrier is removed, can no longer interbreed. This may be due to differences in anatomy, courtship behaviour, or even molecular or genetic changes that otherwise prevent the successful fusion of gametes or embryo formation. They are now **reproductively isolated** from all members of other populations.

There are two models to describe how two populations of the same species can become genetically isolated. These are called **allopatric speciation** and **sympatric speciation**.

Allopatric speciation

Allopatric speciation occurs when two populations are physically separated, in a process called **geographical separation**. The barrier prevents the populations from mixing, and may include mountain ranges, volcanic activity or rising or falling water levels.

Imagine that you could watch what happens to a species of fish over many thousands, or even millions, of years. The species is found in a large lake. Part of a population becomes separated from the rest by retreating water levels, as the lake becomes split up into many smaller lakes, and the land between them dries out and turns to desert. In their new environments, the separated populations do not experience the same environment; the climate, the food supply, the competition could all be significantly different. Some adaptations would be successful in one lake; others would be favoured in another and so the separated populations would evolve in different ways. Because the selection forces are different, the process of natural selection favours different characteristics in different lakes. The fish with phenotypes that favour survival in their local environment will be more likely to pass on the alleles to their offspring. As the allele frequency changes, the phenotypes of the two populations will become more and more different. This is thought to have occurred in Lake Malawi in East Africa during the past million years, which now contains a large number of species of small, brightly-coloured fish called cichlids.

In the Galapagos Islands, we can see that there are different species of finches on the different islands. They look very similar to one another. This could be explained if one species of finch arrived on one of the islands from the mainland. Some spread to different islands, where the selection forces were different. They were separated from other members of their species by the water between the islands, so each evolved along its own path, eventually becoming reproductively isolated from each other. There are now 14 different species of finch on the islands with a wide variety of different beak sizes (Figure 17.10).

Sympatric speciation

Sympatric speciation occurs when populations experience genetic isolation and no gene flow, but still share the same geographical area. How can this happen?

Imagine a field with a species of plant that flowers all day. A new mutation produces an allele that causes some plants to flower only in the morning. These may attract a species of insect that gathers pollen only in the early part of the day. Another mutation in other plants may produce an allele that results in plants that flower only in the evening or at night. These attract a different species of insect for pollination. Both alleles are successful, but the plants form separate populations that are pollinated by one species of insect only. They no longer exchange alleles. Their gene pools are separated. Over time, further changes in phenotype can occur in both groups – slightly altered petals or stigmas could make pollination by the insect concerned even more efficient. There would then be selection for these alleles and, gradually, the two sets of plants could develop into two distinct species. This mechanism of speciation is dependent on a specific type of sympatric speciation called **behavioural isolation**.

Behavioural isolation also occurs in animals. An example of sympatric speciation relates to the apple maggot fly (Figure 17.29). Two hundred years ago, in the United States, the ancestors of these insects laid their eggs only on hawthorns, a plant native to America. Today, however, some flies lay eggs on domestic apples, which were introduced to America by immigrants from Europe. Females generally choose to lay their eggs on the type of fruit that they grew up on, and males tend to look for mates on the type of fruit that they grew up on. So, hawthorn flies generally end up mating with other hawthorn flies and apple flies generally end up mating with other apple flies. This means that gene flow between parts of the population that mate on different types of fruit is reduced. This host shift from hawthorns to apples may be the first step toward sympatric speciation; in less than 200 years, some genetic differences between these two groups of flies have already developed.

Sympatric speciation can also be brought about by a mechanism called **ecological isolation**. The three-spined stickleback (*Gasterosteus aculeatus*) is a species of fish that lives in inland lakes in Western Canada. Scientists have found that populations of this fish occupy different depths of their shared lake habitat, and show adaptations to living in these different zones. Importantly, these do not interbreed with each other – they are genetically isolated. Similar examples exist in plants: some members of the same species become ecologically separated because they grow on soil with different pH values or concentrations of pollutants.

Figure 17.30 summarises the types and causes of speciation.

Figure 17.29 The apple maggot fly, *Rhagoletis pomonella*.

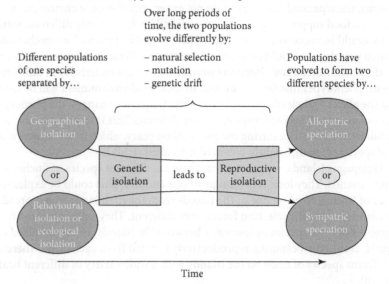

Over long periods of time, the two populations evolve differently by:

Different populations of one species separated by…

– natural selection
– mutation
– genetic drift

Populations have evolved to form two different species by…

Geographical isolation

or

Behavioural isolation or ecological isolation

Genetic isolation — leads to — Reproductive isolation

Allopatric speciation

or

Sympatric speciation

Time

Figure 17.30 The types and causes of speciation.

Worked example

Lake Malawi in East Africa contains around 400 species of cichlids, small, brightly-coloured fish. It is thought that all of these species evolved from a common ancestor, which lived around 5 million years ago.

a. Describe one way in which scientists could find out whether cichlids from two different populations belong to the same species.

Answer

The scientists could enclose two members of each population in the same fish tank and allow them to interbreed. If the offspring they produce are fertile, then the original fish belong to the same species.

b. Many species of cichlids are similar in size and appearance. Suggest how behavioural isolation may help to maintain these groups as separate species.

Answer

The fish may have behavioural patterns, such as courtship rituals or fertility windows, that make breeding possible only between members of the same species.

---

10. Draw a Venn diagram to describe the similarities and differences between natural selection and selective breeding (artificial selection).

11. Copper mining usually kills plants: the heavy metal ion is a metabolic poison. However, near copper mines, the grass *Agrostis tenuis* has evolved copper resistance and can grow well in polluted soils. The plants can transport copper out of their cells, so that it accumulates in the cell wall and does not interfere with their metabolism.

    a. Explain how natural selection could produce a copper-tolerant population near a copper mine.

    b. Copper-tolerant *Agrostis tenuis* plants flower at a different time of the year from those that are not copper tolerant. Explain how this could lead to sympatric speciation in *Agrostis*.

12. Many vegetables have arisen from the selective breeding of the wild mustard, *Brassica oleracea*, to form cultivars. In the original wild population, plants show variation in the size and shape of leaves, stems, terminal buds, lateral buds and flower clusters. Selection for specific features has led to the development of a wide variety of cultivars (Figure 17.31).

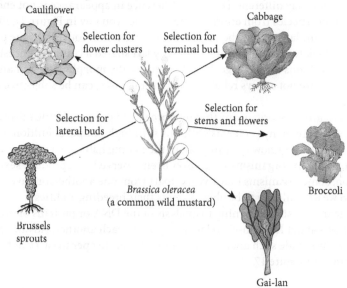

Figure 17.31 The cultivars of *Brassica oleracea*.

a. The gai-lan vegetable has a thick stem and pronounced leaves and sepals. Suggest and explain how farmers used selective breeding to produce the gai-lan vegetable.

b. Outline how a scientist could determine if two different cultivars belong to different species.

Key ideas

→ New species arise by a process called speciation, which requires two populations of a pre-existing species becoming reproductively isolated.

→ Allopatric speciation occurs when two populations are apart. Sympatric speciation occurs when populations become isolated despite occupying the same area.

→ The clearest example of allopatric speciation is seen when a population of organisms becomes geographically separated from the rest of the species. The selection forces on this population may be different from those on the other populations of the species, and so the allele frequencies begin to diverge.

→ Forces that can change allele frequency include mutation, migration, non-random mating, natural selection and genetic drift.

→ A point is reached when individuals are unable to breed successfully with the rest of the species, even if the geographical barrier is removed. When reproductive isolation has been achieved, new species have been formed.

17.5 Molecular comparisons between species

As you will see in Chapter 18, scientists are often interested in classifying organisms into groups. One reason is to determine their evolutionary relationships. However, there are two significant problems that scientists have encountered when they attempt to classify organisms:

- We can recognise a dog, a sheep and a human as being separate species because they look completely different. However, difference in appearance is not enough to distinguish one species from another. The two dogs you saw in Figure 17.20 are far from looking similar, yet they belong to the same species. Conversely, the plants in Figure 17.31 look remarkably similar, but they last shared a common ancestor many hundreds of millions of years ago. Visible features, the **morphology** and **anatomy**, of individuals, are not always reliable, and in some cases, can be subjective and open to debate.

- A species is a group of organisms that interbreed and produce fertile offspring. However, there are many problems with this seemingly simple definition. For example, we may not know anything about the mating habits of a particular organism, and some organisms have never been observed to reproduce sexually. In addition, some organisms may live far apart from one another, or have become extinct, so we are unlikely to be able to assess their breeding abilities.

Increasingly, taxonomists are turning to analysis of the DNA or proteins of organisms to make decisions about how closely related they are to each other. Advances in biotechnology have made such analyses much easier and cheaper to carry out. An example is shown in Figure 17.32.

Link

You will explore the use of databases and DNA and protein sequence comparison tools in Chapter 19.

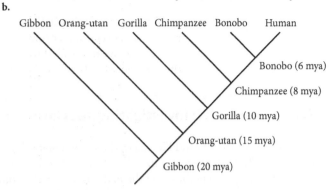

Figure 17.32 Our closest living relative, the pygmy chimp or bonobo **(a)**. The family tree of apes, including humans, has been determined **(b)**. The more differences between the sequences, the further back in time two species shared a common ancestor (mya = millions of years ago).

The DNA base sequence that can be directly compared between chimpanzees and humans is almost 99% identical. On average, the proteins that are made in our cells each differ by only one amino acid from those made in the cells of chimpanzees. To put this into perspective, the genetic difference between humans and chimpanzees is approximately 10 times less than between mice and rats. But the genetic difference between a human and a chimp is about 10 times more than between any two humans. This evidence has been used to estimate that humans and bonobos diverged from a common ancestor about six million years ago.

To make comparisons between many different organisms, a gene or a protein must be studied that is common to all of them. For example, most vertebrates produce the protein haemoglobin. We can therefore use the amino acid sequences of the beta- (ß-) polypeptide chain of haemoglobin to work out the relationships between any two species of vertebrates. The more dissimilar the amino acid sequences, the further back in time their evolutionary lines diverged from a common ancestor. Table 17.6 shows the number of differences in the amino acid sequences in the ß-chain of the haemoglobin of each listed animal and a human.

Animal	Number of differences in the primary sequence
gorilla	1
gibbon	2
rhesus monkey	8
dog	15
horse	25
mouse	27
grey kangaroo	28
chicken	45
frog	67

Table 17.6 The number of differences in the amino acid sequences in the ß-chain of the haemoglobin of each listed animal and a human. Analyses such as this use DNA as a 'molecular clock,' as the rate of mutation over long periods of time is relatively consistent.

While haemoglobin is useful for comparing different species of vertebrates, it does not occur in many other groups of organisms. We need to find a more universal protein to compare a wider range of organisms. Cytochrome c is a protein molecule that forms part of the electron transport chain. Because of its vital role in aerobic respiration, it is found in every eukaryotic cell, including all animals, plants and fungi. It is encoded by the mitochondrial genome, a circular DNA molecule that replicates separately to nuclear DNA in eukaryotic cells. The amino acid sequences of cytochrome c in many species have been determined, and comparing them shows how closely the species are related. Human

Link

You may recall that the electron transport chain, found in the inner mitochondrial membrane, is involved in aerobic respiration. You encountered this mechanism in Chapter 12.

cytochrome c contains 104 amino acids; 37 of these have been found at equivalent positions in every cytochrome c that has been sequenced. Scientists deduce that these molecules have descended from a precursor cytochrome c in a primitive microorganism that existed more than two billion years ago.

Assignment 17.1: Investigating speciation in *Anolis* lizards

The Caribbean Islands and Florida in the United States have many different species of *Anolis* lizards (Figure 17.33). Each species of lizard is found only on one island or a small group of islands. These lizards usually have green or brown bodies, and the males have large, colourful, inflatable dewlaps ('chin flaps'), which they use in display. These displays help the males to defend their territories, and are also used in courtship.

Figure 17.33 From left to right: *Anolis oculatus, Anolis sagrei, Anolis sagrei sagrei, Anolis carolinensis* and *Anolis equestris*.

Researchers are interested in how so many different species of *Anolis* lizards have arisen. Several teams have been working on the DNA sequences of the different species. The species *Anolis carolinensis* has the distinction of being the very first reptile whose DNA was completely sequenced.

Earlier research had suggested that the original home of the *Anolis* lizards was in Cuba, and that they had spread from there. Data collected in previous studies suggested that each of the different lizard species on the various Caribbean islands had developed following separate colonisations from Cuba.

The researchers made a prediction that – if this hypothesis were true – the relationships between the different species of lizards would be closer to the Cuban species than to each other. They tested this hypothesis by analysing the base sequence of the mitochondrial DNA in each of the five species. Table 17.7 shows the researchers' results of an investigation into a number of *Anolis* lizards. The larger the number, the greater the differences between the two species. The smaller the number, the closer the relationship between the two species.

Mean difference between mitochondrial base sequences of species						
Species	Species	Species 1	Species 2	Species 3	Species 4	Species 5
1	A. longiceps		0.125	0.137	0.175	0.119
2	A. maynardi	0.125		0.119	0.168	0.114
3	A. brunneus	0.137	0.119		0.167	0.113
4	A. carolinensis	0.175	0.168	0.167		0.152
5	A. porcatus	0.119	0.114	0.113	0.152	

Table 17.7 Mean pairwise divergence for the mitochondrial DNA of *Anolis* lizards. The lower the number, the greater the similarity between the base sequences of the two species.

QUESTIONS

A1. Use the data in Table 17.7 to determine the species to which each of the following is most closely related:
  a. i. *A. longiceps*   **ii.** *A. maynardi*   **iii.** *A. brunneus*   **iv.** *A. carolinensis*
  b. Explain what this suggests about *A. porcatus*.

A2. Suggest how the lizards on the different islands evolved different features.

A3. Suggest why hybrids between these species of lizard are very rare.

A4. The dewlaps of male lizards of different species may be different colours. Suggest how this could make one population of lizards reproductively isolated from another population.

A5. How could the researchers determine whether the five types of lizards that they tested really do belong to different species?

A6. Suggest further investigations that could test the hypothesis that the various species of *Anolis* lizards arose from separate colonisations of islands by lizards originating from Cuba.

★A7. Mitochondrial DNA has a higher rate of mutation than nuclear DNA, acquiring one mutation on average every 25 000 years. Suggest an advantage of using mitochondrial DNA rather than nuclear DNA to assess the relationships of the lizard species.

Key idea

→ Comparisons of the DNA nucleotide sequences of two species can give information about how closely they are related to each other.

CHAPTER OVERVIEW

Review the mini mind map to link new learning with your prior learning, and to highlight which of the key concepts underpin the chapter.

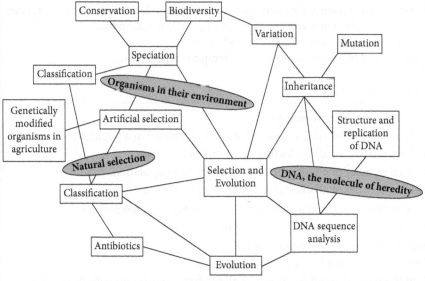

Try copying this mini mind map and expanding upon it. Use your notes from other chapters to help you explore how the essential ideas, theories and principles can be linked further together.

WHAT YOU HAVE LEARNED

Now that you have finished this chapter you should be able to:

- explain, with examples, that phenotypic variation is due to genetic factors or environmental factors or a combination of genetic and environmental factors
- explain what is meant by discontinuous variation and continuous variation
- explain the genetic basis of discontinuous variation and continuous variation
- use the *t*-test to compare the means of two different samples (the formula for the *t*-test will be provided)
- explain that natural selection occurs because populations have the capacity to produce many offspring that compete for resources; in the 'struggle for existence' individuals that are best adapted are most likely to survive to reproduce and pass on their alleles to the next generation
- explain how environmental factors can act as stabilising, disruptive and directional forces of natural selection
- explain how selection, the founder effect and genetic drift, including the bottleneck effect, may affect allele frequencies in populations
- outline how bacteria become resistant to antibiotics as an example of natural selection
- use the Hardy–Weinberg principle to calculate allele and genotype frequencies in populations and state the conditions when this principle can be applied (the two equations for the Hardy–Weinberg principle will be provided)
- describe the principles of selective breeding (artificial selection)
- outline the following examples of selective breeding:
  ○ the introduction of disease resistance to varieties of wheat and rice
  ○ inbreeding and **hybridisation** to produce vigorous, uniform varieties of maize
  ○ improving the milk yield of dairy cattle
- outline the theory of evolution as a process leading to the formation of new species from pre-existing species over time, as a result of changes to gene pools of populations from generation to generation
- discuss how DNA sequence data can show evolutionary relationships between species
- explain how speciation may occur as a result of genetic isolation by:
  ○ geographical separation (allopatric speciation)
  ○ ecological and behavioural separation (sympatric speciation).

CHAPTER REVIEW

1. What is meant by the term 'gene pool'?
   A. The total number of alleles of all genes in all organisms in a population.
   B. The total number of all alleles of all genes in an organism.
   C. The total number of all genes in all organisms in a population.
   D. The total number of all alleles for a gene in a population.

2. Natural selection can act on a population in a number of different ways:
   1. selection favouring a recessive phenotype
   2. selection favouring a dominant phenotype
   3. selection favouring a heterozygous phenotype.
   Which of these could result in the loss of alleles from a large population within a few generations?
   A. 1 only   B. 1 and 2
   C. 2 and 3   D. 3 only

3. Copy the table below. Place a tick or ticks into each row in the table to indicate which mechanism(s) has been observed to change the gene pool of a population and cause an evolutionary change. The first one has been done for you.

Observation	Natural selection	Founder effect	Genetic bottleneck	Genetic drift
The human species has a small gene pool because around 70 000 years ago a volcanic super-eruption may have reduced our ancestral population to only several thousand.			✓	
The northern elephant seal species has a small gene pool because it was hunted to near-extinction in the late 19th century.				
In a small, reproductively isolated population of frogs, all have become green in colour.				
In some isolated communities living in Venezuela, Huntington's disease is very common. These individuals descend from an immigrant with the disease from Spain in the 17th century.				
Extinct in the wild, Przewalski's horse survived in zoos, and was successfully reintroduced into the steppes of Mongolia in 2006.				
In populations that changed from a hunter-gathering lifestyle to farming, lactose tolerance developed as a means to consume milk – a sterile, nutritious food source.				
The native human population of the Americas is thought to have descended from as few as 20 individuals who crossed the land bridge from Siberia to Alaska during the last Ice Age.				
The development of a peacock's tail has been driven by a female's mating preference for elaborate displays.				

4. 'If we evolved from chimpanzees, why do chimpanzees still live today?' Explain what is wrong with this question.

5. Before the work of Darwin and Russel Wallace, the French naturalist Jean-Baptiste Lamarck suggested that living things changed by inheriting acquired characteristics. He suggested that giraffes stretched their necks to reach food, and their offspring inherited stretched necks. This is now known to be wrong. Explain how the theory of natural selection could have led to the development of the long neck of a giraffe.

6. The modern cheetah population is thought to have descended from a much larger population, which experienced a population bottleneck. Scientists have observed that male and female cheetahs are more likely to be promiscuous (have multiple partners during the mating season) than most other cat species. Suggest an explanation for this observation.

7. Warfarin is a poison used to control rat populations. Susceptible rats fed on warfarin die from internal bleeding. Resistant rats have a dominant allele that encodes a modified enzyme that preserves normal blood clotting even after consuming warfarin. However, animals carrying the allele for resistance need large quantities of vitamin K in their diet.

a. In areas where warfarin is used, the percentage of resistant rats is maintained at about 50% of the total rat population. Use this information to estimate the frequency of the recessive allele for susceptibility in this population. The two Hardy–Weinberg equations are $p + q = 1$ and $p^2 + 2pq + q^2 = 1$.

b. Suggest why rats that are homozygous dominant for the resistance allele are unlikely to survive in the wild.

c. Describe how the process of natural selection maintains the proportion of resistant rats at about 50% of the total population.

8. A biologist investigated the evolutionary relationships between different species of apricot and plum tree, which belong to the genus *Prunus* (Figure 17.34). The variation in the sequence of amino acids in a protein common to all *Prunus* species was compared (Table 17.8).

Figure 17.34 Ripe fruit on an apricot tree, *Prunus armeniaca*.

Species	Number of differences in the amino acid sequence of protein compared with *P. armeniaca*
P. brigantina	1
P. mandshurica	4
P. mume	4
P. zhengheensis	7
P. sibirica	8

Table 17.8

a. Suggest the advantages of measuring variation in amino acid sequence in this investigation, compared with the appearance of the fruit.

b. Explain how these results suggest that *P. brigantina* is the most closely related to *Prunus armeniaca*.

c. Explain why it is not a valid assumption that *P. mandshurica* and *P. mume* are the most closely related pair of species of trees.

d. The scientists subsequently analysed the DNA base sequences that encode protein A in the species. Give two reasons to explain why the comparison of DNA base sequences will give more information than the amino acid sequence of protein A.

e. Explain why improvement in some characteristics, such as fruit size, will not continue to improve over many generations of selective breeding using self-pollination.

Classification, biodiversity and conservation

We classify organisms in an attempt to understand the relationships between different groups of living things. However, not all organisms fit within the groups that we create. The picture shows *Euglena*, which is green and carries out photosynthesis like a plant. But *Euglena* also moves around and consumes food like an animal. To be classed as an animal or plant, an organism must be multicellular, but *Euglena* is unicellular. *Euglena* has no cell wall, but it does have membrane-bound organelles. So how do we classify *Euglena*?

Prior understanding

You may recall some information about classification of organisms as animal, plant, fungi or bacteria from previous courses. You may also recall cell structure from Chapter 1 and nucleic acids and protein synthesis from Chapter 6. You may previously have used terms such as species, ecosystem and conservation. Calculation of Pearson's linear correlation coefficient was introduced in Chapter 2.

Learning aims

In this chapter, you will learn about the current domain classification system and how this is based on genetic information rather than appearance of organisms. You will discover how biologists measure biodiversity and learn the meanings of terms such as species, ecosystem and niche. You will describe some methods of measuring the distribution and abundance of organisms. You will also discover what causes organisms to become endangered or extinct and what steps can be taken to prevent this.

18.1 Classification (Syllabus 18.1.1–18.1.6)

18.2 Biodiversity (Syllabus 18.2.1–18.2.6)

18.3 Conservation (Syllabus 18.3.1–18.3.6)

18.1 Classification

Aristotle (384–322BCE) made one of the earliest attempts at classification. He grouped living things according to characteristics. For example, animals with blood and animals without blood; animals that live in water and animals that live on land. This marked the beginning of a trend to use properties to group organisms. However, Aristotle's method presented problems, as some organisms live in water and on land. One present day method of classifying organisms is into species.

SPECIES

As you read in Chapter 17, the term species is used to describe organisms of different types, but what does the term species mean? There are different uses of the term, each with a different meaning.

Biological species concept

The **biological species concept** is based on gene-flow within a group of organisms. Two organisms are said to be in the same species if they can **interbreed** to produce viable, fertile offspring. For example, all races of humans are classed as the same species because they can mate, give birth to healthy offspring and these offspring in turn can produce more offspring.

Problems with the biological concept of classification arise from reproductive isolation. One type of reproductive isolation is pre-zygotic, that is, it occurs before fertilisation. The two dogs in Figure 18.1 are the same biological species as they could, in theory, produce viable fertile offspring, but mating between these two dogs is unlikely.

Another type of reproductive isolation is post-zygotic, that is, it occurs after fertilisation. For example, horses and donkeys will mate in captivity. As you read in Chapter 17, the offspring of a horse and donkey, called a mule (Figure 17.28), is a hybrid. Hybrids, such as the mule, are viable but not fertile, so they cannot interbreed to produce offspring of their own.

Another problem with this species definition is that some organisms are never observed to interbreed. For example, some plants only reproduce asexually, so biologists cannot use the biological species concept for these plants.

Morphological species concept

Darwin introduced the **morphological species concept** based on the appearance of organisms. Where a morphological gap occurs between two organisms, that gap forms the species boundary.

Problems with the morphological species concept include variation within species and similarities between different species. For example, the peppered moth occurs in two forms that are morphologically distinct on the basis of markings, as was shown in Figure 17.11, but are the same biological species.

Sometimes two biologically different species can be morphologically very similar. The chiffchaff and the willow warbler, shown in Figure 18.2, are biologically different species.

Ecological species concept

It has been suggested that species are better defined by the ecology of the habitat to which they are adapted. This is known as the **ecological species concept**.

This can cause problems when two different species are adapted to the same ecological niche, such as the aye-aye and the woodpecker in Madagascar. The aye-aye is a mammal and the woodpecker is a bird, but they are adapted to find the same food in a very similar way. In addition, some species such as the malaria vector mosquito spend the larval stage of its life cycle in water and the rest of its life on land or in the air.

Figure 18.1 These two dogs are the same biological species but are unlikely to mate.

a.

b.

Figure 18.2 These birds are biologically different species yet are morphologically similar.

THE DOMAIN SYSTEM OF CLASSIFICATION

Until 1990, most biologists used a system of classification that divided cellular life into five kingdoms, one of which included prokaryotes. In 1977, Carl Woese and co-workers proposed that there is sufficient difference in the ribosomal RNA genes within this kingdom to merit a further level of classification based on three domains. This, and other differences, led to the three domains used by most biologists today: **Archaea**, **Bacteria** and **Eukarya**.

Link

You may find it helpful to look back at Chapter 1 for the differences between prokaryotic and eukaryotic cells.

Domain Archaea

These are prokaryotic organisms, often living in extreme environments. They are different from bacteria because they have different ribosomal RNA (rRNA) genes than bacteria, phospholipids in their cell surface membranes have different structures, and different cell walls. Their phospholipids differ in structure from that of all other organisms in several ways. These differences in the phospholipids and membranes include:

- the fatty acid and glycerol part are joined by an **ether bond**, rather than an ester bond found in other organisms
- the fatty acid chains have branches and rings in their structure, rather than the unbranched chains found in other organisms
- the three-dimensional shape of the glycerol part is the mirror-image of the type found in other organisms
- some members of domain Archaea have lipid monolayers, as opposed to the lipid bilayers found in other organisms.

Most, but not all, members of domain Archaea have a cell wall. When a cell wall is present, it is formed from surface-layer (S-layer) proteins that form a rigid, cross-linked structure. The cell wall does not contain peptidoglycan that is found in bacterial cell walls.

Examples of domain Archaea include:

- thermoacidophiles, which live in very hot water that is acidic
- halophiles, which live in water with a very high salt concentration
- methanogens, which produce methane gas as a metabolic waste product.

Domain Bacteria

These are prokaryotic organisms that are traditionally referred to as bacteria. They have rRNA genes that are distinct from those of domain Archaea and Eukarya. Their cell walls are made from peptidoglycan and their cell surface membranes have phospholipids that are similar in structure to those of domain Eukarya.

Examples of domain Bacteria include:

- coliforms, which ferment lactose and some of which live in the human intestines
- cyanobacteria, which are photosynthetic and live in water
- spirochaetes, which are helix-shaped and include some human pathogens.

Domain Eukarya

These are eukaryotic organisms that are divided into four **kingdoms**.

Kingdom Animalia contains the animals. Animals are characterised by being multicellular, having no cell walls and no chloroplasts. As animals are not capable of photosynthesis, they must take in nutrients from outside.

Kingdom Plantae contains the plants. Plants are also multicellular, having cell walls made from cellulose, hemicellulose and pectin. Their cells have a large central vacuole and organelles called plastids, which include chloroplasts and other organelles containing photosynthetic pigments.

Kingdom Fungi contains the fungi. Fungal cells are more similar to animal cells than to plant cells, but have differences, such as a cell wall made from chitin and less separation between cells. Any divisions between cells that are present are called septa. The larger fungi that have few septa are effectively one giant multinucleate cell.

Kingdom Protoctista contains the organisms that are eukaryotic but do not fit within the animal, plant or fungal kingdoms. The kingdom includes unicellular animal-like organisms such as amoeba, large multicellular organisms such as kelp (Figure 18.3) and unicellular photosynthetic organisms such as *Euglena*.

Figure 18.3 This underwater kelp forest is comprised of large multicellular protoctists that can grow up to 60 m in length. They are often referred to as seaweeds, but they are not plants.

Each kingdom in domain Eukarya is further divided into **phylum**, **class**, **order**, **family**, **genus** and **species**, with fewer organisms in each level. Table 18.1 shows how this is done using the example of the insect *Coccinella septempunctata*.

Taxon	Description	Example
domain	the largest group, containing three types of organism with very different biochemistry	Eukarya
kingdom	major groups of organisms within each domain	Animalia
phylum	a group of organisms that share a common body plan, for example having a segmented body, an external skeleton made of chitin, and jointed limbs	Arthropoda
class	a major group within a phylum, for example all the arthropods with three pairs of jointed legs	Insecta
order	a subset of a class, for example all the beetle-like insects	Coleoptera (beetles)
family	a group containing organisms with very similar features	Cocconellidae (ladybirds)
genus	a closely related group within a family, for example red or orange ladybirds with black spots	*Coccinella*
species	a group of individuals that can breed with one another to produce fertile offspring, but not with other members of the same genus	*Coccinella septempunctata* (the seven-spot ladybird)

Table 18.1 Complete classification of *Coccinella septempunctata*.

Notice how the organism in Table 18.1 is referred to by the last two names in the classification: *Coccinella septempunctata*. This scientific naming system is called the **binomial** (meaning two name) **system**. The first name has an uppercase letter and denotes the genus. The second name has a lowercase letter and denotes the species.

VIRUS CLASSIFICATION

The domain system is used for cellular life and organisms, but viruses do not have cells and can be classed as non-living so require a separate classification system.

Viruses require a host cell within which to replicate and so are referred to as **obligate intracellular parasites**.

Viruses were discovered in 1901, but had no classification system until David Baltimore, in the early 1970s, suggested that viruses be classified according to their genetic material (genome).

Viruses can have DNA or RNA genomes and this, in turn, can be single stranded (ss) or double stranded (ds). When a virus has a single-stranded RNA genome, the genome is called positive (+) or negative (−) depending whether the RNA can be used as mRNA. A single-stranded RNA genome that is positive is called +ssRNA and this can be used as mRNA directly within the cell. A single-stranded RNA genome that is negative is called −ssRNA and this must be transcribed to another, positive, single strand of RNA within the cell before it can be used as a message. This gives the five genomes of viruses in the Baltimore system.

There are two further classes depending on whether the virus has the enzyme reverse transcriptase (RT) or not. The seven classes are denoted by Roman numerals (Table 18.2).

Tip

You may recall the seven life processes from previous courses. Viruses do not exhibit respiration, excretion, sensitivity or nutrition so on that basis are not classed as living.

Link

Messenger RNA, or mRNA, is the molecule that carries the codons for polypeptide synthesis, as was described in Chapter 6.

Link

Reverse transcriptase, RT, is an enzyme that catalyses the formation of double-stranded DNA using an RNA template. This will be described in Chapter 19.

Baltimore class	genome	example
I	dsDNA	poxviruses, such as the causative agent of smallpox, the first disease to be completely eradicated by vaccination
II	ssDNA	parvoviruses, such as the causative agent of canine parvovirus, which is potentially fatal in young dogs
III	dsRNA	reoviruses, such as rotavirus, which causes severe diarrhoea in children and can be fatal if untreated
IV	+ssRNA	rhinovirus, such as the causative agent of common cold, one of the most frequently occurring infectious diseases in humans
V	−ssRNA	rhabdoviruses, such as the causative agent of rabies, which is a serious disease of animals and humans
VI	dsRNA,RT	retroviruses, such as the causative agent of HIV/AIDS, described in Chapter 10
VII	dsDNA,RT	hepadnaviruses, such as the causative agent of hepatitis B, which is a serious liver disease in humans

Table 18.2 The Baltimore classification of viruses.

1. The lion *Panthera leo* lives in sub-Saharan Africa and the tiger *Panthera tigris* lives in Asia.
   a. Explain the relationship between these two animals using their binomial names.
   b. Explain why *P. leo* and *P. tigris* are unlikely to mate in the wild and state the type of reproductive isolation that occurs.
   c. *P. leo* and *P. tigris* can mate in captivity and produce a sterile hybrid called a tigon. A tigon has physical features of both a lion and a tiger. Explain why the tigon is difficult to classify as a species.
2. Attempts have been made at classification for the last 2300 years. Explain why the domain system of classification was introduced so recently.
3. Describe the similarities and differences between members of domain Archaea and those of domain Bacteria.
4. Fungi were previously classified in kingdom Plantae because some fungi, such as mushrooms, grew in soil and had rigid cell walls. Explain why fungi are now classified in their own kingdom.

5. Nucleic acid and protein sequence comparisons between organisms can be used to help with classification. Cytochrome *c* is a protein involved in the electron transport chain in mitochondria. The amino acid sequences of cytochrome *c* from different organisms are often compared. Suggest why amino acid sequence information from cytochrome *c* is more useful in classification than amino acid sequence information from haemoglobin.

6. a. The zebra and the donkey are two different species. In captivity, a zebra and a donkey can breed to produce a hybrid offspring called a zonkey. The diploid number of a zebra is 32 and the diploid number of a donkey is 62. Use this information to suggest why the zonkey cannot mate to produce offspring of its own.
   b. Suggest why it is difficult to classify the zonkey as a species.

7. a. List the five different types of genome that define separate classes of viruses.
   b. Explain why viruses cannot be included in the domain system of classification.

★ 8. Poliovirus, the causative agent of polio, is in Baltimore class IV. Once poliovirus is inside a cell, it can start making virus protein in a much shorter time than many viruses in other classes. Suggest why.

Key ideas

→ The term species can be defined in biological, morphological and ecological ways, each with advantages and disadvantages.

→ The current classification system for cellular life uses three domains: Archaea, Bacteria and Eukarya.

→ Archaea and bacteria are prokaryotic with differences in RNA, phospholipids and cell walls.

→ Organisms in Eukarya are further divided into kingdom, phylum, class, order, family, genus and species.

→ Each of the four kingdoms in Eukarya, Animalia, Plantae, Fungi and Protoctista, have characteristic features.

→ Viruses are classified separately according to whether their genome is DNA or RNA, double stranded or single stranded.

18.2 Biodiversity

Biodiversity means the variety of life. This meaning is summarised in Figure 18.4.

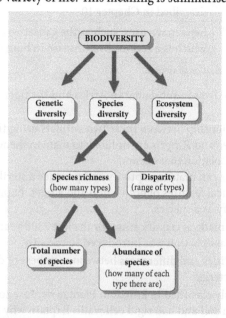

Figure 18.4 The meaning of the term biodiversity.

In order to describe biodiversity, we need to use the terms ecosystem, habitat and niche.

An **ecosystem** is a partly self-contained place containing organisms and their non-living surroundings where the organisms interact with each other and interact with their non-living surroundings. Figure 18.5 shows a tropical rainforest ecosystem. Another example of an ecosystem is a coral reef (Figure 18.6).

Biotic factors are the effects of living things in an ecosystem; **abiotic factors** are the non-living things.

In and around a coral reef there will be the coral itself, plus thousands of species of fish, plants, birds, invertebrates, algae, marine mammals and reptiles and prokaryotes. The interactions between these organisms constitute the biotic factors. There will also be water, dissolved minerals, carbon dioxide and oxygen gases, annual temperature variations and many more abiotic factors with which the living things interact. Natural ecosystems cannot be completely self-contained, as, for example, a forest will have boundaries with other ecosystems, such as a lake or an urban ecosystem. At these boundaries, biotic and abiotic factors overlap between the different ecosystems.

A **habitat** is where an organism lives. The habitat of an earthworm may be only a few cubic metres of soil. Marine habitats are the largest. One example is that of a female leatherback turtle that was tracked travelling from Indonesia to the Pacific coast of the USA. During this journey, the turtle covered over 20 000 km!

The **ecological niche** of an organism is the role of the organism in the ecosystem and the conditions necessary for the survival of the organism. This can be very complex. For example, the steppe is a type of ecosystem found in central Asia and south-eastern Europe. Steppe is characterised by flat, unforested grassland. In a steppe ecosystem the ecological niche of grass is a producer in the food chain for the animals that live there. Grass also removes carbon dioxide from the atmosphere and releases oxygen. Grass takes in minerals and water from the soil and provides organic matter for decomposers in soil when it dies. Grass provides cover for small animals, places for insects to lay eggs and spiders to make webs. Grass also releases water vapour from transpiration, which can affect rainfall elsewhere in the same ecosystem. The roots of grass provide habitats for some fungi and grass can be hosts for viral pathogens.

Figure 18.5 The world's tropical rainforest ecosystems are estimated to contain 50% of global land plant and animal species.

Figure 18.6 Coral reefs are some of the most diverse ecosystems on Earth.

ASSESSING BIODIVERSITY

Biodiversity is important for the natural sustainability of all life on Earth, including human life. Maintaining biodiversity ensures that ecosystems are stable and function correctly. That means fertile soil, water suitable for drinking and air suitable for efficient gas exchange. Biodiversity can be assessed in different ways, including:

- the number and range of different ecosystems and habitats
- the number of species and their relative abundance
- the genetic variation within each species.

To determine the number of a particular species in an ecosystem, **random sampling** must be carried out. Random sampling can be done in different ways depending on the organism and its habitat.

Methods of random sampling
Random sampling is useful when the distribution of organisms is thought to be uniform. (even). For example, small plants may be distributed quite evenly over open grassland, or a certain species of fish may be found in equal numbers at different positions in a lake. Methods of random sampling include:

- frame quadrat
- line transect
- belt transect
- mark-release-recapture.

A **frame quadrat** (Figure 18.7) can be used for small plants and sessile animals. Sessile animals are those that are fixed in position or only move extremely small distances, such as some sea snails that remain attached to rocks in one place for their entire life.

Quadrats can be made from wood, wire or plastic

Larger quadrats, for example, for sampling in woodland, can be laid out with string and pegs

50 cm

50 cm

All the species within a quadrat can be counted, or the abundance of each species estimated

A quadrat with sides of 0.5 m has an area of 0.25 m². Wire fixed at 10 cm intervals gives 25 smaller units, each of 0.01 m², to make counting easier

Figure 18.7 A frame quadrat.

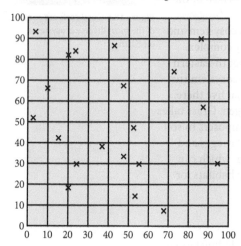

Figure 18.8 The positions for a frame quadrat in a field are shown as randomly generated co-ordinates.

When sampling at random using a frame quadrat, the positions where the quadrat is to be placed are decided using random numbers. These random numbers are then used to generate co-ordinates on a grid (Figure 18.8), and these co-ordinates can be measured out in the area to be studied, such as a field.

Frame quadrats are useful for estimating the number, or **abundance**, of a species of small plant or sessile animal in a particular area. To do this, the total number of organisms counted in all quadrats is divided by the total area of quadrats used. This gives an estimate of the number of organisms per square metre. This can then be multiplied by the total area where the sampling took place to get an estimate of the number of organisms in that area.

Frame quadrats are also useful for estimating **percentage cover**. That is why some quadrats have a smaller-sized grid inside. The method for estimating percentage cover is shown in Figure 18.9.

This plant covers four whole + five half squares = 26% cover

Each square in this 0.25 m² quadrat represents 4% of the total area within the quadrat

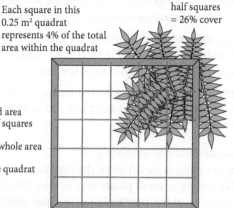

To measure percentage cover:
- Lay a frame quadrat over the selected area
- Count the number of whole and half squares occupied by the species
- Calculate this as a percentage of the whole area of the quadrat
- Repeat for all the other species in the quadrat

Figure 18.9 Use of a frame quadrat to estimate percentage cover.

Frame quadrats on their own can be used to compare different areas. If a gradually changing habitat needs to be sampled then a transect is used.

Examples of these include increasing distance from a river bank, and one study even involved observers counting large birds from a train on a 60 km journey!

There are two basic types of transect: **line transect** and **belt transect**.

A line transect is constructed using a tape measure or marked rope between two poles. The tape or rope forms the line. Organisms that are touching the line are recorded along with the distance from one end of the line. A choice can be made whether to count all the organisms being studied or whether to count those occurring at predetermined regular intervals, such as 5 m.

A belt transect is constructed in a similar way. A frame quadrat is placed along the line, touching one side of the line. Again the quadrat can be placed without leaving any gaps, or at longer distance intervals. If the interval method is used, this is called an interrupted belt transect.

Both types of transect are summarised in Figure 18.10.

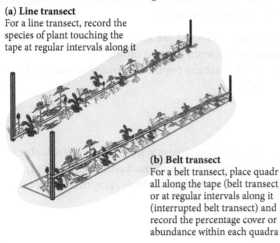

(a) Line transect
For a line transect, record the species of plant touching the tape at regular intervals along it

(b) Belt transect
For a belt transect, place quadr all along the tape (belt transect or at regular intervals along it (interrupted belt transect) and record the percentage cover or abundance within each quadra

Figure 18.10 Line and belt transects.

The **mark-release-recapture** method is used when organisms move quickly or are difficult to see. Also, the organisms should be large enough to mark and live in a reasonably restricted area to ensure no significant immigration or emigration. For example, small mammals such as mice will not remain in one place to be counted within a quadrat and fish are free to move and are not visible in lakes.

Figure 18.11 shows the mark-release-recapture method.

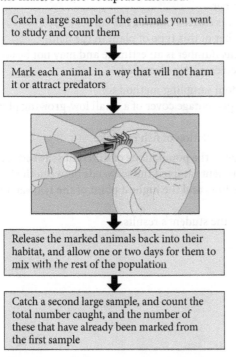

Catch a large sample of the animals you want to study and count them

Mark each animal in a way that will not harm it or attract predators

Release the marked animals back into their habitat, and allow one or two days for them to mix with the rest of the population

Catch a second large sample, and count the total number caught, and the number of these that have already been marked from the first sample

Figure 18.11 The mark-release-recapture method.

Care must be taken:

- that a large enough sample is taken in the first catch
- that the marking method does not harm the animal or make it more or less attractive to predators
- that the mark is waterproof and secure
- that marked animals have the opportunity to mix again with the remaining population
- that enough time is allowed for mixing of marked and unmarked organisms between the first and second sample
- that the time between the first and second sample is not so long as to allow significant numbers of births and deaths to occur.

In the mark-release-recapture method, we assume that there is no significant immigration or emigration between the two samples.

The equation used to estimate the population from the mark-release-recapture method is called the **Lincoln index**:

Tip

When you need to use the Lincoln index equation, it will be provided.

$$N = \frac{n_1 \times n_2}{m_2}$$

where N = estimate of population size, n_1 = number of individuals captured in first sample, n_2 = number of individuals (both marked and unmarked) captured in second sample and m_2 = number of marked individuals recaptured in second sample.

9. Describe what is meant by:
   a. ecosystem
   b. ecological niche.

10. A frame quadrat measures $0.5\,\text{m} \times 0.5\,\text{m}$.
    a. Calculate the area of this quadrat in m^2.
    b. The quadrat was placed at random in 25 positions in a field. Calculate the total area sampled.
    c. The total number of a certain species of plant within all 25 quadrats was 368. Calculate the mean number of these plants per square metre. Give your answer as a whole number.
    d. The sampling was carried out in a field of total area $75\,\text{m}^2$. Calculate the estimated number of this type of plant in the field.
    e. Explain why this number is an estimate and may not be accurate. Assume all plants have been correctly identified.

11. State the best random sampling method to:
    a. determine the percentage cover of a small low-growing plant as distance from a busy road increases
    b. determine the population of fish in a lake.

12. A student investigated the populations of different seaweeds growing on a rocky coast. The student sampled the seaweeds between the lowest point of the sea water (low tide line) and the highest point of the sea water (high tide line) during the day.
    Figure 18.12 shows the student's results.

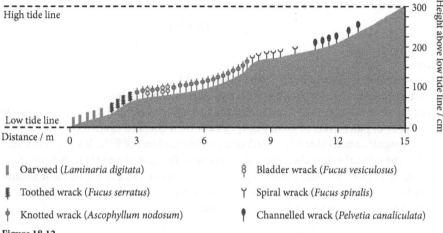

Oarweed (*Laminaria digitata*)

Toothed wrack (*Fucus serratus*)

Knotted wrack (*Ascophyllum nodosum*)

Bladder wrack (*Fucus vesiculosus*)

Spiral wrack (*Fucus spiralis*)

Channelled wrack (*Pelvetia canaliculata*)

Figure 18.12

a. Suggest which sampling method was used to obtain these results.
b. Which two species of seaweeds share the same habitat?
c. Use Figure 18.12 to estimate the range of seashore heights occupied by channelled wrack.
d. Suggest and explain which species of seaweed is **least** tolerant of dry conditions.

13. A student used the mark-release-recapture method to estimate the population of centipedes living on a decaying log. The student captured 50 centipedes, marked them in a harmless way and released them. After 48 hours, the student captured a second sample of 70 centipedes, 17 of which had marks.
a. Use this information and the Lincoln index,

$$N = \frac{n_1 \times n_2}{m_2}$$

to estimate the population of centipedes on the decaying log.
b. Give three reasons why this is only an estimate of the population.
c. Suggest two reasons why the 48-hour time interval between samples was chosen.

ANALYSIS OF RESULTS

When we collect results from random sampling experiments, we may need to analyse these further. For example, when studying areas affected by industrial pollution, we may want to find whether the abundance of a copper-tolerant plant varies with copper content in soil. We may also wish to determine a numerical value for biodiversity.

Spearman's rank correlation coefficient

Monotonic variables are those where an increase in one variable results in either an increase or a decrease in the other variable. These are referred to as positive and negative correlations, respectively. When there are two sets of paired measurements that are monotonic, then **Spearman's rank correlation coefficient** can be used to analyse the relationship between these variables. The variables have to be paired and represent continuous data. Variables do not have to be normally distributed to use Spearman's rank correlation coefficient, so this analysis is the best one to use first when checking for correlation.

Spearman's rank correlation coefficient gives the strength and direction of the correlation. The values of this correlation coefficient range from −1 to +1. A value of −1 means a perfect negative correlation, 0 means no correlation and +1 means a perfect positive correlation. Because ranks rather than actual values are used it looks at trends and asks, 'is there a correlation?'.

Tip

Remember, the null hypothesis will usually be in the form 'There is no correlation between variable 1 and variable 2'.

Tip

The symbol Σ, called sigma, means the sum of the quantities that follow the symbol.

Tip

The value $p = 0.05$ represents a probability of 0.05, or 5%. If a p-value is $\leq$ 0.05, there is less than 5% probability of the result occurring by chance, so the null hypothesis is rejected.

Tip

Soil moisture content is an abiotic factor in the habitat of *Astrantia major*.

The equation to calculate Spearman's rank correlation coefficient (r_s) is:

$$r_s = 1 - \left(\frac{6 \times \sum D^2}{n^3 - n} \right)$$

where

n = the number of pairs of items in the sample, and

D = the difference in rank between each pair of ranked measurements.

The number of degrees of freedom for the test is then calculated, which is 2 less than the number of paired sets of data. The r_s value is then checked against the critical value for r_s at a significance value of $p = 0.05$ or a confidence level of 95%. If r_s is greater than or equal to the critical value, then the null hypothesis is rejected and there is a significant correlation. A positive value means a positive correlation and a negative value means a negative correlation. If r_s is less than the critical value, then the null hypothesis is accepted and there is no significant correlation.

Worked example

A student investigated the relationship between the abundance of masterwort plants, *Astrantia major* and soil moisture content. The student counted the number of *A. major* in 10 randomly placed quadrats and measured the soil water content at each position. The results are shown in Table 18.3.

Sample number	1	2	3	4	5	6	7	8	9	10
Number of *A. major* in one quadrat	10	14	12	8	2	9	15	5	11	14
Soil moisture content / %	22	26	27	12	10	14	23	9	20	24

Table 18.3

Use the results in Table 18.3 and Spearman's rank correlation coefficient to determine whether there is a correlation between the abundance of *A. major* and soil moisture content.

Table 18.4 provides the critical values for Spearman's rank correlation coefficient at $p = 0.05$. As the number of pairs increases, the critical value decreases in order to take random variation into account.

Number of pairs of values (n)	4	5	6	7	8	9	10	11	12
Critical value	1.00	0.90	0.83	0.79	0.74	0.68	0.65	0.61	0.59

Table 18.4 If a calculated value is greater than the critical value for a given number of pairs, then we say that $p \leq 0.05$.

Answer

First write the null hypothesis:

There is no correlation between the abundance of *A. major* and soil moisture content.

Next, rank (order) the values in each data set, so the highest has a value of 1 and the lowest is 10.

Sample number	1	2	3	4	5	6	7	8	9	10
Number of *A. major* in one quadrat	10	14	12	8	2	9	15	5	11	14
Rank	6	2.5	4	8	10	7	1	9	5	2.5
Soil moisture content / %	22	26	27	12	10	14	23	9	20	24
Rank	5	2	1	8	9	7	4	10	6	3

Note that the rank for the value 14 (number of *A. major* in one quadrat in samples 2 and 10) is 22.5. This is because they are the same, so we assign the mean of the next two ranks, which is 2.5. If three are the same we assign the average of the three ranks.

Next, we find *D* the difference between each pair of ranks.

Then, square the *D* values.

Sample number	1	2	3	4	5	6	7	8	9	10
Number of *A. major* in one quadrat	10	14	12	8	2	9	15	5	11	14
Rank	6	2.5	4	8	10	7	1	9	5	2.5
Soil moisture content / %	22	26	27	12	10	14	23	9	20	24
Rank	5	2	1	8	9	7	4	10	6	3
D	1	0.5	3	0	1	0	−3	−1	−1	−0.5
D²	1	0.25	9	0	1	0	9	1	1	0.25

Now calculate the sum of the *D²* values:

$\Sigma D^2 = 22.5$

Then substitute the values into the equation:

$$r_s = 1 - \left(\frac{6 \times \Sigma D^2}{n^3 - n} \right)$$

$$r_s = 1 - \frac{(6 \times 22.5)}{(10^3 - 10)}$$

$$= 0.864$$

Finally, compare this value with the critical value for 10 pairs of results, which is 0.65 (see Table 18.4).

This r_s value is greater than the critical value, so the null hypothesis is rejected and we can conclude that there is a significant positive correlation.

If the r_s value had been negative, we ignore the sign when comparing the value with the critical value, then use the negative sign to conclude that there was a negative correlation.

These data give a sigmoid curve so although there is a correlation it is not a linear one. If you wanted to see how closely the two variables are correlated to a line of best fit, it would be better to use a Pearson coefficient.

---

14. A member of domain Archaea called *Natronococcus* is thought to favour alkaline conditions for growth. A student set up growth media at 10 different pH values and used these to grow *Natronococcus*. After 24 hours incubation, the diameter of the colonies was measured. The larger the size of the colonies, the better the pH for *Natronococcus* growth. The results are shown in Table 18.5.

pH	6.0	7.6	8.0	8.6	9.0	9.4	9.5	9.6	9.9	10.0
Rank										
Mean colony diameter / mm	5.5	6.0	3.8	5.9	6.2	6.5	9.0	12.2	8.5	10.0
Rank										
D										
D²										

Table 18.5

Tip

pH is measured on a logarithmic scale. As this is not a linear scale, Spearman's rank correlation coefficient can be used, but Pearson's linear correlation coefficient would not be suitable.

a. Write a null hypothesis for this investigation.
b. Copy and complete Table 18.5 to calculate Spearman's rank correlation coefficient. Use your answer to determine whether there is a correlation between increasing pH and better growth conditions for *Natronococcus*. You will need to refer to Table 18.4, which gives the critical values for Spearman's rank correlation coefficient at $p = 0.05$, and the equation for Spearman's rank correlation coefficient:

$$r_s = 1 - \left(\frac{6 \times \sum D^2}{n^3 - n} \right)$$

where D = difference in rank between each pair of ranked measurements and n = number of pairs of items in the sample.

Link

Pearson's linear correlation coefficient was introduced in Chapter 2.

Pearson's linear correlation coefficient

Pearson's linear correlation coefficient is similar to Spearman's rank test, but with one important difference. Pearson's linear correlation coefficient tests for a linear correlation. That means when one value increases or decreases in equal-sized intervals, the other value also increases or decreases in equal-sized intervals. This means you could not use it if one variable was discrete, even if you assigned a number to it. Whereas Spearman's coefficient detects trends, and so could indicate the trend of a curved line of best fit, Pearson's coefficient is a measure of how close the values are to a straight line of best fit. Outliers can therefore have a significant effect, changing the coefficient so the strength of the correlation appears much weaker than it really is. This is why outliers should not be included in the calculation. Pearson's linear correlation coefficient is usually used when the data are normally distributed. A comparison of Pearson's and Spearman's tests and their results is shown in Figure 18.13.

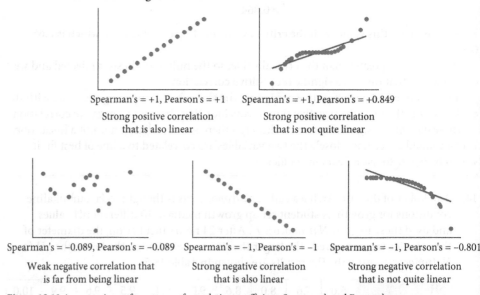

Spearman's = +1, Pearson's = +1
Strong positive correlation that is also linear

Spearman's = +1, Pearson's = +0.849
Strong positive correlation that is not quite linear

Spearman's = −0.089, Pearson's = −0.089
Weak negative correlation that is far from being linear

Spearman's = −1, Pearson's = −1
Strong negative correlation that is also linear

Spearman's = −1, Pearson's = −0.801
Strong negative correlation that is not quite linear

Figure 18.13 A comparison of two types of correlation coefficient: Spearman's and Pearson's.

The equation to calculate Pearson's linear correlation coefficient, r, is given below:

Tip

The number of degrees of freedom for Pearson's linear correlation coefficient is 2 less than the number of pairs of data.

$$r = \frac{\sum xy - n\bar{x}\bar{y}}{(n-1)s_x s_y}$$

where x, y = observations, $\bar{x}$, $\bar{y}$ = means, n = sample size (number of observations) and s = sample standard deviation.

Worked example

Hydrobia is a type of water snail. A student investigated how the density of *Hydrobia* varies with particle size in the sand on which the snail lives. Table 18.6 and Figure 18.14 show the results.

Fine sand particles in size range 20–180 μm / %	Density of *Hydrobia* / 100 per m²
13.4	31
19.8	94
27.8	156
35.2	195
41.5	255
47.2	345

Table 18.6

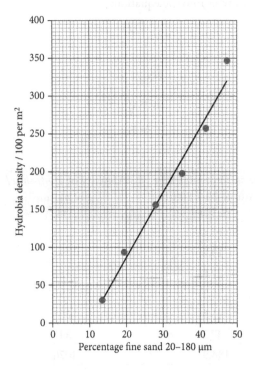

Figure 18.14

Use Pearson's linear correlation coefficient to show the strength of the positive linear correlation between the percentage of fine particles in the sand and the density of *Hydrobia*.

Answer

First, assign one set of values as x and one as y. Next, calculate the mean of each data set and use that to work out the standard deviation of each data set using the equation below. To calculate s you first need to subtract the mean from each value and then square it:

Tip

The answer below shows all the steps needed to calculate the Pearson coefficient; you will not be expected to carry out all of these – you will be given partially completed calculations to finish.

	Fine sand particles in size range 20–180 µm / % (x)	Density of Hydrobia / 100 per m² (y)	$x - \bar{x}$	$(x - \bar{x})^2$	$y - \bar{y}$	$(y - \bar{y})^2$
	13.4	31	−17.4	303.3	−148.3	22002.8
	19.8	94	−11.0	121.4	−85.3	7281.8
	27.8	156	−3.0	9.1	−23.3	544.4
	35.2	195	4.4	19.2	15.7	245.4
	41.5	255	10.7	114.1	75.7	5725.4
	47.2	345	16.4	268.4	165.7	27445.4
Totals	184.9	1076		835.5		63245.2
Mean $(\bar{x}, \bar{y})$	30.8	179.3				

We can now substitute values into the equation:

$$s = \sqrt{\frac{\sum(x - x)^2}{n - 1}}$$

For x:

$$s_x = \sqrt{\frac{835.5}{5}}$$

$$= 12.9$$

For y:

$$s_y = \sqrt{\frac{63245.2}{5}}$$

$$= 112.5$$

Next, calculate xy:

	Fine sand particles in size range 20–180 µm / % (x)	Density of Hydrobia / 100 per m² (y)	xy
	13.4	31	415
	19.8	94	1861
	27.8	156	4337
	35.2	195	6864
	41.5	255	10583
	47.2	345	16284
Totals	184.9	1076	40344
Mean $(\bar{x}, \bar{y})$	30.8	179.3	

Now we can substitute these values into Pearson's equation:

$$r = \frac{\sum xy - n\bar{x}\bar{y}}{(n - 1)s_x s_y}$$

$$r = \frac{40344 - (6 \times 30.8 \times 179.3)}{(5 \times 12.9 \times 112.5)}$$

$$= \frac{40344 - 33135}{7256}$$

$$r = 0.99$$

which indicates a strong positive correlation that is very close to a linear relationship.

15. In an investigation into abiotic factors in a mountain ecosystem, mean annual temperatures were measured at different altitudes. The results are shown in Table 18.7.

Altitude / m	200	400	600	800	1000	1200	1400	1600	1800	2000
Mean annual temperature / °C	21	22	17	16	13	10	9	7	8	7

Table 18.7

a. Plot a scatter graph using the results in Table 18.7.
b. When finding the strength of the correlation between altitude and mean annual temperature, state why Pearson's linear correlation coefficient is suitable.
c. The value for Pearson's linear correlation coefficient, calculated from Table 18.7, is $r = -0.968$.

   State what this value shows about the correlation between altitude and mean annual temperature.
d. Explain how:
   i. the number of degrees of freedom is calculated for this statistical test
   ii. a table of critical values is used to determine whether $r = -0.968$ is significant.

Some relationships show a correlation because there is a cause and effect relationship between the variables. For example, there is a cause and effect relationship in the correlation between pH and the size of colonies of *Natronococcus*. This is because of pH affecting enzyme reaction rates.

Other relationships show correlation but no direct cause and effect relationship. For example, in ancient times people believed that lice were beneficial to human health. This was because fewer lice were seen on people with disease than on healthy people, so the conclusion was that people became ill because the lice had left. We now know that lice are sensitive to body temperature. People with fever will be less attractive to lice.

Another example is the rise in carbon dioxide output of countries being correlated with obesity in those countries. The conclusion would appear to be that carbon dioxide causes obesity, however industrialisation correlates positively with wealth, and wealth correlates positively with over-eating.

Simpson's index of diversity

One way to obtain a numerical value for biodiversity is to use **Simpson's index of diversity**. The calculation takes into account the number of different species, or types, of organisms present and the relative abundance of each.

The equation for calculating Simpson's index of diversity, D is:

$$D = 1 - \left(\Sigma \left(\frac{n}{N} \right)^2 \right)$$

where n = number of individuals of each type present in the sample (types may be species and/or higher taxa such as genera, families, etc.) and N = the total numbers of all individuals of all types present in the sample.

The values of D will be in the range 0–1, with 0 meaning no diversity and 1 meaning infinite diversity. One index value on its own holds very little value; its real value comes when two locations, or the same location at two different times, are compared. Then we can consider why there are differences.

Worked example

A study was carried out to investigate the biodiversity of a temperate grassland. The area was sampled and the populations of various plants were estimated. The results are shown in Table 18.8.

Species	Number of individuals recorded in the sample (n)	n/N	$(n/N)^2$
grass species 1	80	0.345	0.1190
grass species 2	45	0.194	0.0376
clover	9	0.039	0.0015
black medick	22	0.095	0.0090
daisy	13	0.056	0.0031
dandelion	3	0.013	0.00017
speedwell	10	0.043	0.0019
self-heal	14	0.060	0.0036
moss	36	0.155	0.0241
Totals	232		0.2

Table 18.8

a. Using the results in Table 18.7, calculate Simpson's index of diversity for the plant species in this area.

b. Comment on the value of D.

Answer

a. $D = 1 - \left(\Sigma \left(\frac{n}{N} \right)^2 \right)$

$D = 1 - 0.2$

$= 0.8$

b. In the range 0–1, a D value of 0.8 is quite high, indicating high biodiversity.

16. Two lakes, A and B, were sampled randomly for various fish species. Estimates of the populations of each fish species in lake A and lake B are shown in Table 18.9.

Fish species	Population estimate lake A	Population estimate lake B
stickleback	1242	895
brown trout	66	27
carp	108	34
steelhead	18	4
northern pike	7	1
bream	229	88
roach	453	110

Table 18.9

a. Use Simpson's index of diversity to make a comparison of the biodiversity of fish in these two lakes.

$$D = 1 - \left(\Sigma \left(\frac{n}{N} \right)^2 \right)$$

b. The northern pike is the top predator in both these lakes. Explain what would happen to the biodiversity in lake A if the number of pike increased.

Assignment 18.1: Biodiversity in crop farming

Many farms in temperate climates traditionally had fields separated by hedges as shown in Figure 18.15.

Figure 18.15 On this farm, wheat is grown in a field bordered with hedges.

Hedges are rows of woody plants, usually between 1 and 5 m tall, that form boundaries between fields. As mechanisation of farming increased in the 20th century, farmers removed hedges so that fields became larger, and so easier to operate large machines. In addition, the area of ground occupied by the hedges then became available for crop growth.

A study was carried out of the effects of a 2 m tall hedge on various factors at different distances from the hedge. The results are shown in Figure 18.16.

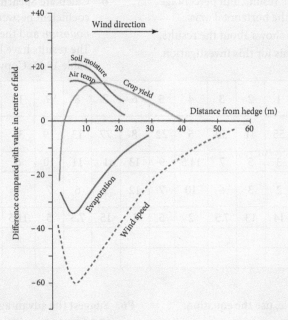

Figure 18.16

QUESTIONS

A1. State the types of factors shown in Figure 18.16.

A2. Suggest how the results for crop yield were collected.

A3. Suggest why soil moisture content decreases further from the hedge.

⭐**A4.** Suggest reasons why:
   a. the crop yield decreases closer to the hedge than about 4 m
   b. there is increased crop yield between 4 and 20 m from the hedge.

A5. Give reasons why hedges will increase biodiversity in the field.

Experimental skills 18.1: Investigating correlations in species abundance

The mollusc *Littorina littorea* is **native** to the northern coasts of the Atlantic ocean. *L. littorea* is omnivorous but grazes extensively on algae that grow on coastal rocks. One species of algae, *Ascophyllum nodosum*, produces a compound that is toxic to *L. littorea* as a defence against grazing.

A student proposed the hypothesis that there would be fewer *L. littorea* on *A. nodosum* than on bare rock or on other species of algae.

The student used a frame quadrat and a transect line parallel to the sea across an area where both species were found.

The quadrats measured 1 m by 1 m and were placed at 2 m intervals along the transect line. The percentage cover of *A. nodosum* was measured and the numbers of *L. littorea* were recorded.

QUESTIONS

P1. State the independent and dependent variables in this investigation.

P2. Name the sampling method used in this investigation.

The results are summarised in Table 18.10.

Quadrat number	1	2	3	4	5	6	7	8	9	10	11	12	13	14	15
Percentage cover of *A. nodosum*	55	41	30	5	22	8	77	13	19	36	27	5	52	12	62
Number of *L. littorea*	2	3	6	10	7	12	1	6	7	4	5	7	4	8	5

Table 18.10

P3. Plot a scatter graph of the results. Put percentage cover of *A. nodosum* on the horizontal axis.

P4. Describe what the graph shows about the results.

P5. a. Write a null hypothesis for this investigation.

b. Calculate Spearman's rank correlation coefficient between the percentage cover of *A. nodosum* and the number of *L. littorea*. The results have been ranked for you in Table 18.11. Copy and complete this table.

Quadrat number	1	2	3	4	5	6	7	8	9	10	11	12	13	14	15
Percentage cover of *A. nodosum*	55	41	30	5	22	8	77	13	19	36	27	5	52	12	62
Rank	3	5	7	14.5	9	13	1	11	10	6	8	14.5	4	12	2
Number of *L. littorea*	2	3	6	10	7	12	1	6	7	4	5	7	4	8	5
Rank	14	13	7.5	2	5	1	15	7.5	5	11.5	9.5	5	11.5	3	9.5
D															
D²															

Table 18.11

After completing the table, use the equation:

$$r_s = 1 - \left(\frac{6 \times \sum D^2}{n^3 - n} \right)$$

c. Comment on the significance of your answer. The critical value for 15 pairs of data at $p = 0.05$ is 0.44.

P6. Suggest the advantage of plotting a scatter graph in addition to calculating Spearman's rank correlation coefficient on the same results.

Key ideas

→ The terms ecosystem, habitat and niche are used in ecology.

→ Biodiversity can be assessed as the number and range of ecosystems and habitats, as the number of different species and their relative abundance, or as the genetic variation within each species.

→ It is important that sampling of an area is carried out randomly.

→ Frame quadrats, line transects, belt transects and the mark-release-recapture method can be used for random sampling of organisms.

→ The Lincoln index is used with the mark-release-recapture method to calculate an estimate of the population.

→ Simpson's index of diversity can be used to calculate the biodiversity of an area.

→ Spearman's rank correlation coefficient and Pearson linear correlation coefficient can be used to analyse the relationships between two variables.

18.3 Conservation

Conservation in biology means working with and caring for organisms and ecosystems. In the biological context, this means maintaining the same level of biodiversity. In order to do this, we need to understand ecological processes. When ecological processes go wrong, then individual species can become endangered. **Endangered** means their numbers fall so low that they have difficulty in finding mates and may then become **extinct**. If this happens, then the ecosystem will be damaged.

One problem with conservation is that the high-profile endangered species tend to be the photogenic mammals and flowering plants. Smaller invertebrates do not attract so much attention and there is currently no information on endangered members of domains Archaea or Bacteria!

REASONS FOR EXTINCTION

Animals and plants have become extinct long before humans appeared on Earth. The best-known example of this is the dinosaurs, the last of which became extinct 65 million years before humans. However, human impacts on biodiversity have been severe over the last 250 years, so many people believe that we have a responsibility to save species from extinction if possible.

There are many, often complex, factors in the causes of extinction of individual species. These factors are:

- hunting by humans, predation and competition by introduced species, introduced diseases
- habitat loss and degradation
- climate change.

Hunting by humans

Hunting of animals for food, for the pet trade or for supposed medicinal use can lead to extinction. One well-documented example of this is the dodo, *Raphus cucullatus,* a large flightless bird that was native to the island of Mauritius. The dodo was endangered by the introduction of other animals to Mauritius, such as dogs and cats. The use for the dodo as a food source by visiting sailors led to its extinction in the 17th century.

Many animals are endangered in the present day by illegal hunting activities: one such example is the sun bear, *Helarctos malayanus*, of Southeast Asia, which is hunted for food and for products that have supposed medicinal benefits to humans.

Predation and competition by introduced species

Invasive alien species are species of animals or plants that have been introduced, accidentally or deliberately, to a habitat and have come to threaten local biodiversity. Reasons for the threat from these species can be due to the lack of natural predators or their ability to compete better for resources than native species.

One of the world's rarest ducks is the Campbell Island teal, *Anas nesiotis*. *A. nesiotis* is a flightless duck native to Campbell Island and some other small islands 700 km south of New Zealand. The Norway rat, *Rattus norvegicus*, was introduced to Campbell Island in the 19th century and the rat population reached an estimated 200 000 by the year 2000. Rats preyed on many species of small animals and birds, but by preying on *A. nesiotis*, rats almost caused the extinction of this species. Breeding pairs of *A. nesiotis* were found on one particular island that had remained free of rats. Captive breeding and reintroduction along with the elimination of rats has allowed the population of *A. nesiotis* to recover.

Introduced diseases

Pathogens can be introduced, usually by accident, and have devastating effects on populations.

Dutch elm disease is a fungal disease affecting elm trees. This disease is believed to have been endemic in Asia but was accidentally introduced to Europe and North America and has now spread to New Zealand. The disease causes the death of elm trees. These large trees are an important part of temperate forest ecosystems. Dutch elm disease is spread between trees by elm bark beetles and kills trees rapidly. Trees respond to the infection by blocking their xylem vessels in an attempt to stop the fungus from spreading. This prevents transpiration and is ultimately fatal. The disease is now controlled by human intervention, but otherwise diseases such as this could lead to the extinction of a species.

The rodents *Rattus nativitatis* and *R. macleari* were native to Christmas Island in the Indian Ocean but were both extinct by the early 20th century. Researchers have since studied DNA samples from preserved remains of these animals. Evidence suggests that a protozoan disease was responsible for their extinction.

Habitat loss and degradation

One of the best-known examples of an endangered species is the giant panda *Ailuropoda melanoleuca*, shown in Figure 18.17.

In 2019, *A. melanoleuca* was classed as vulnerable due to a number of factors. One of these was habitat destruction and habitat fragmentation. Habitat fragmentation is where a large, continuous habitat is changed to smaller, discontinuous habitats. Habitat destruction was caused by logging, which is cutting down trees for wood to use in construction. Habitat fragmentation was caused by the building of roads and railways through the habitat, restricting the movement of the animals. In addition to these problems, *A. melanoleuca* has a very low birth rate, both in the wild and in captivity.

Climate change

Climate change is causing the polar bear, *Ursus maritimus*, to become endangered. *U. maritimus* is a predator of fish and seals, hunting for prey by making holes in ice to reach sea water below. Climate change is the gradual and long-term change in weather patterns on Earth that include the increase in average temperature. This increase in temperature is causing polar ice to become thinner, so is less able to support the weight

Figure 18.17 Giant panda, *A. melanoleuca*, in a breeding research institute in Chengdu, China.

of these large animals. Unable to hunt for prey, fewer *U. maritimus* are able to reach breeding age, causing numbers to fall.

REASONS FOR CONSERVATION

Endangered organisms should be preserved for several different reasons:

- All organisms are part of complex food webs. If one organism becomes extinct, it could endanger others.
- Increased biodiversity in an ecosystem means increased stability in that ecosystem, so by losing species we make the future of established ecosystems less predictable.
- Some important medicines have been made from plants and fungi (Figure 18.18). For example, around 7000 medicines are derived from plant products and many antibiotics have come from fungi. Losing plant and fungal species may mean the loss of important drug discoveries in the future.
- People like to see a variety of animals and plants, and responsible ecotourism generates valuable income for some communities.

METHODS OF CONSERVATION

Species whose numbers are falling can be saved from becoming endangered, and those that are endangered can be assisted to increase again in number.

The **International Union for Conservation of Nature** (IUCN) was founded in 1948. The IUCN produces, and regularly updates, its red list of endangered species. This enables people in different countries to be aware of species that are in danger. The IUCN also helps to prevent illegal trade in endangered species, and tries to expand and introduce more protected areas.

The **Convention on International Trade in Endangered Species of Wild Fauna and Flora** (CITES), is a treaty involving many different countries to protect endangered animals and plants. Key successes have been the reduction in the fur trade and getting some countries to stop importing products such as shells of sea turtles.

Endangered species can be protected by education of local people who may be hunting an endangered animal without realising its endangered status. Laws can be put in place to prevent hunting or habitat destruction, but the problem remains in enforcing these laws. For example, the catching of endangered whale species hundreds of miles from land is very difficult to prevent.

Nature reserves and **conservation areas** can be established where human impact is easier to control. Conserved areas can be on land or at sea. Those on land are sometimes called national parks, and those at sea are called marine parks. Involvement of local people in management of these areas helps to raise awareness about endangered species. Management includes maintaining watch over the areas to prevent illegal hunting or fishing. Money raised from tourism can be used to employ local people as rangers, or be returned to the community through health or school programmes.

One way to increase the numbers of endangered species is through **captive breeding** programmes. Zoos can be used to breed animals in safe surroundings and botanic gardens can be used to do the same for plants. The offspring can then be returned to the wild in a managed way to increase numbers.

Frozen zoos are storage facilities for DNA, eggs, sperm, embryos and tissues of animals. As the name suggests, the samples are stored at very low temperature, usually in liquid nitrogen, which is at −196 °C, to prevent deterioration with time. The first frozen zoo was established in 1972 and now various facilities exist around the world. The aim is to provide a 'library of life' with the possibility in the future of restoring extinct species.

Seed banks are the equivalent of frozen zoos for plants. In 2012, scientists published the results of an experiment to regenerate a plant from a 32 000-year old seed that had

Tip

A stable ecosystem is one that is capable of returning to its original state after some change has been imposed.

Link

Many antibiotics, such as penicillin, were first extracted from fungi, as described in Chapter 10.

Tip

Ecotourism is travel to exotic or threatened natural areas, which provides economic support for conservation in those areas.

Figure 18.18 The common painkiller aspirin was originally made from an extract taken from the bark of the willow tree.

been discovered buried in frozen soil at a depth of almost 40 m. This proves that DNA is stable over long time periods at low temperatures and gives the possibility of future regeneration of extinct plant species from seed banks.

One of the current problems facing frozen zoos and seed banks is knowing what to store. There are millions of plant and animal species, so decisions must be made as to which species are relevant.

Assisted reproduction is another method that can be used to raise numbers of endangered animals. This can take the form of *in vitro* **fertilisation** (IVF), **embryo transfer** and **surrogacy**.

In vitro, literally meaning 'in glass', fertilisation is where eggs and sperm are extracted from the same species and allowed to fertilise in a dish in the laboratory. IVF was initially developed for humans with fertility problems but can also be used with animals. For example, the last surviving male northern white rhino died in 2018. Scientists extracted sperm from the animal that could be used in IVF. Eggs would need to be extracted from females, and the embryo implanted into a much younger member of a closely related species for this to be successful. There is also the problem of a very small gene pool for the restored generation, should this work. The gene pool is the total number of genes in a population. The process of IVF is summarised in Figure 18.19.

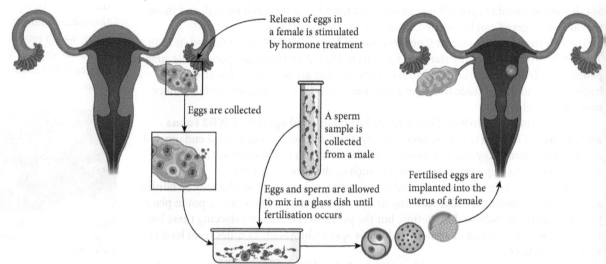

Release of eggs in a female is stimulated by hormone treatment

Eggs are collected

A sperm sample is collected from a male

Eggs and sperm are allowed to mix in a glass dish until fertilisation occurs

Fertilised eggs are implanted into the uterus of a female

Figure 18.19 *In vitro* fertilisation is one method of assisted reproduction.

Embryo transfer is where an embryo from one species is implanted into the uterus of another closely related species. In 2000 a frozen zoo in the USA implanted a previously frozen embryo of a critically endangered African wildcat into a female domestic cat. The domestic cat then gave birth to a healthy African wildcat.

Surrogacy in conservation is where one species is chosen to represent many others. The needs of a surrogate species should reflect the needs of the others that they represent. This facilitates making conservation decisions, such as location and size of reserves required. Surrogacy can have an **umbrella effect** where other species are also protected. For example, reserves designed to protect the northern spotted owl in the USA were also found to be protecting species of salamanders and molluscs. In this case the owl is the surrogate species.

Key ideas

→ Populations and species can become extinct because of climate change, habitat loss, competition and hunting by humans.
→ Biodiversity needs to be maintained for the stability of ecosystems and food chains.
→ Zoos, botanic gardens, conserved areas (national parks and marine parks), frozen zoos and seed banks all play a role in conservation of endangered species.
→ Assisted reproduction methods can be used to raise numbers of endangered species.
→ Invasive alien species need to be controlled as they can outcompete native species and cause them to become endangered.
→ The International Union for the Conservation of Nature (IUCN) and the Convention on International Trade in Endangered Species of Wild Fauna and Flora (CITES) both play important roles in conservation worldwide.

17. Suggest why the IUCN red list has:
    a. many more vertebrates than invertebrates
    b. no prokaryotes.

18. In 2016, the last two known Yangtze giant softshell turtles, a male and a female, were taken into captivity. Both animals were thought to be over 90 years old. The last known female died in 2019 following an attempt to artificially inseminate her. Artificial insemination is a type of assisted reproduction where the sperm from the male is introduced to the female through a thin, flexible tube.
    a. Suggest why the turtles did not breed in captivity.
    b. Freshwater turtles lay eggs in the sand on river banks. Sand is in high demand for building. Removal of the sand leaves nowhere for the turtles to lay eggs. Suggest how this situation could be changed.

19. The tuatara is a reptile that is only found on some small islands close to New Zealand.
    The temperature at which the tuatara eggs are incubated determines the gender of the offspring:

    • higher than 22 °C produces only males
    • lower than 22 °C produces only females.

    a. In the year 2000, the temperature during the egg-laying season was in the range 20–24 °C. Suggest the proportions of male and female tuataras that hatched in the year 2000.
    b. Climate change forecasts have estimated that the average temperature in the habitat of the tuatara could increase by 2.5–4.5 °C in the next 100 years. Suggest the effect this will have on the tuatara population.
    c. Give two reasons why animals such as the tuatara should be protected.

CHAPTER OVERVIEW

Review the mini mind map to link new learning with your prior learning, and to highlight which of the key concepts underpin the chapter.

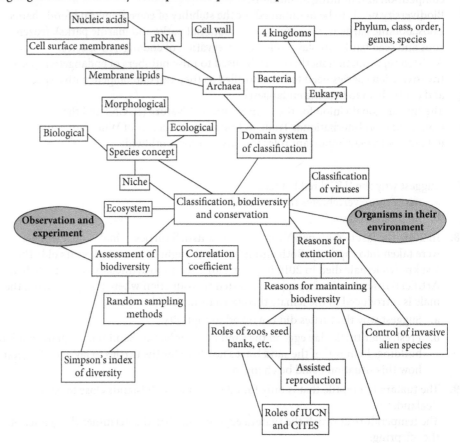

Try copying this mini mind map and expanding upon it. Use your notes from other chapters to help you explore how the essential ideas, theories and principles can be linked further together.

WHAT YOU HAVE LEARNED

Now that you have finished this chapter you should be able to:

- discuss the meaning of the term species, limited to the biological species concept, morphological species concept and ecological species concept
- describe the classification of organisms into three domains: Archaea, Bacteria and Eukarya
- state that Archaea and Bacteria are prokaryotes and that there are differences between them, limited to differences in membrane lipids, ribosomal RNA and composition of cell walls
- describe the classification of organisms in the Eukarya domain into the taxonomic hierarchy of kingdom, phylum, class, order, family, genus and species
- outline the characteristic features of the kingdoms Protoctista, Fungi, Plantae and Animalia
- outline how viruses are classified, limited to the type of nucleic acid (RNA or DNA) and whether this is single stranded or double stranded
- define the terms ecosystem and niche

- explain that biodiversity can be assessed at different levels:
  - o the number and range of different ecosystems and habitats
  - o the number of species and their relative abundance
  - o the genetic variation within each species
- explain the importance of random sampling in determining the biodiversity of an area
- describe and use suitable methods to assess the distribution and abundance of organisms in an area, limited to frame quadrats, line transects, belt transects and mark-release-recapture using the Lincoln index
- use Spearman's rank correlation and Pearson's linear correlation to analyse the relationships between two variables, including how biotic and abiotic factors affect the distribution and abundance of species
- use Simpson's index of diversity (D) to calculate the biodiversity of an area, and state the significance of different values of D
- explain why populations and species become extinct as a result of:
  - o climate change
  - o competition
  - o hunting by humans
  - o degradation and loss of habitats
- outline reasons for maintaining biodiversity
- outline the roles of zoos, botanic gardens, conserved areas (including national parks and marine parks), frozen zoos and seed banks, in the conservation of endangered species
- describe methods of assisted reproduction used in the conservation of endangered mammals, limited to IVF, embryo transfer and surrogacy
- explain reasons for controlling invasive alien species
- outline the role in conservation of the International Union for the Conservation of Nature (IUCN) and the Convention on International Trade in Endangered Species of Wild Fauna and Flora (CITES).

CHAPTER REVIEW

1. Up until the late 20th century, cellular life was classified into five kingdoms: Animalia, Plantae, Fungi, Protoctista and Prokaryotae.
   a. Explain why kingdom Prokaryotae was later split into two domains.
   b. Name the domain that includes the other four kingdoms.
   c. Give two differences between the kingdoms:
      i. Animalia and Plantae
      ii. Plantae and Fungi.
2. a. Describe the biological concept of species.
   b. Discuss whether the biological or the morphological concept of species is more suitable for the domestic dog.
3. a. Outline how viruses are classified.
   b. Explain why viruses require a separate classification system.
4. A public park contains a small pond filled with water. The pond contains fish, amphibia, invertebrates and various microorganisms.
   a. Explain why the pond can be considered as an ecosystem.
   b. The pond contains water beetles, which are arthropods that live in water. Describe a method for estimating the population of water beetles in the pond. Include any considerations that would make the estimate more accurate.

    c. A species of fish, A, that has existed in the pond for many years feeds on a species of amphibian, B, that also lives in the pond. The populations of both animals have been relatively stable for many years. Explain why introducing a new fish species that also feeds on species B could cause numbers of species A and B in the pond to fall dramatically.

5. In many countries people clear areas of forest so they can plant food crops. Discuss the effects this has on the ecosystem of the area.

6. Zoos are places where people pay to see animals, some of which are not native to the local area. Outline the roles that can be played by zoos in conservation of endangered species.

7. The slow loris is a small primate mammal that lives in trees in south and Southeast Asia. The slow loris is threatened by habitat loss and also by the illegal pet trade. The slow loris is captured and sold for use as pets in many countries worldwide. People who buy the pets often know nothing about the slow loris and are only attracted by its appearance.

    a. Explain how the illegal pet trade caused the slow loris to become endangered.

    b. Suggest how an organisation such as the Convention on International Trade in Endangered Species of Wild Fauna and Flora (CITES) can help save the slow loris from becoming extinct.

8. a. The population of land snails was investigated in an enclosed area. First, 320 snails were captured and their shells marked with a small spot of paint before being released. One week later, 400 snails were captured, 26 of which had the marks on their shells. Use this information to estimate the population of land snails in this area. Give your answer to two significant figures.

    b. Give three factors that could make the estimate in (a) inaccurate.

Chapter 19

Genetic technology

C.ETH2.0
785kb

Will it soon be possible to design and make an artificial cell? In 2019, researchers made one more step towards this possibility.

In this small tube is the world's first computer-generated artificial genome. It is a molecule called *Caulobacter ethensis-2.0*, a piece of DNA 785 701 base pairs long (just over 785 kilobases or kb). Researchers at ETH Zürich in Switzerland used a computer algorithm to produce a DNA molecule with a specific sequence. This comprises the full genetic instructions to make a new bacterium, with 680 genes, 580 of which could be transcribed normally.

The power of computers will vastly expand in the 2020s. Bioinformatics holds great potential. Among the possible future applications of this work are synthetic microorganisms that could produce complex medicinal drugs, or break down environmental pollutants. However, as promising as this research is, it demands thoughtful consideration. Many are concerned that such developments may lead to unpredictable and significant ethical, social and environmental dilemmas.

Prior understanding

You may remember that the molecule of inheritance in living organisms is DNA. You may recall the structure of DNA and RNA from Chapter 6 and how these molecules are involved in protein synthesis. You may also remember the basis of inheritance from Chapter 16, including some mechanisms of controlling gene expression.

Learning aims

In this chapter, you will learn how our knowledge of the structure and function of DNA led to the development of a new field of science, genetic technology. You will explore the social and ethical dilemmas associated with these new technologies and the caution with which they are used.

19.1 Principles of genetic technology
    (Syllabus 19.1.1–19.1.11)

19.2 Genetic technology applied to medicine
    (Syllabus 19.2.1–19.2.4)

19.3 Genetically modified organisms in agriculture
    (Syllabus 19.3.1, 19.3.2)

19.1 Principles of genetic technology

Genetics has come a long way since the experiments of Mendel and his pea plants. The fastest pace of development has been in the past 40 years. There now exists an ever-increasing number of gene technologies, from research to medicine, and from agriculture to industry. However, these advances have been associated with moral and ethical dilemmas. As you will see later in this chapter, gene technology has the potential for abuse, and we should think carefully about where it might lead.

Figure 19.1 provides an overview of the molecular 'tools' and other laboratory equipment found in the kit of a gene technologist.

Link

Gene technology is a branch of biotechnology. You will have encountered this before, including the use of immobilised enzymes (Chapter 3) and monoclonal antibodies (Chapter 11).

Link

The genetic code is universal – the same codons (base triplets) code for the same amino acids in all species. You encountered this in Chapter 6.

Tip

Genetic engineering adjusts the content of an organism's **genome**, its complete DNA content. This in turn changes its **proteome**, the complete set of proteins it can produce.

Tip

Do not confuse genetic engineering with selective breeding, in which whole sets of genes are involved and which takes much longer.

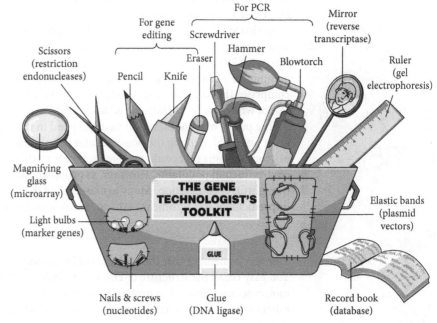

Figure 19.1 The molecular tools that are used by gene technologists. You will encounter all of these during this chapter.

GENETIC ENGINEERING

Scientists have active imaginations. As you saw in Chapter 6, it became clear in the 1960s that all organisms, from bacteria to higher mammals, use the same molecule to control inheritance. Not only that, they read the molecule in the same way. Scientists wondered: could a piece of DNA from one organism be put into another and still code for the same proteins? An exciting new world of possibilities emerged, in which scientists imagined producing organisms with new and valuable functions. The technique of **genetic engineering** was born.

Genetic engineering involves the deliberate manipulation of genetic material to modify specific characteristics of an organism. This usually involves taking DNA from the genome of one species, called **donor DNA**, and adding it to the genome of another, the **recipient** organism. This produces **recombinant DNA** that is expressed in a **transgenic organism** (or a **genetically modified organism**, GMO) to form **recombinant protein**.

One of the earliest transgenic organisms was the bacterium *E. coli*, genetically engineered to produce human insulin in 1978. Recombinant human insulin is still used to treat people with diabetes today. Many other successes followed (Table 19.1).

Year	Transgenic organism
1973	first transgenic bacterium
1974	first transgenic animal (mouse)
1978	first recombinant human protein produced by transgenic bacteria (insulin)
1981	first vaccine produced by genetic engineering (foot and mouth disease)
1983	first transgenic plant
1994	first GM plant approved for food (FlavrSavr tomato)
1995	first insecticide-producing transgenic crop (Bt corn)
2000	first nutrient-enriched transgenic crop (golden rice)
2003	first transgenic pet (Glo-Fish®)
2009	first recombinant human protein produced by a transgenic animal (antithrombin)
2015	first transgenic animal approved for food (AquAdvantage® salmon)

Table 19.1 The development of transgenic organisms. Note that the FlavrSavr tomatoes are not described as transgenic. This is because gene technologists made a change to their genome; they did not contain a donor gene from another organism.

Link

In Chapter 14, you read how insulin is required to treat people with diabetes. Before production of recombinant insulin using yeast, insulin was extracted from the pancreases of slaughtered animals.

So how do gene technologists produce GMOs? The process generally has four steps (Figure 19.2). You will explore these methods and molecules in the following paragraphs.

Isolation

Remember that in most examples of genetic engineering, a donor gene from one organism is introduced into another organism. The first step in genetic engineering must therefore be to obtain (isolate) a section of DNA that contains the gene required. This can be difficult: in all species, there are thousands of genes 'hidden' among other DNA sequences in chromosomes. The gene technologist has three methods to choose from:

- **restriction**
- **reverse transcription**
- **automated polynucleotide synthesis**.

Restriction

This was the first method used to isolate genes from a genome. It takes its name from the enzymes used in the process, called **restriction endonucleases** (or simply 'restriction enzymes'). These occur naturally in bacteria and play a role in a defence system against viruses: they 'restrict' the replication of these invaders, by cutting up their DNA. This property makes them very useful in isolating genes in genetic engineering: they can be used as 'molecular scissors'.

Restriction endonucleases do not cut DNA at random. There are thousands of different restriction endonucleases that have slight differences in the shape of their active site. Each enzyme cuts across the double-stranded DNA molecule by hydrolysing phosphodiester bonds between nucleotides found in a specific nucleotide base sequence, usually 4–6 bases in length, known as the **restriction** (or **recognition**) **site**.

A challenge for the gene technologist is to identify restriction sites that occur on either side of the gene in the genome of the organism, so that the gene can be 'cut out' and isolated. They must also ensure that any enzymes that are used do not have restriction sites within the gene. A further preference is to choose restriction endonuclease(s) that do not cut phosphodiester bonds that are directly opposite in

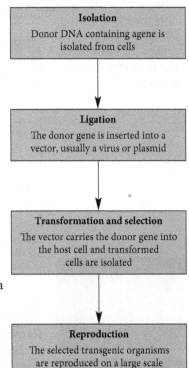

Isolation
Donor DNA containing a gene is isolated from cells

Ligation
The donor gene is inserted into a vector, usually a virus or plasmid

Transformation and selection
The vector carries the donor gene into the host cell and transformed cells are isolated

Reproduction
The selected transgenic organisms are reproduced on a large scale

Figure 19.2 An overview of the process of genetic engineering to make transgenic organisms. To undertake these four steps, the gene technologist requires some of the molecular 'tools' shown in Figure 19.1. These can today be bought commercially.

the DNA molecule. Instead, an enzyme is usually chosen that makes 'staggered cuts' to produce two DNA molecules with 'overhangs'. These are short single-stranded lengths of unpaired bases that are formed at each cut end, called **sticky ends** (Figure 19.3). The advantage of sticky ends over 'blunt ends' is that the gene technologist has greater control over which DNA strands join – a sticky end can only join to another complementary sticky end with a complementary base sequence. Three examples of restriction endonucleases and their restriction sites are shown in Table 19.2.

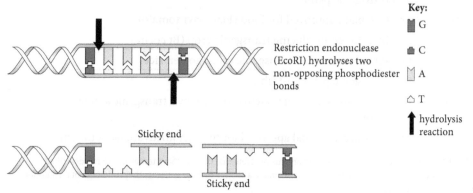

Figure 19.3 The production of 'sticky ends' by the restriction endonuclease EcoRI, which cuts the sugar-phosphate backbone on both strands at non-opposing phosphodiester bonds. Note that restriction sites are usually **palindromes**, which means that the sequence and its complementary sequence on the other strand are the same but reversed (e.g. GAATTC has the complement CTTAAG).

Restriction enzyme	Source of enzyme	Restriction site	Site of cut across DNA	Sticky end
EcoRI	*Escherichia coli*	5′ –G↓AATTC– 3′ 3′ –CTTAA↑G– 5′	–G │ AATTC– –CTTAA │ G–	AATT–3′
BamHI	*Bacillus amyloliquefaciens*	5′ –G↓GATCC– 3′ 3′ –CCTAG↑G– 5′	–G │ GATCC– –CCTAG │ G–	GATC–3′
HaeIII	*Haemophilus aegyptius*	5′ –GG↓CC– 3′ 3′ –CC↑GG– 5′	–GG │ CC– –CC │ GG–	none (blunt ends)

Table 19.2 Three examples of restriction endonucleases and their restriction sites. The enzymes are named after the bacterium from which they were extracted. EcoRI was the first restriction enzyme to be discovered, in the R strain of the bacterium *Escherichia coli*.

Reverse transcription

An alternative method that can be used to isolate a gene uses the enzyme **reverse transcriptase**. This is an enzyme found naturally in some retroviruses, including HIV. In nature, its purpose is to convert viral RNA into a complementary strand of DNA that integrates into the genome of a host T-helper cell. In effect, this is the reverse of transcription, a process that you encountered in Chapter 6. Gene technologists have taken advantage of the function of this enzyme in the laboratory.

To make a molecule of DNA containing a specific gene from mature mRNA, we need to find cells that are actively expressing that gene. Human insulin, for example, is synthesised in the beta cells in the islets of Langerhans of the pancreas. The cytoplasm of these cells contains millions, possibly billions, of copies of this mRNA for insulin. This mature mRNA can be extracted and used to make a strand of **complementary DNA** (**cDNA**) using the enzyme reverse transcriptase. The single-stranded cDNA is then used to make double-stranded DNA using DNA polymerase (Figure 19.4).

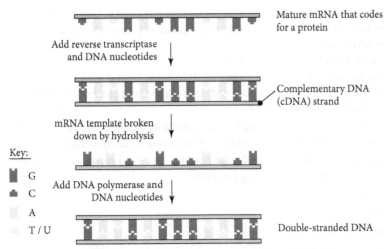

Figure 19.4 A molecule called complementary DNA (cDNA) can be synthesised in the laboratory using two enzymes, reverse transcriptase and DNA polymerase. This process, called reverse transcription, uses a mature mRNA template (without introns) obtained from cells actively expressing the gene to be isolated. A third enzyme is used to destroy the original mRNA template. It is possible during this process to add single-strand sticky ends to the double-stranded cDNA molecule for insertion into a vector later. Note that only 12 bases/base pairs are shown (most mRNA molecules will be much longer than this).

As a method of isolation, reverse transcription is usually preferred to restriction. This is because it avoids the necessity of finding appropriate restriction enzymes, and overcomes the problem posed by the limited number of copies of the DNA genome in a cell. Another advantage is that the DNA molecule produced using this process will not contain introns. As you saw in Chapter 16, prokaryotic DNA does not contain introns, and therefore bacteria do not have the mechanisms for splicing them out. If a gene is inserted into bacteria, and transcribed and translated without splicing out the introns, the protein will almost certainly not function properly.

Automated polynucleotide synthesis

Increasingly, genes are being chemically synthesised from nucleotides by 'working backwards' from the primary sequence of a protein. If the amino acid sequence of the protein is known, special techniques can make a DNA molecule with the encoding base sequence. It is not only genes that can be produced in this way. As you saw at the start of this chapter, whole genomes can now be designed and synthesised using these emerging technologies.

Worked example

Restriction enzymes, reverse transcriptase and DNA polymerase are three enzymes used to isolate DNA sequences to make recombinant DNA. In another method, a piece of DNA can be produced directly from nucleotides.

a. Explain what is meant by recombinant DNA.

Answer

Recombinant DNA is a molecule of DNA that contains nucleotide sequences from two or more species. When introduced into a cell, a transgenic organism is the result.

b. Outline the roles of these three enzymes in isolating DNA sequences.

Answer

Restriction enzymes have specific recognition base sequences in DNA and hydrolyse phosphodiester bonds to cut the double strands. Carefully selected restriction enzymes can therefore be used to remove a molecule of DNA containing a gene from a larger molecule of DNA. Reverse transcriptase converts a molecule of mRNA, produced by the transcription of a gene in a cell, into a molecule of complementary DNA (cDNA). DNA polymerase is then used to produce a complementary strand to make double-stranded DNA comprising the sequence of the expressed gene.

1. Restriction endonucleases are an important tool in genetic engineering.
    a. Use your knowledge of restriction to explain why restriction endonucleases have two active sites.
    b. Explain what is meant by a 'sticky end'.
    c. Refer to Table 19.2. If a long piece of DNA is incubated with HaeIII, more restriction fragments are produced than if the same piece of DNA is incubated with EcoRI. Explain why.
2. Give two reasons why a DNA molecule containing a gene isolated using reverse transcription will be shorter than a DNA molecule isolated using restriction enzymes.
3. State the type or locations of cells that could be used in reverse transcription to isolate the genes encoding the following proteins.
    a. Myosin
    b. ADH
    c. Rubisco

Ligation

Once the gene technologist has isolated the gene of interest, the next process can begin. It must be put into a **vector**, a 'delivery vehicle' that will facilitate its uptake by a host cell. A vector is also needed because a length of DNA containing a gene on its own will not be transcribed inside a host cell or replicated when the cell divides.

In most examples of genetic engineering, vectors are bacterial plasmids. As you saw in Chapter 1, these are small, looped molecules of double-stranded DNA that occur naturally in bacteria (Figure 19.5). Plasmids contain genes such as the gene for antibiotic resistance. They have an origin of replication, which means that they can be copied and passed on to daughter cells when the bacterium divides. Gene technologists once purified these plasmids from bacteria. Now they can purchase them commercially.

To insert the isolated gene into a plasmid, the gene technologist must first cut open the plasmid with the same restriction enzyme used to isolate the gene. Using the same enzyme ensures that the plasmid has complementary sticky ends to the isolated gene. The sticky ends **hybridise** (join) as hydrogen bonds form between the complementary bases. Next, another enzyme, DNA ligase, catalyses the condensation reactions between adjacent nucleotides to form phosphodiester bonds on both DNA strands. This process, called **ligation**, also occurs during lagging strand synthesis of DNA replication (see Chapter 6). Ligation joins the two DNA molecules together to form a **recombinant plasmid** (Figure 19.6).

Most plasmids used in genetic engineering have a range of restriction sites that enable gene technologists to insert donor genes that have a wide variety of sticky ends. In addition, most plasmids contain a promoter region next to these restriction sites. As you saw in Chapter 16, promoters facilitate gene expression by allowing transcription factors and RNA polymerase to bind to the DNA. In this way, gene technologists can ensure that the transgenic organism will express the donor gene.

Transformation and selection

Once a recombinant plasmid has been produced, the next challenge is to encourage cells to absorb it. The process of introducing the new DNA incorporated into living cells is called **transformation**. This is a challenge: most cells have evolved effective barriers such as the surface membrane to keep large, foreign molecules out. Various techniques are used to make cells more permeable using ions or heat, but their success is extremely unreliable. For every cell that takes up a plasmid, tens of thousands do not. Scientists therefore need to be able to identify those few cells that have successfully taken up the foreign DNA and been transformed.

So how can a gene technologist tell which bacteria have been transformed, and which have not? A clever way is to include in the plasmid vector another gene, called a **marker gene**, as well as the donor gene. The product of this second gene makes the bacteria easy to detect – it literally 'marks' those cells that have actually taken up the recombinant DNA.

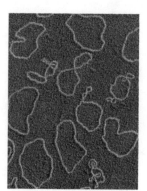

Figure 19.5 A transmission electron micrograph of plasmids extracted from *E. coli*.

Tip

Think of DNA ligase as 'molecular glue'. It has the opposite role to a restriction enzyme.

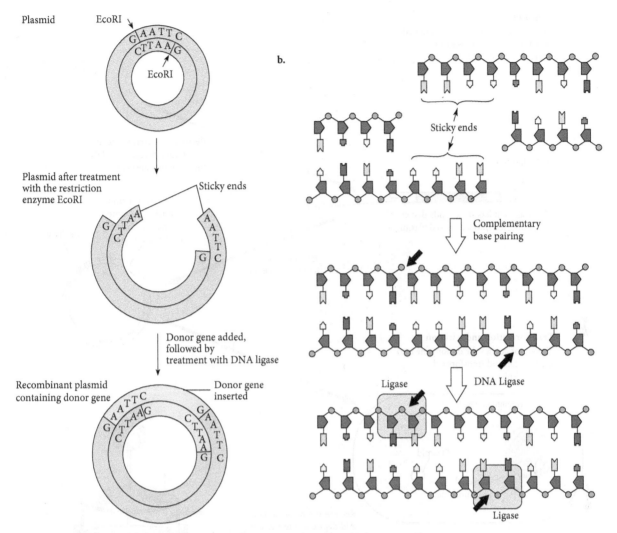

Figure 19.6 A plasmid vector is cut using the same restriction enzyme used to isolate the donor gene. This produces sticky ends that are complementary and will hybridise when they are mixed other, increasing the efficiency of ligation (**a**). DNA ligase catalyses condensation reactions to form phosphodiester bonds between adjacent nucleotides on both strands of the DNA (**b**).

In previous decades, most marker genes provided the transformed cells with antibiotic resistance. The bacteria were then grown on a medium that contains the antibiotic, and only those bacteria that took up the plasmid would be able to survive and grow. However, alternative marker genes are preferred today because of the potential danger of such antibiotic resistance genes being accidentally transferred to other bacteria, including pathogens.

The most common marker gene included in plasmid vectors is the gene that encodes **green fluorescent protein (GFP)**. This is naturally found in the genome of the crystal jellyfish, *Aequorea victoria*, which glows bright green when exposed to blue-violet light. Cells or tissues that have absorbed the recombinant plasmid will fluoresce when exposed to UV light, and can easily be identified.

Reproduction

Once the gene technologist has identified transgenic cells that have been successfully transformed, the next step is to get them to divide. Usually, this is done on a small scale at first in small volumes of nutrient broth. The bacteria are then grown in large industrial fermenters to facilitate the mass-production of the recombinant protein. Figure 19.7 shows a summary of the entire process of genetic engineering. You may wish to refer back to Figure 19.2 to cross-reference the four steps involved.

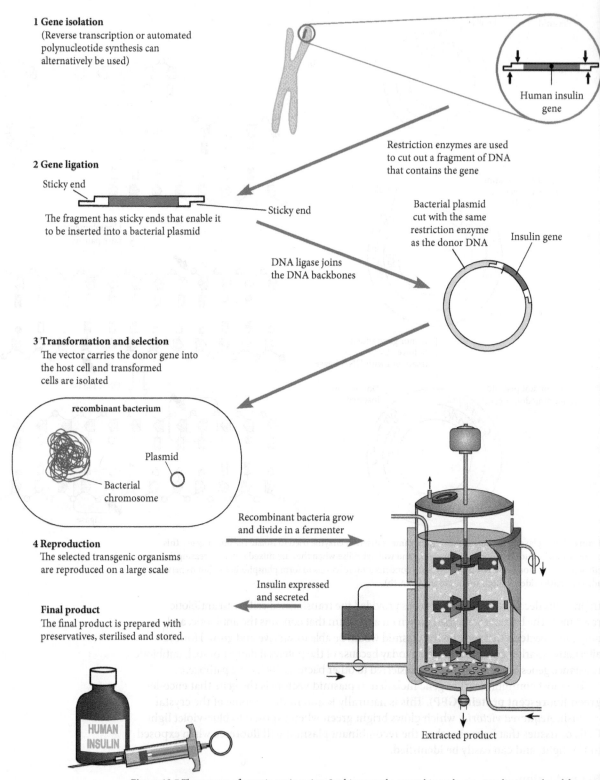

1 Gene isolation
(Reverse transcription or automated polynucleotide synthesis can alternatively be used)

Human insulin gene

Restriction enzymes are used to cut out a fragment of DNA that contains the gene

2 Gene ligation

Sticky end

The fragment has sticky ends that enable it to be inserted into a bacterial plasmid

Sticky end

Bacterial plasmid cut with the same restriction enzyme as the donor DNA

Insulin gene

DNA ligase joins the DNA backbones

3 Transformation and selection
The vector carries the donor gene into the host cell and transformed cells are isolated

recombinant bacterium

Plasmid

Bacterial chromosome

Recombinant bacteria grow and divide in a fermenter

4 Reproduction
The selected transgenic organisms are reproduced on a large scale

Insulin expressed and secreted

Final product
The final product is prepared with preservatives, sterilised and stored.

HUMAN INSULIN

Extracted product

Figure 19.7 The process of genetic engineering. In this example, recombinant human insulin is produced from transgenic bacteria.

Worked example

Recombinant human insulin was first made in transgenic *E. coli*, a bacterium. As well as the human insulin gene, the *lac* operon you encountered in Chapter 16 was also added. This contains a gene that encodes the enzyme β-galactosidase. This metabolises lactose, but it can also hydrolyse a synthetic, colourless substrate called X-gal when added to plates of transformed bacteria into a blue product.

a. Explain why the β-galactosidase gene is described as a marker gene.

Answer

When transcribed by cells that have absorbed them, marker genes produce products that can be easily identified. The presence of β-galactosidase enzyme in a cell will enable it to convert the colourless substrate X-gal into a blue product, allowing the scientist to identify the colonies that developed from this cell as being successfully transformed.

b. Using your knowledge of the *lac* operon, and assuming that the donor gene is controlled by this regulatory sequence, suggest another advantage of using β-galactosidase as a marker gene in bacteria.

Answer

The enzyme β-galactosidase is only produced in the presence of lactose. Therefore, it may be possible to 'turn on' the expression of the donor gene in transgenic organisms by adding lactose to the nutrient medium in which the bacteria reproduce.

---

4. Explain why a plasmid must:
   a. contain a variety of restriction sites
   b. contain a promoter next to these restriction sites
   c. contain an origin of replication
   d. contain a marker gene
   e. have a low molecular mass.

5. Before ligation, when the isolated gene and the plasmid vector are mixed, it is important to heat the mixtures to denature the restriction enzymes. Explain why.

6. Calculate the total number of phosphodiester bonds that must be formed when one cut plasmid and an isolated donor gene are joined together by DNA ligase.

GENE EDITING

New techniques in genetic technology are quickly adopted and adapted by research groups around the world. A recent example is **gene editing** (Figure 19.8).

Like restriction, gene editing is a natural mechanism that all bacteria use to defend themselves against viruses. It relies on base sequences in bacterial genomes called **clustered regularly interspaced short palindromic repeats** (CRISPR, pronounced 'Crisper'). These sequences encode short RNA molecules called **guide RNA** (gRNA), which attach to a bacterial endonuclease called **CRISPR-associated enzyme-9** (Cas9). This enzyme is able to bind to viral DNA molecules in the cytoplasm that have a complementary sequence to the gRNA. Cas9 then makes a double-stranded cut in the sugar-phosphate backbone in defence.

CRISPR sequences are 'molecular memories' of viral infections. In nature, they come from parts of genomes of viruses that infected ancestors of bacteria. When transcribed into gRNA, these sequences direct Cas9 towards viral DNA with complementary sequences. Once the virus is made harmless, the Cas9 enzyme again integrates parts of its DNA into the CRISPR regions of the bacterial genome. These new sequences are then transcribed to make more gRNA molecules, and the cycle of defence continues.

Importantly for gene technology, artificial gRNA molecules can be made in the laboratory. Automated polynucleotide synthesis can easily produce molecules with complementary sequence to any molecule of DNA. Guided by these molecules, scientists can use recombinant Cas9 to accurately cut DNA at a precise location, chosen by the scientist, in target DNA (Figure 19.9).

Figure 19.8 The molecular biologists Emmanuelle Charpentier, from France (left), and Dr Jennifer Doudna, an American (right). In 2012, they realised the potential of CRISPR/Cas9 as a tool for gene editing.

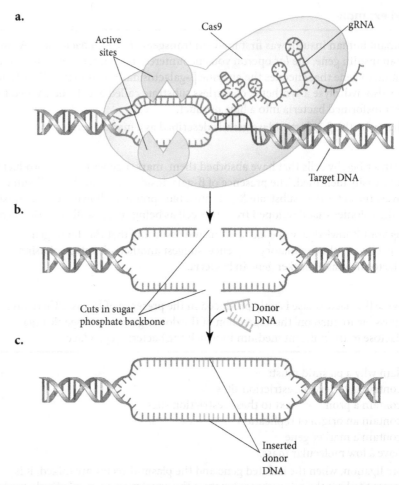

a.

Cas9

gRNA

Active sites

Target DNA

b.

Cuts in sugar phosphate backbone

Donor DNA

c.

Inserted donor DNA

Figure 19.9 The CRISPR-Cas9 system offers a 'programmable' DNA cutting mechanism. Cas9 (blue) is an endonuclease that cuts target DNA (black) on both strands in a specific region (green) determined by a guide RNA (gRNA; red). The gRNA has a sequence of 20 bases that can locate and bind Cas9 to a strand of DNA with the complementary base sequence (**a**). Hydrolysis of two phosphodiester bonds introduces a double-stranded break in the DNA molecule (**b**). Short segments of DNA can be added by the gene technologist that are inserted into the sites of the cut (**c**). Note that two different gRNA molecules would be required to remove ('delete') regions from the target DNA molecule (not shown).

It did not take long for the potential of these new molecular tools to be recognised. After target DNA is cut by Cas9, short lengths of purpose-built DNA can be inserted. This allows scientists to 'cut and paste' or 'rewrite' genes with far greater precision than was possible when using restriction enzymes and vectors. CRISPR-Cas9 technology has already been used to remove mutant alleles from genomes and insert new base sequences to treat genetic disorders. In 2019, a new process, called CRISPR-Prime, was developed that combines CRISPR/Cas9 and reverse transcription to improve the accuracy of the gene editing process. Gene editing promises to revolutionise existing gene therapy technologies, which you will encounter later in this chapter.

Worked example

CRISPR/Cas9 is a new technique that can be used to change the base sequence of DNA molecules. It uses guide RNA (gRNA) molecules with specific sequences of 20 bases to direct an endonuclease to cut DNA.

a. Determine the base sequence of the following guide RNA (gRNA) molecule that could be used to edit the sequence of DNA: CTAGCTCGATAAAGATCGAA.

Answer

GAUCGAGCUAUUUCUAGCUU

b. Outline why the use of the CRISPR-Cas9 enzyme may be preferred to the use of restriction enzymes in genetic engineering.

Answer

Although both restriction enzymes and CRISPR-Cas9 enzymes enable scientists to cut DNA at specific sequences, CRISPR-Cas9 can be directed to cut specific sequences by making a guide RNA (gRNA) complementary to that sequence.

---

7. Refer to Figure 19.9 to answer this question.
   a. Outline the process by which a bacterium produces a guide RNA (gRNA).
   b. Suggest a role for the loops, formed by complementary base pairing, within a gRNA molecule.

8. Suggest possible applications for editing the following genomes using CRISPR-Cas9 technology:
   a. cancer cells growing in the laboratory
   b. crop plants
   c. pathogens, such as *Plasmodium falciparum*.

Assignment 19.1: Genetically modified mosquitoes

A large number of serious diseases, including dengue and yellow fever, are spread by female mosquitoes through biting (Figure 19.10). Mosquitoes breed in bodies of still water, and the incidence of these mosquito-borne diseases has risen as the human population has increased.

QUESTIONS

A1. Suggest **two** reasons to explain why an increase in human population numbers leads to a rise in the mosquito numbers.

Gene technologists have attempted to combat the spread of mosquito-borne diseases by producing genetically modified male mosquitoes. In one study, a donor gene called tTA, which encodes a protein that kills cells in the developing wing muscles of female larvae, was introduced into the genome of these mosquitoes. In addition to the tTA gene, a marker gene that encodes a fluorescent product was also included.

Figure 19.10 The dengue virus is transmitted by the female *Aedes aegypti* mosquito. In Singapore, posters update neighbourhoods on the prevalence of the disease at different times of the year.

A2. With reference to this example, explain the role of marker genes in genetic technology.

A3. Suggest how the scientists could ensure that the tTA gene was only expressed in the developing wing muscles of female larvae.

A4. Suggest **two** reasons to explain why an inability to fly would reduce the spread of mosquito-borne viruses in humans.

★**A5.** Switching on the gene coding for tTA in the mosquito larvae, rather than in the eggs, increases the effectiveness of this method of mosquito control. Suggest why.

More recently, some researchers have explored the possibility of using gene editing to combat the spread of mosquito-borne diseases. In one study, female mosquitoes were produced that develop male characteristics. These females are unable to bite or lay eggs.

A6. Suggest the advantages of using gene editing, compared with genetic engineering using restriction enzymes, in the production of GM mosquitoes.

AMPLIFYING AND SEPARATING DNA

So far in this chapter, you have only considered how gene technologists make changes to DNA. However, in many cases, they use tools to simply analyse DNA, perhaps to detect mutations or to sequence it, and draw conclusions. Two techniques that are vital in this regard are the polymerase chain reaction (PCR) and gel electrophoresis.

The polymerase chain reaction (PCR)

Glance back to Figure 19.1. In the earliest examples of genetic engineering, scientists initially started their work with very few copies of the gene of interest. They extracted donor DNA from large quantities of cells, such as the human pancreas, in order to isolate the gene such as the one encoding insulin using restriction enzymes. Today, however, scientists only need small quantities of the donor DNA. A process of **cloning**, or **amplifying** them, is now possible in the laboratory. This process is called the **polymerase chain reaction (PCR)**.

PCR is an entirely automated process that occurs *in vitro* (outside living cells), in a special machine called a **thermal cycler**. It can be thought of as DNA replication that occurs in small tubes in the laboratory. It is remarkably efficient: it can make billions of copies of a strand of DNA in a few hours. Only a very small amount of DNA is needed at the start of the process – in theory, just a single molecule. It was developed in 1983 by Kary Mullis, for which discovery he won a Nobel Prize 10 years later. The requirements and method of PCR are summarised in Figure 19.11 and a thermal cycler ('PCR machine') is shown in Figure 19.12.

Figure 19.12 A thermal cycler, usually referred to as a 'PCR machine'.

Link

In Chapter 18 you encountered the kingdom Archaea. *Thermus aquaticus* belongs to this kingdom and shares it with many other species that can tolerate extreme environmental conditions.

Link

In Chapter 6 you studied the mechanism of DNA replication in the cell cycle.

Equipment

- A sample of the DNA molecule to be amplified (the 'template DNA')
- Excess (unlimited) nucleotides (with each of the four bases)
- Taq DNA polymerase enzyme
- Primers of the two different sequences
- Heat-conductive plastic tubes
- Thermal cycler and power supply

Method

1. Add the template DNA, nucleotides, Taq DNA polymerase and primers to the plastic tubes and mix contents thoroughly.

2. Place the plastic tubes into the thermal cycler. Switch it on.

3. After 2-3 hours, analyse the contents of the plastic tubes.

Figure 19.11 The equipment and method required to undertake PCR in the laboratory.

An important feature of PCR is that the DNA polymerase used in the process is **thermostable**. This particular DNA polymerase, **Taq DNA polymerase**, was derived from the thermophilic bacterium *Thermus aquaticus*, which grows naturally in hydrothermal springs at high temperatures. Therefore, it is not denatured by the high temperatures used in the process. It works in the same way as DNA polymerase from any other organism: it produces two identical strands from template DNA by moving along the separated strands, forming phosphodiester bonds between complementary nucleotides to make double-stranded DNA.

Another important ingredient in the PCR are the **primers**. These are included to ensure that the process does not copy the entire DNA template, which is in some cases the whole genome of an organism. Instead, a specific region of the template is amplified. To do this, two primers are required. These are single-stranded molecules of DNA of around 20 bases in length. These **anneal** (form complementary base pairs) to complementary sequences on the two DNA strands forming short lengths of double-stranded DNA. PCR primers have two purposes:

- *Taq* DNA polymerase, like all replication enzymes, requires some existing double-stranded DNA to begin the process of DNA replication.
- Only the DNA between the primer sequences is replicated, so by choosing appropriate primers it is possible to ensure that only a specific target sequence is copied. They essentially 'signal' to *Taq* DNA polymerase where to start and stop copying.

The molecular mechanism of PCR is outlined in Figure 19.13.

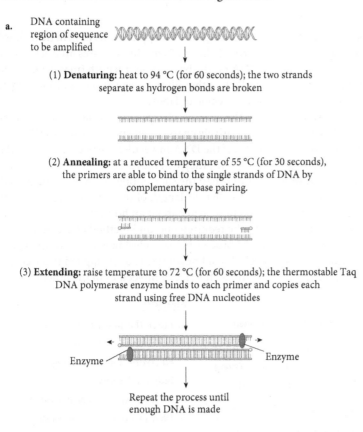

a. DNA containing region of sequence to be amplified

(1) **Denaturing:** heat to 94 °C (for 60 seconds); the two strands separate as hydrogen bonds are broken

(2) **Annealing:** at a reduced temperature of 55 °C (for 30 seconds), the primers are able to bind to the single strands of DNA by complementary base pairing.

(3) **Extending:** raise temperature to 72 °C (for 60 seconds); the thermostable Taq DNA polymerase enzyme binds to each primer and copies each strand using free DNA nucleotides

Enzyme   Enzyme

Repeat the process until enough DNA is made

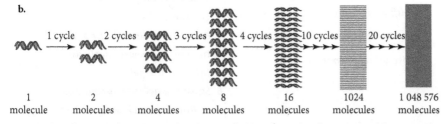

b.

| 1 cycle | 2 cycles | 3 cycles | 4 cycles | 10 cycles | 20 cycles |

| 1 molecule | 2 molecules | 4 molecules | 8 molecules | 16 molecules | 1024 molecules | 1 048 576 molecules |

Figure 19.13 The polymerase chain reaction (PCR), controlled by a thermal cycler, consists of three steps that comprise one cycle, and these 2–3-minute cycles are repeated 20–30 times (a). In a second cycle of reactions, there will be four double-stranded copies for every original, and at the end of the third cycle there will be eight, and so on (after n cycles, there is an amplification factor of 2^n) (b). This is why it is called a chain reaction, as the number of molecules increases exponentially.

Such was its capacity to amplify DNA, the revolutionary procedure of PCR quickly led to further, and sometimes unexpected, applications. Professor Alec Jeffreys famously discovered the technique of DNA fingerprinting by accident in 1984. Developments since have made it possible to amplify the tiniest amounts of DNA, for example, even molecules extracted from the remains of extinct woolly mammoths (see the introduction to Chapter 6). Many crimes have been solved with the help of PCR, using DNA extracted from tiny traces of biological material.

9. Suggest the optimum temperature of *Taq* DNA polymerase.
10. How many cycles of PCR would be needed to produce one million copies of a DNA molecule, starting from just one DNA molecule?
11. a. The DNA template used in PCR is a polynucleotide, whereas the primers used in PCR are referred to as oligonucleotides. Suggest the meaning of the term oligonucleotide.
    b. Explain the importance of primers in PCR.

Gel electrophoresis

Another laboratory technique that has significant applications in genetic technology is **gel electrophoresis**. This is a form of chromatography that uses an electric field to separate different molecules of DNA on the basis of their length (mass). Often these pieces of DNA are obtained from PCR, although they may also be produced by the action of restriction enzymes on a larger molecule of DNA.

Gel electrophoresis depends on the negative charge of DNA molecules. Each nucleotide in a molecule of DNA contains a negatively charged phosphate group, so DNA is attracted to the positive electrode (the anode). If the DNA molecules are placed into a well in the gel near the negative electrode (the cathode), they will move from negative to positive, with longer lengths of DNA moving slower than the shorter ones as they are all 'pulled' through the gel by the electric field. So, the smaller the length of the DNA molecule, the further along the gel it will move in a given time. After about an hour, the current is turned off and the DNA molecules stop moving. The requirements and method of gel electrophoresis are summarised in Figure 19.14 and the process being undertaken is shown in Figure 19.15. Note that a mixture of DNA molecules of known size is often applied to a well at one end of the gel, so that scientists can estimate the size of molecules in the other lanes.

Link

You last encountered chromatography in Chapter 13, when you analysed the photosynthetic pigments in leaf cells on the basis of their molecular mass.

Equipment	Method
• DNA samples to be analysed	1. Place the gel into the tank.
• A thin slab of gel made of agarose or polyacrylamide, containing wells prepared at one end	2. Cover the gel in an electrolytic (conducting) buffer solution.
• Electrophoresis tank	3. Using a micropipette, add the DNA samples into the wells.
• Electrolyte (conducting) buffer solution	4. Switch on the power supply (around 75 V).
• A micropipette	5. After approximately 1 hour, turn off the power supply and take out the gel.
• Dye solution to stain DNA	6. Stain the gel with a coloured or fluorescent chemical or add single-stranded radioactive DNA probes to visualise the DNA fragments as bands.

Figure 19.14 The equipment and method required to undertake PCR in the laboratory.

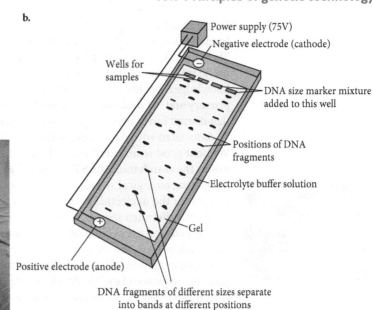

b.

Power supply (75V)

Negative electrode (cathode)

Wells for samples

DNA size marker mixture added to this well

Positions of DNA fragments

Electrolyte buffer solution

Gel

Positive electrode (anode)

DNA fragments of different sizes separate into bands at different positions

a.

Figure 19.15 Using a micropipette to load the wells in agarose gel with samples of DNA mixed with blue tracking dye (**a**). During gel electrophoresis, an electric field is applied between two electrodes, which causes negatively charged DNA molecules to move to the anode (positive electrode). The distance travelled by a molecule is proportional to its length, which can be estimated by comparing the distance travelled by DNA molecules of known length in a DNA size marker mixture (**b**).

As you will see in the remainder of this chapter, gel electrophoresis has useful applications in genetic technology. In all cases, it relies on the ability to determine the different sizes of fragments of DNA.

STORING AND ANALYSING GENETIC INFORMATION

By its very nature, genetic technology involves incredibly large numbers. The human genome, for example, consists of three billion (3×10^9) **base pairs**, and this is present twice in each cell. A human genome sequence was first published in 2003, which concluded a collaborative and highly energetic decade-long project involving research centres in the United States, the United Kingdom, Japan, France, Germany and China. Today, after some great advances in genetic technology, it is possible to sequence a human genome in less than one day. The Wellcome Sanger Institute near Cambridge (UK) is estimated to produce 12 billion bases of data per day. The collection, processing and analysis of biological information using data and computer software is called **bioinformatics**.

An ongoing challenge associated with the development of bioinformatics has been to develop computer hardware that can store and manage vast quantities of data. Open-access databases have been built, which not only store the sequences of genomes, but also proteomes and protein sequences and structures (Table 19.3). Some databases hold very specific information. For example, COSMIC, the Catalogue of Somatic Mutations in Cancer, is a database that holds records of the mutations detected in the genomes of human cancer cells.

Link

More than 98.5% of the human genome sequence has been found not to encode proteins. You have encountered some of this non-coding DNA in Chapter 6 (introns) and Chapter 16 (promoter regions).

Name of database	Type of data stored
The European Molecular Biology Laboratory (EMBL)	nucleotide sequences of genes and genome sequences
Universal Protein Resource (UniProt)	primary sequences of proteins and the functions of many proteins
Protein Data Bank (PDB)	amino acid sequences of proteins and details of their three-dimensional structures

Table 19.3 Three examples of databases used in bioinformatics to store data.

Research centres such as the Wellcome Sanger Institute not only sequence and analyse human DNA and proteins. They also use information from a range of human pathogens. These include the parasite that causes malaria, *Plasmodium*. Studies like this can reveal very useful information about the proteome – the proteins that the organism can make. This knowledge is beneficial in a number of ways. For example, finding genes, and predicting the primary and 3D structure of proteins they code for, could help develop vaccines and drugs that disrupt functions of proteins in pathogens when they infect humans.

Associated with databases are the tools for selection and retrieval of information. Without good search and retrieval tools, all the information stored would be of little value. The search tool **BLAST** (**Basic Local Alignment Search Tool**), is an algorithm for comparing biological sequence information, such as the amino acid sequences of different proteins or the nucleotide base sequences of genes. Researchers use BLAST to find similarities between sequences that they are studying and those already saved in databases.

One practical use of bioinformatics is the DNA **microarray**. A microarray is a small, solid chip made of glass, silicon or nylon (Figure 19.16). To the surface of this chip, thousands of short single-stranded DNA (ssDNA) molecules, called **probes**, are fixed as spots in a regular grid. The sequence of each molecule is unique to a different gene in a genome. Using this chip, it is possible to test for the expression of a very large number of genes simultaneously and very quickly. This is done by adding to the microarray DNA obtained from a specific cell, which hybridises to the fixed probes only if it has a complementary sequence. An example of the use of a DNA microarray is shown in Figure 19.17.

Figure 19.16 Although there are many different designs, most microarray chips are about the size of a microscope slide.

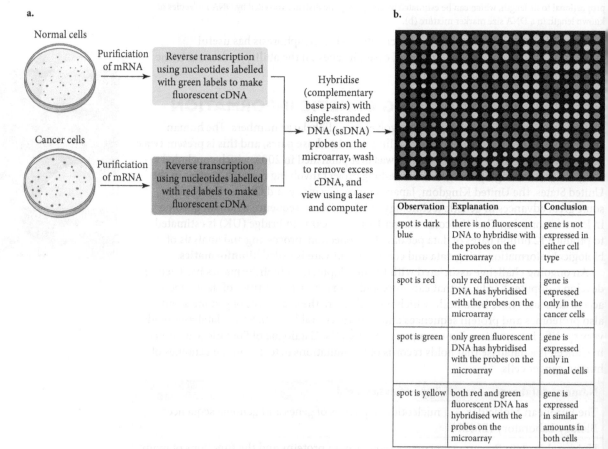

Figure 19.17 The process of microarray analysis enables scientists to analyse patterns of gene expression in a cell. Here, the expression of genes in a cancerous cell compared to that in a healthy (control) cell is investigated. This is carried out by labelling DNA produced by reverse transcription of mRNA extracted from cancer and normal cells with different coloured tags, and applying both samples to the same microarray (**a**). After laser fluorescence, a computer takes a digital image of the microarray, which can then be analysed. Each dot corresponds to a unique DNA sequence found in a gene, where the colour of the spot corresponds to the expression of the gene (**b**).

One hope is that bioinformatics and the use of microarrays may help scientists to develop **personalised medicines** to treat people with diseases caused by a wide variety of different mutations.

12. Which diagram in Figure 19.18 shows the correct arrangement of the gel and the electrodes during electrophoresis? Refer to Figure 19.15 to help you with this question.

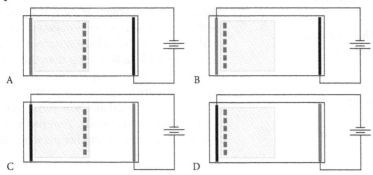

Figure 19.18

13. Suggest and explain why it is more difficult to determine the proteome of a human than it is to determine the proteome of a bacterium from their genomes.

14. Apart from the example described in Figure 19.17, suggest three uses of DNA microarrays.

15. The Wellcome Sanger Institute produces 12 billion bases of data per day.
    a. Estimate the number of bases sequenced in one year. Give your answer in standard form.
    b. Describe the role of databases in the storage and analysis of this data.
    c. Explain some of the benefits of analysing the three-dimensional structures of proteins.

Experimental skills 19.1: Coral bleaching: clues from a microarray study

In recent years, a phenomenon known as coral bleaching has become a growing problem. Colourful algae associated with coral colonies die, resulting in a loss of the coral's photosynthetic nutrient source. To date, bleaching has killed around 50% of the coral in the Great Barrier Reef in Australia and has affected reefs elsewhere from the Caribbean to Japan. Experts warn that the loss of reefs may result in a catastrophic reduction in marine biodiversity and cause devastation to human societies that depend on them.

Some evidence suggests that the increase in coral bleaching may be related to an increase in mean global temperatures. Scientists in the United States used microarray analysis to investigate the effects of water temperature on gene expression patterns in coral.

QUESTIONS

P1. Identify the null hypothesis (H_0) that the scientists should make for this investigation.

The scientists used the following method:
- At 2.00 p.m. each day for 12 consecutive days, a coral branch of length 10–30 mm was cut from each of the same three colonies of different species of coral.
- At the same time, a temperature probe was used to record the temperature of the water 1 metre away from the coral colony.
- The branch was transferred to a tube and was stored at –20 °C.
- At the end of the study, reverse transcription of the mRNA contained in 200 mg of each sample was carried out using fluorescent nucleotides to produce cDNA.
- Each fluorescent cDNA sample was heated to 95 °C and then transferred to separate coral DNA microarrays. A digital image was captured for analysis.

P2. Calculate:

    a. the total mass of coral analysed in this investigation

    b. the number of digital images that were captured in this investigation.

P3. Suggest why the branches were stored at −20 °C until the RNA was extracted.

P4. Before the cDNA is applied to the microarray it must be heated to a high temperature. Explain why.

Figure 19.19 shows the results of this investigation. The scientists published the expression patterns of a specific group of genes. They concluded that temperatures above 30.5 °C are responsible for corals increasing the expression of these genes.

P5. Calculate the percentage increase in gene expression of species C between day 6 and day 8 of the study.

☆**P6.** Evaluate the claim that marine temperature directly affects the expression of the group of genes.

☆**P7.** Another group of scientists wanted to compare the gene expression patterns of two species of coral, one of which has been observed to be more susceptible to bleaching than the other. Describe how microarray analysis could be used to compare the gene expression patterns of two species of coral and how you could draw conclusions from this study. Your method should be detailed enough for another person to follow.

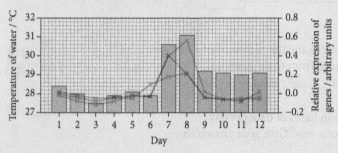

Figure 19.19

Key ideas

→ Genetic technology involves using a variety of techniques to investigate the sequence of nucleotides in DNA and alter an organism's DNA. Genetic engineering involves extracting DNA from one organism and placing it into the DNA of another to form recombinant DNA. Gene(s) must be inserted in such a way that they will be expressed in the genetically modified organism (GMO).

→ Restriction endonucleases cut across DNA at specific restriction sites: staggered cuts give rise to sticky ends; straight cuts give blunt ends. Pieces of DNA with sticky ends that are complementary to each other are able to join together by forming hydrogen bonds. DNA ligase joins together the sugar–phosphate backbones of pieces of DNA by catalysing the formation of phosphodiester bonds.

→ Vectors, such as plasmids, viruses and liposomes, are used to carry pieces of DNA into cells.

→ Plasmids are small circles of double-stranded DNA that can be cut with restriction enzymes and have promoters and marker genes (e.g. genes for antibiotics or GFP) inserted into them alongside the gene(s) to transform the host cell.

→ A promoter must be inserted alongside the gene because organisms will not transcribe and express a gene unless there is a binding site for RNA polymerase.

→ Cells that have taken up plasmids with the desired gene can be identified by detecting fluorescence (GFP) or appropriate staining.

→ Lengths of DNA for genetic modification can be synthesised directly from mRNA by using the enzyme reverse transcriptase. Specific lengths of DNA can also be synthesised from nucleotides using knowledge of the genetic code.

→ Gene editing is a form of genetic engineering in which DNA can be inserted into a genome using the CRISPR-Cas9 system.

→ Electrophoresis is used to separate fragments of DNA of different lengths; the material to be tested is placed in wells in agarose gel and a voltage applied across the gel.

→ In the polymerase chain reaction (PCR) large numbers of copies of DNA are made from very small quantities. Heat-stable DNA polymerases such as *Taq* polymerase are used.

→ A gene probe is a length of single-stranded DNA, which has a known base sequence and is used to hybridise with lengths of DNA that have the complementary sequence; probes are labelled in some way to make them 'visible' (e.g. with radioactive phosphorus). PCR and gene probes are used in forensic investigations.

→ Microarrays containing many thousands of gene probes are used to analyse genes in different genomes and detect the presence of mRNA from cells to detect gene expression at any one time.

→ Databases hold huge amounts of information on DNA and proteins. Bioinformatics deals with the storage and analysis of biological data; researchers use the data to compare genes from different organisms, and to search for mutant alleles that cause genetic diseases.

19.2 Genetic technology applied to medicine

Genetic technology has made some significant progress in the development of the diagnosis and treatment of disease. In addition to the techniques you have seen so far in this chapter, including database analysis and microarray scans, there are many others. These include recombinant protein synthesis, genetic screening and gene therapy, which you will now consider. However, it is important to understand that in all three cases, these technologies are currently applicable to a small number of genetic diseases. These are examples for which scientists know the precise underlying genetic cause, and, in general, those caused by defects in a single gene.

RECOMBINANT PROTEIN SYNTHESIS

You will recall from earlier in this chapter that the first human protein produced by genetic engineering was insulin. This was chosen for a number of reasons: globally, type I diabetes affects millions of people, and the insulin protein is small and produced from a gene with only two introns. But which other genetic disorders have been treated using **recombinant protein synthesis** by transgenic organisms? Table 19.4 summarises the context of recombinant insulin and two other recombinant proteins that are produced today.

Recombinant protein, gene name and locus	Genetic disease for which it is produced to treat	Issues with previous form of treatment that were overcome using recombinant protein
Insulin (*INS*, chromosome 11)	Type I diabetes. Individual cannot produce insulin as the beta cells in the islets of Langerhans have died.	Insulin was previously purified from the pancreas of slaughtered animals. This has cultural and ethical sensitivities.
Factor VIII (*F8*, chromosome X)	Haemophilia. Individuals cannot produce a blood clotting factor. This can mean that minor haemorrhages (cuts and bruises) can be deadly.	Factor VIII used to be extracted from donor blood. This poses the risk of transmission of pathogens including HIV and the hepatitis C virus.
Adenosine deaminase (*ADA*, chromosome 20)	ADA-dependent severe combined immunodeficiency (ADA-SCID). Individuals cannot produce a metabolic enzyme. This results in the death of lymphocytes and an absent immune response.	No traditional method previously available. Babies with ADA-SCID usually died very early due to infection (see later).

Table 19.4 Examples of three recombinant human proteins used to treat genetic diseases caused by recessive alleles.

Link

You have encountered the mechanisms that underlie some of the diseases described in Table 19.4 in Chapter 14 (diabetes) and Chapter 16 (haemophilia).

Another advantage of using recombinant human proteins is that they can be continually produced on an industrial scale by transgenic microorganisms in fermenters. These proteins also have the same primary, and therefore three-dimensional, structure as natural proteins, and are hence biologically active in the human body.

GENETIC SCREENING

As you saw in Chapter 16, a genotype is the combination of alleles that an individual has for a specific gene, of which they will usually have two copies. One of the most common applications of genetic technology is quick and accurate analysis of an individual's genotype. This is called **genetic screening**, or genetic testing. This involves analysing a sample of biological tissue taken from the blood, or cheek cells in adults, or cells obtained from an embryo or the placenta of an unborn child.

Genetic screening is possible because scientists know the function of a large number of genes. In some cases, they know exactly how the base sequence is different in the different versions (or alleles) of the genes that cause genetic diseases. Table 19.5 gives an overview of three diseases for which genetic screening is currently offered.

There are several methods used in genetic screening. One of the simplest involves gel electrophoresis. This can be used to distinguish between different alleles of a gene that have different lengths: DNA strands containing different alleles will be positioned at different places on the gel after electrophoresis. For example, in Huntington's disease, dominant, disease-causing alleles will have more bases than recessive alleles. This, as you saw in Chapter 16, means that DNA fragments containing these alleles are larger and will move less distance on the gel when analysed.

Disease and gene(s)	Characteristics of disease and basis of genetic screening
Breast cancer susceptibility (*BRCA1* and *BRCA2*)	In some countries, breast cancer affects up to one in eight women during their lifetime. There are a great number of different mutations that have been identified in the *BRCA1* and *BRCA2* genes that increase the risk of developing breast cancer. Screening can help to choose the best treatment for a particular patient. For example, the drug Herceptin is only effective at treating a certain type of breast cancer, caused by mutation in the *HER2* gene, so would be ineffective in treating patients with mutations in *BRCA1* or *BRCA2*.
Huntington's disease (*HTT*)	Huntington's disease (HD) was explored in Chapter 16. Because HD is caused by a dominant allele, a positive test means that an individual will definitely develop the disease, and there is at least a 50% chance that they will pass the disease on to any children. The test can also indicate the likely severity of the disease: more CAG repeats (longer alleles) are associated with an earlier age of onset.
Cystic fibrosis (*CFTR*)	If the test reveals that the prospective father and mother both carry one mutant *CFTR* allele (are heterozygous), then there is a one-in-four chance that any child they have will suffer from cystic fibrosis.

Table 19.5 Three examples of genetic diseases for which genetic screening is available.

Individuals who undergo genetic screening can be grouped into two categories. There are many advantages associated with undertaking the procedure.

- Genetic screening of an individual for an inherited genetic disease that usually develops in later life, such as Huntington's disease and breast cancer. A positive result for this test may provide the individual with the opportunity to make lifestyle changes to delay the onset of the disease, or seek more regular check-ups or even pre-emptive surgery. The person may also decide to avoid starting a family.
- Genetic screening of an individual or couple to estimate the chance of an inherited genetic disease affecting a child, such as cystic fibrosis. A positive result for such a test will better enable parents or couples planning a family to make some difficult decisions. These include whether or not to have a child, to use donor eggs or sperm, or whether to terminate an affected pregnancy. If they decide to pursue the pregnancy, a positive diagnosis does provide an opportunity to make adjustments to living arrangements and financial planning to prepare to care for a child with a disability.

There are, however, significant **social issues** and **ethical issues** to consider related to genetic screening. It is now possible in many countries to carry out *in vitro* fertilisation (IVF) and an **embryo biopsy** to test whether an embryo has inherited an allele from its parent(s). This technique is called **pre-implantation genetic diagnosis** (PGD) and generally involves the destruction of embryos that are either found to be affected by a genetic disease, or not choosing these to be implanted for pregnancy.

This has raised some significant questions. Do all embryos and unborn children have a right to life? Could this lead to a trend in which people are selecting the sex of the embryo that they choose to implant? Even a genetic screen of a child or adult raises some challenging questions. Would an individual be able to deal with the anxiety of knowing that symptoms for an incurable disease would begin in the next few years? Usually, the results of a genetic screen are communicated to the individual very sensitively by a highly trained genetic counsellor. He or she will explain the meaning of the test results, discuss the implications for the patient and their family, and advise on the next course of action.

> **Tip**
>
> Social considerations are issues that relate to an individual and his or her relationships with other people. Ethical considerations relate to a set of agreed standards, to determine what is acceptable in society.

One final issue related to the social and ethical nature of genetic screening concerns the confidentiality of the results. Clearly, the individual who has been screened must be informed, but what of their parents, siblings and children, some of whom are also likely to have the allele? Some people are also very concerned about the possibility of this information being obtained by employers, and mortgage or insurance providers. Many countries have already passed laws to protect individuals against discrimination on the basis of their genotype.

GENE THERAPY

Link

Liposomes are another type of vector that can be used to deliver alleles in gene therapy. You encountered these spherical structures, which have a phospholipid bilayer envelope, in Chapter 4.

Throughout the late 1980s and the 1990s, scientific journals published the loci (position) and sequences of a range of genes associated with serious recessive diseases. This caught the imagination of researchers. Might it be possible to insert a working copy of an allele into an individual who has a recessive genetic disorder? They hoped that this would compensate for the lack of a functional protein in their cells. **Gene therapy** is the name of this approach.

Gene therapy sounds simple, but achieving success is easier said than done. Scientists can easily isolate donor DNA containing the healthy allele, but a significant problem is getting the genes into the exact cells that need them, and then making sure that they are expressed. To be effective, an appropriate vector must be used to deliver the allele to the cells that need it. Here you will consider how viruses have been used to deliver alleles to treat ADA-dependent severe combined immunodeficiency (ADA-SCID) and some types of inherited eye disease.

Using gene therapy to treat ADA-SCID

In 1990, Raj and Van DeSilva were desperate. Their young daughter, Ashanthi, was dying. For much of her early childhood growing up in the United States she was confined to a 'plastic bubble' – an isolated room in a hospital, in which the air supply was kept sterile. Ashanti had ADA-SCID, which is caused by having a homozygous recessive genotype for a mutant allele for the adenosine deaminase (ADA) enzyme gene. In the absence of this enzyme, Ashanthi's lymphocyte count had depleted. This is because the substrate of the missing enzyme accumulated in cells and lymphocytes are particularly sensitive to its toxic effects.

Link

Viruses need host cells to express their DNA to make new virus particles. (Chapters 1 and 10). Viruses used in gene therapy have been modified to prevent them causing disease.

Ashanthi DeSilva was the first patient to be treated using gene therapy. Scientists removed ADA-deficient lymphocytes from her bone marrow and grew them in the laboratory. These cells were then infected with a genetically modified retrovirus containing a dominant, functional allele of the *ADA* gene. Ashanthi was given four doses of these cells, and over the four months of the treatment, her condition improved. Today, she is a healthy woman leading a normal life.

Following on from the success of Ashanthi's treatment, many more gene therapy trials were commissioned in the 1990s. Unfortunately, in France in 2000, four children who had received gene therapy for another form of SCID developed leukaemia as a result of using retrovirus as a vector. Retroviruses insert their genes into the host's genome (Figure 19.20a). However, this is at random. Sometimes, another gene or promoter of a gene is disrupted by the inserted DNA, and the cell becomes cancerous. Scientists began to explore the use of other viruses, such as the **adenovirus** (Figure 19.20b), which does not insert its genes into the host genome.

Using gene therapy to treat hereditary eye diseases

The adenovirus and another virus (vaccinia) have been used to treat a type of hereditary blindness, Leber's congenital amaurosis (LCA). Patients with LCA are homozygous recessive for mutant alleles for the *RPE65* gene and lack a light-sensitive protein pigment in the retina. In 2007, the retina of patients was injected with a suspension of virus containing the normal allele of this gene. People who had been blind from a young age have been able to see again. The success of these trials is due to a number of reasons. The eye is considered to be a good organ for developing gene therapies because it is small and easy to target. Also, there is little activity of immune cells inside the eye, so there is a low risk of harmful immune responses to the vector.

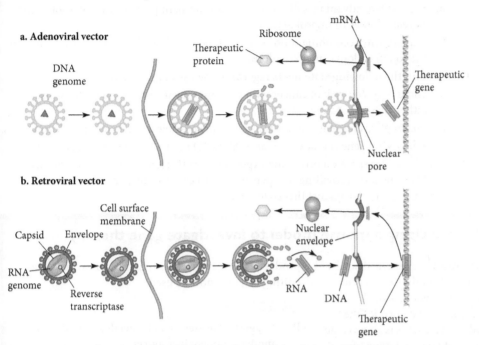

Figure 19.20 Treating disease using gene therapy with adenoviral (**a**) and retroviral (**b**) vectors.

Similar to genetic screening, there are many social and ethical considerations associated with the process of gene therapy. It is important here to make a distinction between **somatic gene therapy** and **germline gene therapy**. All attempts of gene therapy so far have involved placing the allele in body cells, known as **somatic cells**. Therefore, the eggs or sperm of the patient would still carry the defective alleles which could be passed on to future generations. Inserting the allele into **germ cells**, defined as the gametes or cells of the very early embryo, would result in all of these cells inheriting the working copy of the allele, and the donor gene being passed on to subsequent generations. Many people are opposed to this. They suggest that the long-term effects of such genetic modifications are not known. Also, they suggest that we might find it difficult to decide when to stop – and this may result in people paying money to create 'designer babies', with modified characteristics such as intelligence, height or hair colour. Germline gene therapy, and more recently germline gene editing, is therefore currently illegal in most countries.

Worked example

List some of the advantages of genetic screening an individual for breast cancer susceptibility caused by mutations in the *BRCA1* and *BRCA2* genes.

Answer

If an individual receives a positive test for increased breast cancer susceptibility, they may decide to have pre-emptive surgery (breast removal, called a mastectomy), or at least go to the doctor for more regular check-ups to identify any early signs of tumours. Knowing if a patient has a mutation in one of these genes may allow a doctor to better personalise medicine if the disease develops, by prescribing medicines that are known to be successful at treating cancers caused by mutations in one of these genes.

★ **16.** Recombinant human insulin is normally produced in yeast, factor VIII in hamster cells, and adenosine deaminase in the cells of an insect larva grown in culture. In all cases, the human gene is combined with a sequence of DNA that allows it to be translated by ribosomes attached to the rough endoplasmic reticulum, rather than free ribosomes in the cytoplasm.

    a. Suggest two advantages of producing recombinant proteins in eukaryotes, such as yeast, rather than prokaryotic cells.

    b. Explain why recombinant proteins must be translated by ribosomes attached to the rough endoplasmic reticulum.

17. Explain the following statements regarding the use of gene therapy:

    a. There is greater risk of cancer with gene therapy if a retroviral vector is used rather than an adenoviral vector.

    b. Treatment using gene therapy is usually only possible for recessive genetic disorders.

    c. Adenoviral gene therapy to treat ADA-SCID has to be repeated every few years.

🔎**18.** 'The resources spent on providing expensive genetic technology to treat the rich could be spent on providing inexpensive health care for the poor.' Discuss the arguments for and against this statement.

Experimental skills 19.2: Using a mouse model to investigate gene therapy

Phenylketonuria (PKU) is a recessive genetic disorder in which dangerously high levels of the amino acid phenylalanine accumulate in the blood. This is because of a mutation in the gene that encodes the protein phenylalanine hydroxylase. This is an enzyme that catalyses the conversion of phenylalanine into another amino acid, tyrosine.

    A number of investigations have been carried out into the use of gene therapy to treat mice that have phenylketonuria. In one study, genetically modified mice containing two mutant alleles for the gene encoding phenylalanine hydroxylase were produced. Mouse models of disease have traditionally been

produced using restriction enzymes. However, gene editing can also now be used.

QUESTIONS

★**P1.** Suggest what steps will be needed to make mouse models of phenylketonuria.

★**P2.** Outline how gene editing could be used to produce mouse models of phenylketonuria.

Gene therapy involves introducing working (non-mutated) versions of a gene into an organism's cells. The graph in Figure 19.21 shows the results of one investigation using a mouse model of PKU.

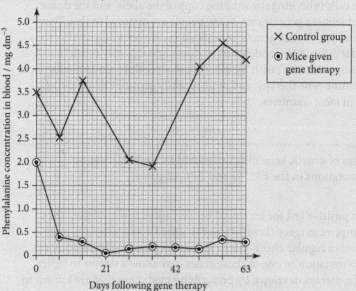

Figure 19.21 The effect of gene therapy on the phenylalanine concentration of the blood of mice with PKU.

P3. Describe the effect that gene therapy has on the phenylalanine concentration in the blood of these mice.

P4. Explain the effect that gene therapy has on the phenylalanine concentration in the blood of these mice.

P5. Suggest reasons for the variation in the concentration of phenylalanine over the course of the study in the control mice.

Tip

The purpose of a control experiment in an investigation is to remove the effect of the independent variable, to prove that it is this which has an effect on the dependent variable.

P6. Suggest what treatment the control group of mice may have been given in this investigation.

Key ideas

→ Many recombinant human proteins are now produced by GMOs, such as bacteria and yeasts. This makes available drugs, such as recombinant adenosine deaminase to treat SCID and recombinant factor VIII to treat haemophilia. This production also boosts the supply of others that were previously available from other sources, such as insulin.

→ Genetic screening involves testing people to find out if they carry any faulty alleles for genes that can cause disease; there are genetic tests for many genetic diseases including breast cancer associated with *BRCA1* and *BRCA2*, Huntington's disease and cystic fibrosis. Genetic counsellors may help people who find that they or their unborn child have a disease-causing allele, to make a decision about how to act on this information.

→ Gene therapy involves the addition to human or animal cells of functional alleles of genes that can cure, or reduce the symptoms of genetic diseases, such as SCID and some inherited eye diseases. Successful gene therapy involves selecting a suitable vector, such as viruses or liposomes, or inserting DNA directly into cells.

19.3 Genetically modified organisms in agriculture

Worldwide, nearly a billion people are malnourished. The United Nations predicts that by 2050, 70% more food must be produced in order to adequately feed the growing global population. Meanwhile, farmland is increasingly under threat from industrialisation and harvests are failing more frequently due to extreme weather patterns. In this context, some have suggested that genetic technology could help to solve the global demand for food. Genetically modified (GM) crop plants and animals offer promise to enhance pest and environmental resistance, and even their nutritional content (Figure 19.22). However, this suggestion is fraught with controversy. You will consider these concerns later.

It is worth remembering that humans have been changing the genetic makeup of species by practising selective breeding for millennia. Our ancestors kept seeds from the best crops and planted them in subsequent years, and restricted which animals were able to mate. This has altered the genetic makeup of many types of plant and animals. However, as you saw earlier in this chapter, gene technologists can now take a donor gene from any living organism and insert it into any other organism. This has been done in plants for a variety of purposes and has produced GM crop varieties that are now grown on land all over the world (Figure 19.23).

Figure 19.22 Growing genetically modified cotton (*Gossypium hirsutum*) has enabled farmers to harvest higher yields and enhance their standard of living.

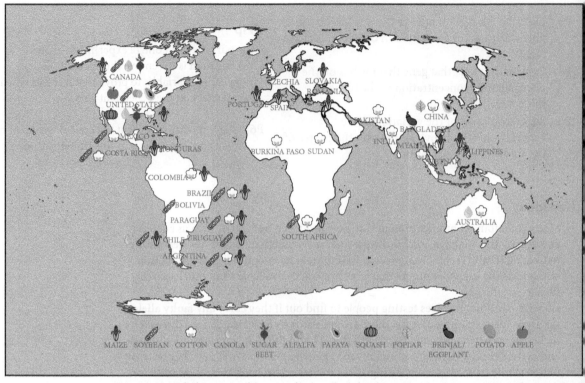

Figure 19.23 By the beginning of the 2020s, land in all inhabited continents has been used to grow GM crops.

To explore the reasons why genetically modified organisms have been developed for food production, you will consider three case studies. These are insect resistance in cotton and herbicide resistance in soybean plants, and GM salmon.

INSECT RESISTANCE IN COTTON

Take a look back at Figure 19.22. This is **Bt cotton**. *Bt* is short for *Bacillus thuringiensis*, a soil bacterium. This is the donor organism of the *cry* gene, which encodes a protein that acts as a very effective **insecticide**. *Bt* cotton expresses the *cry* gene and therefore kills any insect that feeds on it. This only affects the digestive system of insects, so it is considered harmless to other organisms, including humans. *Bt* cotton has significantly reduced the yield of cotton lost to pests such as boll worm and the boll weevil bore.

HERBICIDE RESISTANCE IN SOYBEAN

Glyphosate is a powerful synthetic **herbicide**, which can be applied to soil to kill weeds. Weeds compete with crop plants for space, light, water and nutrients from the soil, and so reduce yield. Glyphosate is absorbed by roots and is translocated to the growing tips of plants, where it inhibits an enzyme involved in the production of amino acids. Protein synthesis therefore cannot occur, and the plant dies. In some varieties of GM soybean (*Glycine max*), a bacterial gene coding for an enzyme that inactivates glyphosate has been introduced. These plants are now 'herbicide resistant' because they are able to degrade the glyphosate molecules before they can inhibit the production of amino acids. Spraying glyphosate on fields of these crops will therefore ensure that weed competition is kept to a low level.

There are many opponents of the development and use of genetically modified crop plants such as insect resistant cotton and herbicide resistant soybean. Many argue that the solution to global hunger and malnutrition lies in redistributing existing food supplies and solving political issues. They claim that transgenic products are being rushed to market before their implications are fully understood. Figure 19.24 summarises some of the most common arguments for and against the use of genetically modified organisms (GMOs) in crop farming.

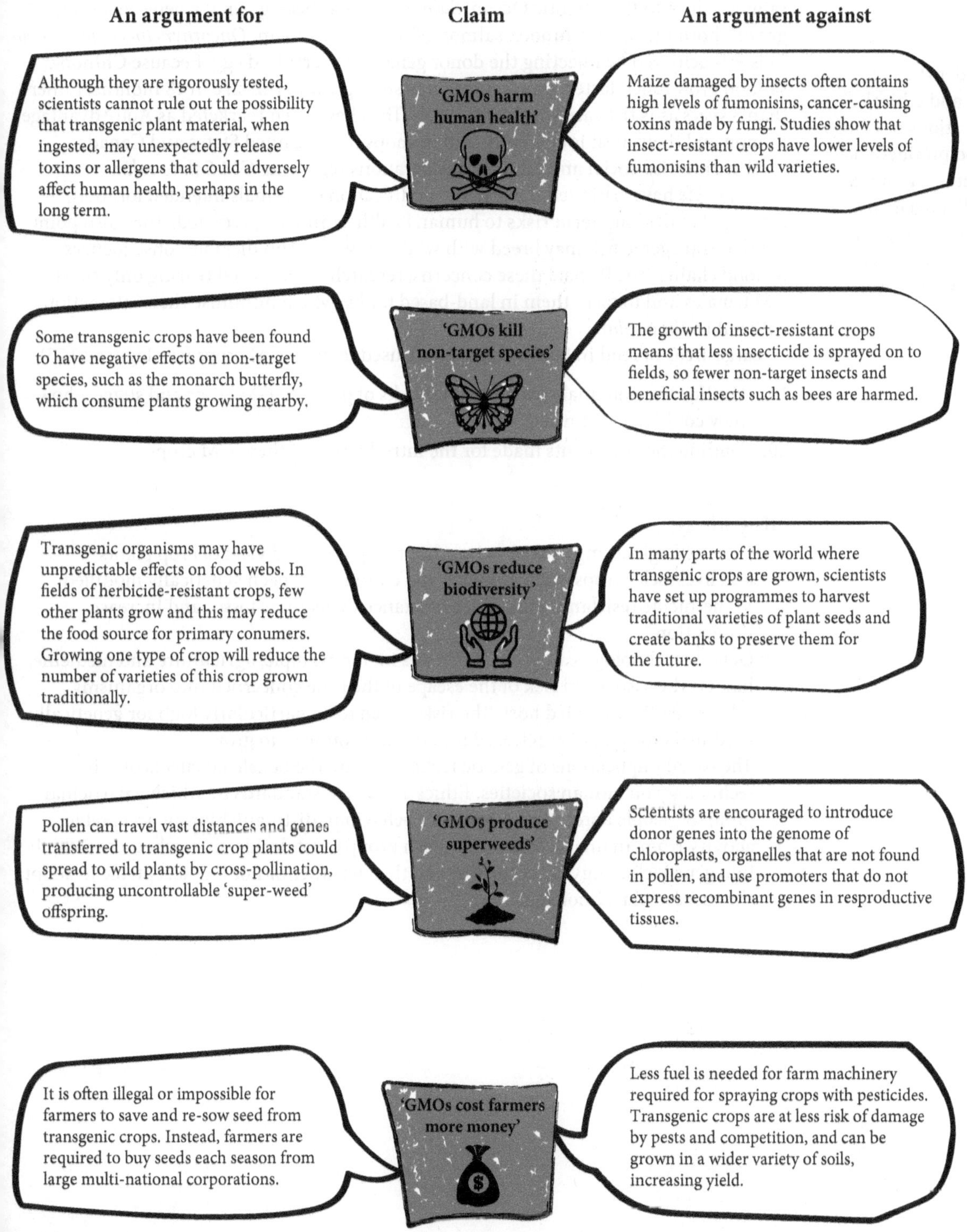

An argument for	Claim	An argument against

An argument for

Although they are rigorously tested, scientists cannot rule out the possibility that transgenic plant material, when ingested, may unexpectedly release toxins or allergens that could adversely affect human health, perhaps in the long term.

Claim: 'GMOs harm human health'

An argument against

Maize damaged by insects often contains high levels of fumonisins, cancer-causing toxins made by fungi. Studies show that insect-resistant crops have lower levels of fumonisins than wild varieties.

Some transgenic crops have been found to have negative effects on non-target species, such as the monarch butterfly, which consume plants growing nearby.

Claim: 'GMOs kill non-target species'

The growth of insect-resistant crops means that less insecticide is sprayed on to fields, so fewer non-target insects and beneficial insects such as bees are harmed.

Transgenic organisms may have unpredictable effects on food webs. In fields of herbicide-resistant crops, few other plants grow and this may reduce the food source for primary conumers. Growing one type of crop will reduce the number of varieties of this crop grown traditionally.

Claim: 'GMOs reduce biodiversity'

In many parts of the world where transgenic crops are grown, scientists have set up programmes to harvest traditional varieties of plant seeds and create banks to preserve them for the future.

Pollen can travel vast distances and genes transferred to transgenic crop plants could spread to wild plants by cross-pollination, producing uncontrollable 'super-weed' offspring.

Claim: 'GMOs produce superweeds'

Scientists are encouraged to introduce donor genes into the genome of chloroplasts, organelles that are not found in pollen, and use promoters that do not express recombinant genes in resproductive tissues.

It is often illegal or impossible for farmers to save and re-sow seed from transgenic crops. Instead, farmers are required to buy seeds each season from large multi-national corporations.

Claim: 'GMOs cost farmers more money'

Less fuel is needed for farm machinery required for spraying crops with pesticides. Transgenic crops are at less risk of damage by pests and competition, and can be grown in a wider variety of soils, increasing yield.

Figure 19.24 The ethical and social implications of growing genetically modified crop plants used in food production.

GM SALMON

Crop plants are not the only transgenic organisms produced for human consumption. Some animals intended for food have also been genetically modified. A variety of wild salmon native to the Atlantic Ocean, *Salmo salar*, has been given the gene encoding growth hormone from Chinook salmon of the Pacific Ocean, *Oncorhynchus tshawytscha*. This was achieved by injecting the donor gene into a fertilised egg. Because Chinook salmon grow much faster, transgenic *Salmo salar* grow to full size in 18 months, rather than 3 years as is the case with normal fish. They have been marketed as AquAdvantage® salmon and their meat is currently sold in shops in the United States and Canada.

There are also social and ethical considerations regarding GM *Salmo salar*. Some people have criticised the genetic modification of animals intended for food, arguing that the long-term risks to human health cannot be predicted. They also point out that transgenic fish may breed with wild fish, with unpredictable consequences on food chains. To alleviate these concerns, researchers proposed rearing only sterile GM females and to farm them in land-based tanks, which minimise their interaction with wild *Salmo salar*.

Large areas of land in South America are used to grow GM crops (see Figure 19.23).

19. Outline the potential environmental risks of growing GM crops and suggest how they could be minimised.
20. Outline the arguments made for the introduction of these GM crops.

Tip

Consider the social and ethical implications of GMOs in food production in context. To make overall judgements, each consideration needs to be assessed on a case-by-case basis.

Key ideas

→ Genetic engineering is used to improve the quality and yield of crop plants and livestock. Crops, such as maize and cotton, have been genetically modified for herbicide resistance and insect resistance to decrease losses and increase production.

→ Genetic technology can provide benefits in, for example, agriculture and medicine, but has the associated risk of the escape of the gene concerned into organisms other than the intended host. The risk is seen to be particularly high for genetically modified crops that are released into the environment to grow.

→ The social implications of genetic technology are the beneficial effects of this technology on human societies. Ethics are sets of standards by which a particular group of people agree to regulate their behaviour, distinguishing an acceptable activity from an unacceptable one. Each group must decide, first, whether research into gene technology is acceptable, and then whether or not it is acceptable to adopt the successful technologies.

CHAPTER OVERVIEW

Review the mini mind map to link new learning with your prior learning, and to highlight which of the key concepts underpin the chapter.

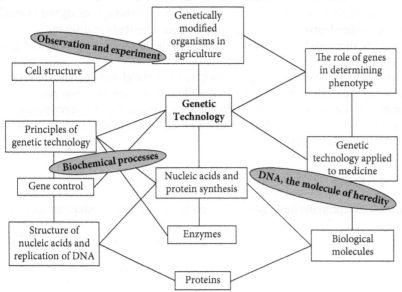

Try copying this mini mind map and expanding upon it. Use your notes from other chapters to help you explore how the essential ideas, theories and principles can be linked further together.

WHAT YOU HAVE LEARNED

Now that you have finished this chapter you should be able to:

- define the term recombinant DNA
- explain that genetic engineering is the deliberate manipulation of genetic material to modify specific characteristics of an organism and that this may involve transferring a gene into an organism so that the gene is expressed
- explain that genes to be transferred into an organism may be:
  - extracted from the DNA of a donor organism
  - synthesised from the mRNA of a donor organism
  - synthesised chemically from nucleotides
- explain the roles of restriction endonucleases, DNA ligase, plasmids, DNA polymerase and reverse transcriptase in the transfer of a gene into an organism
- explain why a promoter may have to be transferred into an organism as well as the desired gene
- explain how gene expression may be confirmed by the use of marker genes coding for fluorescent products
- explain that gene editing is a form of genetic engineering involving the insertion, deletion or replacement of DNA at specific sites in the genome
- describe and explain the steps involved in the polymerase chain reaction (PCR) to clone and amplify DNA, including the role of *Taq* polymerase
- describe and explain how gel electrophoresis is used to separate DNA fragments of different lengths
- outline how microarrays are used in the analysis of genomes and in detecting mRNA in studies of gene expression
- outline the benefits of using databases that provide information about nucleotide sequences of genes, genome sequences, amino acid sequences of proteins and protein structures

- explain the advantages of using recombinant human proteins to treat disease, using the examples insulin, factor VIII and adenosine deaminase
- outline the advantages of genetic screening, using the examples breast cancer (*BRCA1* and *BRCA2*), Huntington's disease and cystic fibrosis
- outline how genetic diseases can be treated with gene therapy, using the examples severe combined immunodeficiency (SCID) and inherited eye diseases
- discuss the social and ethical considerations of using genetic screening and gene therapy in medicine
- explain that genetic engineering may help to solve the global demand for food by improving the quality and productivity of farmed animals and crop plants, using the examples of GM salmon, herbicide resistance in soybean and insect resistance in cotton
- discuss the ethical and social implications of using genetically modified organisms (GMOs) in food production.

CHAPTER REVIEW

1. Figure 19.25 represents the stages in the production of insulin using genetic technology.

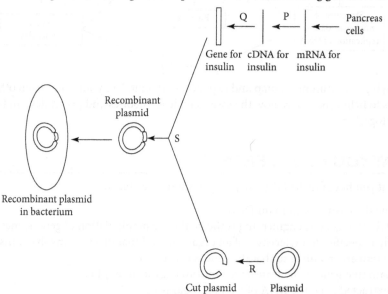

Figure 19.25

State the letters that indicate stages at which the following processes occur:
a. hydrolysis
b. condensation
c. polymerisation
d. complementary base pairing.

2. Draw a table to compare PCR with the process of DNA replication in a human cell during the cell cycle.

3. Using the key terms 'store', 'compare' and 'predict' in your answer, outline the roles of databases in genetic technology.

4. Identify the 'odd one out' from these series of terms. Explain your choice. The first one has been done for you.
   a. gene editing, gene therapy, genetic screening. **Answer:** the odd one out is genetic screening, because it does not modify the genome of an individual (unlike the other two).
   b. genetic engineering, selective breeding, gene therapy
   c. adenosine deaminase, factor VIII, insulin
   d. breast cancer susceptibility, Huntington's disease, cystic fibrosis

5. Summarise the challenges posed by gene therapy that have had to be overcome by researchers. Decide which procedures are ethical and which are not.

6. Cotton plants have been genetically engineered to be resistant to the cotton boll weevil. The *cry* gene from the bacterium *Bacillus thuringiensis* that codes for a toxin is inserted into plants.
   a. Explain why a promoter is attached to the gene for *Bt* toxin.
   b. Explain the role of DNA ligase in producing genetically engineered cotton plants.
   c. Suggest why the effectiveness of *Bt* toxin is decreasing in areas that grow *Bt* cotton plants.

7. Look back at Figure 19.1 at the start of this chapter. Choose three items labelled in green from the gene technologist's 'toolbox'. Explain how they represent accurate analogies for the molecule or laboratory process.

8. At the start of this chapter, you read about the production of the world's first genome, belonging to a theoretical organism called *Caulobacter ethensis-2.0*. Refer back to the story to help you answer the following questions.
   a. Explain what is meant by the term genome.
   b. Calculate the mean number of base pairs in one gene of *Caulobacter ethensis-2.0*.
   c. Suggest why 100 genes failed to be expressed when the scientists attempted to transcribe the genome of *Caulobacter ethensis-2.0*.
   d. Apart from the possibilities mentioned in the introduction, suggest reasons to explain why researchers may use gene editing to make changes to the sequence of genomes of wild bacteria.

Clues to selected problems

CHAPTER 1

Q4a. Think about how many nanometres are in one micrometre.

Q14. Think about the kind of cells that require a lot of energy.

CHAPTER 2

Q3c. Does Benedict's solution test only for glucose?

Q7. Consider the following properties of starch in your answer: its insolubility in water, the branched-like nature of amylopectin, and the tightly packed nature of amylose.

Q11b. How many hydrolysis reactions are involved in breaking down triglycerides and glycogen? Also, remember that only glucose can be used directly in respiration to release energy.

Q15. Don't forget that the folding of the polypeptide chain to form a tertiary structure in proteins can result in amino acid residues with different properties to be positioned on the inside, or the outside, of the protein.

Assignment 2.1 A2a. Remember that when a peptide bond is formed, the carboxyl group of one amino acid (–COOH) undergoes a condensation reaction with the amine group of the other amino acid (–NH2). Although these amino acids have slightly different structures from those you are familiar with, try to identify a key carbon atom in one amino acid, and a nitrogen atom in the other.

CHAPTER 3

Q5. Consider the bonds between the amino acids that make up the polypeptide chain.

Q8c. Anomalous data means any measurement that is clearly outside of the range of data for the other measurements. Look for any readings in the table that appear to be significantly higher or lower than the others.

Q10ci. Look at the similar central square and pentagon shapes in the molecules. How might their similarity in shape result in an effect on beta-lactamase?

Q11. Carefully look at the diagrams. Which methods would reduce the access of the enzyme to the substrate, for example? Which methods would carry a risk that the enzyme might be lost from the immobilisation material?

Experimental skills 3.2 P2. A null hypothesis consists of the following statement: 'A change in <independent variable> has no significant effect on the change in <dependent variable>.'

Experimental skills 3.2 P3b. Testing the untreated milk with a glucose test strip before pouring it into the reaction vessel would avoid false-positive results. What do you think this means?

CHAPTER 4

Q3. Think carefully about how vesicles are formed, and how they can be used to release substances from the cell. What does this suggest about the material from which the membrane of the vesicle is made?

Q7. Consider the position of the protein – is it found on the inside or outside of the cell?

Q8. Ask yourself which molecules in the cell surface membrane are required to respond to hormones. How might muscle cells be different from other cells with regard to these molecules?

Experimental skills 4.3 P1a. Summarise the overall trend, then talk about the shape of the plotted line in more detail. It can help to break up the graph into distinct sections and describe the shape of each section with reference to temperature and light intensity. You should include key data points, for example, temperature in this case.

Experimental skills 4.3 P1d. Your method must include finding the point at which the plotted line intersects the appropriate axis (interpolation) to estimate an appropriate concentration.

Experimental skills 4.3 P4. In your plan, have you:

1. Identified an appropriate range of concentrations of ethanol?

2. Identified an appropriate number of different concentrations of ethanol, separated by appropriate intervals?

3. Described how different concentrations of ethanol will be prepared from the stock solutions?

4. Described how other variables that could affect membrane integrity will be controlled?

5. Described a control experiment for the investigation and explained why this is important?

6. Described how the intensity of the red colour in the solution will be objectively and precisely measured to minimise the effect of random and systematic errors on the accuracy of the data?

7. Identified a hazard in the experiment and how a safety precaution will be followed to minimise the risk?

CHAPTER 5

Q4b. Think carefully about which other cells in the human body share the characteristics of the cells in the germinal epithelium of the testes. Why might telomerase have a similar effect in these cells?

Q12. Try to use the following key terms in your answer: 'nucleotides', 'ATP', 'enzymes', 'replication', 'organelles', 'microtubules'.

Q13. Remember from Chapter 1 that mitochondria are the site of aerobic respiration in a cell. What do they produce and why might this be important at these specific points in mitosis?

Experimental skills 5.1 P3h. Use extrapolation to estimate a figure, which involves extending the line from the measurement closest to the tip of the root to the reading that would be the case when the mitotic index reached zero.

Assignment 5.1 A2. Include the terms 'range,' 'intervals' and 'trend' in your answer.

CHAPTER 6

Q5. Consider the differences between the purine and pyrimidine nitrogenous bases.

Experimental skills 6.1 P3. Remember that 'validity' refers not only to how suitable an experiment is to achieving its aims, but also whether the data obtained can be generalised to other situations. The scientists used poly-U RNA in their first experiment. Could they change this for something else?

Experimental skills 6.1 P7. This data is continuous, so you must draw a graph that allows you to estimate the values between plotted points.

Q14b. Consider the roles of RNA in a cell – what part does it play during protein synthesis, and for how long is it needed?

CHAPTER 7

Q4. Think about what the main function of a leaf is.

Q9. Think about what could obstruct the flow of materials.

CHAPTER 8

Q11. There are 1000 cm^3 in 1 dm^3.

Q24. The time taken for one cardiac cycle is the time at which the pressure lines return to the same values they had at time zero.

CHAPTER 9

Q8. Think about how the smooth muscles being contracted or relaxed affects the diameter of the lumens of the airways.

Q12. Use Figure 9.9 to help you. Remember that crossing one layer of squamous epithelium requires crossing two membranes: inner and outer.

Assignment 9.1 A4c. Think about concentration gradients for diffusion and also the haemoglobin dissociation curve from Chapter 8.

CHAPTER 10

Q3. Think about factors such as levels of public sanitation and education about the mode of transmission.

Q6. The question asks for biological issues, so ensure your answer looks at these rather than factors such as funding or political problems.

Experimental skills 10.1 P3. A dilution series is where a substance is diluted by the same fraction a number of times. The fraction is usually ½ or $\frac{1}{10}$ for each successive dilution. 10-fold is $\frac{1}{10}$.

CHAPTER 11

Q9a. Do all pathogens have a cell surface membrane? Refer to Chapter 10 for more information.

Q9c. Cancer cells form from normal body cells. Refer to Chapter 5 for more information.

Q12. The spleen and lymph nodes act as the site of differentiation and storage of B-lymphocytes.

Experimental skills 11.1 P2b. Consider factors that may cause a difference, if they were to vary, in the intensity of the line on the pregnancy test strip. These must be standardised.

Assignment 11.1 A2. Consider passive immunity in your answer.

CHAPTER 12

Q3. The greater the proportion of hydrogen atoms in a molecule the more energy is available from that molecule in respiration.

Q18. Think about how oxygen gets to respiring cells and what will happen when this supply of oxygen stops.

CHAPTER 13

Q7. Think about how rubisco contributes to plant growth.

Q8. Calculate the number of TP molecules made per second, but remember two 3-carbon TP molecules are needed to make one 6-carbon glucose molecule.

Assignment 13.1 A4a. Remember that respiration will also be occurring at the same time as photosynthesis.

CHAPTER 14

Q15a. Consider the contents of the three structures.

Q15b. Consider the size of the three molecules.

Q15c. Consider the word 'pressure'.

Q18a. Remember that an inability to secrete ADH means that the permeability of the collecting duct remains low.

Experimental skills 14.1 P8. Think about some of the factors that were not, or could not, be standardised in this investigation. These could include: pre-treatment, body mass, sex and age, and so on.

Experimental skills 14.2 P1. Your statement should be in the following form: 'There is no significant difference between the _____ and the _____.'

Experimental skills 14.2 P3g. Remember that 10 volunteers consumed MSG and 10 volunteers consumed sodium chloride, and the values of the dependent variable are means of their measurements.

CHAPTER 15

Assignment 15.1 A1. You will need to compare the trends shown in the graphs and give reasons for the differences. You should also give a comparative data quote to support your answer and remember to give values from both axes.

Q13. Think about how molecules bind to receptors and what influences the specificity of receptors.

CHAPTER 16

Q4c. Remember from your work in Chapter 5 that prophase involves the condensation of chromosomes, which makes them appear shorter and thicker.

Q8cii. Use the terms 'probability,' 'chance' and 'random' in your answer.

Q9. For a child to be born with blood group O, he or she must inherit the recessive Io allele from each parent.

Q13. A girl with red-green colourblindness must have a father who is red-green colourblind. Why?

Experimental skills 16.1 P2b. The red pigment is the product of an enzyme-catalysed reaction.

Assignment 16.1 A1. Include the word 'affinity' in your answer.

CHAPTER 17

Q1. Look for traits that you can measure quantitatively (discontinuous data) and qualitatively (continuous data).

Q8c. Consider how release and recapture may be used.

Experimental skills 17.1 P1. Seeds are often ingested with the fleshy fruit of the durian.

Assignment 17.1 A6. Take a look at a map of the Caribbean Islands and the south-eastern United States. Where is Cuba in relation to these?

CHAPTER 18

Q1a. Compare the information that is given by the first parts of the names and also by the second parts of the names.

Q5. Think about the range of organisms that will have cytochrome c compared with those that will have haemoglobin.

Q16. Calculate the index of diversity for each lake separately, then compare the two results.

CHAPTER 19

Q8. You will see later in this chapter that:

    a. Mutations in genes such as *BRCA1* and *BRCA2* increase a person's susceptibility to developing breast cancer.

    b. Rice and soybean plants have been modified to produce proteins that allow them to resist pests or herbicides.

    c. Databases hold the sequence of genes that encode antigens and other proteins produced by many pathogens.

Q18. Consider some of the diseases that are common in parts of the world affected by natural disasters and poor healthcare – refer to Chapter 10 for more information.

Practice exam-style questions

Exam-style questions have been written by the authors. In examinations, the way marks are awarded may be different. References to assessment and/or assessment preparation are the publisher's interpretation of the syllabus requirements and may not fully reflect the approach of Cambridge Assessment International Education.

CHAPTER I

E1. Which is the same length as 0.1 µm?
    A. 0.000 01 mm   **B.** 1.0×10^{-3} mm
    C. 100 nm   **D.** 1.0×10^{3} nm   [1 mark]

E2. Which structure is not surrounded by a double membrane?
    A. Chloroplast   **B.** Golgi body
    C. Mitochondrion **D.** Nucleus   [1 mark]

E3. **a.** Describe and explain how to make a temporary preparation for observing plant cells with a light microscope. Include any safety precautions you should take.   [6 marks]

    b. A student observed a cell using an eyepiece graticule and stage micrometer scale, as shown in Figure 1.31.

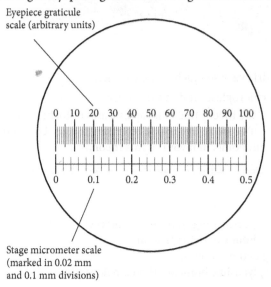

Figure 1.31

    The diameter of the cell measured 24 divisions on the graticule scale. Calculate the diameter of the cell in micrometres. Show your working.   [3 marks]

    c. A student used a camera attached to a microscope to produce a photomicrograph of a plant cell. The length of the cell on the micrograph was 20 cm. The total magnification of the image was ×1000. Calculate the actual length of the cell in nanometres, showing your working. Give your final answer in standard form.   [3 marks]

E4. Figure 1.32 shows the structure of *Amoeba*, a single-celled organism.

    a. Name the structures X, Y and Z. Give a reason for each of your answers. [6 marks]

    b. Is the *Amoeba* cell prokaryotic or eukaryotic? Explain your answer. [3 marks]

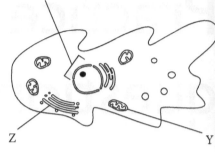

Figure 1.32

E5. **a.** Figure 1.33 shows an electron micrograph of a white blood cell.

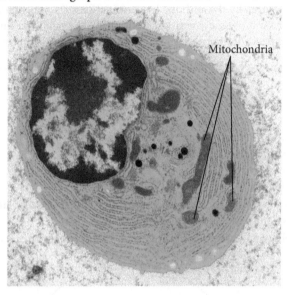

Figure 1.33

    Suggest why the mitochondria vary so much in appearance. [2 marks]

    b. Compare the structures of chloroplasts and mitochondria. [6 marks]

    c. The endosymbiotic theory is the theory that chloroplasts and mitochondria are descended from once separate prokaryotic cells that began to live inside other larger cells. Use your knowledge of these organelles to suggest the main points in favour of this theory. [3 marks]

CHAPTER 2

E1. Which statements about amylopectin and glycogen are correct?

    1. Both have $\alpha(1,4)$ glycosidic bonds in their structure.

    2. Amylopectin contains β-glucose residues.

    3. Glycogen has more $\alpha(1,6)$ glycosidic branches than amylopectin.

    A. 1 only

    B. 1 and 2

    C. 1 and 3

    D. 2 and 3 [1 mark]

E2. What describes the quaternary structure of collagen fibres?

    A. The position and number of α-helices and β-pleated sheets

    B. Its fibrous structure

    C. The specific order of amino acids in the polypeptide chains

    D. Three polypeptide chains joined together [1 mark]

E3. Chitin is a macromolecule found in a number of different organisms, for example in fungal cell walls and the hard outer skeletons of insects. It is a polysaccharide consisting of glucosamine residues joined together by $\alpha(1,4)$ glycosidic bonds. Glucosamine is a monosaccharide similar in structure to glucose, but has an amine group ($-NH_2$) in place of a hydroxyl ($-OH$) group. Figure 2.46 shows a glucosamine molecule.

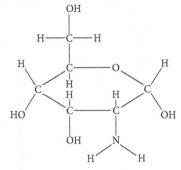

Figure 2.46 The structure of glucosamine, the monomer of the polysaccharide chitin.

   a. When two molecules of glucosamine are joined in a condensation reaction, a (1,4) glycosidic bond forms to produce a disaccharide called chitobiose. Draw a diagram to show a molecule of chitobiose. Note that the carbons are numbered in the same order as in α-glucose. [3 marks]

   b. Glucosamine is a reducing sugar, but chitobiose is not. Explain how you could conduct a biochemical test to distinguish these two samples, and how you would draw a conclusion from your observations. [4 marks]

E4. Haemoglobin is an important protein in the red blood cells of mammals. Haemoglobin has a quaternary structure in which four polypeptide chains bind to a haem group, which each contain an iron ion. In the functioning protein, two amino acids bind to the haem group. These two amino acids, histidine and aspartate, are found at positions 53 and 108 in the primary structure of the protein.

   a. Explain what is meant by 'primary structure'. [1 mark]

   b. Explain how the structure of the protein brings these two amino acids close to each other. [2 marks]

   c. Explain in detail how the tertiary and quaternary levels of protein structure of the haemoglobin molecule contribute to its role in the transport of oxygen. [6 marks]

E5. Many species of rodent hibernate over the winter to conserve energy and increase the chance of survival. They often have brown adipose tissue, or brown fat, as well as white fat. The fat cells in brown fat tissue have many more mitochondria per cell than white fat and the tissue has more capillaries. The cells have a greater metabolic rate and generate heat, helping to maintain the hibernating animal's body temperature just high enough so that it survives the winter (Figure 2.47).

Figure 2.47 Hibernating animals can lower their metabolic rate and body temperature to quite surprising levels for many months at a time.

   a. Referring to your knowledge from Chapter 1, describe why cells in brown fat tissue are able to release more energy than cells in white fat tissue. [1 mark]

   b. The percentage body mass of brown fat in rodents decreases during the non-feeding season and increases during the feeding season. Suggest explanations for this observation. [3 marks]

   c. Glycerol can be converted into glucose for use in respiration. Explain in detail how glycerol can be produced from a triglyceride. [2 marks]

   d. Outline a biological test that could be carried out to show the presence of triglyceride in a liquid mixture and describe the positive result for that test. [3 marks]

E6. The silk that spiders use to make their webs is one of the strongest natural materials known. It is composed mainly of two macromolecules called spidroin 1 and spidroin 2.

Figure 2.48 shows a section of spidroin 1.

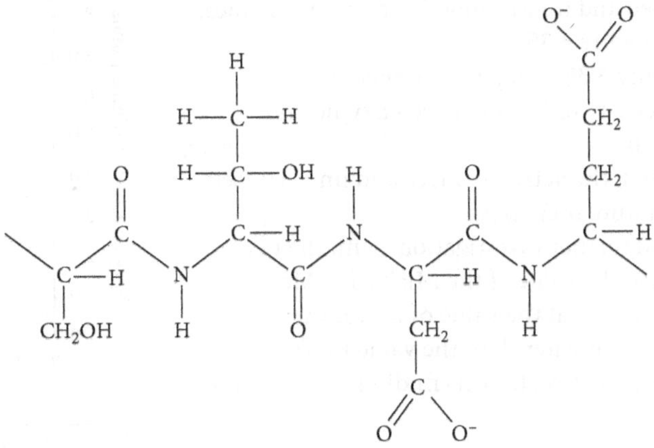

Figure 2.48 A short section of spidroin 1.

a. State two features shown in Figure 2.48 that indicate that spidroins are proteins. [2 marks]

b. The loss of a hydrogen atom from the −COOH group to become −COO⁻ is called deprotonation. Explain why the deprotonation of some R groups can be important for the tertiary structure in proteins. [3 marks]

Figure 2.49 shows the results of an investigation into the tensile strength (ability to stretch before breaking) of spidroin 1 and 2; a scientist applied increasing forces to lengths of the two proteins.

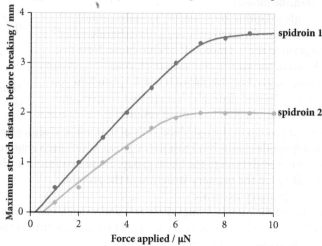

Figure 2.49 The results of an investigation into the tensile strength of spidroin 1 and 2.

i. State the independent variable in this investigation. [1 mark]

ii. Suggest which factors were standardised in this investigation. [2 marks]

iii. Compare in detail the tensile strength of spidroin 1 and spidroin 2. [5 marks]

c. Spidroin proteins are surrounded by an outer layer, or 'skin', of triglycerides in spider silk. Suggest **one** benefit of this arrangement. [1 mark]

CHAPTER 3

E1. In a graph showing the relationship between temperature and the rate of a reaction catalysed by a human enzyme, there is a sharp fall after a certain temperature. Which of the following statements best explain this? [1 mark]

1. The tertiary structure of the enzyme is changed.
2. There are more collisions between enzyme and substrate molecules.
3. The shapes of the active site and the substrate are no longer complementary.
4. Bonds are broken between the variable (R) groups of the amino acids in the polypeptide chain(s) of the enzyme.

A. 1, 2 and 3   **B.** 1, 2 and 4   **C.** 1, 3 and 4   **D.** 2, 3 and 4

E2. An investigation was carried out into the effect of temperature on the activity of immobilised and non-immobilised (free) enzymes. The results are shown in Figure 3.35.

a. With reference to Figure 3.35, compare the effect of temperature on the activity of immobilised enzyme and enzyme free in solution. [4 marks]

b. Explain the difference in the activity of free and immobilised enzyme as the temperature increases. [4 marks]

★ **c.** The student conducted further investigations with the two enzymes and calculated the value of the Michaelis–Menten constant (K_m). They found that the value of K_m for the immobilised enzyme was greater than the value for free enzyme. Suggest an explanation for this finding. [2 marks]

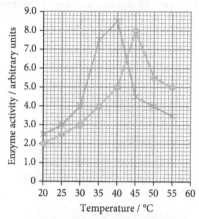

Figure 3.35 The effect of temperature on free and immobilised enzyme activity in an investigation.

E3. One way in which cells control a sequence of biochemical reactions (a metabolic pathway) is to use the end-product of a chain of reactions as a non-competitive, reversible inhibitor. As the enzyme converts substrate to product, it is slowed down because the end-product binds to another part of the enzyme and prevents more substrate binding. This is shown in Figure 3.36.

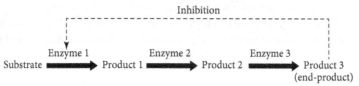

Figure 3.36 The mechanism of end-product inhibition

  a. Describe what is meant by a non-competitive, reversible inhibitor. [3 marks]

  b. Assume that the metabolic pathway shown in Figure 3.36 occurs in the cytoplasm. Describe and explain the effect of the following events on the rate of the metabolic pathway:

    i. the immediate packaging of the end-product into vesicles and export from the cell by exocytosis [2 marks]

    ii. the polymerisation of the end-product into a macromolecule for storage inside the cell [2 marks]

    iii. the accumulation of the end-product in the cytoplasm of the cell. [2 marks]

  c. End-product inhibition is common in metabolic pathways that make products that need to be maintained in the cell within a narrow range of concentrations. Explain how end-product inhibition achieves this. [2 marks]

  d. Use the information in the following passage to draw a diagram, similar to Figure 3.36, to show how this metabolic pathway illustrates end-product inhibition. [2 marks]

    'In a metabolic pathway called the Krebs cycle, succinate dehydrogenase converts a substrate called succinic acid into fumaric acid. Fumaric acid is then converted into malonic acid by an enzyme called fumarase. Fumaric acid has a similar shape to the structure of succinic acid.'

E4. Most 'biological' washing powders contain a range of enzymes that remove stains from dirty clothes.

  a. Explain why a biological washing powder should contain more than one enzyme in order to wash clothes that have stains such as dried egg and blood. [2 marks]

  b. Many biological washing powders work best at temperatures between 40 °C and 50 °C. Explain why these biological washing powders are not as effective in a very hot wash at 90 °C. [2 marks]

  c. Some washing powders contain enzymes extracted from bacteria found in hot water springs. Suggest one advantage of using enzymes from these bacteria in washing powders. [1 mark]

  d. Many washing powders are able to wash clothes effectively at 25 °C. Suggest one advantage of this. [1 mark]

  e. Suggest reasons why enzymes in biological washing powders are now added to the washing machine in an encapsulated form instead of a powdered form. [2 marks]

E5. The mass production of textiles such as clothes causes pollution to waterways. One major pollutant is a family of molecules called azo dyes, which are used to add colour to textiles. Azo-dyes increase the risk of developing cancer.

Environmental scientists found that a species of fungus called *Pleurotus ostreatus* can be used to treat water pollution. This is a fungus that carries out decomposition using extracellular enzymes called laccases. These enzymes are able to break down azo dyes into harmless products.

  a. Explain what is meant by the term 'extracellular enzyme'. [1 mark]

  b. The ability of *Pleurotus ostreatus* to break down azo dyes was investigated. A suspension of live cells was added to water contaminated with various concentrations of an azo dye. The results are shown in Table 3.7.

Azo dye concentration / mg dm^{-3}	Percentage breakdown of azo dye / %	
	After 3 days	After 7 days
20	38.5	100.0
40	35.7	96.5
60	32.5	92.1
80	30.2	88.3
100	10.6	63.5

Table 3.7 The relationship between concentration of azo dye and the percentage breakdown by laccase enzymes after 3 days and 7 days.

Describe the results shown in Table 3.7. [2 marks]

  c. When a similar experiment was performed using the free laccases from *Pleurotus ostreatus* cells, all
  concentrations of azo dye shown in Table 3.7 were broken down within a few hours. Suggest why free
  laccase from cell lysates breaks down the azo-dye more quickly than live cells.   [2 marks]
  d. Laccases may be immobilised. Evaluate the use of immobilised enzymes in the treatment of waste water from
  the textile industry, compared with the use of non-immobilised (free) enzymes.   [4 marks]
  e. Suggest why researchers added protease inhibitors to the cell lysates used in this investigation.   [1 mark]
  f. The researchers wanted to design a simple method to find out the concentration of laccase in the extract from
  one fungus.

  They obtained:

  • 1.0 g dm^{-3} stock solution of laccase
  • azo dye agar plates with wells into which enzyme solutions can be placed (azo dye agar plates are Petri
    dishes containing azo dye mixed with starch).

  Figure 3.37 shows how the scientist used the plates to find the concentration of laccase. The scientist thought
  that the area of the decolourised area was proportional to the concentration of laccase.

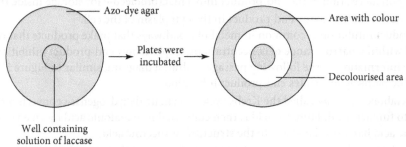

Figure 3.37 A method used to determine the ability of an enzyme to digest its substrate.

  i. Identify the independent and dependent variables in this investigation.   [2 marks]
  ii. Describe how the students could use the method outlined in Figure 3.37 to find out the concentration of
  the enzyme laccase in their extract.
  Your method should be set out in a logical way and be detailed enough to let another person follow it. [8 marks]

CHAPTER 4

E1. Viruses infect and reproduce inside cells. When new virus particles leave the host cell, they are enclosed in a
phospholipid bilayer acquired from its cell surface membrane. This is called a viral envelope.
  a. Explain why phospholipids spontaneously arrange into a phospholipid bilayer.   [2 marks]
  b. Suggest how the presence of a viral envelope containing phospholipids enhances the ability of viruses to
  infect other cells.   [2 marks]
  c. Scientists have found that the ratio of cholesterol to phospholipids in some viral envelopes is higher than
  in the cell surface membrane of the cells they infect. Suggest **one** difference between the properties of
  these structures.   [1 mark]
E2. Omeprazole is a class of medicinal drug called a proton pump inhibitor (PPI). It binds to and inhibits pump
proteins from moving hydrogen ions against a concentration gradient. One use is to reduce the acidity inside the
stomach. It is prescribed to people who have persistent heartburn, or indigestion. This is caused by excess acid in
the stomach.
  a. i. Explain what is meant by the term 'pump protein'.   [2 marks]
  ii. Using the information provided, explain why omeprazole inhibits the process of active transport. [2 marks]
  b. Suggest why one of the common side effects of PPI drugs taken orally is diarrhoea.   [2 marks]
  c. Omeprazole diffuses into cells and binds to the cytoplasmic portion of the proton pump. Explain what this
  indicates about the chemical nature of the drug.   [2 marks]

E3. A student carried out an experiment to estimate the water potential of the cells in epidermis tissue obtained from an onion.

A range of sodium chloride solutions of different concentrations between 0% and 50% were prepared. Three small pieces of onion epidermis were immersed in each solution, and also in distilled water. All tissue was left to incubate at room temperature for 20 minutes.

The student mounted each piece of tissue on a slide. They then counted 25 cells and recorded the number of cells they could see that were plasmolysed. They moved the slide a little and continued to count until they had recorded results for 10 fields of view for each tissue sample, so the number of counted cells was 250 for each solution.

The student collated all the data in order to calculate the mean, median and mode of the number of plasmolysed cells, as shown in Table 4.5.

Concentration of sodium chloride solution/ %	Number of plasmolysed cells counted out of 25 cells (10 repeats)										Mean	Median	Mode
0.0	0	0	0	0	1	1	0	0	1	0	0.3	0.0	0
0.1	2	1	2	4	2	3	2	2	1	2	2.1	2.0	2
0.2	5	5	3	24	6	6	5	5	5	6	7.0	5.0	5
0.3	14	12	15	15	14	14	12	12	18	17	14.3	14.0	14
0.4	22	23	22	21	21	23	23	23	22	23	22.3	22.5	23
0.5	24	25	25	25	23	25	25	25	25	25			

Table 4.5

a. Explain why it was important to count the same number of cells for each solution. [1 mark]
b. Calculate the missing values of the mean, median and mode for the plasmolysed cells counted in the solution of 0.5% sodium chloride concentration. [1 mark]
c. Suggest **one** reason why the student identified the fourth reading obtained for the 0.2% solution of sodium chloride as anomalous. [1 mark]
d. Explain why plotting the median or mode on a graph would more accurately display the effect of concentration of sodium chloride on onion cell plasmolysis. [2 marks]
e. Explain why the values obtained for the median and mode could be misleading. [2 marks]
f. The student wanted to obtain a more accurate value for the water potential of these cells. Which one of the following steps would be most appropriate to achieve this? [1 mark]
   A. Repeat the experiment using glucose instead of sodium chloride.
   B. Repeat the experiment using a wider range of sodium chloride concentrations.
   C. Repeat the experiment at a higher temperature, to speed up the rate of osmosis.
   D. Repeat the experiment using smaller intervals of sodium chloride concentration between the range 0.3 and 0.4 mol dm^{-3}.
   The student noticed that different plant cells showed different degrees of plasmolysis. The teacher suggested that measuring the time taken for a plant cell to show incipient plasmolysis would be a more accurate way of estimating the effect of solutions of different concentrations. Incipient plasmolysis is the point at which part of the cell surface membrane initially pulls away from the cell wall.
g. The following questions will guide you through a method that could be used to investigate the effect of sodium chloride concentration on the time taken for incipient plasmolysis to occur in onion cells.
   i. Identify the independent and the dependent variable in this investigation. [2 marks]
   ii. Explain why the student should watch and record the time taken for incipient plasmolysis in at least five cells in each solution, rather than just one. [2 marks]
   iii. Explain why it was important to add the same volume of sodium chloride solution to each microscope slide. [2 marks]
   iv. Identify a source of error in this investigation and suggest how this could be avoided. [2 marks]

E4. Figure 4.33 shows the structures of two types of lipid called stearin and phosphatidylcholine.

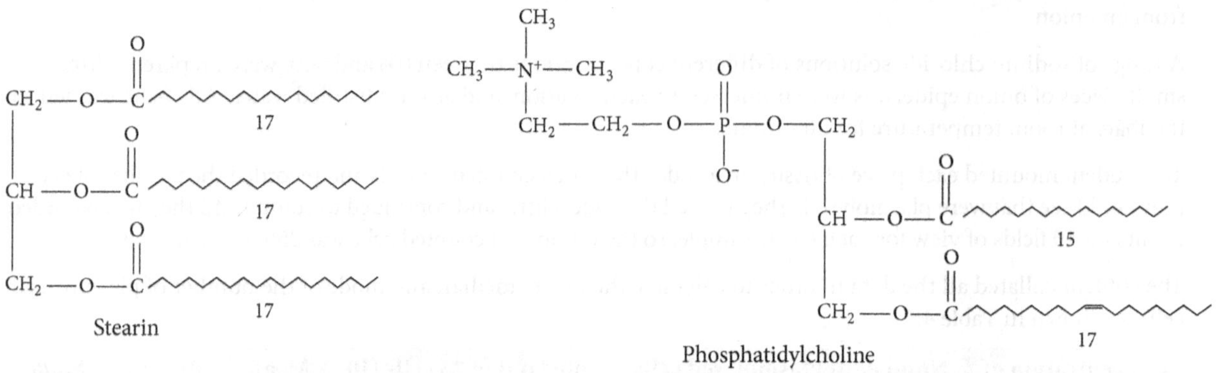

Stearin

Phosphatidylcholine

Figure 4.33

    a. Using your knowledge of the structure of lipids, identify which lipid is described in each statement below and give reasons for your answers.

      i. This lipid is unsaturated. [2 marks]

      ii. This lipid is amphipathic. [2 marks]

      iii. This lipid gives more ATP when used in respiration to release energy. [2 marks]

      iv. This lipid has hydrocarbon tails of different lengths. [2 marks]

    b. Phosphatidylcholine is a phospholipid found in the cell surface membrane of eukaryotic cells. Suggest and explain how it is arranged in a cell surface membrane. [3 marks]

    c. Describe in detail how a molecule of stearin is formed from its component molecules. [4 marks]

E5. The cell surface membranes of cells found lining the walls of the small intestine contain channels through which chloride ions, Cl⁻, are moved out of the cells by active transport. When the channels are open, chloride ions accumulate outside the cells. This causes water and sodium ions, Na⁺, to also move out of the cells and into the lumen of the intestine. This causes diarrhoea.

    a. Explain how an increase in the concentration of chloride ions outside a cell results in water moving out of that cell. Use the term 'water potential' in your answer. [3 marks]

    b. The bacterium *Vibrio cholerae* causes cholera. When this bacterium enters the small intestine, it produces a toxin that binds to receptors on the cell surface membranes of the cells in the intestinal lining. The cells then take up the toxin by endocytosis.

    The uptake of the toxin triggers a cell signalling mechanism that results in the Cl⁻ channels remaining permanently open. The cells therefore lose a large volume of water, resulting in diarrhoea and dehydration.

      i. Describe how a cell would take up the toxin by endocytosis. [2 marks]

      ii. Explain in detail what is meant by the term 'cell signalling mechanism'. [4 marks]

    The cell surface membranes also contain glucose-sodium cotransporter proteins. These use active transport to move glucose molecules and sodium ions into the cells from the lumen of the intestine. Individuals suffering from cholera can be treated by providing a drink that contains glucose and sodium ions. This causes the cells lining the small intestine to reabsorb water.

    c. Suggest how a drink containing glucose and sodium ions can reduce diarrhoea. [3 marks]

CHAPTER 5

E1. A new chemotherapy drug was trialled on a specific type of liver cancer. Figure 5.30 shows the results of a study into the effects of the drug on the division of cancer cells growing in a laboratory. During a period of 8 days, the same volume of the chemotherapy drug was applied to a sample of the cells growing in the laboratory at two points in the study. After each application, the number of cells growing 24 hours later was counted.

a. The chromosomes of cancer cells usually have longer telomeres than chromosomes taken from healthy tissue from the same person. Suggest why. [3 marks]

b. Explain why a control experiment should be used in this study. [2 marks]

c. Calculate the percentage change in the number of cancer cells over the course of the study. Give your answer to one decimal place. [1 mark]

d. Identify the two time points at which the drug was applied to the cells during the study. Explain your answer. [3 marks]

e. Further research showed that applying the chemotherapy drug to cells cultured in the laboratory at more regular intervals reduced the number of cancer cells to zero. Using your knowledge of the location of stem cells in the body, suggest why such a method should not be used when treating cancer cells growing in tumours in the body. [2 marks]

f. Mitosis stops if the spindle fibres are not attached correctly to the centromeres. Several proteins, known as spindle checkpoint proteins, are involved in this process. Explain why cancer cells may contain more or fewer chromosomes than normal cells. [3 marks]

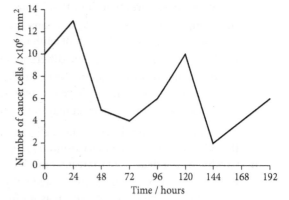

Figure 5.30 The effect of a chemotherapy drug on the division of liver cancer cells over a period of 8 days.

E2. Sunburn occurs when skin is exposed to high levels of ultraviolet (UV) light. Skin tissue contains a small proportion of stem cells called epidermal stem cells. These cells are thought to be able to produce new cells to replace damaged skin tissue, for example after sunburn.

a. Describe the characteristics of a stem cell. [2 marks]

b. Researchers collected data about telomere length in epidermal stem cells of people who had a history of sunburn and people who did not. The results are shown in Figure 5.31.

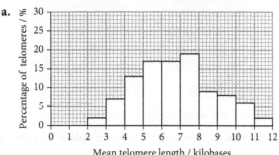

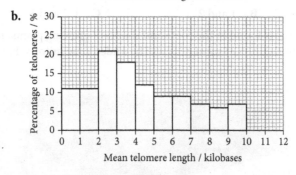

Figure 5.31 The average length of telomeres in epidermal stem cells from people who have no history of sunburn (a) and from people with a history of sunburn (b).

     i. State two features that show that Figure 5.31 is a histogram, rather than a bar chart. [2 marks]

     ii. Compare the mean lengths of telomeres in epidermal stem cells from people with a history of sunburn and those without a history of sunburn. [4 marks]

     iii. Suggest explanations for the differences you have described in your answer to ii. [3 marks]

     iv. Based on this investigation, the researchers concluded that: 'People with a history of sunburn have epidermal stem cells that have shorter telomeres than individuals with no history of sunburn.' Suggest why this conclusion may not be valid. [3 marks]

  c. UV light can damage DNA in skin cells, and this can lead to skin cancer. Explain how a primary tumour in the skin can lead to secondary tumours in other organs. [3 marks]

  d. Colchicine is a drug used to treat stomach cancer. It prevents the formation of microtubules.

     i. In a sample of dividing cells treated with colchicine, all cells are seen to be in the same stage of mitosis. Suggest which stage, and give reasons for your answer. [3 marks]

     ii. Suggest why cells such as stomach cancer cells are often seen to have more mitochondria and rough endoplasmic reticulum compared with the cells from which they formed. [2 marks]

⚗ E3. Mitosis occurs in apical meristematic tissue in plants.

  a. Below are four statements that apply to many plant cells.

     1. They have chromosomes that are visible with a light microscope.

     2. They contain chromosomes that consist of pairs of sister chromatids.

     3. They have a nuclear envelope that encloses the chromosomes.

     4. They are differentiated.

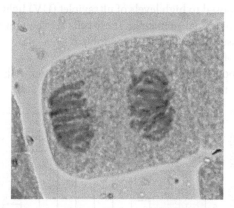

Figure 5.32 Photomicrograph of a plant cell undergoing anaphase. Magnification ×400.

Cell A in Figure 5.32 shows the root tip meristematic tissue of a plant. Identify which statements apply to cell A. [1 mark]

  A. 1 only   **B.** 1 and 2   **C.** 1, 2 and 3   **D.** 4 only

  b. Apical meristematic tissue is found in the roots and shoots of growing plants. Identify in which region of the root tip (A, B, C or D) in Figure 5.33 most cells will be undergoing mitosis. Explain your answer. [2 marks]

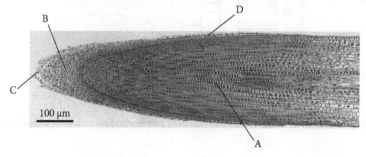

Figure 5.33 Photomicrograph of a longitudinal section through a plant root.

c. Use the scale bar to calculate the magnification of the root tip in this photomicrograph. [2 marks]
d. Describe how an eyepiece graticule and stage micrometer were used to calculate the length of the scale bar. [4 marks]
e. Draw a labelled scientific diagram of the root tip of the plant, showing the separate layers of tissue. [3 marks]

E4. A scientist investigated the relationship between the distance of cells from the tip of a root, their mean volume and the percentage of cells that were undergoing mitosis. Figure 5.34 shows the results of the study.

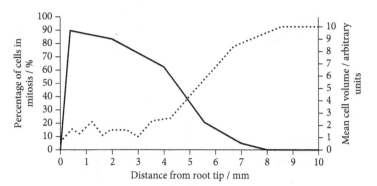

Figure 5.34 The relationship between the distance of cells from the end of a root tip, and the percentage of cells in mitosis and their volume.

Use Figure 5.34 to help you answer the following questions.

a. Describe how the percentage of cells undergoing mitosis varies as cells are sampled from increasing distances from the tip of the root. [2 marks]
b. With reference to both plotted lines, explain the results of the study. [4 marks]
c. Chromosomes must undergo certain changes in order for cell division to take place. Explain the reasons for two of these changes. [2 marks]

CHAPTER 6

E1. Figure 6.31 shows the DNA structure as determined by Watson and Crick in 1953.

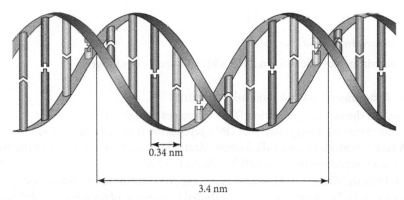

Figure 6.31 The structure of DNA. This diagram shows the distance between two base pairs (0.34 nm) and the distance occupied by one complete turn of the double helix (3.4 nm).

If the double helix takes 3.4 nm to make one complete turn and base pairs are 0.34 nm apart, how many base pairs are found in five complete turns? [1 mark]

A. 5   **B.** 10   **C.** 50   **D.** 100

E2. Figure 6.32 shows an experiment that led to the discovery of the semi-conservative nature of DNA replication. In this investigation, a radioactive isotope of nitrogen called ^{15}N was supplied in the culture medium and was incorporated into the DNA of bacteria. The bacteria were abruptly transferred into culture medium containing the less dense ^{14}N isotope and this was incorporated into the DNA of bacteria of the first and second generations.

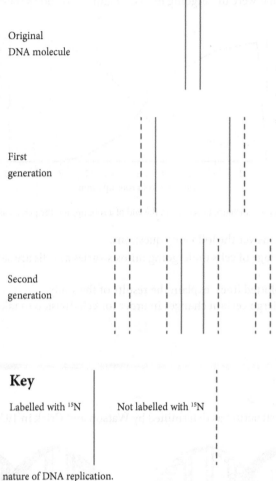

Original
DNA molecule

First
generation

Second
generation

Key

Labelled with ^{15}N Not labelled with ^{15}N

Figure 6.32 The semi-conservative nature of DNA replication.

a. Explain why bacteria are not able to form DNA without a source of nitrogen. [2 marks]

b. Explain:
  i. why all the first-generation DNA molecules contained ^{15}N [2 marks]
  ii. why only half of the second-generation DNA molecules contained ^{15}N. [2 marks]

c. Explain how this experiment suggested that DNA replication is semi-conservative. [2 marks]

d. During DNA replication, molecules called single-strand binding proteins attach to the separated strands after they have been separated by DNA helicase. Suggest why. [1 mark]

E3. Alfred Hershey and Martha Chase conducted a series of experiments in the 1950s to investigate whether inherited genetic material is DNA or protein. They infected two groups of bacteria with different types of virus. One type of virus had been labelled with a radioactive isotope of phosphorus (^{32}P), while the other had been labelled with radioactive sulfur (^{35}S). Figure 6.33 illustrates how they conducted this experiment and shows their results.

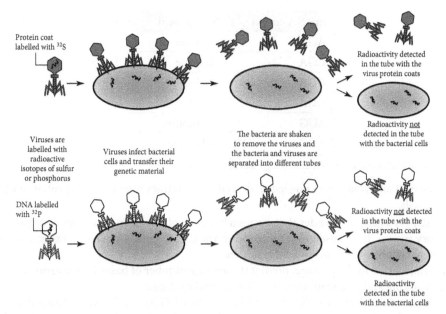

Figure 6.33 The investigation by Hershey and Chase.

a. Explain how the results of the investigation by Hershey and Chase proved that genetic material is DNA. [2 marks]

b. Before the work of Hershey and Chase, many scientists believed that proteins held genetic information. Using your knowledge of the structures of chromosomes and proteins, suggest why. [3 marks]

c. The genetic code is common to all organisms, including many DNA viruses. State the term used to describe this feature of the genetic code. [1 mark]

E4. In the early 1940s, two scientists named George Beadle and Edward Tatum concluded that each gene codes for a different polypeptide. They showed that when they mutated a single gene in the bread mould *Neurospora crassa* the offspring of the organism could not grow on medium lacking a given nutrient. This showed that the fungus lacked the enzyme (a protein) for the synthesis of that nutrient.

a. Define the term 'mutation'. [1 mark]

b. Describe one way in which Beadle and Tatum could make their findings reliable. [1 mark]

c. Suggest **two** reasons why *Neurospora* was chosen as the model organism in this investigation. [2 marks]

d. Suggest a factor that would have been standardised in this investigation. [1 mark]

E5. Bacteria undertake protein synthesis in a similar process to that in eukaryotic cells.

a. It takes eukaryotic cells longer than prokaryotic cells to make a protein. Using your knowledge of the structure of chromosomes and genes in these cells, suggest why. [2 marks]

b. Explain how the degeneracy of the genetic code reduces the effects of mutation on cells. [4 marks]

Tetracycline is an antibiotic that kills many species of bacteria. During protein synthesis, it attaches to the prokaryotic ribosome at the position shown in Figure 6.34. The diagram also shows some molecules involved in translation.

c. Use Table 6.8 to identify the amino acids X and Y. Do not forget that the table shows the mRNA codons, not DNA triplets. [2 marks]

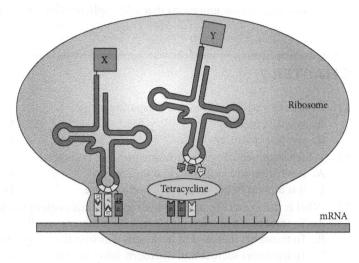

Figure 6.34 The action of the antibiotic tetracycline on bacterial ribosomes.

mRNA codon	Amino acid
AGA	arginine
CGA	arginine
GGA	glycine
UGA	STOP
AUG	methionine
AUC	isoleucine

Table 6.8 A selection of mRNA codons and the amino acids that they encode.

    d. Using Figure 6.34 and your own knowledge, describe how tetracycline prevents protein synthesis in prokaryotic cells. [4 marks]

E6. Cystic fibrosis (CF) is a genetic disorder that causes a build-up of mucus in a number of organs, especially those of the respiratory system. It is caused by a mutation in the CFTR gene. This has 27 exons and 26 introns and encodes a protein that has 1480 amino acid residues.

    a. Using the information in the passage, predict the average number of bases in one exon in the CFTR gene. Show your working and give your answer to three significant figures. [2 marks]

Different people with CF often have different mutations in the CFTR gene. One very common mutation is a deletion of three nucleotides, resulting in the deletion of a phenylalanine residue at position 508 (this mutation has the code ΔF508). The text below shows a section of the base sequence of the normal CFTR gene (a) and a section of the CFTR gene with this mutation (b). The amino acids encoded by this section of the gene are also shown.

(a) Normal CFTR sequence:

Nucleotide:	ATC	ATC	TTT	GGT	GTT
Amino acid:	Ile	Ile	Phe	Gly	Val

(b) ΔF508 CFTR sequence:

Nucleotide:	ATC	ATT	GGT	GTT	AAA
Amino acid:	Ile	Ile	Gly	Val	Leu

    b. i. Copy the normal CFTR base sequence. Draw a box around the three nucleotide bases that have been deleted in the ΔF508 mutation. [1 mark]

      ii. Explain how this mutation illustrates the degeneracy of the genetic code. [2 marks]

Another common mutation is a nucleotide substitution. This causes a problem during splicing of the primary transcript of CFTR.

    c. i. Explain what is meant by the term 'primary transcript'. [1 mark]

      ii. Suggest two chemical reactions that are involved in the processing of the primary transcript into an mRNA molecule. [2 marks]

      iii. Suggest how mutations that affect splicing can result in disease. [4 marks]

CHAPTER 7

E1. How does water travel past the Casparian strip? [1 mark]
    A. apoplast pathway   **C.** both apoplast and symplast pathways
    B. symplast pathway   **D.** neither apoplast nor symplast pathways

E2. Which statement about hydrostatic pressure is correct? [1 mark]
    A. It is higher in a sink than in a source.   **C.** It is lower in a sink than in a source.
    B. It is highest in between a sink and a source.   **D.** It is lowest in between a sink and a source.

E3. What is the role of the sucrose–hydrogen cotransporter proteins in companion cells? [1 mark]
    A. To transport hydrogen ions out of the companion cell.
    B. To transport sucrose and hydrogen ions into the companion cell.
    C. To transport sucrose and hydrogen ions out of the companion cell.
    D. To transport sucrose into the phloem sieve tube element.

E4. Figure 7.25 shows changes in the circumference of the trunks of three tree species during a hot day in summer.

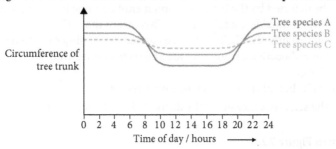

Circumference of
tree trunk

Time of day / hours

Figure 7.25

Suggest explanations for the similarities and differences shown for the three species. [6 marks]

E5. a. Figure 7.26 shows a micrograph of a transverse section through a leaf of the tea plant *Camellia sinensis*. The centre of the image contains the main 'vein' (vascular bundle) of the leaf.

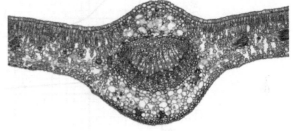

   i. Explain what type of microscope was used to produce the image in Figure 7.26. [2 marks]

   ii. Explain how the image in Figure 7.26 would look different if it was a longitudinal section through the central 'vein'. [2 marks]

   iii. Draw a labelled plan diagram of the section shown in Figure 7.26 to include the xylem and phloem tissues. [3 marks] **Figure 7.26**

b. *Camellia sinensis* is adapted to living in warm environments with a low rainfall. One of these adaptations is in its pattern of stomatal opening, as shown in Figure 7.27.

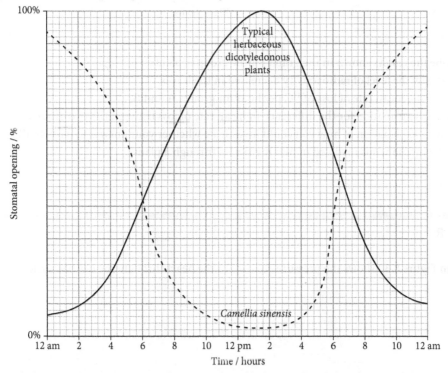

Figure 7.27 A graph showing daily changes in stomatal opening in typical herbaceous dicotyledonous plants and *Camellia sinensis*.

   i. Describe how the pattern of stomatal opening in *Camellia sinensis* differs from that of typical herbaceous dicotyledonous plants. [2 marks]

   ii. Explain the benefit of this pattern of stomatal opening for *Camellia sinensis*. [2 marks]

E6. A group of scientists investigated the transport of sugar in plants using the radioactive isotope ^{14}C (carbon-14). The presence of ^{14}C can be detected by the beta radiation it emits. In one investigation, the leaves of tomato plants were exposed to carbon dioxide containing ^{14}C. This meant that ^{14}C would be assimilated into the sugars the leaves produced in photosynthesis. The plants each carried developing tomato fruits. The plants were then removed from exposure to radioactive carbon dioxide and were divided into two groups:

- group A was kept at 20 °C
- group B was kept at 20 °C but all the tomato fruits were removed.

Over the next 11 hours the scientists measured the decrease in radioactivity in the leaves as a measure of the amount of ^{14}C present.

Their results are shown in Figure 7.28.

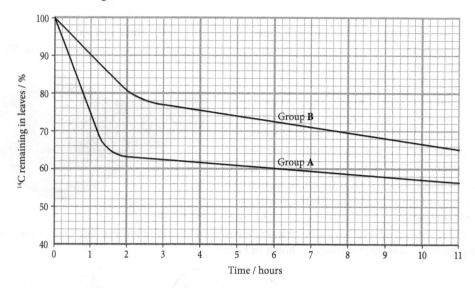

Figure 7.28

a. **i.** Describe the results shown in the graph for the group A plants. [2 marks]

 ii. Suggest why the level of radioactivity decreased in the leaves. [1 mark]

b. **i.** Calculate the mean rate of loss of ^{14}C per hour for groups A and B. [2 marks]

 ii. Suggest an explanation for the difference in mean rate of loss between the two groups. [2 marks]

 iii. Suggest an explanation of how you could investigate whether your answer to part (b)(ii) is correct. [2 marks]

c. Describe the mechanisms involved in moving sugars through a plant. [6 marks]

CHAPTER 8

E1. The graph in Figure 8.30 shows the pressure changes in the heart through one cardiac cycle.

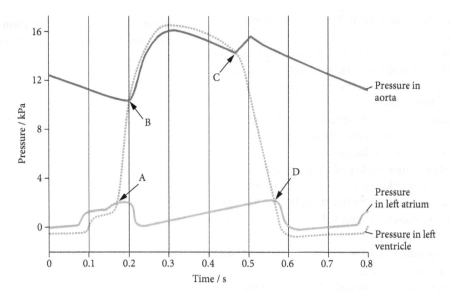

Figure 8.30

Which letter in Figure 8.30 shows the point at which the atrioventricular valve closes? [1 mark]

E2. Name the part of the circulatory system described in each statement.
   a. The blood vessel that transports oxygenated blood from the heart. [1 mark]
   b. The cell with a large, lobed nucleus. [1 mark]
   c. Structure that prevents blood flowing from the pulmonary artery into the right ventricle. [1 mark]
   d. Type of cell that forms the capillary wall. [1 mark]

E3. Red blood cells contain haemoglobin.

Each red blood cell contains approximately 2.4×10^8 molecules of haemoglobin.
   a. A certain individual has 6.5×10^6 red blood cells per mm^3 of blood.
   Calculate the number of haemoglobin molecules in $1\,mm^3$ of this person's blood.
   Show your working. [2 marks]
   b. Describe the role of haemoglobin in the transport of:
   i. oxygen [2 marks]
   ii. carbon dioxide. [2 marks]

E4. The cardiac cycle is initiated by the sinoatrial node.
   a. State the location of the sinoatrial node in the heart. [1 mark]
   b. Explain how events in the cardiac cycle ensure that the atria and ventricles do **not** contract at the same time. [4 marks]
   c. Diagrams (a) and (b) in Figure 8.31 show two stages in the cardiac cycle.

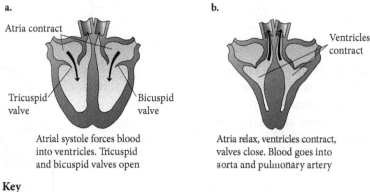

Key
→ Direction of blood flow

Figure 8.31

Describe the changes that take place in the cardiac cycle between (a) and (b) in Figure 8.31. [5 marks]

E5. A student investigated the ability of haemoglobin to bind oxygen.

The student obtained a sample of haemoglobin that was extracted from red blood cells. The haemoglobin was diluted in a buffer solution that was the same pH as that of red blood cells.

Equal volumes of the haemoglobin in the buffer solution were each exposed to different partial pressures of oxygen in solution.

Haemoglobin changes colour slightly when binding with oxygen, so a colorimeter was used to measure the percentage saturation of the haemoglobin with oxygen.

The results are shown in Table 8.3.

a. In this investigation, state the:
   i. independent variable [1 mark]
   ii. dependent variable. [1 mark]
b. The buffer solution ensured that the pH of the haemoglobin remained constant during the investigation.
   State **two other** variables that should be controlled in this investigation. [2 marks]
c. Plot a graph of the results. [4 marks]
d. The partial pressure of oxygen in a particular respiring tissue is 2.7 kPa. The partial pressure of oxygen at the lungs is 14 kPa.
   i. Use your graph to determine the difference in percentage saturation of haemoglobin between the lungs and this respiring tissue. [2 marks]
   ii. Explain why the actual change in percentage saturation of haemoglobin in the circulatory system will be greater than the value you determined in (d) (i). [3 marks]

Partial pressure of oxygen / kPa	Saturation of haemoglobin with oxygen / %
0	0.0
1	8.0
2	24.5
3	42.5
4	58.0
5	71.0
6	80.5
7	85.0
8	88.5
9	92.0
10	93.5
11	96.0
12	97.0
13	97.5
14	98.0

Table 8.3

CHAPTER 9

E1. Which row in the table shows features of the walls of the bronchi? [1 mark]

	Cartilage	Goblet cells	Smooth muscle	Ciliated epithelium
A	✓	✗	✗	✗
B	✓	✓	✗	✗
C	✗	✗	✓	✓
D	✓	✓	✓	✓

E2. Figure 9.12 shows some of the structures in the human gas exchange system.
a. Describe how structure A is adapted for its function. [2 marks]
b. **i.** Give the letters of the structures that contain goblet cells. [1 mark]
   ii. Goblet cells produce mucus. Describe how mucus and cilia work together to maintain the health of the gas exchange system. [4 marks]

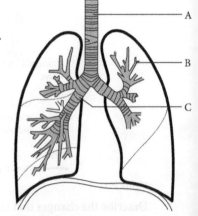

Figure 9.12

E3. A student viewed a transverse section of a structure from the gas exchange system using a light microscope at low power.

The student made a plan diagram of the section (Figure 9.13).

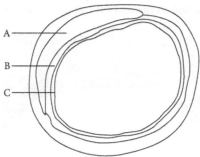

Figure 9.13

   a. Identify the structure of the gas exchange system shown in the plan diagram.   [1 mark]

   b. Identify:

      i. the type of tissue labelled A   [1 mark]

      ii. the type of tissue labelled B   [1 mark]

      iii the type of epithelium labelled C.   [1 mark]

   c. The bacteria *Bordatella pertussis* causes a disease called whooping cough. *B. pertussis* infects the lining of the upper airways and causes loss of cilia.

      Explain why infection with *B. pertussis* causes coughing.   [3 marks]

E4. The surfaces of the gas exchange system are protected by mucus and macrophages.

Mucus contains a glycoprotein called mucin and an aqueous solution of ions.

   a. **i.** Name two structures in the gas exchange system that synthesise and secrete mucus.   [2 marks]

      ii. Explain why mucus contains ions.   [3 marks]

   b. Describe the role of macrophages in protecting the surfaces of the gas exchange system.   [2 marks]

E5. A student researches the gas exchange system.

   a. The student finds a comparison between the efficiency of gas exchange in healthy lungs and in diseased lungs.

      The results are shown in Table 9.3.

Time in alveolar capillary / s	Partial pressure of oxygen in alveolar capillary / kPa	
	Healthy lungs	Diseased lungs
0.0	5.3	5.3
0.1	7.0	6.5
0.2	10.5	7.5
0.3	12.7	8.5
0.4	14.3	10.0
0.5	15.0	11.0
0.6	15.0	12.5

Table 9.3

      i. Draw a graph of partial pressure of oxygen in the alveolar capillary against time in the alveolar capillary. Put healthy and diseased lungs on the same graph. Label the lines clearly.   [4 marks]

      ii. Blood takes 0.6 s to pass though the alveolar capillary that was used in this study.

         Explain the trend in the results for the healthy lungs between 0.5 and 0.6 s.   [2 marks]

b. A student is provided with a microscope slide with a stained section of a mammalian bronchus.

    i. The student places the slide on the microscope stage and shines a light through the slide.
Explain how the student would use the microscope to view this slide at low power. [2 marks]

    The photograph (Figure 9.14) shows the section when viewed at low power.

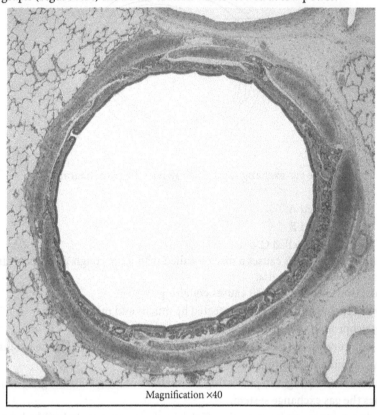

Magnification ×40

Figure 9.14

    ii. Make a large, labelled, plan diagram of the bronchus section from the photograph. [6 marks]

    iii. Calculate the internal diameter of this bronchus. [3 marks]

CHAPTER 10

E1. Which steps could be taken as precautions against these diseases?

    1. use of reverse transcriptase inhibitors

    2. boiling water before drinking

    3. vaccination

    4. covering water storage tanks

	Cholera	Malaria	HIV/AIDS
A	4	2	1
B	1	2	3
C	3	1	4
D	2	4	1

E2. Which of these show correct routes of transmission for each disease? [1 mark]

	Cholera	HIV/AIDS	Tuberculosis
A	blood transfusion	contaminated water	insect vector
B	contaminated water	blood transfusion	unpasteurised milk
C	insect vector	unpasteurised milk	blood transfusion
D	unpasteurised milk	insect vector	contaminated water

E3. Many people who have HIV/AIDS also have tuberculosis (TB).

   a. HIV/AIDS is a sexually transmitted disease. List two other ways that HIV/AIDS can be transmitted. [2 marks]

   b. Table 10.4 shows the results of a study on survival rates of patients who developed TB who already had HIV/AIDS.

    The study shows survival rates for patients who were given antiviral drugs and those who were **not** given antiviral drugs. All patients were given the same antibiotics to treat the TB.

	Survival rate with TB / %	
Time after TB diagnosis / years	Those given antiviral drugs	Those not given antiviral drugs
1	96	44
2	94	19
3	88	9

Table 10.4

    Explain the results shown in Table 10.4. [5 marks]

E4. **a.** Explain why penicillin can kill bacteria but does **not** harm human cells. [3 marks]

   b. The penicillin molecule contains a beta lactam ring structure.

    Many bacteria have evolved to produce an enzyme called beta-lactamase that breaks the beta lactam part of the structure.

    i. Explain how the production of beta-lactamase makes these bacteria resistant to penicillin. [3 marks]

    ii. Clavulanic acid is another molecule containing a beta lactam ring structure. Clavulanic acid has a similar structure to penicillin but is **not** broken down by the bacterial beta-lactamase enzyme.

     Clavulanic acid on its own has no antibiotic activity.

     Suggest how giving patients penicillin together with clavulanic acid can be an effective treatment for penicillin resistant bacterial infections. [2 marks]

E5. A student is researching the global effect of tuberculosis (TB).

   The student finds the graph in Figure 10.18 that shows the numbers of new cases of TB in some countries in 2012.

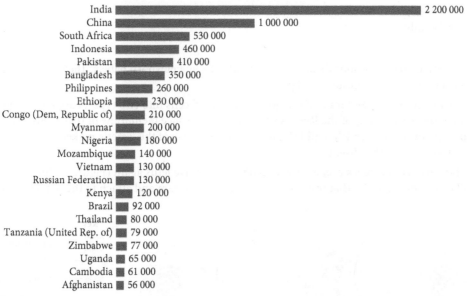

Figure 10.18

   a. The results shown in Figure 10.18 are from World Health Organization estimates.

    Explain why the exact number of people with TB in each country cannot be known in any year. [3 marks]

   b. The prevalence of a disease in a country is the proportion of people in the country who have the disease.

    Explain why the results in Figure 10.18 cannot be used to compare the prevalence of TB in different countries.

c. The populations of four of the countries in 2012 are given in Table 10.5.

Country	Population in 2012
India	1.3×10^9
South Africa	5.3×10^7
Indonesia	2.5×10^8
Vietnam	9.0×10^7

Table 10.5

    i. Use the results in Figure 10.18 and the information in Table 10.5 to calculate the new cases of TB in 2012 as a percentage of the population of each country. Give your answers to an appropriate number of significant figures.
[3 marks]

    ii. Draw a bar graph of the new cases of TB in 2012 as a percentage of the population of each country in Table 10.5.
[3 marks]

d. Explain how the graph in Figure 10.18 and your graph from (c) (ii) can both be used to gain information about the global effect of TB.
[4 marks]

CHAPTER 11

E1. Which statements correctly describe lymphocytes and their functions?
[1 mark]

    1. A small number of B-lymphocytes and T-lymphocytes become memory cells.

    2. Plasma cells develop from B-lymphocytes and secrete antibodies.

    3. Each B-lymphocyte has the ability to make several types of antibody molecule.

    4. Some T-lymphocytes stimulate macrophages to kill infected cells.

        A. 1, 2, 3 and 4

        B. 1, 2 and 3 only

        C. 1, 2 and 4 only

        D. 1 and 4 only

E2. Erythroblastosis fetalis is a serious disorder that can affect some newborn babies. The development of this disorder is explained by the Rhesus blood group.

The Rhesus blood group is a system that categorises red blood cells into two types, based on the presence (positive) or absence (negative) of the Rhesus antigen on the surface membrane of these cells. The disease occurs when a Rh⁻ woman conceives a Rh⁺ child (Figure 11.30a), because of the genetic makeup of the father. Her first baby will not suffer from this disease.

However, if the woman becomes pregnant with a Rhesus-positive fetus again (Figure 11.30c), her immune system will produce antibodies that travel into the fetal bloodstream and attach to the red blood cells in the fetus.

Key

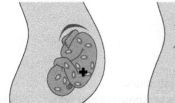

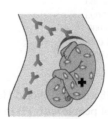
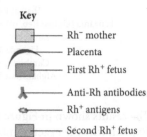

▭—	Rh⁻ mother
⌒—	Placenta
▬—	First Rh⁺ fetus
⅄—	Anti-Rh antibodies
⬭—	Rh⁺ antigens
▬—	Second Rh⁺ fetus

Figure 11.30 The structures shown in this diagram are not to scale.

a. Suggest the cause of the symptoms of erythroblastosis fetalis. [2 marks]

b. Explain why:

   i. erythroblastosis fetalis does not affect the first Rh⁺ child of a Rh⁻ mother [2 marks]

   ii. erythroblastosis fetalis affects the second Rh⁺ child of a Rh⁻ mother. [3 marks]

c. During birth, fetal blood can enter the mother's body. If a mother is injected with antibodies specific to the Rhesus antigen immediately after she gives birth, future immune responses against later fetuses may be avoided. Suggest how the injection of these antibodies would prevent a future immune response to a subsequent Rh⁺ fetus. [2 marks]

E3. Koalas (*Phascolarctos cinereus*) are a species native to Australia (Figure 11.31).

Figure 11.31

In recent decades, the number of koalas in Australia has declined sharply. One reason for this is the spread of the sexually transmitted disease chlamydia, caused by a pathogenic bacterium. Symptoms of chlamydia include blindness and infertility.

One method that has been suggested to conserve koala numbers is to trap and treat sample populations of koalas with antibiotics.

During a trial to investigate the potential effects of this 'trap-and-treat' method, research was carried out to find the percentage of koalas that were immune to chlamydia. At the start of the investigation, the trial area had an estimated population of 2000 koalas.

The research suggested several steps;

- Humane steel traps, all of the same size, were used. Once caught, blood samples were taken from the trapped koalas.
- A standard test for antibodies to chlamydia was used to screen the blood samples.
- The percentage testing positive for antibodies to chlamydia was calculated for each sample.
- The trapped koalas were then provided with antibiotics until symptoms were no longer obvious, and released without harm.

Table 11.6 shows the results of this research.

Year	Number of koalas tested	Percentage positive for chlamydia antibodies
2010	42	65
2012	32	58
2014	59	49
2016	18	19
2018	41	17
2018 (control)	34	55

Table 11.6

a. Explain why the identification of antibodies for chlamydia was a positive test for immunity to chlamydia. [1 mark]

b. You will need to refer to Table 11.6 to help you answer these questions.

  i. Estimate the number of koalas in the original population (in 2010) that suffered from chlamydia. [1 mark]

  ii. State in which year the most reliable data was obtained. Explain your answer. [2 marks]

  iii. Suggest one reason why the number you have calculated may be an underestimate. [1 mark]

  iv. Suggest one reason why the number you have calculated may be an overestimate. [1 mark]

  v. The researchers concluded that the 'trap-and-treat' method could help to reduce the spread of chlamydia among the koala population. Give two reasons to explain why this conclusion may not be valid. [2 marks]

In 2014, Australian scientists developed a vaccine for the type of chlamydia that affects koalas.

c. To achieve a successful vaccination programme, it would not be necessary for all koalas in a population to be vaccinated. Explain why. [2 marks]

d. It normally takes 1–2 weeks for immunity to develop after receiving a vaccine. Explain why vaccines do not offer immediate protection against an infectious disease. [3 marks]

E4. Human immunodeficiency virus (HIV) infects T-helper cells. Once it has entered a cell, its genetic material is incorporated into the chromosomes of the host cell, and is replicated if the cell divides. The immune system will produce antibodies against the virus, and these can be detected in the blood. A person who has anti-HIV antibodies is said to be HIV-positive.

Eventually, after many years, large numbers of new viruses are formed, ready to infect more T-helper cells. Gradually, the number of functioning T-helper cells in the person's body decreases and the person develops AIDS.

a. Diagnosis of HIV is sometimes carried out by identifying antibodies for HIV in a patient's blood. Explain why this method may be ineffective as an HIV test on a newborn baby of an HIV-positive mother.

b. Figure 11.32 shows how the number of copies of HIV and antibodies complementary to HIV vary over time in the blood of a person who has been infected with HIV.

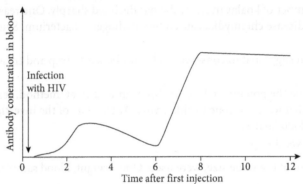

Figure 11.32 The course of an HIV infection.

  i. Describe the process that results in the rise in the number of copies of antibodies complementary to HIV in the blood of the individual over the first three months of infection.

  ii. The number of copies of antibodies complementary to HIV falls between years 1 to 8. Explain why.

  iii. Suggest why a patient suffering from AIDS might die from a usually non-fatal infection or cancer.

CHAPTER 12

E1. The respiratory quotient (RQ) can be used to gain information about respiration in organisms.

  a. State the equation used to calculate RQ. [1 mark]

  b. The fatty acid stearic acid is used for aerobic respiration; the equation is:

  $$C_{18}H_{36}O_2 + \ldots\ldots\ldots O_2 \rightarrow \ldots\ldots\ldots CO_2 + 18H_2O$$

  i. Calculate the number of molecules of oxygen and carbon dioxide that are required for the aerobic respiration of one molecule of stearic acid. [2 marks]

  ii. Calculate the RQ for stearic acid. [1 mark]

c. A study of respiration was carried out using mice.
Mice were given food that was high in starch content.
The mice were then given no food for 4 hours.
Figure 12.17 shows how the RQ of the mice varied with time after feeding.

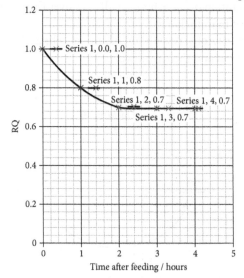

Figure 12.17

Explain the results shown in the graph. [4 marks]

d. Yeast cells were supplied with glucose in a suspension that contained very little oxygen.
The RQ value of the respiring yeast cells increased to a value much greater than 1.0.
Suggest an explanation for this RQ value. [2 marks]

E2. Moths are flying insects. A study was carried out on the rate of oxygen consumption of moths of different body mass when at rest and when flying.

Moths of the family *Sphingidae* were contained within 500 cm³ glass jars at a constant temperature of 22 °C.
The change in volume of oxygen in the air in the jar was measured and used to determine the rate of oxygen consumption.
Figure 12.18 shows the results.

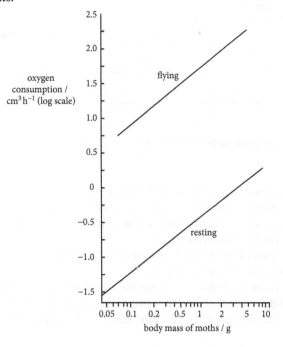

Figure 12.18

    a. Explain the results shown in the graph. [4 marks]

    b. Moths do not regulate their body temperature. The temperature of the body tissues in the moth is approximately the same as the surroundings.

       The temperature range in the habitat of these moths is 22–32 °C.

       Explain how the rate of respiration of these moths will vary over this temperature range. [3 marks]

E3. **a.** Describe how ATP is produced in mitochondria. [9 marks]

    b. Describe the role of NAD in anaerobic respiration in animal cells and in yeast cells. [7 marks]

E4. Figure 12.19 shows the link reaction and the Krebs cycle.

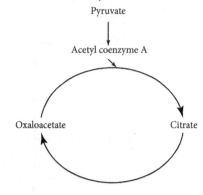

Figure 12.19

    a. State precisely where the link reaction and Krebs cycle occur in cells. [1 mark]

    b. Copy the diagram and add:

       i. the number of carbon atoms in each of the four named substances [2 marks]

       ii. arrows to show the positions and products of decarboxylation [2 marks]

       iii. arrows to show the positions and products of dehydrogenation. [3 marks]

⚠ E5. A student set up a simple respirometer to compare the rates of respiration of:

- yeast
- germinating peas
- blowfly larvae.

The respirometer is shown in Figure 12.20.

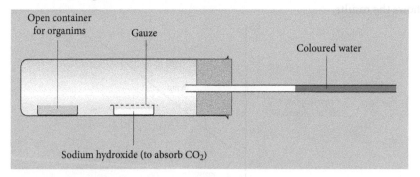

Figure 12.20

The student planned to place each organism into the open container in the respirometer. The rate of oxygen uptake would be shown by the movement of the coloured water in the tube.

    a. Suggest how the respirometer could be maintained at constant temperature. [1 mark]

    b. Describe a suitable control for this investigation. [2 marks]

    c. State the dependent and independent variables in this investigation. [2 marks]

    d. Write a suitable hypothesis for this investigation. [1 mark]

    e. State what measurements are needed to determine the rate of uptake of oxygen in $cm^3 h^{-1}$. [3 marks]

    f. Describe a method that the student could use, with this respirometer, to compare the rates of respiration of these three organisms. The method should include sufficient detail for someone else to follow. [7 marks]

CHAPTER 13

E1. Chloroplasts were isolated from a plant. Figure 13.22 is a transmission electron micrograph of part of a chloroplast.

  a. Write the letter from Figure 13.22 where:

    i. photosynthetic pigments are found   [1 mark]

    ii. the enzyme rubisco is found.   [1 mark]

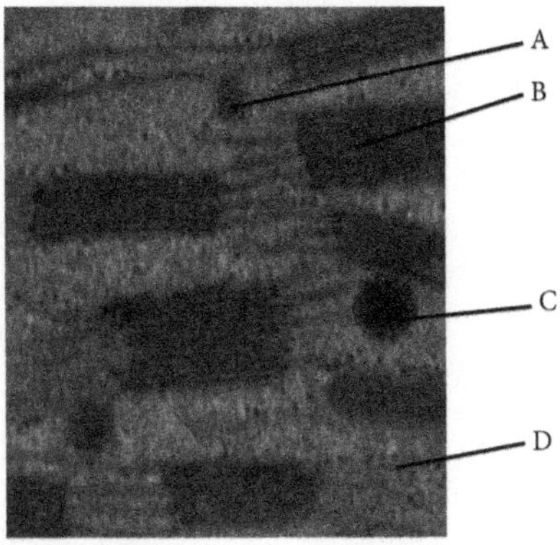

  b. These chloroplasts only contain two photosynthetic pigments.

    i. Name the procedure that can be used to separate photosynthetic pigments.   [1 mark]

    Figure 13.23 shows the absorption spectra of chlorophyll *a* and chlorophyll *b* that were extracted from the chloroplasts.

Figure 13.22

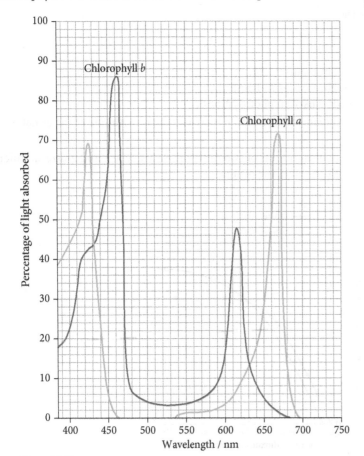

Figure 13.23

    ii. Describe how the absorption spectra of chlorophyll *a* and chlorophyll *b* compare.   [3 marks]

    iii. Suggest the ranges of wavelengths that give the minimum and maximum rates of photosynthesis in the action spectrum of these chloroplasts.   [2 marks]

E2. A plant grower produces tomato plants in a greenhouse.

Artificial lights are used to maintain high light intensity.

  a. Figure 13.24 summarises the steps that occur in the light-dependent stages of photosynthesis.

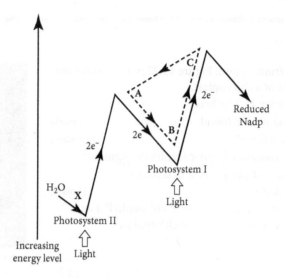

Figure 13.24

    i. Describe what happens at step X in Figure 13.24.   [3 marks]

    ii. Name the process shown by the three steps A, B and C.   [1 mark]

    iii. Describe how the reduced NADP is used in photosynthesis.   [2 marks]

  b. Explain why increasing the carbon dioxide concentration increases the growth rate of the tomatoes at high light intensity.   [6 marks]

E3. A study was carried out on the rates of photosynthesis of two different plants, wheat and maize, at different carbon dioxide concentrations. Figure 13.25 shows the results.

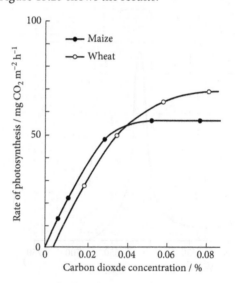

Figure 13.25

  a. The rate of photosynthesis in the graph was measured as the rate of uptake of carbon dioxide.

    i. Name the enzyme that uses carbon dioxide as a substrate in photosynthesis.   [1 mark]

    ii. Name the other substrate of this enzyme when the product of the enzyme is glycerate 3 phosphate.   [1 mark]

  b. Explain the results in Figure 13.25.   [6 marks]

E4. *Elodea* is a plant that lives in water (Figure 13.26). An investigation on the effect of temperature on the rate of photosynthesis of *Elodea* was carried out.

Figure 13.26

Each experiment was carried out using a fresh sample of *Elodea* in water that had the same initial carbon dioxide concentration. The light intensity was constant for all experiments. Each experiment was carried out for 60 s.

The results are shown in Table 13.2.

Temperature / °C	Uptake of CO_2 in 60 s / nmol	Rate of photosynthesis / nmol s^{-1}
0	48	0.8
5	126	
10	204	
15	318	
20	450	

Table 13.2

a. Copy and complete Table 13.2. [1 mark]
b. Plot a graph of the results in Table 13.2 to show the effect of temperature on the rate of photosynthesis. [4 marks]
c. State the limiting factor for the rate of photosynthesis in all five experiments. [1 marks]
d. Describe and explain the trend shown in these results. [3 marks]

E5. *Chlorella* is a single-celled photosynthetic eukaryote that lives in water.

Figure 13.27 shows apparatus that can be used to measure the rate of photosynthesis of *Chlorella*.

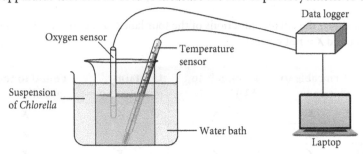

Figure 13.27

a. Describe a method where the apparatus in Figure 13.27 could be used to investigate the effect of temperature on the rate of photosynthesis of *Chlorella*. [8 marks]
b. i. Sketch a graph of the results you would predict for this experiment. [3 marks]
ii. Explain the shape of your graph. [3 marks]

CHAPTER 14

E1. Figure 14.23 shows a simple summary of the activities of the kidney. Identify the processes and fluids labelled in Figure 14.23.

	Ultrafiltration	Selective reabsorption	Filtered Blood	Urine
A	P	Q	R	S
B	Q	P	R	S
C	P	Q	S	R
D	Q	P	S	R

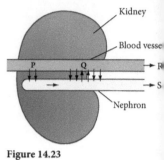

Figure 14.23

E2. Figure 14.24 shows a simple diagram of a nephron.

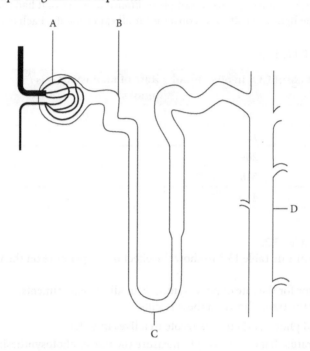

Figure 14.24

Identify the row that shows the correct functions of the four labelled sections of the nephron in the table below. Note that ✓ = correct and ✗ = not correct. [1 mark]

	Permeable to protein	Responds to ADH	Contains glucose	Found in the medulla
A	✗	✗	✗	✗
B	✗	✗	✓	✓
C	✓	✓	✗	✗
D	✓	✓	✓	✓

E3. Urea is the molecule of nitrogenous waste in mammals. Outline the formation of urea from excess amino acids by liver cells. [2 marks]

E4. Abscisic acid (ABA) is a plant stress hormone that causes stomata to close during high temperatures or dry conditions. A student investigated the effect of ABA on stomata aperture (diameter of the open pore). The student was provided with:

- 100 cm³ of 1.0 mmol dm⁻³ ABA solution
- Freshly picked leaves from a plant that had been kept in light for 3 hours to ensure that all of its stomata were fully open.

a. i. Identify the independent variable and the dependent variable in this investigation. [2 marks]

ii. The student used serial dilution to prepare 50 cm³ of four solutions of ABA of concentrations 0.5 mmol dm⁻³, 0.25 mmol dm⁻³, 0.125 mmol dm⁻³ and 0.0625 mmol dm⁻³. Describe how they were prepared. [3 marks]

b. State a null hypothesis that the student could test about the effect of ABA concentration on the stomatal aperture. [1 mark]

c. To ensure that a valid comparison can be made of the results, the student considered what else to include in their method.

i. Identify three variables that the student should standardise in this investigation. [3 marks]

ii. Describe how the student could have standardised two of these variables. [2 marks]

iii. The stomata chosen for measurement were selected using a grid and random number generator. Explain why. [1 mark]

d. Explain why a control is necessary in this investigation. [1 mark]

e. Explain how the student could estimate the minimum concentration of ABA solution that has a measurable effect on the stomatal aperture. [2 marks]

f. State one hazard in this investigation and explain how the risk can be minimised. [1 mark]

E5. A study was carried out into the effect of eating a meal on blood glucose concentration and the concentration of two hormones in the blood, insulin and glucagon. The results are shown in Figure 14.25.

a. Calculate the percentage increase in the concentration of insulin between the time when a meal was eaten and 2 hours. Show your working.

b. Explain why the person was told not to eat or drink anything other than water for 12 hours before the meal. [2 marks]

c. Predict the shapes of the lines for another hour after this study ended, assuming that the individual did not take another meal. [3 marks]

d. Sketch a graph to show the relative amounts of cAMP in liver cells during the course of this study. You do not need to include values on your axes. [2 marks]

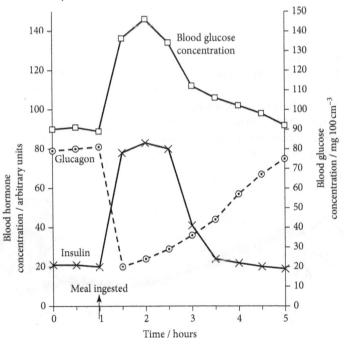

Figure 14.25

Practice exam-style questions

△ **E6.** The pH of urine is very variable. After consuming certain foods or as the result of specific disorders, the pH can be as low as 4.5. In other scenarios, it can rise to a value of 8. On average, the pH of urine is around 6.

A pharmaceutical company aimed to develop a more effective glucose biosensor, able to display accurate numerical results over a wider pH range.

a. State whether the glucose biosensor is a qualitative, quantitative or semi-quantitative test. Explain your answer. [1 mark]

b. Explain why the glucose biosensor may show a difference in activity when exposed to urine of different pH values. [2 marks]

The pharmaceutical company prepared a graph to show how the Michaelis–Menten constant (K_m) of a fungal extract varied over a pH range (Figure 14.26).

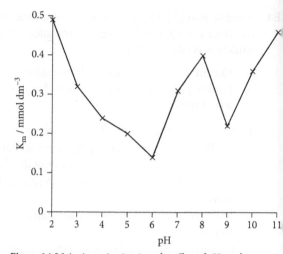

Figure 14.26 An investigation into the effect of pH on the affinity of glucose oxidase in a fungal extract for its substrate.

c. Suggest what was added to each of the test-tube to maintain the pH. [1 mark]

d. Identify a hazard in this investigation and state a means by which risk can be minimised. [1 mark]

e. In this investigation, temperature is a standardised variable. Explain what this means, and describe how temperature could be standardised. [2 marks]

f. Suggest a suitable control for this investigation and explain why this is necessary. [2 marks]

g. Further tests showed that the fungal extract contained two slightly different glucose oxidase enzymes. Explain how Figure 14.26 illustrates this. [3 marks]

CHAPTER 15

E1. Figure 15.30 shows a cholinergic synapse.
  a. Name the substance labelled:
    i. A [1 mark]
    ii. B [1 mark]
    iii. C [1 mark]
  b. Describe the effect of the movement of C. [2 marks]
  c. With reference to Figure 15.30, explain the role of acetylcholinesterase. [3 marks]
  d. Atropine is a drug that can be used to decrease saliva production by the salivary glands during surgery. Atropine has a similar shape to acetylcholine but does not cause ion channels to open. Suggest how atropine works. [4 marks]

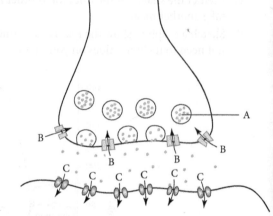

Figure 15.30

E2. Figure 15.31 shows a sensory neurone.

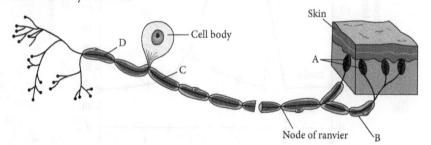

Figure 15.31

a. Name the parts labelled:

    i. A   [1 mark]

    ii. B   [1 mark]

    iii. C   [1 mark]

    iv. D   [1 mark]

b. Explain the role of the cell body in maintaining the resting potential in the sensory neurone.   [3 marks]

c. Figure 15.32 shows how the membrane potential in the sensory neurone varies with the stimulus strength.

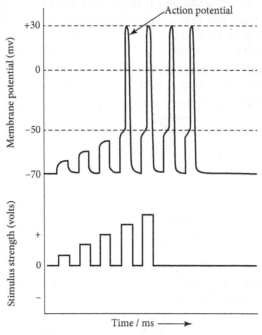

Figure 15.32

    i. State the term used for the membrane potential of $^-70$ mV in Figure 15.32.   [1 mark]

    ii. Describe what the information in Figure 15.32 shows about this sensory neurone.   [3 marks]

E3. The sliding filament model explains how striated muscle contracts.

Figure 15.33 is a diagram of a sarcomere in the relaxed position.

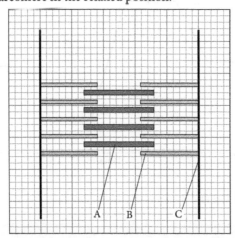

Figure 15.33

a. State what is represented by:

    i. A   [1 mark]

    ii. B   [1 mark]

    iii. C   [1 mark]

  b. Draw the same sarcomere in the contracted position. [2 marks]
  c. Outline the role of each of the following in the sliding filament model:
      i. calcium ions [2 marks]
      ii. myosin heads. [2 marks]

E4. a. Describe the need for control and co-ordination systems in mammals and in plants. [5 marks]
  b. Outline the differences between control systems in mammals and in plants. [5 marks]

△ E5. A student carried out an investigation into the effect of auxin on the pH of plant cell walls. Auxin was added and the pH measured every 10 minutes. The results are shown in Table 15.6.

Time after adding auxin / minutes	Cell wall pH
0	5.9
10	5.8
20	5.2
30	4.9
40	4.8
50	4.7
60	4.7

Table 15.6

  a. i. Plot a graph of the results in Table 15.6. [3 marks]
      ii. Complete the graph with a smooth curve. [1 mark]
  b. Use your graph to determine the rate of change of pH at 15 minutes. Show your working and give your answer in min^{-1}. [3 marks]
  c. Figure 15.34 shows a method for applying auxin to a plant stem.

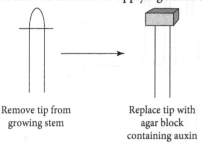

Remove tip from growing stem   Replace tip with agar block containing auxin

Figure 15.34

Describe an experiment, using the method shown in Figure 15.34, to show the effect of varying concentration of auxin on rate of elongation of the stem. [8 marks]

CHAPTER 16

E1. Figure 16.32 shows the DNA content of a cell that is undertaking a series of processes over a period of time.

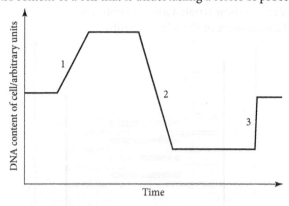

Figure 16.32

Identify the processes numbered 1, 2 and 3. [1 mark]

	1	2	3
A	DNA replication	mitosis	meiosis
B	fertilisation	meiosis	mitosis
C	mitosis	meiosis	DNA replication
D	DNA replication	meiosis	fertilisation

E2. Manx cats do not have tails. Tailless cats are heterozygous. Normal cats with tails have two copies of the dominant allele (T).

    a. Draw a genetic diagram to show how two Manx cats can have a kitten with a tail. [4 marks]

    b. Genetic screening showed that embryos that are homozygous for the recessive allele fail to develop in the womb and are never born. By drawing a Punnett square, predict the actual ratio of phenotypes that would be formed between a cross of two heterozygous parents. [2 marks]

E3. Huntington's disease (HD) is caused by a dominant allele of a gene called *HTT* which contains many repeats of a triplet sequence of nucleotides with the base sequence CAG. An investigation was carried out into the effect of number of CAG triplet repeats in the *HTT* gene and the age at which symptoms of HD first appear. The age at which symptoms of HD first appear (Figure 16.33).

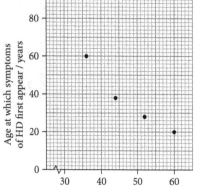

    a. State the age at which symptoms of HD first appear if an individual is found to have 50 CAG triplet repeats. [1 mark]

    b. To make a valid comparison of the variables such as those in Figure 16.33, state two variables which the investigators should standardise. [2 marks]

    c. Identify a statistical test that could be used to determine if the correlation between the number of CAG triplet repeats in the *HTT* gene and the age at which symptoms of HD first appear is significant. [1 mark]

Figure 16.33

    d. What other information is required to determine if this study is reliable? Explain your answer and describe how this information should be illustrated on the graph. [3 marks]

E4. The inheritance of comb shape in chickens can be explained by two genes on different autosomes (Figure 16.34):
- walnut-combed male (AABB, AaBB or AABb, which can be described as A–B–)
- rose-combed male (Aabb or AAbb, which can be described as A–bb)
- pea-combed male (aaBB or aaBb, which can be described as aaB–)
- single-combed male (aabb).

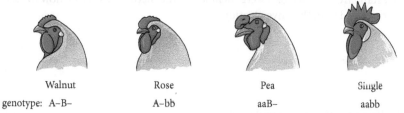

Walnut	Rose	Pea	Single
genotype: A–B–	A–bb	aaB–	aabb

Figure 16.34 Four comb shapes in chickens. The comb is a flap of skin found on top of the head. The genotype of two genes involved in determining comb shape is shown.

A cross was made between two chickens of genotype AaBb.

    a. Explain why the offspring of these two parents are referred to as the F2 generation. [1 mark]

    b. Draw a genetic diagram to show the possible phenotypes of the offspring in the F2 generation. [5 marks]

    c. Draw a Punnett square to identify the ratio of phenotypes of the chickens in the F2 generation. [4 marks]

    d. Using the information provided, explain why the inheritance of comb shape in chickens is an example of epistasis. [2 marks]

E5. A recent study by scientists in China investigated the process by which petunia flowers discolour and wilt. This appears to involve a transcription factor named PhFBH4, which is produced by petal cells in greater amounts as they get older.

    a. Explain what is meant by the term transcription factor. [2 marks]

    b. Predict what the scientists observed when:

        i. they made plants in which the gene encoding PhFBH4 is 'switched off' by hydrolysing the mRNA transcribed from the *PhFBH4* gene [2 marks]

        ii. they made transgenic (genetically modified) plants that produced more PhFBH4 protein than normal. [1 mark]

    c. Figure 16.35 shows the results of an investigation into the effect of a number of environmental factors on the expression of the *PhFBH4* gene in petunia plants. Different plants were exposed to different environmental conditions for 12 hours. At each 3-hour interval, the scientists determined the expression levels of *PhFBH4* in each plant by measuring the concentration of mRNA for this gene in petal cells. 'Standardised conditions' refers to the conditions usually associated with normal room conditions for keeping plants at home, such as a temperature of 25 °C, in plenty of tap water (normal root medium).

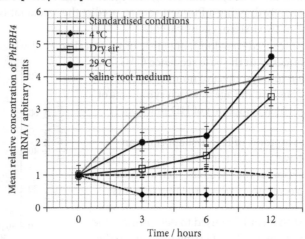

Figure 16.35 The effect of various environmental conditions on the expression of the *PhFBH4* gene in petunia plants.

i. Explain why measuring mRNA concentration provides an accurate indication of the rate of transcription of a gene. [2 marks]

ii. State the null hypothesis that the researchers could test about the effect of dry air on the expression of the *PhFBH4* gene in petunia plants. [1 mark]

iii. Compare the effects of high temperature and saline root medium on *PhFBH4* expression, with standardised conditions. [3 marks]

iv. To ensure that a valid comparison can be made of the effects of environmental factors on *PhFBH4* expression, standardised variables were included in the method. Describe how to standardise two variables that are not referred to in this investigation. [2 marks]

CHAPTER 17

E1. The following all affect variation in a population:
   1. immigration   2. genetic drift
   3. mutation   4. non-random mating.

   Which will reduce the ability of a population to evolve in response to an environmental change? [1 mark]
   A. 1 and 2 only   **B.** 1 and 3 only
   C. 2 and 4 only   **D.** 3 and 4 only

E2. Table 17.9 shows the differences in the frequency of three alleles, found in populations of a species of bat found in three cities in the Middle East and North Africa (MENA). Bats hunt at night and use advanced hearing to locate their prey.

	Amman, Jordan	Cairo, Egypt	Muscat, Oman
Allele A	0.55	0.72	0.10
Allele B	0.43	0.44	0.29
Allele C	0.08	0.01	0.05

Table 17.9

a. Allele A is recessive and is of a gene with only two variants. Calculate the frequency of heterozygotes in the population of bats in Amman. The two Hardy–Weinberg equations are $p + q = 1$ and $p^2 + 2pq + q^2 = 1$. [2 marks]

b. Use Table 17.9 to explain why alleles A and B are likely to be of different genes. [1 mark]

c. Bats that carry allele C have reduced levels of hearing. Explain why the Hardy–Weinberg principle does not apply to this allele. [1 mark]

d. The scientists expanded the investigation to determine the difference in frequency of allele C between bats found in different global regions. The scientists used separate *t*-tests to find out if the differences in allele frequencies were significant. Table 17.10 shows the results of these calculations.

	$P < 0.05$
MENA and Sub-Saharan Africa	not significant
MENA and Southern Asia	not significant
MENA and Southeast Asia	significant

Table 17.10

i. State one reason why the *t*-test is a suitable statistical test to use for the scientists' data. [1 mark]

ii. Using the terms 'probability' and 'chance' in your answer, discuss the significance of the differences between the frequency of allele C in these populations. [2 marks]

E3. Antibiotics are substances that kill bacteria but do not normally harm human cells. An example is penicillin: it stops cell wall formation, so preventing cell reproduction. However, some bacteria are resistant to penicillin, in some cases because they have an allele that codes for the production of an enzyme, penicillinase, which inactivates the antibiotic.

A microbiologist investigated the development of penicillin resistance in bacteria. In this investigation:

- A long rectangular container was divided into six sections (Figure 17.34a).
- Five different concentrations of penicillin were prepared by serial dilution using distilled water from a stock solution of 1.0 mg cm^{-3}. Six solutions of sterile medium containing no antibiotic, or antibiotic of the five different concentrations, were poured into the six different sections. These were then left to solidify.
- At the start of the investigation (time = 0 hours), the microbiologist inoculated a region of section A with 10 cm^3 suspension of genetically identical bacteria that had no resistance to the antibiotic.
- The plate was then incubated for 10 days at a constant temperature. Figure 17.34b shows some of the observations over this time period. The time point at which bacteria began to grow into each new section was recorded. For example, the bacteria took 8 hours to develop sufficient resistance to move into section B (from the time point 24 hours to 32 hours).

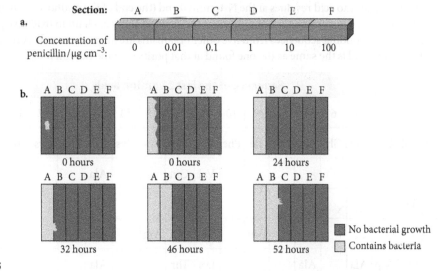

Figure 17.35

a. Identify the independent and the dependent variables in this investigation. [2 marks]

b. Describe how serial dilution could be used to prepare solutions of penicillin of the concentrations shown in Figure 17.34a. [3 marks]

  c. Explain why:

    i. a sterile medium was used [1 mark]

    ii. the temperature was kept constant. [1 mark]

  d. Explain how bacteria evolve to become able to transition into a new medium section after 32 and 52 hours after the experiment started. [3 marks]

The scientist conducted 10 replicates of this investigation. Table 17.11 shows the results.

Concentration of penicillin in section / µg cm^{-3}	Mean time taken for transition ± SE / hours	Confidence intervals / hours	
		Lower limit	Upper limit
0.01	8.0 ± 0.15	7.7	8.3
0.1	6.0 ± 0.30	5.4	6.6
1	5.8 ± 0.10	5.6	6.0
10	5.4 ± 0.15		
100	5.1 ± 0.35	4.4	5.8

SE = standard error

Table 17.11

    i. Explain why the time taken for bacteria to evolve resistance to the same concentration of penicillin differs between the 10 replicates. [1 mark]

    ii. Calculate the difference between the mean time of transition to section B and the mean time of transition to section F. [1 mark]

    iii. The confidence interval = mean ± 2SE. Use this formula to calculate the missing confidence intervals. [1 mark]

  e. One conclusion from this data is that the rate at which bacteria evolve antibiotic resistance increases as the concentration of antibiotic increases.

    i. Identify two successive sections which appear to support this conclusion and give a reason for your choice. [2 marks]

    ii. State which statistical test could have been used to confirm this conclusion and give a reason for your choice. [2 marks]

    iii. State a null hypothesis for this test. [1 mark]

    iv. Use the results of this investigation to discuss whether antibiotic resistance in bacteria demonstrates continuous or discontinuous variation. [2 marks]

E4. Table 17.12 shows the 22 amino acid residues at the N-terminal end (the end of the amino acid chain that finishes with an amine group) of human cytochrome c, a component of the electron transfer chain in oxidative phosphorylation in mitochondria. The corresponding sequences from five other organisms are aligned beneath. A blank cell in the table indicates that the amino acid is the same as the one found at that position in the human molecule.

Species	Sequence of the last 22 amino acids																					
	1	2	3	4	5	6	7	8	9	10	11	12	13	14	15	16	17	18	19	20	21	22
human	Gly	Asp	Val	Glu	Lys	Gly	Lys	Lys	Ile	Phe	Ile	Met	Lys	Cys	Ser	Gln	Cys	His	Thr	Val	Glu	Lys
pig											Val	Gln			Ala							
chicken			Ile								Val	Gln										
fruit fly									Leu		Val	Gln	Arg		Ala							Ala
wheat		Asn	Pro	Asp	Ala		Ala				Lys	Thr			Ala						Asp	Ala
yeast		Ser	Ala	Lys			Ala	Thr	Leu		Lys	Thr	Arg		Glu	Leu						

Table 17.12 Differences in the amino acid sequences of cytochrome c between humans and other species.

1	2	3	4	5	6	7	8	9	10	11	12	13	14	15	16	17	18	19	20	21	22
Gly	Asp	Val	Glu	Lys	Gly	Lys	Lys	Val	Phe	Val	Gin	Lys	Cys	Ala	Gin	Cys	His	Thr	Val	Glu	Asn

Table 17.13 The sequence of amino acids in the cytochrome of dogfish.

    a. List at which positions in the cytochrome c molecule the amino acids are always the same, in every species tested (Table 17.12). [1 mark]

    b. Cys stands for the amino acid cysteine. Disulphide bonds form between cysteine molecules. Explain what the information in Table 17.12 implies about the tertiary and quaternary structure of cytochrome c from all species. [2 marks]

    c. More closely-related species have more similarity in the amino acid sequence of shared proteins. Table 17.13 shows the sequence of amino acids in the cytochrome of dogfish. Explain the position of the dogfish among the species in Table 17.12 and explain your decision. [2 marks]

    d. Suggest how the data in Table 17.13 could be used to make a quantitative estimate of the relative closeness of relationship between each of the species. [3 marks]

E5. The Hawaiian Islands are in the Pacific Ocean. Figure 17.35 shows the islands and their dates of formation, millions of years (My) before the present. Fossil skulls of extinct birds belonging to a family called the moa-nalo have been found on three of the four island groups. Evidence suggests that these were giant, flightless birds with features similar to modern ducks.

Recently, mitochondrial DNA has been extracted from the moa-nalo fossils for analysis. Mitochondrial DNA has a higher rate of mutation than nuclear DNA.

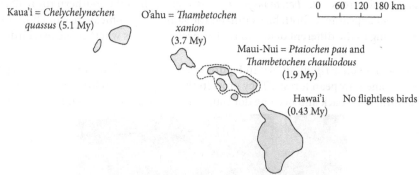

Figure 17.35 The Hawaiian islands with the ages of each island listed (My = millions of years). The dotted line shows the size of Maui-Nui around 1 million years ago.

    a. Describe the advantages of using mitochondrial DNA rather than nuclear DNA in classifying organisms. [2 marks]

    b. Explain why evolution of different species of moa-nalo on each island supports Darwin's theory of natural selection. [4 marks]

    c. Explain why studies of evolution and the formation of new species often feature:

        i. islands [2 marks]

        ii. birds. [2 marks]

    d. The date of an island's formation provides a maximum age for species inhabiting the island. Give reasons to suggest why there were never any giant flightless birds on the island of Hawai'i. [2 marks]

    e. There are more species of plants endemic to Kauai, Oahu and Maui-Nui covered with spines than those endemic to Hawai'i. Explain why. [3 marks]

E6. If their environment is stable, many species of snail will reproduce asexually. However, if a new pathogen or parasite infects a population, some snails are able to switch to sexual reproduction.

    a. Suggest why genetic drift is more likely to affect small populations of snails than larger populations. [2 marks]

    b. Suggest some of the advantages of asexual reproduction while the environment is stable. [2 marks]

    c. Describe why the effects of a genetic bottleneck on a species are similar to those associated with long periods of asexual reproduction. [3 marks]

    d. Suggest why some organisms change their reproduction from asexual to sexual if a new infectious disease or parasite emerges. [3 marks]

CHAPTER 18

E1. The southern royal albatross is one of the largest flying bird species. Table 18.12 shows the taxonomic classification of the southern royal albatross. The information is in the correct order, but some information is missing.

Taxon	Name
kingdom	Animalia
	Chordata
	Aves
	Procellariiformes
	Diomedeidae
genus	*Diomedea*
species	*epomophora*

Table 18.12

    a. State the domain to which the southern royal albatross belongs. [1 mark]
    b. Name the class to which the southern royal albatross belongs. [1 mark]
    c. Give the scientific binomial name of the southern royal albatross. [1 mark]
    d. The light-mantled albatross, *Phoebetria palpebrata*, is smaller than the southern royal albatross and has a very different appearance. Both birds inhabit the same geographical area and hunt for prey in the same way. Referring to the different definitions of a species, discuss whether the two birds are of the same species. [3 marks]

E2. Bluebells (*Hyacinthoides non-scripta*), shown in Figure 18.20, are small flowering plants that grow from bulbs among the trees in some European forests. They grow in the early spring before leaves are on the trees, flower in the late spring, and then die back and become dormant during the summer when it is shady under the trees.

Figure 18.20

In a study of the bluebells, the average number of bluebell flowers per square metre was determined.

    a. Explain how a frame quadrat could be used to gather representative data to determine the average number of bluebell flowers per square metre. [4 marks]
    b. Bluebells flower in the months of April and May. Scientists wanted to know if climatic conditions the previous year affected the flowering of the bluebell plants in the next year. Permanent quadrats were set up the previous summer, and the soil moisture content and average daily soil temperature were recorded in each

one during June and July of one particular year. The bluebell flower density was measured in each quadrat in April and May of the following year. Spearman's rank correlation was calculated for each combination of month and soil condition. The results are shown in Table 18.13.

Month	Soil moisture content	Average daily soil temperature
June	+0.77 ($P < 0.05$)	−0.83 ($P < 0.05$)
July	+0.69 ($P < 0.05$)	−0.71 ($P < 0.05$)

Table 18.13

    i. State the type of factors that include quantity of rainfall and average daily temperature in an ecosystem. [1 mark]

    ii. Explain what the results in Table 18.13 show about ideal conditions in the previous year for flowering of bluebells. [3 marks]

    iii. Scientists have predicted that global warming will make northern European summers become warmer and drier. Use this data to predict the long-term future of the bluebell. [2 marks]

  c. Japanese knotweed is a plant that was introduced to many European countries from Asia in the 19th century. Japanese knotweed is a tall, fast-growing plant, that reaches a height of 2 m. Bluebells grow with their leaves at a maximum height of 0.1 m.

    Suggest how Japanese knotweed in woodlands can cause bluebells to be endangered in those areas. [3 marks]

E3. A study was carried out to compare the numbers of insect species found in a forest and in a field of crops that was close to the forest. Traps were set up in both locations and the numbers of insects of each species caught in the traps were recorded. The results are shown in Table 18.14.

Species of insect	Number of organisms of each species caught in trap	
	Forest	**Field**
bird-cherry oat aphid	0	216
beech aphid	563	0
large white butterfly	20	0
lacewing	12	3
7-spot ladybird	36	0
2-spot ladybird	9	1
total number of organisms of all species	640	220

Table 18.14

  a. The insect species in Table 18.14 are all flying insects. Some of these insects are less than 2 mm in length and can move over a very large area. Explain why the mark-release-recapture method would **not** be suitable for estimating the population of these insects. [2 marks]

  b. i. Use Simpson's index of diversity to calculate the index of diversity for the insect species in the forest. Show your working. [4 marks]

$$D = 1 - \left(\sum \left(\frac{n}{N} \right)^2 \right)$$

    where n = number of individuals of each type present in the sample (types may be species and/or higher taxa such as genera, families, etc.) and N = the total number of all individuals of all types present in the sample.

    ii. Estimate, without calculation, how the diversity index of the insects in the field differs from that of the forest. Give two reasons for this. [3 marks]

  c. Explain whether the forest or the field is the more stable ecosystem. [2 marks]

E4. All of the world's major ecosystems are found on the continent of South America.

  a. Explain how these different ecosystems contribute to biodiversity in South America. [5 marks]

  b. Outline the need to maintain biodiversity. [3 marks]

△ **E5.** The spikemoss fern (*Selaginella involvens*) shown in Figure 18.21, is a low-growing plant found in tropical forests.

Figure 18.21

A student noticed that there were more *S. involvens* plants in some areas of the forest than in others. The student proposed the hypothesis that the abundance of *S. involvens* was positively correlated with light intensity.

  a. Describe a suitable sampling technique for this investigation. [3 marks]

  b. Describe any preliminary work that should be carried out to ensure data are as representative as possible. [3 marks]

  c. Describe how variables should be controlled or monitored in this investigation. [4 marks]

  d. Describe how the results should be analysed and give the expected outcome if the student's hypothesis is correct. [3 marks]

  e. The student measured the percentage cover of *S. involvens* and recorded the light intensity at different locations in the forest. The results are shown in Table 18.15.

Percentage cover of *S. involvens* / %	46	44	50	75	45	40	45	66
Light intensity / arbitrary units	870	920	1120	2050	270	580	800	1590

Table 18.15

The standard deviations for the results are:

- percentage cover of *S. involvens*, 12.4
- light intensity, 565.

  i. Use these values and the results in the table to calculate a value for Pearson's linear correlation coefficient in this study. The equation for Pearson's linear correlation coefficient is:

$$r = \frac{\sum xy - n\bar{x}\bar{y}}{(n-1)s_x s_y}$$

[4 marks]

  ii. Comment on the value of Pearson's linear correlation coefficient. [1 mark]

CHAPTER 19

E1. Babies born with adenosine deaminase (ADA)-associated severe combined immunodeficiency (SCID) can be treated using gene therapy. Some of the steps in the procedure are listed:
  1. Bone marrow cells are removed from baby with ADA-SCID.
  2. Normal allele for adenosine deaminase taken up by stem cells.
  3. Stem cells from bone marrow are obtained.
  4. Stem cells transfused back into the baby.
  5. Stem cells infected with genetically engineered virus containing functional allele for ADA.
  6. Immune system produces T- and B-lymphocytes.

  What is the correct order of the events in this treatment? [1 mark]
  A. 1→3→5→2→4→6
  B. 2→4→3→1→3→6
  C. 5→3→1→6→2→4
  D. 6→1→2→5→3→4

E2. The Japanese chemist Osamu Shimomura purified green fluorescent protein (GFP) from jellyfish. In recent decades, the gene that encodes GFP has become one of the most widely used marker genes in genetic engineering, in preference to previously used marker genes. Marker genes are introduced into transgenic organisms to provide an indication of levels of gene expression.
  a. Name the type of enzyme that would have been used to cut the GFP gene out of the DNA of the jellyfish. [1 mark]
  b. An alternative method to obtain donor DNA is to collect mRNA from cells in the jellyfish. Suggest why isolating the mRNA coding for GFP in a jellyfish cell is easier than isolating the DNA for GFP in a cell. [2 marks]
  c. It is possible for scientists to make GMOs that produce GFP in only one tissue. Suggest and explain how this could be done. [2 marks]
  d. Describe why the use of previously used marker genes has now been discouraged. [2 marks]
  e. List the characteristics of marker gene products that would make them useful in genetic engineering. [3 marks]
  Glo-fish® were the first GM organism to be marketed as pets. These were produced by introducing the GFP gene into the embryo of a zebrafish (*Danio rerio*). Glo-fish® fluoresce under certain wavelengths of light. In 2003, Glo-fish® were approved for sale as pets in the USA.
  f. Outline how the gene construct is inserted into cells of *D. rerio*. [3 marks]
  g. Most GM animals are not approved as easily as Glo-fish®. Suggest reasons to explain why it has proved difficult for GM salmon to receive approval for human consumption. [3 marks]

E3. The polymerase chain reaction (PCR) is a laboratory technique used to amplify DNA. Figure 19.26 shows some of the temperature changes that occur during one cycle of PCR.

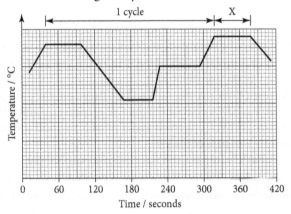

Figure 19.26

  a. Describe and explain the events that occur at the time period indicated by X. [3 marks]

b. Estimate the time taken for *Taq* DNA polymerase to copy the DNA template in this example. [1 mark]

c. Explain why *Taq* DNA polymerase does not need to be added for each cycle. [1 mark]

During PCR, small sections of single-stranded DNA called primers bind to regions of the template DNA. *Taq* DNA polymerase attaches to these regions of double-stranded DNA to begin replication.

A team of scientists undertook a series of investigations in an attempt to improve the efficiency of PCR to amplify DNA. One problem that scientists encounter with PCR is that some of the primer molecules sometimes attach to each other because of regions of complementary base sequence in the primers. The result is the formation of 'primer dimers'.

In this investigation:

- The scientists made five pairs of primers of length 18 nucleotides each, which all had a complementary sequence to each other. The five pairs of primers had different ratios of C+G to A+T bases.
- For each primer pair, they prepared a solution of 0.1 µmol dm^{-3} by carrying out a serial dilution using the initial stock solution of 0.1 µmol dm^{-3}.
- They added 500 ml of each primer pair solution to a PCR reaction tube and heated this mixture slowly from room temperature.
- Using a special colorimeter, the scientists measured the temperature at which the DNA strands dissociated from each other. This is called the melting temperature (T_m).

d. Identify the independent and dependent variables in this investigation. [2 marks]

e. State three factors that were standardised by the investigators. [3 marks]

f. State one reason why the results of the investigation were considered to be reliable. [1 mark]

g. Describe how a solution of 0.1 µmol dm^{-3} could be prepared using serial dilution of the initial stock solution of 0.1 µmol dm^{-3}. [3 marks]

h. Identify a statistical test that could be used to determine whether there is a significant correlation between the ratio of C+G to A+T in the primer dimer and the melting temperature of DNA. Explain your answer. [2 marks]

i. Suggest an explanation for the results of this study. [2 marks]

Table 19.6 shows the results of this investigation.

Ratio of C+G to A+T in primer dimer	Melting temperature of DNA (T_m)/ °C
1.0	53.28
1.5	54.38
2.0	54.42
2.5	55.55
3.0	56.60

Table 19.6

E4. Huntington's disease (HD) is caused by dominant alleles of the *HTT* gene containing a region of CAG repeats. The number of these repeats determines whether or not an allele of this gene will cause the disorder. Generally, an allele with 36 or more CAG repeats will cause HD.

Scientists took DNA samples from three people, A, B and C. They used the polymerase chain reaction (PCR) to produce many copies of the piece of DNA containing the CAG repeats obtained from each person. They then separated the DNA fragments by gel electrophoresis and a suitable method was used to visualise the fragments. Figure 19.27 shows the appearance of part of the gel.

a. Explain why PCR was necessary before genetic screening was carried out. [2 marks]

b. Explain why fragments of DNA of different lengths are separated by gel electrophoresis. [3 marks]

c. One of these three people tested positive for Huntington's disease. Identify this person and explain your answer. [2 marks]

d. The diagram only shows part of the gel. Suggest how the scientists found the number of CAG repeats in the bands shown on the gel. [1 mark]

e. Two bands are usually seen for each person tested. Suggest why only one band was seen for Person C. [2 marks]
f. Suggest one advantage and one disadvantage of screening for faulty alleles of *HTT* before any symptoms occur. [2 marks]
g. Outline how gene editing may in the future be used to correct the genetic fault associated with HD. [3 marks]

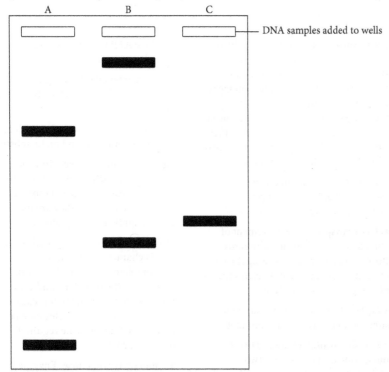

Figure 19.27

E5. Pangolins, native to Africa and Asia, are the only mammals with true scales. Many species are endangered and captive breeding programmes are active in many zoos. However, despite a ban on the trade in pangolin scales, they are illegally transported and sold as ingredients in some traditional medicines.

Scientists analysed three tonnes of pangolin scales seized at a shipping port in Hong Kong. The researchers took small samples of several hundred scales and used PCR to amplify DNA sequences found in specific regions of the pangolin genome. These are called short tandem repeats (STRs). STRs consist of 2–5 bases that are repeated between 10 and 30 times. These STRs therefore have a very variable length between individuals, but there is more similarity between individuals that are closely related.

a. Explain what is meant by the term amplify. [1 mark]
b. Outline how the researchers should design primers used in this PCR. [2 marks]

Using the technique of gel electrophoresis, the researchers confirmed that the majority of the pangolin scales originated from species in Africa, rather than Asia.

c. Suggest how the researchers were able to reach this conclusion. Draw a sketch to help you explain your answer. [3 marks]
d. The seized pangolin scales were very densely packed into sealed containers in an attempt to avoid detection by the authorities. Suggest how this would have made estimating the relative mass of scales that originated in Africa and Asia difficult. [2 marks]
e. Outline how the results of investigations such as this can benefit captive breeding programmes. [2 marks]

Glossary

abiotic factor Any non-living part of an ecosystem. For example, soil pH, light intensity.

abscisic acid (ABA) An inhibitory plant growth regulator that causes closure of stomata in dry conditions.

absorption spectrum In the case of photosynthesis, the proportion of incident light that is absorbed at different wavelengths; usually shown as a line graph.

abundance The relative numbers of a particular species in one location; a measure of how common or rare a species is.

accuracy A measure of the closeness of the measured data to the true value. Accurate data is more likely to be obtained if the equipment and methods used in the investigation were precise.

acetyl coenzyme A A molecule that delivers an acetyl (2-carbon) group to the Krebs cycle.

acetylcholinesterase The enzyme that catalyses the hydrolysis of acetylcholine at synapses.

acid hydrolysis The process by which a disaccharide is heated in the presence of acid, which results in the breakage of the glycosidic bond and release of two monosaccharides.

actin One of the fibrous protein filaments in muscles that has a role in muscle contraction.

actin-myosin cross bridges The structures formed when myosin heads attach to specific binding sites on actin filaments.

action potential The change in electrical potential across a membrane that is associated with a nerve impulse.

action spectrum The variation of the rate of photosynthesis with wavelength of light; usually shown as a line graph.

activation energy The minimum amount of energy needed to trigger a reaction.

active immunity Immunity gained by being exposed to an antigen, so the body makes its own antibodies (and memory cells) complementary to it. It can be natural (for example, by previously developing a disease caused by an infectious organism), or artificial (for example, by obtaining a weakened form of the antigen in a vaccine).

active site A 'dent' on an enzyme molecule that binds to a protein (substrate) during a reaction.

active transport The movement of chemical substances, usually across the cell surface membrane, against a concentration gradient; requires the use of energy from respiration.

adapted Features that are suited to survival in a specific environment.

adenosine diphosphate (ADP) A nucleotide derivative with two phosphate groups that is a product of ATP hydrolysis and also used for ATP synthesis.

adenosine triphosphate (ATP) A nucleotide derivative with three phosphate groups that is the universal energy currency in cells.

adenovirus Viruses containing a double stranded DNA genome that are often used in gene therapy treatments.

adenylyl cyclase An enzyme that catalyses the formation of cyclic AMP from ATP.

adipose tissue Tissue, often found just beneath the skin, consisting of cells that contain a high volume of triglyceride.

adhesion The force that attracts water molecules to the cellulose walls of xylem vessels in a plant.

aerenchyma Plant tissue with many air spaces that allows efficient gas exchange between the air and parts of the plant that are under water or in waterlogged soil.

aerobic respiration Chemical process in living organisms using oxygen to produce ATP from the breakdown of complex organic substances.

afferent arteriole Leading towards; e.g. the afferent blood vessel leads towards a glomerulus.

affinity The attraction of an enzyme for, or its ability to bind to, its substrate.

agglutination A process in which antibodies attach to antigens on the surface of bacteria, causing the bacteria to clump together so they are more easily engulfed by phagocytes.

albinism An inherited genetic disorder that is characterised in humans by the complete or partial absence of pigment in the skin, hair and eyes. In the most common form, it is caused by having two recessive alleles for a gene that encodes an enzyme required for melanin production.

alcohol dehydrogenase Enzyme that converts the alcohol ethanol into the aldehyde ethanal together with the reduction of NAD; the name alcohol dehydrogenase refers to the removal of hydrogen from an alcohol.

aleurone layer A structure within some seeds that produces hydrolytic enzymes as part of germination.

allele Different forms of a gene that control particular characteristics.

allele frequencies The proportion of the total number of alleles that one particular allele accounts for.

all or nothing principle The fact that an action potential will either be produced or not be produced according to whether the stimulus is above or below the threshold; action potentials cannot vary in voltage or in time of duration.

allopatric speciation Speciation that takes place as a result of two populations living in different places and having no contact with each other.

allosteric site The region on the surface of an enzyme to which a non-competitive inhibitor binds.

alpha (α) cell A cell in the islets of Langerhans in the pancreas that detects changes in blood glucose concentration and secretes glucagon if they become too low.

alpha helix A helical structure formed by a polypeptide chain, held in place by hydrogen bonds. It is an example of secondary structure in a protein.

alternative hypothesis A statement relating to an investigation that you are expecting to accept.

alveoli (Singular: alveolus) Small, spherical air sacs at the end of bronchioles where gas exchange occurs between the lungs and the blood.

amino acids One of the building blocks of protein; each amino acid has an amine group $-NH_2$ and an acid carboxyl group $-COOH$, and the general formula $NH_2-CHR-COOH$, where R is one of 20 side groups.

amine group see amino group

amino group A portion of an amino acid molecule consisting of a nitrogen atom attached to two hydrogen atoms.

amphipathic A molecule that has both hydrophilic and hydrophobic parts.

amplification A process during cell signalling in which an initial signal (usually a hormone binding to its receptor) is increased in strength due to a series of enzyme-catalysed reactions in the cytoplasm.

amplify To increase the number of copies of. Usually referred to in terms of PCR, which amplifies DNA.

amylase An enzyme that hydrolyses starch into maltose.

amylopectin A water-insoluble polysaccharide and highly branched polymer of α-glucose units, bonded to each other through α(1,4) and α(1,6) glycosidic bonds, found in plants. It is one of the two components of starch, the other being amylose.

amylose A water-insoluble polysaccharide and unbranched polymer of α-glucose units, bonded to each other through α(1,4) glycosidic bonds, found in plants. It is one of the two components of starch, the other being amylopectin.

anabolic Type of reaction where smaller molecules are joined together to make larger molecules.

anaerobic respiration Respiration without oxygen, forming lactic acid in animals, and ethanol and carbon dioxide in plants and microorganisms.

anaphase The phase of the cell cycle when the spindle fibres drag the sister chromatids to opposite ends of the cell.

anatomy Relating to structural features within an organism.

anneal The process by which primers attach to complementary sequences in the DNA template during PCR.

anomalous data A piece of data that deviates from what is standard, normal, or expected.

antibiotic Compound produced by an organism that has anti-microbial effects, either killing or stopping the growth of bacteria.

antibiotic resistance Phenomenon where bacterial cells that were one susceptible to antibiotics have developed mechanisms in order to survive and continue to multiply in the presence of the antibiotic.

antibody A protein molecule with a quaternary structure, produced by B-lymphocytes. This is complementary with a specific antigen, and combined with that antigen to form an antibody–antigen complex. This has the effect of neutralising the antigen or destroying it.

anticodon The three unpaired bases in tRNA.

antidiuretic hormone (ADH) A hormone produced by the hypothalamus and secreted from the pituitary gland that increases water reabsorption in the kidneys and therefore reduces water loss in urine.

antigen A macromolecule, usually protein or glycoprotein in nature, that stimulates an immune response.

antigen–antibody complex This forms when an antibody, complementary to a specific antigen, binds to that antigen.

antigen-binding site The region on an antibody that is complementary to, and binds, a specific antigen.

antigenic concealment Describes the process by which a cell, usually pathogenic, is able to prevent the immune system from identifying its antigens.

antigenic variability The frequent changing of antigens on the surface of a pathogen, which means that memory cells produced as a result of a prior infection are less likely to recognise it.

antigen presentation The process that occurs after phagocytosis in which antigens obtained from a partially digested pathogen are presented to T-lymphocytes in the cell surface membrane of an antigen-presenting cell (APC).

antigen-presenting cell A cell that displays antigens in a form that T-lymphocytes can recognise.

antiparallel The way in which the two strands in a polynucleotide run in opposite directions.

antiserum A preparation derived from blood that contains antibodies specific to a particular antigen. These are often used to give passive immunity to individuals exposed to a serious disease, e.g. rabies or tetanus.

aorta Artery that carries blood from the left ventricle towards to body.

apical meristem A tissue found in the tips of roots and shoots that contains fast-dividing cells responsible for plant growth.

apoplast (apoplastic) pathway The route through which some water and mineral ions move into a plant root, through the cell walls and intercellular spaces.

aquaporins Channel proteins for water molecules in a cell surface membrane.

Archaea see domain Archaea

arteriole A small branch of an artery leading towards capillaries.

artery A blood vessel that carries blood away from the heart; has thicker, more muscular walls than other blood vessels.

artificial immunity Immunity that is not gained by direct interaction between one organism and another. The pathogen responsible for the disease has not entered the body. It can be active (for example, obtained by vaccination), or passive (for example, obtained by injection of antiserum).

ascending limb One of the structures in the nephron of the kidney; drains urine into the distal convoluted tubule.

asexual reproduction A method of reproduction that involves only one parent.

assimilates Organic compounds produced by the process of assimilation. For example, carbohydrates, proteins and lipids.

assimilation The process within a living organism of producing useful organic compounds from substances originally taken in from the surroundings; for example, plants using carbon dioxide and water to produce carbohydrates during photosynthesis, or animals using amino acids from their diet to synthesise new proteins.

assisted reproduction Methods that use human intervention to help animals produce offspring.

asymptomatic Showing no symptoms.

atoms Tiny particles which cannot be chemically broken down.

ATP hydrolase (ATPase) The enzyme that catalyses the hydrolysis of ATP to ADP and inorganic phosphate with the release of energy.

ATP synthase An enzyme that joins ADP and phosphate to resynthesise ATP.

atria The chambers at the top of the heart that receive blood from veins and pump blood into ventricles.

atrial systole Term given to the contraction of the atria when blood is pumped down into the ventricles.

atrioventricular node Electrically conductive part of the heart that plays a part in co-ordination of the cardiac cycle, ensuring that the atria and ventricles do not contract at the same time.

atrioventricular valves Valves within the heart, located between the atria and ventricles, that allow blood to flow from the atria to the ventricles, but not in the opposite direction.

attenuation The process by which a pathogen is weakened so that it can be given in a vaccine without causing harm.

automated polynucleotide synthesis The production of DNA or RNA molecules from nucleotides, usually from the primary amino acid sequence of a protein using the genetic code.

autosome Any chromosome that is not a sex-determining chromosome.

auxin Group of plant growth regulating substances that work by promoting cell elongation.

axon An extension of the cytoplasm of a neurone that carries nerve impulses away from the cell body.

Bacteria see domain Bacteria

bactericidal antibiotic An antibiotic that kills bacteria. (Compared with a bacteriostatic antibiotic which stops the growth and multiplication of bacteria.)

bacteriostatic antibiotic An antibiotic that stops the growth and multiplication of bacteria. (Compared with a bactericidal antibiotic which kills bacteria.)

basement membrane A thin, fibrous, extracellular matrix of tissue that separates the epithelium, mesothelium and endothelium from underlying connective tissue.

base pairs A pair of complementary bases in a double-stranded nucleic acid molecule, consisting of a purine in one strand linked by hydrogen bonds to a pyrimidine in the other. Cytosine always pairs with guanine, and adenine with thymine (in DNA) or uracil (in RNA).

behavioural isolation The separation of two populations because they have different behaviours which prevent them breeding together.

belt transect A sampling method for species in a narrow rectangular area that records how far the organisms are from a particular place and how many of them are present.

Benedict's solution (or Benedict's reagent) A test for the presence of reducing sugars such as glucose. The unknown substance is heated with Benedict's reagent. A change from a clear blue solution to give a yellow, red or brown precipitate indicates the presence of reducing sugars.

Benedict's test A chemical test which determines the presence of non-reducing sugars in a test solution.

benign A benign tumour is a mass of cells which is slow-growing and does not spread to other organs.

beta (β) cell A cell in the islets of Langerhans in the pancreas that detects changes in blood glucose concentration and secretes insulin if they become too high.

binomial system System of naming organisms with two names: the genus name followed by the species name.

biochemistry The study of chemical processes within and relating to living organisms.

biodiversity The variability of all organisms within an ecosystem.

bioinformatics The collection, processing and analysing of biological information using computerised databases.

biological species concept Two organisms are of the same species if they can reproduce to give viable, fertile offspring.

bioreactor A vessel in which a chemical process is carried out which involves organisms or biochemically active substances derived from such organisms.

biosensor A device that uses an enzyme to measure the concentration of a chemical compound.

biotic factor A factor caused by the interaction of any living thing with any other living thing in an ecosystem. For example, predation, competition

bivalent A pair of homologous chromosomes.

bladder A hollow muscular organ that collects and stores urine from the kidneys before disposal by urination.

BLAST Basic Local Alignment Search Tool. An algorithm that enables scientists to make comparisons between nucleotide sequence data stored on databases.

blood glucose concentration The concentration of glucose dissolved in the blood plasma. It is normally expressed as mg 100 cm^{-3}.

B-lymphocytes (B-cells) Lymphocytes involved in the humoral response. These can differentiate, in response to binding an antigen, to form plasma cells that secrete antibodies.

Bohr shift The lowering in affinity of haemoglobin for oxygen in conditions of increased partial pressure of carbon dioxide.

bone marrow A tissue consisting of stem cells that can divide rapidly to form all types of blood cell.

Bowman's capsule A cup-like sac at the beginning of the tubular component of a nephron in the mammalian kidney that performs the first step in the filtration of blood to form urine.

β-pleated sheet A chain of amino acids in a polypeptide, connected by at least two or three backbone hydrogen bonds, forming a generally twisted, pleated sheet. It is an example of a secondary structure in a protein.

broad spectrum antibiotic An antibiotic that affects a wide range of different types of bacteria.

bronchi (Singular: bronchus) The two large branches of the trachea that continue to divide into smaller airways called bronchioles.

bronchiole Small branches of bronchi that ultimately lead to alveoli. Terminal bronchioles have smooth muscle in their walls, and have a diameter of about 1 mm. Respiratory bronchioles do not have smooth muscle and their diameter is about half that of the terminal bronchioles.

***Bt* cotton** Cotton plants that have been genetically modified to express the *cry* gene, derived from the bacterium *Bacillus thuringensis,* which encodes an insecticide.

calibration Determining a correct value by comparison with a standard.

calibration curve A general method for determining the concentration of a substance in an unknown sample by comparing the unknown to a set of standard samples of known concentration.

Calvin cycle Part of the light-independent reactions in photosynthesis where CO_2 is fixed to RuBP, which is split to TP, then RuBP is subsequently regenerated.

cambium Stem cell tissue found in many plants that can differentiate to form either xylem or phloem tissue.

cancer cell A cell which, owing to the mutation of genes involved in regulating the cell cycle, is able to divide uncontrollably.

capillary The smallest blood vessels that allow exchange of substances between blood and tissue fluid; they have walls only one endothelial cell thick.

capsid Protein coat of a virus.

capsule Protective outer layer found in some bacteria; surrounds the cell wall.

captive breeding Allowing animals to reproduce in artificially protected surroundings, such as a zoo.

carbaminohaemoglobin Substance formed when carbon dioxide binds to haemoglobin.

carbohydrate A chemical containing carbon, hydrogen and oxygen; the hydrogen and oxygen being in the ratio of 2:1.

carbonic acid Product of the reaction between carbon dioxide and water, its formation is catalysed by carbonic anhydrase, its formula is H_2CO_3.

carbonic anhydrase Enzyme in red blood cells that catalyses the reaction between aqueous carbon dioxide and water to form carbonic acid.

carbon fixation The step in the light-independent reactions of photosynthesis where CO_2 reacts with RuBP; catalysed by the enzyme rubisco.

carboxyl group A functional group in amino acids with the formula –COOH.

carcinogen A substance capable of causing cancer in living tissue.

cardiac cycle The series of events that occur within the heart during one complete heartbeat.

cardiac muscle Type of muscle found in the heart which can conduct electrical impulses and contains a structure that initiates the contraction of the muscle.

carotene A group of pigments, some of which are involved in photosynthesis

and transfer energy from light to chlorophyll.

carrier protein Protein that facilitates the diffusion of different molecules across a cell surface membrane. They can use ATP in the process of active transport, but are sometimes independent of ATP in facilitated diffusion.

cartilage A tough connective tissue, more flexible than bone.

Cas9 CRISPR-associated enzyme-9. This enzyme is guided by gRNA molecules to recognise and cut DNA at a specific site.

Casparian strip An impermeable band, composed of the waxy substance suberin, running through the cell walls of the endodermis in plant roots; it provides a means of controlling which substances will enter the vascular tissue at the centre of the root.

catalase An enzyme that catalyses the decomposition (break down) of hydrogen peroxide to water and oxygen.

catalyst Substance added to a chemical reaction to alter the speed of the reaction.

categorical data Data that can be grouped into separate categories.

categorical variation see discontinuous variation

cell The basic unit of living organisms.

cell body The part of a neurone containing the nucleus and being larger in diameter than the axon or dendrites.

cell cycle The sequence of events in a cell that prepares it to undertake DNA replication, mitosis and cytokinesis.

cell plate The structure that forms in the middle of the dividing plant cell during telophase, formed by the joining of Golgi vesicles at the equator of the cell.

cell sap Fluid found inside a plant cell vacuole.

cell surface antigens Cell-specific proteins and glycoproteins in a cell surface membrane.

cell surface membrane The structure that separates the contents of a cell from its surroundings, while still allowing substances to move in and out of the cell.

cell surface receptors Proteins that are embedded in the membranes of cells. They act in cell signalling by receiving extracellular molecules.

cellular response An immune response involving T-cells. The two most important cells involved are T-helper cells and cytotoxic (killer) T-cells.

cellulose Polysaccharide polymer, composed of β-glucose monomers, which is a major component of the cell wall in plants and algae.

cell wall Permeable structure outside the cell surface membrane, which provides support for the cell; found in plants, algae, fungi and bacteria.

central nervous system The collective name for the brain and spinal cord.

centrioles Structures that appear at prophase in animal cells which are composed of short lengths of microtubules lying parallel to one another and arranged around a central cavity to form a cylinder. They form part of the centrosome and are involved in spindle assembly during mitosis.

centromere The place on a chromosome where their sister chromatids are held together.

centrosome An organelle near the nucleus of a cell which contains the centrioles (in animal cells) and from which the spindle fibres develop in cell division.

channel proteins Proteins that facilitate the movement of ions across a cell surface membrane.

chemiosmosis Movement of hydrogen ions down their electrical gradient and chemical concentration gradient by facilitated diffusion through the ATP synthase protein found in the inner mitochondrial membrane.

chiasma (Plural: chiasmata) Breaking and crosswise re-joining of homologous sister chromatids during meiosis.

chi-squared test Statistical test to check if a null hypothesis is true.

chitin A nitrogen-containing polymer that forms part of the cell walls of fungi.

chloride shift Describes when chloride ions (Cl⁻) enter red blood cells when hydrogencarbonate / HCO_3^- leave, to maintain a neutral charge in the cytoplasm. An equal number of chloride ions enter as hydrogencarbonate ions leave.

chlorophyll Green pigment found in chloroplasts which absorbs light.

chlorophyll a Specific type of chlorophyll that absorbs strongly in the violet-blue and orange-red parts of the spectrum, so contributing to the green colour of plants.

chlorophyll *b* Specific type of chlorophyll that absorbs strongly in the blue part of the spectrum, so contributing to the green colour of plants.

chloroplast Organelle in which photosynthesis takes place.

chloroplast envelope Double membrane layer forming the outer surface of a chloroplast.

cholera An infectious disease caused by the bacterial pathogen *Vibrio cholera* that is transmitted through the faecal-oral route and causes severe diarrhoea.

cholesterol A lipid-like substance in cell surface membranes that prevents too much movement of other molecules in the membrane.

cholinergic synapse A gap between two neurones; on the arrival of an action potential the neurotransmitter acetylcholine diffuses across the gap to enable a new action potential to be stimulated on the other side.

chromatin The mixture of DNA and protein in the nucleus that condenses to form chromosomes during cell division.

chromatogram A visible record of the results of a chromatography experiment.

chromatography A laboratory technique for the separation of a mixture of substances that are soluble in a solvent.

chromosome A thread-like structure in the nucleus of a cell, made of a single molecule of DNA coiled many times, held together by protein.

cilia Hair-like structures on the surface of some cells, that are able to move, e.g. on ciliated epithelial cells in the bronchi and trachea.

ciliated epithelial cell A type of columnar epithelial cell that has hair-like extensions on the exposed surface; these hair-like extensions waft mucus across the surface of the epithelium.

class Taxonomic rank in classification between phylum and order.

cleavage furrow The inner folding of the cell surface membrane during cytokinesis in animal cells, which ultimately meets in the middle and causes the cell to split in half.

climate change The long-term change in Earth's weather patterns, including higher average temperatures.

clonal expansion The process by which an activated lymphocyte divides rapidly by mitosis to form many clones that share specificity for the same antigen.

clonal selection A process in which lymphocytes divide by mitosis to form a clone of identical lymphocytes. It is triggered when an antigen binds to the cell surface antibodies (B-cells) or receptors (T-cells).

clone (noun) An organism or cell that is produced by asexual reproduction or mitosis and that is genetically identical to its parent.

clone (verb) To increase the number of copies of. Usually referred to in terms of PCR, which clones DNA.

codominant A form of inheritance pattern wherein the alleles of a gene pair in a heterozygote are fully expressed and produce a blended phenotype.

codon A sequence of three bases on an mRNA molecule that codes for one amino acid.

coenzyme Substance that is required for an enzyme-catalysed reaction to be completed.

cohesion An intermolecular attractive force that holds together, for example, water molecules as they pass through plant xylem vessels.

cohesion-tension The mechanism by which water passes up plant xylem vessels.

collagen The most abundant protein in the human body; a major component of connective tissues that make up tendons, ligaments, skin and muscles.

collagen fibres Formed when three collagen polypeptides interact by hydrogen bonding.

collagen polypeptide A very long chain of amino acid residues, joined by peptide bonds, of a recurring sequence. Three of these join together by hydrogen bonding to form a collagen fibre.

collecting duct Part of the renal tubule in a kidney in which water absorption takes place under the control of ADH, producing urine of variable concentration depending on overall water levels in the body.

colorimeter An item of scientific equipment that measures the proportion of light that passes through a sample of liquid. This provides a numerical value for the colour of a solution.

colostrum A high-protein, low-fat liquid produced by a mother's breasts soon after childbirth.

colour standards In biochemical tests, a set of solutions of distinguishable colour and known concentration, which can be used to determine the concentration of a solution.

companion cell A cell closely associated with a phloem sieve tube element; its function is to keep its associated sieve tube element alive and allow it to function; companion cells are connected to sieve tube elements by many plasmodesmata.

competition Interaction between organisms that results from limited resources and where both organisms will be harmed.

competitive inhibitors Molecules that bind to the active site in an enzyme but no reaction takes place.

complementary The shape of the active site is complementary to the shape of the substrate which simply means it fits exactly.

complementary base pairing Describes the consistent pairing of bases in a polynucleotide (A with T and C with G).

complementary DNA (cDNA) A short piece of double-stranded DNA, usually containing the sequence of a gene, produced by the action of reverse transcriptase and DNA polymerase.

compound tissue A tissue containing more than one type of cell; for example, phloem tissue contains both phloem sieve tube elements and companion cells.

conclusion A judgement or decision that has been made based on results collected in an experiment.

condensation reaction A reaction in which two compounds are joined together by removing the elements of a water molecule.

condense The process by which chromosomes become shorter and thicker during prophase, in readiness for the process of mitosis.

conformational change A change in the shape of a macromolecule, often induced by environmental factors

conservation Taking care of and working with organisms, habitats or ecosystems.

conservation areas Areas within ecosystems that are protected from human impacts.

constant region The region of an antibody molecule that is common in structure between different antibodies.

continuous replication The synthesis of a new strand of a replicating DNA molecule in a process that does not stop until the whole molecule is copied.

continuous variation Differences between individuals of a species in which each one can lie at any point in the range between the highest and lowest values.

control experiment An experiment designed to check or correct the results of another experiment by removing the variable or variables operating in that other experiment.

controlled variable Variable that must remain constant during an investigation, to ensure that changing the independent variable is responsible for any change in the dependent variable.

Convention on International Trade in Endangered Species of Wild Fauna and Flora (CITES) An international agreement between governments aiming to ensure that trade in animals and plants does not endanger these organisms

co-ordination system An interconnected group of cells, tissues or organs that work together to control the regulation of a physiological (bodily) factor.

correlation A relationship between two factors.

cortex The outer region of an organ or tissue, especially when it is distinct from the inner region or medulla. In plants, the outer layer of cells in a plant root or stem, just behind the epidermis.

cotransport A process in which two substances are transported at the same time across a membrane by one protein.

cotransporter protein A protein that transports two substances across a membrane in the process of cotransport; for example, the sucrose-hydrogen cotransporter protein found in plant phloem companion cells.

cotyledon Embryonic leaf formed in a seed.

counter-current multiplier An arrangement in which fluid in adjacent tubes flows in opposite directions, which allows a large concentration gradient to develop.

covalent bond Very strong bond formed when two atoms share electrons with each other.

CRISPR Clustered Regularly Interspaced Short Palindromic Repeats. Regions of the bacterial genome that contain sequences derived from viral infections, and which form the basis of CRISPR-Cas9 gene editing technology.

cristae Folds of inner membrane in mitochondria; hold many enzymes involved in respiration.

critical value A cut-off area outside which a test statistic is unlikely to lie.

crossing over The exchange of genetic material between non-sister chromatids of homologous chromosomes during meiosis, which results in new allelic combinations in the daughter cells.

cyclic AMP (cAMP) A derivative of adenosine triphosphate used as a second messenger during intracellular signal transduction.

cyclic photophosphorylation Part of the light-dependent stage of photosynthesis where the excited electron from photosystem I is returned to the photosystem resulting in synthesis of ATP, so NADP is not reduced.

cytokine A class of chemicals, released by T-helper cells, which stimulate the activity of macrophages, cytotoxic T-cells and B-cells in the immune response.

cytokinesis The process by which the cell splits in half after mitosis.

cytoplasm The cytosol and all the organelles in a cell (apart from the nucleus).

cytoskeleton The lattice or internal framework of a cell composed of protein filaments and microtubules in the cytoplasm; it has a role in controlling cell shape, maintaining intracellular organisation, and in cell movement.

cytosol Jelly-like material within a cell, in which the organelles are suspended.

daughter cell One of the two cells that forms during mitosis after a parent cell divides by mitosis and cytokinesis.

deamination The breakdown of excess amino acids in the liver, by the removal of the amine group. This forms ammonia, which ultimately is made into urea.

decarboxylation Removal of a carboxyl (–COOH) group from a carboxylic acid with the release of CO_2 and the reduction of a coenzyme.

decondense The process by which the chromosomes become longer and thicker at the end of mitosis such that protein synthesis is possible in gap phase 1.

degenerate Describes a triplet that codes for the same amino acid as other triplets.

degrees of freedom A measure of how many values can change in a statistical calculation. With the t-test, the number of degrees of freedom can be calculated by using the formula $v = n_1 + n_2 - 2$,

where v = degrees of freedom and n = sample size (number of observations). With the chi-squared test, the number of degrees of freedom can be calculated by using the formula $v = c - 1$, where c = number of classes.

dehydrogenation The removal of hydrogen from a compound using an enzyme; dehydrogenation will result in the oxidation of the compound.

DELLA protein A family of factors that inhibit seed germination but which are destroyed by a mechanism involving gibberellin.

denatured Describes an enzyme that has been 'inactivated' when hydrogen bonds are broken through heating or by extreme values of pH.

dendrite An extension of the cytoplasm of a neurone that carries nerve impulses towards the cell body.

deoxyribonucleotide The monomer, or single unit, of DNA, deoxyribonucleic acid. Each deoxyribonucleotide has three parts: a nitrogenous base, a deoxyribose sugar, and one phosphate group.

deoxyribose The sugar found in the nucleotides of DNA.

dependent variable The variable that is measured by the experimenter during an investigation.

depolarisation Change in the distribution of charge across a membrane where the difference in charge becomes smaller.

descending limb The portion of the renal tubule constituting the first part of the loop of Henlé.

descriptive statistics Mathematical processes that can be used to provide more information on data collected in an investigation. For example, a mean, a sample standard deviation and a standard error of the mean can be calculated, to assess the reliability of a specific given dataset.

diabetes A disorder in which there is a deficiency of the pancreatic hormone insulin, which results in a failure to metabolise sugars and starch.

dialysis tubing Artificial, partially permeable membrane tubing. Also called Visking tubing.

differentiate The structural and functional specialisation of cells during an organism's development.

differentiation The process by which a stem cell specialises in terms of structure and function to form a cell with a specific purpose.

diffusion The spread of particles from regions of higher concentration to regions of lower concentration.

digestion The process in which larger molecules are broken down into smaller molecules to allow nutrients to be absorbed.

dipeptide A molecule made up of two amino acid molecules joined by a condensation reaction.

diploid Containing two sets of chromosomes.

dipole Describes a molecule that has a small positive charge in some areas and a small negative charge in others.

directional forces Selective forces that cause a population of organisms to change by natural selection such that more individuals have a phenotype corresponding to one extreme of a range of values. Individuals with values representing intermediate phenotypes or at the other extreme are not selected for.

directional selection Natural selection that causes a gradual change in allele frequency over many generations.

disaccharide A sugar molecule consisting of two monosaccharides joined together by a glycosidic bond.

discontinuous replication The synthesis of a new strand of a replicating DNA molecule as a series of short fragments that are later joined together.

discontinuous variation Differences between individuals of a species in which each one belongs to one of a small number of distinct categories, with no intermediates.

disruptive forces Selective forces that cause a population of organisms to change by natural selection such that more individuals have a phenotype corresponding to both extremes of a range of values. Individuals with values representing intermediate phenotypes are not selected for.

disruptive selection Natural selection that maintains relatively high frequencies of two different sets of alleles; individuals with intermediate features and allele sets are not selected for.

dissociation Chemical term for breaking any type of bonds in a substance to produce two or more smaller substances.

dissociation curve The graph showing how the percentage saturation of haemoglobin with oxygen varies with partial pressure of oxygen.

disulfide bonds A covalent bond that has formed between the R groups of two amino acid residues that contain the element sulfur.

distal (or second) convoluted tubule (DCT) A portion of kidney nephron between the loop of Henlé and the collecting duct system.

DNA Deoxyribonucleic acid – a nucleic acid that stores genetic information in every living cell.

DNA helicase An enzyme that 'unwinds' a DNA molecule and breaks the hydrogen bonds that connect the bases.

DNA ligase An enzyme that helps the joining of DNA strands together by catalysing the formation of a phosphodiester bond.

DNA polymerase An enzyme that catalyses reactions that join nucleotides to form DNA.

domain Archaea Group of prokaryotic organisms having molecular characteristics separating them from domain Bacteria.

domain Bacteria Group of prokaryotic organisms having molecular characteristics separating them from domain Archaea.

domain Eukarya Group of organisms whose cells contain a true, membrane-bound nucleus.

dominant An allele that needs to be present only once in the genotype to have an effect on the phenotype.

donor DNA A DNA molecule taken from the cell of one organism that will

be introduced into the cell of another organism.

double circulation system Where blood passes twice through the heart in one complete circulation of the body: once for the pulmonary circuit and once for the systemic circuit.

double helix A pair of parallel helices intertwined about a common axis, especially that in the structure of the DNA molecule.

double-stranded The makeup of a DNA molecule, consisting of two long chains of nucleotides twisted into a double helix and joined by hydrogen bonds between the complementary bases adenine and thymine or cytosine and guanine.

ecological isolation The separation of two populations because they live in different environments in the same area and so cannot breed together.

ecological niche The role and position of an organism in its environment.

ecological species concept A group of organisms that are adapted to the same ecological niche are of the same species.

ecosystem The interaction of all living and non-living things in a particular area.

effector An organ or cell, usually a muscle of gland, that carries out a response to a nerve impulse.

efferent arteriole Leading away from; e.g. the efferent blood vessel leads away from a glomerulus.

elastic fibre Bundles of protein (elastin) that can stretch up to 1.5 times its original length and recoil to its original size.

elastin A flexible protein found in elastic tissues such as the walls of alveoli.

electrochemical gradient The combination of the differences in electrical charge and chemical concentration across a membrane due to the greater numbers of a charged substance on one side of the membrane.

electron micrograph Image taken through an electron microscope.

electron microscope High-powered microscope that uses electrons, not light, to form an image.

electron transport chain A series of reduction-oxidation (redox) reactions that occur on the inner membrane of mitochondria as part of aerobic respiration.

embryo biopsy The removal of a cell or small number of cells from an early embryo for the purposes of genetic screening.

embryonic stem cell A stem cell found in the early embryo which is able to differentiate into any cell type found in the human body.

embryo transfer A method of assisted reproduction where an embryo is placed within the uterus of another animal.

emulsion A mixture of micelles and water.

endangered An organism becomes endangered when its population falls so low that it is difficult for the numbers to recover.

endemic A disease is said to be endemic when it is continually and usually present in a particular geographical region.

endocrine gland Organ that secretes hormones directly into the blood.

endocrine system The glands that produce hormones.

endocytosis The process by which a cell surface membrane surrounds a particle and encloses the particle in a vesicle to bring the particle into the cell.

endodermis A single layer of cells in a plant root containing the Casparian strip.

endoplasmic reticulum (ER) Network of membranes in the cytoplasm used to transport substances within the cell. See also smooth endoplasmic reticulum and rough endoplasmic reticulum.

endosperm The tissue found in many seeds that surrounds the embryo and provides the respiratory substrate for germination.

endothelium A layer of cells lining the inner surface of blood vessels, formed from squamous epithelial cells.

energy A quantity that must be changed or transferred in order for something to happen; glucose and ATP each contain a store of chemical energy.

energy stores In a biological context these are molecules that can be used to release chemical energy through respiration, such as carbohydrates and lipids.

environment The surroundings or conditions in which a person, animal or plant lives.

environmental factor A feature of the environment of an organism that affects its survival.

enzyme Catalyst (usually a protein) produced by cells, which controls the rate of chemical reactions in cells.

enzyme cascade A series of reactions involving enzymes, which occur one by one, that amplify (increase) an initial signal. The product of each reaction is required for the next reaction.

enzyme–substrate complex The combination of an enzyme and a substrate.

epistasis The interaction of alleles found at more than one locus, to have a combined effect on the phenotype of an organism.

equator The middle of the cell, at which position chromosomes line up during metaphase.

ester bonds The bonds that join fatty acids to glycerol molecules in triglycerides.

ethanol fermentation Anaerobic respiration in yeast and plants that uses glucose and results in the formation of ethanol and carbon dioxide.

ether bond Bond between two parts of a molecule that has an oxygen atom between two carbon atoms, R – O – R.

ethical issues Considerations that relate to a set of agreed standards, to determine what is acceptable in society.

Eukarya see domain Eukarya

eukaryotic cell Cell with a nucleus and other membrane-bound organelles; found in all eukaryotic organisms, e.g. animals, plants and fungi.

evolution The change of characteristics of species over time. It can lead to the formation of new species from pre-existing ones, as a result of changes to the gene pools of populations over many generations.

excretion The removal of toxic or waste products of metabolism from the body.

exocytosis The process by which a substance is released from the cell through a vesicle that transports the substance to the cell surface and then fuses with the membrane to let the substance out.

exon The sections of a gene that code for amino acid sequences in a protein.

exothermic A reaction that releases thermal (heat) energy.

extinct An organism or species that has no living members.

extracellular enzyme An enzyme that is secreted by a cell and functions outside of that cell.

eyepiece graticule Small transparent scale placed on top of a microscope eyepiece lens.

facilitated diffusion A passive diffusion process in which a channel protein makes it easy for ions to move through cell surface membranes.

family Taxonomic rank in classification between order and genus.

fatty acids Acids containing a long hydrocarbon chain attached to a – COOH group.

fenestrations Small pores found between the endothelial cells in the wall of the glomerulus.

fertilisation The fusion of gametes to form a new organism of the same species.

fibrous capsule A tough fibrous layer surrounding the kidney and covered in a layer of adipose capsule. It provides some protection from trauma and physical damage.

fibrous proteins Proteins that form long, thin molecules.

first filial, F1, generation The first generation of a cross between two individuals.

fitness The ability of an organism to survive and reproduce.

flaccid Describing a plant cell that has collapsed and lost its usual shape due to water loss.

flagella (Singular: flagellum) Protein fibrils that allow bacterium cells to 'swim'.

fluid mosaic model A theory that describes the unfixed structure of cell surface membranes in phospholipids.

foreign Describes an antigen that is not usually found in the host body.

founder effect The reduction in a gene pool compared with the main populations of a species, resulting from only two or three individuals (with only a selection of the alleles in the gene pool) starting off a new population.

frame quadrat A square used to outline a sample area.

frameshift A mutation in which an insertion or deletion of base pairs occurs that involves a number of base pairs not divisible by three. This consequently disrupts the triplet reading frame of a DNA sequence.

frozen zoo Storage facility containing sperm, eggs, DNA and tissue samples from organisms kept at –196 °C in liquid nitrogen.

fructose A simple sugar (monosaccharide).

fusogen Any chemical that is used in the fusing of two cells.

galactose A sugar that joins with glucose to form the disaccharide lactose.

β-galactosidase A hydrolytic enzyme that splits lactose into glucose and galactose.

gamete A haploid cell specialised for fertilisation.

gap phase 0 (G_0) Some cells can exit, temporarily or permanently, the cell cycle and enter this phase; often this occurs as they irreversibly differentiate.

gap phase 1 (G_1) The time period during the cell cycle that is defined by a cell rapidly growing after mitosis, and preparing for DNA replication in S phase.

gap phase 2 (G_2) The time period during the cell cycle that is defined by a cell preparing for mitosis by synthesising the subunits of microtubules and ATP.

gas exchange Also called gaseous exchange. The movement of gases, usually oxygen and carbon dioxide, from areas of high concentration to areas of lower concentration by simple diffusion.

gated ion channels Channels in cells that can be opened or closed to allow ions to move through them.

gel electrophoresis A method of separating lengths of DNA or proteins according to their relative mass or charge.

gene A length of DNA that codes for a particular polypeptide.

gene control The molecular process that determine whether or not a gene is expressed to produce a protein.

gene editing A genetic technology that uses CRISPR-Cas9 to make precise cuts in DNA using molecules called guide RNA to determine the cutting site.

gene flow The transfer of alleles, by mating, from one population to another.

gene mutation A change in the base sequence that affects the gene in that part of the DNA sequence.

gene pool The complete range of DNA base sequences in all the organisms in a species or population.

gene therapy The introduction of DNA containing functional alleles into cells of individuals homozygous for a recessive genetic disease.

genetically modified organism (GMO) An organism that has undergone genetic engineering. This normally, but not always, involves accepting DNA from another organism.

genetically unique An individual that is different from other individuals with regard to the alleles that it possesses.

genetic bottleneck A period when the numbers of a species fall to a very low level, resulting in the loss of a large number of alleles and therefore a reduction in the gene pool of the species.

genetic code The way in which a sequence of DNA bases represents a sequence of amino acids.

genetic diagram Conventional way of showing a cross in genetics.

genetic drift The gradual change in allele frequencies in a small population, where some alleles are lost or favoured just by chance and not by natural selection.

genetic engineering The laboratory procedures used to produce genetically modified organisms.

genetic isolation Populations that are no longer able to breed together; there is no exchange of genes.

genetic screening Taking a small sample of tissue from a living organism, which can later be tested, e.g. to determine whether the organism is homozygous for a recessive allele of a gene.

genetics The study of inheritance patterns between different generations.

genome The entirety of the DNA that is found in a cell of an organism.

genotype A description of the alleles that an individual has for a given gene (on its two homologous chromosomes).

genotype frequencies The proportion of individuals with a specific genotype with regard to the proportion of individuals with all other genotypes.

genus Taxonomic rank in classification between family and species.

geographical separation The separation of two populations of the same species by a geographic barrier that prevents them from meeting.

germ cells Gametes or cells of the very early embryo.

germination The breaking of dormancy in a seed.

germline gene therapy The introduction of a therapeutic allele into the gametes or early embryo of individuals homozygous for a recessive genetic disease.

gibberellin receptor A protein that is complementary to the shape of gibberellin and binds to gibberellin to form a complex with biological functions.

gibberellins Group of substances in plants that regulate processes including growth, dormancy, germination, flowering, and leaf and fruit senescence.

globular protein A protein that forms a ball-shaped molecule.

glomerular filtrate Fluid that results from ultrafiltration and which passes through the nephron of the kidney.

glomerulus A cluster of capillaries found within the 'cup' of a Bowman's capsule in the cortex of the kidney.

glucagon A small peptide hormone secreted by the α-cells in the islets of Langerhans in the pancreas that brings about an increase in the blood glucose level.

gluconeogenesis The process by which glucose is generated from certain non-carbohydrate carbon substrates.

α-glucose Alpha glucose – a form of the simple sugar (monosaccharide) glucose.

β-glucose Beta glucose – a form of the simple sugar (monosaccharide) glucose.

glucose A monosaccharide that exists in two forms: α-glucose and β-glucose.

glucose oxidase An enzyme that converts glucose and oxygen to gluconic acid and hydrogen peroxide.

glucose-sodium cotransporter protein A family of glucose transporter found in the intestinal epithelial cells that contribute to glucose reabsorption in the small intestine.

glucose tolerance test A medical procedure in which a person is provided with a sample of glucose to ingest, and the concentration of glucose in the blood is measured at regular intervals to determine the body's response.

glycerate 3 phosphate (GP) A three-carbon phosphorylated sugar that is an intermediate in the Calvin cycle; also called triose phosphate (TP).

glycerol A chemical that combines with three fatty acid molecules to form a lipid.

glycogen A carbohydrate composed of many α-glucose molecules joined together.

glycogen phosphorylase An enzyme that catalyses the hydrolysis of glycogen into glucose-1-phosphate.

glycogen synthase An enzyme that catalyses the condensation of glucose into glycogen.

glycogenesis Production of glycogen from glucose.

glycogenolysis Breakdown of glycogen into glucose.

glycolipid Composite molecule that is part lipid and part carbohydrate.

glycolysis A series of enzyme-catalysed reactions that occur in the cytoplasm of cells where glucose is split into two molecules of pyruvate; glycolysis is common to aerobic and anaerobic respiration.

glycoprotein Composite molecule that is part protein and part carbohydrate.

glycosidic bond A bond formed between two sugar molecules.

glyphosate A herbicide that prevents the formation of some amino acids by plants. Some plants have been genetically engineered to break down this molecule and can grow in its presence.

goblet cell A single-celled mucous gland, which produces and releases mucus. Goblet cells are found in the epithelium of the trachea and bronchi.

Golgi body Also known as Golgi apparatus or Golgi complex. A cell organelle that modifies and packages substances such as proteins before they are transported elsewhere in the cell, or released outside the cell.

goodness of fit The extent to which observed data matches the values expected by theory, which can be determined by the use of the chi-squared test.

G protein A protein associated with the inner surface of the cell surface membrane that acts as molecular switch inside cells; it is involved in transmitting signals from a variety of stimuli outside a cell to its interior.

grana (Singular: granum) A stack of thylakoid membranes; found in chloroplasts.

granules Small particles found in plant cells that contain starch and act as energy stores.

gravitropism The response of plants to gravity; the response can be positive where the direction of plant growth changes towards the Earth or negative where the change in direction is away from the Earth.

green fluorescent protein (GFP) The most common marker protein used in genetic engineering, encoded by a marker gene.

growth The process by which a cell or organism increases in size and usually complexity.

guard cell A long epidermal cell found in pairs that control the opening and closing of a stoma.

guide RNA (gRNA) A short length of RNA produced by CRISPR regions found in a bacterial genome. These can be made in the laboratory to facilitate gene editing using CRISPR-Cas9.

habitat A place where an organism lives; part of an ecosystem.

haem group A chemical substance containing an iron ion that is an important component of haemoglobin.

haemoglobin A red pigment that transports oxygen around the body.

haemoglobinic acid A weak acid formed when hydrogen ions bind to haemoglobin in red blood cells, the formation of which promotes the dissociation of oxyhaemoglobin.

haemophilia An inherited genetic disorder that impairs the body's ability to make blood clots, a process needed to stop bleeding. In the most common form, it is caused by having recessive allele(s) for a gene that is found on the X chromosome.

haploid Cells that contain only one set of chromosomes.

Hardy–Weinberg equations Equations used to predict allele and genotype frequencies in a population: $p^2 + 2pq + q^2 = 1$ and $p + q = 1$.

Hardy–Weinberg principle The statement that the allele frequencies in a population will remain the same over successive generations, provided that a number of assumptions are made.

hepatocyte A cell from the liver.

herbaceous Non-woody plants, for example, strawberry plants and buttercups.

herbaceous dicotyledonous plants The main group of flowering plants, including, for example, strawberry plants and buttercups.

herbicide A chemical that kills plants.

herd immunity Immunity that occurs when a large enough proportion of the population are immune so that a disease is unlikely to spread to those who are not.

heterozygous An individual having two different alleles of a particular gene.

histone proteins Proteins present in chromatin, around which are coiled DNA molecules.

HIV / AIDS An infectious disease caused by the viral pathogen human immunodeficiency virus (HIV) that is transmitted through the sexual and blood-to-blood contact routes and can cause weakening of the immune system, leaving patients susceptible to other diseases.

homeostasis Keeping a constant internal environment.

homologous Chromosomes that have the same structural features and pattern of genes.

homozygosity The possession of two identical alleles of a particular gene or genes by an individual.

homozygous An individual having two identical alleles of a particular gene.

hormone A chemical substance secreted by an endocrine gland that is carried in the blood plasma to another part of the body to have an effect on a target cell.

human chorionic gonadotrophin (hCG) A hormone that is secreted by the placenta and that stimulates ovulation and secretion of progesterone by the ovary. It is the hormone which is detected by the pregnancy test.

humoral response An immune response mediated by antibodies released by plasma cells, which themselves differentiated from a single B-lymphocyte by clonal selection.

Huntington's disease A fatal genetic disorder that causes the progressive breakdown of nerve cells in the brain.

hybrid Organism resulting from a mating between two different species; hybrids are usually sterile.

hybrid chromosomes A chromosome that consists of genetic material from both parents, produced by crossing over during prophase I of meiosis.

hybridise see hybridisation

hybridisation The attachment, by hydrogen-bonding, of two DNA or RNA molecules with complementary base sequences.

hybridoma cell A cell formed by the fusion of a plasma cell and a cancer cell; it can both secrete antibodies and divide by mitosis to form other cells like itself.

hybrid vigour An increased ability to survive and grow well, as a result of outbreeding and therefore increased heterozygosity.

hydrogen bond A weak force of electrostatic attraction between two atoms. In biology, these usually form between a hydrogen atom and another atom.

hydrogencarbonate ions The negatively charged product that is formed when carbonic acid dissociates in aqueous solution, formula HCO_3^-; the other product is a positively charged hydrogen ion.

hydrolysis Reactions in which compounds are broken down by reacting with water.

hydrolysis of ATP A reaction in which chemical energy stored in ATP is released; essential for the process of active transport.

hydrophilic A molecule that is attracted to water molecules and tends to be dissolved by water. Water-loving.

hydrophobic A molecule that is repelled by water molecules and tends to be dissolved by oil. Water-hating.

hydrostatic pressure The pressure exerted by a liquid enclosed in a space. For example, that of plant sap on the walls of phloem sieve tubes, or of fluid on the capillary walls of the glomerulus or the walls of the Bowman's capsule.

hyperpolarisation When the membrane potential becomes more polarised than in its usual resting state.

hypertonic Describes a solution whose solute concentration is higher than the solute concentration inside a cell.

hypothalamus A small region of the brain located at the base, near the pituitary gland. It plays a crucial role in many important functions, including the synthesis of hormones such as ADH.

hypotonic Describes a solution whose solute concentration is lower than the solute concentration inside a cell.

immobilised enzyme An enzyme attached to an inert, insoluble material, which restricts its movement.

immune A state in which the immune system produces so many antibodies so quickly that the antigen is destroyed before it can make us ill.

immune response The response of the immune system to invasion by foreign cells.

immune system This consists of a number of lymphoid organs, for example the thymus gland, spleen, tonsils, and lymph nodes, linked by lymphatic vessels and capillaries. The lymphoid organs contain billions of lymphocytes that are responsible for identifying and eliminating pathogens.

immunity The ability to resist an infectious disease, owing to the presence of antibodies and/or memory cells.

immunoglobulin Protein molecules produced by B-lymphocytes. They are also known as antibodies.

inbreeding Breeding that occurs between members of a species that are very closely related.

inbreeding depression A loss of the ability to survive and grow well, due to breeding between close relatives; this increases the chance of harmful recessive alleles coming together in an individual, and being expressed.

independent assortment The random alignment of homologous chromosomes on either side of the equator during prophase and metaphase I, which means that they are inherited by the gametes in any order.

independent variable The variable that is changed by the experimenter during an investigation.

induced-fit hypothesis A type of enzyme action in which the substrate binds to the active site of the protein.

inducer A substrate whose presence results in the synthesis of an inducible enzyme.

inducible enzyme Enzymes that are synthesised only in the presence of their substrate.

infectious disease A disease that is caused by a pathogen and that can be transmitted from an infected person to an uninfected person.

inheritance The study of inheritance patterns between different generations.

inhibition To limit, prevent or block the action or function of; as in to inhibit an enzyme, or to inhibit a chemical reaction.

insecticide A chemical that kills insects.

insulin Hormone produced by the pancreas that lowers the blood sugar level.

integral protein A membrane protein that is permanently attached to the biological membrane, usually emerging on both sides of the structure.

interbreed Mating between two organisms of different species.

intermediate neurone Type of neurone that has synapses with sensory and motor neurones; intermediate neurones in the spinal cord are part of the reflex arc; also called relay neurones and interneurones.

International Union for Conservation of Nature (IUCN) A global environmental organisation with the aim of conserving the integrity and diversity of ecosystems.

interphase The phase in the cell cycle when DNA replication takes place.

intracellular enzyme An enzyme that functions within the cell in which it was produced.

intracellular signaling A series of chemical reactions that occur inside a cell, in which the product of each reaction causes the next reaction to occur.

intramolecular hydrogen bonds Weak chemical bonds involving hydrogen atoms that occur between atoms within a molecule itself.

intron Non-coding section of DNA.

invasive alien species Organisms that are introduced to a new habitat resulting in negative consequences for organisms within the new habitat.

in vitro fertilisation (IVF) Method of assisted reproduction where sperm and egg are allowed to fertilise in laboratory conditions.

ion channel An integral membrane protein that traverses the cell surface membrane and forms a channel to facilitate the movement of ions through the membrane according to their electrochemical gradient.

iodine–potassium iodide solution Iodine dissolved in potassium iodide solution, this is a straw-coloured solution that turns blue/black in the presence of starch.

ionising radiation Radiation consisting of particles, X-rays, or gamma rays with enough energy to cause ionisation in the medium through which it passes.

islets of Langerhans Regions of the pancreas consisting of α-cells that secrete the hormone glucagon, and β-cells that secrete the hormone insulin.

isomers Each of two or more compounds with the same formula but a different arrangement of atoms in the molecule and different properties.

isotonic Describes a solution whose solute concentration is equal to the solute concentration inside a cell.

K_m constant see Michaelis–Menten constant

kidney A bean-shaped organ found in vertebrates that is an organ of excretion. It filters the blood to produce urine.

kinetochore A protein structure found at the centre of a sister chromatid to which microtubules attach during mitosis.

kingdom Taxonomic rank in classification between domain and phylum.

kingdom Animalia Group of organisms in domain Eukarya whose cells lack cell walls.

kingdom Fungi Group of organisms in domain Eukarya whose cells have chitin cell walls.

kingdom Plantae Group of organisms in domain Eukarya whose cells have cellulose cell walls.

kingdom Protoctista Group of organisms in domain Eukarya that cannot be classified within the other kingdoms.

Krebs cycle A series of enzyme-catalysed reactions that occur in the mitochondrial matrix as part of aerobic respiration; the purpose of the Krebs cycle is to reduce the coenzymes NAD and FAD.

lacA A structural gene in the *lac* operon that encodes lactose permease.

lac operon A length of DNA containing genes that code for one or more proteins, and also base sequences which control whether or not the genes will be expressed.

lactate The dissociated form of lactic acid that is the product of anaerobic respiration in animal cells.

lactate fermentation Anaerobic respiration in animals that uses glucose and results in the formation of lactate.

lactose Disaccharide composed of glucose and galactose.

lactose permease A channel protein specific to lactose, which allows lactose to move into bacterial cells by facilitated diffusion.

lactose transacetylase An enzyme that modifies lactose, which plays a role in lactose absorption in bacteria.

lacY A structural gene in the *lac* operon that encodes lactose transacetylase.

lacZ A structural gene in the *lac* operon that encodes β-galactosidase.

lagging strand A lagging strand of DNA requires a slight delay before it is copied, and it must be replicated discontinuously in small fragments.

latent heat of vaporisation The energy that is lost from liquid water as it evaporates.

Le A gene whose protein product is an enzyme that catalyses the final reaction in a metabolic pathway that produces gibberellin.

le A gene whose protein product is a non-functional enzyme that catalyses the final reaction in a metabolic pathway that produces gibberellin.

leading strand Strand of DNA being replicated continuously. In DNA replication, the strand that is made in the 5' to 3' direction by continuous replication at the 3' growing tip.

leucocyte A colourless cell, also called a white blood cell, which circulates in the blood and body fluids and is involved in counteracting foreign substances and disease. There are several types, including lymphocytes, neutrophils and macrophages.

ligand A molecule that binds to a protein receptor either in the cell surface membrane (if the ligand is hydrophilic) or inside the cytoplasm (if the ligand is hydrophobic).

ligation The formation of a phosphodiester bond between two nucleotides, catalysed by the enzyme DNA ligase.

light-dependent stage Series of reactions in photosynthesis that occur in the grana of chloroplasts; light is used to split water, synthesise ATP and reduce NADP; oxygen is a waste product.

light-independent stage Series of reactions in photosynthesis that occur in the stroma of chloroplasts; CO_2 is used to synthesise carbohydrate and ATP and reduced NADP are used in the Calvin cycle.

light microscope Optical instrument for producing an enlarged image of a small object.

lignin A strong, impermeable substance found in the walls of xylem vessels, which acts as waterproofing and also provides structural support.

limiting factor Any variable that, when changed, affects the rate of a process. For example, factors such as enzyme concentration or substrate concentration that limit any further increase in the rate of an enzyme-catalysed reaction.

Lincoln index A method for estimating population sizes using the mark-release-recapture method.

line transect Sampling method where frame quadrats are placed along a line, either continuously or at intervals.

linked genes Genes that are inherited together with the other gene(s) as they are located on the same chromosome.

link reaction The decarboxylation of pyruvate that occurs in the mitochondrial matrix and connects glycolysis with the Krebs cycle.

lipase An enzyme in the pancreas that catalyses the breakdown of fats

lipid Fat made up of molecules of carbon, hydrogen and oxygen.

liposome A small structure consisting of a phospholipid bilayer that encloses a small volume of water or water-soluble substances. Liposomes can be used in gene therapy to transport DNA into a cell.

liver An organ in vertebrates, which detoxifies various metabolites, synthesises proteins, and produces biochemicals necessary for digestion.

lock-and-key hypothesis In this analogy, the lock is the enzyme and the key is the substrate. Only the correctly sized key (substrate) fits into the key hole (active site) of the lock (enzyme).

locus (Plural: loci) A fixed position on a chromosome, like the position of a gene.

longitudinal section (LS) A view along the length of a specimen, as opposed to a transverse section (TS) which is a view across the length.

loop of Henlé The part of a kidney tubule which forms a long loop in the medulla of the kidney, from which water and salts are reabsorbed into the blood.

lumen The hollow channel though which a liquid or gas passes.

lymphatic vessel Vessel through which a fluid called lymph circulates, which carries some immune cells.

lymph node Small swelling in the lymphatic system where lymph is filtered and lymphocytes are formed.

lymphocyte Large white blood cell, derived from stem cells in the bone marrow, that helps to defend the body against infection. There are two types of lymphocyte, B-lymphocytes that make antibodies (humoral response) and T-lymphocytes which are involved in the cellular response.

lyse To burst. Animal cells, which lack a cell wall, will lyse when they are immersed in a solution of very high water potential.

lysosome Vesicle that contains the digestive enzyme lysozyme.

lysozyme A digestive enzyme, found in lysosomes, tears and saliva, which breaks down bacterial cell walls and unwanted organelles within a cell.

macrofibril A fine fibre-like strand, consisting of cellulose microfibrils packed together.

macromolecule A large molecule built by joining together a small number of simple molecules with covalent bonds.

macrophages Leucocytes that are the largest type of phagocyte, which engulf pathogens. They exist outside of the bloodstream and engage in phagocytosis in the lymph, tissue fluid and lungs.

magnification How many times larger an image is than an object.

malaria An infectious disease caused by four different species of the *Plasmodium* parasite that is transmitted by the female *Anopheles* mosquito acting as a vector and that causes fever, headaches, nausea and vomiting.

malignant A malignant tumour is a mass of cells which is fast-growing and can spread to other organs.

maltose A disaccharide composed of two glucose molecules.

marker gene A gene that encodes an easily-identifiable protein product, used in genetic engineering to determine which cells have successfully absorbed recombinant DNA.

mark-release-recapture Method used in estimation of population size where organisms are captured, marked, released and captured again. The proportion of marked organisms in the second capture is then used to estimate the population.

mass flow The movement of a liquid through transport vessels, for example, the movement of blood through blood vessels, of water through xylem vessels, and of dissolved organic substances in solution through phloem vessels.

matrix Fluid inside mitochondria.

maximum velocity (V_{max}) The maximum rate of an enzyme-catalysed reaction at high substrate concentrations.

medulla The inner region of an organ such as the kidney, especially when it is distinct from the outer region or cortex.

meiosis Cell division in which the number of chromosomes is halved.

membrane repolarisation The change in distribution of charge across a membrane where the charge difference increases to its original value.

memory cell Lymphocyte that responds rapidly to reinfection and triggers a secondary immune response.

meristem (meristematic tissue) A tissue found in the tips of roots and shoots (apical meristem) and around the edges of stems and trunks (lateral meristem) that contains fast-dividing cells responsible for plant growth.

mesophyll The inner cells of a plant leaf, made up of the palisade mesophyll layer and the spongy mesophyll layer.

messenger RNA (mRNA) Messenger ribonucleic acid – a molecule of RNA with the code for a protein.

metabolic waste Substances left over from metabolic processes (such as cellular respiration) which cannot be used by the organism (they are surplus or toxic), and must therefore be excreted.

metabolism The sum of all of the chemical reactions that occur in a cell.

metaphase The phase in the cell cycle when two spindle fibres attach to the centromeres and pull the chromosomes into a line across the centre of the cell.

metastasis The process by which cancer cells are released from a malignant tumour and travel in the bloodstream and lymphatic system.

micelles Tiny droplets that are formed when lipids do not fully dissolve in water.

Michaelis–Menten constant (K_m) The substrate concentration at which the reaction rate is half of V_{max}, used as a measure of the efficiency of an enzyme; the lower the value of K_m, the more efficient the enzyme.

microarray A small piece of glass or plastic on which thousands of tiny pieces of single-stranded DNA are attached in a regular arrangement; the DNA pieces act as probes and allow rapid analysis of DNA of unknown composition.

microfibril A fine fibre-like strand, consisting of cellulose.

micrograph Image taken through a microscope.

micrometre (μm) A measure of distance equal to one thousandth of a millimetre or one millionth of a metre.

microtubules Fine protein filaments that help provide structure and shape to cell cytoplasm; make up the spindle, and attach to the centromeres of chromosomes during mitosis.

microvilli Small finger-like projections found on the surface of some cells to increase surface area, e.g. on the surface of villi in the small intestine to aid food absorption.

middle lamella Layer of pectin, found between the cell walls of adjacent plant cells, which holds plant cells together.

migration The movement of individuals into (immigration) or out from (emigration) a population.

millimetre (mm) A measure of distance equal to one thousandth of a metre.

missense mutation A mutation in which a single nucleotide (base) change results in a codon that codes for a different amino acid.

mitochondria (Singular: mitochondrion) Organelle that is the site of aerobic respiration.

mitosis A type of eukaryotic cell division in which the new cells have exactly the same number and type of chromosomes as the parent cell.

mitotic index Expressed as a percentage, this is the proportion of cells that are undertaking mitosis in a given sample of tissue.

molecular biology The study of the structure and interactions of cellular molecules, such as nucleic acids and proteins, which carry out the biological processes essential for cellular functions and maintenance.

molecular formula A means of presenting information about the chemical proportions of atoms that make up a particular chemical compound or molecule, using chemical element symbols and numbers.

molecules Small particles made up of atoms that are linked by bonds.

monoclonal antibody Antibodies produced by a single clone of B-lymphocytes that have differentiated to form plasma cells.

monocytes Leucocytes that have a large, kidney-shaped nucleus, which are found in the bloodstream and engage in phagocytosis to engulf pathogens. They can leave the blood and become macrophages.

monomers Small molecules that can be chemically bonded to form polymers.

monosaccharides Simple sugars – the monomers from which all larger carbohydrates are made.

monounsaturated Monounsaturated fatty acids are fatty acids that have one double bond in the fatty acid chain with the rest of the carbon atoms being single-bonded.

morphological species concept Organisms that have similar body shape and physical features are of the same species; used for organisms that reproduce asexually.

morphology Relating to structural features of an organism.

motif A recurring structural feature of a protein, such as an α-helix or β-pleated sheet (both are examples of secondary structures).

motor end plate The terminal of a motor neurone that occurs at a muscle cell.

motor neurone Cell within the peripheral nervous system that carries nerve impulses from the central nervous system to effectors, which are usually muscles.

mucous glands Mostly multicellular structures that release mucus onto some internal body surfaces such as the upper airways (nose and trachea).

mucus A sticky aqueous solution of mineral salts and glycoproteins covering some internal surfaces in the body; mucus is produced by goblet cells and mucous glands.

multicellular Organism made of many different cells.

multiple alleles Several alleles that affect a characteristic, for example, height, blood group.

multipotent cells see pluripotent cells

mutagen An agent, such as radiation or a chemical substance, which causes genetic mutation.

mutagenic agent Factor that increases the risk of gene mutation, such as ionising radiation and X-rays.

mutation A change in the base sequence in a DNA molecule that may result in an altered polypeptide.

myelinated neurone Cell in the nervous system that has insulation in the form of Schwann cells that surround the axon and/or dendrites; transmission of nerve impulses is significantly faster than in non-myelinated neurones.

myelin sheath Layer of fatty insulation surrounding the axon and/or dendrites of a myelinated neurone; formed from Schwann cells.

myeloma A cancer of the plasma cells.

myofibril The cylinder-shaped subunit of a muscle cell that forms a long fibre with contractile properties.

myosin One of the fibrous proteins in muscle cells that works together with actin in the sliding filament model of muscle contraction.

nanometre (nm) A measure of distance equal to one thousandth of a micrometre or one millionth of a millimetre or one billionth of a metre.

narrow spectrum antibiotic An antibiotic that is only effective against specific types of bacteria.

native Of a species, one that has historically lived in a particular ecosystem.

natural immunity An immunity gained by direct interaction between one organism and another. It can be active (for example, by previously developing a disease caused by an infectious organism), or passive (for example, by obtaining antibodies from the mother via the placenta or in breast milk).

natural selection The way in which individuals with particular characteristics have a greater chance of survival than individuals without those characteristics, and are therefore more likely to breed and pass on the genes for these characteristics to their offspring.

negative feedback A process in which the change in a physiological factor brings about processes which move its level back towards the set point.

nephron A kidney tubule.

nervous system The part of an animal that co-ordinates its actions by transmitting signals to and from different parts of its body.

neuromuscular junction The type of synapse found between a motor neurone and a muscle cell.

neurotransmitter Chemical that is released, on arrival of an action potential, from the presynaptic knob of a neurone at a synapse, diffusing across the synaptic cleft to stimulate a new action potential on the post-synaptic side.

neutrophil Leucocytes that have a large, kidney-shaped nucleus, which are found in the bloodstream and engage in phagocytosis to engulf pathogens. They can leave the blood and become macrophages.

nitrogenous base A nitrogen-containing molecule that is an important component of nucleotides found in DNA and RNA: adenine, guanine, cytosine, thymine and uracil.

nitrogenous waste Metabolic waste that contains nitrogen atoms, which must be excreted from the body due to its toxicity.

node of Ranvier Gap on an axon or dendrite between two Schwann cells where an action potential can occur.

non-competitive inhibitors Molecules that inhibit enzyme reactions by altering the shape of the enzyme molecule.

non-cyclic photophosphorylation Part of the light-dependent stage of photosynthesis involving both photosystems I and II; photolysis, the reduction of NADP and ATP synthesis; NADP is reduced because the electrons from photosystem II are not returned.

non-homologous Chromosomes that have different structural features and pattern of genes. Chromosomes 3 and 19, for example.

non-overlapping Describes the triplets in a DNA code that are read sequentially.

non-polar A state in which the distribution of charge in a molecule is uneven.

non-reducing sugar Sugar such as sucrose that is not readily reduced in other substances.

non-self antigen Antigen that is not usually found on cells of the host organism.

nonsense mutation A mutation in which a codon that corresponds to one of the twenty amino acids specified by the genetic code is changed to a STOP codon.

non-specific immune response Carried out by non-specific immune cells, this is a response to one of many antigens, not just one. An example is phagocytosis by monocytes, neutrophils and macrophages.

non-transcribed strand The strand in DNA that is not used as a code for making proteins.

non-vascular Describes an organism, or tissue, that does not have specialised transport vessels, for example, the plant moss.

normal distribution The distribution of data that tends to cluster around an intermediate value with no bias to either extreme value.

nuclear envelope Two membranes that enclose the nucleus of a cell.

nucleic acid A complex organic substance present in living cells whose molecules consist of many nucleotides linked in a long chain: DNA or RNA.

nucleolus Dark region within a nucleus; the site of RNA production.

nucleotides Molecules that combine to form strands of DNA and RNA.

nucleotide addition A type of mutation that occurs when one or more extra nucleotides are inserted, so extra bases are added to the DNA sequence.

nucleotide deletion A type of mutation that occurs when one or more nucleotides are removed, so the DNA sequence has fewer bases.

nucleotide substitution A type of mutation that occurs when a nucleotide is replaced by a nucleotide with a different DNA base.

nucleus The largest organelle in a cell, containing most of the cell's DNA.

null hypothesis A statement relating to an investigation that you are expecting to reject.

obligate intracellular parasites Description used for viruses because they require a host cell within which to replicate.

observation A statement based on something one has seen or otherwise noticed.

oestrogen Female sex hormone that stimulates the development of female secondary sexual characteristics and has a role in the menstrual cycle.

Okazaki fragment A newly formed DNA fragment that forms part of the lagging strand during replication and which is linked by DNA ligase to produce a continuous strand.

operon A functioning unit of DNA containing a cluster of genes under the control of a single promoter.

opportunistic disease A disease caused by a microorganism that is normally non-pathogenic, but can be pathogenic in people whose immune system is weakened, for example with HIV / AIDS.

optimum pH The pH in which enzymes function best.

optimum temperature The temperature at which enzymes work fastest (about 40 °C).

order Taxonomic rank in classification between class and family.

organelle A specialised structure within a cell carrying out a particular function. Mitochondria and chloroplasts are examples of organelles.

organic compounds Compounds containing carbon, hydrogen and often other elements such as oxygen or nitrogen; they are usually associated with living organisms.

origin (in chromatography) The position where the sample to be separated is placed at the start, before solvent is applied.

osmoreceptor cells Cells in the hypothalamus that are able to detect changes in the water potential of the blood.

osmoregulation The maintenance of constant osmotic pressure in the fluids of an organism by the control of water and salt concentrations.

osmosis The movement of water from a less concentrated solution to a more concentrated solution through a selectively permeable membrane.

outbreeding Breeding between individuals that are not closely related.

ovaries The organs in females that produce ova.

oxidative phosphorylation The process in aerobic respiration that occurs in mitochondria and uses reduction-oxidation (redox) reactions to release the energy necessary for ATP synthesis from ADP and inorganic phosphate.

oxidising agent A reactant that becomes reduced causing another reactant to become oxidised.

oxyhaemoglobin The product formed when oxygen binds with haemoglobin; each haemoglobin molecule can bind a maximum of 4 oxygen molecules (8 oxygen atoms).

palindrome A sequence of bases that reads the same in the 5′ to 3′ direction on one strand as in the 3′ to 5′ direction on the other strand.

pancreas An organ lying close to the stomach that functions both as an exocrine gland (secreting pancreatic juice) and an endocrine gland (secreting insulin and glucagon).

pandemic The worldwide spread of a disease caused by a new pathogen or by a pathogen to which people have no immunity.

parasite An organism that gets benefit from another organism (the host) but which gives no benefit to the host; viruses are also classed as parasites because viruses require a host cell in which to replicate.

parent cell The cell that divides by mitosis and cytokinesis to form two daughter cells.

partially permeable membrane A membrane that contains pores large enough for water molecules to pass through but not large enough for the solute molecules to pass through.

partial pressure The proportion of total pressure in a mixture of gases that is due to one particular gas.

passive Describes a biological process that does not require a cell to do anything to make it happen.

passive immunity Immunity that is obtained without exposure to the antigen. It can be artificial (by giving an individual an injection containing the appropriate antibodies) or natural (by obtaining antibodies from the mother via the placenta or breast milk).

pasteurisation Process by which foods are heated to 100 ˚C or lower for a short time sufficient to kill pathogenic microorganisms without significant effect on the flavour of the foods; pasteurisation is different from sterilisation as not all microorganisms are killed in pasteurisation.

pathogen An infectious agent, such as a virus or some types of bacteria.

Pearson's linear correlation A measure of the strength of the linear relationship between two variables that consist of continuous data with a normal distribution.

pectin A carbohydrate found in the middle lamella between plant cell walls.

pedigree diagram A diagram that shows the occurrence and appearance or phenotypes of a particular gene or organism and its ancestors from one generation to the next.

penicillin The first antibiotic to be mass-produced and commercially available, its mode of action is bactericidal by inhibiting bacterial cell wall synthesis.

peptide bond A bond between two amino acids.

peptidoglycan A polymer made of polysaccharide and peptide chains; the substance that prokaryotic cell walls are made of.

percentage cover A measure of abundance used for plants or sessile animals.

percentage error A measure of the accuracy of the measurement made during an investigation.

peripheral nervous system The parts of the nervous system that are in the outer parts of the body and not within the brain or spinal cord.

peripheral protein A membrane protein that is permanently attached to the biological membrane, usually emerging on only one side of the structure.

peristalsis Wavelike muscular contractions in tubular structures, including the ureter.

peroxidase An enzyme that catalyses the oxidation of a particular substrate by hydrogen peroxide.

personalised medicines Medicinal drugs that are targeted to an individual, usually as a result of genetic screening.

phagocyte A cell that ingests (engulfs) and digests foreign matter or microorganisms. There are a number of different types of phagocyte, which differ in their location, for example, monocytes, neutrophils and macrophages.

phagocytosis The process by which a phagocyte engulfs large particles or whole cells, as a defence mechanism.

phagolysosome A cytoplasmic body formed by the fusion of a phagosome with a lysosome in a process that occurs during phagocytosis.

phagosome A vacuole in the cytoplasm of a cell, containing a phagocytosed particle enclosed within a part of the cell surface membrane.

phase The specific time window in the cell cycle during which specific events occur.

pH buffer An aqueous solution consisting of a mixture of a weak acid and its conjugate base, or vice versa. Its pH changes very little when a small amount of strong acid or base is added to it.

phenotype The physical effects of the genotype on the individual.

phloem A tissue found in many plants that transports dissolved organic compounds, such as sucrose and amino acids, from regions where the substances are made or stored, to where they are used or stored, in the process of translocation. Phloem tissue contains different types of cell, including phloem sieve tube elements and companion cells.

phloem sieve tube element The individual transport cells within phloem tissue; they get their name because their end walls form sieve plates.

phosphodiester bonds The bonds that join sugar and phosphate molecules in a polynucleotide.

phospholipid Lipid containing two fatty acids, attached to glycerol via an ester bond.

phospholipid bilayer A two-layered arrangement of phospholipid molecules that form a cell surface membrane, the hydrophobic lipid tails facing inward and the hydrophilic phosphate heads facing outward.

phosphorylation The transfer of a phosphate group to an organic compound.

photoexcitation The process by which energy from light causes electrons to be removed from orbitals in photosynthetic pigments.

photoionisation The process by which electrons that have been removed by photoexcitation are taken by electron acceptor molecules.

photolysis The process of splitting water in the presence of light; part of the light-dependent stage of photosynthesis producing oxygen gas, hydrogen ions and electrons and catalysed by the oxygen-evolving complex.

photomicrograph Image (photograph) taken through a light microscope.

photophosphorylation The general name for synthesising ATP from ADP and inorganic phosphate using energy from light.

photosynthesis Process in which energy from light is transferred to produce complex organic molecules.

photosystem I A group of membrane proteins that use light energy for synthesis of ATP and reduction of NADP; involved in cyclic photophosphorylation and the second photosystem involved in non-cyclic photophosphorylation.

photosystem II A group of membrane proteins that use electrons from photolysis to provide energy for ATP synthesis and to reduce NADP; photosystem II is the first photosystem involved in non-cyclic photophosphorylation.

phototropism The response of plants to the direction of a light source; the response can be positive where the direction of plant growth changes towards the light source or negative where the change in direction is away from the light source.

phylum Phylum is a taxonomic rank in classification between kingdom and class. Plural is phyla.

phytochrome-interacting protein (PIF) A transcription factor that initiates the expression of genes including the amylase gene in plants.

pits Microscopic holes in the secondary cell wall of xylem vessel elements, which allow water and dissolved mineral ions to pass in and out of the xylem vessels.

plan diagram A diagram showing the outlines of different tissues but not individual cells.

plasma The liquid part of blood that does not include red blood cells, white blood cells or platelets.

plasma cell B-lymphocyte that has differentiated to form a cell that secretes large numbers of antibodies.

plasmid Small loop of DNA found in bacteria. Used in genetic technology as a vector to transfer genes into organisms.

plasmodesmata (Singular: plasmodesma) Cytoplasmic connections between adjacent plant cells through tiny gaps in their cell walls.

plasmolysis The shrinking of cytoplasm away from the cell wall of a plant or bacterium due to water loss from osmosis, thereby resulting in gaps between the cell wall and cell surface membrane.

pluripotent cells Types of cell that can differentiate into several different kinds of cells, for example, different types of blood cells. Also called multipotent cells.

podocytes Cells in Bowman's capsule in the kidneys that wrap around capillaries of the glomerulus.

polar Describes a molecule with partial positive and partial negative charges across the molecule.

pole The ends of the cell, from which position the spindle fibres radiate during prophase to attach to the centromeres of the chromosomes.

polygenes A number of different genes at different loci that all contribute to a particular aspect of phenotype.

polymer A long chain of chemically bonded smaller molecules called monomers.

polymerase chain reaction (PCR) A (usually fully automated) method of rapidly producing many copies of a small DNA sample.

polynucleotide Polymer whose molecules comprise many nucleotides, such as DNA and RNA.

polypeptide chain Long chain of amino acids.

polysaccharide a polymer whose subunits are monosaccharides joined together by glycosidic bonds.

polysome (polyribosome) A cluster of ribosomes held together by a strand of messenger RNA which each is translating.

polyunsaturated Polyunsaturated fats are fats in which the constituent hydrocarbon chain possesses two or more carbon–carbon double bonds.

population All of the organisms of the same species present in the same place and at the same time that can interbreed with one another.

positive feedback A process in which the change in a physiological factor brings about processes which move its level even further in the direction of the initial change.

posterior pituitary gland The posterior lobe of the pituitary gland, found at the base of the brain, which is part of the endocrine system.

post-exposure prophylaxis (PEP) Intensive course of drugs given to people who have been infected with human immunodeficiency virus (HIV) with the aim of preventing disease development, but without guaranteed success.

potassium iodide solution Used to dissolve iodine for the test for starch.

power stroke Stage in the sliding filament model of muscle contraction where myosin heads that are bound to actin bend, pulling the actin filaments.

pre-implantation genetic diagnosis (PGD) The genetic screening of an embryo, produced by IVF. The results of the test will determine whether or not the embryo is implanted into the uterus.

primary cell wall The first cell wall that forms during the development of xylem vessel elements; made of cellulose.

primary immune response The stimulation of the immune system that occurs when the body is first exposed to an antigen. It is usually slow to occur and produces a small number of antibodies, which means that symptoms of disease can develop.

primary structure The unique sequence of amino acids in a polypeptide.

primary transcript The molecule of RNA that is formed by the transcription of a gene. It contains the base sequence also present in the introns of that gene, which must be removed before it is translated.

primer A short length of DNA with a base sequence complementary to the part of a DNA strand that is to be copied during the PCR.

probe A short region of single-stranded DNA with a specific sequence complementary to a sequence of DNA to be identified on a gel after electrophoresis.

products The molecules produced by a reaction.

prokaryotic cell Cell that does not have a nucleus or other membrane-bound organelles; found in Bacteria and Archaea.

promoter A region of DNA that initiates transcription of a particular gene, usually found just upstream of that gene.

prophase The first stage of mitosis, when the two membranes of the nuclear envelope break down.

prophylactic A course of drugs designed to prevent a disease.

proportional (simple) dilution A method by which a unit volume of a liquid material of interest is combined with an appropriate volume of a solvent liquid to achieve the desired concentration.

prosthetic group A non-protein component in a protein molecule that is vital for the function of the protein. An example is the iron-containing haem group in haemoglobin.

protein A molecule made up of long chains of amino acids.

protein kinase A kinase enzyme that modifies other proteins by chemically adding phosphate groups to them (phosphorylation).

protein synthesis A process of creating protein molecules, the amino acid sequence of which is determined by the DNA base sequence.

proteome The complete set of proteins produced by a cell.

proton pump A carrier proton that actively transports hydrogen ions (protons) across a cell surface membrane.

proximal (or first) convoluted tubule (PCT) The portion of the duct system of the nephron of the kidney which leads from Bowman's capsule to the loop of Henlé.

pseudopodia A temporary protrusion of the surface of a cell, such as a phagocyte, which is used for movement and feeding.

pulmonary artery Blood vessel that carries blood away from the right ventricle towards the lungs.

pulmonary circulation system The part of the circulation system that carries blood to and from the lungs.

pulmonary vein Blood vessel that carries blood away from the lungs towards the left ventricle.

pump proteins Proteins that use ATP in the process of active transport to allow the transport of different molecules across a cell surface membrane.

Punnett square A diagram that is used to predict the genotypes of a particular cross or breeding experiment.

purine The nitrogenous bases adenine and guanine present in DNA.

Purkyne tissue Electrically conductive fibres located in the ventricle walls of the heart that co-ordinate the ventricles contracting at the same time.

pyrimidine The nitrogenous bases cytosine and thymine present in DNA, and uracil, present in RNA.

pyruvate The dissociated form of pyruvic acid; two molecules of pyruvate are produced from one molecule of glucose in glycolysis.

qualitative In experiments, this relates to measuring a factor in a way that does not use numbers.

quantitative In experiments, this relates to measuring a factor using numerical values.

quaternary structure The structure of proteins that contain more than one polypeptide chain.

random error Fluctuations in the measured data owing to the precision limitations of the measurement device.

randomly orientated The random alignment of homologous chromosomes on either side of the equator during prophase and metaphase I, which means that they are inherited by the gametes in any order.

random mating The situation in which an individual is equally likely to mate with any other individual in a population.

random sampling Method used to collect unbiased information about numbers or distribution of organisms.

receptor A cell that is sensitive to a stimulus, a change in the environment. It usually generates an action potential as a result of the change.

recessive An allele that needs to be present twice in the genotype to have an effect on the phenotype.

recipient The cell into which a donor DNA molecule is introduced during genetic engineering.

recombinant An organism that shows two or more phenotypes that are different to those in either of its two parents.

recombinant chromosomes Chromosomes that are formed by the exchange of genetic material between non-sister chromatids of homologous pairs.

recombinant DNA A DNA molecule that consists of genetic material that originates from two different species.

recombinant plasmid A circular DNA molecule, usually extracted from a bacterium, which contains genetic material that originates from another species.

recombinant protein A protein produced by a transgenic organism.

recombinant protein synthesis The production of proteins from recombinant DNA in a transgenic organism.

recovery period Part of the refractory period following an action potential where sodium channels become active after being opened.

red blood cell Specialised cell in the blood that carries haemoglobin for the transport of oxygen around the body.

redox indicator A substance that changes colour when converted between its oxidised state and its reduced state.

reducing agent A reactant that becomes oxidised causing another reactant to become reduced.

reducing sugar Sugar such as glucose and fructose that readily lose electrons to another substance.

refractory period Period of time following the start of depolarisation where a second stimulus will not generate an action potential.

regulatory gene Genes that code for proteins that regulate the expression of other genes (e.g. repressors).

reliability A measure of the repeatability of data. A reliable set of data will be consistently obtained when the experiment is repeated, and is free from or not influenced much by anomalous results.

renal artery One of the pair of large blood vessels that branch off from

the abdominal aorta (the abdominal portion of the major artery leading from the heart) and enter into each kidney.

renal vein The short thick veins which return blood from the kidneys to the vena cava.

replicable In an experiment, something that can be copied or recreated.

repressible enzyme Enzymes that are synthesised only in the absence of a repressor.

repressor Proteins that inhibit transcription by binding to a section of DNA near the promoter and interfering with the ability of RNA polymerase to transcribe the gene.

reproductive isolation The inability of two groups of organisms of the same species to breed with one another, e.g. Because of geographical separation or because of behavioural differences.

residue A single unit that makes up a polymer, such as an amino acid in a polypeptide or protein.

resolution Also known as resolving power. The shortest distance between two points that can still be seen as separate objects.

respiration The production of ATP for use in energy-requiring processes, e.g. active transport and muscle contraction.

respiratory quotient The ratio of the number of molecules (or volume) of carbon dioxide produced in respiration to the number of molecules (or volume) of oxygen used in respiration.

respiratory substrate A compound that can be used as a starting substance in the reactions occurring in respiration.

respirometer A piece of equipment used to measure the rate of exchange of oxygen and carbon dioxide when an organism is respiring, the measurements then can be used to calculate the rate of respiration of the organism.

resting potential The potential, or voltage, inside a neurone relative to outside the neurone when there is no action potential or hyperpolarisation.

restriction The process by which a restriction enzyme makes a double-stranded cut in DNA at a specific (restriction) site.

restriction endonuclease An enzyme used to cut DNA at a specific base sequence.

restriction site The recognition sequence in DNA at which restriction enzymes cut. Also called a recognition site.

reverse transcriptase An enzyme derived from retroviruses that produces a single strand of complementary DNA from a molecule of mRNA.

reverse transcription The process, catalysed by the enzyme reverse transcriptase, by which a single strand of complementary DNA from a molecule of mRNA.

R_f value Ratio of the distance travelled by a pigment in chromatography to the distance travelled by the solvent.

ribonucleotide A nucleotide containing ribose as its pentose component.

ribose A pentose sugar that occurs widely in nature as a constituent of ribonucleotides, the monomer that when joined together makes RNA.

ribosome Very small organelles formed from RNA; the site of protein synthesis where amino acids are joined together to form proteins.

70S ribosome Ribosomes found in prokaryotic cells, and in the mitochondria and chloroplasts of eukaryotic cells.

80S ribosome Ribosomes found in the cytoplasm of eukaryotic cells.

ribulose bisphosphate (RuBP) Five-carbon phosphorylated sugar used as a substrate for rubisco in the Calvin cycle for the carbon-fixation step.

RNA Ribonucleic acid – a nucleic acid involved in protein synthesis; different types include messenger RNA (mRNA), transfer RNA (tRNA) and ribosomal RNA (rRNA).

RNA polymerase An enzyme that links RNA nucleotides by forming phosphodiester bonds between their ribose and phosphate groups.

root hair cell Cell on the surface of a plant root; it has a very large surface area to volume ratio for the absorption of water and mineral ions.

rough endoplasmic reticulum (RER) Endoplasmic reticulum with ribosomes attached to its outer surface, giving it a 'bumpy' appearance; proteins produced at the ribosomes move into the channels formed by the membranes of the endoplasmic reticulum to travel to other parts of the cell.

rubisco The enzyme in the Calvin cycle that catalyses the carbon-fixation step.

saltatory Mode of conduction of action potentials in myelinated neurones; means 'jumping' and refers to the action potentials jumping between nodes of Ranvier resulting in faster transmission.

sarcolemma The membrane surrounding a muscle cell.

sarcomere The repeating unit in a myofibril between two Z discs.

sarcoplasmic reticulum System of membrane-bound tubules within a muscle cell that store Ca^{2+} for release when the muscle needs to contract.

saturated fatty acid A fatty acid in which all the carbon atoms use two of their bonds to join to other carbon atoms and two to join to hydrogen atoms.

scanning electron microscope (SEM) An electron microscope that uses electrons reflected from the object to produce a three-dimensional image.

Schwann cell Flat cell that grows around the axon or dendrite of a myelinated neurone to form the myelin sheath.

secondary cell wall A second cell wall inside the first (primary) cell wall of xylem vessel elements; it is impregnated with impermeable lignin to provide waterproofing and structural support.

secondary immune response The stimulation of the immune system that occurs when the body is exposed to an antigen for a second time. It is usually much faster than the primary immune response and produces a large number of antibodies, which means that symptoms of disease do not develop.

secondary structure The three-dimensional structure of α-helices and β-pleated-sheets adopted by a polypeptide chain, owing to hydrogen bonding between neighbouring residues.

second filial, F2, generation The second generation of a cross between two individuals.

second messenger Intracellular signalling molecules released by the cell in response to exposure to extracellular signalling molecules. Examples include cyclic AMP (cAMP) in the control of blood glucose concentration and calcium ions in the control of stomata closure.

seedbank Storage facility in which seeds of plants are kept to preserve biodiversity.

selection The process by which an individual becomes more likely to survive and reproduce to pass on its alleles.

selection advantage (or selective advantage) The benefit that a particular feature brings to an individual to increase the likelihood of it being able to survive and reproduce in a specific environment.

selection force (selection pressure) An environmental factor that confers greater chances of survival and reproduction on some individuals than on others in a population. Sometimes called selective pressure or selection agent.

selective breeding (artificial selection) The selection by humans of organisms with desired characteristics, from which to breed.

selective reabsorption The absorption of some of the components of the glomerular filtrate back into the blood as the filtrate flows through the nephrons of the kidney.

self antigen Considerations that relate to an individual and his or her on cells of the host organism.

semi-conservative replication Describes the process of copying DNA in which one strand is retained from the original and one new one is made.

semilunar valve Type of valve in the heart that is similar to valves in veins,

each consisting of a pair of cusp-shaped flaps that open and close due to pressure differences; found between the ventricles and the arteries.

semi-quantitative In experiments, this relates to measuring a factor on the basis of an estimate of numerical values, e.g. shades of a colour.

sense strand The strand of DNA that is used as a code for making proteins.

sensory neurone Cell within the peripheral nervous system that carries nerve impulses from receptors to the central nervous system.

septum Dividing wall in the middle of the heart separating the left and right atria and the left and right ventricles; the septum prevents the mixing of oxygenated and deoxygenated blood.

serial dilution The stepwise dilution of a substance in solution.

set point The ideal value of a physiological factor that the body controls in homeostasis.

sex chromosomes The chromosomes that determine sex in animals. In humans, these are the X and Y chromosomes.

sexually-reproducing Organisms that reproduce by producing gametes, which then fuse during fertilisation to form a zygote.

sickle cell anaemia A disease that results in the formation of a defective type of haemoglobin that causes sickle-shaped red blood cells to form. It is caused by the inheritance of two recessive alleles for the beta-globin gene.

sieve plates The perforated end walls of phloem sieve tube elements; the perforations are called sieve pores and allow the movement of materials, such as sucrose, from one sieve tube element to another along the phloem vessel.

silent mutation A change in the sequence of nucleotide bases which make up DNA, without a subsequent change in the amino acid or the function of the overall protein.

signalling molecules Molecules that interact with a target cell to bring about a change inside that cell, by

binding to cell surface receptors, and/ or by entering into the cell through its membrane or endocytosis to bind to cytoplasmic receptors.

signal transduction The transmission of molecular signals from a cell's exterior to its interior.

simple tissue A tissue made of only one type of cell, for example, spongy mesophyll.

Simpson's index of diversity A method of expressing the biodiversity of a habitat taking into account the numbers of each species and the abundance of each species.

sink A part of a plant to which assimilates are moved by translocation, for example, a developing fruit.

sinoatrial node The pacemaker of the heart that releases regular electrical impulses to initiate the cardiac cycle; located in the wall of the right atrium.

sister chromatids The two strands into which a chromosome splits during cell division. They are held together by a centromere.

smooth endoplasmic reticulum (SER) Endoplasmic reticulum with no ribosomes attached to its outer surface; involved in several functions including the synthesis of hormones and lipids.

smooth muscle Type of muscle that does not appear striated under the light microscope and is under involuntary control; found in the walls of hollow organs such as some blood vessels and some airways.

social issues Considerations that relate to an individual and his or her relationships with other people.

sodium alginate A polysaccharide which, in the presence of calcium, forms a gel.

sodium-potassium-ATPase pump Active transport carrier protein in the membrane of neurones, and some other cells, that uses energy from ATP hydrolysis to move sodium ions out of the cell and potassium ions into the cell.

solute Substance that can be dissolved in a solvent.

solution A mixture in which the particles of one or more substances (the solute) are distributed uniformly throughout another substance (the solvent).

solvent Liquid in which other substances (solutes) can dissolve.

solvent front The distance travelled by the solvent in chromatography; measured from the origin.

somatic cells Body cells.

somatic gene therapy The introduction of a therapeutic allele into the body cells of individuals homozygous for a recessive genetic disease.

source A part of a plant from which assimilates are moved by translocation, for example, a leaf.

spatial Related to space or location.

Spearman's rank correlation coefficient Method of expressing the strength and direction (positive or negative) of the relationship between two variables that consist of data that is not normally distributed.

speciation The production of new species.

species A group of organisms with similar morphology and physiology, which can breed together to produce fertile o spring and which is reproductively isolated from other species. A species can be defined biologically, ecologically or morphologically

specific Of one type, or able to interact with one type.

specific heat capacity The amount of heat energy that has to be transferred to an object to increase its temperature.

specific immune response Carried out by specific immune cells such as lymphocytes, this is a response to just one antigen.

spindle A structure composed of spindle fibres that develops during mitosis and meiosis and is responsible for separation and migration of chromosomes.

spindle fibres Microtubules which attach to centromeres during meiosis and mitosis.

spleen The largest lymphatic organ in the body. It serves as a blood reservoir, disintegrates old red blood cells, and is the site of leucocyte storage.

splicing The process in which introns are removed from the primary transcript and exons and stuck together to form mRNA.

spongy mesophyll A layer of cells inside a leaf, next to the lower epidermis, characterised by many air spaces which allow free movement of gases and water vapour.

squamous epithelial cell A type of epithelial cell that is large and flattened, giving squamous epithelial tissue the appearance of scales on a fish; found in alveolar walls and capillary walls.

stabilising forces Selective forces that cause a population of organisms to change by natural selection such that more individuals have a phenotype corresponding to intermediate values. Individuals with values representing either extreme of the phenotypes are not selected for.

stabilising selection Natural selection that tends to keep allele frequencies relatively constant over many generations.

stage micrometer scale Microscope slide marked with a finely divided scale.

standard deviation A measure of the spread of data collected in an investigation around the mean value of that data. More reliable data will have a smaller standard deviation.

standard error A measure of the reliability of the mean value. This is a measure of how close the sample mean is to the population mean.

standard form A way of writing down very large or small numbers written in the form $N \times 10^n$ or $N \times 10^{-n}$, where $1 > N < 10$.

START codon The first codon of a primary RNA transcript translated by a ribosome. The start codon always codes for methionine in eukaryotes and is of the base sequence AUG.

statistical test A mathematical process which provides an indication of the strength of a series of the results of an investigation.

stem cell Undifferentiated cells that retain the ability to become specialised and divide indefinitely.

stem cell therapy The use of stem cells to treat or prevent a disease or condition.

sterilisation Process of killing all microorganisms; usually carried out by heating to over 120 °C in steam, by use of dry heat, or by radiation.

steroid Class of chemical substance including cholesterol and many hormones that has a structure consisting of four rings.

sticky ends Short, single-stranded overhangs that form after restriction.

stimulus A change in the environment, either internal or external, that is detected by a receptor, which usually leads to a response.

stomata (Singular: stoma) Microscopic openings or pores in plant tissue (usually leaves) that allow for gas exchange with the atmosphere.

STOP codon In the genetic code, a stop codon is a nucleotide triplet within messenger RNA that signals a termination of translation into proteins.

stress hormone A hormone released in response to conditions of stress, which are away from the optimum conditions.

striated muscle Skeletal muscle that has a banded appearance under the light microscope.

stroma Fluid inside chloroplasts.

structural formula A formula that shows the arrangement of atoms in the molecule of a compound.

structural gene Genes that code for proteins required by a cell for its normal function. Examples include globular proteins such as enzymes and haemoglobin, and fibrous proteins including keratin and collagen.

struggle for existence In natural selection, the competition for resources, mates or other factors needed for individuals in a species to survive and reproduce.

suberin A waxy substance found in the cell walls of the endodermis layer in plant roots; forms the Casparian strip.

subjectivity The bias of a judgement due to personal feelings, taste or opinions. It can reduce the validity of an investigation.

substrate The molecule that an enzyme allows to react.

substrate-linked reactions Enzyme-catalysed reactions that occur in respiration where an inorganic phosphate group is added to ADP directly from another phosphorylated substance.

sucrose A disaccharide composed of a glucose molecule and a fructose molecule.

sucrose-hydrogen cotransporter protein A membrane transport protein (cotransporter) that couples the movement of hydrogen ions (protons) with its concentration gradient and movement of sucrose against its concentration gradient.

sugar–phosphate backbone The portion of a DNA molecule that consists of 5-carbon deoxyribose sugars and phosphate groups. These sugars are linked together by a phosphodiester bond, between carbon 4 of their chain, and a CH_2 group that is attached to a phosphate ion.

supercoiled The process by which chromatin is further packaged by winding upon itself to form a chromosome.

(ecological) surrogacy Where something, such as a species, an ecosystem or an abiotic factor is used to represent some other factor in an ecosystem.

survival of the fittest In natural selection, the tendency for better adapted individuals to survive and reproduce.

sympatric speciation The emergence of a new species from another species where the two are living in the same place; it can happen, for example, as a result of polyploidy.

symplast (symplastic) pathway The route that some water and mineral ions move into a plant root, through the cytoplasm of adjacent cells.

symptomless carriers People who are infected with a potentially pathogenic microorganism but who do not show any signs or symptoms of the disease.

synapse A gap between two neurones; on the arrival of an action potential a neurotransmitter diffuses across the gap to enable a new action potential to be stimulated on the other side.

synaptic cleft The gap between two neurones at a synapse.

synaptic knob / synaptic button The slightly enlarged end of a neurone at a synapse.

synthesis (S) phase The time period during the cell cycle that is defined by a cell copying its DNA by semi-conservative replication.

systemic circulation system The part of the circulation system that carries blood to and from the body organs, except the lungs.

systematic error Systematic errors are usually due to a problem that is present throughout the experiment; they cause a consistent inaccuracy in the measured data.

***Taq* DNA polymerase** An enzyme derived from bacteria that are adapted to live in hot water springs; the enzyme has a very high optimum temperature, and is used to synthesise new strands of DNA during PCR.

target cell A cell which has receptors for a hormone, drug or other signalling molecule.

target organ The specific organ where a hormone, or other substance, has its effect.

telomere The very end of the arms of a chromosome, which protects the loss of genes found in the DNA further down the chromosome arms which would occur as the result of cell division.

telophase The final stage of mitosis, when the sister chromatids unravel to become strands of DNA again and the nuclear envelope reforms.

temporal Related to time.

tension A pulling force, for example that exerted downwards by gravity on the columns of water in plant xylem vessels.

tertiary structure The three-dimensional shape of a polypeptide molecule.

test cross Testing a suspected heterozygote by crossing it with a known recessive homozygote.

testes The organs in males that produce sperm.

testosterone Male sex hormone that stimulates the development of male secondary sexual characteristics.

test strips A small instrument with an absorbent pad that is placed down into urine or blood, which changes colour to give an estimate of the glucose concentration.

T-helper cell A differentiated T-lymphocyte that helps other cells in the immune response by recognising foreign antigens.

thermal cycler A 'PCR machine'. This heats and cools the mixture of molecules during a PCR reaction in a series of cycles.

thermostable Able to withstand high temperatures.

threshold In the nervous system, the minimum stimulus, or minimum amount of membrane depolarisation, required to produce an action potential.

thylakoid Inner membrane of chloroplast; embedded with chlorophyll molecules.

thymus A small gland in the top part of the chest, between the lungs. It is the site of T-lymphocyte maturation.

tissue fluid Fluid that surrounds most cells in the body, formed by liquid components of blood being forced out of capillaries and removed by the lymphatic system.

tissue repair The replacement of cells in tissues that are predisposed to lose cells over time, for example the skin and the intestinal epithelium.

T-killer (cytotoxic) cells A differentiated T-lymphocyte that destroys cells that have been infected by the pathogen that carried the specific antigen to which it can respond.

T-lymphocytes (T-cells) Lymphocytes involved in the cellular response.

tonoplast Membrane surrounding the large permanent vacuole found in plant cells.

totipotent cells Cells that can mature into any type of body cell or cell type that is associated with embryo support and development, or even a whole organism.

trachea Also called the windpipe. The organ that allows air to flow from the back of the throat (larynx) to the lungs.

transfer RNA (tRNA) A molecule of RNA that transports amino acids to ribosomes.

transcribed/ template strand The strand in DNA that is used as a code for making proteins.

transcription The process of using the coded information in DNA to form mRNA.

transcriptional mechanisms Mechanisms in gene expression that prevent or promote transcription, and thereby turn off or turn on the synthesis of RNA.

transcription factor A protein that controls the rate of transcription of genetic information from DNA to mRNA, by binding to a specific DNA sequence called a promoter.

transformation The process by which recipient cells are encouraged to absorb recombinant DNA in genetic engineering.

transgenic organism An adjective used to describe DNA, or an organism, in which the DNA has been modified so that it contains base sequences from more than one organism.

translation The process in which the mRNA code is used to make a polypeptide.

translocation The process by which dissolved organic compounds, such as sucrose and amino acids, are transported in many plants, from regions where the substances are made or stored (sources), to where they are used or stored (sinks).

transmission cycle In human diseases, the process by which a pathogen infects a person, leaves that infected person and infects another, previously uninfected, person.

transmission electron microscope (TEM) An electron microscope that uses electrons transmitted through (passed through) the object to produce a very detailed image.

transpiration The process by which water and dissolved mineral ions are transported in many plants, from the roots, up the stem, and to the leaves where the water evaporates and diffuses into the atmosphere.

transpiration pull The force pulling water (and dissolved mineral ions) up through xylem vessels from roots to leaves.

transpiration stream The flow of water and dissolved mineral ions through a plant from the roots to the leaves through the xylem vessels.

transverse section (TS) A view across the length of a specimen, as opposed to a longitudinal section (LS) which is a view along the length.

trichome Fine, hair-like outgrowths from plants.

triglyceride Molecule made up of a glycerol molecule bonded to three fatty acid molecules.

triose phosphate (TP) A three-carbon phosphorylated sugar that is an intermediate in the Calvin cycle; also called glycerate 3 phosphate (GP).

triplet A sequence of three bases on a DNA molecule coding for one amino acid.

tropomyosin Protein in muscles that blocks the positions on actin filaments where myosin heads would bind when muscles are relaxed.

troponin Calcium-binding protein in muscles that causes tropomyosin to move, exposing myosin binding sites on actin, when calcium is present in order for muscles to contract.

trypsin An endopeptidase found in pancreatic juice.

T-tubule system Extensions of the cell surface membranes of muscle cells that extend into the cell to allow faster transmission of the action potential into the cell.

tuberculosis (TB) An infectious disease caused by the bacterial pathogens *Mycobacterium tuberculosis* that is transmitted through the aerosol route and *Mycobacterium bovis* that is transmitted from infected cattle through unpasteurised dairy products. TB causes a severe persistent cough although can affect other organs besides the lungs such as kidneys.

tumour A mass of cancer cells that forms in the body.

turgid Describes a plant cell that has become firm owing to water retention.

ultrafiltration Filtration under pressure.

umbrella effect Protection given to other organisms by the presence of one, usually easily observable and well-studied, species.

unicellular Organism made of a single cell; e.g. *Amoeba* or bacteria.

universal Describes the way that the genetic code is almost identical in all living organisms.

unlinked genes Genes found on different chromosomes.

unsaturated fatty acid A fatty acid in which some of the carbon atoms have a double bond linking them to a neighbouring carbon atom.

uracil A nitrogenous base found in RNA. It is complementary to the base adenine, much like the base thymine in DNA.

urea A compound which is the main nitrogenous waste and breakdown product of protein metabolism in mammals and is excreted in urine.

ureter The duct by which urine passes from the kidney to the bladder.

urine A liquid by-product of metabolism in humans and in many other animals.

vaccination The administration of attenuated (weakened) pathogens or antigens from a pathogen into humans to induce an immune response.

vacuole Fluid-filled space inside a cell.

validity A measure of the ability to draw conclusions from the data of an investigation. In a valid investigation, the only variable that had an effect on the dependent variable was the independent variable.

variable region The region of an antibody molecule that differs in structure between different antibodies.

variation Differences between the DNA base sequences of individuals within a species.

vascular Organism or tissue containing specialised, tube-like transport vessels, such as xylem and phloem in many plants, and blood vessels in many animals.

vascular bundle A strand of vascular tissue found in many plants, consisting mainly of xylem and phloem tissue, as well as cambium; found close to the surface in plant stems and as 'veins' in leaves: these transport vessels also provide structural support.

vector An organism, usually an insect, that transmits a disease but does not contract the disease; a vector acts as a host for a pathogen and the pathogen often completes part of its life cycle in the vector.

vector (gene technology) In genetic technology, a molecule such as a bacterial plasmid that can be used to transfer a donor gene into a recipient organism, or a virus that can be used to transfer a therapeutic allele into a cell in gene therapy.

vein Type of blood vessel that carries blood back to the heart under low pressure; a vein has a relatively thinner wall and wider lumen than an artery.

vena cava Vein that returns blood to the right atrium in the heart from the systemic circulation.

ventricle One of the lower chambers of the heart that receives blood from the atria and pumps blood out into arteries.

ventricular diastole State in the cardiac cycle where the ventricles are relaxed so they can receive blood from the atria.

ventricular systole The contraction of the ventricles where blood will be forced out into arteries.

venule A small branch of a vein carrying blood from capillaries to larger veins on its way back to the heart.

vertical transmission Passing a pathogen from mother to child either in the uterus or during childbirth.

vesicle Small, fluid-filled, membrane-bound organelle; functions include transporting materials to, from and through the cell surface membrane.

virus A non-living infective particle composed of a nucleic acid (RNA or DNA) core surrounded by a capsid (protein coat).

Visking tubing Artificial, partially permeable membrane tubing. Also called dialysis tubing.

voltage-gated channel Type of ion channel that is activated to open or close depending on changes in the membrane potential close to the channel.

water potential A measurement of the proportion of a solution that consists of water molecules. It determines the tendency of water molecules to move out of a solution.

water stress A limited supply of water.

white blood cell A colourless cell, also called a leucocyte, which circulates in the blood and body fluids and is involved in counteracting foreign substances and disease. There are several types, including lymphocytes, neutrophils and macrophages.

xanthophyll Group of pigments, some of which are involved in photosynthesis, that absorb blue-violet light and appear yellow.

xerophyte A type of plant adapted to live in very dry conditions with little liquid water – these environments include hot dry deserts as well as cold ice- and snow-covered conditions.

xylem A tissue found in many plants that transports water and dissolved mineral ions from the roots, up the stem, to the leaves, in the process of transpiration; xylem tissue contains different types of cell, including xylem vessel elements.

xylem vessel A hollow tube made of xylem vessel elements; found in many plants to transport water and dissolved mineral ions.

xylem vessel element One of the cell types found within xylem tissue; forms xylem vessels.

zygote The fertilised ovum; the first cell that forms when gametes fuse.

Syllabus and assessment summary

INTRODUCTION

The information in this introductory section is taken from the Cambridge International AS & A Level Biology syllabus (9700) for examination from 2022. You should always refer to the appropriate syllabus document for the year of your examination to confirm the details and for more information. The syllabus document is available on the Cambridge International website at www.cambridgeinternational.org.

There are five papers to be taken for the full A Level in this subject, and three papers to be taken for AS. These are summarised in Table 1.

Paper	AS or A Level	Format	Time	Number of marks	Content assessed	Assessment objectives
1	AS (31%) & A Level (15.5%)	40 multiple choice questions	1 hour 15 minutes	40	AS syllabus	AO1 (50%) AO2 (50%)
2	AS (46%) & A Level (23%)	structured questions	1 hour 15 minutes	60	AS syllabus	AO1 (50%) AO2 (50%)
3	AS (23%) & A Level (11.5%)	structured questions in a laboratory setting	2 hours	40	experimental and practical skills	AO3 (100%)
4	A Level only (38.5%)	structured questions	2 hours	100	A Level syllabus, assuming AS understanding	AO1 (50%) AO2 (50%)
5	A Level only (11.5%)	two compulsory questions	1 hour 15 minutes	30	practical skills of planning, analysis and evaluation	AO3 (100%)

Table 1

KEY CONCEPTS

Look out for the key concepts which underpin the Cambridge International AS & A Level Biology syllabus to strengthen your understanding. You will find them in the mini mind maps at the end of each chapter in this book. The key concepts given on the syllabus are:

- **Cells as the units of life**
  A cell is the basic unit of life and all organisms are composed of one or more cells. There are two fundamental types of cell: prokaryotic and eukaryotic. Understanding how cells work provides an insight into the fundamental processes of all living organisms.
- **Biochemical processes**
  Cells are dynamic structures within which the chemistry of life takes place. Biochemistry and molecular biology help to explain how and why cells function as they do.
- **DNA, the molecule of heredity**
  Cells contain the molecule of heredity, DNA. DNA is essential for the continuity and evolution of life by allowing genetic information to be stored accurately, to be copied to daughter cells, to be passed from one generation to the next and for the controlled production of proteins. Rare errors in the accurate copying of DNA known as mutations result in genetic variation and are essential for evolution.

- **Natural selection**

  Natural selection acts on genetic variation and is the major mechanism in evolution, including speciation. Natural selection results in the accumulation of beneficial genetic mutations within populations and explains how populations can adapt to meet the demands of changing environments.

- **Organisms in their environment**

  All organisms interact with their biotic and abiotic environment. Studying these interactions allows biologists to understand better the effect of human activities on ecosystems, to develop more effective strategies to conserve biodiversity and to predict more accurately the future implications for humans of changes in the natural world.

- **Observation and experiment**

  The different fields of biology are intertwined and cannot be studied in isolation. Observation, enquiry, experimentation and fieldwork are fundamental to biology, allowing relevant evidence to be collected and considered as a basis on which to build new models and theories. Such models and theories are further tested by experimentation and observation in a cyclical process of feedback and refinement, allowing the development of robust and evidence-based conceptual understandings.

ASSESSMENT OBJECTIVES FROM THE CAMBRIDGE INTERNATIONAL SYLLABUS

The assessment objectives of the syllabus are shown below, these are what will be tested in your Cambridge examinations.

AO1: KNOWLEDGE AND UNDERSTANDING

(40% of both AS & A Level syllabus)

Candidates should be able to demonstrate knowledge and understanding of:
- scientific phenomena, facts, laws, definitions, concepts and theories
- scientific vocabulary, terminology and conventions (including symbols, quantities and units)
- scientific instruments and apparatus, including techniques of operation and aspects of safety
- scientific quantities and their determination
- scientific and technological applications with their social, economic and environmental implications.

AO2: HANDLING, APPLYING AND EVALUATING INFORMATION

(40% of both AS & A Level syllabus)

Candidates should be able to handle, apply and evaluate information, in words or using other forms of presentation (e.g. symbols, graphical or numerical) to:
- locate, select, organise and present information from a variety of sources
- translate information from one form to another
- manipulate numerical and other data
- use information to identify patterns, report trends and draw conclusions
- give reasoned explanations for phenomena, patterns and relationships
- make predictions and construct arguments to support hypotheses
- apply knowledge, including principles, to new situations
- evaluate information and hypotheses
- demonstrate an awareness of the limitations of biological theories and models
- solve problems.

AO3: EXPERIMENTAL SKILLS AND INVESTIGATIONS

(20% of both AS & A Level syllabus)

Candidates should be able to:
- plan experiments and investigations
- collect, record and present observations, measurements and estimates
- analyse and interpret experimental data to reach conclusions
- evaluate methods and quality of experimental data, and suggest improvements to experiments.

PREPARING FOR EXAMINATIONS

It is a good idea to become familiar with the format of the examination papers you will be taking for your course. Information on the different formats you will encounter is given below. You may also find it helpful to look at past examination papers for your course and to practise these under timed conditions.

Multiple choice question papers

For your own time management, you may find it helpful to allocate a sensible amount of time for checking at the end of your examination, and divide the remaining time allocation by the number of questions to obtain the approximate time you should spend on each question. Try to be aware of the approximate time allocation you have worked out as you go through a paper in order to allow yourself sufficient time.

One common misconception that students have about multiple choice questions is that they are easy because the correct answer is provided. However, common incorrect answers may also be options. Hence, just because you find your answer in the four options does not automatically mean that you are correct. This is especially true in calculations and where the understanding of a concept is trickier. When checking answers, go through each of the options and check whether those that you have not chosen can definitely be eliminated.

Structured question papers

Structured question papers assess both AO1 and AO2 skills. Structured means that the questions are split into parts, such as (a), (b), (c) etc. and sub-parts (i), (ii), (iii) etc.

For time management purposes you may wish to work out approximate time allocations for questions, making sure to also allow time for looking through the paper at the beginning and going back through to check your answers at the end.

Questions are often not actually questions at all, but command statements. Most of these start with a command word, such as state, calculate, explain, suggest etc. The command word is used to tell you what type of answer is expected.

The table below lists out command words and is taken from the Cambridge International AS & A Level Biology syllabus (9700) for examination from 2022.

Command word	What it means
Assess	make an informed judgement
Calculate	work out from given facts, figures or information
Comment	give an informed opinion
Compare	identify/comment on similarities and/or differences
Contrast	identify/comment on differences
Define	give precise meaning
Describe	state the points of a topic / give characteristics and main features
Discuss	write about issue(s) or topic(s) in depth in a structured way
Explain	set out purposes or reasons/make the relationships between things evident/provide why and/or how and support with relevant evidence
Give	produce an answer from a given source or recall/memory
Identify	name/select/recognise
Outline	Set out main points
Predict	suggest what may happen based on available information
Sketch	make a simple drawing showing the key features
State	express in clear terms
Suggest	apply knowledge and understanding to situations where there are a range of valid responses in order to make proposals / put forward considerations

In addition to a command word, there may be some additional guidance or context. For example, 'Explain with reference to Fig. 1.1 / Table 1.1…' In this case, specific information from that photograph, diagram, graph or table is expected, and you should ensure your answer includes information extracted from the figure or table.

Pay attention to the mark allocation for each question part or sub-part. A mark allocation shows the number of marks that can be awarded. A mark allocation of [2] for example, may indicate that more than one thing needs to be shown or stated.

In calculations, it is always important to show working, even when marks can be allocated for a correct answer with no working. If a final answer is wrong, or wrongly rounded, then marks can be awarded for the working. If the working is not there, then no mark can be allocated.

The number of significant figures required in an answer may be given in the question. If not, the numerical data used for the calculation should be used as a guide. Your answer should be based on the value that has the least number of significant figures. For example, if a calculation involves the values 12 and 78.6, then the first of these has two significant figures and the second has three. The answer in this case should have two significant figures. You may not be awarded the full mark allocation for an answer to an inappropriate number of significant figures.

The number of answer lines or the depth of answer space has also been carefully considered. With average size writing, this is the space that a full answer that can score all of the allocated marks is expected to occupy. Some very well written answers can score maximum marks in significantly less space, but be aware of the line allocation. If you have written two lines when seven are allocated, are you sure you have included everything relevant?

In written answers, take care to use key syllabus terms correctly. In general, some spelling mistakes are acceptable but take care with spellings of scientific words that sound similar. If a spelling mistake makes your intended word have a different meaning, then marks may not be awarded. There are no mark allocations for grammar or for writing in full sentences, so if you find it easier to structure your ideas into bullet points, then do so.

Also, marks are awarded for correct science, so it may be possible to supplement your answer with a small diagram or sketch graph that is correctly annotated, even where there is no blank space allocated for a drawing.

If you really do not know an answer, then writing some key information about the topic may just result in the award of a mark. Repeating the question in many and varied ways will not.

PRACTICAL SKILLS PAPERS

Examinations in laboratory settings

Some papers, such as Paper 3 of the Cambridge International syllabus, are taken in the laboratory and actual experiments are carried out. There can be separate laboratories for each question. For example, half the students may be doing question 1 in one laboratory while the other half are doing question 2 in a separate laboratory. At the appropriate time, and after equipment has been reset and refreshed, students will swap rooms and do the other question.

When you arrive, you will be allocated some bench space and all the necessary equipment and resources will be there ready for you. When you are instructed to start, it is advisable to read through the whole question before starting. That means you can get an overview of what is required and have time to check resources and organise equipment.

There will be a science specialist, called the supervisor, present in the room or in an adjoining room. This person is not watching your methods and is certainly not allocating marks for you based on your working. Their role is to carry out the same experiment, out of sight of the students, so their results can be included in the pack of papers that are sent for marking. They are also available if some piece of equipment is not working, or if you run out of some resource.

Make sure you follow the instructions. This may sound obvious, but you can be presented with equipment that you have used in another experiment previously and assume, under exam pressure, that you are doing the same thing as before. This may not be the case, so only do what you are asked to do in the question paper. There can be a lot of reading, particularly in question 1. Take your time with this and make sure not to miss anything. The time allocation for the paper includes this reading time. While the Experimental Skills sections in this book are based on the syllabus content of each chapter, the practical context in an examination may not be taken from the syllabus. If the context of a question is unfamiliar, do not worry. Instead, continue to follow the instructions given.

Making measurements and recording results

Repeat readings should be made and an average calculated from all results as you would normally do in a class practical.

If not told otherwise, at least six values of the independent variable should be taken if the trend is expected to be linear, and at least nine for a curved trend.

Sometimes a reaction will take time to happen, such as 15 or 20 minutes. If this is part of the procedure, the instructions in the question will tell you what to do in that time, such as continue with another part of the question. You will not be expected to do nothing for up to 20 minutes.

All results should be recorded in a single table. You should plan the table before you use the apparatus to take results. The table should have columns for all the raw data that is going to be measured and columns for any calculated values. You should label each column with the quantity and an appropriate unit. Standard scientific convention is that there should be a distinguishing mark between the quantity and unit. For example, if the time measured in seconds is measured using a stop-watch, then acceptable column headings could be:

time/seconds	t/s	time/s

Other examples include

concentration/mol dm^{-3}	Appearance of solution	pH

Measurements taken in an experiment should be recorded to the same precision as the measuring instrument. For example, when using a ruler to measure length, measurements are made to the nearest millimetre, so a length of 10 centimetres should be recorded as 10.0 cm.

d/cm
12.1
10.0

In the second data row of the table, 10 would be incorrect because it indicates that d has only been recorded to the nearest centimetre not the nearest millimetre and d could have a value between 9.5 cm and 10.5 cm. All raw readings of a quantity should be recorded to the same degree of precision, i.e. all quantities should be recorded to the same number of decimal places. In timing experiments, using a stopwatch, times can either be recorded to the nearest 0.01 s or rounded to the nearest 0.1 s. Since the column heading has both a quantity and a unit, units should not be included with the data values in the table.

When calculating values, the number of significant figures in the calculated value will depend on the number of significant figures of the measured quantity with the least number of significant figures. The number of significant figures in a calculated value should be the same as the smallest number of significant figures in the data used for the calculation. For example, a radius r is measured and r^2 is calculated. Since r is given to three significant figures (sf), then r^2 should be recorded to three significant figures.

Graph plotting and analysis

$\dfrac{r}{cm}$	Full calculation of r^2 $\dfrac{}{cm^2}$	$\dfrac{r^2}{cm^2}$
24.2	585.64	586
24.1	580.81	581
24.3	590.49	590

The table shows that r has been recorded to the nearest millimetre and is given to three significant figures.

It is good practice to draw graphs with a sharp pencil and a clear plastic ruler. Graph paper provided will have 2 mm squares with bold lines every 2 cm, forming large squares.

Each axis should correspond to one of the columns from the table of results and have the same label as the column heading. The x-axis should be labelled with the independent variable and the y-axis with the dependent variable.

A scale also needs to be added to each axis. Numerical values should be added at every large square of the graph paper. When deciding the scales, you should aim to use as much of the graph paper as possible so that the plotted data points occupy at least half the graph paper in each direction. It may not be necessary to start the graph at the origin. For example, for data in the range 350–600, the axis could start at 300. The scales should be simple – this will help you plot the data points accurately and help you read data from the graph. Typical simple scales are those that increase in multiples of 1, 2 or 5, e.g. 0.2, 0.4, 0.6 (increasing by 0.2) or 50, 100, 150 (increasing by 50).

It is good practice to plot the data points as a fine cross or a dot in a circle using a sharp pencil. The plotting should be accurate to within one millimetre and the diameter of the point should be less than one millimetre. All the data points from the table of results should be plotted.

Having plotted the points, the line should then be drawn. Pay attention to any instructions given in the question paper about the type of line to draw. If the trend appears to be linear, then a straight line of best fit may be drawn. If there are any points that do not appear to follow the trend, it is worth checking that the points have been plotted correctly. Any point that still does not follow the trend should be circled and labelled as an anomalous point. This point should then be ignored when drawing the line of best fit. If the points indicate a linear trend, a clear ruler should be used so that there is a balance of points about each side of the line. A best fit line should only ever be drawn with reference to the plotted points. Do not try to force a best fit line to pass through the origin because you think that, in theory, it should. The thickness of the line should be less than one millimetre.

If the points indicate a curved trend, then a smooth single curve should be drawn. You may need to draw a tangent to the curve. A tangent is a straight line that touches the curve at one point only. You may encounter a curved trend when measuring the rate of an enzyme-catalysed reaction. See Chapter 3 for more details.

One of the questions, usually question 2, involves observing a prepared slide under a light microscope and making one or two drawings from your observation. There will be blank space on the question paper to make the drawing. Your drawing should occupy most of this space.

A plan diagram is made from a low magnification view. It should only have lines at tissue boundaries and no shading or addition of texture anywhere. All lines should be clear and continuous with no breaks and not ragged. There should be no overlapping lines. Individual cells should not be drawn. Tissue outlines should be approximately the correct shape and proportions should also be approximately correct. No drawing aids, such as a ruler or a compass, should be used for drawing these lines. If labels are to be added, then a ruler should be used for the label lines only. Label lines should not cross each other. Label lines should end by touching, or just entering, the part you are labelling.

If you are asked to draw a diagram of a specific part of the specimen, then this will be in the instructions. For example, you may be asked to draw four cells from a named tissue. In this case, a higher magnification should be used. Only draw the number of cells given in the instructions. If these are plant cells, then draw the cell walls with two lines, if the thickness of the cell wall is visible. Any visible details can be added, although nuclei and darker-stained structures should not be shaded. Do not add features that you assume are there but you cannot see. As with the plan drawing, there should be no overlapping lines.

Papers assessing planning, analysis and evaluation

The skills of planning, analysis and evaluation may be in different questions, so can be in different experimental contexts, and may not be in that order. For example, question 1 may ask about analysis of data and question 2 may ask you to plan an experiment in a totally different context, or the other way around.

The analysis and evaluation of data will be structured and you will be told what is expected in each question part. The context of the question may be from an experiment that you do not recognise, but sufficient information will always be given to answer any questions that are not directly drawn from the syllabus content. You will not be asked to carry out a full statistical analysis such as a Spearman's rank or Pearson's linear correlation coefficient, but you may be required to complete part of these calculations. Any of the statistical analysis equations that are required will be provided. Other calculations, such as percentage changes, ratios and means can be expected.

Questions can be asked about naming independent and dependent variables in an experiment and sometimes identifying those that cannot be standardised.

You may be asked to make conclusions from results that are provided in a table or in some other form. In this case, you should use the information provided rather than your theoretical knowledge.

Planning

The planning skill requires you to incorporate all the practical skills that you have practised throughout the course. To plan an experiment or investigation you will need to consider:

- identifying variables and quantities that need to be kept constant (standardised)
- planning a workable method
- explaining how the data will be collected
- explaining how the results will be analysed
- explaining any safety precautions.

The initial stage of planning any investigation is to consider the quantities that are involved and to decide how to make the investigation valid. Your plan should include the apparatus you need and the measuring equipment. Initially, you will need to decide which variable to change: the independent variable. You need to describe the method of changing the independent variable and some suggested values for this. You will then need to decide which variable to measure: the dependent variable. The independent variable is usually plotted on the x-axis and the dependent variable is usually plotted on the y-axis of a graph. To make the experiment fair, you will then need to consider the variable or variables that need to be kept constant: standardised variables, and details of how this is to be achieved.

Include a diagram of how the apparatus is to be assembled. Your plan should also include any further details on techniques used to ensure that measurements are as accurate as possible.

Some safety information should always be included. Standard laboratory rules are assumed, so these do not need to be stated. Safety information should be specific to the investigation, even if you only state that your plan is for a low risk procedure. Specific safety information may include the use of a water bath so that a test-tube is not heated directly, eye protection if caustic solutions are to be used, immediate washing if a stain is splashed on skin and wearing of gloves if a digestive enzyme is used.

Evaluation

After an experiment, you may be asked for an evaluation. This can be in terms of sources of uncertainty (sometimes called errors) or the limitations of the procedure. You can also be asked to suggest improvements or modifications to the method, possibly using other, named, pieces of equipment. Modifications should be suggested on the basis of giving greater confidence in the results. These should be specific and not refer to working more carefully or using better equipment. For example, where a colour change is gradual and judged by eye, then a modification could be to use a colorimeter and measure absorbance of samples at regular time intervals.

When considering sources of error, think about any difficulties you had in making accurate measurements. Always give reasons for these difficulties.

Limitations of the procedure can often arise from having too few measurements. In this case, the reason why too few measurements is a limitation should be explained. Limitations can also be linked to the equipment and experimental design.

Answers

CHAPTER I – CELL STRUCTURE

In-text questions

1. A photograph is an image formed through the action of light. A photomicrograph is a type of photograph, but taken through a microscope.

2. The image is three-dimensional.

3. (a) $50 \, \mu m = \frac{50}{1000} \, mm = 0.05 \, mm = 5 \times 10^{-2} \, mm$

   (b) $50 \, \mu m = 50 \times 1000 \, nm = 50\,000 \, nm = 5 \times 10^4 \, nm$

4. (a) $6 \, \mu m = 6 \times 1000 \, nm = 6000 \, nm$ or $6 \times 10^3 \, nm$
   (b) Size of image = magnification × size of object
   $$= 4000 \times 6 \, \mu m$$
   $$= 24\,000 \, \mu m$$
   $$= 24 \, mm$$

5. Resolution is the ability to distinguish between two objects that are close together. Magnification is how many times bigger the image is than the object.

6. Advantage: greater resolution/greater magnification. Disadvantage: not in colour/specimen cannot be living.

7. Eukaryotic cell because it contains a nucleus / other organelles.

8. Some substances are used up by a cell, such as oxygen, and so constantly need to be taken in. Other substances are produced as waste and need to be removed because they are not needed and may be harmful, such as carbon dioxide.

9. When DNA is unravelled, the individual genes can be accessed to be 'read'. However, during cell division, it is easier to divide the DNA between the daughter cells if it is condensed into chromosomes.

10. Rough endoplasmic reticulum. Rough ER has ribosomes on it (they appear as small black dots). The smooth ER has no ribosomes.

11. It is more efficient as the substances can travel directly to the places where they are needed; it may also be quicker as some substances may not be (very) soluble.

12. C

13. A secretory cell, e.g. one that secretes digestive enzymes.

14. Muscle cells, as they need to release a lot of ATP for use in the energy-requiring process of muscle contraction. (Or any other valid example.)

15. Otherwise they would damage the cell by breaking down needed organelles.

16. The cilia do not move to remove mucus from the lungs so the mucus is removed instead by heavy coughing **or** the coughing is an involuntary response to the build-up of mucus/irritants in the lungs.

17. As digested food substances cannot be absorbed properly, the condition leads to malnutrition and dehydration.

18. When a plant cell loses water the volume of the cytoplasm decreases and the cell surface membrane no longer exerts pressure against the cell wall. This causes the whole cell to lose rigidity.

19. Animals either have a rigid skeleton (like humans for example), or they live in water which supports the body (like jellyfish for example).

20. Not all plant cells are exposed to light, for example those in roots, so there is no benefit in them containing chloroplasts.

21. Makes the plant less likely to be eaten by herbivores.

22. DNA in prokaryotic cells is circular and not associated with histone proteins, whereas DNA in eukaryotic cells is in single lengths, and is associated with histone proteins to form chromatin. Prokaryotic cells contain DNA plasmids, unlike eukaryotic cells. There is much less DNA in a prokaryotic cell compared with a eukaryotic cell.

23. Prokaryotic cell = 5 μm = 5000 nm
    5000 nm ÷ 25 nm = 200 times bigger

Assignment 1.1

A1. (a) Size of object = size of image ÷ magnification
    $$= 68 \div 12\,000 \, mm$$
    $$= 0.005667 \, mm$$
    $$= 5.7 \, \mu m \text{ (to two significant figures)}$$
    (b) Measure the diameter several times in different directions. Calculate the mean average.

A2. (a) Diagram **a** is too simple and inaccurate, and is not labelled. Diagram **b** may be a more accurate representation of the photomicrograph but a biological drawing is meant to be a clear and simple representation without shading and unnecessary detail. Labels should not be on the diagram itself as they may obscure some details, and without label lines it might not be clear which part a label is referring to. Diagram **c** is best as it shows the main parts of the cell, without any confusing shading, and is clearly labelled.
    (b) Include a title and the magnification. Make the drawing of the nucleus more circular.

Experimental skills 1.1

P1. (a) Otherwise some cells will overlie other cells making it very difficult to distinguish separate cells.
    (b) To make the cells and their structures more visible.
    (c) To prevent the specimen from drying out.

P2. (a) To avoid the lens hitting the slide, which could damage both.
    (b) Turning the coarse focusing dial moves the lens much more than does turning the fine focusing dial. This means that it is easier and quicker to bring the image into view with the coarse focusing dial. However, to bring the image into sharp focus, which only requires slight adjustment to the position of the lens, it is easier to use the fine focusing dial.
    (c) You can examine a much larger area of the specimen with a lower power lens, and then move the slide so the parts that you want to examine in more detail are in the centre of your field of view. When you change to a higher power lens to see in more detail, the parts you want to observe will still be in the centre of your field of view.

P3. Clear biological diagram, labelled to show the cell wall, nucleus and cytoplasm. Suitable title.

P4. (a) 40.

    (b) 0.1mm ÷ 40 = 0.0025 mm
                $= 2.5 \times 10^{-3}$ mm

    (c) 12 divisions = $12 \times 2.5 \times 10^{-3}$ mm
                     $= 0.03$ mm
                     $= 30 \ \mu m$

    (d) The graticule will need recalibrating for the new objective lens.

P5. (a) Total number of cells in the five shaded 1 mm² squares = 17
         Average number of cells per mm² = 17 ÷ 5 = 3.4
         Total area occupied by the blood on the slide = 4 cm² = 400 mm²

         So total number of cells on slide = 3.4 × 400 = 1360

    (b) Use a greater number of squares when calculating the mean; use a method of randomly choosing which squares to include in the sample (for example, allocate co-ordinates to the grid and use a random number generator to select which squares to include).

Chapter review

1. B

2. D

3. 15 mm = 15 000 μm
    Magnification = image size ÷ object size = 15 000 ÷ 75 = ×200

4. × 10 × 25 = × 250

5. Scanning electron microscope / SEM, because the image is three-dimensional.

6. Any three from: cell wall / chloroplast / large permanent vacuole / tonoplast / plasmodesmata.

7. Mitochondria, which are the site of aerobic respiration, which produces most ATP.

8. Contains lysozymes, to break down / digest, unwanted organelles / cells.

9. Any three from: smaller size / peptidoglycan cell wall / circular DNA / DNA not associated with histone proteins / 70S ribosomes / capsule / flagella / plasmids / do not have double-layered membrane-bound organelles.

10. Not cellular, do not exhibit all seven characteristics of life.

Practice exam-style questions

E1. C [1]

E2. B [1]

E3. (a) Any six from:

Obtain a one-cell-thick sample of tissue **or** take a single layer of cells if using an onion **or** squash sample to separate cells. [1]

Explanation: to allow single cells to be seen / to avoid overlapping cells. [1]

Place the tissue on a microscope slide, and add a suitable stain e.g. iodine–potassium iodide solution. [1]

Explanation: stain will make certain structures more visible. [1]

Add a coverslip, avoiding trapping air bubbles. [1]

Explanation: coverslip helps stop sample drying out / air bubbles obscure view of sample. [1]

Safety precaution: wear eye protection in case of spills of stain. [1]

    (b) 40 divisions on the graticule scale = 0.2 mm [1]
         1 division on graticule scale = 0.2 mm ÷ 40 = 0.005 mm [1]
         24 divisions = 12 × 0.005 mm = 0.06 mm = 60 μm [1]

    (c) Size of image = 20 cm = 2.0×10^8 nm
         Size of object = size of image ÷ magnification
                       $= 2.0 \times 10^8 \div 1000$ nm
                       $= 2.0 \times 10^5$ nm [3]

E4. (a) X = nucleus, contains nucleolus / nuclear pores [2]
         Y = mitochondrion, folded inner membrane / cristae [2]
         Z = Golgi body, producing vesicles [2]

    (b) Eukaryotic [no mark]
         Any three from:
         Presence of double-layer membrane-bound organelles / nucleus / mitochondria / Golgi body / lysosomes. [3]

E5. (a) The cell has a three-dimensional structure, mitochondria are lying in different orientations. [2]

    (b) Similarities:
         Any three from: double membrane, internal membranes with large surface area, 70S ribosomes, circular DNA. [3]
         Differences:
         Any three from: chloroplasts contain chlorophyll, mitochondria do not; chloroplasts contain stacked internal membranes / thylakoids / grana, mitochondria contain folded inner membrane / cristae; chloroplasts are site of photosynthesis, mitochondria are site of aerobic respiration; chloroplasts contain stroma, mitochondria contain matrix. [3]

    (c) Mitochondria and chloroplasts have many similarities to prokaryotic cells: small size, 70S ribosomes, circular DNA. [3]

CHAPTER 2 – BIOLOGICAL MOLECULES

In-text questions

1. The negative result for the iodine test indicates that starch is not present in the sample. The positive results for the biuret and emulsion tests indicate that protein and lipids are respectively present in the sample.

2. When disaccharides are heated strongly in the presence of acid, they are hydrolysed into monosaccharides (the glycosidic bond is broken). Because all monosaccharides are reducing sugars, a positive result will be seen when the Benedict's test is undertaken.

3. (a) When heated, Benedict's solution changes colour from blue in the presence of reducing sugar. The exact colour that is seen depends on the concentration of reducing sugar in the sample. The order of colours, red-orange-yellow-green, indicates solutions of decreasing concentration.

    (b) Factors that must be standardised in a comparison of reducing sugar concentration include:
       • The volume of fruit juice used in the test.
       • The volume/concentration of Benedict's solution used in the test.
       • The temperature at which the mixture of fruit juice and Benedict's solution was heated.
       • The time that the mixture of fruit juice and Benedict's solution was heated for.

    (c) This investigation attempted to estimate the glucose concentration of a solution using the Benedict's test. However, the Benedict's test will give a positive result – a colour change – for any reducing sugar. Therefore, it is possible that one solution with a glucose concentration lower than another may have given a colour in the Benedict's test that suggested a greater concentration of reducing sugars (which includes fructose – a sugar commonly found in fruit juices).

4. A molecule of water reacts with sucrose to break the glycosidic bond between the carbon-1 of α-glucose and the carbon-4 of fructose.

5. (a)

Concentration of glucose / %	Volume of 1% glucose solution / cm³	Volume of water / cm³
0.8	8	2
0.6	6	4
0.4	4	6
0.2	2	8

(b) (i)   A 1 dm³ solution of glucose of concentration 0.2% could be produced by mixing 2 ml of the 1% glucose solution with 998 ml distilled water.

(ii)   A 1 dm³ solution of glucose of concentration 0.05% could be produced by mixing 0.5 ml of the 1% glucose solution with 999.5 ml distilled water.

(iii)   A 1 dm³ solution of glucose of concentration 0.001% could be produced by mixing 0.01 ml of the 1% glucose solution with 999.99 ml distilled water.

6. An example answer presented as a table is shown below. Each correct comparison (row) is given.

Starch	Cellulose
formed by joining α-glucose monomers	formed by joining β-glucose monomers
residues are connected by α(1,4) glycosidic bonds	residues are connected by β(1,4) glycosidic bonds
contains two types of polysaccharide: amylose and amylopectin	contains one type of polysaccharide
amylopectin (but not amylose) is a branched molecule	an unbranched molecule
easily digested by animals	not easily digested by animals
found in starch granules inside plant cells	found in plant cell walls packed into microfibrils
energy storage role	structural role - provides the cell wall with tensile strength

7. Starch is an effective energy storage carbohydrate in plants because:
   - It is insoluble, which means that it does not affect the water potential of the cells in which it is stored. The advantage of this is that it has no effect on the amount of water that leaves or enters the cell by osmosis.
   - Long-term energy storage is achieved by the use of amylose, a tightly packed unbranched polysaccharide. Shorter-term energy storage is provided by amylopectin, which has a branched structure that has more 'ends' from which enzymes can remove glucose molecules for use in respiration.

8. The diagram should show how an α(1,1) glycosidic bond forms between two trehalose residues as follows. The second residue

must be rotated through 180° in order to allow the two carbons to be adjacent to each other.

Carbon atoms do not need to be numbered.

9. Three condensation reactions occur when the triglyceride is formed, each releasing one molecule of water. Triglycerides consist of one molecule of glycerol which is attached to three fatty acids.

10. (a)   Increasing the number of C=C bonds causes more 'kinks' in the fatty acid chains. This prevents their packing together in triglycerides, which means that they are more easily separated and the material changes from a solid to a liquid at lower temperatures.

(b)   A triglyceride consists of one molecule of glycerol which is attached to three fatty acids. In a polyunsaturated triglyceride, at least one of the hydrocarbon chains derived from a fatty acid has two or more C=C double bonds.

11. (a)   Calculation shows how the energy value for lipid is divided by the energy value for carbohydrate (39.4 ÷ 15.8). The correct answer is 2.5 (times) to one decimal place.

(b)   Triglycerides take longer to break down, so cannot be used to release energy rapidly. They are also insoluble, which means that it does not cause a large volume of water to move into the cells in which it is stored by osmosis. The advantage of this is that it has no effect on the amount of water that leaves or enters the cell by osmosis.

(c)   The burning biological sample will release heat to the air and the tongs holding it, meaning that not all of the energy is transferred to the water. Better insulating the apparatus may help to minimise heat loss and give a more accurate reading.

(d)   Although the table suggests that lipids yield nearly 2.5 times the energy of a carbohydrate, this is likely to be an average value. There are many types of carbohydrate and lipid, some of which may give different energy values. Also, energy is not released by combustion (burning) inside cells, so it is hard to generalise these findings.

12. A peptide bond forms between two amino acids in a condensation reaction. The hydroxyl (–OH) group on the carboxylic acid group (–COOH) on one amino acid reacts with the amine group (–NH₂) on the other amino acid. The result is a peptide bond, which is a direct link between a carbon atom and a nitrogen atom (C–N).

13. Depending on the order in which the amino acids have been joined together, the diagram will look like this:

Or like this:

H—N—C—C—N—C—C
with H, O, H, H, OH and H—C—H, H groups (amino acid / peptide structural diagram)

14. A polypeptide is a linear chain of amino acids of a specific sequence. A protein is a polypeptide that has developed a three-dimensional structure, as a result of the interactions between chemical groups of different amino acid residues.

15. Globular proteins will form a ball-shaped molecule in aqueous solution. This is the case with all enzymes. This occurs because the folding of the polypeptide chain results in residues with hydrophilic R groups being placed on the outer surface of the molecule, in order to interact with water, and hydrophobic molecules being brought onto the inside of the molecule to avoid water.

16. (a) The sequence of R groups and how they interact with each other through hydrogen and disulfide bonds will determine the secondary structure (of α-helices and β-pleated sheets), which will help determine the tertiary structure (the overall shape of that individual protein), and thus its function.
    (b) Cysteine and methionine can form disulfide bonds because they have free –SH groups.
    (c) Carbohydrates usually consist of polymers of monomers that are identical - for example, alpha-glucose in the case of starch. Proteins, on the other hand, are made from a variety of monomers – amino acids – that have a range of structures and chemical properties. Combining these in different orders will result in polypeptides that take on different structures and therefore can serve different functions.

17. β-mercaptoethanol will only affect proteins whose tertiary structure contains disulfide bridges. Although these are very common, not all proteins will have this type of bond, so the functions of some will not be affected by the addition of this chemical.

18. (a) (i) This is important for the transport of solutes by the blood in animals, and in the xylem and phloem of plants.
        (ii) This is important to maintain body temperature of endotherms, and in the maintenance of the temperature of bodies of water that are the habitat for aquatic organisms.
        (iii) This plays a very important role in cooling of organisms, through sweating and evaporation (animals) and transpiration (plants).
    (b) • Water is an excellent solvent because the δ– and δ+ charges on the water molecules form hydrogen bonds with charged particles and the polar regions of some larger molecules, which become surrounded by a 'shell' of water molecules. This causes solids to spread out among the water molecules and dissolve into a solution.
        • Water has a high specific heat capacity because a lot of heat energy is required to break the hydrogen bonds between water molecules, separating them from each other.
        • Water has a high latent heat of vaporisation because a lot of heat energy is required to break the hydrogen bonds between water molecules, separating them from each other and turning the material into a gas.

Assignment 2.1

A1. (a) Collagen contains the chemical elements: carbon, oxygen, hydrogen, nitrogen and sulfur. In addition to these, haemoglobin also contains the element iron.
    (b) Fibrous proteins are usually insoluble in water and have structural roles.
    (c) A model answer presented below. Each correct comparison is given:
    • Similarities
       ° Both are proteins with a quaternary level of structure (comprising three polypeptides in collagen, and four in haemoglobin).
    • Differences
       ° Collagen is a fibrous protein with a structural role (increasing the tensile strength of body structures); haemoglobin is a globular protein with a physiological role (carriage of oxygen).
       ° Only haemoglobin has a prosthetic (haem) group.

A2. (a) The proline and hydroxyproline are joined by a peptide bond as shown, between the carboxyl group of one residue and the amino group of the next:

(structural diagram of dipeptide of proline and hydroxyproline with peptide bond)

The peptide bond should be indicated.

    (b) The formation of this dipeptide requires the loss of a molecule of water.

A3. Hydroxyproline, which can only be made from proline in the presence of vitamin C, is required for the production of collagen. In its absence, collagen is not formed properly. Structures that are normally provided with higher tensile strength (e.g. cartilage and skin in the mouth) are weaker and are easily damaged.

A4. (a) Percentage change = ((size of population in 1925 – size of population in 1905) ÷ size of population in 1905) × 100. Answer = 43.5 % (increase).
    (b) Percentage change = ((percentage of mothers in 1965 – percentage of mothers in 1905) ÷ percentage of mothers in 1905) × 100. Answer = –60.8 (decrease).
    (c) The evidence to support this claim is debatable. The fall in the number of deaths due to scurvy does not coincide perfectly with the reduction in the number of women who breastfeed their babies. Also, the fact that there has been an increased size of the British population in the 20th century means that the number of cases of scurvy may have actually increased. In addition, the number of deaths is for individuals of all ages, but the effect of breastfeeding would only influence babies.

Experimental skills 2.1

P1. To make the 0.25% glucose solution, 10 cm³ of 0.5% glucose solution is added to 10 cm³ of distilled water in the third beaker. To make the 0.125% glucose solution, 10 cm³ of 0.25% glucose solution is added to 10 cm³ of distilled water in the fourth beaker. To make the 0.0625% glucose solution, 10 cm³ of 0.125%

glucose solution is added to 10 cm³ of distilled water in the fifth beaker.

P2. It can be difficult to measure volumes accurately using a syringe because of parallax error. This occurs because the position of the meniscus (the upper surface of the liquid) may appear in an incorrect position to the observer, if they take the measurement at any position other than at eye level.

P3. Uncertainty is ± half the smallest division on syringe, so 0.1 cm³. Therefore, the percentage error is found by using the equation percentage error = (uncertainty ÷ value being measured) x 100. Percentage error = $\frac{0.1}{10}$ × 100 = 1.00% to 3 significant figures.

P4. (a) The colour of Benedict's solution at 60 seconds with the glucose concentration of 0.125% and the glucose concentration of 0.0625% is blue.

(b) Using the hot water bath carries the risk of burns. A test-tube holder should be used to place the test-tubes into and take them out of the hot water bath.

(c) Washing the test-tubes thoroughly with distilled water between each experiment is important to remove the solutions from the previous experiment. This could cross contaminate the next Benedict's test. This could make forming observations difficult and cause false-positive or false-negative results, which are inaccurate.

(d) A control is required in an investigation to remove the effect of the independent variable – the factor that is varied by the investigator. In this experiment, the scientist must show that there is a negative result in the absence of reducing sugar. Therefore, the Benedict's test must be carried out on a sample of distilled water, which should remain blue in colour.

P5. (a) The graph should have an origin (0) at the intersection of the x- and y-axes. The units should be shown on both axes (seconds for the y-axis and % for the x-axis).

(b) The same volume of urine should be used as the volume of glucose solutions in this investigation. The same volume of Benedict's solution should be added to the urine, which is then heated at the same temperature. The time taken for the first appearance of a colour change should be recorded and this value should be used to read off the concentration of glucose solution on the x-axis, by finding where the value intersects the calibration curve.

P6. The scientist may not be able to make an estimate of the glucose concentration of urine using colour standards because: (i) the concentration of glucose in urine may be too small to cause the Benedict's solution to change colour, and (ii) making a decision based on the colour of a solution is subjective and open to interpretation/bias. To improve the accuracy, use a colorimeter or carry out experiments using more concentrations within the range of the estimate.

P7. Both of the glucose tests undertaken by the scientist are semi-quantitative because they do not provide an exact value for the concentration of the glucose. Instead, they illustrate whether one sample has more or less of a particular substance than another, allowing comparisons to be made of the solutions.

Experimental skills 2.2

P1. The advantage of storing starch in plant cells, rather than glucose, is because starch is insoluble. This means that it does not cause water to move into the cell by osmosis. In addition, glucose is less reactive, so using starch as an energy store means that it will last for longer in the cell.

P2. By referring to the other values for amylose:amylopectin ratios, we can see that they have been calculated by dividing the value for amylose percentage by the value for amylopectin percentage. The missing values are therefore 0.449 for mung bean, and 0.266 for rice.

P3. Amylopectin is a highly branched storage carbohydrate with lots of 'ends' from which enzymes can release glucose residues by catalysing hydrolysis reactions. Therefore, the potato, given its starch content has the highest percentage of amylopectin (the smallest amylose:amylopectin ratio) will be able to mobilise its starch energy stores most rapidly.

P4. Amylose consists of a tightly packed spiral of glucose molecules without branches. Amylopectin is a highly branched molecule. The greater the proportion of amylose (the higher the amylose:amylopectin ratio), the less the space required to package starch, and the smaller the starch granules in cells.

P5. Temperature (use a thermostatically controlled water bath) / pH (use buffers) / concentration of starch from each plant tissue (use a similar mass of tissue) / time for hydrolysis and/ or to run the molecules on a chromatogram (use a stopwatch).

P6. A sample of the mixture of starch-enzyme can be extracted and tested for the presence of starch using iodine–potassium iodide solution. If the orange-brown solution does not change colour to blue–black, then the hydrolysis of the starch is complete.

P7. The amylases in fungal extract 3 hydrolyse only amylopectin, and only at the α(1,6) glycosidic bonds where the branches link to the main chain. If a starch sample consists of a greater proportion of amylopectin (has a lower amylose:amylopectin ratio), then a greater number of spots will be seen on the chromatogram. If there are fewer branches, or if the starch sample consists mainly of amylose (has a higher amylose:amylopectin ratio), then there will be fewer, larger spots. The plant tissue corresponding to chromatogram B has the greatest amylose:amylopectin ratio, whereas the plant tissue corresponding to chromatogram C has the lowest.

P8. (a) Only one spot will be seen, at the very bottom of the chromatogram. This is because cellulose has no α(1,6) glycosidic bonds (there are no branches), and the only spot will be close to the origin as this undigested molecule will be so large.

(b) More spots will be seen, which are smaller (and therefore at the top of the chromatogram). This is because glycogen is a polysaccharide similar to amylopectin in its structure, but it has a greater number of branches.

P9. A similar number of spots will be seen for all chromatograms, because α-amylase hydrolyses α(1,4) glycosidic bonds, which are present in both amylose and amylopectin.

P10.(a) The Pearson's linear correlation test can be used to investigate the data in this study because:
- continuous data has been collected
- the data is from a population that is approximately normally distributed
- the scatter graph (Figure P1) indicates the possibility of a linear relationship
- the number of paired observations is seven, which is more than the minimum of five.

(b) Two factors may appear to be correlated but it could be a coincidence that they increase together, or decrease together. There may be no direct link between the two.

(c) The positive figure suggests that there is a positive, linear correlation (as one variable, amylose:amylopectin ratio, increases, so does the other (mean granule size)). The value of 0.775 is relatively close to 1, suggesting that the correlation is relatively strong – the data points cluster around a straight line of best fit.

(d) The total number of pairs of data (n) should be identified, which is seven. Then, 2 should be subtracted from this value. This can be written as degrees of freedom (df) = (n–2) = (7–2) = 5.

(e) The student should use the probability table at 0.05 (5%). They should compare the calculated r value to the critical value, which is 0.754. The correlation is significant, because the calculated r value (0.775) is higher than the critical value. This means that there is less than a 0.05 (5%) probability that the r value is due to chance.

(f) The starch granules from only one cell from each of the species were measured. It may be that the cell which was chosen was abnormal compared to other cells from the species, which may have given an anomalous amylose:amylopectin ratio.

Chapter review

1. Hydrolysis of a disaccharide will produce two monosaccharides, which are reducing sugars. Some disaccharides, such as sucrose and lactose, are non-reducing sugars and would not give a positive result with the Benedict's test. Only after hydrolysing them into their monosaccharide products will a positive Benedict's result be obtained.

2. A model answer presented as a table is provided below. Each correct row is given

similarities	differences
both contain the elements carbon, oxygen and hydrogen	only amino acids can contain the elements nitrogen and sulfur
both are monomers which can be joined to each other to form polymers called macromolecules	monosaccharides are joined by glycosidic bonds, while amino acids are joined by peptide bonds
both can be joined by condensation reactions with the loss of a molecule of water	only amino acids can differ in their structure by having a different chemical group for their R group

3.

found in:	hydrogen bond	disulphide bond	peptide bond	α (1,4) glycosidic bond
Primary structure of a protein	✗	✗	✓	✗
Tertiary structure of a protein	✓	✓	✓	✗
Amylopectin	✓	✗	✗	✓
Cellulose	✓	✗	✗	✗

4. Although starch and cellulose are formed from glucose, the isomer of the glucose monomers that have been joined differs (in starch, the glucose is α-glucose, whereas in cellulose it is β-glucose). Enzymes that digest starch, for example amylase, will recognise and catalyse the hydrolysis of α(1,4) glycosidic bonds. Very few enzymes will be able to recognise and catalyse the hydrolysis of β(1,4) glycosidic bonds. Note that some bacteria can hydrolyse cellulose using the enzyme cellulase, these bacteria are found in ruminants (e.g. cows and sheep).

5. Keratin consists of a protein with a tertiary structure in which hydrophobic molecules are found on the outer surface of the molecule. These make the protein insoluble in water, so that it does not dissolve when hair is washed.

6. This statement is usually true of globular proteins – for example, haemoglobin and insulin, – which consist of more than one polypeptide chain that has taken on a tertiary structure before being joined in the final protein molecule. In fibrous proteins, such as collagen and keratin, the polypeptide chains can remain as a linear molecule coiled into a helix that is reminiscent of the α-helices found in secondary structures.

7. (a) Insulin consists of two polypeptide chains in which a number of amino acids with different R groups have been joined in a precise sequence, by peptide bonds. The exact order of these amino acids means that some R groups are placed in precise positions relative to each other, allowing for the formation of hydrogen bonds between them and the development of secondary structures such as α-helices and β-pleated sheets. The secondary structures then twist and fold around each other to form a complex, three-dimensional shape. In the case of insulin, this level of organisation occurs for two polypeptides: these are then joined together in a quaternary structure which allows the protein to carry out its functions.

(b) Key properties listed in the passage that indicate that insulin is a globular protein include: (i) it is a hormone, (ii) it travels dissolved in the blood (it is soluble).

(c) The synthetic analogues of insulin are likely to have a similar tertiary and quaternary structure, as they need to bind to receptors in the body, and would not be able to do this if their structure was too different.

Practice exam-style questions

E1. D. [1]

E2. A [1]

E3. (a) The diagram should show how a 1,4 glycosidic bond forms between two glucosamine residues. [2]

Carbon atoms do not need to be numbered.

(b) Only glucosamine, and not chitobiose, will give a positive result when the Benedict's test is carried out. [1] First, a small volume of Benedict's solution is mixed with a solution of the sugar, which is then heated in a water bath set at 80 °C. [1] After a period of incubation, a colour change (from blue to green, yellow, orange or red) will indicate the presence of a reducing sugar, glucosamine. [1] To prove that chitobiose is a non-reducing sugar, the disaccharide is first hydrolysed, using hot acid, before being subject to a Benedict's test. [1] A positive result will indicate the presence of a reducing sugar (glucosamine monomers).

E4. (a) The primary structure of a protein is the sequence of amino acids in the polypeptide chain. [1]

(b) The polypeptide chain folds on itself in order to take on a tertiary structure, [1] which can have the effect of bringing amino acids from distant points in the chain close together. [1]

(c) Any [6] marks from:

Haemoglobin consists of four polypeptide chains in which a number of amino acids with different R groups have been joined in a precise sequence, [1] by peptide bonds. [1] The exact order of these amino acids means that some R groups are placed in precise positions relative to each other, [1] allowing for the formation of hydrogen bonds between them [1] and the development of secondary structures such as α-helices and β-pleated sheets. [1] The secondary structures then twist and fold around each other to form a complex, three-dimensional shape. [1] In the case of haemoglobin, this level of organisation occurs for four polypeptides: these are then joined together in a quaternary structure. [1] The presence of a prosthetic group called a haem group, [1] which contains an iron ion, allows the molecule to transport four molecules of oxygen. [1] The tertiary structure of haemoglobin can change shape when oxygen is bound or released, [1] which further enables it to take up or unload oxygen molecules.

E5. (a) Cells in brown fat tissue have more mitochondria, which are the site of aerobic respiration. [1]

(b) In the non-feeding season, the triglycerides found in fat cells are mobilised and respired, and converted into fatty acids and glycerol that are used in respiration. [1] The energy from fat mobilisation is then used, possibly for movement and to maintain body temperature. [1] In the feeding season, the food eaten is converted into fat for storage, which is also used for heat insulation to maintain body temperature. [1]

(c) Glycerol can be produced from a triglyceride by the process of hydrolysis. [1] Triglycerides consist of three fatty acids joined to a molecule of glycerol. Therefore, three hydrolysis reactions must take place in order to release glycerol. [1]

(d) Mix the liquid mixture containing triglyceride with ethanol in a clean test-tube. [1] Pour the mixture into another clean test-tube containing distilled water. [1] If a triglyceride was present in the original mixture, then a cloudy white emulsion is seen. [1]

E6. (a) Peptide bonds (C–N) are present in the molecule at each third residue. [1] Each third residue has a variable R group attached to the (alpha) carbon. [1]

(b) Dissociation of some chemical groups can result in the formation of an ionised (charged) atom called an ion, including examples such as $-COO^-$ and NH_3^+. [1] Ions can attract each other to form ionic bonds, [1] which can be important in holding together the tertiary and quaternary structures of proteins. [1]

(i) The independent variable in this investigation was the force applied to the protein. [1]

(ii) Variables that should be standardised in this investigation include: the mass of protein used; [1] the temperature at which the experiments were conducted. [1]

(iii) When a force of any size is applied, spidroin 1 stretches a greater distance than spidroin 2. [1] Spidroin 1 is able to stretch a longer distance before breaking than spidroin 2. [1] The maximum distance that spidroin can stretch without breaking is 3.6 mm, whereas this value for spidroin 2 is 2.0 mm. Spidroin 1 can also withstand a greater force before breaking than spidroin 2. [1] The value of 3.6 mm is seen when a force of 9 µN is applied with spidroin 1, [1] whereas a value of 2.0 mm is seen when a force of 7 µN is applied with spidroin 2. [1]

(c) Coating the spidroin proteins with a 'skin' of triglyceride makes the material waterproof. [1]

CHAPTER 3 – ENZYMES

In-text questions

1. Intracellular enzymes are enzymes that have their effects on their substrates inside a cell. Extracellular enzymes have their effects on their substrates in the environment outside of a cell.

2. A change in the primary structure of a protein means that its amino acid sequence will change. This will alter the R groups found along the molecule and, therefore, the number and position of hydrogen and ionic bonds, and hydrophobic interactions. This will cause a change in the overall three-dimensional shape of the enzyme (its tertiary structure) and the shape of its active site. The active site of the enzyme may no longer be complementary to its substrate, meaning that an enzyme-substrate complex cannot be formed and the enzyme cannot function.

3. (a) To transfer a phosphate group to a hexose sugar, energy must be supplied. This is called the activation energy. Hexokinase works as an enzyme by lowering the value of this activation energy, meaning that the reaction is more likely to occur. The effect is an increase in the rate of the reaction.

(b) The lock-and-key hypothesis of enzyme action states that an enzyme is specific to only one substrate – the shape of the active site of the enzyme is precisely matched (complementary) to the shape of the substrate, and cannot change. In contrast, the induced-fit model of enzyme action suggests that the shape of the active site is slightly flexible, i.e. it can change slightly in order to accommodate substrates of slightly different shapes.

(c) Glucose, fructose and galactose are all hexose (6-carbon) sugars of a very similar – but not identical – shape. The fact that hexokinase can add a phosphate group to any of these hexose sugars suggests that its active site can bind them, and that the active site is slightly flexible in shape.

4. (a) The bacterial enzyme that digests plastic is likely to be an extracellular enzyme, because the plastic substrate on which it acts is found in the organism's environment – outside the cell.

(b) All biochemical reactions occur at a faster rate as temperature is increased. If bacteria are produced that secrete this enzyme in an environment with a higher temperature, then the rate of reaction will be higher (because molecules have greater kinetic energy), and so plastic will be broken down more rapidly.

(c) The advantages of breaking down waste plastic with enzymes include the fact that the objects will no longer pose some threats to wildlife (e.g. plastic bags would be destroyed, meaning that they cannot be caught around larger organisms), and the fact that less would be disposed of in landfill or burned, which can pollute the environment. The disadvantages include the fact that it may not be completely broken down, leaving residual small particles of plastic, which could still present a hazard to organisms. Another disadvantage may be to also discourage the recycling of plastics and therefore the consumption of crude oil.

5. When an enzyme is denatured, it is only the hydrogen and ionic bonds, and hydrophobic interactions, that are broken in its secondary, tertiary and quaternary structures. The peptide bonds, which hold together amino acids in the primary structure, are not broken.

6. (a) This is necessary because otherwise the two solutions may not be of the temperature at which the experiment will be conducted. It will take some time for the solutions to reach the required temperature. To allow both solutions time to reach the correct temperature before starting the reaction. This is called pre-incubation. Temperature is the independent variable in this experiment. If one or other solution were not at the same temperature when mixed, the reaction would take place at a different temperature and give an inaccurate result.

(b) This is to increase the validity of comparisons of the results. If different volumes of iodine–potassium iodide solution were added, then the colour changes observed may not be entirely due to the difference in starch concentration between the samples. The volume of iodine–potassium iodide solution must be a standardised variable. The colour change from orange–brown to blue–black would occur at different times if different volumes of reagent were used. Hence for the results at each temperature to be comparable, the volume of reagent used must be the same every time.

(c) Iodine reagent may act as an enzyme inhibitor so cannot be added to the main reaction mixture; it must be used to test samples drawn from the main reaction mixture. Adding iodine–potassium iodide solution will reduce the rate of reaction, as it interferes with the formation of enzyme–substrate complexes. In these cases, samples have to be drawn from the reaction mixture for testing, and so it is not possible to continuously monitor.

(d) Suggested table as follows:

temperature/ °C	rate of reaction (1/time)min⁻¹
5 °C	0.125
25 °C	0.250
40 °C	1.000
55 °C	0.167
70 °C	0.000

Table values: rate of reaction $(1/\text{time})\,\text{min}^{-1}$

(e) Either:
- Repeat the iodine test at closer temperature intervals, for example every 2 °C, over 10 °C either side of the peak and re-plot the graph. Then draw a vertical line down to the temperature axis from the new position of the peak

Or:
- Plot a line graph of the time taken for a colour change against temperature. Draw a smooth curve and then draw a vertical line down to the temperature axis from the peak. This would give a more accurate value for the optimum temperature

Or:
- Record the colour at more regular intervals (e.g. every 30 seconds)

Or:
- Reduce subjectivity by using a method that does not give qualitative data, i.e. use a colorimeter to measure colour changes and give a numerical reading.

(f) At 70 °C, the kinetic energy of molecules is very high and intramolecular vibrations will disrupt the hydrogen and ionic bonds, as well as hydrophobic interactions, holding the tertiary structure of the enzyme together. This may have the effect of changing the shape of the active site of the enzyme, which will no longer be complementary to the substrate. This means that fewer enzyme–substrate complexes will form, resulting in a slower rate of reaction.

(g) At very low temperatures, molecules have little kinetic energy and so will be less likely to collide and form enzyme–substrate complexes.

7. (a) (i) Q_{10} at 20 °C to 30 °C = $\frac{87}{76}$ = 1.14; Q_{10} at 25 °C to 35 °C = $\frac{90}{82}$ = 1.10;

(ii) The Q_{10} values tell us that the effect of an increase in temperature on reaction rate is slightly greater at lower temperatures (20 °C to 30 °C) than at higher temperatures (25 °C to 35 °C)

(b) • There is an increase from 15 °C to 36 °C, and then a decrease from 36 °C to 60 °C (the peak activity of the enzyme / optimum temperature is 36 °C).
- There is a steeper decrease from 36 °C to 39 °C than at subsequent temperatures.
- The increase between 15 °C and 29 °C is steeper than the increase from 29 °C to 36 °C.
- The activity between temperatures 15 °C and 39 °C is higher than at temperatures higher than 39 °C.
- Manipulated data to support the marking points, which needs a comparison of values (for temperature in °C and percentage activity in %). For example, the highest activity of the enzyme, 91%, is seen when the temperature is 36 °C, while the lowest is at 60 °C (40% activity).

(c) Although the shape of the active site of the protease is exactly complementary to the shape of the substrate at the optimum temperature, the enzyme is still able to catalyse the breakdown of substrate at temperatures slightly higher or less than this value, because it does still have an ability to bind to the substrate.

8. (a) The estimate for optimum pH based on the data in Table 3.3 would be more accurate than that calculated based on the data in Table 3.1 because a greater number of values for the independent variable (pH) were used, and the intervals between them were therefore narrower. The data in Table 3.1 only allows you to say that the optimum temperature lies between 25 °C and 65 °C. This is because the temperature intervals of 20 °C are quite large and so the time taken for a colour change might be even quicker at some temperatures between 25 °C and 65 °C, than at 45 °C.

(b) The estimate for optimum pH based on the data in Table 3.3 would be more reliable than the optimum temperature calculated based on the data in Table 3.1 because five trials were conducted, and the mean value for the rate of reaction was found by taking a mean of the five readings obtained. The mean value would therefore be more reliable as compared to the data in Table 3.1, which shows only three trials.

(c) The anomalous data point in Table 3.1 is the third trial for the experiments conducted at 5 °C (130 seconds). This value is very different from the other two values, and should be excluded from the calculation of the mean value.

9. The initial rate of an enzyme-catalysed reaction should be measured when investigating the effect of a competitive inhibitor because the inhibitor will reduce the ability of the enzyme to bind to its substrate over time.

10. (a) Lysozyme has an active site that is complementary to a component of bacterial cell walls, but not plant cell walls. This suggests that the cell walls of these organisms contain molecules of different types.

    (b) K_m is a measure of the affinity of an enzyme for its substrate, where the greater value, the lower the affinity. Because lysozyme has a K_m value which is less than that of penicillinase, it would suggest that the rate of a reaction catalysed by lysozyme reaches the maximum rate (V_{max}) at a lower substrate concentration.

    (c) (i) The shapes of penicillin and sulbactam are similar. This means that if sulbactam is applied to bacteria that are producing the enzyme penicillinase, some enzyme molecules will attach to sulbactam rather than penicillin. By inhibiting some enzyme molecules, this will allow penicillin to take its effect on the bacterial cells.

       (ii) The value of K_m for an enzyme will increase in the presence of a competitive inhibitor because its affinity for its substrate will reduce. This is because the inhibitor will attach to the active site of the enzyme, limiting its ability to attach to the substrate.

11. The three methods of enzyme immobilisation – cross-linkage, encapsulation and adsorption – have many advantages and disadvantages:

    - Cross-linkage: due to chemical modification, this is likely to affect the affinity of the enzyme for its substrate. However, it is possible with this method to release the enzyme from its reaction column, if necessary, by using an enzyme or chemical that breaks the bonds between enzymes.
    - Encapsulation: depending on the material used to immobilise the enzyme, it is likely to affect the affinity of the enzyme for its substrate. However, there will be a lesser chance of the enzyme being lost from the reaction column over time.
    - Adsorption: unlikely to affect the affinity of the enzyme for its substrate, but there will be a greater chance of the enzyme being lost from the reaction column over time.

12. The graph shows that as the temperature of the reaction vessel increases from 20 °C to 40 °C, there is an increase in the volume of clear juice extracted with all three types of enzyme. Between these temperatures, free enzyme has a higher activity at lower temperatures, because there are more successful collisions between enzymes and substrate. In addition, the active site of the enzyme is completely exposed to the substrate. This contrasts with the active site of the immobilised enzymes, which is partially obscured from the substrate by the material used to immobilise them. At temperatures above 40 °C, free enzyme begins to denature and is completely denatured by 60 °C, because the shape of its active site has deformed and is unable to bind to the substrate. The immobilised enzymes, however remain active, because their tertiary structure is strengthened by the material used to immobilise them, and hence their active site remains complementary to the shape of the substrate as hydrogen bonds in the enzyme are less easily broken. The greater volume of clear fruit juice extracted using membrane-bound enzyme compared with enzyme encapsulated inside beads reflects the fact that the former enzymes are more accessible to the substrate; in the latter case, the substrate needs to diffuse into the beads to reach the enzyme.

Assignment 3.1

A1. Applying a high temperature to food will kill microorganisms and denature enzymes. This will prevent these agents from digesting the material and will keep it fresh.

A2. Because a weakened immune system may be owing to an increased activity of the enzyme 5-lipoxygenase, you could treat space researchers with an inhibitor of this enzyme (either competitive or non-competitive), which would reduce its activity and boost the activity of white blood cells.

A3. (a) A urease bioreactor would mean that the space researcher would not need to go to the lavatory as often, as the toxic products of metabolism in urine (including urea) would be removed from the fluid they produce. In addition, it may be possible to recycle the water produced by the bioreactor to enable its direct consumption by the researcher.

    (b) Human urease and urease from other organisms all have a common substrate – urea. This has the same chemical structure in all organisms, and therefore urease from any organism should have an active site that is complementary to this common substrate.

    (c) The scientist should use a Mn^{2+} concentration of 0.05 mol dm^{-3}

    (b) • The affinity of the enzyme for its substrate is increased because Mn^{2+} ions may prevent the attachment of non-competitive inhibitors to the enzyme. This is because manganese ions may occupy allosteric sites on the enzyme, preventing the attachment of non-competitive inhibitors which would otherwise reduce its affinity for its substrate. This means that there is a greater probability of enzyme–substrate complex formation.
    • The affinity of the enzyme for its substrate is increased because Mn^{2+} ions may stabilise the tertiary structure and limit the breaking of hydrogen bonds, upon fluctuations in pH and temperature. This is because manganese ions may help stabilise the tertiary structure of the enzyme, meaning that the active site remains complementary to the shape of the active site even if there are changes in temperature or pH, which means that there is a greater probability of enzyme–substrate complex formation.
    • The affinity of the enzyme for its substrate is increased) because Mn^{2+} ions may make the active site more complementary to the substrate (by the induced-fit mechanism. This is because manganese ions may make the active site of the enzyme more complimentary to the shape of the substrate, which means that there is a greater probability of enzyme–substrate complex formation.

    (e) The biuret test could be used to assess the protein content of the product solution. If a positive result is identified (the biuret reagent will turn from blue to lilac), then the enzyme has been lost from the beads.

    (f) This new method would not require the attachment of enzyme molecules to a chemical substance, which would reduce the likelihood of effects on the affinity for the substrate. It would also enable the use of free enzymes in solution, which would gain kinetic energy more readily if higher temperatures were used, and would result in a higher rate of reaction.

    (g) This method could not be used with enzymes that make macromolecules because they would be larger than the monomers that produce them. The monomers would enter the matrix and after polymerisation would be too large to diffuse out of the material.

Experimental skills 3.1

P1. The independent variable is the factor that is changed by the experimenter. In this investigation, the factor is the species of plant tissue. The dependent variable is the factor that is measured. Here, the student measured the volume of oxygen collected in 30 seconds. The dependent variable must be the factor that is directly measured. Some further processing of data is required here to calculate the rate of reaction (which is not the dependent variable).

P2. The hazard with the greatest level of risk in this investigation is the hydrogen peroxide solution. This chemical is corrosive and a strong oxidising agent, so eye protection must be worn.

P3. Both tables should place the independent variable in the left-most column, with units present in only the headings. The data should be rank-ordered in descending order, and the number of decimal places should be consistent. The tables should be ruled with a full border, for example:

Student 1

Plant tissue	Volume of oxygen gas collected in 30 seconds/cm³
potato	21.4
carrot	19.5
celery	13.2
apple	10.2

and

Student 2

Plant tissue	Volume of oxygen gas collected in 30 seconds/cm³
potato	23.7
carrot	20.5
celery	13.4
apple	11.0

P4. A bar chart is a more appropriate means to display this data because the independent variable consists of discontinuous, qualitative data.

P5. Data with low reliability consists of measurements that have not been repeated, and therefore no mean value was calculated. It is possible that some or all of the measurements are inaccurate (anomalous). If another person were to conduct the investigation, it is likely that they would not obtain similar readings.

P6. Student 2 obtained greater volumes of gas than Student 1. Possible reasons could include: more oxygen was lost to the air using the displacement method as the bung would need to be replaced after adding the hydrogen peroxide or plant tissue (the gas syringe has a second opening in its bung for the direct addition of hydrogen peroxide), or that some of the oxygen dissolved in the water in method (b). In the flasks containing the most vigorous reactions (i.e. the plant tissue with the most catalase), more heat would be released than in the flasks containing the least vigorous reactions. The difference in temperature gain between the two flasks will have introduced another variable to the investigation.

P7. (a) The mass of the contents of the conical flask would reduce over time because oxygen gas is released as a product into the atmosphere.

(b) The main reason is because the loss in mass is measured every five seconds, then the initial rate of reaction can be calculated. In the previous experiments, the total volume of gas released in 30 seconds was recorded, and it is possible that the reaction might end before this time interval elapses.

(c) In the first second after adding the plant tissue, there is a steep loss of mass. However, from two seconds to five seconds, the rate of the loss of mass reduces, and after five seconds there is no further loss of mass.

(d) The benefit of calculating and plotting the mass of contents as a percentage of original mass means that any differences in the mass of plant tissue between experiments is standardised. This would allow the student to make valid comparisons of the data.

(e) Data logging allows data gathered electronically to be recorded and graphs of the variable against time generated by a computer. This is particularly useful for reactions that happen very quickly, which would be true of this investigation, or very slowly.

Experimental skills 3.2

P1. The independent variable in this investigation would be the diameter of beads. The dependent variable would be the concentration of glucose in the product solution.

P2. The null hypothesis should be that there is no significant difference between the concentration of glucose (mmol dm⁻³) in lactase-treated milk produced using spherical enzyme–alginate beads with different diameters.

P3. (a) This ensures that lactase is operating at a temperature closer to its optimum value.

(b) This confirms that the untreated milk does not contain glucose and therefore give a false-positive result if glucose is detected in the product solution.

(c) This allows enough time for the substrate (lactose in milk) and enzyme (immobilised in beads) to be in contact with each other, form enzyme–substrate complexes, and react to form the products.

(d) If the enzyme solution was mixed with the calcium chloride solution, much of the enzyme would be wasted as the majority of the calcium chloride is discarded. In addition, the beads would not be as effective in encapsulating the enzyme. Instead of the enzyme being spread throughout the bead, it would likely be found mainly on the surface layer of the bead.

P4. 5 ml starch solution + 5 ml boiled enzyme should be included and tested for the presence/ absence of starch, at each temperature used. (The most appropriate control for an enzyme-focused investigation is a sample that contains denatured enzyme. This is because it will confirm it is the active site of the enzyme, rather than the presence of the enzyme per se, that is responsible for any changes in the dependent variable.)

P5. A digital glucose detector (sometimes called a glucose biosensor) could be applied to the solution in order to give a quantitative measure of glucose concentration.

P6. (a) As galactose has a similar shape to lactose, it will compete with lactose for access to the active site of the enzyme and act as a competitive inhibitor, limiting the number of enzyme–substrate complexes that are formed. The rate of hydrolysis of lactose will hence decrease.

(b) If the flow rate is too slow, then the treated milk will be slower to leave the vessel containing the beads and there will be a higher concentration of product galactose in the reaction vessel. This will inhibit the lactase, reducing the efficiency of further lactose digestion. If the flow rate is too fast, then the probability of lactose and lactase successfully colliding in the column is reduced, which would limit the rate of this enzyme-catalysed reaction.

(c) To make a valid comparison of the data obtained in this investigation, the student should ensure that there is only one variable that changes (bead diameter). All other variables should be standardised (kept the same) between experiments using different beads.

P7. An advantage of immobilising enzymes over live cells is that the tertiary structure of the enzyme is strengthened and the enzyme is able to operate at temperature and pH values further from its optimum. However, immobilised enzymes have a reduced affinity for the substrate, owing to the material used to immobilise the enzyme acting to obstruct the active site from the substrate.

An advantage of immobilising live cells over enzymes is that they are self-replenishing over time and do not need to be replaced if damaged. However, using immobilised cells means that the chance of the product being contaminated with enzyme (or other substances from the cells) is likely to be greater and therefore will need a supply of nutrients, which reduces maintenance costs.

Chapter review

1. An active site is a region of an enzyme which is specific to one substrate, to which its shape is complementary.

2. (a) Heating egg 'white' causes the proteins in the material to gain kinetic energy. This increases the intramolecular variations, resulting in hydrogen bonds and ionic bonds being overcome, which changes the tertiary structure of the protein and causes the hydrophobic R groups to become exposed to the aqueous environment.

(b) The egg 'white' would change from colourless to white, and its texture would change from slippery/slimy to become hardened. The low pH of vinegar would cause the proteins in the egg 'white' to denature. This is because the increase in hydrogen ion concentration affects the ability of hydrogen bonds to form between different parts of the protein. This will change the tertiary structure of the protein and causes the hydrophobic R groups to become exposed to the aqueous environment.

3. Suggested table as follows:

How enzymes are denatured by extremes of temperature	How enzymes are denatured by extremes of pH
hydrogen bonds, ionic bonds and hydrophobic interactions are affected	hydrogen bonds, ionic bonds and hydrophobic interactions are affected
enzyme gains kinetic energy, causing increased molecular vibration	charges on amino acid R groups are affected
active site changes shape	active site changes shape
is always irreversible	is sometimes reversible

4. (a) Enzyme – (v) a catalyst (usually a protein) produced by cells, which controls the rate of chemical reactions in cells.
(b) Substrate – (iii) molecule that an enzyme acts upon.
(c) Product – (ii) molecule produced by a reaction.
(d) Active site – (i) a 'dent' on an enzyme molecule that binds to a protein during a reaction.
(e) Competitive inhibitor – (vi) molecule that binds to the active site in an enzyme but no reaction takes place.
(f) Non-competitive inhibitor – (iv) molecule that inhibits enzyme reactions by altering the shape of the enzyme molecule without binding to the active site.

5. (a)(i) Competitive inhibition (ethanol competes with ethylene glycol for the active site of alcohol dehydrogenase).

(ii) Competitive inhibition (oxygen competes with carbon dioxide for the active site of ribulose bisphosphate carboxylase).

(iii) Non-competitive (arsenic binds to a region of cyclooxygenase enzyme away from its active site, but changes the shape of the active site).

(b) Answer: The researcher should carry out investigations into the effect of changing substrate concentration on the initial rate of reaction, but using different concentrations of thiourea the inhibitor. All other factors, such as temperature, enzyme concentration and pH, should be kept the same between experiments. Graphs should then be drawn to identify the value of K_m by finding the substrate concentration that gives the maximum initial rate of reaction (V_{max}). If the inhibitor is a competitive inhibitor of the enzyme, then there will be an increase in the value of K_m, but no change in the value of V_{max}. If the inhibitor is a non-competitive inhibitor of the enzyme, then there will be no change in the value of K_m, but a decrease in the value of V_{max}.

6. (a) (i) X   (ii) Y   (iii) X and Y
   (iv) X and Y   (v) X and Y

(b) (i) A competitive inhibitor would increase Km but not affect Vmax, because it decreases the affinity of the enzyme for its substrate. The substrate is competing with the inhibitor for the active site, and the inhibition is overcome by increasing the substrate concentration.

(ii) A non-competitive inhibitor would not affect Km but would reduce Vmax, because it did not affect the affinity of the enzyme for its substrate. The substrate is not competitive with the inhibitor for the active site and therefore the inhibition cannot be overcome by increasing the substrate concentration.

7.

I would enhance the validity of this investigation by . . .	I would enhance the accuracy of my estimate by . . .	I would improve the reliability of the result by . . .
Standardising other variables, including pH, enzyme concentration, substrate concentration, that would also affect enzyme activity and could affect the measurement of the effect of the independent variable (temperature) on the dependent variable (colour change).	Carrying out the experiment at a wide range of temperatures, separated by small intervals. If possible, a very accurate estimate could be obtained by carrying out the experiment at temperatures just above or below the value at which the enzyme denatures most rapidly.	Repeating the investigation for each temperature at least three times, and then calculating a mean value. If any reading appears to be anomalous, then this will be excluded from the calculation to find the mean, and ideally another reading would be obtained to replace it.

Practice exam-style questions

E1. C [1]

E2. (a) Suggested points (maximum of 4 marks) include:
- Immobilised enzyme has a higher optimum temperature than free enzyme: the highest activity of immobilised enzyme occurs at a higher temperature than that of free enzyme. [1]
- The optimum temperature for free enzyme is 40 °C and immobilised enzyme is 45 °C. [1]

- Free enzyme has a higher maximum activity than immobilised enzyme: the highest activity of free enzyme is greater than that of immobilised enzyme. [1]
- The maximum activity of free enzyme is 8.5 units and immobilised enzyme is 8.0 units. [1]
- Between 20 °C and 40 °C the activity of free enzyme is higher. [1]
- Between 40 °C and 55 °C the activity of immobilised enzyme is higher. [1]
- At the lowest temperature (20 °C), free enzyme has a higher activity. [1]

At the highest temperature (55 °C), immobilised enzyme has a higher activity. [1]

(b) As the temperature increases, the activity of the free and immobilised enzymes increases. [1] However, the activity of the free enzyme increases more steeply to its optimum temperature, because the molecules are free to move and collide with molecules of the substrate. [1] This is unlike the immobilised enzymes, which are held in position, cannot move, and therefore are less likely to collide with substrates to form enzyme–substrate complexes. [1] The immobilised enzyme, however, is stabilised and more resistant to higher temperatures than the free enzyme – its tertiary structure is less likely to be affected by increasing temperatures – so it has a greater optimum temperature and is also more active as temperatures rise higher. [1]

(c) The value of K_m is greater if an enzyme has a lower affinity for its substrate. [1] A possible explanation that the immobilised enzyme has a lower affinity for its substrate could be because the material used to immobilise the enzyme has reduced its ability to form enzyme–substrate complexes: its active site has been blocked or prevented from binding to its substrate. [1]

E3. (a) A non-competitive inhibitor reduces the ability of an enzyme to catalyse its specific chemical reaction by binding to a site on the enzyme away from the active site, [1] but by changing the shape of the active site and reducing its ability to bind to its substrate. [1]

(b) (i) If the final product (product 3) is removed from the cell immediately after it is produced, then it will not be able to inhibit enzyme 1. [1] Enzyme 1 will therefore continue to catalyse the conversion of substrate to product 2, which will then be converted into product 3. The reaction pathway will therefore continue to produce more and more molecules of product 3. [1]

(ii) If the final product (product 3) is polymerised to form a macromolecule stored in the cell, then it will not be able to inhibit enzyme 1. [1] Enzyme 1 will therefore continue to catalyse the conversion of substrate to product 2, which will then be converted into product 3. The reaction pathway will therefore continue to produce more and more molecules of product 3. [1]

(iii) If the final product (product 3) accumulates in the cell, then it will inhibit enzyme 1. [1] Enzyme 1 will therefore not be able to catalyse the conversion of substrate to product 2, so the rate of synthesis of product 3 will therefore start to reduce. [1]

(c) In the environment of a cell, the production of too much product 3 could be toxic, so end-product inhibition prevents the accumulation of this molecule. [1] However, when the concentration of product 3 begins to fall, the reaction pathway will begin to produce product 3 at a faster rate. [1]

(d) The diagram should look like this:

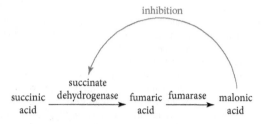

[1 mark for all substrates, products and enzymes in the correct order; 1 mark for the line from malonic acid to succinate dehydrogenase showing inhibition]

E4. (a) Different stains contain different types of biological molecules, [1] or different types of biological molecule. [1] For example, dried egg will contain albumin protein, which can be digested by different proteases from the haemoglobin found in blood stains.

(b) In a very hot wash at 90 °C, these biological washing powders will not be effective because the enzymes they contain will be denatured [1] and will no longer have active sites that are complimentary to their substrates / will no longer be able to form enzyme–substrate complexes. [1]

(c) These enzymes will be adapted to work well (have an active site that is complementary to their substrates) at high temperatures – they will not denature in environments of high temperature. [1]

(d) Reducing the temperature of the conditions in which washing occurs will reduce the energy required and therefore will be more ecologically acceptable. [1]

(e) Reasons will include: [1 mark each]
- Encapsulated enzymes will have a greater resistance to high temperatures (they will not be denatured until the temperature rises very high). [1]
- There is less chance of the enzymes being washed away in the waste water, so they may remain in the washing machine and can therefore be recycled between washes. [1]

E5. (a) An extracellular enzyme is an enzyme that digests its substrates outside of cells. [1]

(b) In both experiments, the percentage breakdown of azo-dye reduces as the concentration increases. [1] In addition, in both experiments, there is a steeper decrease in the percentage breakdown from 80 to 100 mg dm^{-3} as compared with increasing concentrations from 20 to 80 mg dm^{-3}. [1]

(c) For live cells to break down azo-dyes, the substrate must diffuse into the cell (which can take time). [1] In addition, free laccase from cell lysates is more likely to be at a greater concentration than that found in cells. [1]

(d) Any four of the following points:

Advantages:
- An immobilised enzyme can be reused in a continuous process. Because the enzyme is tightly attached to a gel or encapsulated in beads, significant quantities of it are not lost from the reaction column during use and it can therefore be recycled, saving money. [1]
- An immobilised enzyme can withstand a wider range of pH and temperature values. The material used to bind or encapsulate the enzyme stabilises its tertiary structure, which means that hydrogen bonds are less easily broken and its active site retains a shape complementary to its substrate at temperature or pH values far from its optimum. [1]

Disadvantages:

- The product could be retained within the immobilisation material: not all product molecules would pass through the reaction vessel to be collected, because they may get 'stuck' in the immobilisation material. [1]
- The substrate needs to diffuse into the immobilisation material to encounter the enzyme; the substrate would need to diffuse into the immobilisation material to reach the enzyme, which means that it is less likely that enzyme–substrate complexes will be formed.
- The enzyme cannot move, so collisions with substrate are less likely: the enzyme would be unable to move if immobilised, which means that it is less likely that enzyme–substrate complexes will be formed. [1]
- The affinity of an immobilised enzyme for its substrate is reduced. The material used to bind or encapsulate the enzyme acts to slightly block the access of the substrate to the active site, meaning that the formation of enzyme–substrate complexes is less likely and hence its affinity is reduced. [1]
- The active site is less accessible to the substrate as it is masked by the immobilisation material AND the active site is distorted by the immobilisation process or material: the active site of the enzyme may be 'blocked' by the material used to immobilise it, which means that it is less likely that enzyme–substrate complexes will be formed AND the active site of the enzyme may change shape owing to the material used to immobilise it, which means that it is less likely that enzyme–substrate complexes will be formed. [1]

(e) Protease inhibitors will reduce the rate of hydrolysis of proteins, including the extracted laccase enzyme. [1]

(f) (i) The independent variable in this investigation is the concentration of laccase enzyme. [1] The dependent variable is the area/ diameter/ radius of the circle that is clear/ colourless (does not contain azo-dye). [1]

   (ii) The plan should include the following points: [1 mark each, to a maximum of 8 marks]

- Regarding the independent variable:
  ○ A reference is made to diluting the 0.5 g dm⁻³ laccase / stock amylase solution to a minimum of 5 dilutions.
  ○ A reference is made to concentrations from 0.5 g dm⁻³ downwards with correct units.
  ○ Reference is made to the use of a control with example (e.g. boiled enzyme to denature).
- Regarding the dependent variable:
  ○ A reference is made to a suitable method of measuring diameters / width / radius / area of clear zones, e.g. using (suitable) ruler, or use a (transparent) grid/graph paper and count the number of squares.
  ○ Reference is made to testing the (fungal) extract / fungal laccase / fungal enzyme.
  ○ Plot a calibration curve of known concentrations and use it to determine extract concentration.
- Regarding the standardising of variables: [max. 3 marks]
  ○ ref. to suitable stated volume / same volume of laccase (in each well), maximum 1 cm³
  ○ leave (all plates) for same period of time (maximum 24 hours)
  ○ method of maintaining at same / constant / optimum / stated temperature, e.g. incubator / constant temp. room
  ○ use a buffer to keep the pH of the agar same, e.g. making or adding buffer to agar or starch (solutions)
  ○ same concentration of azo-dye (in the agar plates)
  ○ same depth / volume of agar in Petri dish
  ○ cover to prevent contamination / evaporation

- Reference to safety: medium to high risk investigation / hazard and suitable safety precaution. The use of azo-dyes is potentially very harmful (they are carcinogenic), so use of protective eyewear and gloves is important.
- Reference to reliability: a minimum of three replicates and calculate a mean or identify / eliminate / remove anomalies.

CHAPTER 4 – CELL MEMBRANES AND TRANSPORT

In-text questions

1. A phospholipid molecule consists of a phosphate head attached, via a glycerol molecule, to two long hydrocarbon tails. It is, in effect, a triglyceride which has a phosphate group in place of one of the fatty acid tails.

2. Phospholipids are amphipathic, which means that one part of the molecule is polar and hydrophilic (the phosphate head), whereas other parts of the molecule are non-polar and hydrophobic (the two hydrocarbon tails). When mixed with water or an aqueous solution, a small number of phospholipids will arrange themselves into a sphere consisting of a bilayer, surrounding a small pocket of water. Within this bilayer, their hydrophobic tails are facing inwards, towards each other and away from the surrounding water. The phosphate heads of the phospholipids on either side of the bilayer will be associated with water.

3. (a) Vesicles are produced by the Golgi body and are used to move substances around the cell, and sometimes outside of the cell by exocytosis. To be 'pinched off' from the Golgi body and to 'merge' with the cell surface membrane, the material used to make the membrane of the vesicle must be flexible and fluid. Both properties are explained if we assume that this material consists of a phospholipid bilayer.

   (b) Liposomes can be produced that contain molecules of a medicinal drug. When these make contact the cell surface membrane, they may merge with it, and this can result in the delivery of the contents into the cell cytoplasm.

4. Unsaturated phospholipids will be less able to pack closely together, which would have the effect of increasing the fluidity (reducing the stability) of the cell surface membrane. The same effect also occurs if the concentration of cholesterol in the phospholipid bilayer is increased. If the ambient temperature of an organism drops, then this would have the effect of reducing the fluidity of its cell surface membranes as the molecules inside them have less kinetic energy. By increasing the proportion of unsaturated phospholipids and concentration of cholesterol in the membrane, this will have the opposite effect and ensure that the properties of the cell remain unchanged. Increasing the proportion of saturated phospholipids, or decreasing the concentration of cholesterol, would have the opposite effect, if room temperature were to rise.

5. Despite having a hydrophilic –OH group that associates with the phosphate heads of phospholipids, the majority of the cholesterol molecule is hydrophobic. Hence, this will have a similar effect to the hydrocarbon tails of fatty acids and repel hydrophilic molecules and ions to prevent them from diffusing directly across the phospholipid bilayer.

6. A: phospholipid; B: cholesterol; C: integral membrane protein; D: peripheral membrane protein; E: glycoprotein.

7. Possible functions of peripheral proteins include: (i) as an (immobilised) enzyme; (ii) for cell-to-cell adhesion; (iii) as a receptor to bind to hormones or other signalling molecules.

8. (a) (i) A particle, sometimes called a signalling molecule, that attaches to a receptor, in the cell surface membrane or in the cytoplasm.

    (ii) A series of biochemical reactions that are initiated by a ligand and ultimately trigger a response.

    (iii) A change in the target cell, brought about by the ligand activating an intracellular signalling cascade.

    (b) Cell division is often controlled by signals, and these may be defective in the development of cancer/ hormones can act as signals, and understanding these mechanisms may benefit people with disorders related to hormones (e.g. diabetes) or neurotransmitters.

    (c) Muscle cells are likely to have a greater number of cell surface receptor (proteins) in their surface membranes compared with other cells. This means that, per unit time, they will be able to bind to more molecules of adrenaline and therefore react to stimuli more rapidly.

9. All organs in animals with cells that respire, including lungs, kidney, liver, small intestine; in plants, leaves and roots.

10. (a) As the surface area to volume ratio of a cell increases, the rate of diffusion of particles into all parts of the cell across the cell surface membrane increases.

    (b) Temperature, because a higher temperature causes faster diffusion, because dissolved substances have more kinetic energy (or converse for lower temperatures). Concentration of acid, because higher concentration causes faster diffusion, because the concentration gradient would be greater (or converse).

    (c) D

    (d) Any two of the following: (i) cells are not cube-shaped; (ii) Exchange of substances across a membrane involves active transport as well as diffusion; (iii) The agar block does not have an outer membrane / surface / border.

11. Active transport requires the energy released by ATP hydrolysis, so cells engaged in the process will have a high number of mitochondria, as well as the protein carriers (pump proteins) required to transport the molecule required.

12. Channel proteins, which provide a hydrophilic pore through which molecules can passively diffuse, are involved in facilitated diffusion, but not active transport. Carrier proteins, which change shape to transfer molecules from one side of a membrane to another, are involved in both. However, the carrier proteins involved in active transport require energy from ATP hydrolysis and are sometimes called pump proteins. Channel, carrier and pump proteins are all integral membrane proteins that are highly specific for the molecules or ions that they transport.

13. (a) The proteins involved in facilitated diffusion and active transport in this study are specific to glucose; sucrose will not be able to bind to them and so cannot be transported.

    (b) In both graphs, the intracellular concentration of glucose rises but eventually plateaus and will reach a maximum. This is because there is a finite number of membrane proteins specific to glucose, and ultimately, they will all be actively transporting the molecule. Therefore, protein number is a limiting factor, and at the maximum intracellular concentration all the carrier proteins are working at full

capacity. In the case of active transport, the line extends beyond the extracellular concentration, as after this point the glucose molecules are moving against their concentration gradient. However, this does not occur with facilitated diffusion; this stops rising once the intracellular concentration reaches the extracellular concentration as the concentrations of glucose on either side of the membrane have reached equilibrium.

    (c) If a respiratory poison such as cyanide was used, the graph for active transport (a) would show that no glucose would be transported – it would show a line parallel with the x-axis. This is because ATP, released by mitochondria during aerobic respiration, is absent, and this is required for active transport to take place. In the case of facilitated diffusion (b), there should be no change to the shape of the graph as energy, provided by ATP hydrolysis, is not required for the process to occur.

14. The erythrocytes have been placed in a solution of concentrated sodium chloride, which will have a lower water potential than the contents of their cytoplasm. Therefore, there is a net loss of water to the surrounding solution, causing the cells to shrink and take on a 'spiky' (crenated) appearance.

15. The observation that the flame cells are required to actively transport ions from the cytoplasm to the surrounding environment suggests that these ions are at a greater concentration outside of the flatworm and constantly move into the cells by (facilitated) diffusion. It would suggest that this species of flatworm lives in a habitat which has a water potential lower than the contents of its cells, i.e.salt water.

16. The water potential of the fruit preserves or syrup is likely to be very high. Therefore, if a bacterium lands on this solution, it will rapidly lose water by osmosis from its cytoplasm. The result will be that this bacterium will die due to dehydration.

17. The concentration of ions in sports drinks must not be too high, because otherwise it will have a water potential that is lower than the water potential found in epithelial cells that line the organs of the digestive system. This will cause water to move out of these cells, by osmosis, into the alimentary canal, which would have the effect of dehydrating the sportsperson.

Assignment 4.1

A1. (a) The surface area of the cubes is as follows:
    5 cm cube: $6 \times 5 \times 5 = 150$ cm^2
    4 cm cube: $6 \times 4 \times 4 = 96$ cm^2
    3 cm cube: $6 \times 3 \times 3 = 54$ cm^2
    2 cm cube: $6 \times 2 \times 2 = 24$ cm^2
    1 cm cube: $6 \times 1 \times 1 = 6$ cm^2

    (b) The volumes of the cubes are as follows:
    5 cm cube: $5 \times 5 \times 5 = 125$ cm^3
    4 cm cube: $4 \times 4 \times 4 = 64$ cm^3
    3 cm cube: $3 \times 3 \times 3 = 27$ cm^3
    2 cm cube: $2 \times 2 \times 2 = 8$ cm^3
    1 cm cube: $1 \times 1 \times 1 = 1$ cm^3

    (c) The surface area to volume ratio of the cubes is as follows:
    5 cm cube: $150{:}125 = 6{:}5 = 1.2$
    4 cm cube: $96{:}64 = 3{:}2 = 1.5$
    3 cm cube: $54{:}27 = 2{:}1 = 2$
    2 cm cube: $24{:}8 = 3{:}1 = 3$
    1 cm cube: $6{:}1 = 6$

Answers

A2. (a) The table of results should look like this:

length of side of cube/ cm	surface area: volume ratio/ no units	time taken for whole cube to become colourless/ seconds			
		trial 1	trial 2	trial 3	mean
1.0	6	75	55	70	67
2.0	3	210	210	220	213
3.0	2	340	330	365	345
4.0	1.5	530	565	530	542
5.0	1.2	855	770	790	805

(b)

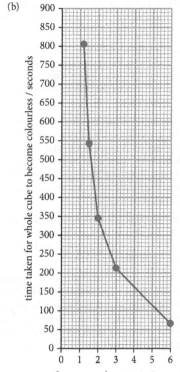

surface area: volume ratio / no units

(c) As surface area: volume ratio increases, the time taken for diffusion to the centre of the block decreases.

(d) This statement is correct. The rate of movement of the molecules in the acid would be the same, no matter what the size of block it was moving through.

A3. (1) This is a random error. Although it partly depends on the measuring instrument (ruler), which would give a systematic error, the greatest difficulty is in human judgement in cutting the blocks.

(2) This is a random error, as the temperature could have changed either up or down at different times during the investigation.

(3) This is a systematic error, as it is the same for each block, and would always reduce the rate of entry of the sodium hydroxide to the block.

Assignment 4.2

A1. A hydrophilic molecule is one that is readily able to dissolve in water. It is not able to mix with hydrophobic, or lipid-soluble, substances.

A2. (a) A ligand is a molecule that binds to a protein receptor either in the cell surface membrane (if the ligand is hydrophilic) or inside the cytoplasm (if the ligand is hydrophobic).

(b) The completed diagram should look like this: <insert new artwork as per sketch below

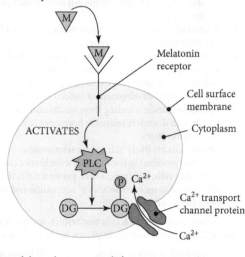

DG : Diacylglycerol   PLC : Phospholipase C

M : Melatonin

Ca^{2+} : Calcium ion

- Melatonin should be shown attaching to its receptor, which is an integral membrane protein.
- The receptor, when bound to melatonin, activates phospholipase C, which is found in the cytoplasm.
- Phospholipase C then phosphorylates diacyl glycerol.

(c) Many ligands will stimulate their target cells to activate or repress the expression of genes, which can have an effect on the cell. Other methods include opening ion channels to cause a change in membrane potential (see Chapter 15) and acting directly as a membrane-bound enzyme (see Chapter 14).

A3. (a) This process is called exocytosis. This involves a vesicle containing the molecules to be released fusing with the cell surface membrane (this is achieved by the fluidity of the phospholipid bilayer). As the two phospholipid bilayers combine, the contents of the vesicle are released into the extracellular environment of the cell.

(b) Insulin acts to reduce blood glucose levels, so in its absence, blood glucose levels rise. Before waking, melatonin is released in lesser quantities by the pineal gland, which means that it stimulates release of less insulin by pancreatic cells. The result is a rise in blood glucose concentration.

Experimental skills 4.1

P1. Red blood cells do not contain organelles, which means that calculations and analysis of the structure of the cell surface membrane will not be affected by the presence of membranes found around other organelles. In addition, red blood cells can be easily obtained by taking blood samples from animals.

P2. The biuret test could be used to test for the presence of protein in a purified sample of erythrocyte ghosts. Mixing the sample with a small volume of blue biuret solution will result in a colour change from blue to violet, if protein is present.

P3. Suggested features could include: (i) the ability of small or hydrophobic molecules to pass directly through the membrane, or (ii) the fact that molecules are able to spread out rapidly across the cell surface membrane.

P4. The pattern of freeze-fracture electron micrographs supports the idea that the cell surface membrane has a mosaic structure because there were 'bumps' and 'dents' in the surface of the interior of the phospholipid bilayer, which could only be present if proteins extended through the surface. In addition, the direction of the fracture lines was not uniform, suggesting that there are molecules within the structure that occupy a random, non-uniform shape.

P5. The method of fluorescent antibody tagging supports the idea that the cell surface membrane has a fluid structure because the differentially-coloured markers are able to spread through the fused cell over time. This showed that membrane proteins are free to move within the membrane, rather than being fixed in peripheral layers.

P6. (a) 95% confidence interval error bars are a measure of the closeness of the sample mean to the actual (population) mean. As seen in the graph, the error bars are wider for the data collected for the Golgi body. This means that there is less certainty about the actual value of time taken for the fluorescent markers to diffuse completely. For the other two organelles, the plotted error bars are narrow and do not overlap, so there is likely to be a significant difference between these mean values. However, the fact that the error bars overlap for the Golgi body and the chloroplast suggests that the significance of the potential difference between their mean values of the data points is questionable.

(b) An appropriate statistical test that could be used to assess the significance of the difference between the rate of diffusion in mitochondria and the Golgi body would be the t-test, as the data is continuous and the mean values are being compared.

(c) The rate of diffusion of proteins in the membranes of the Golgi body is the highest, because this produces vesicles, small membrane-bound structures that are required to transport substances around the cell. Some vesicles are involved in exocytosis, during which process they fuse with the cell surface membrane in order to release their contents outside of the cell. In both cases, the material that forms the membrane must be very fluid, which agrees with this study. The membranes of chloroplasts and mitochondria do not need to be as flexible and fluid as those of the Golgi body, as they do not physically and directly associate with other organelles.

Experimental skills 4.2

P1. The students should measure appropriate volumes of the stock solution of sucrose with appropriate volumes of distilled water. To produce volumes of 20 cm³ of each of these solutions, the following ratios should be used:

– To prepare 20 cm³ of a solution of 0.8 mol dm⁻³, 16 cm³ of the stock sucrose solution should be mixed with 4 cm³ of distilled water.

– To prepare 20 cm³ of a solution of 0.6 mol dm⁻³, 12 cm³ of the stock sucrose solution should be mixed with 8 cm³ of distilled water.

– To prepare 20 cm³ of a solution of 0.4 mol dm⁻³, 8 cm³ of the stock sucrose solution should be mixed with 12 cm³ of distilled water.

– To prepare 20 cm³ of a solution of 0.2 mol dm⁻³, 4 cm³ of the stock sucrose solution should be mixed with 16 cm³ of distilled water.

P2. Potato tissue that was placed into a solution with a higher water potential (hypotonic) than its cells will have increased in size (positive value), whereas potato tissue that was placed into a solution with a lower water potential (hypertonic) than its cells will have reduced in size (negative value).

P3. This would allow a valid comparison to be made of the proportional change from the original mass; the initial mass of the potato cylinders may not have been exactly equal.

P4. Plot a line graph of percentage change in mass of the potato tissue (y-axis) against the concentration of the solution (x-axis). Add a calibration curve. The point at which this curve crosses the x-axis is the point at which the plant tissue has zero change in mass. This represents the point at which the external solution has a concentration with a water potential equal to that of the plant tissue.

P5. Reasons would include: (i) the results were obtained for potato tissue only, whereas other plant cells will have different values for water potential; (ii) only six solutions were used, meaning that the accuracy of the estimate (as based on the point at which the line crosses the x-axis) was limited – further solutions could be used around this value.

P6. Regarding the use of a knife, cutting should be done away from hands and using a tile. Regarding the use of plant material, gloves should be worn in case of allergy.

Experimental skills 4.3

P1. (a) There is a general trend of increasing absorbance with increasing temperature. At temperatures below 30 °C, the absorbance values are very low (less than 0.1), suggesting little or no leakage of pigment from the cells. Between temperatures 25 °C and 40 °C, there is a gradual increase in absorbance, then the absorbance increases more steeply after 40 °C as shown by an increased gradient of the curve. The absorbance increases even more sharply after 65 °C. For temperatures greater than 80 °C, the change in absorbance is increasing sharply, as shown by the steep gradient.

(b) The results of this investigation confirm the fluidity of the membrane, and its protein components, which explains why at high temperatures the structure loses its integrity and allows cell contents to escape into its surroundings, including the red pigment in the beetroot. The changes in absorbance from low to high can be observed by the naked eye as changes in intensity of red-purple colour of the solution, from very light at lower temperatures to dark at higher temperatures. The absorbance (and therefore colour intensity) of the surrounding liquid is proportional to the concentration of pigment that has been able to leave the beetroot cells, which correlates directly with the permeability of the beetroot cell surface membranes. Only a small concentration of pigment is able to leave the cells at low temperatures. This is because the kinetic energy of the pigment molecules and the molecules within the cell surface membrane is relatively low and the cell surface membrane acts as a barrier to contain the contents of the beetroot cells. At higher temperatures, the kinetic energy of the molecules in the cell surface membrane is greater and they move more relative to each other, increasing the size of the gaps between them. There is also an increase in the kinetic energy of the pigment molecules, so their rate of diffusion out of the cell increases and they begin to leave the cell at a greater rate, increasing the concentration of the pigment cells in the surrounding solution and therefore increasing the absorbance and colour intensity. Beyond a particular temperature (approximately 65 °C), the kinetic energy of membrane proteins is so great that they denature, which, owing to changes in their tertiary structure, results in large gaps forming in the membrane. At this point, large quantities of the pigment are able to diffuse out of the solution increasing the concentration of pigment in the surrounding solution and thus increasing the absorbance and colour intensity of the solution.

(c) A line graph is used to show the relationship between two variables where the independent variable is continuous, as is the case in this investigation. It has the advantage of allowing the estimation of the effect of temperatures that were not tested during the investigation.

(d) The value can be found by drawing a vertical line from the value of 65 °C on the x-axis to the curve, and then a horizontal line from this point on the curve to the y-axis, then reading off the absorbance value on the y-axis. To improve the accuracy of the estimate, the experiment could be repeated using a series of temperatures around this value, and another graph could be drawn.

P2. (1) The precision of a 10 cm³ measuring cylinder is greater than a 50 cm³ cylinder, allowing for a more precise measurement of the water. This enhances the accuracy of the data collected.

(2) The concentration of pigment in cells and/or the density of cells may be different at different depths of the vegetable, both of which are factors that need to be controlled. This enhances the validity of the data collected.

(3) This removes any pigment that remains on the surface of the beetroot pieces without releasing any further pigment from undamaged cells. This enhances the validity of the data collected.

(4) This light frequency will move effectively through the red-coloured solution and can be detected by the colorimeter. This enhances the accuracy of the data collected.

(5) Measuring the baseline with distilled water ensures that any change in the colour of the solution will provide a measurable reading.

P3. (a) Possible systematic errors that could have arisen in this investigation could include:
- [P2. (1)] Systematic errors due to inaccurate measurement of volumes as a result of parallax error (the imprecise measurement of the meniscus of a liquid) when measuring the volume of distilled water required for the separate boiling tubes. Be sure to read the meniscus of the liquid at eye level.
- [P2. (4)] Systematic errors due to incorrect calibration of the colorimeter. This could be avoided by measuring the absorbance of distilled water before, and between, the measurement of absorbance values for the different solutions. This is to make sure the colorimeter is always calibrated correctly.

(b) During the experiment, the pigment may concentrate around the surface of the beetroot cylinder, particularly at higher temperatures. This would mean that the pigment fails to diffuse fully into the water in which it is suspended. This would have the effect of limiting the diffusion gradient between the tissue and the water. By agitating the boiling tube every 5 minutes or so, a steep diffusion gradient would be maintained, allowing further pigment to leave the tissue

(c) A control experiment is required to show that it is the change in temperature, rather than another factor, that has the measured effect on membrane permeability. In this investigation, the control investigation would involve placing a cylinder of beetroot, of identical mass/volume to the others, distilled water at room temperature. There should be no detectable release of pigment in the control experiment relative to the equivalent experiment at the same temperature.

P4. The plan should:
- Use a range of concentrations, appropriate to the method of dilution that is chosen, that would give a series of results from which trends can be identified and on which predictions can be made. For example, the 100% ethanol solution can be diluted using either the serial or simple (proportional) dilution method. In the serial dilution method, 9 cm³ distilled water is topped up to 10 cm³ with 100% ethanol to give a 10% solution. Then, 9 cm³ distilled water is topped up to 10 cm³ with the 10% ethanol solution to give a 1% ethanol solution. This can be repeated to give 0.1% and 0.01% concentrations. In the simple (proportional) dilution method, a series of dilutions of ethanol of concentrations 20%, 40%, 60%, and 80% are prepared by diluting a specific volume of 100% ethanol with distilled water. For example, adding 2 cm³ water to 8 cm³ 100% ethanol will give a 10 cm³ solution of 80% concentration.
- Decide on a number of concentrations that would enable a trend line to be drawn and on which predictions can be made.
- Describe how the temperature of the different ethanol concentrations should be consistent, using a thermostatically controlled water bath (set to remain constant at no more than 25 °C). The mass/volume of the beetroot cylinders placed into the different concentrations should also be the same, as well as the time that they were allowed to incubate. If any of these factors were to also vary in addition to the concentration of the ethanol between the tubes, then no valid comparison could be made of the collected data.
- Have a control experiment to show that it is the ethanol, rather than another factor, that has the measured effect on membrane permeability. In this investigation, the control investigation would involve placing a cylinder of beetroot, of identical mass/volume to the others, into a solution of 0% ethanol (100% distilled water) for the same amount of time to incubate, and at the same temperature. The plan should predict that there should be no detectable release of pigment in the control experiment.
- Describe how the intensity of the red colour in the solution will be measured in a similar way to the method that they employed in the practical lesson, by way of using a colorimeter set to transmit blue-green light and measure absorbance. They should explain how distilled water is used to provide a baseline level of absorbance between measured solutions, and carry out three replicates of each absorbance reading, to minimise the chance of systematic and random errors, respectively and improve data reliability.
- Recognise that the use of a sharp knife represents a hazard and how risk can be minimised. In addition, ethanol is highly flammable, and so no naked flames should be brought near the laboratory equipment.

Explain how an estimate could be made by finding the lowest value of ethanol concentration on the x-axis that corresponds to the value of zero absorbance on the y-axis. Another investigation could then be carried out using a narrower range of ethanol values that concentrate on this section of the line, in order to make a more accurate estimate of the actual concentration of ethanol that does not cause pigment leakage.

P5. (a) The completed table should look like this:

Temperature / °C	Mean absorbance / arbitrary units	Standard deviation (s) / arbitrary units	Standard error (SE)/ arbitrary units	Upper limit of 95% confidence interval	Lower limit of 95% confidence interval
5	0.034	0.011	0.006	0.046	0.022
25	0.041	0.004	0.002	0.045	0.037
40	0.290	0.046	0.026	0.293	0.187
60	0.747	0.257	0.148	1.043	0.450
75	1.113	0.086	0.050	1.213	1.014
90	1.690	0.066	0.038	1.766	1.614

To calculate the value of standard error (SE) for the mean absorbance value obtained at 5 °C, the following formula is used:

$$SE = \frac{s}{\sqrt{n}}$$

Where:

SE = standard error,
s = sample standard deviation,
n = sample size

For 5 °C, three measurements were taken (n = 3) and the sample standard deviation is 0.011

Therefore,

$SE = 0.011 \div \sqrt{3}$
  = 0.006

To calculate the 95% confidence limit:
95% confidence limit = mean ± 2 × SE

Therefore, the upper limit of 95% confidence interval is 0.034 + (2 × 0.006) = 0.046.

The lower limit of 95% confidence interval is 0.034 - (2 × 0.006) = 0.022.

(b) The upper confidence interval of the mean value for the experiment conducted at 60 °C (1.043) is higher than the lower confidence interval of the mean value for the experiment conducted at 75 °C (1.014). These confidence intervals therefore overlap, and we would therefore not expect the difference between the two mean values to be significant if a statistical test were to be carried out.

(c) An appropriate statistical test that could be used to assess the significance of the difference between the mean values of absorbance obtained at different temperatures would be the t-test, as the data is continuous and the mean values are being compared.

Chapter review

1.

	Simple diffusion	Facilitated diffusion	Active transport
type of membrane molecule involved	lipids	proteins	proteins
force driving the process	concentration gradient	concentration gradient	ATP hydrolysis
direction of transport	with concentration gradient	with concentration gradient	against concentration gradient
specificity	non-specific	specific	specific
saturation at high concentration of transported molecules	no	yes	yes

2. The water potential of a solution is a measure of the ability of water molecules to move by diffusion out of this solution into another solution separated by a partially permeable membrane.

3.

Fluid	Mosaic
a	b
d	c
f	e

4. The analogy is accurate in that the proteins are less common than the phospholipid molecules, and that they are all able to move relative to one another. The analogy is inaccurate because some proteins are limited to just one side of the cell surface membrane and can have polysaccharides attached to them.

5. Active transport and facilitated diffusion are performed by different membrane proteins, with different specificities and binding sites for molecules to be transported.

6. Aquaporins are likely to have a very narrow, hydrophilic pore through which water molecules can pass.

7. The phospholipids, which are found in the cell surface membrane, are required to maintain the integrity of the cell. If they are broken down, then the cell will lyse (burst) and the contents, including haemoglobin, is released.

8. (a) There will be no significant difference between the net water entry into the bag in experiments using different concentrations of glucose solution.

(b)

independent variable	dependent variable	standardised variables
concentration of glucose solution	volume of solution in dialysis (Visking) tubing bag	temperature, time period for experiment

(c) • Temperature: a thermostatically controlled water bath.
   • Time period for experiment: a stopwatch.

Practice exam-style questions

E1. (a) In the cell surface membrane, phospholipid molecules are arranged in a bilayer in which the hydrophilic phosphate heads point away from each other [1] (and into the cytoplasm or extracellular environment) and the hydrophobic hydrocarbon tails make contact with each other in the middle of the structure. [1]

(b) The phospholipid bilayer is fluid. [1] Therefore, when the viral envelope and cell surface membrane make contact, they are more likely to attach and fuse together. [1] This allows the virus particle to enter the cell. [1]

(c) The greater proportion of cholesterol will reduce the fluidity / increase the stability of the envelope, relative to the cell surface membrane. [1]

E2. (a) (i) This is an integral protein found in the cell surface membrane, which uses active transport to move particles from one side of the cell surface membrane to the other. [1]

(ii) ATP supplies energy for active processes in the cell, [1] suggesting that the movement of membrane proteins was passive and due to simple diffusion. [1]

(b) The loss of hydrogen ions into the stomach lumen will cause the water potential to fall inside the stomach, [1] which will cause water to move out of stomach cells by osmosis and into the stomach lumen. [1] The increased volume of water in the alimentary canal will result in diarrhoea.

(c) This would suggest that the drug is hydrophobic (non-polar) in nature, [1] as it is able to diffuse directly across the phospholipid bilayer and into the cytoplasm to bind to the inner portion of the integral protein carrier. [1]

E3. (a) Counting the same number of cells for each concentration means that a valid comparison can be made between the counted values. [1]

(b) For the 0.5% sodium chloride solution, the mean is 24.7 cells, the median is 25 cells and the mode is 25 cells. [1]

(c) The number of plasmolysed cells counted in the 0.2% sodium chloride solution (fourth reading) is anomalous because it is much higher than the other two trials (does not fit a general correlation / does not follow trend). [1]

(d) Plotting the median or the mode would not take into account the anomalous data reading (the fourth reading counted in the 0.2% sodium chloride solution). [1] The value for the mean has been affected by this anomalous result (it is 7.0 rather than 5.0). [1]

(e) The mean of 24.7 indicates that there were some readings in which there were some cells that were not plasmolysed. [1] The median and mode would incorrectly suggest that in all readings every cell had been plasmolysed. [1]

(f) D. [1]

(g) (i) The independent variable (which is varied by the investigator) is the concentration of sodium chloride solution. [1] The dependent variable is the time taken until incipient plasmolysis occurs. [1]

(ii) This would improve the reliability of each measurement. [1] By calculating an average of the five different cells, the effect of anomalous results would be reduced. [1]

(iii) This ensures that another variable that could have an effect on the time taken for incipient plasmolysis to occur has been standardised. [1] This will improve the validity of the conclusion of the study. [1]

(iv) A possible source of error in this investigation would be the difficulty in judging the point at which incipient plasmolysis occurs. [1] This can be minimised by taking a video of the experiment, through the microscope, so that when this occurs can be accurately determined and the time checked for accuracy. [1]

E4. (a) (i) Phosphatidylcholine is unsaturated. [1] It has a double line in one of its hydrocarbon tails, which indicates the presence of a C=C double bond. [1]

(ii) Phosphatidylcholine is amphipathic. [1] It has a region that is hydrophobic (the fatty acid tails) and a region that is hydrophilic (the charged 'head' that contains phosphate).

(iii) Stearin [1] would release more energy when used in respiration because there is a greater number of carbon and hydrogen atoms present (it has three hydrocarbon 'tails,' rather than two as found in phosphatidylcholine). [1]

(iv) Phosphatidylcholine has hydrocarbon tails of different lengths. [1] This is obvious from the different sizes of the lines that point in a vertical direction, compared with those in stearin, which are of the same length. [1]

(b) In the cell surface membrane, the hydrocarbon tails of phosphatidylcholine will associate with each other in a bilayer [1] such that the heads, containing phosphate, will face the external and internal environment [1] and associate with water. [1]

(c) Stearin is a triglyceride, which is made in condensation reactions [1] in which one glycerol molecule [1] and three long-chain fatty acids [1] react. Three water molecules are removed. [1 marl]

E5. (a) The accumulation of chloride ions outside a cell lowers the water potential, [1] creating a water potential gradient from inside the cell to outside. [1] Water therefore moves out of the cell by osmosis. [1]

(b) (i) To absorb molecules from the environment, the cell surface membrane can extend around a toxin, [1] before fusing together to form a vesicle. [1]

(ii) A cell signalling mechanism occurs when a ligand (signalling molecule/hormone) [1] attaches to a specific protein receptor found at the cell surface membrane, [1] and sets off a series of events inside the cell. [1] These could involve the triggering of a secondary messenger/ activation of enzymes or an enzyme cascade/ phosphorylation events. [1]

(c) Glucose and sodium ions from the drink will enter the small intestine and move into the cells [1] through the glucose-sodium cotransporter protein. [1] This decreases water potential inside the cells, so water moves in by osmosis down the water potential gradient and reduces water loss in diarrhoea. [1]

CHAPTER 5 – THE MITOTIC CELL CYCLE

In-text questions

1. Histones.

2. Chromosomes are made of DNA, which is wound around histone proteins. This material is packaged into a structure with two arms, which each have telomeres at their ends. The arms are held together by a construction called a centromere (before S-phase).

3. Stem cells and cancer cells are similar in that they are able to continuously divide by mitosis. They are different because cancer cells are usually differentiated (specialised) and their division cannot be controlled, unlike stem cells.

4. (a) Telomerase adds DNA to the ends of chromosomes to rebuild their losses after cell division.

(b) Cells that express telomerase in the human body will be those that retain the ability to divide for their whole lifetime. These include stem cells in the skin/ bone marrow/ liver.

5. (a) B (DNA synthesis, and the production of sister chromatids, occurs in S phase).
    (b) A (stem cells are not differentiated).

6. The nuclear envelope disintegrates at the start of prophase, but reassembles at the end of telophase. This is important, because during the process of mitosis (nuclear division), sister chromatids will be separated into two daughter cells. The presence of a nuclear envelope would prevent this process from happening.

7. (a) During gap phase 1 (G_1), cells rapidly grow after mitosis, and prepare for DNA replication in S phase. During gap phase 2 (G_2), cells prepare for mitosis by synthesising the subunits of microtubules and ATP.
    (b) Chromosomes and sister chromatids are effectively synonyms (different words that describe the same thing). However, (sister) chromatids only exist after DNA replication has occurred in S phase, and are held together by the centromere. After the sister chromatids are separated at anaphase, they are again referred to as chromosomes.
    (c) Mitosis is the division of the nucleus and its contents, and has four phases called prophase, metaphase, anaphase and telophase. Cytokinesis is the division of the cytoplasm and its contents to make two independent cells, and happens at the very end of mitosis, before the next cell cycle begins.
    (d) A benign tumour is a mass of cells that is slow-growing and does not spread to other organs, whereas a malignant tumour is a mass of cells that is fast-growing and can spread to other organs in the blood and lymphatic system.
    (e) During metaphase, the chromosomes are aligned at the equator of the cell, whereas during anaphase, the sister chromatids that comprised each chromosome are pulled apart to opposite poles of the cell.
    (f) A telomere is a cap of DNA found at the end of the chromosome arms, whereas a centromere is a constriction in the chromosome which is usually found near the middle. The length of telomeres has been associated with cellular ageing, whereas the role of centromeres relates to cell division, as it is here where the spindle fibres attach during mitosis. Both structures are made of DNA and histone proteins.

8. (a) (i) The quantity of DNA remains at 2 units from 0 to 1 hour, but then rises to reach 4 units at 2 hours. It then remains at 4 units for another 2 hours until 4 hours, when it falls steeply over 30 minutes to 2 units. Its DNA content remains at this level until the 6-hour point, when the same changes in DNA are repeated.
        (ii) The replication of DNA (in the S phase of interphase) takes 1 hour. During this time, the quantity of DNA in the cell doubles as chromosomes become a pair of sister chromatids. Between S phase and mitosis, the DNA content of the cell is twice that of the parent cell. The division of this DNA between two daughter cells, at time = 4 hours and 9 hours, is very rapid as the phases of anaphase, telophase and cytokinesis occur very quickly.
    (b) (i) The cell does not have a nucleus and is somewhat rectangular in shape. There are about four sister chromatids visible, which appear as strands that meet at the middle (equator) of the cell. The sister chromatids are fully attached to each other and the arms of the structures are spread out away from the equator of the cell.
        (ii) Spindle fibres have attached to the centromeres of the chromosomes and pulled them to align at the middle (equator) of the cell. This cell is undertaking metaphase.

9. (E)→D→C→A→G→F→B→(E).
    E – In interphase/early prophase, the nucleus becomes darkly stained.

D – In prophase, the chromosomes begin to condense and start to become visible in the cell. The nuclear envelope begins to disintegrate.

C – In metaphase, the chromosomes line up along the equator of the cell. Spindle fibres radiate from the poles of the cell and attach to the centromeres, the structures that hold together the two sister chromatids.

A – In anaphase, the spindle fibres contract and pull the sister chromatids apart and to opposite poles of the cell.

G – In early telophase, a nuclear envelope begins to form around each set of separated sister chromatids, now called chromosomes.

F – In late telophase, the cytoplasm begins to divide between each set of separated sister chromatids in the process of cytokinesis. A cell plate also forms in order to lay down a new cell wall.

B – The cycle begins again as the cells enter interphase/early prophase.

10. (a) Chromosomes will be pulled to opposite poles of the cell during anaphase. If they were decondensed, rather than tightly packaged, then there would be a greater opportunity for the DNA to be damaged or broken. There may be enzymes or other molecules in the cytoplasm that would make contact with the DNA during mitosis (in the absence of the nuclear envelope), which could also damage it.
    (b) After S phase, each chromosome consists of a pair of sister chromatids joined at the centromere. In prophase, this can be easily seen, meaning that the cell contains twice the required DNA content. After cytokinesis, the cell will have half of this quantity, because the events of anaphase, telophase and cytokinesis have resulted in the sister chromatids being separated into two separate cells.
    (c) The drawing should show a cell without a nucleus, with all six of the chromosomes lined up along the equator of the cell. Spindle fibres, radiating from the poles of the cell, should be attached to the centromeres of each chromosome on both sides.

For example:

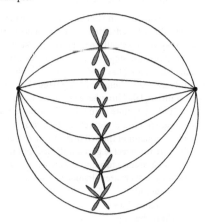

11. The period of interphase is a very busy time for a cell. During this period, a cell must perform its role, which will require the production of specific proteins and other molecules. In addition, the cell must prepare for the next time it undertakes mitosis and cytokinesis. In gap phase 1, a cell will prepare for the DNA replication that will occur in the next stage, by synthesising the raw materials needed to produce DNA, including nucleotides, ATP and key enzymes. In synthesis (S) phase, a cell will replicate its DNA so that each chromosome consists of two sister chromatids. In gap phase 2, a cell will produce more organelles

that will be shared between the daughter cells during cytokinesis, ATP that will provide energy for the processes that take place during mitosis, and the proteins that will form microtubules that play a key role in mitosis.

12. Mitochondria are the site of aerobic respiration and they produce ATP. ATP hydrolysis is used to release energy, which is needed during the most active periods of mitosis. These include the assembly of the spindle, the separation of sister chromatids at anaphase, the reassembly of the nuclear envelope at telophase, and the development of a cleavage furrow during the process of cytokinesis. Mitochondria will be found at these sites because they will be able to provide ATP more quickly to the structures that require it.

13. Myocytes will likely have many nuclei. This is because they undertake mitosis (nuclear division) in the absence of cytokinesis (cytoplasm division). They are also likely to be very large, because in the absence of cytoplasmic division and the sharing of organelles these nuclei will remain in one cell.

14. (a) An eyepiece graticule is a transparent ruler with numbers, but no units. It is positioned inside the eyepiece. A stage micrometer is a microscope slide with a scale on its surface. It is required to calculate the length of the divisions on the eyepiece graticule at a particular magnification.

   (b) If one division on the stage micrometer measures five divisions on the eyepiece graticule at a magnification of ×100, then each division on the stage micrometer is 0.1 mm in length. Therefore, there are five divisions on the eyepiece graticule in 0.1 mm at this magnification. Therefore, when the stage micrometer is replaced with the slide containing the root tip squash, the size of the cells at this magnification can be calculated. For example, if a cell is 5 eyepiece divisions in length, then 1 eyepiece graticule division = 20 µm, so the length of the cell = 5 × 20 µm =100 µm.

   (c) Length of actual scale bar = 10 mm
   Length of magnified scale bar = 0.025 mm
   Magnification = 10 / 0.025 = x400

15. (a) In the case of Student 1, the lines are broken rather than continuous, the sister chromatids are not in proportion to the cell – they are too small (inaccurate representation), and key labels are missing (e.g. cell wall, sister chromatids). For Student 2, shading has been used and not all sister chromatids are drawn (one is missing from the right-hand side). In addition, label lines have not been drawn using a ruler. For Student 3, more than one cell is drawn and sister chromatids are not in the correct position relative to each other and the equator of the cell (inaccurate representation). Some labels are incorrect (e.g. cell wall label points to the cell surface membrane; chromosomes should be sister chromatids) and others are missing, e.g. cell wall.

   (b) The diagram should:
   - use continuous lines
   - not include shading
   - include only the focus cell (one cell)
   - show all contents of the cell in proportion to the cell
   - include labels for all structures
   - include label lines that are drawn using a ruler
   - include all components (in this case, that refers to four sister chromatids on each side of the cell)
   - show the contents of cell in the correct positions relative to each other
   - use as much of the available space as possible.

For example:

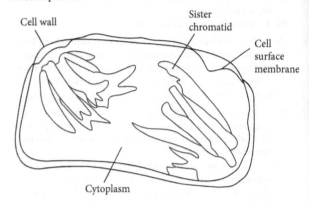

A2. Preparing a series of solutions that have a wider range of concentrations, together with smaller intervals between them, would enable a graph to be plotted that will enable an accurate trend between the relationship between the independent and dependent variables to be identified.

A3. (a) A greater number of readings for different values of the independent variable would be required to draw a line graph.

   (b) A greater number and range of values for the independent variable means that any relationship between the independent and dependent variables will be accurate and allow for a more statistically significant conclusion to be drawn. This would not be possible with a bar chart. In addition, it is possible to use a line graph to make estimates of the effect of untested concentrations of paclitaxel solutions on the number of cells in mitosis (by reading off from the plotted graph).

   (c) The temperature is kept constant for the duration of the investigation, but the time for incubation and how temperature will be maintained (e.g. by using a thermostatically controlled water bath) is not described. The two garlic cloves that are placed into each paclitaxel solution are from different bulbs; it would make the investigation more valid if all cloves used were from the same bulb to minimise any genetic variation that could influence the mitotic index between different garlic plants. In addition, some key variables have not been controlled to make the investigation valid, for example the duration of the incubation period and the length of the basal plate of the clove that is immersed. This means that we cannot be sure if any observed effect is due entirely to the independent variable.

   (d) Only one clove of garlic is used for each paclitaxel solution and (we assume) only one count of cells is performed for each microscope slide. This means that any anomalous data could not be identified, and a mean value could not be calculated. The reliability of the data yielded from this study would be very poor. To improve the investigation, the researcher should incubate more than one clove of garlic in each paclitaxel solution and conduct a number of counts of cells for each slide. Calculating a mean value would reduce the likelihood of anomalous data having a significant effect on the dependent variable.

Assignment 5.1

A1. The dependent variable is the number of cells that are undertaking mitosis. This will be expressed as a proportion of the total number of cells that are counted, and multiplied by 100 to give a percentage figure.

(e) A control (garlic clove placed into distilled water) is not used. This is needed in order to remove the effect of the independent variable, paclitaxel concentration, to confirm its effect. The researcher attempted to include a control in this investigation; however, tap water is not pure water and may contain substances that have an effect on the rate of mitosis in plants.

A4. Null hypothesis: There is no significant correlation between the duration of incubation using different chemotherapy drugs and the percentage of cells undertaking mitosis.
Alternative hypothesis: There is a significant correlation between the duration of incubation using different chemotherapy drugs and the percentage of cells undertaking mitosis.

A5. The Spearman's rank statistical test can be used in this study to test for a correlation between two sets of paired data that are not distributed normally, because it is assumed that:
- The data points within the samples are independent of each other.
- Ordinal data has been collected: this data can be converted to an ordinal scale using ranking.
- If plotted, a scatter diagram would indicate the possibility of an increasing or decreasing relationship.
- There are at least 5 sets of paired observations.
- All cells were selected at random, with each cell having an equal chance of being selected.

A6. Two from vincristine, podophyllotoxin and camptothecin. This is because the calculated value of r_s is higher than the critical value of r_s at $p=0.05$ ($r_s = 0.362$).

A7. If data obtained in a scientific investigation is significant at a probability level of 0.01 rather than 0.05, this means that the results would be seen by chance only one in every 100 times that experiment is conducted. The conclusion would be more convincing than if the data is significant at a probability level of 0.05, which means that the results would be seen by chance only one in every 20 times that experiment is conducted.

A8. (a) The value of $r = -0.40$ indicates that there is a clear negative association between the age of an individual and the mean length of their telomeres.
(b) The Pearson's linear correlation coefficient is used to determine the strength of a correlation between data that is normally distributed, which is likely the case with mean telomere lengths.

Experimental skills 5.1

P1. (a) Hot hydrochloric acid breaks down the pectin in the middle lamella between plant cell walls. This allows the stain to better penetrate the tissue in the next stage, and also ensures that it will be possible later to squash the tissue to a thickness of one cell.
(b) Rinsing the roots removes the hydrochloric acid, which is important because any residue of hydrochloric acid remaining on the roots could limit the effectiveness of the stain.
(c) The period of incubation for the roots to remain in the stain allows for the stain to fully penetrate the plant tissue.
(d) Lowering the coverslip onto the specimen is important, otherwise air bubbles would be trapped, which would obscure the view of chromosomes.
(e) Squashing the root tissue is important because if one layer of cells is positioned on top of another, the view of the chromosomes will be obscured.

P2. The absence of the middle lamella and the cell wall in animal cells means that they may be less able to withstand the force applied to the slide in order to squash them into one layer. The cells may burst and would not show chromosomes in the positions they were when fixed.

P3. (a) Mitotic index = $(12 \div 16) \times 100 = 75\%$
(b) Mitotic index = $(8 \div 16) \times 100 = 50\%$
(c) Only 16 cells were counted from each onion. To improve the reliability, the scientist should select, at random, a number of fields of view (minimum of three for each treatment) in which to identify cells and calculate their mitotic index. By calculating the mean of these values, the effect of anomalous results will be minimised and the reliability of the data improved.
(d) To make a valid comparison between two sets of data, any variables apart from the independent variable must have been standardised during the investigation. The independent variable in this case was the presence or absence of vincristine, but it can be clearly seen from the diagram that one onion is larger than the other. Possibly, the smaller onion (in the vincristine solution) would have had roots with a reduced mitotic index because of the difference in size. In addition, other variables should be controlled, including temperature (by using a thermostatically controlled water bath or incubator), and time for the onions to be immersed in the water or solution.
(e) Vincristine is likely to be at least harmful, and substances in the onion could be allergens. Wearing gloves and eye protection during the investigation is advisable for this reason.
(f) Spindle fibres attach to the centromeres of chromosomes during metaphase and contract to pull the sister chromatids apart to opposite poles of the cell during anaphase.
(g) There are many cells in the tissue treated with vincristine whose cell cycles have been stopped during prophase. This is in keeping with the absence of spindle fibre formation, as the chromosomes cannot align on the equator and cannot subsequently be pulled apart into separate sister chromatids in anaphase.
(h) A straight line of best fit is drawn and extrapolated to the x-axis. The value of (expected) 1.35 mm, which is where the line intersects the x-axis, is recorded.

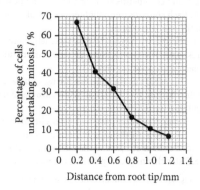

Distance from root tip/ mm	Percentage of cells undertaking mitosis/ %
0.2	67
0.4	41
0.6	32
0.8	17
1.0	11
1.2	7

Chapter review

1. Mitosis produces genetically identical cells, for tissue repair / replacing damaged cells/ replacing old cells, and for growth.

2. (a) G2
   (b) 120 minutes, which is the duration of S-phase
   (c) During G2 phase, which is any time between 5.00 a.m. and 11.00 a.m.
   (d) mitosis
   (e) Cytokinesis will occur towards the end of mitosis, so any time between 11.30 and 12.00 would be acceptable.

3. (a) Centromere – a constriction in the middle of a chromosome that holds together the sister chromatids. (iv)
   (b) Sister chromatid – consists of a structure attached at the centromere to another structure which is genetically identical with it. (viii)
   (c) Telomere – the end of a chromosome arm. Its shortening has been implicated in cellular ageing. (v)
   (d) DNA – the molecule of heredity; contains genes that encode proteins. (iii)
   (e) Histone – a type of protein around which DNA is wound during the process of supercoiling of chromosomes. (i)
   (f) Spindle – made of microtubules, this attaches to the centromeres of all chromosomes at prophase. (vii)
   (g) Equator – the middle of the cell, where the chromosomes align during metaphase. (vi)
   (h) Poles – the ends of the cell, to which position the sister chromatids are pulled during anaphase. (ii)

4. (a) True. Centrosomes replicate during interphase, before mitosis begins.
   (b) True. Sister chromatids are formed by the replication of DNA. Each contains one daughter DNA molecule identical with the parent molecule.
   (c) False. Microtubules extend from the kinetochore to the nearest pole. The kinetochores in sister chromatids are connected to opposite poles.
   (d) False. This occurs during mitosis during spindle manufacture (polymerisation, assembly) and sister chromatid movement (depolymerisation, disassembly).
   (e) False. Kinetochores are found on sister chromatids.
   (f) False. Telomeres are the caps at the ends of chromosomes. Microtubules are attached at the centromeres (kinetochores).
   (g) True. Sister chromatids separate at the start of anaphase.

5. Cancer cells will only spread around the body when a tumour becomes malignant. In plants, cells are kept in place by their surrounding cell wall, so are unable to move around the vessels. There is no equivalent to the bloodstream or lymphatic system in plants, which is responsible for transporting cancer cells during metastasis in animals.

6. The figure is found by dividing the total number of cells undertaking mitosis (19) by the total cells counted (100). Multiplying the resulting figure by 24 hours (the average time for a cell cycle in this tissue type), then by 60 minutes provides the correct answer of 274 minutes.

7. (a) (i) 60 sister chromatids would be present.
       (ii) 30 centromeres would be present.
       (iii) 120 telomeres would be present.
   (b) (i) False. During metaphase, microtubules attach to the centromeres of chromosomes.
       (ii) False. The nuclear envelope reforms during telophase.
       (iii) False. During cytokinesis, chromosomes consist of only one chromatid (sister chromatids were pulled apart at anaphase).
   (c) In organisms that have a greater number of chromosomes, it will likely take longer for these chromosomes to condense (become shorter and thicker) at the start of mitosis (prophase).

8. (a) Histones.
   (b) Centromeres have two functions: to hold together sister chromatids after DNA replication in S phase, and to provide a site of attachment for the spindle during mitosis.
   (c) During interphase, chromosomes decondense – their DNA is unwound from histone proteins and is being read by the cell in the process of gene expression. Chromosomes can only be seen with a light microscope when they condense, which occurs at the start of mitosis.
   (d) A sketch should show a structure with two sister chromatids, connected by a constriction (or small circular structure), labelled centromere. The telomeres, found at the ends of the four arms of the sister chromatids, should also be labelled. For example,

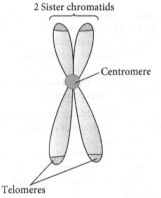

2 Sister chromatids

Centromere

Telomeres

   (e) Suggestions should include:
   - Chromosomes are found in the nucleus.
   - Each chromosome contains many genes / controls many characteristics.
   - Chromosomes must be inherited from both parents for normal growth / development / functioning of the embryo.

Practice exam-style questions

E1. (a) Cancer cells are immortal (they do not stop dividing). [1] This may be because the length of their telomeres is sustained, [1] preventing the loss of genes/ ageing and death. [1]
   (b) The purpose of a control experiment is to prove that the independent variable – the application of the chemotherapy drug – is responsible for any effects that are seen on the dependent variable – the number of cells. [1] In the absence of a control experiment, the researcher cannot be sure that the drug is responsible. [1]
   (c) The number of cells in the sample reduces from 10×10^6 to 6×10^6 during the study. Therefore, the percentage change is $(6\times10^6 - 10\times10^6) \div 10\times10^6 \times 100 = -40.0\%$ [1]

(d) 24 hours [1] and 120 hours [1] because after these time points, the number of cells reduces steeply. [1]

(e) Applying the chemotherapy drug at more regular intervals would likely reduce the number of cancer cells more effectively, but this would also reduce the number of healthy stem cells [1] found in the body in places such as the bone marrow, hair follicles and testes/ ovaries and cause side effects. [1]

(f) In normal cells, mitosis will stop at metaphase if spindle fibres are not correctly attached, because the separation of sister chromatids is prevented. [1] In cancer cells, anaphase will take place even when spindle fibres not correctly attached, [1] resulting in incorrect distribution of sister chromatids into daughter cells. [1]

E2. (a) Stem cells retain the ability to divide indefinitely [1] and are undifferentiated. [1]

(b) (i) This chart is a histogram, rather than a bar chart, because (max 2 marks):
- The data on the *x*-axis is quantitative in nature and continuous. [1]
- The bars are touching/ the values for the *x*-axis are placed at the beginning and the ends of the bars. [1]
- The width of the bars is not the same. [1]

(ii) On average, the mean length of telomeres in individuals with no history of sunburn was greater than the length of individuals with a history of sunburn. [1] The most common (modal) length of telomeres in cells taken from individuals with a history of sunburn is 3 kilobases (21% of mean telomere lengths), compared with 8 kilobases (19% of telomere lengths) in individuals with no history of sunburn. [1] There were no individuals with no history of sunburn who had telomeres of mean length 2 kilobases or less, unlike those individuals with a history of sunburn, 11% of which had lengths of this length. [1] There were no individuals with a history of sunburn who had telomeres of mean length more than 10 kilobases, unlike those individuals who had no history of sunburn, 16% of which had telomeres of this length. [1] In conclusion, the distribution of mean telomere lengths is skewed towards the smaller sizes in the stem cells from individuals with a history of sunburn.

(iii) Sunburn damages/ kills skin cells [1] which will need to be replaced by the division of epidermal stem cells. [1] With each stem cell division, telomeres become shorter, so the greater damage and replacement of skin cells in individuals with a history of sunburn would mean that their epidermal stem cells have shorter telomeres. [1]

(iv) Maximum of 3 marks: A correlation only shows there is a relationship not a cause/effect. [1] The people chosen for the study may have been unrepresentative of the whole population [1] and other factors may be contributing to the shortened telomeres in people with a history of sunburn (e.g. these people may have been older, may have been smokers, and so on). [1] There are some individuals who have a history of sunburn who have mean telomere lengths which are greater than those without a history of sunburn and vice versa. [1]

(c) A primary tumour (for example, in the skin) will grow to the extent that some cancer cells will be released from it and into the blood or lymphatic system. [1] These will be transported around the body and can become attached to tissues elsewhere, [1] where they will begin to divide and form a secondary tumour. [1]

(d) (i) Cells would remain in prophase. [1] This is because the spindle cannot form in the presence of colchicine, [1] and therefore metaphase cannot begin. [1]

(ii) Cancer cells divide more rapidly than normal cells, and therefore will require more energy (from mitochondria) [1] and proteins (from rough endoplasmic reticulum) for growth. [1]

E3. (a) Answer: B. [1] This cell has visible chromosomes (when viewed using a light microscope) during metaphase, in which stage the chromosomes consist of sister chromatids (joined at the centromere) and there is no nuclear envelope. It is a meristem cell, so it is undifferentiated.

(b) Answer: B. [1] The most terminal layer comprises the root cap (beneath the epidermis), which loses cells as the root tip pushes into the soil. Meristem cells rapidly replicate behind the root cap, replacing these cells, and producing cells that differentiate and elongate behind them. [1]

(c) Magnification = actual size of bar in image / magnified size of bar in image.
Magnification = 10 000 μm / 100 μm
Magnification = x100 [2]

(d) Place a stage micrometer on the stage of the microscope. [1] Line up one of the divisions on the eyepiece graticule with a fixed point on the stage micrometer. [1] Count the number of divisions on the eyepiece graticule that correspond with a set measurement on the stage micrometer. [1] Calculate the distance in micrometres of one division on the eyepiece graticule. [1]

(e) The diagram should:
- use continuous lines
- not include shading
- only draw the tissue layers (not every cell)
- have labels to identify structures, with lines drawn straight
- include all components (in this case, that refers to the three different tissue layers)
- include the contents of root, namely the different tissue layers, in the correct position relative to each other
- use as much of the available space as possible. [3]

For example:

E4. (a) At the very tip of the root, in the root cap, very few cells are undergoing mitosis. However, just behind the cap, the majority of cells are dividing in the meristematic tissue. [1] As distance from the tip increases further, the number of cells undergoing mitosis falls gradually, and then more suddenly, until the value reaches zero. [1]

(b) At the root tip behind the cap, meristematic tissue exists which is the site of cells undergoing rapid mitosis. [1] This produces daughter cells that move backwards, further away from the root tip as the root grows. [1] As these daughter cells move further from the tip, they stop dividing, and begin to elongate, [1] as their vacuoles gain water by osmosis. This results in their volume increasing. [1]

(c) First, chromosomes must have their DNA replicated in order to form two sister chromatids. This occurs in synthesis (S) phase and ensures that each daughter cell inherits a copy of each chromosome. [1] Second, chromosomes must shorten and thicken (condense) at the start of mitosis to ensure that the events of the process, in which they move around the cell, do not damage them. [1]

CHAPTER 6 – NUCLEIC ACIDS AND PROTEIN SYNTHESIS

In-text questions

1. The diagram should show all three components of a nucleotide – the pentose sugar as a pentagon, with a phosphate group attached to the fifth carbon and shown as a circle, and a base as a rectangle attached to the first carbon.

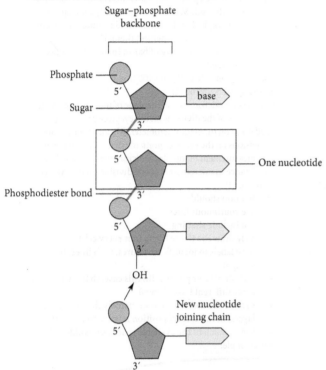

2. DNA (a polynucleotide) is a polymer because it consists of many monomers joined together in a long chain. It is a macromolecule because it is a giant molecule, consisting of many thousands of atoms.

3. (a) Guanine is the odd one out, because it is a purine base (unlike the pyrimidines cytosine and thymine). Purine bases have a single-ring structure, while pyrimidine bases have a double-ring structure.
   (b) Phosphate is the odd one out, because it is not the name of a bond found in DNA. This is unlike phosphodiester (bonds), which link together adjacent nucleotides, and hydrogen (bonds), which form between complementary base pairs between the two polynucleotide strands.
   (c) Nucleotide is the odd one out, because this is a monomer. A polynucleotide is a polymer of many nucleotides, and can also be referred to as a nucleic acid.

4. DNA consists of two polynucleotide strands, held together by hydrogen bonds, that form a double helix. They run in opposite directions, meaning that they are antiparallel in nature. Both polynucleotide strands are polymers, because they consist of nucleotides that have been joined together by phosphodiester bonds.

5. Two reasons to explain why base pairing must be complementary in a DNA molecule are: (i) purines are larger than pyrimidines and if a purine was to bind with another purine, or a pyrimidine was

to bind with another pyrimidine, then there would be areas of the double helix in which the width of the molecule varies; and (ii) the number of hydrogen atoms on a purine and the nitrogen/oxygen atoms on the other correspond in the same matching positions.

6. (a) The structure of RNA differs from the structure of DNA as follows:
   - the pentose sugar is ribose, not deoxyribose.
   - the base thymine is replaced by a different base, called uracil. So, the four bases in RNA are A, U, C and G.
   - most forms of RNA are made up of a single strand, rather than two antiparallel strands joined together. RNA can, however, fold into three-dimensional structures.
   - RNA molecules are usually shorter than DNA molecules: they contain fewer nucleotides joined together. A DNA molecule can consist of over a billion nucleotides, whereas RNA usually consists of a few hundred, which varies depending on the precise role it has in the cell. RNA molecules are also less stable and this is appropriate given their short-term functions in the cell.
   (b) The structure of adenosine triphosphate (ATP) differs from the structure of a nucleotide found in DNA as follows:
   - ATP contains the pentose sugar ribose, whereas nucleotides in DNA contain the pentose sugar deoxyribose.
   - the base in ATP is adenine, whereas the bases found in the nucleotides in DNA can also be guanine, cytosine and thymine.
   - three phosphate groups are attached to ATP, whereas only one is attached to the nucleotides in DNA.

7. (a) (i) After one generation, the DNA produced is neither heavy nor light – it is exactly halfway between the two. This excludes the conservative replication theory.
       (ii) After two generations, the DNA is either light or half-and-half. This rules out dispersive replication.
   (b) After 10 generations, there will still be two bands – one half-and-half and one light – but the light band will be much darker.
   (c) The graph will have generation number on the x-axis, and proportion of DNA containing N isotopes on the y-axis. The graph should show three lines. The first line, labelled ^{15}N, will drop abruptly to half of its original value at the first generation, and then to zero at the second generation. The second line, labelled $\frac{^{15}N}{^{14}N}$, will be zero at the start of the graph, but will rise to half at the first generation, and then decline gradually to zero by the end of the graph. The third line, labelled ^{15}N, will be zero at the start of the graph and at generation 1, but would increase to half at generation 2, and then rise gradually thereafter.

8. (a) There are 12 nucleotides shown in the diagram (6 on each polynucleotide strand).
   (b) The bond labelled X is a phosphodiester bond. This connects two adjacent nucleotides.
   (c) The enzymes involved in DNA replication are: DNA helicase, DNA polymerase and DNA ligase. DNA ligase is required for the replication of the lagging strand only.
   (d) DNA is a molecule in which the two polynucleotide strands run in opposite directions relative to each other (the strands are antiparallel). Therefore, although DNA polymerase can extend the leading strand in a 5' to 3' direction in a continuous process, it must replicate the lagging strand in short fragments to achieve an extension in the same 5' to 3' direction. This may be because DNA polymerase has an active site that is complementary to the 3' end of the growing polynucleotide chain.

9. A

10. The intron and exon in a gene differ with respect to the role of the base sequences they contain. Base sequences in the intron do not encode amino acids, whereas base sequences in the exon do encode amino acids. However, introns are thought to be involved in controlling gene expression (the stimulation of transcription).

11. The structure of a molecule of mRNA differs from the structure of a molecule of tRNA as follows:
    - mRNA is a linear (relatively straight) molecule, whereas tRNA has a three-dimensional 'clover-leaf' shape.
    - mRNA does not contain intramolecular hydrogen bonds (owing to complementary base pairing), whereas tRNA does.
    - mRNA molecules have codons, triplets of bases, arranged along its length. tRNA molecules have one anticodon, which binds to one codon during translation.
    - tRNA molecules carry an amino acid, whereas mRNA molecules do not.

12. (a) Valine-Valine-Serine-Threonine-Leucine.
    (b) B

13. (a) The correct order of protein synthesis is as follows:
       C. The two strands of a DNA molecule separate
       F. RNA nucleotides join with the exposed DNA bases and form a molecule of mRNA
       A. The mRNA molecule leaves the nucleus
       G. The primary transcript undergoes splicing and intron sequences are removed
       E. A ribosome attaches to the mRNA molecule
       B. tRNA molecules bring specific amino acids to the mRNA molecule
       D. Peptide bonds form between the amino acids
    (b) Condensation reactions happen in stage D of the sequence.
    (c) G, E, B and D.
    (d) Translation begins with stage E.

14. (a) Two features that could be recognised by nucleases that digest RNA could be:
       - the pentose sugar is ribose, not deoxyribose
       - the base thymine is replaced by a different base, called uracil. So, the four bases in RNA are A, U, C and G.
    (b) Molecules of mRNA are produced when a gene is expressed and, once they have been translated, they may no longer be needed, and should be broken down in case their encoded protein is over-produced.

Assignment 6.1

A1. Purine percentage = 47.3%; pyrimidine percentage = 52.7%.

A2. The percentage of purine bases is nearly equal to the percentage of pyrimidine bases in all three organisms. This would suggest that DNA has a double-stranded structure in which purines on one strand attach to pyrimidines on the other strand.

A3. Bases are not paired in the DNA molecule, suggesting that it consists of only one polynucleotide chain / the DNA is not double-stranded.

A4. (a) $p<0.05$ indicates that in only 1 of 20 times of carrying out the investigation would the observed difference be seen, if it were simply due to chance or an external factor.
    (b) If Chargaff's rule is true for this virus, then in this sample the expected number of purines and pyrimidines should be $(451+525) \div 2 = 488$.

    So, in this example:

    $\chi^2 = ((451-488)^2 + (525-488)^2) \div 488$

    $\chi^2 = 5.611$

    (c) The χ^2 calculation for the virus gives a value of 5.611. This is greater than the critical value of 3.841, suggesting that there is a significant difference between the observed and expected values for the purine:pyrimidine ratios. Therefore, the null hypothesis must be rejected, as there is less than a probability of 0.05 that this difference was due to chance alone. There is a biological reason for this difference: virus DNA is often single-stranded, and does not have base pairs.

Experimental skills 6.1

P1. Without mRNA, the tRNA molecules holding amino acids will not be brought to the ribosome – translation will not occur.

P2. A molecule that contains nucleotides with only uracil bases will not form intramolecular hydrogen bonds because there will be no sections that are complementary to each other. This is in contrast with tRNA molecules, which have regions that contain bases that are complementary to bases in other regions of the molecule. These will form complementary hydrogen bonds with each other.

P3. The scientists could repeat the experiment but instead add a poly-A mRNA (which produced a protein that consisted of just lysine amino acids), or a poly-C mRNA (which produced a protein that consisted of just proline amino acids), or a poly-G mRNA (which produced a protein that consisted of just glycine amino acids).

P4. The two independent variables in this experiment were the length of the poly-U RNA molecule (how many uracil bases were present) and the concentration of the poly-U RNA molecule.

P5. Factors that should be standardised in this investigation include: concentration of ribosomes in the sample, volume of tRNA solution applied to the membrane, time permitted for measuring radioactivity.

P6. The purpose of a control in any experiment is to remove the effect of the independent variable(s), in order to see if any other factors have an effect on the dependent variable (in this case, the radioactivity that was measured). An appropriate control experiment for this investigation would be to carry out the same experiment but in the absence of the poly-U molecule. The expected result would be that no radioactivity would be measured.

P7. A graph of this data should include the first independent variable on the x-axis (concentration of poly-U RNA) and the dependent variable on the y-axis (radioactivity). The five different RNA lengths (U_2, U_3 U_4, U_5 and U_6) would then be plotted as separate, straight lines, crosses joined, which are labelled.
    - display the independent variable on the x-axis and the dependent variable on the y-axis

- use a small cross to mark each data point
- make sure the intersection of the crosses is exactly on the required point
- make sure the plotted points are connected with a clear, sharp and unbroken smooth curve passing through all the points (assuming there are no anomalous points, in which case a line of best fit should be drawn)
- do not extrapolate (extend) the curve beyond the plotted points
- draw the peak where it naturally falls on the curve rather than at the highest point.

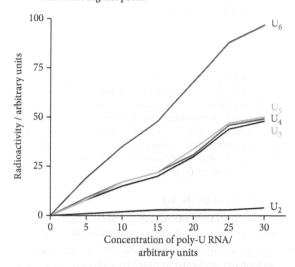

P8. Nirenberg and Leder predicted that three mRNA bases comprise a codon. Their evidence suggests that this is the case: very little radioactivity was detected on the nitrocellulose filter with poly-U RNA molecules when just two bases were used, suggesting that these molecules did not attach to the ribosomes. When poly-U RNA molecules with more than two bases were used, radioactivity was detected on the filter, suggesting that these molecules attached to the ribosomes, which did not pass through the material. When the length of the poly-U RNA reaches six bases, there is a sharp rise in the radioactivity, which would suggest that more tRNA molecules are attaching to the ribosome. This is in agreement with the prediction, because a poly-U RNA of six bases will be able to bind to two tRNA molecules, which would cause the nitrocellulose filter to have twice the amount of radioactivity.

Chapter review

1. (a) A = deoxyribose sugar; B = phosphate group; C = nitrogenous bases.
   (b) The structure of DNA suits its function because:
   - The bases are protected on the inside of the molecule and the two strands are held together by numerous hydrogen bonds, so DNA is a very stable molecule and is not easily damaged.
   - There are four different bases, which can appear in any order, so their sequence can encode information, like writing with a four-letter alphabet.
   - DNA is a very long molecule, so it stores a great deal of information (human DNA has three billion base pairs).
   - The two complementary strands mean there are two copies of the information, which is useful for repair, copying and error checking.
   (c) The DNA double helix is unzipped as DNA helicase breaks the hydrogen bonds between complementary bases on the

separate polynucleotide strands. DNA polymerase catalyses condensation reactions between adjacent nucleotides, which become joined by phosphodiester bonds, to form two daughter strands that have a base sequence complementary to that of the parental strands.

2.

	replication of leading strand	replication of lagging strand
Uses free DNA nucleotides	✓	✓
Is a continuous process	✓	✗
Requires DNA polymerase	✓	✓
Requires DNA ligase	✗	✓
Takes place in the 5′ to 3′ direction	✓	✓

3. Gemcitabine has a very similar structure to the DNA nucleotide containing the base cytosine. Therefore, gemcitabine would be incorporated into the synthesised DNA molecules by DNA polymerase, which has an active site that is complementary to it. However, gemcitabine does not have a phosphate group at the equivalent position to the nucleotide that contains cytosine. This means that it will not be able to form phosphodiester bonds with other nucleotides when it is incorporated into the growing nucleotide chain.

4. (a) Eight amino acids is the maximum that could be coded for by this section of mRNA.
   (b) Because this section of mRNA contains two AUG codons and two GCC codons, only six different tRNA molecules would be needed to translate this molecule.

5. First, the gene for each tRNA (molecule) is transcribed. This occurs when hydrogen bonds in DNA are broken, by the unzipping of the DNA double helix by DNA helicase. One strand of the DNA is then transcribed by RNA polymerase, which involves joining (free RNA) nucleotides together by the formation of phosphodiester bonds. The free RNA nucleotides are arranged in the correct order owing to complementary base pairing. Note that tRNA is not translated to form a polypeptide.

6. DNA replication and transcription share a number of similarities and a number of differences, which should be arranged into a table; for example:

DNA replication	DNA transcription
involves unzipping the DNA double helix to expose the bases (by DNA helicase)	involves unzipping the DNA double helix to expose the bases (by DNA helicase)
uses free DNA nucleotides to produce a second strand	uses free RNA nucleotides to make the mRNA
requires the enzyme DNA polymerase	involves RNA replication, which requires the enzyme RNA polymerase
Is a discontinuous process that requires the enzyme DNA ligase to join together the Okazaki fragments of the lagging strand	Is a continuous process

7. (a) DNA replication – DNA makes a copy of itself (x)
   (b) mRNA – carries DNA code from nucleus to cytoplasm (vii)
   (c) anticodon – triplets of bases on one end of tRNA (v)
   (d) ribosome – where translation takes place (ix)
   (e) DNA – material of which genes are made (iv)
   (f) codon – mRNA triplet of bases that determines which amino acid will be brought in (viii)
   (g) mitochondrion – the site of synthesis of ATP, which is required as a source of energy for protein synthesis (iii)
   (h) transcription – the synthesis of mRNA by RNA polymerase, which happens in the nucleus (vi)
   (i) translation – mRNA is decoded and directs the sequencing of amino acids to make a polypeptide chain (i)
   (j) mutation – change in base sequence (ii)

8. (a) (i) The library in this analogy represents the nucleus of a cell.
       (ii) The library door in this analogy represents the nuclear pore.
       (iii) The photocopier in this analogy represents the enzyme RNA polymerase.
       (iv) The photocopied instruction sheet in this analogy represents the primary transcript/ mRNA.
       (v) The workbench in this analogy represents the ribosome.
       (vi) The pieces of wood and plastic that make the model car in this analogy represents the amino acid subunits.
   (b) Transfer RNA (tRNA) molecules bring amino acids to the ribosome during translation. The amino acid they carry is determined by the sequence of their anticodon. This enables the ribosome to produce a polypeptide of an amino acid sequence that corresponds to the sequence of codons in the mRNA.
   (c) This process is called splicing. It involves the removal of introns from the primary RNA transcript and the joining of exons to make the mRNA, ready for translation.

9. Genetic diseases are caused by a change to the nucleotide/base, sequence of, the DNA in a gene. This causes the formation of a new allele owing to base substitution, insertion or deletion. This means that, in the primary sequence of the polypeptide, the sequence of amino acids may change, because a tRNA with a different anticodon brings a different amino acid to the ribosome. The polypeptide will then take on a modified secondary, tertiary, and sometimes quaternary structure, which may influence its ability to function in the usual way in the cell. This may affect the ability of the cell to carry out its normal processes, resulting in genetic disease.

Practice exam-style questions

E1. B: 10 [1]

E2. (a) Nitrogen is required for the synthesis of nitrogenous bases. [1] These are key components of the nucleotide subunits that make up DNA.
   (b) (i) All the first-generation DNA molecules contained ^{15}N because this was the only source of nitrogen available to these bacteria in the culture medium. [1] Therefore, when DNA was replicated semi-conservatively by the parent to make the DNA of the offspring of the first generation, all DNA contained this isotope. [1]
       (ii) Only half of the first-generation DNA molecules contained ^{15}N because when the half-^{14}N and half-^{15}N molecules produced in the first generation were replicated, [1] half of them obtained a new strand containing the ^{15}N nucleotides. [1]

   (c) This investigation provided strong evidence that DNA replication is semi-conservative because with each new generation, only half of the DNA strands containing ^{14}N were retained in the new DNA molecules. [1] In semi-conservative replication, each strand of DNA acts as a template (for the synthesis of a complementary strand). This means that each new DNA molecule has one old / parental / original, strand and one new / daughter strand. [1]
   (d) The purpose of single-strand binding proteins is to prevent the complementary bases on the two strands of the separated DNA molecule from pairing again. [1]

E3. (a) Sulfur is present in proteins but not DNA, [1] and phosphorus is present in nucleic acids (such as DNA) and not proteins. [1] Therefore, only the material containing phosphorus (i.e. DNA) is transferred to the bacteria. [1]
   (b) This was because proteins, with their almost infinite variety of complex structures, [1] many of which were known about in the 1940s, were thought to be able to store the huge amount of information required to construct and maintain a living organism. It was also known that proteins had a great variety in structure because of the 20 naturally occurring monomers, [1] as opposed to the four nucleotide monomers in DNA.
   (c) The genetic code is universal. [1]

E4. (a) A mutation is a change in the base sequence of a gene. [1]
   (b) The scientists could repeat their investigation and calculate a mean value for the numerical data they obtained. [1]
   (c) *Neurospora* was a good model organism to use in this investigation because it has a short generation time, requires very little attention (nutrition or incubation) and its growth is rapid and obvious. [2 marks maximum]
   (d) Factors that must be standardised in this investigation include:
       • mass of *Neurospora* organism applied to the food source
       • mass of food source provided to *Neurospora*
       • temperature of incubation
       • incubation time. [1 mark maximum]

E5. (a) To transcribe a eukaryotic gene, the DNA must first be dissociated from histone proteins, [1] and the primary RNA transcript will need to undergo splicing to remove introns to make mRNA. [1] Neither of these processes need to occur in prokaryotic transcription.
   (b) There are more possible combinations of base triplets than there are amino acids, [1] so some amino acids are carried by more than one type of tRNA molecule. [1] Therefore, some base substitutions will not change the identity of the amino acid incorporated into the growing polypeptide chain during translation [1] and so a polypeptide with the normal primary structure, able to fold into the correct shape, will be formed. [1]
   (c) X = Methionine; [1] and Y = Glycine. [1]
   (d) The tetracycline molecule prevents translation from taking place. [1] This is because it prevents the anticodon of a tRNA molecule attaching to a codon on the mRNA. [1] This means that the tRNA cannot bring an amino acid to the ribosome [1] and the polypeptide chain cannot be extended. [1]

E6. (a) The polypeptide chain that forms the protein encoded by CFTR is 1,480 amino acids in length. To encode this number of amino acids, $(1480 \times 3) = 4,440$ nucleotide bases would be required in the mRNA sequence. This means that the average number of bases in an exon in the CFTR gene is $(4440 \div 27) = 164$, to 3 significant figures.

(b) (i) The box should be drawn around the bases CTT, as shown:

ATC	ATC	TTT	GGT	GTT
Ile	Ile	Phe	Gly	Val

   (ii) The deletion of CTT, which spans two triplets in the DNA sequence, does not change the identity of other amino acids. This is because the triplet that encodes isoleucine in the normal gene (ATC) is changed to ATT, [1] which also encodes isoleucine, illustrating the degeneracy of the genetic code. [1]

(c) (i) The primary transcript is the molecule of RNA produced during transcription in the nucleus. [1] It is an exact complimentary copy of the base sequence of a gene, so it contains both introns and exons.

   (ii) Introns are removed by the hydrolysis of phosphodiester bonds between the introns and exons, [1] and then the mRNA is produced by joining the exons together by catalysing the formation of new phosphodiester bonds. [1]

   (iii) 4 marks maximum from:

Mutations could cause problems during splicing by preventing the removal of introns from the primary transcript, [1] which would result in base sequences being translated by the ribosome that should not be, [1] or the removal of the exons, [1] which would mean that important base sequences that should be translated are not. [1] The results of these errors would be that polypeptide synthesis would be affected, with effects on the cell and hence disease would result. [1]

CHAPTER 7 – TRANSPORT IN PLANTS

In-text questions

1. C

2. Substances need to be moved from one part of a plant to another. It is usually quicker if the substances move through transport vessels rather than by other processes such as diffusion or osmosis from cell to cell. However, for some small plants, such as mosses, the distances substances need to move are short enough that processes such diffusion or osmosis are sufficient, and specialised transport systems are not needed.

3. On one side cambium cells differentiate to form xylem, and on the other side they differentiate to form phloem.

4. This allows maximum light absorption for photosynthesis by the upper surface which would be reduced by the presence of vascular bundles.

5. The vascular bundles are more prominent on the lower surface and so are more accessible to the aphids here. The sap containing the food that the aphids want flows through the phloem tissue which is on the lower side of the vascular bundles.

6. The xylem tissue has been stained. This is because water travels up the xylem vessels during transpiration. As the water travels up it carries with it the food dye.

7. Lignin strengthens the wall of the xylem vessel and makes it mostly impermeable so water can be transported effectively. The pits allow some water and dissolved mineral ions to leave the xylem vessels to supply the surrounding cells.

8. Phloem sieve tube elements lack many of the usual organelles, which means that they need the companion cells to keep them alive. Xylem vessel elements are no longer living, so have no need for companion cells.

Or:

Companion cells move sucrose into the sieve tube elements and so are essential for the translocation process, whereas movement through xylem vessels is driven by transpiration from the leaves.

9. Substances will travel more quickly through xylem vessels because they are hollow tubes, whereas in phloem sieve tube elements the presence of sieve plates and some organelles will slow down substance movement.

10. B

11. The Casparian strip provides an impermeable barrier to water travelling along the apoplast pathway. This forces the water and its dissolved mineral ions to enter the symplast pathway. As it does so, it passes through the cell surface membrane, which is therefore able to exert some control over what enters.

12. The attraction of water molecules to each other is the cohesion force. The attraction of water molecules to the cellulose cell wall of the xylem vessel elements is the adhesion force. The tension force within the water column is due to the column being pulled up by evaporation while gravity acts downwards.

13. Water enters the root by osmosis passing from an area of higher water potential in the soil to an area of lower water potential in the root. Water moves deeper into the root to the xylem because it continues to move by osmosis from areas of a higher water potential (because those areas are continually gaining water) to areas of lower water potential (because those areas are continually losing water). Later, water moves out of the xylem by osmosis into the leaf mesophyll cells as these have a lower water potential because they have lost water by evaporation.

14. Most plant cells are made rigid by the cell wall and the cell surface membrane (which surrounds all the other cell contents) pressing against each other. This happens in phloem sieve tube elements, but cannot happen in xylem vessel elements, as these contain no cell surface membrane nor other internal cell structures. The secondary cell wall therefore provides additional support to maintain the shape of the xylem vessel element. As part of this maintenance of shape, xylem vessel elements also have to be able to withstand the inward force caused by adhesion to the water molecules carried within the xylem. If the xylem vessel walls were not strong enough, the inward force could cause the xylem vessels to close inwards, as the column of water becomes thinner due to tension. This could prevent or impede movement of water up the xylem vessels.

15. Clear plan diagram of transverse section showing tissues and not individual cells; labelled/annotated to show visible features, e.g. leaf rolled up – to reduce air movement over stomata and create humid conditions to reduce transpiration, pits and hairs – to reduce air movement and create humid conditions, thick cuticle – to reduce evaporation; correct use of pencil and ruler.

16. (a) Carbon dioxide is not an assimilate because it has not been produced in the plant to be used by the plant.
(b) Although carbon dioxide contains carbon, it does not also contain hydrogen.

17. Transpiration is driven by the evaporation of water from leaves which pulls more water up the stem – it cannot push water back down the stem. Translocation moves substances from sources to sinks and can occur in any direction.

18. B

19. It will have no effect on transpiration as this is a passive process. It will, however, stop translocation as this relies on the active transport of hydrogen ions, a process which requires energy from respiration.

20. There will be sucrose present above the ring but not below. As the tree is photosynthesising, sucrose will be transported downwards from the leaves, but will not be able to get past the ring.

Assignment 7.1

A1. As water is lost from the leaves by transpiration, more water is taken up by the shoot. This causes the air bubble to move through the capillary tubing. The quicker the rate of transpiration the quicker the air bubble moves.

A2. If a certain mass of water is lost from the plant by transpiration, the overall mass of the plant and container combined will decrease by the same amount.

A3. The fewer the number of remaining leaves, the lower the distance moved by the air bubble (in an hour). This is because with fewer leaves, less water is lost by transpiration and so less water is taken in by the shoot to replace what has been lost.

A4. They could have repeated the investigation with other shoots and taken mean results.

A5. With 5 leaves the bubble moved 15 mm in 1 hour.

The mean distance moved per leaf was $\frac{15}{5}$ = 3 mm per hour.
A distance of 3 mm is equivalent to a volume of water of
$\pi r^2 h = \pi \times 1^2 \times 3$ mm^3
     = 9.426 mm^3
Rate of water uptake per minute = 9.426 mm^3 ÷ 60
                  = 0.157 mm^3 per minute

A6. The volume of water taken in might not be exactly the same as that lost in transpiration. This could be, for example, because some is used in photosynthesis, some might be used in growth (cell elongation), and some might be produced in respiration.

Experimental skills 7.1

P1. (a) Clear biological diagram, labelled to show visible features, e.g. cell sieve plate, sieve pores, phloem sieve tube elements, cytoplasm. Correct use of pencil and ruler.
   (b) A longitudinal section because it goes along the length of the stem. If it was a transverse section, the sieve plate would be seen as approximately circular with all the pores visible.
   (c) The apparent diameter of a pore will depend on where the section is taken: a section through the centre of a pore will appear wider than a section which only just goes through the edge of a pore.

P2. (a) Clear biological diagram, labelled to show visible features, e.g. secondary cell wall, / lignin pits. Correct use of pencil and ruler.
   (b) A scanning electron microscope because the image is 'three-dimensional'.

P3. The plan diagram shows eight vascular bundles but the photomicrograph shows more. The plan diagram shows a perfectly circular cross section, unlike the photomicrograph. The sizes and shapes of the vascular bundles in the plan diagram are not the same as in the photomicrograph. The labels for phloem and xylem are the wrong way around.

P4. Clear plan diagram showing tissues and not individual cells, labelled to show visible features, e.g. xylem, phloem, endodermis, epidermis. Correct use of pencil and ruler.

Chapter review

1. A

2. D

3. B

4. When a liquid moves along a vessel.

5. (a) The xylem and phloem are at the centre of the root, with the xylem at the very centre.
   (b) Xylem and phloem are together in vascular bundles arranged in a 'cylindrical' pattern close to the surface. The phloem is closer to the surface than the xylem.
   (c) Xylem and phloem are together in vascular bundles or 'veins' which are usually closer to the lower surface of the leaf. The phloem is closer to the lower surface than the xylem.

6. (a) Any two from:

   Lignified secondary cell wall – to maintain tube structure

   Pits – to allow passage of some water/mineral ions through the cell wall

   Hollow – to allow unrestricted movement of water/mineral ions.
   (b) Sieve plates – to allow movement from one sieve tube element to the next.
   Lack of organelles – to reduce restrictions to movement of organic compounds.

7. To connect the cytoplasm of one cell to the next, to allow movement of substances along the symplast pathway.

8. Transpiration pull moves water columns along xylem vessels. Cohesion-tension mechanism by which cohesion and adhesion forces overcome the tension caused by gravity. Water moves by osmosis from xylem into leaf mesophyll cells. Transpiration pull caused by evaporation of water from mesophyll cells into internal air spaces. Diffusion of water vapour along a concentration gradient into the atmosphere.

Practice exam-style questions

E1. B

E2. C

E3. B

E4. All trees show a decrease in trunk circumference during the middle of the day. [1]

The middle of the day is when transpiration is at its greatest. [1]

As water moves rapidly through the xylem vessels adhesion forces cause xylem vessels to narrow. [1]

The greatest decrease is in A and the least change in C. [1]

This is because transpiration occurs most rapidly (in the middle of the day) in A and least in C. [1]

C is best able to reduce the rate of transpiration (in the middle of the day) / A is least able to reduce the rate of transpiration. [1]

E5. (a) (i) Light microscope. [1]
          Because if it was an electron microscope the magnification would be much greater. [1]
       (ii) The 'vein' / vascular bundle / vessels would appear as (a) long narrow structure(s). [1]

          Because a longitudinal section would run along the length of the leaf / 'vein'. [1]
       (iii) Clear plan diagram showing tissues and not individual cells. [1]
          Xylem and phloem correctly identified and labelled. [1]
          Suitable title, correct use of pencil and ruler. [1]
   (b) (i) For *Camellia sinensis* the percentage of stomatal opening is highest in the middle of the night and lowest in the middle of the day. [1]
          This is the opposite of the typical herbaceous dicotyledonous plants. [1]

(ii) (The pattern of opening) reduces water losses by transpiration. [1]
Because stomata are mostly open when it is cooler / are mostly closed when it is warmer. [1]

E6. (a) (i) The percentage of ^{14}C decreased most rapidly in the first (approx.) 1.4 hours, from 100% to (approx.) 66%. [1]
It then decreased steadily at a slower rate for the remaining time, to (approx.) 56% after 11 hours. [1]

(ii) The ^{14}C was being assimilated into sugars and transported away from the leaves. [1]

(b) (i) Group A: (100 − 56) ÷ 11 = 4.0% per hour [1]
Group B: (100 − 65) ÷ 11 = 3.2% per hour [1]

(ii) In group B the fruit had been removed, but in the group A plants the fruit provide an additional sink. [1]
So, in the group A plants sugars are moved away from the leaves more quickly than in group B. [1]

(iii) Investigate for the presence of radioactivity in the fruit in group A. [1]
If the fruit are a sink, then there will be radioactivity present. [1]

(c) Proton pumps actively transport hydrogen ions out of companion cells. [1]
Hydrogen ions together with sugar / sucrose diffuse into companion cells through cotransporter proteins. [1]
Sugar / sucrose diffuses into phloem sieve tube elements, lowering water potential, causing water to enter phloem sieve tube elements. [1]
Hydrostatic pressure increases in source causing movement through phloem sieve tubes away from source. [1]
Sugar / sucrose moves by mass flow from source to sink. [1]
At the sink, sugar / sucrose leaves the sieve tube, causing water to leave, reducing hydrostatic pressure. [1]

CHAPTER 8 – TRANSPORT IN MAMMALS

In-text questions

1. (a) Closed means that blood circulates within blood vessels (arteries capillaries and veins).
(b) Double means there are two circuits – pulmonary and systemic; blood passes twice through the heart in one complete circulation.

2. aorta and pulmonary artery

3. (a) capillaries
(b) Arteries have a pulse whereas veins do not. Blood in arteries surges because of ventricle contractions, putting the blood under fluctuating pressure.

4. aorta, artery supplying the kidney, capillary in the lungs, vena cava

5. In a mammal, blood flows from the heart to the lungs then back to the heart. The path, in a mammal, is heart lungs heart body heart

6. Arteries carry blood at higher pressure (than veins) so their walls are thicker to withstand the higher pressure. Arteries contain more elastic fibres and smooth muscle (than veins)).

7. Capillaries have thin walls to bring blood close to tissues to minimise the diffusion distance. They have a narrow lumen to slow blood flow and to bring red blood cells close to tissues; and fenestrations to speed up diffusion and to allow formation of tissue fluid.

8. Valves are needed in veins not arteries because pressure is much lower in veins than arteries; arteries are closer to the heart which provides pressure for forward flow; blood is more likely to flow backwards in veins.

9. B

10. (a) biconcave disc
(b) The biconcave disc shape increases the surface area to volume ratio of the red blood cell, allowing for faster diffusion of oxygen.

11. number in $1dm^3 = \dfrac{(3 \times 10^{13})}{5}$
$= 6 \times 10^{12}$

number in $1cm^3 = \dfrac{(6 \times 10^{12})}{1000}$
$= 6 \times 10^9$

or $\dfrac{(30 \times 10^{12})}{(5 \times 10^3)}$

$= 6 \times 10^9$

12. (a) Monocyte should be drawn as round but with a slightly irregular shape and have a large kidney-shaped nucleus.
(b) Neutrophil should be drawn as round but with a slightly irregular shape and have a large nucleus with 3, 4 or 5 lobes.
(c) Lymphocyte should be drawn with a rounded shape and have a large round nucleus almost filling the cytoplasm.

13. Any two from: plasma is part of blood and tissue fluid surrounds cells; plasma is contained within blood vessels and tissue fluid is outside blood vessels; plasma has a higher protein content than tissue fluid.

14. (a) Any two from: glucose, amino acids, fatty acids, lipids, glycerol.
(b) carbon dioxide and urea

15. Tissue fluid is formed from plasma. Liquid is forced out of capillaries under hydrostatic pressure. The movement of liquid out of capillaries is greater at the arterial end. The liquid moves through intercellular gaps/clefts; water and small solutes move out but red blood cells and large proteins remain in the blood and reduce its water potential. As hydrostatic pressure drops across the capillary bed the osmotic pressure becomes more significant and water moves down the water potential gradient at the venous end, from the tissue into the plasma.

16. D

17. The first oxygen molecule binding causes a conformational change in haemoglobin; this make it easier for the next oxygen molecule to bind. The same occurs for subsequent oxygen molecules. The graph would be a straight line if haemoglobin had equal affinity for each oxygen / if it was equally easy for each oxygen to bind.

18. (a) oxyhaemoglobin
(b) haemoglobinic acid

19. (a) (i) 57.5% (in range 57 – 58%)
(ii) 25%
(b) The affinity for oxygen is lower at higher partial pressure of carbon dioxide; oxygen is dissociated more easily at higher carbon dioxide partial pressures; oxygen is dissociated at respiring tissues.

20. (a) carbon dioxide + water → carbonic acid
accept symbols if correct and accept reversible reaction symbol
(b) The chloride shift describes when chloride ions (Cl⁻) enter red blood cells when hydrogencarbonate / HCO_3^- leave, to maintain a neutral charge in the cytoplasm. An equal number of chloride ions enter as hydrogencarbonate ions leave.

21. (a) atria / left atrium and right atrium
(b) septum
(c) atrioventricular valves / bicuspid and tricuspid valves
(d) pulmonary artery

22. A

23. The sinoatrial node (SAN) initiates the cardiac cycle; it sends out regular electrical impulses.

24. time for one cycle = 0.70 – 0.75s

$$\frac{60}{0.70} = 85.7$$

$$\frac{60}{0.75} = 80$$

answer in range 80 – 85.7 beats per minute

25. (a) $\frac{70}{1000} = 0.070 dm^3$ in each beat

0.070 x 95

= 6.65 *or* 6.7dm³

(b) $\frac{5.1}{0.070}$

= 73

Assignment 8.1

A1. a – pulmonary artery
c – right atrium
e – right ventricle
g – left atrioventricular valve
I – right ventricle
k – right atrium
m - aorta
o – left atrium

b – aorta
d – semilunar valve
f – left ventricle
h – left ventricle
j – right atrioventricular valve
l – vena cava
n – pulmonary veins

A2. $\frac{120}{0.7}$

= 171cm³

A3. (a) stroke volume is greater for the trained person than the untrained person at all heart rates; stroke volume increases with heart rate for the trained person; stroke volume remains almost constant with increasing heart rate for the untrained person; comparative data quote using both lines and values from both axes.

(b) (i) trained person:
CO = SV x HR
= 185 x 200
= 37 000 cm³ min⁻¹
= 37 (l min⁻¹)

(ii) untrained person:
CO = SV x HR
= 122 x 200
= 24 400 cm³ min⁻¹
= 24.4 (l min⁻¹)

(iii) more blood is pumped per minute when exercising; more oxygen can be delivered to muscles per minute; more carbon dioxide can be removed from muscles per minute; aerobic respiration can keep going for longer

Experimental skills 8.1

P1. $\frac{(9+6+11+8+9)}{5}$

= 9 *answer to be given as a whole number*

P2. (a) 0.05 x 0.05 x 0.1
= 2.5 x 10⁻⁴ mm³

(b) $\frac{2.5 \times 10^{-4}}{1000}$
= 2.5 x 10⁻⁷ ml

P3. (value from P1 / (value from P2(b))

$\frac{8}{2.5 \times 10^{-7}}$

= 3.2 x 10⁷

P4. (value from P3) x 200
3.2 x 10⁷ x 200
= 6.4 x 10⁹

P5. (a) there would be too many cells to count if not diluted

(b) Distilled water has a water potential of (close to) zero. The water potential of a red blood cell cytoplasm is lower than that of distilled water, so water would enter the red blood cells by osmosis causing bursting / lysis; therefore, cells would not be visible to count.

P6. The cells would be transparent (without a stain)). The shape of the nucleus is used to identify the white blood cells; the nucleus would only be clearly visible with a stain.

P7. Any one from: wash hands after use; wear gloves; cover cuts; wipe spills with disinfectant.

Chapter review

1. closed means blood flows within vessels (arteries capillaries and veins); double means *either* blood passes twice though the heart in one complete circulation *or* blood is pumped to the pulmonary and systemic parts / lungs and body at the same time

2. (a) The aorta carries a much larger volume of blood to the whole body; as arteries divide, their internal diameter decreases.

(b) The aorta carries blood at a much higher pressure; pressure decreases with distance from the heart and radial artery further away; the aorta needs more elastic fibres to withstand higher pressure.

3. D

4. 3 x 10¹³ need replaced in 120 days
120 days = 120 x 24 x 60 x 60 = 1.0 x 10⁷ s

$\frac{3 \times 10^{13}}{1.0 \times 10^{7}}$

= 3 x 10⁶ or 3 million

5. There is a higher concentration of proteins in plasma, so plasma has a lower water potential (compared with tissue fluid). Water therefore moves (from tissue fluid to plasma) by osmosis through capillary walls / through capillary endothelial cells. Water also moves through gaps/clefts between endothelial cells.

6. (a) sinoatrial node
(b) Purkyne fibres

7. (a) atria contract; blood pressure in atria increases; as pressure becomes greater than in ventricles; atrioventricular valves open; blood flows from atria into ventricles

(b) ventricles contract; blood pressure in ventricles increases; as pressure becomes greater than in atria; atrioventricular valves close as pressure in ventricles become greater than in arteries; semilunar valves open; blood flows from ventricles into arteries

8. In VSD, oxygenated and deoxygenated blood will mix in the ventricles. Blood flowing to the body will have a lower oxygen content; blood flowing to the lungs will be partly oxygenated. Blood pressure flowing to the body (in the aorta will be lower); blood pressure flowing to the lungs (in the pulmonary artery) will be higher.

Practice exam-style questions

E1. A [1]

E2. (a) aorta [1]
(b) neutrophil [1]
(c) (right) semilunar valve [1]
(d) endothelium / endothelial [1]

E3. (a) 1.56 x 10¹⁵ [2]
if answer incorrect or missing, award 1 mark for $2.4 \times 10^8 \times 6.5 \times 10^6$

(b) (i) Any two from: binds oxygen at the lungs; [1] can bind 4 oxygen molecules / 8 oxygen atoms per molecule; [1] forms oxyhaemoglobin; [1] releases / dissociates oxygen at high(er) partial pressure of CO_2 [1]

(ii) Any two from: binds carbon dioxide at respiring tissues; [1] to form carboxy-haemoglobin; [1] CO_2 binds at N-terminal / at $-NH_2$ of polypeptide chain; [1] CO_2 dissociates at lungs / at region of low partial pressure of CO_2 [1]

E4. (a) in the <u>wall</u> of the <u>right</u> atrium [1]

(b) Any four from: wave of electrical activity passing across atria is blocked by insulating tissue from passing to ventricles; [1] atrioventricular node is stimulated and after a short delay; [1] conducts electrical impulse to Purkyne tissue; [1] causing contraction of ventricles; [1] impulse released from Purkyne tissue is delayed after contraction of atria [1]

(c) Any five from: at *a* ventricles are relaxed <u>and</u> fill with blood, increasing pressure; [1] (then) ventricular systole/ ventricles contract; [1] blood pressure in ventricles becomes greater than that in atria; [1] atrioventricular valves forced shut by higher ventricular pressure; [1] blood pressure in ventricles becomes greater than that in arteries / aorta and pulmonary artery; [1] semilunar valves open; [1] blood flows from ventricles into arteries / into aorta and pulmonary artery [1]

E5. (a) (i) partial pressure of oxygen [1]

(ii) percentage saturation of haemoglobin with oxygen [1]

(b) Any two from: temperature; [1] time that each sample is exposed to oxygen; [1] partial pressure of carbon dioxide [1]

(c) graph with partial pressure of oxygen in kPa on horizontal axis and percentage saturation of haemoglobin with oxygen on vertical axis; [1] linear scales with zeros on both axes scaled so points cover at least half the grid in both directions; [1] all points plotted to within 1mm; [1] smooth curve drawn through all points [1]

(d) (i) value read correctly from graph, in range 36 – 38%; [1] value subtracted from 98.0%, in range 60 – 62% [1]

(ii) Any three from: respiring tissue will raise the partial pressure of CO_2; [1] Bohr shift will occur; [1] curve will shift to the left; [1] percentage saturation at 2.7kPa will be lower than on this graph. [1]

CHAPTER 9 – GAS EXCHANGE

In-text questions

1. 11cm x 10
   = 110mm
   time = $\dfrac{distance}{speed}$
   = $\dfrac{110mm}{5mms^{-1}}$
   = 22s

2. Drawings should include the following: unbroken lines and no shading or ruled lines anywhere; cells should not have parallel vertical sides; nuclei to be large and at the bottom; cilia should not be straight; labels for nucleus and cilia; additional organelles should not to be drawn or labelled.

3. trachea, bronchi, bronchioles (terminal bronchioles then respiratory bronchioles), alveoli

4. (a) trachea, bronchi
   (b) trachea, bronchi, (terminal) bronchioles

5. Possible reasons why gas exchange does not occur in the trachea are: the walls are too thick or the diffusion distance is too great; there are not enough capillaries or there is insufficient blood supply.

6. Mucus traps solid particles, dust or pathogens. Cilia beat to move the mucus containing the trapped particles out of airways.

7. Thicker, stickier mucus would be more difficult for cilia to move; if mucus is not moved, bacteria could reproduce in the mucus; mucus could block gas exchange.

8. It widens the airways or increases the diameter of the lumen in the trachea, bronchi and bronchioles. This reduces the resistance to air flow allowing a greater volume of air to pass and the person to breathe more easily.

9. $\dfrac{70m^2}{5\times10^8}$
   = 1.4×10^{-7} m^2

10. (a) Bar chart (vertical or horizontal) with three gases on the same axes. Inhaled and exhaled in pairs for each gas. Bars should not touch, and the gaps between the different gases should be larger than the gaps between each inhaled-exhaled pair. Inhaled and exhaled can be differently shaded and a key used.

    (b) Bar chart is suitable because data are not continuous. A line graph or histogram would be used for continuous data. The percentages in inhaled and exhaled separately add up to 100%, but two pie charts would be required and the question asks for one graph.

11. C

    The percentages of nitrogen in inhaled and exhaled air are the same meaning there is no net movement. Some nitrogen molecules will diffuse, but the rate of diffusion will be equal in both directions.

12. D

    The two cell surface membranes of the squamous epithelial cells lining the alveolus, plus the two cell surface membranes of the capillary epithelial cell, plus the cell surface membrane of the red blood cell, which gives a total of 5.

13. Gas exchange happens by (simple) diffusion. Oxygen from the air diffuses into the blood and carbon dioxide from the blood diffuses into the air. Both gases move down their concentration gradients.

14. The concentration gradients of oxygen and carbon dioxide are maintained by the following: good ventilation; air with a high (21%) concentration of oxygen and a low (0.04%) concentration of carbon dioxide is continually being brought in; air with a low (16%) concentration of oxygen and a high (4%) concentration of carbon dioxide is continually being removed; deoxygenated blood is continually being brought to the alveoli; oxygenated blood is continually being removed from the alveoli.

15. $\dfrac{(21-16)}{21}$ x 100
    = 24%

Assignment 9.1

A1. If cilia are paralysed, mucus cannot be moved (out of the airways), so mucus will sink down to lower airways.

A2. Smokers may cough more frequently to remove the mucus, as the cilia are paralysed and so cannot move the mucus.

A3. Smokers are more likely to have infectious diseases of the lungs because pathogens/bacteria/viruses become trapped in mucus, which then cannot be removed. This could lead to bacteria multiplying in the mucus. Also, pathogens have longer to infect cells or pass into the blood or tissues.

A4. (a) Without elastic fibres, alveoli will not be able to recoil after expanding. Air is therefore not expelled from alveoli or excess air remains in alveoli during exhalation. This will cause a reduced concentration gradient of oxygen and carbon dioxide, which will reduce the rate of diffusion of gases.

(b) Fewer, larger air spaces will have a lower surface area for gas exchange. This will reduce the overall rate of diffusion and gas exchange will become less efficient.

(c) Supplying air with 80% oxygen will increase the concentration gradient of oxygen between alveoli and blood, which will speed up diffusion. In people with severe emphysema, diffusion is too slow with normal (21% oxygen) air due to reduced efficiency of gas exchange. The higher percentage of oxygen will also increase the partial pressure of oxygen in the lungs, meaning an increased chance that haemoglobin is (almost) fully saturated.

A5. (a) Those who stopped smoking had survival rates similar to those who had never smoked across all age ranges. The smokers had lower survival rates, with the biggest difference in survival rates being between the ages of 60 and 90. A comparative data quote should always be included. For example, at age 70 the survival rates for smokers was 55%, while that for non-smokers and those who had stopped smoking was 82%.

(b) The study only involved men from one profession and in one country. Other factors that could have contributed to lower survival rates, such as diet and genetics, were not taken into account.

Experimental skills 9.1

P1. (a) Diagrams should include the following: unbroken lines at tissue boundaries and no shading or stippling anywhere; no individual cells and no ruled lines; four clear layers; cartilage as the thickest layer and smooth muscle approximately the same thickness as the epithelium.

(b) Labels added as follows: epithelium label pointing to top layer; cartilage label pointing to thickest layer; smooth muscle label pointing to bottom layer.

(c) < I can't give the answer to this until we know the final image size. magnification is x25 when image is 10cm tall>

(d) 1 soak the tissue in preservative to prevent decay

2 soak the tissue in paraffin wax to make the tissue easier to cut into sections

3 cut a thin section from the tissue

4 soak the tissue in a suitable stain

5 place the section onto a slide and add a coverslip

(e) To make cells or tissues visible as they would be almost transparent otherwise.
To differentiate between different tissue types, as in this image cartilage is blue and smooth muscle is pink.

Chapter review

1. B – only the trachea and bronchi contain cartilage in their walls.

2. A – goblet cells do not have cilia.

3. (respiratory bronchioles then terminal bronchioles) bronchioles, bronchi, trachea

4. (a) goblet cells, mucous glands
(b) Mucus traps dust/pathogens and moistens inhaled air.
(c) The regular/co-ordinated wafting/waving movement of cilia pushes mucus upwards.

5. When relaxed, the smooth muscles increase the diameter of the lumen of the bronchioles. This increases the air flow (or decreases

the air resistance) so that more oxygen can be brought to alveoli and more carbon dioxide can be removed. More oxygen will be required by muscles and more carbon dioxide will be produced by muscles.

6. The plan drawing should have: unbroken lines at tissue boundaries and no shading or stippling anywhere; no individual cells and no ruled lines; three clear layers including cartilage; the thickness of each layer approximately in proportion to the photograph; ciliated epithelium, lumen and smooth muscle all labelled.

7. A sample drawing is given below. Drawings should show: outline of the cell with width less than height; outline of cilia rather than cilia as single lines; large nucleus in lower part of the cell,

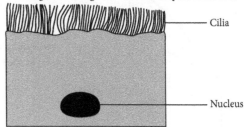

8. (a) Elastic fibres stretch to allow alveoli to expand; this expanding increases their surface area for gas exchange. The elastic fibres prevent the alveoli from bursting and recoil to help expel air.

(b) Answers may include the following adaptations of alveoli: large surface area; squamous epithelial cells, which provide a thin wall for a short diffusion distance; alveoli are in close contact with capillaries; a large number of capillaries surround alveoli.

Practice exam-style questions

E1. D [1]

E2. (a) Answers may include the following (1 mark for each structure linked with a function; maximum 2 marks):
wide lumen or wide diameter – for unobstructed air flow [1]
rings of cartilage – to prevent walls collapsing or to prevent trachea vessel from bursting [1]
contains smooth muscle – to relax and increase lumen diameter during exercise [1]

(b) (i) A and C [1]
(ii) Answers may include the following (maximum of 3 marks for mucus; maximum of 4 marks in total):

Mucus is a sticky liquid; [1] that traps dust/particulates and microorganisms/pathogens. [1] Mucus forms a barrier for pathogens [1] and increases the distance pathogens must travel to reach epithelial cells. [1] Mucus production increases during infection. [1] Cilia move mucus [1] in a co-ordinated way; [1] mucus is pushed to throat to be swallowed. [1]

E3. (a) trachea   [1]
(b) (i) cartilage   [1]
(ii) smooth muscle   [1]
(iii) ciliated (epithelium)   [1]
(c) Answers may include the following (maximum of 3 marks):
cilia move mucus; [1] mucus cannot be moved (when cilia have been lost); [1]
presence of bacteria or infection will cause excess mucus to be secreted; [1]
mucus will sink into lower airways/bronchioles/alveoli;   [1]
coughing needed to remove mucus or bring mucus up out of airways. [1]

E4. (a) (i) goblet cells [1]
mucous glands [1]
(ii) so mucus has the same water potential as cells; [1] so there is no <u>net</u> movement of water in or out of cells (due to mucus) [1] by osmosis. [1]

(b) Any two from: macrophages are phagocytes / phagocytic; [1] they engulf [1] pathogens/bacteria/fungal spores; [1] they engulf dust/particulates. [1]

E5. (a) (i)

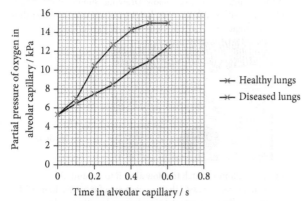

The graph should have: axes with linear scales and labelled with quantities and units; [1] scales on both axes allow points to cover at least half the available space; [1] all points plotted correctly for both lines; [1] smooth curves of best fit *or* ruled lines between points *and* both lines labelled *or* a key used. [1]

(ii) <u>Net</u> movement of oxygen stops (note that it is not correct to say that diffusion stops). [1] Partial pressure / concentration of oxygen is equal in alveolar air and blood. [1]

(b) (i) Any two from: They should select lowest power or lowest magnification objective (lens) [1] and bring the slide into focus [1] by moving the stage (or slide) away from the objective (lens). [1]

(ii) The drawing should: have unbroken lines with no shading anywhere and diameter at least 6cm; [1] show no individual cells and outlines should not be perfect circles; [1] include clear layers and clear separate blocks of cartilage; [1] smooth muscle and cartilage layers to be approximately equal thickness. [1] Labels (maximum 2 marks for any two or three correct labels from): cartilage; [1] smooth muscle; [1] ciliated epithelium; [1] lumen. [1]

(iii) Formula: image size = actual size × magnification OR I = AM [1]
values can be given when final image size is known [1]
final answer can be given when final image size is known [1]

CHAPTER 10 – INFECTIOUS DISEASE

In-text questions

1. *Vibrio cholerae* (not *cholera*)

2. Diarrhoea causes dehydration.

3. Reasons why cholera is most prevalent in less-industrialised countries: lack of sewage treatment; lack of drinking water treatment; higher chance of drinking water being contaminated with the pathogen; lack of education about mode of transmission.

4. Mostly transmitted in the tropics. Almost no transmission outside of the tropics or in temperate climates. Malaria is referred to as a tropical disease because it has the highest incidence or is endemic in tropical countries; the vector – the *Anopheles* mosquito – is only found in the tropics.

5. Precautions include: covering the skin when mosquitoes are active; using insect repellent; sleeping under a mosquito net or using insect screens; taking malaria prophylaxis / preventative drugs.

6. Difficulties include: there are many stages of the parasite life-cycle; many antigens; antigenic concealment.

7. An infectious disease is a disease caused by a pathogen that can be transmitted from an infected person to uninfected people.

8. TB is transmitted by an infected person coughing, spitting, sneezing or exhaling the bacteria / pathogen *Mycobacterium tuberculosis* in the droplets. The droplets (containing the pathogen) are then inhaled by an uninfected person.

9. Many cases of TB will not be diagnosed and many cases will not be reported, making the numbers much higher than published numbers.

10. Death rate from TB in UK steadily decreased from 1900, possibly due to less overcrowding in living conditions. First treatment discovered in early 1940s. Around 4 years after this, death rate decreased more steeply. Immunisation introduced in the early 1950s but did not cause death rate to fall any more steeply than treatment. By late 1950s death rate was starting to fall more slowly and levelled off at about 2000 per year in late 1980s. Death rate started to rise in 1990s due to antibiotic resistance.

11. human immunodeficiency virus

12. $\text{magnification of image} = \dfrac{\text{length of magnified scale bar}}{\text{length of actual scale bar}}$
$= \dfrac{9\,\text{mm}}{120\,\text{nm}}$
$= \dfrac{9 \times 10^6\,\text{nm}}{120\,\text{nm}}$
$= \text{x75000}$

13. Suggestions may include: using barrier contraceptives / condoms / femidoms; only have one sexual partner / refraining from sex; using sterile needles / refraining from injecting drugs; having regular HIV tests.

14. HIV infects T-helper cells; people with HIV / AIDS have low numbers of T-helper cells; people with HIV / AIDS are therefore more susceptible to infection by other pathogens making them vulnerable to opportunistic infection / pathogens that do not normally cause serious disease.

15. (a) An antibiotic is a substance produced by an organism / fungus that kills or inhibits the growth of microorganisms / bacteria.
(b) Viruses have no metabolic processes to inhibit; viruses use the metabolic processes and machinery of the cell to replicate.

16. B

17. Some bacteria in a population will undergo random mutations or there will be genetic variation in bacteria. This is part of evolution. Some of these bacteria will become naturally resistant to an antibiotic. Through natural selection the resistant bacteria survive and multiply leading to a strain that is resistant to the antibiotic.

18. Bacterial diseases are becoming difficult or impossible to treat, which is leading to increased death rates from bacterial diseases. This is a burden on the healthcare system and results in increased cost of healthcare. It also means there is a requirement for use of more powerful antibiotics with more serious side effects.

Assignment 10.1

A1. An opportunistic pathogen does not normally cause disease but it will cause disease if the conditions are correct e.g. weakened immune system.

A2. The graph shows the figures as a proportion / similar to a percentage / similar to a fraction, which makes the figures comparable (between the different years). Also, the population could change over the period shown.

A3. There is an overall increase between 2001 and 2008 in the number of people with *C. difficile* being treated with antibiotic A; the number is roughly constant between 2001 and 2003; there's a decrease in 2004, followed by a sharp increase after 2004 / from 2005 onwards.

A4. Between 2001 and 2002 / 2003 death rates increase as antibiotic A use remains constant; this suggests development of antibiotic resistance in *C. difficile*. Between 2003 and 2004 death rates increase as antibiotic A use decreases; this suggests deaths related to lack of effective antibiotic treatment. Between 2005 and 2008 death rates increase as antibiotic A use also increases; this suggests an increase in *C. difficile* cases.

A5. (a) $\left[\dfrac{(9.80 - 3.38)}{3.38}\right] \times 100$
$= 190\%$

(b)

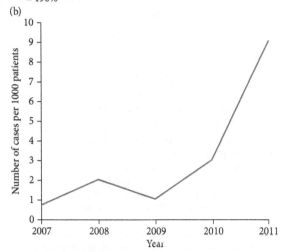

year	number of cases per 1000 patients
2007	0.75
2008	2.35
2009	1.61
2010	3.38
2011	9.80

(c) There is an overall increase in cases between 2007 and 2011; a slight decrease in 2009; a sharp increase between 2010 and 2011.

Experimental skills 10.1

P1. (a) Disc W is a control to demonstrate that neither the water used to dissolve the antibiotics nor the filter paper is not causing a clear zone

(b) diameter measured from Figure 00 to be $\dfrac{1.0\text{cm}}{10\text{mm}}$

magnification of image $= \dfrac{0.5}{\text{image is} 1 / 2 \text{size}}$

so actual diameter $= \dfrac{2.0\text{cm}}{20\text{mm}}$

(c) Antibiotic C is the most effective against this type of bacteria because the concentration of antibiotic (in the agar) decreases further from the disc / the clear zone is largest showing antibiotic is active at low concentrations.

(d) The clear zone would appear in both cases. The clear zone would look the same whether bacteria were killed or inhibited from growing.

(e) Variables that should be kept the same include: same concentration of antibiotics in the discs; same size / diameter / thickness of discs; same incubation temperature; same incubation time; same type of agar.

P2. (a) so the concentration is not affected by contamination from the previous solution

(b) Method:

1. Remove 5cm³ of the 0.025% solution and add to 5cm³ of ethanol to make a 0.00125% solution = 1.25 x 10⁻³%

2. Use separate pipettes for the antibiotic solution and for the ethanol. Mix the new dilution thoroughly.

3. Remove 5cm³ of the 0.00125% solution and add to 5cm³ of ethanol to make a 0.000625% solution, again using separate pipettes. = 6.25 x 10⁻⁴%

P3. Remove 1cm³ from the 1% solution and add to 9cm³ (distilled) water. Mix thoroughly. Use a pipette to measure the volumes. Use a separate pipette for the water and for the antibiotic solution. Mix thoroughly. This dilution is 0.1% (w / v). Remove 1cm³ of this solution using a new / clean pipette, add to another 9cm³ water and mix. This dilution is 0.01% (w / v). Repeat this procedure to make a solution of 0.001% (w / v) concentration.

Chapter review

1. C

2. (a) In refugee camps there is likely to be poor sanitation / no proper sewage facility; lack of drinking water treatment, leading to a high likelihood of food or drinking water becoming contaminated with human waste containing the cholera pathogen.

(b) Wash hands thoroughly after passing faeces and before handling food; avoid raw or undercooked or unwashed foods; drink bottled, boiled or treated water; ensure cooking and eating utensils are washed thoroughly.

3. (a) *Mycobacterium tuberculosis* and *Mycobacterium bovis*

(b) Drink and eat only sterilised or pasteurised milk and other dairy products; perform regular checks on cattle to detect infection; treat or remove any infected cattle.

4. (a) Any two from: *Plasmodium vivax*, *Plasmodium malariae*, *Plasmodium falciparum*, *Plasmodium ovale*

(b) A female *Anopheles* mosquito acts as a vector and takes a blood meal / feeds on blood from an infected person; the mosquito takes in the gametes of the parasite which then reproduces in the mosquito; the parasite is then present in the mosquito saliva and is introduced into an uninfected person when the mosquito takes blood meal.

(c) Either drain or fill in the lake; coat the water surface with oil so that mosquito larvae cannot access oxygen; stock the lake with fish that eat mosquito larvae; spray the area with insecticide.

5. (a) human immunodeficiency virus

(b) HIV / AIDS is caused by a virus. Antibiotics are only effective against bacteria – they are not effective against viruses.

Viruses have no metabolic pathways or no specific machinery to target.

6. C

7. (a) Penicillin contains beta lactam ring structure, which acts as a competitive inhibitor of (DD) transpeptidase enzyme that makes cross-links in peptidoglycan. The rate of breaking cross-links in peptidoglycan is greater than the rate of making cross-links, leading to the bacterial cell walls becoming weakened and the bacterial cell is killed.

   (b) In the mid-20th century, antibiotics were discovered and made widely available. Many bacterial diseases that could previously not be cured could then easily be cured with antibiotics.

   (c) Antibiotic resistance is increasing and affecting more bacterial diseases. Therefore, diseases that were easily cured could become impossible to cure again.

8. (a) By passing the resistance gene to other non-resistant bacterial cells via a plasmid; by non-resistant bacteria undergoing random mutation or the same mutation as the resistant ones; by continued use of the antibiotic as the non-resistant ones will be killed leaving only resistant ones to multiply.

   (b) To reduce antibiotic resistance: use antibiotics more sparingly or only as a last resort or only under medical supervision; use antibiotics in combinations; ensure people complete the course of antibiotics; isolate patients infected with antibiotic resistant forms of bacteria; stop routine use of antibiotics in healthy farm animals.

Practice exam-style questions

E1. D [1]

E2. B [1]

E3. (a) Via organ transplant / tissue transplant / blood transfusion; [1] from mother to unborn baby / through breastfeeding / during childbirth [1]

   (b) Any five from: survival rates with antiviral drugs are higher; [1] comparative data quote to support; [1] TB is an opportunistic disease; [1] HIV / AIDS weakens the immune system; [1] patients with HIV / AIDS more susceptible to TB; [1] antiviral drugs slow the progression of HIV / AIDS; [1] those receiving antiviral drugs have more T-helper cells / CD4+ lymphocytes [1]

E4. (a) Any three from: (penicillin) is a competitive inhibitor of an enzyme / transpeptidase; [1] inhibits peptidoglycan synthesis / cross-linking; [1]
   enzyme that is inhibited is not found in human cells; [1] peptidoglycan cross-links are not formed in human cells. [1]

   (b) (i) It changes the three-dimensional shape of the penicillin molecule; [1] so it no longer fits in the active site of the target enzyme; [1] and therefore no longer inhibits cell wall synthesis. [1]

      (ii) clavulanic acid acts as an inhibitor of beta-lactamase; [1] it occupies the active site of beta-lactamase so penicillin cannot bind. [1]

E5. (a) Any three from: the number of cases will vary throughout the year; [1] due to effect of immigration / emigration / deaths / new cases; [1] some cases not diagnosed; [1] some cases not reported. [1]

   (b) Any two from: the populations of the countries are different; [1] an example given e.g. population of India much higher than South Africa; [1] the results are not proportions / percentages / fractions. [1]

(c) (i) India: $\left(\dfrac{2.2 \times 10^6}{1.3 \times 10^9}\right) \times 100 = 0.17\%$

   South Africa: $\dfrac{5.3 \times 10^5}{5.3 \times 10^7} \times 100 = 1.0\%$

   Indonesia: $\left(\dfrac{4.6 \times 10^5}{2.5 \times 10^8}\right) \times 100 = 0.18\%$

   Vietnam: $\left(\dfrac{1.3 \times 10^5}{9.0 \times 10^7}\right) \times 100 = 0.14\%$

   all 4 correct to 2 s.f. = 3 marks; 3 correct to 2 s.f. = 2 marks; 2 correct to 2 s.f. = 1 mark; all 4 correct to any other no. of s.f. = 2 marks; 3 correct to any other no. of s.f. = 1 mark

   (ii) bar graph with axes scaled so more than half the grid is used; [1] bars can be horizontal or vertical but should be to correct values (as calculated in (i)) and not touching; [1] countries labelled on axis in centre of bars or inside or at ends of bars [1]

(d) Any four from: Figure E1 shows total numbers of cases; [1] graph from (c) (ii) shows prevalence; [1] total number of cases per country could be used for e.g. quantities of antibiotics required in that country; [1] prevalence could be used for e.g. to gauge relative seriousness of the disease in country; [1] comparison of results e.g. India has higher total number of cases than South Africa but South Africa has much higher prevalence / India has higher number of cases than Indonesia but the prevalence is about the same. [1]

CHAPTER 11 – IMMUNITY

In-text questions

1. The origin of all leucocytes is the bone marrow.

2. Non-self antigens are protein or glycoprotein molecules that are found in the cell surface membrane of foreign cells (cells that are not usually found in the body).

3. (a) Neutrophils are cells that have a lobed nucleus, whereas macrophages form from monocytes that have a large, kidney-shaped nucleus. Neutrophils are found mainly in the blood, but can leave the capillaries and enter tissue fluid. Macrophages are found in lymph, tissue fluid and lungs, where they kill microbes before they enter the blood.

   (b) After being engulfed by a phagocyte, a pathogen will be enclosed in a phagosome, which is a small vacuole consisting of a phospholipid bilayer surrounding the pathogen. Soon afterwards, a lysosome containing hydrolytic enzymes such as lysozyme will fuse with the phagosome, to form a phagolysosome. The pathogen is then digested.

   (c) A lysosome is a small vacuole consisting of a phospholipid bilayer surrounding hydrolytic enzymes such as lysozyme.

4. (a) A

   (b)

5. The typhoid bacteria may have developed a mechanism to prevent phagosome formation, or the fusion of a lysosome with the phagosome. Alternatively, they may be able to withstand attack by hydrolytic enzymes, perhaps through the use of enzyme inhibitors or a thick slime capsule.

6. (a) Actual size of cell = measurement of size of image of cell / magnification (3000)
   Actual size of cell is estimated to be 5-7 μm, depending on the measurement chosen.
   (b) Plasma cells produce large numbers of antibodies, which are proteins. These are made by ribosomes, which are often found attached to the rough endoplasmic reticulum. Mitochondria, which produce ATP, are also required to produce high numbers of proteins, because protein synthesis is ATP-dependent.

7. The chemicals used to destroy bone marrow stem cells will reduce the number of white blood cells (leucocytes) that are formed. Therefore, a patient being treated for leukaemia would have fewer leucocytes to start an immune response and would be susceptible to a pathogen causing disease.

8. When the receptor of a specific T-cell attaches to an antigen, it is activated and undergoes rapid mitosis. The daughter cells of this process will be clones, and will produce the same receptors that are placed in the cell surface membrane.

9. (a) Some pathogens, such as viruses, do not have cell surface membranes, and would therefore not be affected by the cytotoxic T-cells. In addition, some other pathogens may have a cell wall or capsule that reduces the effectiveness of this mechanism.
   (b) If holes are made in the cell surface membrane, there will be a rapid intake of water from the surroundings by osmosis. This is because the water potential inside the cell is usually lower than outside.
   (c) Cancer cells are formed from normal body cells. Because they are likely to have similar or identical antigens to normal body cells, immune cells are unlikely to be able to recognise them.

10. (a) In the humoral response, macrophages will ingest pathogens by phagocytosis. They then place fragments of the partially digested pathogens into their cell surface membranes. These may be recognised as foreign by B-cells, some of which will become plasma cells and secrete antibodies.
    (b) In the cellular response, macrophages will again ingest pathogens by phagocytosis. They then place fragments of the partially digested pathogens into their cell surface membranes. These may be recognised as foreign by T-cells, some of which will become T-helper cells and secrete cytokines (which stimulate other immune cells) and T-killer cells, which will directly kill other pathogens carrying the same antigens.

11. (a) D
    (b) In the diagram, the antigen-presenting cell (APC), which is usually a macrophage, places fragments of the pathogen in its cell surface membrane. Next, the antigen is attached to by the receptor found on the cell surface membrane of a T-helper lymphocyte, which is then stimulated to undergo clonal selection and rapid mitosis to form many cells carrying the same receptors.

12. The B-lymphocytes are obtained from the spleen or lymph nodes as these are the organs in which they are found at greatest number. To obtain B-lymphocytes from the blood, a high volume of blood would be required, and it would be more difficult to obtain the required cells as there are a large number of other cell types present.

13. (a) This duration gives enough time for the primary immune response to occur, during which antibodies complementary to the introduced antigen are produced.
    (b) Some cells may not produce the required antibody, namely the antibody capable of attaching to the antigen that was introduced. Screening for these cells will therefore ensure that only those cells able to bind to the antigen will be cultured, and will not waste time or money in culturing cells unable to do so.
    (c) Growing cells in a fermenter increases their number rapidly, and therefore the number of antibodies they are able to produce.

14. Some monoclonal antibodies are able to specifically attach to cells with particular antigens. If these antigens are only found on a diseased or cancer cell, then medicinal drugs can be attached to these antibodies. These will then expose only those cells to the drug, minimising the effect of the drug on other, non-target cells.

15. (a) Monoclonal antibodies are produced by plasma cells that derived from one B-lymphocyte by clonal selection (by mitosis), and are therefore complementary to only one antigen. However, polyclonal antibodies are produced by plasma cells that derived from more than one B-lymphocyte, and are therefore complementary to more than one antigen.
    (b) An advantage of using monoclonal antibodies would be that it would be very clear that a specific pathogen is present.
    (c) An advantage of using polyclonal antibodies would be that the presence of a number of strains/ varieties of pathogen could be tested for at once.

16. Vaccines usually contain molecules that are easily hydrolysed by the digestive system, including proteins and glycoproteins. Injecting these molecules into the bloodstream would mean that they are able to come into contact with cells of the immune system.

17. Several childhood vaccines are provided in more than one dose, and the doses are separated by a number of years. On each occasion that the vaccines are administered, there is an immune response. The primary immune response produces memory cells, but on each occasion thereafter a greater number of memory cells is produced upon each exposure. This will result in the building up of a high number of memory cells. This means that there are more cells specific for this pathogen in the immune system, which would not need a high quantity of pathogen cells to cause an activation of the secondary immune response / it increases the probability of encountering pathogens more quickly. Hence, the pathogen is killed faster, meaning that it is unlikely that symptoms would develop if the individual is exposed to this pathogen in the future.

18. (a) 1.5 weeks
    (b) 56 arbitrary units
    (c) This is because memory cells specific to this antigen are still present in the body, and will be activated fast enough if the pathogen were to be met again to avoid the development of symptoms.

19. (a) (i) Percentage decrease during 2016 week 5 and week 12 = ((value at week 12 – value at week 5) ÷ value at week 5) × 100. The percentage decrease in the number of cases of Zika virus disease is therefore ((390 – 1750) ÷ 1750) × 100 = 77.7%
       (ii) The Zika virus is transmitted by mosquitoes, so perhaps pregnant women took more care to ensure they were covered by netting when they slept or wore longer garments to avoid being bitten.

(b) The highest number of babies and fetuses being identified with microcephaly was identified in Week 28 of 2016. The highest number of pregnant women with Zika virus disease was identified in Week 5 of 2016. The difference between these dates is 23 weeks, which is less than the duration of pregnancy and suggests that the infection may be related to the physical defects of the fetus. In addition, the rise and fall in the number of women with Zika virus disease appears to be replicated by a rise and fall in the number of babies identified with microcephaly, but around 26 weeks later. However, there are rather large deviations in the number of babies with microcephaly during the study.

(c) This would suggest that the antibodies complementary to (antigens on) the Zika virus may, in some individuals, be of a similar tertiary structure to antigens found in the cell surface membranes of neurones.

Assignment 11.1

A1. Some lymph nodes are found in the region of the neck. After infection with the bacterium that causes diphtheria, the lymph nodes will swell as the leucocytes inside them undergo rapid mitosis.

A2. (a) A

(b) The antibodies could block the toxin from reaching cell surface membranes, which could prevent it from binding to membrane proteins. It may also be unable to enter cells, which would otherwise cause cell damage.

(c) Using modified toxins in vaccines, rather than weakened pathogens, has the benefit that there is no possibility that the vaccine could change and become pathogenic. It is also usually much more stable/long-lasting, because it is not living, unable to divide or spread around the body.

(d) Adjuvants may play a number of roles in toxoid vaccines. Both relate to the importance of maintaining the specific shape of the antigen to ensure a specific immune response, and the production of memory cells required for immunity.
- Modify the tertiary structure of the toxoid, so that it more closely resembles the real toxin.
- Stabilise the tertiary structure of the toxoid, so that it has a longer 'shelf life' and expiry date.
- Protect the antigen against low pH or enzymes in the digestive system (if administered orally (by mouth)).

A3. (a) From 1980 until 1992, there is a general trend that shows a decrease in the number of cases of diphtheria, while the global percentage of children immunised increases during this period. The number of cases falls from 100 000 to 24 000 during this period, while the percentage of children immunised rises from 30% to 82%, and it remains at around this level until the end of the duration shown. Between 1993 and 1995, there is a sharp increase in the number of cases of diphtheria, which peaks at 59 000 in 1995. Thereafter, the number of cases falls to around 10 000–12 000 per year.

(b) If enough people are vaccinated, usually above 90–95%, then the population is said to have herd immunity. This means that there are too few individuals susceptible to the disease for the pathogen to be passed from person to person.

Experimental skills 11.1

P1. The reduction in phagocytosis by macrophages would mean that he removal of pathogens, such as many bacteria, from the lungs would be inefficient. Bacteria are therefore able to divide and form greater numbers, which could lead to reduced lung function and even death.

P2. Two factors that the scientists must standardise in order to make a valid comparison of the data obtained in the three experiments would include:
- The mass / number of phagocyte cells in each sample.
- The mass / number of E. coli cells added to each sample.
- The temperature / other physical factors at which each sample was incubated.

P3. (a) The difference in the mass of E. coli that were phagocytosed in non-smokers was 5.1 µg and in patients with COPD was 2.7 µg. The difference between these values is 2.4 µg.

(b) 2.4 µg is equal to 2400 ng. Assuming there are 1000 bacterial cells in a nanogram, this would suggest that the phagocytes of non-smokers will ingest 2 400 000 extra bacterial cells in this investigation than the phagocytes of patients with COPD. Expressed as standard form, this number is 2.4×10^6.

(c) C

P4. The chemicals in cigarette smoke may attach to the proteins on the surface of cells. This may reduce the ability of phagocytes to recognise pathogens, and therefore engulf them, or they may prevent effective antigen presentation to lymphocytes.

P5. The most reliable data was obtained for the data for patients with COPD. This is shown by the fact that the 95% confidence interval error bar is the narrowest: it extends above and below the mean the smallest distance of the three bars.

P6. The fact that the difference between these two sets of data is significant at $p<0.05$ indicates that there is less than a 0.05 probability that the difference between them is owing to chance alone (only once in 20 occasions). Therefore, the difference is likely to have an underlying scientific basis.

Experimental skills 11.2

P1. (a) Two bands are seen if a woman is pregnant because molecules of hCG are present in her urine. These are attached to by the free antibodies (with coloured substance attached) complementary to hCG. When the antigen–antibody complexes move along the test strip, immobilised antibodies complementary to hCG will attach to them in the first window, which results in the appearance of a coloured strip. Not all of the antigen–antibody complexes will be attached here, and some will move along the strip to the second window, where antibodies complementary to the anti-hCG antibodies are found. These will attach to the antibody portion of the antigen–antibody complex, resulting in the appearance of a second coloured strip.

(b) Only one band is seen if a woman is not pregnant because there are no molecules of hCG present in her urine. The free antibodies move along the test strip, but will not be attached to by the immobilised antibodies complementary to hCG in the first window. Therefore, no coloured strip will appear. The free antibodies will move along the strip to the second window, where antibodies complementary to the anti-hCG antibodies are found. These will attach to the antibody portion of the antigen–antibody complex, resulting in the appearance of a coloured strip.

(c) The immobilised antibodies are used in the second window to show that the test is functional (that the free antibodies have been able to move). If no lines appear after the test is used, then the woman will know that the test is faulty, and she should attempt another one.

P2. (a) The investigation was based on 25 kits of each type. This means that the data obtained would be reliable, because anomalies could be identified.

(b) Factors that the pharmaceutical company would need to standardise in this investigation include:
- The incubation time (the period of time that the test kit was left in order to develop the coloured band).
- The position on the test strip to which the hCG solution was applied.
- The temperature of the surroundings.

(c) This solution was a control. The purpose of the control was to confirm that a solution containing no hCG gives a negative result, confirming that the presence of hCG is responsible for the positive results observed.

P3. (a) Actual error (uncertainty) is equal to ± half the smallest division on the measuring instrument (micropipette). This $(10 \, \mu l \div 2) = 5 \, \mu l$ (0.005 ml).

(b) The percentage error can be calculated using the following formula: percentage error = (uncertainty ÷ value being measured) × 100. In this investigation, the volume being measured is 1 ml (1 cm³). Therefore, percentage error = (0.005 ÷ 1.000) × 100 = 0.5%

P4. (a)

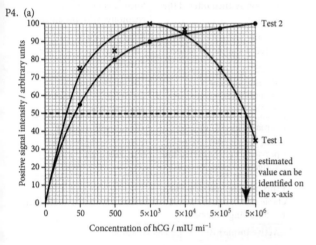

Concentration of hCG / mIU ml⁻¹

(b) As the concentration of hCG in the solution applied to test kit 1 increases, the positive signal intensity of the kit rises steeply from 35.0 arbitrary units at 50 mIU/ml to reach a peak at 5×10^3 miU ml⁻¹ at 98.5%. At concentrations higher than this, the positive signal intensity falls, with an increasing gradient, to reach 69 arbitrary units when the concentration of hCG in the solution is 5×10^6 miU ml⁻¹.

(c) By drawing a line from the 50% value of positive signal intensity and then finding where this line intersects with the plotted line on the x-axis (concentration of hCG), an estimate can be made.

P5. (a) As shown in Figure P3b, if a solution with a very high concentration of hCG (5×10^6 mIU ml⁻¹) is applied to test kit 1, some hCG molecules will bind to the free (anti-hCG) antibodies. However, other hCG molecules will move up the strip and attach to the immobilised antibodies in the first window. This will prevent the immobilised antibodies from attaching to the free antibodies (via hCG) and this will mean that a coloured band does not appear in the first window.

(b) The band in the second window may appear more intense (darker) because the free antibodies that would normally be bound by the immobilised antibodies pass through to the second window, and are attached to by the immobilised antibodies that are complementary to these free antibodies.

P6. The value for the third replicate (79.0 mIU ml⁻¹) appears to be very different to the others, suggesting that it is an anomalous reading. The researchers should exclude this value from the calculation of the mean and/ or carry out another repeat to obtain an 11th replicate.

P7. The table should be completed as follows:

replicate	x	$\bar{x}$	$(x - \bar{x})^2$
1	98.5	95.5	9.00
2	96.0	95.5	0.25
3	88.0	95.5	56.25
4	96.0	95.5	0.25
5	96.0	95.5	0.25
6	95.5	95.5	0.00
7	95.5	95.5	0.00
8	95.0	95.5	0.25
9	99.0	95.5	12.25
10	95.5	95.5	0.00
$\bar{x}$	95.5		
$\sum$			78.5

Next, the formula should be used to calculate the standard deviation,

$$s: s = \sqrt{\frac{\sum (x - \bar{x})^2}{n-1}}$$

So, $s = \sqrt{(78.5 \div (10 - 1))}$ = 2.95 (to 2 decimal places).

P8. (a) The t-test is an appropriate statistical test to use because the data is continuous and the mean values are being compared. The standard deviations are also very similar.

(b) The t-test may not be an appropriate statistical test to use because the collected data does not have a normal distribution.

(c) The t-value can be found by using the following equation:

$$t = \frac{|\bar{x}_1 - \bar{x}_2|}{\sqrt{\left(\frac{s_1^2}{n_1} + \frac{s_2^2}{n_2}\right)}}$$

where $\bar{x}_1 = 35.0$; $\bar{x}_2 = 95.5$;
$s_1 = 1.89$; $s_2 = 2.95$; $n_1 = 10$; $n_2 = 10$.
Therefore, $t = 54.6$

(d) In the t-test, the value for degrees of freedom is the sum of the sample sizes of both groups minus 2. Therefore, because

10 replicates were used to test kit 1, and 10 to test kit 3, df = $(20 - 2) = 18$.

(e) A probability table should be used at a probability level of 0.05 to see if the t-value is higher or lower than the critical value using 18 degrees of freedom. If the t-value is higher than the critical value, then the difference between the mean values is significant.

Chapter review

1. (a) Phagocytosis.
   (b) Antigen-presenting cells (APC) will hydrolyse (digest) molecules found in the pathogen, in order to prevent its spread throughout the body. They will then place fragments of the pathogen in their cell surface membrane (this is called antigen display). These will then be exposed to other cells, including B- and T-lymphocytes, some of which will be selected to divide rapidly by mitosis in a process called clonal selection. This activation is due to the binding of the antigens to membrane antibodies (B) or receptors (T) on their cell surface membrane.
   (c) T-helper cells play a vital role in the immune response by secreting cytokines. These chemicals stimulate other immune cells such as macrophages and T-killer cells, which act to directly kill pathogens, and increase the rate of antibody secretion by plasma cells.
   (d) Plasma cells.

2. (a) The structure of an antibody plays a key role in its function in the humoral immune response.
   - A: the variable region is highly specific for another molecule, an antigen, to which its shape is complementary. This allows the antibody to bind to the antigen, and this can reduce the ability of the pathogen to divide and spread around the body.
   - B: The hinge region allows for flexibility in the structure of the molecule. It allows for the antibody to 'bend' in order to attach to two antigens at once (using both variable regions), and therefore to aid agglutination.
   - C: The constant region of the antibody can be embedded in the cell surface membrane of the B-cells, which allows them to use the molecules as receptors to attach to antigens prior to clonal selection. The constant region can also be attached to by other antibodies, which plays a role in agglutination.
   (b) An antigen–antibody complex is formed when the variable region of the antibody, complementary to an antigen, binds to that antigen. This is achieved by hydrogen bonds that form between the two molecules.
   (c) C   (d) Disulfide.

3. (a) Antigen-presenting cells may mistakenly recognise molecules or small particles found in tree nuts as harmful, and will ingest them by phagocytosis. They may then display these molecules in their cell surface membrane. If B- and T-cells have antibodies/ receptors that are complementary to these antigens, then they will be stimulated to divide by mitosis in the process of clonal selection and an immune response will occur.
   (b) There may be some people who have never been exposed to antigens found in peanuts, who may nevertheless be allergic to them. In addition, it may be that a person with an allergy to peanuts last encountered the antigens in peanuts many years ago, and therefore have very few (if any) antibodies complementary to these antigens in their bloodstream (even though they would have memory cells that would be stimulated by exposure).

4. (a) A table that compares the structure and functions of B- and T-lymphocytes would include the following statements:

B-lymphocytes	T-lymphocytes
Have antibodies in their cell surface membrane, able to bind to antigens	Have receptor proteins in their cell surface membrane, able to bind to antigens
Differentiate into plasma cells (which secrete antibodies) or memory B-cells	Differentiate into cytotoxic T-cells, helper T-cells or memory T-cells
Involved in the humoral immune response	Involved in the cellular immune response

(This list is not exhaustive)

(b) A table that compares the primary and secondary immune responses would include the following statements:

Primary immune response	Secondary immune response
Initiated by a B- or T-lymphocyte that has not previously encountered the antigen	Initiated by a B- or T-memory cell that was formed after a previous encounter with the antigen
Usually associated with the onset of symptoms of the disease	Rarely associated with the onset of symptoms of the disease
The rate and quantity of antibodies secreted is relatively low	The rate and quantity of antibodies secreted is relatively high
Initiated by a B- or T-lymphocyte that has not previously encountered the antigen	The rate and quantity of antibodies secreted is relatively high

(This list is not exhaustive)

(c) A table that compares active and passive immunity would include the following statements:

Active immunity	Passive immunity
Obtained when the individual encounters the antigen and initiates an immune response	Obtained without encountering the antigen; no immune response required
Antibodies produced in response to exposure to the pathogen/ vaccination	Antibodies acquired from another individual from breast milk or injection of antibodies
Forms memory cells	No memory cells are formed
Is permanent	Is temporary

(This list is not exhaustive)

5. (a) Cells of the immune system recognise cells as foreign by identifying antigens on their cell surface membrane as non-self. Most cancer cells are formed from normal body cells, and therefore have antigens that are similar, if not identical, to those found in normal body cells.
   (b) T-killer lymphocytes play a role in directly killing pathogens, cells that are infected with pathogens, and even some cancer

cells. They do this by interacting with pathogens or infected body cells and secreting chemicals such as hydrogen peroxide and the protein perforin by exocytosis. These cause holes to develop in the cell surface membrane of the cell, which gains water rapidly by osmosis and dies.

(c) Immunosuppressants will reduce the activity of the immune system and immune cells will be less likely to recognise any differences that arise in the antigens of cancer cells that form from normal body cells. These unrecognised cancer cells will therefore not be killed, and may continue to divide to form a tumour.

6. The pathogens that are responsible for some infectious diseases regularly change their antigens. This may be due to mutation. Therefore, memory cells that may have been formed as the result of a previous infection would no longer recognise the pathogen, and it would not cause a primary immune response if it infected the body.

7. The processes and definitions match as follows:
   - (a) Phagocytosis – (vii) the process by which a cell engulfs large particles or whole cells, as a defence mechanism.
   - (b) Humoral response – (iv) an immune response mediated by an antibody.
   - (c) Cellular response – (v) an immune response involving T-cells.
   - (d) Primary immune response – (ii) the stimulation of the immune system that occurs when the body is first exposed to an antigen.
   - (e) Secondary immune response – (vi) the stimulation of the immune system that occurs when the body is exposed to an antigen for a second time.
   - (f) Antigen presentation – (i) the display of antigens in the cell surface membrane by macrophages, which allows lymphocytes to more easily recognise the antigen.
   - (g) Clonal selection – (iii) a process in which lymphocytes divide by mitosis to form a clone of identical lymphocytes.

Practice exam-style questions

E1. C [1]

E2. (a) The symptoms of erythroblastosis fetalis would be caused by the lysis (bursting) of red blood cells in the fetus, [1] which causes them to lose their contents (e.g. haemoglobin) into the blood plasma. [1]

(b) (i) In the first pregnancy of a Rh⁻ mother who is carrying a Rh⁺ child, antibodies complementary to the rhesus antigens are not present in the bloodstream of the mother. [1] Therefore, no antibodies are present to attach to this antigen on the red blood cells of the fetus, and erythroblastosis fetalis does not occur. [1]

(ii) During the first pregnancy/ childbirth of a Rh⁻ mother who is carrying a Rh⁺ child, antibodies complementary to the rhesus antigens are made by the mother's plasma cells, and memory cells are produced that are specific to this antigen. [1] During her second pregnancy with a Rh⁺ child, the memory cells will start a secondary immune response [1] and antibodies will be produced that will cross the placenta to bind to the rhesus antigens on the red blood cells of the child. [1]

(c) The injected antibodies will attach to the Rhesus antigen on the surface of any red blood cells that have entered the mother's blood, which means they are not recognised by her immune system. [1] This will prevent the mother from initiating a primary immune response to these antigens, which reduces the likelihood that she will produce antibodies

complementary to the Rhesus antigen if she carries another Rh⁺ fetus later. [1]

E3. (a) The presence of antibodies for chlamydia in the blood of a koala indicates that the animal had previously been infected by this pathogen as it had initiated a primary immune response. [1]

(b) (i) There was an estimated population of 2000 koalas in the trial area at the start of the investigation (in 2010), and 65% of the sample of koalas tested were positive for chlamydia antibodies. Therefore, it would be expected that $(2000 \times 0.65) = 1300$ of the koalas in the initial population were positive for chlamydia.

(ii) The greatest number of koalas were caught and tested in 2014, [1] so the data for this year was the most reliable. [1]

(iii) Some koalas may have been infected long ago with chlamydia, and so the number of antibodies in their blood could have fallen to very low levels (even though memory cells would remain). [1]

(iv) It is possible that koalas infected with chlamydia are more likely to be trapped, possibly because of poor health status, than uninfected koalas. [1]

(v) There are a number of reasons that would suggest this conclusion may not be valid: [1 mark each]
   - Very few koalas in some years were tested (e.g. 2016), so the reliability of the calculated percentages may have been low.
   - The koalas that were treated for chlamydia may have died soon after their release / may not have come into contact with other koalas.
   - Another factor may be responsible for the reduction in the number of koalas with chlamydia (correlation does not necessarily indicate causation).

(c) If a very high percentage of koalas are vaccinated, then herd immunity is achieved. [1] This means that there are too few individuals to carry the pathogen and pass it onto others. [1]

(d) Vaccines will introduce an antigen into the blood that triggers a primary immune response. [1] This ultimately results in the production of memory cells, [1] which are required for a secondary immune response which could occur many months or years later if the pathogen infects the body. [1]

E4. (a) An antigen is a macromolecule, usually a protein or glycoprotein, [1] which triggers an immune response. [1] When the hCG antigen is injected into the bloodstream of the mammal, it will attach to the membrane-bound antibodies in a small number of B-lymphocytes. [1] These will be stimulated to divide rapidly by mitosis [1] in a process called clonal selection. [1] Some of these B-lymphocytes will differentiate to form plasma cells, [1] which are specialised cells able to produce and secrete large numbers of antibodies complementary to the hCG antigen.

(b) (i) The increase in antibodies complementary to hCG is faster after the second injection than the first. [1] The number of antibodies produced is also greater after the second injection than the first. [1]

(ii) This difference (in the number of antibodies and the rate at which they are released) is explained by the fact that after the first exposure (the primary immune response), some of the B-lymphocytes differentiate to form memory B-cells. [1] These remain in the body and will recognise the hCG antigen more rapidly after the second exposure (the secondary immune response). [1]

(iii) Reasons that could explain this include the following. [1]
- The tertiary structure of the equivalent protein in mammals is not complementary to the anti-hCG antibodies.
- The embryos of other mammals do not release a protein like hCG.
- The embryos of other mammals release much smaller amounts of a protein like hCG.

CHAPTER 12 – ENERGY AND RESPIRATION

In-text questions

1. C

2. (a) ATP $\rightarrow$ ADP + P$_i$
   (b) ADP + P$_i$ $\rightarrow$ ATP
   (reversible reaction symbol can be used in place of $\rightarrow$ in each case)

3. olive oil, uncooked lean meat, pasta

4. (a) Number of oxygen atoms on r.h.s. = (16 x 2) + 16 = 48

   There are 2 oxygen atoms in palmitic acid, so on l.h.s. we need 48 – 2 = 46
   Each oxygen molecule has 2 atoms, so $\frac{46}{2}$ = 23

   (b) RQ = $\frac{\text{volume of CO}_2 \text{ produced}}{\text{volume of oxygen used}}$

   = $\frac{16}{23}$

   = 0.70

5. (a) Number of carbon atoms on l.h.s. = 4 x 4 = 16
   Number of carbon atoms required on r.h.s. = 16.
   Each carbon dioxide molecule has only one carbon atom, so 16.

   (b) RQ = $\frac{\text{volume of CO}_2 \text{ produced}}{\text{volume of oxygen used}}$

   = $\frac{16}{19}$

   = 0.84

6. (a) cytoplasm
   (b) mitochondrial matrix
   (c) mitochondrial matrix
   (d) inner membrane of mitochondria (and intermembrane space)

7. (a) 6   (b) 6   (c) 3   (d) 2   (e) 4   (f) 6

8. reduced NAD / NADH

9. A

10. C

11. to increase the surface area for the electron transport chain / ATP synthesis

12. (a) pyruvate and ADP (or inorganic phosphate)
    (b) oxygen

13. Mitochondria may appear circular if the section cuts across the mitochondrion i.e. if it is a transverse section of a mitochondrion.

14. There are more hydrogen atoms in a molecule of fatty acid than in a molecule of glucose. More reduced NAD and reduced FAD can be made using lipid. More electrons released from oxidation of lipid, so more oxidative phosphorylation can occur.

15. C

16. In anaerobic respiration, glucose is not completely oxidised; only glycolysis occurs; there is only substrate-level phosphorylation / no oxidative phosphorylation.

17. glucose ($\rightarrow$ ethanal) $\rightarrow$ ethanol + carbon dioxide

18. The organ / tissues have no blood supply, so will not be receiving oxygen. Anaerobic respiration in cells causes lactate to build up. Lactate is acidic and will cause proteins to denature, eventually killing the cells.

19. Any three from: hollow stem; aerenchyma tissue in roots; ethanol-tolerance of root cells; ability to elongate stem rapidly.

Assignment 12.1

A1. By placing the respirometer (or the tube) with the germinating peas in a water bath

A2. Respiration uses oxygen and produces carbon dioxide. Any carbon dioxide produced is absorbed by the potassium hydroxide, so the volume of air in tube B decreases.

A3. $\frac{0.21 \text{cm}^3}{15}$ minutes
    = 0.014 (cm^{3}min^{-1})

A4. An equal volume of distilled water.

A5. RQ = $\frac{V_1 + V_2}{V_1}$

    = $\frac{0.21 + 0.0}{0.21}$

    = 1.0

A6. Germinating peas are using carbohydrate as a respiratory substrate because RQ = 1.0 for carbohydrate / explanation in terms of volume / use of respiration equation with glucose.

A7. With potassium hydroxide solution the volume of air in tube B would decrease / change the same way as with the peas (accept description in terms of the liquid in the U-tube); without potassium hydroxide solution the volume of air in tube B (or level of liquid in the U-tube) would change: volume in tube B would decrease (or the level of liquid in U-tube on that side would go up). Explanation: RQ for lipid is less than 1.0 / is 0.7; there is a larger volume of oxygen taken in than carbon dioxide given out. This is because lipids / fatty acids are less oxidised / contain less oxygen than carbohydrates.

Experimental skills 12.1

P1. Tube 1 – a control to show whether yeast will change the colour of the indicator when there is no glucose / no respiratory substrate.
    Tube 2 – the investigation tube with yeast, a respiratory substrate and the indicator; there should be a colour change in this tube.
    Tube 3 – a control to show whether killed yeast will change the colour of the indicator in the presence of glucose, and to show whether glucose changes the colour of the indicator.

P2.

temperature / °C	rate of respiration / s^{-1}
25	1.85×10^{-3}
30	2.35×10^{-3}
35	4.26×10^{-3}
40	3.64×10^{-3}
45	3.08×10^{-3}
50	2.67×10^{-3}

P3.

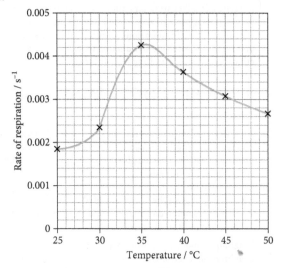

P4. Repeat the experiment with more temperatures in this range – for example 30, 31, 32, 33 up to 40°C; then determine which temperature gives the shortest time to go colourless (the highest rate of respiration).

P5. (a) Either yeast, glucose solution, 2 drops of distilled water; or yeast, glucose solution, 2 drops of reduced methylene blue / colourless methylene blue.

(b) no human judgement required; defined numerical value can be used as end point

P6. A plan for an investigation should include the following points:
- use the optimum temperature / given temperature in range 30–40°C
- use the same mass of yeast in each
- use the same number of drops of methylene blue in each
- vary the concentrations of glucose in each
- suggested concentrations of glucose, e.g. 5 – 50% in steps of 5%
- includes 0% glucose / equal volume of distilled water
- method for determining end point time e.g. judge by eye or colorimeter method
- use of equation to convert times to rates
- any additional information e.g. repeats to determine means etc.

Chapter review

1. Any three from: active transport; protein synthesis; DNA replication; any example of an anabolic reaction / synthesis of a named substance; any example of movement within cells e.g. nuclear division.

2. (a) The ratio (or fraction) of volume of CO_2 taken in to volume of O_2 used.

(b) Lipid requires more oxygen to completely oxidise / lipid is less oxidised than carbohydrate because fatty acids contain so many C–H bonds, so the volume of CO_2 produced is less than the volume of O_2 taken in, so RQ is less than 1.0 (RQ is 0.7 for lipid and 1.0 for carbohydrate).

3. (a) pyruvate

(b) When 1 glucose molecule enters glycolysis, **2** ATP molecules are used. Later in glycolysis **4** ATP molecules are produced. In addition, **2** molecules of NAD are reduced in glycolysis.

4. (a) FAD is a coenzyme; it becomes reduced / accepts hydrogen in the Krebs cycle. FAD takes hydrogen to the electron transport chain / to oxidative phosphorylation and becomes oxidised by giving up H+ and e– (to carrier protein in electron transport chain); FAD gets reused in this process.

(b) NAD is reused / regenerated; NAD is reduced to form reduced NAD and then oxidised again; only a small quantity needs to be synthesised / synthesis is only required to replace lost NAD molecules.

5. (a) citrate

(b) Oxaloacetate is regenerated in decarboxylation reactions: 2 carboxyl groups to be removed / 2 decarboxylation steps; carbon dioxide is released at each decarboxylation step; hydrogen is also removed; hydrogen is accepted by NAD and FAD, so the number of carbon atoms is decreased by 2 / from 6 to 4

(c) substrate-level phosphorylation or phosphorylation of ADP during substrate-linked reactions

6. (a) In chemiosmosis, protons (H+) are pumped into the intermembrane space using energy from redox reactions / from the electron transport chain; concentration and electrochemical gradients are built up; protons are allowed to flow down this gradient into the mitochondrial matrix; they flow through ATP synthase; the electrical potential energy / energy from protons is used to phosphorylate ADP.

(b) oxygen

(c) water

7. (a) glycolysis occurs: glucose forms two molecules of pyruvate; 2 molecules of ATP are produced (per glucose molecule); 2 molecules of NAD are reduced (per glucose molecule); pyruvate is converted to ethanal; ethanal is reduced (by reduced NAD) to ethanol and carbon dioxide; NAD is regenerated during ethanol production.

(b) Cells in roots of the rice plant can perform (more) aerobic respiration due to presence of aerenchyma tissue; these air spaces allow sufficient oxygen supply / efficient gas exchange; anaerobic respiration produces ethanol (and carbon dioxide), which is toxic to plant cells so kills other plant roots; rice has high tolerance of ethanol.

8. (a) DCPIP, which is blue when oxidised and colourless when reduced

(b) respirometer

(c) to absorb carbon dioxide

Practice exam-style questions

E1. (a) $RQ = \dfrac{\text{number of molecules of } CO_2 \text{released}}{\text{number of molecules of } O_2 \text{used}}$ [1]

(Note that volume is acceptable in place of 'number of molecules')

(b) (i) 26 O_2 [1] 18 CO_2 [1]

(ii) $\dfrac{18}{26} = 0.69$ or 0.7 <to be written as fraction> [1]

(c) Any four from: RQ starts at 1.0 as carbohydrate / starch / glucose is used for respiration; [1] when using carbohydrate volume of CO_2 released equals volume of O_2 used; [1] then RQ value falls to 0.7 (and remains at 0.7); [1] fat / lipid / fatty acids are being used for respiration; [1] when using fats the volume of CO_2 released is less than volume of O_2 used; [1] carbohydrate has all been used up. [1]

(d) The yeast cells were undergoing anaerobic respiration; [1] so volume of CO_2 given out is (much) greater than volume of O_2 taken in. [1]

E2. (a) Any four from: oxygen consumption increases with body mass for both resting and flying; [1] moths with greater body mass have more respiring cells / respiring tissue; [1] moths with greater body mass have a higher requirement for ATP; [1] moths are using aerobic respiration; [1] (most) ATP is produced using oxidative phosphorylation; [1] ATP / energy requirement for flying is greater than resting. [1]

(b) Respiration rate will increase with temperature (over this range); [1] (Note that, reference to decrease in rate at the higher end of this temperature range would not be accepted here.) reactions in respiration are temperature-dependent; [1] increased kinetic energy of molecules / increased frequency of collisions of molecules / increased frequency of collisions between enzymes and substrates. [1]

E3. (a) ATP is produced by substrate-level phosphorylation in the Krebs cycle. [1] Plus, any eight from: reduced NAD / reduced FAD; [1] release hydrogen / proton and electron / H^+ and e^-; [1] to carrier proteins / to electron transport chain; [1] energy from redox reactions used to pump H^+ into intermembrane space; [1] chemiosmosis; [1] protons / H^+ flow across membrane down electrochemical gradient; [1] through ATP synthase; [1] energy used to phosphorylate ADP / for the reaction ADP + P_i ATP; [1] oxidative phosphorylation; [1] final electron acceptor is oxygen; [1] oxygen is reduced to produce water. [1]

(b) Any seven from:
In both animal cells and yeast cells: glycolysis occurs; [1] 2 molecules of NAD are reduced per glucose (in glycolysis); [1] NAD is regenerated in next steps of anaerobic respiration. [1]
In animal cells: pyruvate is converted to lactate; [1] using reduced NAD. [1]
In yeast cells: pyruvate is converted to ethanal and carbon dioxide; [1] ethanal is converted to ethanol; [1] using reduced NAD. [1]

E4. (a) mitochondrial matrix [1]
(b) (i) pyruvate – 3 and acetyl coenzyme A – 2; [1] oxaloacetate – 4 and citrate – 6 [1]
(ii) between pyruvate and acetyl coenzyme A and two between citrate and oxaloacetate; [1] carbon dioxide shown at each of these three positions [1]
(iii) between pyruvate and acetyl coenzyme A and two between citrate and oxaloacetate; [1] NAD becoming reduced between pyruvate and acetyl coenzyme A and between citrate and oxaloacetate; [1] FAD becoming reduced between citrate and oxaloacetate [1]

E5. (a) By placing the respirometer in a water bath [1]
(b) An equal mass of any non-living material could be used as a control; [1] everything else would need to be identical. [1]
(c) Dependent variable: rate of respiration / rate of movement of coloured liquid [1]

Independent variable: type of organism [1]
(d) Either: rates of respiration will be equal / different for the three organisms; [1] Or rate of respiration for one named organism will be fastest / slowest [1]
(e) diameter / radius / cross-sectional area of the tube (containing coloured liquid); [1] distance moved by the liquid; [1] time taken for the liquid to move that distance [1]
(f) Any seven from: use an equal mass of each organism; [1] equal mass / concentration / volume of sodium hydroxide; [1] block light from the reaching respirometer (to ensure that the germinating peas cannot be photosynthesising); [1] ensure respirometer is sealed / no gas can enter or leave; [1] suggested temperature in range 20 – 35°C; [1] moving

coloured liquid back along the tube before starting with next organism; [1] allow fresh air / allow oxygen into respirometer between organisms; [1] repeat and calculate mean for each organism / identify anomalous results; [1] suitable safety precaution for handling sodium hydroxide; [1] suitable ethical practice for handling blowfly larvae. [1]

CHAPTER 13 – PHOTOSYNTHESIS
In-text questions

1. (a) An action spectrum shows the rate of photosynthesis at different wavelengths. Isolated chloroplasts can carry out photosynthesis but isolated photosynthetic pigments cannot carry out photosynthesis.
(b) (i) 430–450nm
(ii) green light gives the lowest rate of photosynthesis (note that this is not to do with absorption); green is probably reflected most from the leaf

2. (a) Diagram should include: oval shape with double outer membrane; grana drawn as stacks of double lines; other detail, e.g. loops of DNA or starch grains or lipid droplets; labels for stroma and grana / thylakoids.
(b) grana identified as site of light-dependent stage; stroma identified as site of light-independent stage

3. (a) photolysis is the splitting of a water molecule using energy from light
(b) photoactivation is the removal of an electron from a photosynthetic pigment using energy from light
(c) photophosphorylation is the use of light energy to phosphorylate ADP / to synthesise ATP
(d) photosystem means a group of photosynthetic pigments in the thylakoid membrane

4. (a) H^+ are used to reduce NADP.
(b) The electrons (e^-) are given to photosystem II
(c) Oxygen is released as a waste product

5. In the reaction catalysed by rubisco, the substrates are RuBP and CO_2; the product is (an unstable 6-carbon compound that breaks down to) two molecules of GP.

6. GP is reduced to TP; the reduction uses reduced NADP and ATP; TP is converted to RuBP, using ATP; $\frac{5}{6}$ of TP molecules are used in the regeneration.

7. Improving the efficiency of rubisco would make it work faster / fix more carbon; this would increase the rate of photosynthesis, which would increase the rate of plant growth; this would increase crop yields and the ability to produce more food / feed more people / reduce famine.

8. In 20 seconds $20 \times 3 = 60$ reactions occur

60 reactions make 120 molecules of GP

120 molecules of GP make 120 molecules of TP

$\frac{1}{6}$ of 120 = 20 molecules of TP available to make glucose

TP has 3 carbons and glucose has 6 carbons, so 2 TP needed to make 1 glucose

$\frac{20}{2}$ = 10 glucose molecules

9. D – carbon dioxide concentration

10. The oil-burning heater also emits carbon dioxide, which is a rate-limiting factor in photosynthesis. Therefore, increasing carbon dioxide concentration increases the rate of photosynthesis.

11. (a)

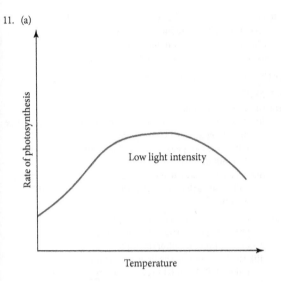

Temperature

(b)

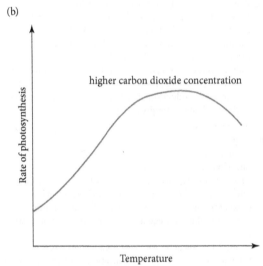

Temperature

Assignment 13.1

A1. light intensity – use the same lamp / same bulb; keep the *Elodea* at the same distance from the lamp; use the same level of background lighting temperature – check the temperature of the water in the boiling tube with a thermometer; the tank of water between the lamp and the *Elodea* is to stop heat from the lamp warming the *Elodea* plant used in the investigation – use the same type / age of plant / cuttings from the same plant; use the same length of plant each time

A2. boiling removes any carbon dioxide that was already dissolved; to start from zero concentration of carbon dioxide / so that all the carbon dioxide in the water comes from the potassium hydrogen carbonate

A3. sketch of a graph with rate of photosynthesis on vertical axis and carbon dioxide concentration *OR* mass of potassium hydrogencarbonate on horizontal axis; line with constant gradient sloping up from the origin curving to a plateau

A4. (a) bubbles may not all have equal volume; bubbles may not be only oxygen / bubbles may also come from respiration
    (b) measure the volume of gas; some method of measuring oxygen content of gas / ensuring all the gas collected is from photosynthesis

A5. (a) changed – distance of *Elodea* from the lamp / brightness of the lamp e.g. using a dimmer, bulbs of different power rating, semi-transparent filters
    controlled – temperature – use the tank between the lamp and the *Elodea*; check the temperature of the water is constant each time light intensity is change
    controlled - plant used in the investigation – use the same, type / age of plant / cuttings from the same plant / length of plant each time
    (b) changed – put the boiling tube in a transparent water bath / in a beaker of water at known temperature; allow time for the temperature of the water in the boiling tube (and the *Elodea*) to come to the desired temperature
    controlled – light intensity by, using the same lamp;/ same distance between lamp and *Elodea*
    controlled - plant used in the investigation – use the same, type / age of plant / cuttings from the same plant / length of plant each time

A6. measuring effect of photosynthesis only and not respiration; colour change of indicator will be more precise than counting bubbles; other (named) effects of the independent variable on the plant will not interfere with results

A7. (a) to prevent osmosis from bursting / causing water loss from chloroplasts
    (b) to ensure the pH remains the same / remains optimum; to ensure enzymes / proteins do not denature

A8. rate = $\frac{1}{time}$

    unit = s^{-1}

A9. The method should include: use of coloured filters to cover the visible spectrum / to cover 400–700nm wavelengths; at least 5 named colours e.g. red, orange, yellow, green, blue, violet; *a method for changing independent variable* e.g. use of one lamp with filter between the lamp and the extract; *a method for blocking unfiltered light from lamp* e.g. lamp is in a sealed box with the filter at the opening; *a method for blocking unfiltered background light* e.g. carry out investigation in a dark room; *methods for controlling other variables:* temperature – place tube with extract in glass beaker of water at known temperature, light intensity – use the same lamp, keep the extract the same distance from the lamp; extract – use an equal volume each time, take sample from the same extract each time; end point – use of a reference colour or colorimeter to judge end point each time.

Experimental skills 13.1

P1. (a) so substances from skin do not interfere with the chromatography / do not affect the results
    (b) pencil / graphite will not dissolve whereas ink from a pen will dissolve; pigments in the ink from the pen will separate and interfere with the results
    (c) If the origin were lower down and touching the solvent, then the leaf pigments would dissolve into the liquid; the leaf pigments would come off the chromatography paper / would not travel up the paper.

P2. The position of the pigments is being compared in order to identify them; the quantity of the pigments is not being compared; the pigments may be present at different concentrations in each sample.

P3. (a) A and E   (b) B and D
        C and F   (c) G and H

P4. (a) A and E – $R_f = 0.99$
C and F – $R_f = 0.45$
B – $R_f = 0.74$
D – $R_f = 0.38$
G – $R_f = 0.32$
H – $R_f = 0.01$

(b) A and E are beta carotene
C and F are chlorophyll *a*
B is xanthophyll
D is chlorophyll *b*
G is fucoxanthin
H is chlorophyll *c*

P5. It is difficult knowing where to take the measurement as the spot is large / elongated / faint.

Chapter review

1. (a) B
   (b) (i) light-dependent (ii) light-independent

2. An absorption spectrum shows the (relative) absorption of different wavelengths of light by a photosynthetic pigment / photosystem / chloroplast / leaf. An action spectrum shows the rates of photosynthesis at different wavelengths of light by a chloroplast / leaf.

3. photoactivation / photoionisation; involves (only) photosystem I; light causes electron to be emitted from chlorophyll *a*; electron passed through electron transport chain / through protein carriers; energy used to pump protons / H^+ across thylakoid membrane into grana; chemiosmosis; energy from protons flowing back into stroma used to synthesise ATP; phosphorylation of ADP using inorganic phosphate / $ADP + P_i \rightarrow ATP$

4. photolysis: $H_2O \rightarrow 2H^+ + 2e^- + \frac{1}{2}O_2$ (or reaction described in words); electrons given to photosystem II; hydrogen used to reduce NADP; oxygen is a waste product

5. (a) ATP is used to supply energy for conversion of GP to TP; ATP is used to supply energy for regeneration of RuBP from TP; reduced NADP is used for conversion of GP to TP
   (b) Carbon-fixation is catalysed by the enzyme rubisco; reactants / substrates are carbon dioxide and ribulose bisphosphate; CO_2 has 1 carbon and RuBP has 5 carbons; product is two 3-carbon molecules called glycerate 3 phosphate.
   (c)

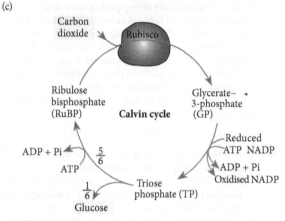

6. Some oxygen is used in respiration and therefore is taken into mitochondria.

7. (a) As temperature increases the rate of photosynthesis increases; enzymes and substrates gain more kinetic energy; enzyme–substrate complexes are formed more quickly; products are formed more quickly; up to the optimum temperature when rate of photosynthesis is maximum; with further rise in temperature rate of photosynthesis decreases; enzymes become denatured
   (b) rate of photosynthesis is higher (at all temperatures)); carbon dioxide is a rate-limiting factor; when the availability of a rate-limiting factor increases the rate can increase; optimum / maximum rate occurs at the same temperature

8. DCPIP is blue when oxidised and becomes colourless when reduced; the DCPIP is reduced when it accepts the electrons that have been released from photosynthetic pigments; energy for release of electrons / photoexcitation comes from light

Practice exam-style questions

E1. (a) (i) B [1] (ii) D [1]
   (b) (i) chromatography [1]
       (ii) Any three from: chlorophyll *a* absorbs over a wider range of wavelengths; [1] chlorophyll *b* has a higher peak; [1] both have two peaks; [1] both have minimum in range 500–600nm. [1]
       (iii) maximum rates at 400–460nm <u>and</u> 600 – 700nm (accept ±10nm at each end of ranges)); [1] minimum at 510 – 580nm [1]

E2. (a) (i) photolysis; [1] water is split to produce hydrogen ions / protons <u>and</u> electrons; [1] and oxygen [1]
       (ii) cyclic photophosphorylation [1]
       (iii) Any two from: in the light-independent stage; [1] to reduce GP to TP; [1] in regeneration of RuBP. [1]
   (b) Any six from: light intensity no longer limiting; [1] carbon dioxide now limiting; [1] carbon dioxide concentration in air is very low; [1] increased rate of ATP synthesis; [1] carbon fixation can occur faster; [1] rate of rubisco-catalysed reaction will increase / CO_2 is substrate for rubisco; [1] Calvin cycle occurs faster; [1] rate of production of GP / TP increases; [1] rate of production of glucose increases; [1] link to increased production of cellulose for cell walls / amino acids / nucleotides for growth. [1]

E3. (a) (i) rubisco [1] (ii) ribulose bisphosphate [1]
   (b) Any six from:
   for maize and wheat
   rate of photosynthesis increases with increasing CO_2 concentration; [1] this happens when CO_2 concentration is rate-limiting; [1] rate of photosynthesis levels off at higher CO_2 concentrations; [1] when other factors become limiting; [1] rates of photosynthesis are equal when CO_2 concentration = 0.04% [1]
   for maize
   at zero CO_2 concentration rate of photosynthesis is zero; [1] rate becomes constant at CO_2 concentrations greater than 0.05% [1]
   for wheat
   at very low CO_2 concentrations rate of photosynthesis is zero; [1] rate becomes constant at CO_2 concentrations greater than 0.08% [1]
   comparison points
   rate is higher for maize up to CO_2 concentration 0.04% / ora for wheat; [1]
   rate is higher for wheat at CO_2 concentration greater than 0.04% / ora for maize [1]

E4. (a)

temperature / °C	uptake of CO_2 in 60s / nmol	rate of photosynthesis / nmol s^{-1}
0	48	0.8
5	126	2.1
10	204	3.4
15	318	5.3
20	450	7.5

all four values = 1 mark

(b) graph with rate of photosynthesis in nmol $CO_2\ s^{-1}$ on vertical axis and temperature in °C on horizontal axis; [1] axes with linear scales to include zero and covering more than half the grid in both directions; [1] all points plotted to within 1mm; [1] smooth curve drawn through all points. [1]

(c) temperature [1]

(d) As temperature increases, rate of photosynthesis increases; [1] rate of enzyme-catalysed reactions increase with temperature; [1] enzyme and substrate molecules have more energy; [1] and so enzyme–substrate complexes form more frequently [1]

E5. (a) Any eight from:

independent variable: method of varying the temperature of the water bath e.g. Bunsen / spirit burner / ice / adding pre-heated water; [1] checking the temperature using the temperature sensor <u>and</u> adjusting if necessary; [1] at least five temperatures at equal intervals given [1]

control variables: method for ensuring light intensity is high and constant; [1] using fresh suspension of *Chlorella* each time; [1] same density of suspension each time; [1] fresh water each time and taken from the same source / equal concentration of CO_2 / same concentration of hydrogencarbonate added [1]

measuring dependent variable: measuring change in O_2 concentration in water; [1] over the same time for each temperature [1]

use of equipment: measuring cylinder to measure *Chlorella* suspension; [1] colorimeter to check density of *Chlorella* suspension [1]

safety: low risk procedure / keep water away from electrical equipment / wash hands after handling *Chlorella* [1]

(b) (i) sketch with rate of photosynthesis on vertical axis <u>and</u> temperature on horizontal axis; [1] line showing increase in rate with temperature; [1] line reaching a peak and then rate decreasing [1]

(ii) rate increases with temperature as frequency of collisions between enzymes and substrates increases / rate of formation of enzyme–substrate complexes increases; [1] reaches an optimum / maximum; [1] rate decreases as higher temperature causes enzymes to denature [1]

CHAPTER 14 – HOMEOSTASIS

In-text questions

1. This is because when a change in a physiological (bodily) factor is detected, mechanisms are put in place that bring that factor back (correct it) to a normal value.

2. Blood will travel to the liver and pancreas containing glucose absorbed from the small intestine. The liver and pancreas are involved in detecting, and modifying, the glucose concentration of the blood, so this means that any large deviations from the set point can be corrected quickly.

3. (a)

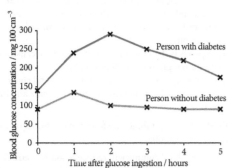

(b) The person with diabetes has a blood glucose concentration that is higher than the person without diabetes at the point at which the glucose solution is ingested (around 50 mg 100 cm^{-3} more). The blood glucose concentration rises more rapidly, and to a higher

peak value, in the person with diabetes than the person without diabetes (290 mg 100 cm^{-3} versus 135 mg 100 cm^{-3}). While the blood glucose concentration in the person with diabetes begins to fall after one hour, the blood glucose concentration in the person with diabetes continues to rise until 2 hours.

(c) This is because of the lack of insulin in the bloodstream of the person with diabetes. Insulin stimulates the absorption of glucose from the blood by cells in the liver and muscles, and its conversion into glycogen.

(d) The individuals should have the same body mass, age and gender. It should also be ensured that they engage in the same level of activity during the course of the study. The volume of glucose solution that was ingested should be the same for both people.

(e) Insulin stimulates the absorption of glucose from the blood by cells in the liver and muscles. If too much glucose leaves the blood, then cells will not be able to obtain enough for respiration, and will die.

4. The mechanism has in common with exocytosis the fact that vesicles in the cytoplasm merge with the cell surface membrane. However, these vesicles contain the glucose carrier proteins in their phospholipid bilayer, and they are not released from the cell when this occurs. Instead, they are placed into the cell surface membrane.

5. **B.** This is incorrect because β cells are the effectors.

6. (a) Glucose is a monosaccharide which is the respiratory substrate. Glycogen is a polysaccharide that consists of many glucose molecules joined together, and which acts as a storage carbohydrate in mammals.

(b) Both processes increase blood glucose concentration. In gluconeogenesis, glucose molecules are synthesised from other organic molecules, such as fatty acids, glycerol and pyruvate. In glycogenolysis, glucose molecules are released when glycogen is hydrolysed.

(c) A hormone is a substance that is released from an endocrine gland, which travels in the blood and affects the activity of target cells that have receptors specific to that hormone in their cell surface membrane or cytoplasm.

(d) The process of signal transduction involves the conversion of a stimulus outside of a cell into a response inside the cell, usually achieved by a hormone or other signalling molecule interacting with a receptor. Amplification refers to the increase in the size of the stimulus to produce a larger response, and is often achieved through signal transduction using second messenger molecules.

(e) Glycogen phosphorylase is an enzyme that removes phosphate groups from molecules and plays an important role in glycogenolysis. Specifically, it is necessary in the process of glycogen hydrolysis and release of glucose. Phosphorylase kinase activates glycogen phosphorylase in the signalling cascade that is initiated by glucagon.

7. (a) Insulin is a protein, which would be hydrolysed by the proteases in the acidic environment of the stomach if consumed by ingestion.

(b) Starch is a polysaccharide which can be hydrolysed to release glucose. This takes time, and occurs gradually, meaning that the release of glucose would be slow and would not result in a large peak of blood glucose concentration.

(c) (i) The high concentration of glucose in the bloodstream causes water to move out of cells by osmosis, dehydrating them. This is detected by osmoreceptors responsible for the feeling of thirst.

(ii) The inability to store glucose as glycogen and its loss in the urine means that the body mass of the individual is unable to increase significantly. In addition, cells respire stores of lipid, which contributes to weight loss.

(iii) The inability of cells to absorb glucose from the blood means that they are unable to engage in a high rate of respiration.

(iv) In the absence of sufficient glucose for respiration, cells can metabolise other substrates, including lipids.

(v) The high concentration of glucose in the bloodstream causes water to move out of cells by osmosis, dehydrating them. This means that more water is lost as urine. This is also caused by excessive drinking.

(d) (i) Facilitated diffusion

(ii) Insulin promotes the fusion of vesicles containing glucose carrier proteins with the cell surface membrane. This means that glucose can move into the cells at an increased rate. In the absence of insulin, there will be fewer carrier proteins to transport, so the rate of glucose intake into cells will reach a plateau at a lower concentration of glucose in the blood plasma than if insulin is present.

(iii) This means that the biosensor is able to measure accurately to the nearest 0.5 mg.

8. The passage should read as follows. The underlined words are those that should be inserted:

'People with diabetes can use a biosensor that is sensitive to glucose. The biosensor contains an immobilised enzyme called glucose oxidase. The substrate for this enzyme is glucose and this is converted into two products: hydrogen peroxide and gluconic acid. The synthesis of these products generates a very small electrical charge, which is proportional in size to the concentration of glucose in the blood. The current is then read by a meter, which displays a quantitative (numerical) reading for blood glucose concentration.'

9. The loop of Henlé and the collecting duct.

10. (a) Every hour, 1.2 dm³ x 60 = 72 dm³. Therefore, each hour, the total volume of blood in the body passes through the kidneys 72 ÷ 5 = 14.4 times.

(b) 0.125 dm³ of 0.7 dm³ plasma passes into the Bowman's capsules. Expressed as a percentage, this is 17.9 % (to 1 dp).

11. (a) The Bowman's capsule is the first part of the nephron, which appears as a cup-like structure. It almost fully surrounds the glomerulus, which is a ball of capillaries. Ultrafiltration occurs between the glomerulus and the Bowman's capsule.

(b) The afferent arteriole is the blood vessel that leads into the glomerulus, and the efferent arteriole leads away. Because the diameter of the lumen of the afferent arteriole is wider than that of the efferent arteriole, the pressure of the blood in the glomerulus is high.

(c) A podocyte is a cell in the epithelium of the Bowman's capsule, which wraps around the capillaries of the glomerulus. The basement membrane is a protein matrix that forms a selective barrier between the endothelial cells in the wall of the glomerular capillary and the cell surface membrane of the podocytes that surround them.

12. (a) The glomerular filtrate contains the same concentration of urea and glucose as the blood. However, the concentration of protein in the blood plasma is much greater than the concentration of protein in the glomerular filtrate.

(b) Urea and glucose are small molecules that are forced out of the glomerular capillaries and into the Bowman's capsule during ultrafiltration. Proteins are much larger and cannot fit through the fenestrations between the podocytes, meaning that they remain in the blood and cannot enter the filtrate.

(c) As the filtrate moves from the Bowman's capsule to the proximal convoluted tubule, plasma proteins maintain a

low water potential in the blood. This means that water (much of which was also removed from the blood by ultrafiltration) moves, by osmosis, from the filtrate back into the blood.

13. The loop of Henlé contains fluid that moves in opposite directions in the descending and the ascending loops. Through its action, it multiplies the negative water potential in the fluid outside the nephron in the medulla.

14. Graph should resemble the following:

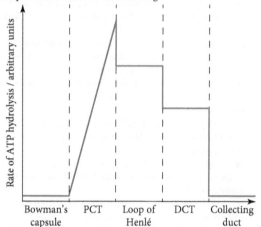

No ATP hydrolysis is required for the process of ultrafiltration between the glomerulus and the Bowman's capsule. In the proximal convoluted tubule, active transport is required to reabsorb glucose from the filtrate and back into the blood. Sodium ion transport between the loop of Henlé and the surrounding fluid in the medulla requires ATP hydrolysis, and the reabsorption of some ions from the filtrate in the distal convoluted tubule also requires active transport and therefore ATP.

15. (a) The proximal convoluted tubule is the most likely 'odd one out' because it is not associated with the process of ultrafiltration.

(b) Protein is the most likely 'odd one out' because it does not leave the blood in the glomerulus during ultrafiltration. Glucose and urea are small enough to fit through the fenestrations and pass into the Bowman's capsule.

(c) Ultrafiltration is the most likely 'odd one out' because it involves the movement of molecules due to hydrostatic pressure – small molecules and ions are forced out of the glomerulus and into the Bowman's capsule. Selective reabsorption and osmosis, which occur between the nephron and the blood elsewhere, occur due to the difference in concentration of ions and molecules between two solutions.

(d) Renal pelvis is the most likely 'odd one out' because it does not contain regions of the nephron. The cortex contains the Bowman's capsule and the proximal and distal convoluted tubules; the medulla contains the loop of Henlé and the collecting duct.

16. The distal convoluted tubules are found in the cortex of the kidney, not the medulla.

17. (a) A reduced blood volume will reduce blood pressure. This will mean that the hydrostatic pressure required for ultrafiltration may not be sufficient to force the small molecules and water into the Bowman's capsule.

(b) Seawater has a very low water potential. When the ions and molecules are forced into the nephron after ultrafiltration, they will draw more water by osmosis than usual from the blood and into the nephron. This may cause the person to become dehydrated.

(c) Protein remains in the efferent arteriole of the glomerular capillaries after ultrafiltration. This means that the water potential of the blood is lower than normal, and hence will result in water moving from the nephron and into the blood more readily by osmosis.

18. (a) A lack of ADH secretion will result in a high volume of dilute urine, which will cause dehydration and intense feelings of thirst.

(b) Injections of ADH during periods of time when the person has not taken in large volumes of water.

19. (a) ADH is secreted by the pituitary gland and causes more aquaporin protein channels to enter the surface membrane of the epithelial cells lining the collecting duct. This means that water moves, by osmosis, from the filtrate to the blood at an increased rate. This decreases the water potential of the filtrate. In the absence of ADH, the number of aquaporins in the surface membrane of the epithelial cells lining the collecting duct is not increased, which means that the movement of water molecules from the filtrate into the blood is slower. The water potential of the filtrate does not, therefore, decrease as much.

(b) (i–iii)

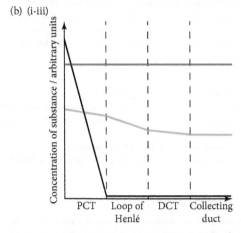

20. (a) Proton pump proteins in the guard cell surface membrane actively transport hydrogen ions, H+ (protons) into neighbouring epidermal cells. The decrease in the hydrogen ion concentration inside the guard cells causes channel proteins specific for potassium (K+) ions in the cell surface membrane to open. These ions move into the guard cells down an electrochemical gradient.

(b) The efflux of hydrogen ions and influx of potassium ions across the cell surface membrane of guard cells decrease their water potential so water enters by osmosis; this causes the guard cells to become turgid and the stoma to open.

(c) Abscisic acid (ABA) is a plant hormone that is secreted by plant cells in stress conditions. ABA inhibits the proton pumps, so hydrogen ions are no longer pumped out. This allows the concentration of hydrogen ions to increase, causing a high positive charge inside the cell. To balance this, potassium ions leave the cell down the electrochemical gradient by facilitated diffusion. In addition, ABA activates calcium ion channel proteins in the tonoplast, the vacuolar membrane. The release of calcium ions from the vacuole and influx into the cytoplasm occurs by facilitated diffusion. Calcium ions are a second messenger. Their rising concentration in the cytoplasm triggers a signalling cascade that results in

potassium ion efflux (potassium ions leaving the cell) by facilitated diffusion via specific protein channels.

The movement of potassium ions out of the cytoplasm rapidly increases the water potential of the contents of the guard cells, causing water to move out by osmosis. This reduces the volume of the guard cells, which become flaccid, causing the stoma to open.

Assignment 14.1

A1. $PM_{2.5}$ may enter the kidneys from the blood, and it may enter the blood in the lungs: the particles are so small that they can diffuse from the air in the alveoli, across the squamous epithelium, and through the endothelial cells of the capillary walls.

A2. (a) 45 000 : 2500 =
450 : 25 =
18 : 1 (ratio of new cases of CKD : new cases of kidney failure)

(b) These results are reliable because a total of 2.5 million people were studied.

(c) The negative control provides evidence that high atmospheric concentrations of $PM_{2.5}$, rather than other factors such as sodium ions, are responsible for the increased numbers of new cases of CKD.

A3. The regions of the United States that have an increased incidence of CKD almost perfectly overlap with the regions in which the atmospheric concentrations of $PM_{2.5}$ are highest. However, there are some regions where the atmospheric concentrations of $PM_{2.5}$ are low, but CKD numbers are high.

(a) Actual width = image width ÷ magnification. Therefore, actual width = [insert at proofs]

(b) The presence of the Bowman's capsules and glomeruli indicate that this section is of the cortex of the kidney, which is where ultrafiltration takes place.

(c) The three defects common in people with CKD are shown on the diagram below.

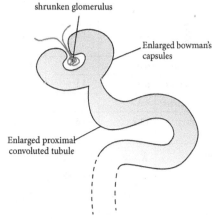

(d) The epithelial cells of the proximal convoluted tubule are vital in the process of selective reabsorption. These cells have many mitochondria to release energy for active transport of glucose. If these cells die, the kidney would be unable to reabsorb glucose from the filtrate back into the blood, which would mean that most of the glucose would be lost in the urine.

(e) A longitudinal section of the cortex would show the length of these tubules and the lumens would be shown in a vertical section.

Experimental skills 14.1

P1. The kangaroo rat lives in a desert (an environment that is very dry and where water is in short supply), so it must conserve as much water in its body as possible. It does this by producing of urine with a very low water potential.

Answers

P2.

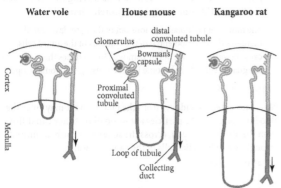

Water vole   **House mouse**   **Kangaroo rat**

P3. The method has been standardised because the rodents were kept at the same temperature, for the same time, in order to collect urine. The same volume of urine was used in the experiment to assess its freezing point.

P4. Urine with a greater solute concentration has a lower water potential. The more solute added to urine, the greater the disruption of hydrogen bonds between the water molecules and the lower the freezing point.

P5. (a) A negative correlation between two factors indicates that as one factor (RMT) increases, another factor (freezing point of urine) decreases. A strong correlation indicates that the gradient of the line, if these factors were plotted on a graph, is high (steep).

(b)

rodent species	mean relative medullar thickness/ no units, x	mean freezing point of urine/ °C, y	xy
Arvicola amphibius	1.5	−1.4	−2.10
Myocastor coypus	2.8	−2.9	−8.12
Rattus rattus	6.0	−3.8	−22.80
Mus musculus	6.3	−4.1	−25.83
Jaculus jaculus	9.8	−4.2	−41.16
Dipodomys deserti	12.2	−5.6	−68.32
mean	$\bar{X} = 6.43$	$\bar{Y} = -3.67$ $= -3.67$	
$n\bar{X}\bar{Y}$	$= -141.59$		$\sum XY = -168.33$
standard deviation	$s_x = 4.06$	$s_y = 1.41$	$r = -0.93$

Note that an answer of $r = -0.94$ is also acceptable if figures have not been rounded in the calculation.

(c) The degrees of freedom in this investigation is found by subtracting 2 from the total number of measurements, which is 12. The degrees of freedom are therefore 10. Using the table of critical values for the Pearson's linear correlation test (r) in Table 2.10, we can see that − 0.93 is higher than the value of − 0.576 (at p = 0.05) and higher than − 0.708 (at p = 0.01). Therefore, this suggests that there is a statistically significant correlation between these two factors.

(d) The researchers should plot a scatter graph and draw a line-of-best-fit. They should extrapolate (extend) this line either side of the two points at the top and lower point of the range. They should then find the RMT value of this animal and draw a line from this axis to intersect with the plotted line (this is called interpolation), and then read off the estimated value of the freezing point of the urine on the other axis.

(e) Spearman's rank is a statistical test to assess the strength of a correlation between two factors whose data is not normally distributed. In this investigation, the freezing points of different samples of urine from the same species of rodent were measured, which is likely to represent a normally distributed dataset.

P6. The larger the RMT, the longer the loop of Henlé. In kidneys with a longer loop of Henlé, the water potential in the medulla is reduced to a much lower value. This will enable the animal to produce urine with far lower water potential, which has a lower freezing point.

P7. C

P8. The conclusion drawn in this study may not be valid because other factors may have varied between the rodents, which could have had an effect on the volume (and therefore water potential) of the urine produced.
- The animals were of different body masses.
- The rodents may be of different ages, or genders, or health status.

Experimental skills 14.2

P1. There is no significant difference between the mean feeling of thirst and hydration levels in individuals who have consumed monosodium glutamate, and individuals who have consumed sodium chloride.

P2. Knowing whether they consumed one food additive or the other could influence the feelings of the individuals, which may reduce the validity of conclusions drawn in this study.

P3. (a) The range of the data is from 1 hour to 5 hours, which is likely to be sufficiently wide to accurately identify a trend in the data. The interval is every hour, which will allow for a graph to be plotted to further indicate this trend.

(b)

time after ingestion of meal/ hours	mean feeling of thirst after ingesting MSG/ arbitrary units	standard error/ arbitrary units	95% confidence interval/ arbitrary units	Upper limit of error bar	Lower limit of error bar	mean feeling of thirst after eating NaCl/ arbitrary units	standard error/ arbitrary units	95% confidence interval/ arbitrary units	Upper limit of error bar	Lower limit of error bar
1	3.3	0.10	0.20	3.50	3.10	2.8	0.21	0.42	3.22	2.38
2	5.7	0.21	0.42	6.12	5.28	4.9	0.26	0.52	5.42	4.38
3	7.5	0.15	0.30	7.80	7.20	7.1	0.14	0.28	7.38	6.82
4	8.9	0.12	0.24	9.14	8.66	7.6	0.07	0.14	7.74	7.46
5	9.1	0.13	0.26	9.36	8.84	9.3	0.16	0.32	9.62	8.98

(c) The confidence limits are the range/interval in which the true value of the mean lies, with 95% probability/ chance (we are 95% confident that the true mean lies within this range).

(d)

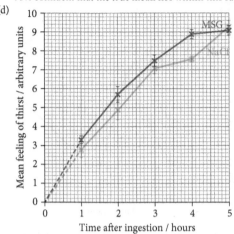

(e) The most reliable mean value is that for the individuals who consumed sodium chloride, 4 hours after ingestion, because this has the smallest value of standard error (SE) of 0.07 arbitrary units, and therefore the narrowest 95% confidence interval error bars.

(f) There may be a significant difference between the mean feelings of thirst of volunteers that ingested MSG and those who ingested sodium chloride at 4 hours after eating. This is because the 95% confidence interval error bars (see P 3 d) do not overlap.

(g) The calculated t-value is <u>less than</u> the critical value for <u>eighteen</u> degrees of freedom. This means there is a probability of ><u>0.05</u> that the difference in reaction time is due to chance. This means there is no significant difference and the null hypothesis is <u>accepted</u>.

P4. The values for both dependent variables in this investigation are judged by an individual, and are compared to a 'scale' of values that have been previously established.

P5. A colorimeter could be used to determine the colour of the urine. This would give a quantitative measure of the concentration (water potential) of the solution, which is not influenced by the experimenter's judgement. Alternatively, measuring the concentration of ADH would give a numerical (quantitative) reading, which reduces the chance that subjectivity in judgement would make the results inaccurate.

Chapter review

1. The maintenance of a constant internal environment.

2. (a) β cells in the islets of Langerhans are receptors, which detect the stimulus of higher blood glucose concentration. They also act as effectors, as they secrete insulin in response. α cells in the islets of Langerhans are also receptors, which detect the stimulus of lower blood glucose concentration. They also act as effectors, as they secrete glucagon in response.
   (b) Osmoreceptor cells in the hypothalamus are receptors, which detect the stimulus of lower blood water potential. They activate effector cells in the posterior pituitary gland to secrete antidiuretic hormone (ADH) in response.
   (c) Epidermal cells in the leaf are receptors, which detect the stimulus of water stress. They activate effector cells in the leaf called guard cells by secreting abscisic acid (ABA) in response, which causes stomata to close.

3.

Insulin	Glucagon
stimulates glycogenesis to reduce blood glucose concentration	stimulates glycogenolysis and gluconeogenesis to increase blood glucose concentration
secreted by β cells in islets of Langerhans in the pancreas	secreted by α-cells in islets of Langerhans in the pancreas
secreted when blood glucose concentration increases	secreted when blood glucose concentration decreases
binds to receptors in surface membrane of target cells in liver/muscle	binds to receptors in surface membrane of target cells in liver
increases permeability of cells to glucose by stimulating vesicles containing glucose channel proteins to fuse with surface membrane	acts via a second messenger system using cAMP

4. The absorbent pad is lowered into a blood sample. In the presence of the enzyme glucose oxidase glucose is converted to gluconic acid and hydrogen peroxide. In the presence of the enzyme peroxidase hydrogen peroxide generates a small electrical current. The current is then detected by an electrode and converted into a digital reading.

5. (a) Osmoregulation is the process by which the water potential of the blood is maintained at near-constant levels.
   (b) Negative feedback describes the processes by which any deviation in the value of a physiological (bodily) factor stimulates processes that correct that change and bring it back to a set point. In the case of osmoregulation, a reduction in the water potential of the blood is detected by the hypothalamus, which stimulates the posterior pituitary gland to secrete antidiuretic hormone to bring about actions that reduce water loss from the body.
   (c) The nervous system is involved in osmoregulation because the osmoreceptor cells in the hypothalamus, when they detect that the water potential of the blood has decreased, initiate nervous impulses that stimulate the posterior pituitary gland. When stimulated, these effector cells secrete antidiuretic hormone, which travels in the blood to the kidneys as part of the endocrine system.
   (d) ADH binds to receptors in membrane of collecting duct epithelial cells. Phosphorylase enzyme is activated. Vesicles containing aquaporins move to the cell surface membrane and fuse with it. Water then passes through aquaporins and into the lumen of the collecting duct.

6. (a) Sometimes true. ABA will bind to the receptors on the surface membrane of guard cells only during times of water stress.
   (b) Sometimes true. Potassium ions will move out of the guard cell by facilitated diffusion only during times of water stress.
   (c) Always true. Regulation of stomatal opening and closure is controlled by negative feedback.
   (d) Sometimes true. Water will move into guard cells by osmosis when water stress is not occurring.
   (e) Sometimes true. Calcium ions will move out of the vacuole by facilitated diffusion when water stress is not occurring.
   (f) False. The density of stomata is always greater on the lower surface of a leaf than the upper surface of a leaf.

Practice exam-style questions

E1. **A** [1]

E2. **A** [1]

E3. Enzymes in the liver remove the amine (NH_2) group from amino acids to form ammonia (this is called deamination). [1] This is then immediately combined with carbon dioxide to make urea. [1]

E4. (a) (i) Independent variable: concentration of ABA. [1]

Dependent variable: diameter of stomatal aperture. [1]

(ii) The student would need to take 50 cm^3 of the 1.0 mmol dm^{-3} ABA solution [1] and add 50 cm^3 distilled water. [1] This will produce 100 cm^3 of a solution of 0.5 mmol dm^{-3}. Repeating this process [1] using the same method would then produce a series of solutions, each half the concentration of the last.

(b) There is no significant correlation between the concentration of ABA and the stomatal aperture. [1]

(c) (i) Standardised variables should include: volume of the ABA solution applied to the leaf; [1] temperature at which the investigation is undertaken; [1] the light intensity of the room in which the investigation is undertaken; [1] the concentration of carbon dioxide in the atmosphere surrounding the epidermis peels; [1] the time of incubation; [1] and the species/ age of the leaf that is used. [1] [3 max.]

(ii) Volume of ABA solution: a pipette with high accuracy can be used to apply the solution to the leaf. [1] Temperature: the investigation could be carried out in a temperature-controlled room or chamber. [1] Concentration of carbon dioxide: experiment conducted in a controlled atmosphere. [1] Light intensity: experiment conducted in a room with constant lighting. [1]

(iii) Random sampling of stomata is required in order to avoid bias in the selection of stomata to measure. [1]

(d) A control is necessary in this investigation to show that it is the ABA, rather than the application of a volume of liquid or any other treatment associated with the investigation, that is responsible for the change in stomatal aperture. [1]

(e) A calibration curve can be plotted with the concentration of ABA on the x-axis and the stomatal aperture on the y-axis. [1] A line is extrapolated from the line-of-best fit to see where it intersects with the x-axis. [1] This represents an estimate of the minimum concentration that results in the change of the stomatal aperture.

(f) Although this is a low risk investigation, the plant material may be an allergen, so gloves must be worn throughout. [1] When preparing the epidermal peels, care should be taken when using the scalpel – it should be used with a tile and cutting should take place away from the body. [1] [1 mark max.]

E5. (a) The concentration of insulin in the blood before a meal was taken was 21 arbitrary units, and after 2 hours was 83 arbitrary units. The percentage increase is therefore (83 – 21) ÷ 21 = 295% [1]

(b) This is because the test had to ensure that the concentration of insulin and glucagon in the bloodstream was affected by only the glucose that was ingested at the start of the study. [1]

(c) The plotted line showing the concentration of insulin in the blood continuing to fall. [1] The plotted line showing the concentration of glucagon in the blood continuing to rise. [1] The plotted line showing the concentration of glucose in the blood continuing to fall. [1]

(d) Because glucagon stimulates the formation of cyclic AMP (cAMP) from ATP as a second messenger, the concentration of cAMP in liver cells will be similar to that of glucagon.

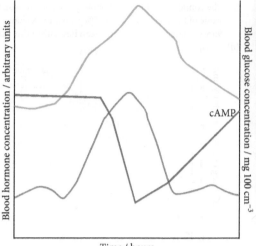

E6. (a) The glucose biosensor is a quantitative test for glucose, because it provides a numerical reading corresponding to the concentration of the sugar. [1]

(b) The glucose biosensor contains an enzyme (glucose oxidase). [1] The activity of all enzymes is variable and can depend on the surrounding pH, which will affect the shape of the active site and influence the likelihood of formation of enzyme–substrate complexes. [1]

(c) pH buffer(s). [1]

(d) The use of solutions of pH 2–3 and pH 10–11 presents a hazard, because these could be corrosive. Safety eyewear and gloves should be worn to minimise the likelihood that they come into contact with the eyes or skin. [1]

(e) Temperature would need to be standardised in this investigation because it would also be expected to affect the affinity of an enzyme for its substrate (and therefore K_m). [1] Temperature can be standardised by incubating the test-tubes containing the reacting molecules in a thermostatically controlled water bath. [1]

(f) The purpose of a control in an investigation is to ensure that the changes that are made to the independent variable (pH) are responsible for the changes observed in the dependent variable (K_m). [1] If K_m does not change in the investigation involving the control, then we can be more confident that it is pH that is responsible for the effects that are seen. A suitable control in this investigation would be to use a sample of boiled (denatured) enzyme.

(g) The graph shows that there are two troughs (low values). [1] When K_m is lowest, an enzyme has its greatest affinity for its substrate. [1] It is unlikely that an enzyme has two optimum pH values as the active sites would be different. [1]

CHAPTER 15 – CONTROL AND CO-ORDINATION

In-text questions

1. D

2. (a) An organ that produces a hormone; the hormone is secreted internally and is released directly into blood (in capillaries).

(b) A chemical messenger that travels in the blood.

(c) A specific organ or cells that the hormone acts upon. The target organ contains receptors specific for that hormone.

3. (a) Insulin is released from the β-cells of the pancreas and acts on liver and muscle cells.

(b) Glucagon is released from the α-cells of the pancreas and acts on liver cells.

(c) Antidiuretic hormone is released from the (posterior) pituitary gland and acts on the nephron / distal convoluted tubule / collecting duct.

4. (a) endocrine system – long duration of months or years and needs to be systemic
    (b) nervous system – response needs to be rapid, specific and only short duration needed
    (c) endocrine system – long duration of days or weeks and needs to be systemic
    (d) nervous system – uses sensory receptors involved in smell, hearing and vision

5. (a) Myelin sheath is made from Schwann cells. Schwann cells are formed of many layers of lipid / myelin around the axon; there are gaps called nodes of Ranvier between Schwann cells.
    (b) The sensory neurone carries the action potential / impulse for a longer distance and transmission / speed of impulse needs to be rapid (myelin speeds up transmission).

6. Antibodies will bind to myelin in the myelin sheath / Schwann cells; electrical insulation around neurones / axons / dendrites will be lost.

7. B

8. (a) active transport   (b) facilitated diffusion

9. (a) the potential outside the cell
    (b) (i) resting potential
        (ii) depolarisation
        (iii) action potential
        (iv) repolarisation
        (v) hyperpolarisation
    (c) in range 2.6–3.0ms
    (d) $\frac{1000}{3.0}$ to $\frac{1000}{2.6}$
        = answer in range 333–385

10. (a) sodium channels open; sodium ions (Na^+) flood in (to axon); charge difference across membrane becomes less
    (b) from -70 to +40mV
        change = 110mV

11. The action potential jumps along an axon between nodes of Ranvier. No conduction can occur where Schwann cells are located.

12. (a) calcium ions / Ca^{2+}
    (b) (i) exocytosis   (ii) diffusion
    (c) Acetylcholinesterase breaks down acetylcholine; this ensures action potentials are not stimulated on the post-synaptic side longer than required.
    (d) If acetylcholinesterase is inhibited, there will be reduced breakdown of acetylcholine, which compensates for reduced levels of acetylcholine; acetylcholine will therefore have more time to bind to receptors.

13. MSG increases the levels of glutamate in the brain. Glutamate (from MSG) binds to post-synaptic receptors; therefore, in presence of MSG, action potentials are stimulated at post-synaptic side when no action potential has arrived at the pre-synaptic side, or if depolarisation is not sufficient to stimulate another action potential.

14. (a) Calcium ions are released from the sarcoplasmic reticulum and bind to troponin, causing tropomyosin to move, exposing myosin binding sites on actin for the myofibril to begin contraction.
    (b) Ca^{2+} binds to troponin, causing it to change shape. This causes the troponin to pull the tropomyosin away from myosin's binding sites so actin can attach.
    (c) Tropomyosin covers myosin binding sites on actin in relaxed myofibrils; this

prevents myosin heads from binding; needs to be displaced to initiate contraction.
    (d) ATP hydrolysis provides the energy to move the myosin head towards the binding position with actin.

15. (a) myosin; and actin at either end
    (b) actin only   (c) myosin only

16. (a) The dark band contains myosin; myosin filaments do not change in length.
    (b) The light band is where actin is not overlapping with myosin. During contraction, the region of overlap at each side grows longer shortening the region that has actin only.

17. Any suggestion relating to storing or releasing calcium ions being less efficient, as the sarcoplasmic reticulum structure will be affected. Any suggestion of muscle weakness or actin fibres coming apart from each other or apart from the rest of the muscle fibre.

18. (a) $\frac{(7.8 \times 10^{-2})}{(5 \times 10^{-9})}$
        = 1.56 x 10^7
    (b) $\frac{(1.56 \times 10^7)}{0.250}$
        = 6.24 x 10^7

19. (a) leaves / leaf lobes
    (b) touch / pressure / mechanical stimulation
    (c) depolarisation
    (d) Any two from: sodium (Na^+); calcium (Ca^{2+}); chloride (Cl^-); potassium (K^+).

20. Auxin regulates the growth of stems and roots, and promotes or inhibits cell elongation. Plants do not have endocrine glands / endocrine systems.

21. (a) In plant stems, auxin stimulates cell elongation; it controls growth of the stem and bending of the stem.
    (b) In plant roots, auxin inhibits cell elongation; it controls growth of the root and bending of the root.

22. protons / hydrogen ions / H^+ and expansins

23. (a) gibberellin(s)
    (b) uptake of water into the embryo
    (c) It stimulates transcription of the amylase gene; amylase breaks down starch into maltose in the endosperm; glucose is produced; glucose is used in respiration / used to make ATP.

24. Ions can freely move across the cell surface membrane when made completely permeable. The concentration of the ions in the cytoplasm will become the same as that of any external solution due to diffusion. This would not happen in untreated cells, as the hydrophobic part of the lipid bilayer does not allow ions to move across by simple diffusion. Ions in untreated cells move through channels, which the cell can close to prevent movement of ions.

Assignment 15.1

A1. Speed of transmission increases with axon diameter for both myelinated and non-myelinated neurones. The relationship between speed and diameter is not linear in both / is linear for smaller diameters and not linear for larger diameters. The rate of increase of speed with diameter of myelinated axons is greater. The speed of transmission for the same diameter axon is / would be much faster in myelinated than non-myelinated. Saltatory conduction in myelinated axons causes the speed of transmission to be greater, and salutatory conduction cannot occur in non-myelinated neurones. Comparative data e.g. speed is 20ms^{-1} in 2.6μm diameter myelinated neurone and 20ms^{-1} in 600–700μm diameter non-myelinated neurone.

A2. (a) 88ms^{-1}

(b) time = $\dfrac{\text{distance}}{\text{speed}}$

$= \dfrac{1.76}{88}$

$= 20$ (ms)

A3. faster than 30ms^{-1}

A4. The difference makes withdrawal reflex of head faster; head can be removed from danger / protected faster; more likely to survive damage to tail.

Experimental skills 15.1

P1. (a) Independent variables: presence or absence of gibberellin; presence or absence of light; temperature

(b) percentage of seeds germinated

P2. Bar chart with bars in pairs *either* water and GA *or* light and dark; percentage germination on vertical axis; temperature on horizontal axis (as temperatures are not equally spaced, horizontal axis can have non-linear scale); bars drawn to correct height and labelled or key used.

P3. Gibberellin is necessary for germination because only a few seeds germinated without it. GA exposure to light increases percentage of germination in both GA and water treated seeds. About $\dfrac{20°C}{15}$ – 20°C gives the highest percentage germination / best temperature for germination. Temperature makes more difference in percentage germination in 0 GA and light and in GA and dark than in GA and light; (any comparative data quote).

P4. (a) For example, chi-squared test; compare dark with light at each temperature; null hypothesis: there is no difference in the results between light and dark; either dark or light can be observed and the other expected; check chi-squared result for significance using table of p values; degrees of freedom = 1; if result is greater than the critical value at p = 0.05 then null hypothesis is rejected and results are significant.

(b) Any three from: use more concentrations of GA; use temperatures below 15°C; use smaller temperature intervals; repeat with more batches of 100 and calculate means.

Chapter review

1. (a) Insulin acts as a chemical messenger; it travels in the blood; is produced in an endocrine gland; and has specific target cells / target organ(s).

(b) Many different types of cells can have receptors for adrenaline. Adrenaline travels in the blood so will reach all parts of the body.

2. Any two from: the endocrine system is used for longer-term effects than the nervous system; the endocrine system is used for more systemic effects / effects on many organs, whereas the nervous system is for more specific effects; the endocrine system uses no physical connections whereas neurones connect cells / organs in nervous system; the endocrine system uses the blood whereas the nervous system uses neurones; the endocrine system uses hormones whereas the nervous system uses action potentials.

3. (a) Mitochondria synthesise ATP for active transport of Na$^+$ and K$^+$ / for the sodium-potassium pump.

(b) Dendrites form connections with other neurones.

(c) Voltage-gated sodium channels close to maintain resting potential; they open to create an action potential.

4. This is a period in which another stimulus will not produce an action potential / period of time that limits the frequency of action potentials. The axon membrane is depolarised; Na$^+$ channels are already open at the start of this period; the axon membrane is repolarising; the axon membrane is hyperpolarised; both Na$^+$ and K$^+$ channels are closed towards the end of this period.

5. Arrival of an action potential at the presynaptic side causes Ca^{2+} channels to open; the presence of Ca^{2+} causes vesicles to fuse with the presynaptic membrane and releasing acetylcholine. Acetylcholine / neurotransmitter diffuses across the synaptic cleft and binds to receptors on the post-synaptic side. This causes Na$^+$ channels to open and new action potential is stimulated.

6. (a) Ca^{2+} binds to troponin, causing it to move tropomyosin away from the binding sites for the myosin heads on actin; the binding sites for myosin heads on actin become exposed so that myosin heads now bind with actin. Myosin heads rotate pulling the actin filament; this is the power stroke. ATP attaching to myosin head makes the myosin head release actin. ATP hydrolysis bends the myosin heads towards the next set of binding sites, further along the actin filament.

(b) Respiration stops so ATP synthesis stops. ATP is required to release myosin heads from actin, so if myosin heads are still attached to actin, the muscle cannot relax.

7. Auxin causes active transport of protons / hydrogen ions / H$^+$ across the cell surface membrane from the cytoplasm into the cell wall. This increases the acidity of the cell wall, which causes it to become flexible. Auxin also stimulates release of expansins into the cell wall, which further increase its flexibility. The presence of protons in the cell wall also causes breakdown of cellulose cross-links. Auxin causes channels for potassium ions to open; K$^+$ moves into the cytoplasm causing the water potential of the cytoplasm to be lowered. Water therefore moves into the cell by osmosis; this increases turgor pressure / causes the cytoplasm to swell and the flexible cell wall expands.

8. Gibberellin speeds up germination; makes seeds germinate at the same time. This will increase percentage germination.

Practice exam-style questions

E1. (a) (i) neurotransmitter / acetylcholine [1]

(ii) calcium ions / Ca^{2+} [1]

(iii) sodium ions / Na$^+$ [1]

(b) depolarises the membrane; [1] sets up a new action potential [1]

(c) Any three from: breaks down acetylcholine / neurotransmitter / substance A; [1] removes substance that can bind to receptors on post-synaptic side; [1] prevents excess action potentials at post-synaptic side; [1] products (of enzyme) enable acetylcholine / neurotransmitter to be resynthesised [1]

(d) Any four from: attaches to acetylcholine receptor; [1] blocks binding of acetylcholine to receptor; [1] inhibits action potential on post-synaptic side; [1] action potentials arriving at pre-synaptic side will have no effect; [1] impulses sent to salivary gland will not reach destination. [1]

E2. (a) (i) receptor cell / named receptor e.g. Pacinian corpuscle [1]

(ii) myelin sheath / Schwann cell [1]

(iii) dendrite [1]   (iv) axon [1]

(b) contains mitochondria; [1] ATP production; [1] for sodium-potassium pump / active transport of Na$^+$ and K$^+$ [1]

(c) (i) resting potential [1]

(ii) The threshold potential is -50mV; [1] stimuli less than the threshold produce no action potential; [1] action potentials do not increase with stimulus strength / all or nothing principle above threshold. [1]

E3. (a) (i) myosin [1]   (ii) actin [1]   (iii) Z line or Z disc [1]

(b) Four lines similar to those labelled A in the central position and two sets of alternate horizontal lines similar to those labelled B that are closer together and have a smaller gap between them; [1] two vertical lines similar to those labelled C closer together [1]

(c) (i) Calcium ions are released from the sarcoplasmic reticulum to trigger contraction; [1] binds to troponin to move tropomyosin away from myosin binding sites [1]

(ii) Myosin heads bind to actin; [1] they move to pull actin during power stroke. [1]

E4. (a) Any five from: respond to changes in (external) environment; [1] adapting to increase chance of survival; [1] respond to internal changes; [1] rapid responses in mammals e.g. withdrawal from danger; [1] rapid responses in plants e.g. Venus fly trap closure; [1] longer-term control in mammals e.g. control of blood glucose; [1] longer-term control in plants e.g. control of growth. [1]

(b) Any five from: mammals have a nervous system; [1] mammals have brain / central nervous system / central point of co-ordination; [1] mammals have direct connections between distant cells / organs; [1] mammals have an endocrine system; [1] both have cells that can produce action potentials; [1] both produce growth regulating substances; [1] both have fast and slow response systems. [1]

E5. (a) (i) pH on vertical axis and time in minutes on horizontal axis; [1] both axes with linear scales and zero on time axis so points cover at least half the grid in both directions; [1] all points plotted to within 1mm. [1]

(ii) smooth curve passing through all points [1]

(b) tangent drawn at 15 minutes; [1] change in y and change in x determined correctly (can be shown on graph or in equation for gradient); [1] gradient correctly determined from their values, ignore negative sign (gradient will be in region of 0.06 or 6.0×10^{-2} if plotted correctly). [1]

(c) Any eight from:
independent variable
use of at least 5 different concentrations of auxin (ignore values) [1]
method of getting auxin into agar, e.g. soak agar in auxin / make agar containing auxin [1]
range includes zero / includes agar soaked in distilled water [1]
dependent variable
measure change in height of stems / growth of stems in a specified time [1]
processing of results

$$\text{growth rate} = \frac{(\text{Change in height})}{(\text{time taken})} \quad [1]$$

control variables
use same type / species / variety of plant [1]
use plants of same age [1]
same soil / same minerals / same watering [1]
same temperature [1]
same light intensity [1]
agar blocks applied same length of time after tip removed [1]
procedures
detail of how change in height is recorded, e.g. photographing / marking start height on a scale [1]
making repeat measurements and calculation of mean [1]
safety
low risk procedure / wear gloves when handling auxin / care with sharp blade for cutting tips [1]

CHAPTER 16 – INHERITANCE

In-text questions

1. This is because it reduces the number of chromosomes in a cell by half (to make haploid gametes).

2.

Mitosis	Meiosis
one division	two divisions
two daughter cells produced	four daughter cells produced
daughter cells have the same number of chromosomes as the parent cell	daughter cells have half the number of chromosomes compared with the parent cell
homologous chromosomes do not pair up	homologous chromosomes pair up as bivalents in prophase I

no crossing over takes place	crossing over takes place during prophase I
daughter cells are genetically identical to the parent cell	daughter cells are not genetically identical to the parent cell – they are unique

3. BOX 1 joins with BOX E (At the end of meiosis I, there are two cells, each containing one copy of each chromosome from a homologous pair consisting of two sister chromatids).
BOX 2 joins with BOX B (At the end of meiosis II, there are four cells, each containing one copy of each chromosome from a homologous pair consisting of one chromatid).
BOX 3 joins with BOX A (Random orientation is due to random alignment of bivalents on either side of the equator in metaphase I).
BOX 4 joins with BOX D (Crossing over is due to the close proximity of bivalents on either side of the equator in metaphase I)
BOX 5 joins with Box C (Random fertilisation means that any paternal gamete can fertilise any maternal gamete).

4. (a) 37

(b) (i) The possible number of unique gametes produced by random orientation is found by the expression 2^n (where n = the haploid number of chromosomes of that species). Therefore, in the case of bears, $2^n = 2^{37} = 1.374 \times 10^{11}$ (nearly 137.5 billion).

(ii) During prophase I, non-sister chromatids of homologous chromosomes in a bivalent come into contact and form chiasmata. At these points, genetic material is swapped between them, resulting in the formation of unique hybrid chromosomes that consist of a mixture of maternal and paternal alleles.

(c) During prophase I, chromosomes condense (shorten and thicken). Because bears have more chromosomes than humans (74 versus 46), this process is likely to take a longer period of time.

5. (a) A variant of a gene.

(b) An allele which, if present just once in the genotype, has an effect on the phenotype.

(c) An allele which has an effect on the phenotype of a heterozygous individual when present with another allele in the genotype of an individual.

6. (a) The allele for black fur (B) is dominant, because only those individuals that have a recessive homozygous(genotype (bb) are brown.

(b) Inbreeding involves the production of offspring between two closely-related individuals. If in a small population there are some individuals that carry the recessive (b) allele, then it is more likely that offspring will be produced that have two copies (genotype bb) and are therefore born with brown fur.

7. (a) These alleles are codominant. This is because a person with one normal allele and one mutant allele has a mild form of the disease, indicating that both alleles have an effect on the phenotype in individuals who are heterozygous for these alleles.

(b)

Genotype	Phenotype
AA or Hb^A Hb^A	unaffected
AS or Hb^A Hb^S	sickle cell trait
SS or Hb^S Hb^S	sickle cell anaemia

Genotype	Phenotype
FF	purple flowers
Ff	purple flowers
ff	white flowers

8. (a)
   (b) A heterozygous plant with purple flowers has the genotype Ff. A white-flowered plant has the genotype ff. A genetic diagram can be used to predict the ratio of phenotypes in the offspring:

Phenotypes of parents:	PURPLE FLOWERS × WHITE FLOWERS
Genotypes of parents:	Ff × ff
Gametes:	F and f and f
Genotypes of offspring:	Ff and ff
Phenotypes of offspring:	PURPLE FLOWERS and WHITE FLOWERS

The ratio of purple flowers : white flowers is predicted to be 1 : 1.

   (c) (i) The ratio would be 952 : 1020 = 14 : 15. This can be further simplified to 1: 1.07
       (ii) The ratio of 14 : 15 is very close to the ratio 1 : 1 predicted in the genetic diagram. It is not exactly equal, however, because the ratios of phenotypes are predictions based on probabilities. The numbers of offspring are high; however, they are not high enough to provide an exact ratio of 1 : 1 due to chance effects and sampling variation.

9. The four ABO blood group phenotypes are as follows: A, B, AB and O.

   Therefore, the only combination of parental genotypes that would give these phenotypes in the offspring are $I^A I^o$ and $I^B I^o$, as shown in the Punnett square below:

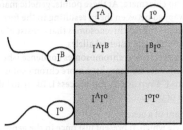

	I^A	I^o
I^B	$I^A I^B$	$I^B I^o$
I^o	$I^A I^o$	$I^o I^o$

10. Because these parents (who are supertasters) have at least one child who is not a supertaster, this would suggest that both parents are heterozygous for the dominant allele of the *TAS2R38* gene. A Punnett square can be used to predict the probability of these parents having a child who is a supertaster as follows:

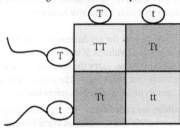

	T	t
T	TT	Tt
t	Tt	tt

A child who is a supertaster would have the genotype TT or Tt – the probability of this is 0.75.

11.
phenotype of parents:	Black male × orange female
genotype of parents:	$X^B Y$ × $X^o X^o$
gametes:	X^B and Y × X^o
genotypes of offspring:	$X^B X^o$, $X^o Y$
phenotypes of offspring:	Tortoiseshell female & orange male

phenotype of parents:	orange male × black female
genotypes of parents:	$X^o Y^o$ × $X^B X^B$
gametes:	X^o and X^o × $X^B Y$
genotypes of offspring:	$X^o Y^B$ × $X^o Y$
phenotypes of offspring:	Tortoiseshell female & orange male

and

phenotype of parents:	orange female × black male
genotype of parents:	$X^o Y$ × $X^B Y^B$
gametes:	X^o and Y × X^B
genotypes of offspring:	$X^o X^B$ × X^o
phenotypes of offspring:	tortoiseshell female orange male

12. (a) A red snapdragon plant has the genotype $C^R C^R$ and a white plant has the genotype $C^W C^W$.

   A genetic diagram can be used to show the expected phenotypes of offspring in the F1 generation:

Phenotypes of parents:	Red × White
Genotypes of parents:	$C^R C^R$ × $C^W C^W$
Gametes:	C^R and C^W
Genotypes of offspring:	$C^R C^W$
Phenotypes of offspring:	pink

   (b) Crossing the offspring of the first cross (the F1 generation), which have the genotype $C^R C^W$, gives the ratios of phenotypes in the F2 generation as shown:

Phenotypes of parents:	PINK × PINK
Genotypes of parents:	$C^R C^W$ × $C^R C^W$
Gametes:	C^R and C^W C^R and C^W
Genotypes of offspring:	$C^R C^R$ $C^R C^W$ $C^W C^W$
Phenotypes of offspring:	RED PINK WHITE

Ratio of phenotypes: 1: 2: 1

   (c) A Punnett square can be used to predict the ratio of phenotypes in the F2 generation:

	C^R	C^W
C^R	$C^R C^R$	$C^R C^W$
C^W	$C^R C^W$	$C^W C^W$

This predicts that the offspring of this cross would occur in a 1 red : 2 pink : 1 white ratio.

13. For a girl to be born with red-green colour-blindness, she would need to inherit two X chromosomes that each have a copy of the mutant recessive allele. Her father must therefore have the condition (as he has only one X chromosome), and would therefore have the genotype $X^r Y$; her mother can either be a carrier of the condition ($X^R X^r$; heterozygous) or also affected ($X^r X^r$; homozygous recessive).

14. (a) It is not possible for a father to pass on the allele that causes haemophilia to his son. This is because a son inherits the Y sex chromosome from his father, which does not carry the allele responsible.

(b) It is always the case that a father passes on the allele that causes haemophilia to his daughter. This is because a daughter inherits the X sex chromosome from her father, which carries the allele responsible.

(c) Assuming that the man with haemophilia has a daughter who then passes the X-chromosome she inherited from him onto her son, it is possible for a grandson to be affected with haemophilia.

15. (a) In this example, the allele for tall stem can be represented by the letter T, and the allele for short stem is t. The allele for purple flowers can be represented by the letter F, and the allele for white flowers is f. Plants that are heterozygous for both genes therefore have the genotype TtFf. A genetic diagram can therefore be drawn that shows the expected genotypes and phenotypes in the offspring as follows.

Phenotypes of parents:	Tall stem, purple flowers × Tall stem, purple flowers			
Genotypes of parents:	Tt Ff	Tt Ff		
Gametes:	TF tF Tf tf × TF tF Tf tf			
Genotypes of offspring:	TTFF TtFF TTFf TtFf, TTff Ttff, ttFF ttFf, ttff			
Phenotypes of offspring:	Tall stem, purple flowers	Tall stem, white flowers	dwarf stem, purple flowers	dwarf stem, white flowers

(b) A Punnett square can be used to predict the ratios of phenotypes in the cross of two plants with genotype TtFf:

	TF	Tf	tF	tf
TF	TTFF *	TTFf *	TtFF *	TtFf *
Tf	TTFf *	TTff **	TtFf *	Ttff **
tF	TtFF *	TtFf *	ttFF ***	ttFf ***
tf	TtFf *	Ttff **	ttFf ***	ttff ****

Tall, purple: 9 *

Tall, white: 3 **

Short, purple: 3 ***

Short, white: 1 ****

16. (a) (i) An example could be LLGG (this fly is homozygous dominant for both genes).

(ii) An example could be AAee (this fly is homozygous dominant for aristae and homozygous recessive for eye colour).

(iii) An example could be wwbb (this fly is homozygous recessive for both genes).

(b) (i) Linked genes have loci that are on the same chromosome.

(ii) If genes are linked, then random orientation and independent segregation will not separate the alleles of an individual during meiosis. Unless crossing over occurs to separate them onto different chromosomes in a bivalent during prophase I, they will not be inherited independently. Therefore, the predicted ratios of phenotypes may not be seen in the offspring.

(iii) Linked genes that are closer together on a chromosome are less likely to be exchanged between homologous chromosomes during prophase I, as the probability of a chiasma forming between them is less. Therefore, the percentage of recombinant flies for eye colour and aristae length would be larger because these genes are further apart than the other linked genes.

(iv) The possible genotypes of a fly with long legs and a grey body (which are both dominant traits) would be LLBB, LLBb, LlBB and LlBb. To identify which is the most likely genotype of this fly, a test cross could be performed with a pure-breeding fly that is homozygous recessive for both of these genes (genotype llbb), which produces gametes of genotype lb. The following ratios of offspring would be expected if these crosses were performed:

If the genotype of the fly is LLBB:

	LB
lb	LlBb

So, all offspring would have long legs and a grey body.

If the genotype of the fly is LLBb:

	LB	Lb
lb	LlBb	Llbb

So, the ratio of the offspring would be 1 long legs, grey body : 1 short legs, grey body

If the genotype of the fly is LlBB:

	LB	lB
lb	LlBb	llBb

So, the ratio of the offspring would be 1 long legs, grey body : 1 short legs, grey body.

If the genotype of the fly is LlBb:

	LB	Lb	lB	lb
lb	LlBb	Llbb	llBb	llbb

So, the ratio of the offspring would be 1 long legs, grey body : 1 long legs, black body : 1 short legs, grey body : 1 short legs, black body.

17. (a) Different alleles of a gene arise due to mutation.

(b) This is an example of epistasis, because one step in the pathway of pigment production is blocked. Allele L encodes a linamarase enzyme that has an active site complementary to its substrate, while allele l does not. In the absence of allele L, no substrate (cyanogenic glucoside) can be converted into cyanide.

(c) The genotype of a plant heterozygous for both genes is LlGg. A Punnett square can be used to predict the ratio of phenotypes in the F1 generation:

	LG	Lg	lG	lg
LG	LLGG *	LLGg *	LlGG *	LlGg *
Lg	LLGg *	LLgg **	LlGg *	Llgg **
lG	LlGG *	LlGg *	llGG ***	llGg ***
lg	LlGg *	Llgg **	llGg ***	llgg ***

Therefore, the phenotypes would be seen in the following ratios:

Rapid cyanide release (L–G–) = 9 *

Slow cyanide release (L–gg) = 3 **

No cyanide release (ll––) = 4 ***

(d) (i) Based on the Punnett square diagram in question 16c, the expected ratio of offspring of the three phenotypes would be:

Rapid cyanide release (L–G–) = 9

Slow cyanide release (L–gg) = 3

No cyanide release (ll––) = 4

Because the total number of offspring in this cross is 240, the expected numbers of plants with these three phenotypes are 135, 45 and 60, respectively.

(ii) The table should be completed as follows:

Phenotype	Observed number (O)	Expected number (E)	(O – E)	(O – E)²	$\frac{(O-E)^2}{E}$
Rapid cyanide release	145	135	10	100	0.74
Slow cyanide release	43	45	–2	4	0.09
No cyanide release	52	60	–8	64	1.07
Total number	240	240			$\sum\frac{(O-E)^2}{E}$ 1.90 (χ^2)

The chi-squared value for this analysis is 1.90. Comparing this with the table of critical values at a probability level of $P < 0.05$ (chi-squared = 5.991), and using two degrees of freedom, shows that the calculated value is lower than the critical value, and therefore the null hypothesis should be accepted. That is, there is no significant difference between the observed and expected numbers of offspring with these phenotypes.

18. (a) The ratio of phenotypes in the offspring of a black and gold dog does not significantly differ from 9 black : 4 gold : 3 brown.

(b) A table should be drawn that is similar to that below:

Phenotype	Observed number (O)	Expected number (E)	(O – E)	(O – E)²	$\frac{(O-E)^2}{E}$
black	53	45	8	64	1.42
gold	10	20	–10	100	5
brown	17	15	2	4	0.27
Total number	80	80			$\sum\frac{(O-E)^2}{E}$ 6.69 (χ^2)

The chi-squared value for this analysis is 6.69. Comparing this with the table of critical values at a probability level of $P < 0.05$ (chi-squared = 5.991), and using two degrees of freedom, shows that the calculated value is higher than the critical value, and therefore the null hypothesis should be rejected. That is, there is a significant difference between the observed and expected numbers of offspring with these phenotypes.

(c) The chi-squared test uses actual numbers, and the number of offspring in these crosses (total: 80) may not be sufficient in order to draw a conclusion. In addition, there may be other genes that interact in these different pairs of dogs, which could make their comparison invalid.

19.

enzyme name	encoded by a gene which is		acts as an enzyme which is	
	regulatory?	structural?	repressible?	inducible?
β galactosidase	✗	✓	✗	✓
lactose transacetylase	✗	✓	✗	✓
RNA polymerase	✗	✓	✗	✗

20.

In the absence of gibberellin

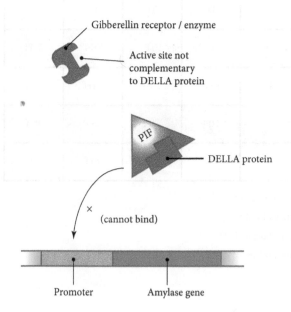

Gibberellin receptor / enzyme

Active site not complementary to DELLA protein

PIF

DELLA protein

× (cannot bind)

Promoter   Amylase gene

In the presence of gibberellin

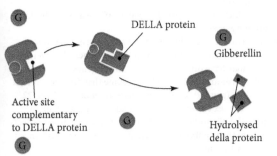

DELLA protein

Gibberellin

Active site complementary to DELLA protein

Hydrolysed della protein

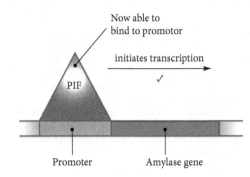

Now able to bind to promotor

initiates transcription
✓

PIF

Promoter

Amylase gene

21. (a) (i) Long stem (tall).
     (ii) Long stem (tall).
     (iii) Short stem (short).
   (b) This observation indicates that the protein encoded by the *le* allele is non-functional and is required for the production of the gibberellin hormone.

Assignment 16.1

A1. During pregnancy, the embryo / fetus will need to obtain oxygen from its mother. Therefore, the affinity of embryonic / fetal haemoglobin will need to be higher than adult haemoglobin.

A2. (a) False. At birth, the percentage of haemoglobin that consists of beta-globin is 14%. At six months, it is 46%. The percentage increase is therefore $(46 - 14) \div 14 = 228.6\%$.
   (b) False. Although 50% of the haemoglobin consists of alpha-globin at birth, only 14% of the haemoglobin consists of beta-globin. The remaining 36% consists of other globin proteins.
   (c) True. Embryonic haemoglobin consists of epsilon and zeta-globin molecules. As shown in the graph, at 10 weeks after fertilisation, the percentage of these globin molecules in haemoglobin has fallen to zero.
   (d) False. Alpha-globin and gamma globin are not expressed until 2 weeks after fertilisation. Epsilon and zeta-globin molecules appear almost immediately after fertilisation.

A3. Hydroxyurea stimulates the expression of the gamma globin gene to increase the production of the gamma globin protein. This combines with alpha-globin to form fetal haemoglobin. This is able to form haemoglobin that does not result in the problems seen with adult haemoglobin, because in these people (with sickle cell anaemia) the beta-globin protein has the amino acid substitution associated with the disease.

A4. (a) The GATA transcription factor will be attached to the promoter upstream of the zeta gene when zeta-globin protein is detectable in the embryo, between 1 and 10 weeks after fertilisation.
   (b) An operon consists of a group of regulatory and structural genes. The alpha-globin gene cluster has two structural genes (zeta-globin and alpha-globin), but no genes that produce proteins that regulate their expression.

(c)

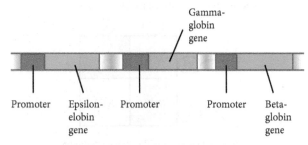

Gamma-globin gene

Promoter   Epsilon-elobin gene   Promoter   Promoter   Beta-globin gene

A5. (a) This is because newborn boys have only one X-chromosome, which means that if the recessive allele for GATA1 is present, he will not be able to produce a functional transcription factor. In girls, two X chromosomes are present, meaning that even if she carries a recessive allele for this gene, she is likely to have an allele on her other X chromosome that produces a functional protein.
   (b) A mutation in a gene encoding a transcription factor may produce a protein that does not recognise the promoter sequence, or is unable to recruit RNA polymerase to initiate transcription.

Experimental skills 16.1

P1. (a) (i) Pure-bred organisms are those that are homozygous for a particular allele of a particular gene.
     (ii) The F1 generation is the first generation of organisms produced by crossing two pure-bred parents. They are usually heterozygous for the gene that is being analysed.
   (b) A test cross occurs when an individual of unknown genotype is crossed with an individual of known genotype, which is usually homozygous recessive. In this case, the cross occurred with red-eyed females, which are homozygous dominant.

P2. (a) A dominant allele is one that has an effect on the phenotype when it is present just once in the genotype.
   (b) The red pigment may be produced by a series of enzyme-controlled reactions. The allele that gives rise to white eyes may produce an enzyme (protein) that is non-functional.
   (c) The genetic diagram should be completed as follows:

phenotypes of parents:	red-eyed female	×	red-eyed male
genotypes of parents:	$X^R X^r$	×	$X^R Y$
gametes:	$\underline{X^R}$ and $\underline{X^r}$	×	$\underline{X^R}$ and $\underline{Y}$
genotypes of offspring:	$\underline{X^R X^R}$ and $\underline{X^R X^r}$ and $\underline{X^R Y}$ and $\underline{X^r Y}$		
phenotypes of offspring:	red-eyed female and	red-eyed female and	red-eyed male and white-eyed male

   (d) As is the case in humans, male fruit flies have only one X chromosome. Therefore, if a recessive allele is found on the chromosome, it will have an effect on the phenotype as there is never a dominant allele present to hide its effect.

P3. (a) The ratio of phenotypes in the offspring produced as a result of this cross will not significantly differ from 1 red : 2 orange : 1 yellow.
   (b) There is a probability of less than 0.05 of this ratio being seen by chance.

(c) (i) A Punnett square can be used to predict the ratios of phenotypes of the offspring of this cross:

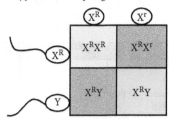

The expected numbers of flies can be found by multiplying the expected fractions by the total number of flies (61+101+58 = 220). Because we would expect half of the flies to be red-eyed females (genotypes X^RX^R and X^RX^r), then $0.5 \times 220 = 110$ would be the expected number. A quarter of the flies would be expected to be red-eyed males (genotype X^RY: $0.25 \times 220 = 55$ flies), and a quarter of the flies would be expected to be white-eyed males (genotype X^rY: $0.25 \times 220 = 55$ flies).

(ii) The table should be drawn that is similar to that below:

Phenotype	Observed number (O)	Expected number (E)	(O − E)	(O − E)²	$\frac{(O-E)^2}{E}$
red-eyed male	61	55	6	36	0.65
red-eyed female	101	110	−9	81	0.74
white-eyed male	58	55	3	9	0.16
total number	220	220		$\sum\frac{(O-E)^2}{E}$	1.55 (χ^2)

(iii) The chi-squared value for this analysis is 1.55. Comparing this with the table of critical values at a probability level of $P < 0.05$ (chi-squared = 5.991), and using two degrees of freedom, shows that the calculated value is lower than the critical value, and therefore the null hypothesis should be accepted. That is, there is no significant difference between the observed and expected numbers of offspring with these phenotypes.

P4. Any two from: they are small and require little care, so can be kept in high numbers in the lab; like humans, male fruit flies have the sex chromosomes XY and the females have XX; males and females are easily distinguishable; they have a short life cycle and many eggs are produced by one female.

Chapter review

1. Meiosis, which involves two consecutive cell divisions, is responsible for halving the chromosome number by reduction, which produces haploid gametes. During meiosis, variation is introduced to the gametes by the processes of crossing over and random orientation during prophase I.

2. (a) Chickens will have 39 bivalents (pairs of homologous chromosomes).
(b) Chickens will have 78 centrosomes (which will be at opposite poles of the cell) during metaphase I. These will attach to the separate centromeres of each member of the bivalents.
(c) Chickens will have 78 centrosomes (which will be at opposite poles of the cell, but at 90 degrees to those that form in metaphase I) during metaphase II. These will attach to the

centromeres of each chromosome and will separate the sister chromatids.
(d) 2^n in chickens is 2^{39}, which is 5.498×10^{11} (nearly 550 billion unique gametes).

3. (a) (i) Homologous chromosomes are the same size and contain the same genes, at the same loci. However, they may have different alleles for these genes.
(ii) An autosome is any chromosome that does not control sex (gender). In humans, these are numbered 1 to 22. The sex chromosomes (non-autosomes) are pair 23.
(iii) The phenotype of an organism is the measurable effects of the genotype. The phenotype can also be influenced by the environment.
(b) During metaphase and anaphase of meiosis, it is possible that the spindle fibres attach to the centromeres of both homologous chromosome 21 members (in meiosis I, or sister chromatids of chromosome 21 in meiosis II), and pull both structures to the same pole of the cell. If this occurs, one of the gametes that is produced will have two copies of chromosome 21, which would then result in a zygote with three copies of the chromosome, if fertilised.
(c) For an individual with Down's syndrome to be affected by a genetic disorder caused by a recessive mutant allele of a gene found in chromosome 21, they would need three copies of that allele, rather than two.

4. (a) Anaphase II (b) Metaphase I
(c) Prophase I (d) Anaphase II
(e) Anaphase II (f) Telophase II

5.

MEIOSIS I
- Separates homologous chromosomes.
- crossing over occurs.
- Takes place in diploid cells

(overlap)
- Have 4 sub-stages (prophase, metaphase, anaphase, telophase).
- Requires cytokinesis.
- Spindle is required to separate genetic material.

MEIOSIS II
- Separates sister chromatids.
- Takes place in haploid cells
- Produces cells that are fully-formed gametes.

6. (a) Never true: although human males have 22 pairs of homologous chromosomes (chromosomes of the same size, with the same genes at the same loci), their 23rd pair of chromosomes are X and Y, which do not have these features.
(b) Never true: crossing over occurs only during prophase I, when non-sister chromatids of homologous chromosomes exchange genetic material.
(c) Sometimes true: if the woman (who has two X chromosomes that carry the mutant recessive allele) has a son, then he will definitely suffer from haemophilia. This is because sons inherit an X chromosome from their mother. However, assuming that the father of her child does not have haemophilia, then her daughters will be carriers of the mutant allele and they will be healthy. This is because daughters inherit one of their X chromosomes from their father.
(d) Always true: a test cross involves mating an individual of unknown genotype with an individual known to be homozygous recessive for that allele.

(e) Never true: because girls have two X chromosomes, the probability of them inheriting a dominant allele that causes disease is greater.

7. (a) Chromosome X is the odd one out as it does not have a homologous pair (the Y chromosome is smaller and has different genes at different loci), and because it is a sex chromosome and not an autosome.

    (b) Albinism in humans is the odd one out because it is a trait that is not controlled by codominant alleles. The alleles that determine the presence of melanin in the skin are either dominant or recessive. In the case of ABO blood group and petal colour in snapdragons, alleles exist that give a codominant phenotype (blood group AB and pink colour, respectively).

    (c) Dominant is the odd one out as it describes an allele; heterozygous and homozygous describe genotypes.

    (d) Genotype is the odd one out because it describes the combination of alleles that an individual has for a given gene, found at the same loci on different homologous chromosomes. Genes and alleles are found at one locus only.

    (e) Anaphase II is the odd one out because at this stage the homologous pairs of chromosomes have been separated into different cells.

8. (a) A 3 : 1 ratio would be seen in the phenotypes of offspring of a monohybrid cross between two heterozygotes that have a gene with a dominant and a recessive allele.

    (b) A 1 : 1 ratio would be seen in the phenotypes of offspring of a monohybrid cross between an individual with heterozygous genotype and an individual with a homozygous recessive genotype that have a gene with a dominant and a recessive allele. This is commonly known as a test cross (although it is also the case in the inheritance of Huntington's disease).

    (c) A 1 : 2 : 1 ratio would be seen in the phenotypes of offspring of a monohybrid cross between two heterozygous individuals that have a gene with a codominant allele.

    (d) A 9 : 3 : 3 : 1 ratio would be seen in the phenotype of offspring of a dihybrid cross between two individuals that are heterozygous for two unlinked genes that each have a dominant and a recessive allele.

    (e) A 9 : 4 : 3 ratio would be seen in the phenotype of offspring of a dihybrid cross between two individuals that are heterozygous for two genes that each have a dominant and a recessive allele that show epistasis.

9. The correct order would be:
    3. lactose binds to repressor
    2. repressor undergoes a conformational change
    4. repressor detaches from the operator
    1. RNA polymerase transcribes structural genes

10.

	Regulation of:	
	lactase synthesis in bacteria	amylase synthesis in plants
involves the expression of structural genes	✓	✓
involves the expression of regulatory genes	✓	✓
requires repressible enzymes	✓	✗
requires inducible enzymes	✓	✗
involves negative feedback	✓	✗

Practice exam-style questions

E1. Answer: D. [1] During DNA replication in S-phase of the cell cycle (1), the DNA content of a cell doubles and each chromosome consists of two sister chromatids. During meiosis (2), two cell divisions take place that reduces the DNA content to a quarter of what it was before, in order to make haploid gametes. At fertilisation (3), the DNA content of the cell doubles as homologous chromosomes are brought together.

E2. (a) The two tailless cats are heterozygous (genotype Tt). Therefore:

parents phenotypes:	Tailless × Tailless
parents genotypes:	Tt × Tt
gametes:	T and t × T and t
offspring genotypes:	TT and Tt and tt
offspring phenotypes:	normal (tailed) taillers dies before birth

Kittens with the genotype TT will have tails.

    (b) In the absence of cats that have the genotype tt, the ratio of phenotypes in the offspring will be $\frac{1}{3}$ with tails (genotype TT) to $\frac{2}{3}$ tailless (genotype Tt). A Punnett square can be used to prove this [1 mark for the parents with the correct genotypes; 1 mark for showing the correct genotypes of the offspring]:

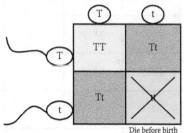

Die before birth

E3. (a) 28 years of age. [1]

    (b) Other variables that the scientists should standardise include gender, ethnicity, and other health status of the patients. [2 max.]

    (c) The Pearson's correlation coefficient could be calculated for this data in order to determine if the relationship between the age at which symptoms of HD first appear and the number of CAG repeats in the *HTT* gene show a statistically significant negative correlation. [1]

    (d) To determine if this study is reliable, there should be a large number of patients [1] who are analysed, with the mean values for the age of onset at each number of CAG repeats shown on the graph. [1] 95% confidence interval error bars should be shown in order to illustrate the degree of similarity between the sample mean and the population mean. [1]

E4. (a) The F2 generation is the offspring of two heterozygotes for at least one gene, [1] which were themselves produced as offspring of a cross between two pure-breeding homozygous individuals of both the dominant allele and the recessive allele. [1]

(b) The expected phenotypes of the offspring in a cross between two chickens with genotype AaBb is as follows. [1 for each correct line in the genetic diagram, with the exception of parental genotypes which are provided]

phenotypes of offspring	genotypes of offspring:	gametes:	genotype of parents:	phenotypes of parents:
genotype of parents:	RrPp	×	RrPp	
phenotype of parents:	walnut	×	walnut [1 mark]	
gametes:	RP Rp rP rp	×	RP Rp rP rp	
genotypes of offspring:	RRPP, RrPP, RRPp, RrPp;	RRpp, Rrpp;	rrPP, rrPp and	rrPP
phenotypes of offspring:	Walnut	rose	pea	single

(c) A Punnett square can be used to predict the ratios of phenotypes in the cross of two chickens with genotype AaBb. [1 mark for each correct ratio of phenotypes identified]

	AB	Ab	aB	ab
AB	AABB*	AABb*	AaBB*	AaBb*
Ab	AABb*	AAbb**	AaBb*	Aabb**
aB	AaBB*	AaBb*	aaBB***	aaBb***
ab	AaBb*	Aabb**	aaBb***	aabb****

	RP	Rp	rP	rp
RP	RRPP *	RRPp *	RrPP *	RrPp *
Rp	RRPp *	RRpp **	RrPp *	Rrpp **
rP	RrPP *	RrPp *	RRPP **	rrPp **
rp	RrPp *	Rrpp **	rrPp **	rrpp ****

The ratio of phenotypes is as follows: walnut: 9; pea: 3; rose: 3; single: 1.

(d) Epistasis indicates that there is an interaction between different genes at different loci. [1] That is, the protein product of one gene determines whether the protein product of another gene is able to affect the phenotype. [1]

E5. (a) A transcription factor is a protein [1] that binds to DNA at specific regions called promoters to influence gene transcription. [1]

(b) (i) Hydrolysing the *PhFBH4* mRNA will prevent the formation of PhFBH4 protein [1] and therefore the plants will have flowers that take longer to discolour and wilt. [1]

(ii) Making more PhFBH4 protein will cause the flowers to discolour and wilt much faster. [1]

(c) (i) In the process of transcription, mRNA is produced. [1] Therefore, if the rate of transcription is higher,

more mRNA will be produced by a cell and a greater concentration of this particular mRNA will be found. [1]

(ii) There is no significant difference between the concentration of *PhFBH4* mRNA found in plants that have been exposed to conditions of dry air and those that have not been exposed to conditions of dry air. [1]

(iii) In both environments, the relative expression of *PhFBH4* (production of *PhFBH4* mRNA) increases over a period of 12 hours, relative to plants grown in standardised conditions. [1] However, the rate of increase of *PhFBH4* expression is higher in the case of plants grown in saline root medium until 6 hours (the rise is from 1 to 2.8 versus from 1 to 2.1) [1] although plants grown at higher temperatures for more than 6 hours have a more rapid increase in *PhFBH4* expression (final value is 4.5 versus 4.0 relative expression). [1]

(iv) Any two from: the pH of the growth medium (which can be kept constant using pH buffers); the light intensity/ time period of light exposure (which can be kept constant by using a light box); the genetic makeup of the plants (which can be kept constant by using cloned plants produced from each other by asexual reproduction). [2 max]

CHAPTER 17 – SELECTION AND EVOLUTION

In-text questions

1. Characteristics that show continuous variation in cheetahs include: height, mass, length of limbs, length of tail, colour of fur, number of spots, size of spots, and so on.

2. While this statement is largely accurate, there are some characteristics that show continuous variation that are not quantitative (numerical) in nature, such as the colour of skin or fur, and this would be difficult to plot on a histogram. Likewise, there are some characteristics that show discontinuous variation that are not qualitative (categorical) in nature, such as number of digits on the hand (it is not possible to have 4.5 fingers, for example).

3. Variation caused by the environment cannot be inherited because there is no molecular basis for passing on this information to the next generation.

4. Some dominant alleles, such as the mutant allele that causes Huntington's disease, *HTT*, are very rare in a population because those who carry them may die before they reproduce. Almost every person in a population is homozygous recessive for the *HTT* gene.

5. The frequency of the dominant allele would be equal to 1 minus the frequency of the recessive allele. Therefore, $p = (1 - q) = 0.91$.

6. (a) If the frequency of individuals who are homozygous recessive is 1 in 10 000 $(1 \div 10\ 000 = 1.0 \times 10^{-4})$, then the frequency of the recessive allele, q, in this population is $\sqrt{(1.0 \times 10^{-4})}$ = 0.01. Therefore, the frequency of the dominant allele, p, is $(1 - 0.01) = 0.99$. To find the frequency of individuals who are heterozygote carriers of the PKU allele in this population, we use the expression 2pq, where $(2 \times 0.99 \times 0.01) = 0.0198 = 0.02$ to 2 dp. Therefore, 2% of individuals are carriers of the recessive PKU allele in this European population.

(b) If the frequency of the recessive allele is 3.0×10^{-3} (0.003), then the frequency of the dominant allele, p, is $(1 - 0.003) = 0.997$. To find the frequency of individuals who are heterozygote carriers of the PKU allele in this population, we use the expression 2pq, where $(2 \times 0.997 \times 0.003) = 0.006$ to 3 dp. Therefore, 0.6% of individuals are carriers of the recessive PKU allele in this sub-Saharan population.

(c) The Hardy–Weinberg Principle states that the frequency of alleles in the gene pool of a population will not change from one generation to the next, if certain conditions are met. A resistance among heterozygotes to ochratoxin A may have provided a selective advantage to Europeans, allowing individuals to consume infected grain without developing renal cancer. However, this trait was of no great advantage in sub-Saharan Africa, where grain stores were not stored over the winter months. While individuals with two copies of the recessive allele (recessive homozygous) died early due to PKU, those who are heterozygote carriers of the PKU allele are more likely to survive, and so the allele responsible has increased in frequency in the European population over many generations.

7. (a) The percentage of vancomycin-resistant bacteria is zero in all years up to 1988. From 1989 to 1994, the percentage increases to 14%, before decreasing in 1995 to 12%. From 1995 to 2003 the percentage increased to 27%.

(b) (i) The mechanism of natural selection can only apply to characteristics that are genetic in their basis, because it depends on those individuals with a competitive advantage to have a greater chance of surviving and reproducing, and passing on the alleles responsible for this advantage. In the case of variation due to environment, this cannot be inherited, so natural selection cannot cause the advantageous trait to increase in the subsequent generations.

(ii) In organisms with faster life cycles, the rate at which the generations occur is greater. Therefore, the selective advantage conferred by alleles in one generation, which make it more likely that an individual will survive, will be passed onto the next generation sooner.

(iii) Disruptive selection is very unusual in nature because it suggests that the phenotypic range of the ancestral form of a species, which is clustered around the mean value, suddenly becomes the least fit. Although this happens in the case of directional selection, disruptive selection also involves the selection for two extremes in phenotype, either of which is advantageous, which is also unlikely.

8. (a) List of points or sketches to include the following: In the original population of moths, most were pale and speckled, but a few were melanic (dark)   A reduction in air quality resulted in the silver-white lichens that usually grow on trees in their habitat to die, changing the colour of their resting places to brown-black   Melanic moths became more likely to survive as they were more camouflaged against this background and were predated less by birds   The melanic moths became more likely to survive and reproduce, passing on the allele responsible for this characteristic to the next generation   Over time, the frequency of the allele increased in the population and more moths were melanic.

(b) (i) In these surveys, 515 moths were captured in total. Therefore, the percentage of melanic moths captured in this study is (395 ÷ 515) × 100 = 77%.

(ii) Rearranging the formula for the standard deviation (s) gives $s = SE \times \sqrt{n}$. For the pale, speckled moth, the value of $s = 11.2 \times \sqrt{20} = 50.1$. For the dark (melanic) moth, the value of $s = 14.9 \times \sqrt{20} = 66.6$.

(iii) The 95% confidence interval error bars are calculated using the formula 95 % CI = 2 SE. The standard error of the mean, SE, is calculated using the standard deviation

(s) and the sample size (n): $SE = \frac{s}{\sqrt{n}}$. For each plotted mean value on the graph, the equation 95% confidence interval error bars (95 % CI) = ±2 SE should be used to calculate the confidence intervals (the upper and lower limits of the bars). Error bars should be drawn so that they extend 2 x SE above and 2 x SE below the bars.

(iv) The 95% CI error bars will not overlap (the upper limit of the error bar for the pale, speckled mean value is 142.4, and the lower limit of the melanic (dark) error bar for the melanic (dark) mean value is 365.2), suggesting that there is a significant difference between the mean values. However, to confirm this, and determine whether the observed difference is real or otherwise due to chance, a *t*-test would need to be performed.

(c) Experiments that could be conducted to provide evidence for the evolution of the peppered moth by natural selection include:
- Releasing equal numbers of normal speckled (pale) and melanic moths into a closed environment with predators (birds) and determining how many moths of each phenotype are recaptured after a certain time.
- Placing normal speckled (pale) and melanic moths onto lichen-free trees and determining how quickly they are seen and eaten by predators.

9. Answer: **C**. The fourth statement is incorrect.

10.

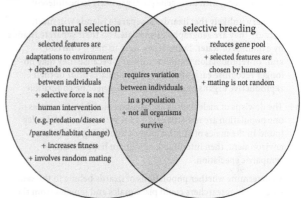

11. (a) In the regions of soil polluted with copper ions, very few other plants will grow. Therefore, if a population of *Agrostis* colonises this area, it will have very little competition for nutrients, water and light. Those individuals with copper tolerance will therefore be more likely to survive, reproduce and pass on the alleles responsible to the next generation, meaning that subsequent generations of *Agrostis* will be more likely to be tolerant to copper.

(b) This could lead to sympatric speciation. If copper-tolerant plants flower at a different time of year to non-copper-tolerant plants, then cross-pollination between the two populations will not happen (genetic isolation). With no gene flow between them, each population will begin to gather genetic differences due to natural selection, mutation, and genetic drift. Over time, this could result in an inability to reproduce (reproductive isolation), even if cross-pollination happens again in the future.

12. (a) Farmers could have self- or cross-pollinated *B. oleracea* plants with the largest leaves. After sowing these seeds and allowing the plants in the next generation to reach maturity, the farmer would then again self- or cross-pollinate *B. oleracea* plants with the largest leaves. Over time, the plants in successive generations will have larger and larger leaves.

(b) A species is defined as a group of individuals that can reproduce to produce offspring that are themselves fertile. To determine if two cultivars belong to the same species, a scientist could cross pollinate the plants, produce seeds, plant the seeds and then attempt to cross-pollinate two individuals from the next generation. If they are able to produce seeds, then the two original cultivars are members of the same species.

Assignment 17.1

A1. (a) (i-iv) Species 5. All species of *Anolis* are more closely related to *A. porcatus* than they are to each other.

(b) This suggests that *A. porcatus* is the most recent common ancestor, this gives evidence that the Cuban population gave rise to the populations on the other islands due to the founder effect.

A2. Separated from other populations by expanses of water, a population of lizards on an island is reproductively isolated. In their different environments, there will have been different selective forces that led to the evolution of distinct characteristics due to natural selection. Mutations in the different populations will not have been exchanged between the two populations, and the effects of genetic drift on allele frequencies would have been unique to each group. If one of the populations of lizards descended from a small number of individuals that left the larger population (a founder effect), then their gene pools are likely to be different.

A3. It is very unlikely that lizards on separate islands will meet, because they are geographically isolated on their separate islands by expanses of water. Even if they were to meet, courtship and/ or reproduction would be unlikely, because they may no longer recognise each other as breeding mates and/ or their reproductive organs or gametes may no longer be compatible.

A4. The dewlaps of male lizards are used in courtship. If females of one population are not attracted to a specific colour of dewlap found in the males of another population living in the same environment, then interbreeding will not happen. This is sympatric speciation.

A5. To determine whether populations of lizards belong to the same species, the researchers could place males and females from the two populations together to encourage them to interbreed. If the resulting offspring are fertile, then the original populations are members of the same species.

A6. Comparing the DNA sequence of lizards on the islands will offer further clues. For example, if the ancestral population of *Anolis* lizards did indeed originate in Cuba, and then spread to islands in one particular direction, then the sequence of DNA of lizards living on islands closer to Cuba should be more similar to the Cuban population than other populations.

A7. Mitochondrial DNA contains genes (and produces proteins) common to all species, given its vital role in aerobic respiration.

Experimental skills 17.1

P1. New species arise through reproductive isolation of two or more populations of the same species. Perhaps some members of an ancestral population of durian trees produced durian fruits with an odour or taste that appealed to carnivores. These individuals may have had their seeds dispersed further away or more effectively than those with the usual odour or taste. Over time, these trees will become reproductively isolated from the original population and as a result, there is no gene flow between these populations and each of them is able to evolve independently from the others, depending on the different selection pressures acting on them, as well as chance events such as genetic drift, bottlenecks and founder effects that may change the gene pool of each population. Natural selection acts on the populations causing the individuals with advantageous alleles to survive and reproduce more successfully than the rest. Over many generations this results in an increased proportion of the population having these alleles. Chance events such as the bottleneck effect can also cause unpredictable changes in isolate gene pools. Eventually, the two populations may independently evolve to the extent that they can no longer interbreed, even if they are given the opportunity to. This is sympatric speciation.

P2. (a) All three characteristics can be measured quantitatively, which is a strong indication that they show continuous variation.

(b) Either genetic (some alleles of certain genes will determine the value of each characteristic) or environmental (some abiotic and biotic factors will determine the value of each characteristic, such as the efficiency of pollination of the flower that produced the seeds).

(c) The study investigated 25 durian fruits for each tree, which suggests that the reliability of data is high. However, all of the fruits from each sample were taken from only one tree, which reduces the reliability.

P3. (a)

number of seeds in durian fruit	
tree A	tree B
19	17
18	16
18	16
17	14
17	14
17	14
17	13
17	13
17	13
16	13
15	13
15	11
14	11
14	11
13	11
13	11
12	11
12	10
12	10
11	10
10	9
10	9
10	8
10	8
9	7
mean: 14.1	11.7
median: 14	11
mode: 17	11

(b) The standard deviation for the number of seeds in fruit from tree A is 3.02. The standard deviation for the number of seeds in fruit from tree B is 2.57.

(c) Standard error of the mean (SE) = $\frac{s}{\sqrt{n}}$, where s = the standard deviation and n = the number of samples. Therefore, SE for tree A is $3.02 \div \sqrt{25} = 0.605$. and SE for tree B is $2.57 \div \sqrt{25} = 0.514$.

(d) The values for SE can be used to calculate the 95% confidence intervals (which can be used to draw error bars on a bar chart for each mean). If these confidence intervals do not overlap, then there is possibly a significant difference between them, although a t-test would be required to determine the significance of this.

P4. (a) Histograms are useful when continuous data is to be plotted on a graph that falls into categories or bands. If this data was plotted as a bar chart, then the number of bars would be very large (more than 50) because each seed mass would have to be counted up separately rather than grouping them as in this example.

(b) It is not possible to tell from these histograms whether the data in any given bar is skewed towards one extreme of the bar. The histograms appear to show a difference in the mass of seeds between the two trees; however, this difference may not exist in reality.

P5. (a) Between 1900 and 2000, the mean mass of durian has steadily increased. This is probably because of selective breeding (artificial selection): growers have selected the trees that produce the largest durians for reproduction, which then transfer the alleles responsible for this trait to the next generation.

(b) The standard deviation of the three samples reduces from 1900 to 2000. This can be concluded from the shape of the curves: the curves become narrower. This means that the sample in 2000 contains more fruits of a similar mass than the fruits in 1900. The standard deviation has reduced from 1900 to 2000. This observation may be due to the selective breeding (artificial selection) and increased homozygosity of the durian trees over the past century, as farmers select for fruits of a larger, more uniform mass.

P6. (a) t = 9.90

(b) The number of degrees of freedom is 18. This is because there are two normally distributed populations (experiment and control) consisting of 10 replicates each. Therefore, $v = c - 2$ and $v = 20 - 2 = 18$.

(c) Using the probability table at the level of 0.05 (5% probability), we can see that the t-value is higher than the critical value (6.314) (2.10), so the difference is significant. This means that there is less than a 5% probability that the difference shown in the table is due to chance alone. In fact, the t-value is higher than the critical value at the level of 0.01 (1% probability), which is 2.88, which means that there is less than a 1% probability that the difference shown in the table is due to chance alone.

(d) The evidence strongly suggests that durian extract (DE) does have an inhibitory effect on aldehyde dehydrogenase activity. There is a difference between the mean values of enzyme activity in the presence and absence of DE, even when the 95 % confidence intervals are accounted for (95 % CI = ±2 SE). Moreover, despite the measurement of some enzyme activity in the presence of denatured enzyme, subtracting this value from active enzyme activity still shows a difference between the comparison.

Chapter review

1. A

2. B

3.

Observation	Natural selection	Founder effect	Genetic bottleneck	Genetic drift
The human species has a small gene pool because around 70 000 years ago a volcanic super-eruption may have reduced our ancestral population to only several thousand.			✓	
The northern elephant seal species has a small gene pool because was hunted to near-extinction in the late 19th century.			✓	
In a small, reproductively isolated population of frogs, all have become green in colour.	✓			✓
In some isolated communities living in Venezuela, Huntington's disease is very common. These individuals descend from an immigrant with the disease from Spain in the 17th century.		✓		
Extinct in the wild, Przewalski's horse survived in zoos, and was successfully reintroduced into the steppes of Mongolia in 2006.		✓	✓	
In populations that changed from a hunter-gathering lifestyle to farming, lactose tolerance developed as a means to consume milk – a sterile, nutritious food source.	✓			
The native human population of the Americas is thought to have descended from as few as 20 individuals who crossed the land bridge from Siberia to Alaska during the last Ice Age.		✓		
The development of a peacock's tail has been driven by a female's mating preference for elaborate displays.	✓			

Answers

4. This question is based on the incorrect assumption that humans evolved on the same pathway as chimpanzees. Instead, humans and chimpanzees are modern species that both evolved from a common ancestor, which no longer exists.

5. In the ancestral species of giraffes, some had longer necks than others. A reduction in food availability at ground level, perhaps due to a prolonged drought, resulted in those with shorter necks dying. Individuals with longer necks became more likely to survive as they were able to reach leaves at greater heights. The animals with longer necks became more likely to survive and reproduce, passing on the allele responsible for this characteristic to the next generation. Over time, the frequency of the allele increases in the population and the giraffe developed long necks as a result.

6. Promiscuity (the taking of multiple mating partners) will increase the genetic diversity in the offspring. This behavioural characteristic may be more common in cheetahs because they have a very small gene pool, due to a genetic bottleneck. Animals that engaged in such behaviour were more likely to have offspring that had greater genetic diversity, and were therefore more likely to survive and reproduce.

7. (a) Warfarin resistance is caused by a dominant allele, so individuals with the dominant homozygous and heterozygous genotypes are resistant. The genotype frequency is therefore represented by $p^2 + 2pq$, which is 0.5. Because $p^2 + 2pq + q^2 = 1$, $q^2 = 0.5$. Therefore, the frequency of the recessive allele for susceptibility in this population is $q = \sqrt{0.5} = 0.71$.
   (b) Rats that are homozygous dominant for the resistance allele require a large amount of vitamin K, which they are unlikely to be able to obtain on a normal diet in the wild.
   (c) In areas that use rat poison, individuals that carry at least one copy of the dominant allele for resistance to warfarin are more likely to survive and reproduce, passing on this allele to the next generation in which it becomes more common. At the same time, in areas that do not use rat poison, rats that are homozygous recessive and therefore carry no alleles for resistance are more likely to survive and reproduce, because they are not dependent on large dietary amounts of vitamin K to survive.

8. (a) Apricot fruits from different species of tree may look similar, and distinguishing them on the basis of size, shape, colour and taste, can be subjective (open to bias). Comparing the amino acid sequence of these species provides objective (no bias) evidence on which conclusions regarding evolutionary relationships can be based. This also gives quantitative data whereas the other features are more qualitative so not so useful for calculations.
   (b) In the protein sequence of *P. brigantina*, there is only one amino acid residue that is different from the protein in *P. armeniaca*. This suggests that there has been the shortest time since the last common ancestor of these two species, and fewest mutations have occurred in the gene sequence that encodes this protein.
   (c) Although both of these species differ by only four amino acid residue differences from *P. armeniaca*, the position of these residues may differ between them.
   (d) DNA base sequences of genes include regions of non-coding DNA such as introns that may also provide information as mutations may have occurred in this sequence. In addition, some mutations in the DNA can change codons but not the encoded amino acid, due to the degeneracy of the genetic code, which will not be reflected in the amino acid sequence of the protein.

(e) This is due to inbreeding depression. If closely-related individuals are crossed over many generations, homozygosity will occur at an increased number of loci. This can reduce the vigour (health) of the individuals, and hence the size of the fruit.

Practice exam-style questions

E1. Answer C: 2 and 4. [1] These phenomena reduce the gene pool of a population, meaning that it will have less phenotypic variation.

E2. (a) According to the Hardy–Weinberg equations, the proportion of heterozygotes in a population is equal to $2pq$. [1] Therefore, if allele A is recessive, $q = 0.55$ in Amman. Therefore, the frequency of the dominant allele, p, is 0.45. So, the frequency of heterozygotes is equal to $2 \times 0.45 \times 0.55 = 0.495$ [1]
   (b) Alleles A and B are likely to be alleles of different genes because in bats in Cairo, the frequencies of these alleles add up to more than 1. [1]
   (c) The Hardy–Weinberg Principle states that the frequency of alleles should not change in individuals from one generation to the next. [1] However, if allele C reduces the hearing of these bats, carriers are less likely to be able to hunt successfully, survive, and reproduce. [1] Therefore, due to natural selection, the frequency of allele C is likely to decrease over time and become very rare. [1]
   (d) (i) A *t*-test can be used with this data, because the scientists are comparing the mean values of two samples, each of which has been calculated from continuous data/ data that has a normal distribution. [1]
       (ii) The reduction in the allele frequencies between Amman and Cairo, and between Cairo and Muscat, are significant: there is less than a 5% probability that these observed differences are due to chance. [1] However, between Amman and Muscat, there is more than 5% probability that the observed difference is due to chance and therefore it is not significant. [1]

E3. (a) The independent variable is the concentration of penicillin. The dependent variable is the time taken for bacterial growth to transition from one section to the next. [1]
   (b) The stock solution of penicillin has a concentration of 1.0 $mg\ cm^{-3}$. To produce solutions of the concentrations used in this study, this solution needs to be diluted. To prepare a solution of 100 $\mu g\ cm^{-3}$ (0.1 $mg\ cm^{-3}$), 1 cm^3 of the original stock solution [1] is mixed with 9 cm^3 distilled water. [1] To prepare subsequent solutions of concentrations that are each successively 10 times more dilute, 1 cm^3 of the prepared stock solution is mixed with 9 cm^3 distilled water. [1]
   (c) (i) The medium used to make the agar blocks must be sterile, because only the growth of bacteria applied to Section A at time = 0 hours should be allowed to grow. The growth of bacteria other than these will interfere with the measurement of the dependent variable. [1]
       (ii) Temperature affects the rate of bacterial cell division. Therefore, if the temperature were to vary during the course of the investigation, the rate at which bacteria transition from one section to the next will be dependent on another factor in addition to penicillin concentration – making the comparison of the data invalid. [1]
   (d) The original population of bacteria were genetically identical (clones). None of them had the ability to grow in medium containing penicillin – they were all susceptible. There may be some with a mutation in a gene that enables it to grow on medium containing penicillin [1] and this enabled this bacterium to grow outside of an area with high competition for resources. [1] Over time, bacteria carrying this mutation spread to fill the section containing this antibiotic. [1]
       (i) Mutation is a random process. [1]

(ii) The mean time taken for bacteria to grow in order to transition into section B was 8.0 hours. The mean time taken for bacteria to grow in order to transition into section F was 5.1 hours. The difference between these two values is 2.9 hours. [1]

(iii) The missing confidence limits for the 100 μg cm⁻³ penicillin solution are 5.10 hours (lower limit) and 5.70 hours (upper limit). [1]

(e) (i) Sections B and C. [1] This is because their confidence intervals do not overlap. [1]

(ii) The *t*-test could be used to determine whether the observed difference between two mean time values is significant. This is because time is a factor that shows continuous variation.

(iii) There is no significant difference between the time it takes for a bacterium to evolve resistance to higher concentrations of antibiotic compared to lower concentrations of antibiotic. [1]

(iv) Although the ability to resist the effects of penicillin is a discontinuous characteristic (a bacterium is either susceptible or resistant), [1] the ability of a bacterium to resist different concentrations of penicillin is a form of continuous data, as these values would likely fall on a normal distribution curve between two extreme values. [1]

E4. (a) Positions 1, 6, 10, 14, 17–20. [1]

(b) Because the cysteine residues are found in the same positions in all proteins, this suggests that strong disulphide bonds form between the same regions of the polypeptide chain. [1] Therefore, the tertiary and quaternary structure of the cytochrome c proteins from all species are likely to be very similar. [1]

(c) The dogfish should be placed between the chicken and the fruit fly. [1] This is because it has five amino acid residues that are different from the human, two more than the chicken, and one fewer than the fruit fly. [1]

(d) The greater the number of amino acid residue differences between two species, the further back in time they shared a common ancestor. [1] This is because amino acid sequences are determined by DNA sequence, [1] which can be changed over time by mutation. [1]

E5. (a) All eukaryotes have mitochondrial DNA, which contains genes common to species as diverse as yeast to humans. [1] Mitochondrial DNA also has a greater mutation rate, meaning that there is more information to determine evolutionary relationships. [1]

(b) An ancestral species of bird, possibly a flying duck, probably arrived on Kauai not long after it formed 5.1 million years ago. [1] As new islands formed, this ancestral species colonised them, and on each island, was reproductively isolated from the others, [1] therefore evolving different characteristics as a result of natural selection, mutation and genetic drift in the absence of gene flow. [1] The birds on each island probably had no predators and therefore over many generations was able to increase in population, and evolve to become larger and lost their ability to fly. [1]

(c) (i) Islands are geographically isolated from others by bodies of water, separated by other populations from expanses of water than most organisms cannot cross. [1] There is therefore no gene flow between the organisms of one island and another. The colonisation of an island by a small group of founding individuals often causes the resultant populations on the island to have a gene pool that is smaller than the original population that the founding individuals came from, thus contributing to the formation of a new species. [1]

(ii) Birds are often the subject of evolutionary studies because for a terrestrial animal to arrive on an island, it usually must be able to fly there. [1] Birds may be blown off course on migratory journeys and to an uninhabited island, where they occupy a niche and evolve to take advantage of their new habitat. [1]

(d) Hawai'i is a relatively young island, formed 430 000 years ago, so there has not been enough time for evolution of giant flightless birds to occur. [1] Also, the giant flightless birds would have been unable to cross the expanse of sea between the islands. [1]

(e) The diet of the moa-nalo may have included plants. To protect themselves from being eaten, these plants may have evolved spines: individuals in an ancestral population that had pointed protrusions will have been more likely to survive, reproduce, and pass on the alleles responsible for this characteristic. [1] On the island of Hawai'i, where no moa-nalo lived, predation would not have been a selective force, [1] and therefore the presence of spines would not have been an advantageous characteristic. [1]

E6. (a) Genetic drift is the change in allele frequency due to chance, and this is more likely to have significant effects on small populations. [1]

(b) Asexual reproduction, the formation of offspring by just one parent, is likely to be faster than sexual reproduction. [1] An individual which is successful due to having a well-adapted set of alleles to its current environment will reproduce more by asexual than one which is not so well adapted. This means it can more rapidly monopolise the habitat where it is. [1]

(c) In a species that reproduces by asexual reproduction, more members of the next generation are likely to be genetically identical. This reduces the gene pool of the species. [1] This effect is similar to a genetic bottleneck, in which a large proportion of individuals of a species will die, [1] leaving only a small number to repopulate the species afterwards. [1]

(d) Sexual reproduction involves processes such as meiosis and random fertilisation of gametes, which increase variation in a species and therefore increase the gene pool. [1] During a period of environmental change, such as the emergence of a new disease, it is an advantage for there to be a greater variation between individuals. This is because there is a higher chance that there will be some individuals that have a resistance to the new disease [1] and these will survive and reproduce to maintain that species and prevent its extinction. [1]

CHAPTER 18 – CLASSIFICATION, BIODIVERSITY AND CONSERVATION

In-text questions

1. (a) They have the same first name (*Panthera*), which means they are of the same genus, but they have different second names so they are a different species.

(b) They are unlikely to mate as they are geographically isolated (live far apart). This is pre-zygotic isolation.

(c) It is sterile so cannot produce offspring meaning the biological species concept does not apply. The tigon is morphologically similar so could be classed in the same species as a lion or a tiger. The tigon is not naturally adapted to any ecological niche so the ecological concept of species does not apply.

2. The domain system uses information from DNA / RNA / genetics, and molecules / molecular structures / biochemistry; technology

for this was only available recently (in 20th century). Previous systems used visible features / visible characteristics to classify organisms.

3. Similarities: unicellular; prokaryotic; no nuclear envelope. Differences: Archaea have no peptidoglycan in their cell walls; both have different rRNA / rRNA genes; Archaea have different structures of phospholipids.

4. Fungi are non-photosynthetic. The cell wall material in fungi is made from chitin whereas plant cell walls are composed of cellulose.

5. Cytochrome c will be found in all organisms with mitochondria – all members of domain Eukaryota, whereas haemoglobin is only found in some animals / some members of kingdom Animalia.

6. (a) The haploid number of the zebra is 16 and the haploid number of the donkey is 31, which would give a diploid number of 47 in the zonkey. An odd number of chromosomes means meiosis cannot occur; chromosomes will not be homologous.
   (b) Because the zonkey cannot produce offspring; is not adapted to any particular habitat; and will have morphological features of two different species.

7. (a) double-stranded DNA (dsDNA); single-stranded DNA (ssDNA); double-stranded RNA (dsRNA); positive single-stranded RNA (+ssRNA); negative single-stranded RNA (–ssRNA)
   (b) Viruses are non-living. They do not respire, excrete etc.; they have no cells; they are not organisms and the domain system of classification is for cellular life. Viruses are obligate intracellular parasites so they require a host cell to replicate – they cannot replicate independently.

8. Poliovirus is in Baltimore class IV so has positive single-stranded RNA (+ssRNA). Therefore its genome can act as mRNA and so can be translated immediately; there is no need for transcription (to produce mRNA).

9. (a) An ecosystem is a self-contained area with biotic and abiotic factors, where the biotic factors interact with each other and with abiotic factors.
   (b) An ecological niche is an organism's role and position in the ecosystem.

10. (a) 0.5×0.5
    $= 0.25m^2$
    (b) $0.25m^2 \times 25$
    $= 6.25m^2$
    (c) $\dfrac{368}{6.25}$
    $= 58.88$
    rounded to 59
    (d) 59×75
    $= 4425$
    (e) It is an estimate because not all plants will have been counted – there may have been more or fewer in the areas sampled (or in the areas not sampled); the original value was rounded up, so when multiplied will give a higher result.

11. (a) belt transect
    (b) mark-release-recapture

12. (a) belt transect / line transect
    (b) knotted wrack (*Ascophyllum nodosum*) and bladder wrack (*Fucus vesiculosus*)
    (c) 200 – 250cm
    (d) oarweed as it is exposed for only a very short time each tide

13. (a) $\dfrac{(50 \times 70)}{17}$
    $= 206$
    (b) Any three from: marked centipedes may not have mixed randomly after release; other centipedes may have arrived or been born between samples; centipedes may have died or been eaten or have moved away between samples; some marks may have come off.
    (c) Any two from: so second sample is taken at same time of day; not too long for centipede life cycle / life span; long enough to allow marked centipedes to mix randomly; less likely that significant immigration/emigration had occurred.

14. (a) there is no correlation between increasing pH and *Natronococcus* growth conditions
    (b)

pH	6.0	7.6	8.0	8.6	9.0	9.4	9.5	9.6	9.9	10.0
rank	10	9	8	7	6	5	4	3	2	1
mean colony diameter / mm	5.5	6.0	3.8	5.9	6.2	6.5	9.0	12.2	8.5	10.0
rank	9	7	10	8	6	5	3	1	4	2
d	1	2	-2	-1	0	0	1	2	-2	-1
d²	1	4	4	1	0	0	1	4	4	1

sum of $d^2 = 20$

$r_s = 1 - \left[\dfrac{(6 \times 20)}{(10^3 - 10)} \right]$

$= 0.88$

this is greater than the critical value of 0.65, which means that the null hypothesis is rejected and there is a correlation between increasing pH and *Natronococcus* growth conditions. The positive sign means that the correlation is positive.

15. (a) graph with altitude / m on the x-axis and mean annual temperature / °C on the y-axis, all points plotted correctly and no line drawn
    (b) two sets of paired data / (paired) data are continuous / scatter graph suggests linear correlation
    (c) strong negative correlation that is almost linear
    (d) (i) $10 - 2 = 8$ or 2 less than the number of paired data sets
        (ii) use a probability table at $p = \dfrac{5\%}{0.05}$; compare r value with critical value; if r value is greater than critical value, then result is significant

16. (a) lake A

fish species	population estimate	$\dfrac{n}{N}$	$\left(\dfrac{n}{N}\right)^2$
stickleback	1242	0.585	0.342
brown trout	66	0.0311	0.000964
carp	108	0.0509	0.00259
steelhead	18	0.00848	0.0000719
northern pike	7	0.00330	0.0000109
bream	229	0.108	0.01160117
roach	453	0.213	0.0455
total	2123		0.40

$D = 1 - 0.40$
$= 0.60$

lake B

fish species	population estimate	$\frac{n}{N}$	$\left(\frac{n}{N}\right)^2$
stickleback	895	0.772	0.596
brown trout	27	0.0233	0.00054
carp	34	0.0293	0.00086
steelhead	4	0.00345	0.0
northern pike	1	0.00086	0.0
bream	88	0.0759	0.0058
roach	110	0.0949	0.009
total	**1159**		**0.61**

$D = 1 - 0.61$
  $= 0.39$

   This value indicates that the biodiversity of fish is much greater in lake A.
   (b) biodiversity would decrease as n would decrease as individuals of other species are eaten; N could also decrease if any one species becomes extinct in that lake

17. (a) Possible reasons may include: vertebrates are more photogenic / appealing to the public / likely to generate more interest; vertebrate populations are better studied; many invertebrate species are still undiscovered.
   (b) It is very difficult to study populations of prokaryotes; if prokaryotes are endangered then it is likely we have not discovered them as their numbers will be so low; the public are less likely to care about prokaryotes.

18. (a) too old / stress of captivity / infertile
   (b) protected areas where removal of sand is not permitted

19. (a) roughly equal proportions
   (b) There will be increasingly more males than females so the males will not be able to mate / fewer females for mating; therefore the species could become extinct.
   (c) Any two from: because they are part of food webs / so other animals do not become endangered; to maintain biodiversity; to maintain ecosystem stability.

Assignment 18.1

A1. crop yield is a biotic factor and the others are abiotic factors

A2. belt transect

A3. more evaporation (because) wind speed increases less shade (from the hedge)

A4. (a) Closer to the hedge there is competition for minerals and water, and less sunlight in the shade of hedge.
   (b) warmer so plant's enzymes more active leading to increased growth
   Moister soil so more water available for growth
   Much less evaporation so less water stress in plant
   Much reduced wind speed so less damage to leaves and roots by wind

A5. Hedges increase biodiversity because: different plants grow in the hedge
   – such as woody plants and smaller flowering plants; hedges give cover for animals or birds nesting and provide food for birds / mammals / insects.

Experimental skills 18.1

P1. *independent* percentage cover of *A. nodosum*
   dependent number of *L. littorea*

P2. belt transect / interrupted transect

P3.

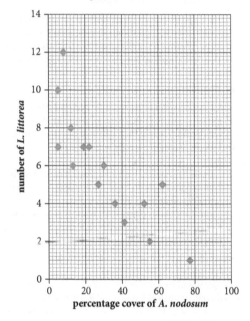

percentage cover of A. nodosum	55	41	30	5	22	8	77	13	19	36	27	5	52	12	62
number of L. littorea	2	3	6	10	7	12	1	6	7	4	5	7	4	8	5

axes labelled with linear scales including zero
scale on axes allows plotted points to cover at least half the grid in both directions
all points plotted to within 1mm

P4. The graph shows a negative correlation.

P5. (a) there is no correlation between the number of *L. littorea* and the percentage cover of *A. nodosum*

Answers

(b)

quadrat number	1	2	3	4	5	6	7	8	9	10	11	12	13	14	15
percentage cover of A. nodosum	55	41	30	5	22	8	77	13	19	36	27	5	52	12	62
rank	3	5	7	14.5	9	13	1	11	10	6	8	14.5	4	12	2
number of L. littorea	2	3	6	10	7	12	1	6	7	4	5	7	4	8	5
rank	14	13	7.5	2	5	1	15	7.5	5	11.5	9.5	5	11.5	3	9.5
D	-11	-8	-0.5	12.5	4	12	-14	3.5	5	-5.5	-1.5	9.5	-7.5	9	-7.5
D^2	121	64	0.25	156.25	16	144	196	12.25	25	30.25	2.25	90.25	56.25	81	56.25

sum of $D^2 = 1051$

$$r_s = 1 - \left[\frac{(6 \times 1051)}{(15^3 - 15)}\right]$$

$$= -0.88$$

(c) 0.88 is greater than the critical value of 0.44 so therefore this is a significant result at the 5% level. The negative sign means that there is a negative correlation between the percentage cover of *A. nodosum* and the number of *L. littorea*.

P6. The scatter graph gives a visual representation that indicates a negative correlation but it does not show whether the correlation is significant. Spearman's rank correlation coefficient can be used to show the strength of the correlation and its significance.

Chapter review

1. (a) differences between prokaryotes in their membrane lipids; differences between prokaryotes in their rRNA; differences between prokaryotes in their cell wall compositions; domain Archaea and domain Bacteria were created
   (b) Eukarya
   (c) (i) animals have no cell wall and plants have cell walls (made from cellulose); plants have chloroplasts / large central vacuoles and animals do not
       (ii) plants have chloroplasts / can photosynthesise and fungi do not; plants have cell walls made from cellulose and fungi have cell walls made from chitin

2. (a) Organisms are of the same species if they can breed to produce viable and fertile offspring.
   (b) Biological is more suitable because they can all theoretically interbreed. Biological is less suitable because there are physical pre-zygotic barriers to reproduction between some varieties on the basis of size.

   Morphological is more suitable because they all share many of the same physical features. Morphological is less suitable because dogs share many of the same physical features with other mammals of different species such as foxes / wolves / hyenas / dingos

3. (a) Viruses are classified according to their genome / genetic material (and whether or not they have the enzyme reverse transcriptase) into: dsDNA; ssDNA; +ssRNA; -ssRNA
   (b) The domain system of classification is for cellular life. Viruses do not have cells and can be argued to be non-living / are non-living as they are not able to independently carry out most characteristics of living things

4. (a) It is self-contained; contains biotic factors that interact, and abiotic factors with which the biotic factors interact.
   (b) By random sampling using mark-release-recapture method.

For example, use a net for capturing the beetles, mark the first sample; ensure that the mark is waterproof and does not harm the beetles (e.g. it does not make the beetles more visible to predators / does not make the beetles less likely to be prey); release the marked sample; allow specified period of time e.g. $\frac{24}{48}$ hours; capture the second sample; second sample taken at same time of day / in same place / with same net; use Lincoln index to calculate estimate

(c) The new species of fish would be an invasive alien species. The new species could be a better competitor than species A and so more of amphibia (species B) are eaten so their population will fall. This reduces the food available for fish species A so their numbers will also fall.

5. *negative effects*
   habitat destruction
   reduces species diversity
   makes ecosystem less stable
   crop plants may be invasive alien species
   some species could become endangered / extinct in that area
   positive effects
   can produce more food for human population
   other native species may flourish in new conditions
   clearance of small areas can increase species diversity

6. Zoos offer protection from hunting / human impacts. Zoos can carry out breeding of the endangered species in captivity, then release the offspring into the wild. Zoos also provide education of people in the need for protection of species and raise funds for conservation projects.

7. (a) they are being removed from their habitat so it becomes more difficult for slow lorises in the wild to reproduce (more difficult to find mates). People who buy them as pets will not breed them /(they will likely only have one); early death eg during transport to market
   if they are bred by pet owners, the offspring will not be released responsibly.
   (b) Educate people that the slow loris is endangered or could become extinct; perform checks on markets where animals are sold (or ports or airports); release animals destined for sale back into the correct habitat; have rangers in forests to protect the animals; provide funds for conservation areas.

8. (a) $N = \dfrac{n_1 \times n_2}{m_2}$

   $= \dfrac{320 \times 400}{26}$

   $= 4900$
   (b) *any three from* some may have died; some may have been born; some marks may have come off; small sample size

Practice exam-style questions

E1. (a) Eukarya [1]　　　(b) Aves [1]

(c) *Diomedea epomophora*　[1]

(d) Using the biological concept they have different species names so are different; [1] using the morphological concept they have different appearances so are different species; [1] using the ecological concept they have the same habitat and same adaptations / occupy same ecological niche so are the same species. [1]

E2. (a) Place the frame quadrat randomly (e.g. using co-ordinates generated from random numbers); [1] use a large number of quadrat positions; [1] count the number of bluebells in each quadrat; [1] the total number of bluebells in all quadrats divided by total area of quadrat sampling (gives average number per square metre)). [1]

(b) (i) abiotic　[1]

(ii) these conditions refer to the previous year / previous summer; [1] higher moisture content is better as there is a positive correlation between soil moisture content the previous year with flower density; [1] lower temperature is better as there is a negative correlation; correlation is weaker as bluebell becomes more dormant in July. [1]

(iii) as summers become warmer and drier this is likely to have a detrimental effect on flowering of bluebells; [1] since they need damp cool conditions in June and July; [1] fewer flowers means less seed

(c) Any three from: the Japanese knotweed is an invasive alien species. [1] It is a better competitor than bluebells / outcompetes bluebells; [1] it blocks the sunlight and; [1] takes up more minerals/water. [1]

E3. (a) *two from* too small to mark [1]; moving over a large area means immigration / emigration will occur [1]; suitable time period between release and recapture would be difficult to determine [1]

(b) (i) $\frac{n}{N}$ values: 0, 0.880, 0.0313, 0.0188, 0.0563, 0.0141 [1]

$\left(\frac{n}{N}\right)^2$ values: 0, 0.774, 0.000980, 0.000353, 0.00317, 0.000199 [1]

sum of $\left(\frac{n}{N}\right)^2$ values = 0.779 [1]

$D = 1 - 0.779$
$\quad = 0.221$ [1]

(ii) lower [1]

Fewer species in field [1]

One (oat aphid) is extremely dominant [1]

(c) the forest, because increased biodiversity gives greater ecosystem stability; [1] greater sustainability / food webs less vulnerable to change. [1]

E4. (a) Any five from: biodiversity can be assessed in terms of habitats and ecosystems; [1] many different habitats; [1] many different niches; [1] many different species; [1] high genetic diversity; [1] many different adaptations., [1] including any two from: different temperatures / rainfall / soil types / altitudes / other named conditions [2]

(b) Any three from: stability of food webs; [1] so animals / plants do not become extinct ; [1] stability of ecosystems; [1] future medicines; [1] human economy eg the demise of insect populations will detrimentally affect many crops such as all fruits and many vegetables [1]

E5. (a) frame quadrat; [1] placed at random; [1] light intensity measurements / use of light meter. [1] percentage cover of spikle moss measured

(b) mapping the area / determining the area to be sampled; [1] generating co-ordinates for the quadrat / use of random numbers for placements; [1] ensuring samples cover a range of light intensities. [1]

(c) Any four from: take light intensity measurements at same time of day; [1] take light intensity measurements in same weather conditions; [1] measure soil moisture; [1] measure soil pH; [1] measure soil temperature; [1] measure distance from other plants; [1] whether the area is grazed. [1]

(d) Any three from: plot a scatter graph; [1] if correlation appears linear then use Pearson's linear correlation coefficient; [1] otherwise use Spearman's rank correlation coefficient; [1] between abundance / percentage cover and light intensity; [1] positive <u>and</u> significant value if hypothesis is correct [1]

(e) (i) means calculated as 51.4 (% cover) and 1025 (light intensity) [1]

sum of xy calculated as 466 540 [1]

$(n - 1) s_x s_y$ calculated as 48 828　[1]

$r = 0.927$ or 0.93　　[1]

(ii) strong positive linear correlation [1]

CHAPTER 19 – GENETIC TECHNOLOGY

In-text questions

1. (a) A restriction enzyme makes two 'cuts' in a DNA molecule by catalysing the hydrolysis of two phosphodiester bonds, one in each polynucleotide strand. Each reaction requires the activity of an active site.

(b) A sticky end is a region of single-stranded DNA with unpaired bases, produced by the action of a restriction enzyme. This 'overhang' can form hydrogen bonds with another sticky end with complementary bases.

(c) HaeIII has a restriction (recognition) site of only four bases in length, whereas the restriction site of EcoRI is six bases in length. A four-base restriction site will occur, on average, each 4^4 (= 256) bases, whereas a six-base restriction site will occur, on average, each 4^6 (= 4 096) bases.

2. A DNA molecule containing a gene isolated using reverse transcription will not have intron sequences, because these will have been spliced out of the molecule after transcription. In addition, it is unlikely that the restriction sites of restriction enzymes used to cut out the DNA sequence would have been on either side of the gene, so there will be some flanking (surrounding) sequence also present in the DNA molecule isolated using restriction enzymes.

3. (a) Muscle cell (myocyte).

(b) Cells of the hypothalamus.

(c) Photosynthetic cells, e.g. palisade mesophyll cells of plant leaves.

4. (a) The presence of a variety of restriction sites in a plasmid will allow the gene technologist to choose from a range of restriction enzymes to cut the plasmid and produce sticky ends that have a complementary base sequence to those of the sticky ends found on the DNA molecule containing the gene to be inserted.

(b) The presence of a promoter next to the restriction sites in a plasmid will activate the transcription of the donor gene. Promoters are regions of DNA to which transcription factors can bind, allowing RNA polymerase to begin transcription of the downstream gene.

(c) The presence of an origin of replication in a plasmid will mean that the DNA will be replicated and passed on to daughter cells when the recombinant bacterium reproduces.

(d) The presence of a marker gene in a plasmid will allow the gene technologist to identify which of the bacteria have successfully absorbed the plasmid and are expressing the donor DNA.

(e) The low molecular mass of a plasmid will allow bacterial host cells to absorb the DNA molecule.

5. If the restriction enzymes used to isolate the gene and cut open the plasmid are different, and if they have restriction sites within the DNA molecule that they did not previously act on, then the recombinant DNA may be cut into pieces.

6. To join two pieces of DNA, four phosphodiester bonds will need to be made by condensation. This is because DNA consists of two polynucleotide strands, each of which must be joined to another DNA molecule at each end.

7. (a) Bacterial genomes have regions called CRISPR sequences. These contain sections of DNA derived from viruses that are transcribed. These transcribed molecules are called guide RNA (gRNA) molecules.

(b) The loops take on a secondary structure due to base-pairing between regions of complementary bases. This characteristic structure allows the Cas9 endonuclease to recognise and bind to them, and position the region of complementary RNA within the active sites.

8. (a) Cells can become cancerous due to mutations in genes that control the cell cycle. Deleting or inserting sections of DNA to remove or replace the sequence of bases responsible may return cancer cells to a normal state.

(b) The genetic modification of crop plants to have useful properties, such as insect or herbicide resistance, could be achieved by deleting or inserting sections of DNA.

(c) The ability of pathogens to cause disease in their hosts, including their attachment or entry into cells, could be prevented by deleting or inserting sections of DNA into genes that encode antigens or toxins.

9. 72 °C is the temperature at which *Taq* DNA polymerase has maximum activity, which is why this temperature was chosen.

10. After 20 2^{20} cycles, over 1 million copies of a DNA molecule will be produced by PCR.

11. (a) PCR primers are short sections of single-stranded DNA of about 15–20 nucleotides in length. This number of monomers is not considered to constitute a polymer, so they are referred to instead as oligonucleotides. 'Oligo' is a prefix from a Greek word meaning 'few'.

(b) PCR primers bind (anneal) close to target sequences by complementary base pairing in the template DNA strand. This initiates DNA replication by providing a 3' end for *Taq* DNA polymerase to attach to and extend the complementary strand. Normally, scientists make two primers for each region of DNA that they wish to amplify. This is because each strand of the DNA must be replicated. They are often referred to as forward and reverse primers. It is also thought that primers reduce the re-annealing of separated strands.

12. D. The DNA molecules are placed into the wells in the gel, which are next to the cathode (negative electrode, black). When the

power supply is switched on, the DNA molecules will move towards the anode (positive electrode, red), through the entire length of the gel.

13. Bacterial (prokaryotic) genes do not have introns, whereas human (eukaryotic) genomes do. The presence of introns in genes allows a cell to produce a range of proteins from one gene.

14. Suggestions of the uses of microarrays include: the comparison of DNA of different species/ individuals; the screening for genetic disease; and investigating the change in gene expression patterns in response to environmental stimuli (see Experimental skills for an example of the last application).

15. (a) If 12 billion (1.2×10^{10}) bases are sequenced every day, and there are 365 days in a year, then the total number is ($1.2 \times 10^{10} \times 365$) = 4.38×10^{12} bases.

(b) Databases enable scientists to store and analyse data in a number of ways, including:
   - Comparison of sequences between different organisms to analyse evolutionary relationships.
   - Identification of differences in base sequence or amino acid sequence that are associated with specific phenotypes.

(c) For example, analysing the three-dimensional structure of proteins will allow scientists to carry out research into:
   - The shape of antigens and toxins that allow pathogens to enter cells and cause disease, so that medicinal drugs can be designed to inhibit them.
   - The shape of proteins that cause dominant disorders, such as HTT, which may provide some information that leads to the development of new medicinal drugs.

16. (a) Eukaryotic cells are preferred to prokaryotic cells as host organisms of recombinant DNA for two reasons. First, they are able to splice mRNA molecules to remove introns, which would otherwise be translated by prokaryotes into non-functional polypeptides. Second, yeast cells have rough endoplasmic reticulum and the Golgi body two membrane-bound organelles that allow for the processing and packaging of polypeptides, which often contributes to their formation of the expected three-dimensional structure as a protein.

(b) If recombinant mRNA is translated by free ribosomes in the cytoplasm, then the proteins produced would remain in the cytoplasm and would not be secreted by exocytosis. Translation by ribosomes attached to the rough endoplasmic reticulum allows for the processing and packaging of proteins into vesicles, and export by exocytosis.

17. (a) Adenoviruses do not activate a mechanism that inserts their genetic material into the cell's genome, unlike retroviruses. Retroviral material may insert into, or nearby, a gene involved in controlling the cell cycle, with effects on the regulation of cell division.

(b) Gene therapy can only usually be used to treat genetic disorders caused by having two copies of a recessive mutant allele. This is because gene therapy involves introducing a piece of DNA containing a gene encoding a functional protein, which is missing in the cells of these people. In dominant genetic diseases, a protein is normally produced, but this defective protein is the cause of the symptoms (for example, in the case of HTT in HD).

(c) Lymphocytes, which are lacking in individuals with ADA-SCID, are produced by stem cells in the bone marrow. It is likely that, over many years, stem cells that were successfully modified by gene therapy may die or fail to divide any more, and therefore will no longer differentiate into functional lymphocytes.

18. Arguments for: the treatment of disease such as cholera and cataracts is very inexpensive, and many people who suffer from these diseases can be cured with great ease, at a fraction of the cost of genetic technologies.

Arguments against: the costs of treating individuals with genetic disease by genetic technology have been reduced considerably by continued research and development, and it may be possible to treat all in the future.

19. The possible environmental risks of growing GM crops include:
• There is a possibility that pollen from GM crops could spread to wild plants and could be consumed by insects or other organisms, with negative effects on these organisms and the food chain to which they belong.
• There is a possibility that pollen from GM crops could spread to wild plants and this could result in the formation of 'superweeds' that have traits that have been introduced to the GM plants (e.g. insect resistance, herbicide resistance).

These risks could be minimised by:
• Planting GM crops a great distance away from other plants with which they can breed.
• Ensuring that GM crop plants are infertile, possibly by inserting the introduced genes into chloroplast DNA, which is not found in pollen.

20. GM crops are grown for a number of reasons, including:
• To produce their own insecticide, which reduces the damage due to insect pests and increases yield.
• To be resistant to herbicide, which means that toxic substances can be sprayed onto plants that compete with the crop for space, light and nutrients.
• To improve their nutritional value. An example is 'golden rice,' which contains additional vitamin A.

Assignment 19.1

A1. Increasing human populations will be associated with an increase in the bodies of still water (see Figure A1 for examples), which are breeding sites for the species. More hosts so more reservoirs for mosquito-borne pathogens.

A2. Marker genes usually encode proteins whose presence is easily determined, e.g. by fluorescence and are attached to the gene for the relevant characteristic. In this example, the gene encoded a fluorescent protein has been inserted next to the *tTA* gene. This enables researchers to determine whether cells are expressing the recombinant DNA by being able to see cells which express the gene in microscopic preparations.

A3. Scientists could insert a promoter sequence next to the tTA gene that usually activates transcription of genes in the developing wing muscle cells of female larvae. Promoters are regions of DNA to which transcription factors can bind, allowing RNA polymerase to begin transcription of the downstream gene.

A4. When larvae develop into mosquitoes, they will be unable to escape their underwater sites of development and will die due to drowning. Also, being unable to fly, female mosquitoes will not be able to find and feed on the blood of target species, and will be less likely to mate with males due to impaired courtship behaviour.

A5. GM larvae do not die immediately, meaning that they will be able to compete with non-GM larvae which will also reduce the numbers of female mosquitoes that survive to adulthood.

A6. Gene editing provides scientists with the ability to cut DNA and insert DNA with precision. This means that they will not be limited by the variety of restriction enzymes (with specific restriction sites) while they are constructing the recombinant DNA molecule to introduce to the host mosquito. Also they can simply remove genes (eg those associated with female characteristics.)

Experimental skills 19.1

P1. There is no relationship between temperature and the gene expression patterns of coral.

P2. (a) The researchers used 36 branches of coral and extracted 200 mg of tissue from each. Therefore, the total mass of coral used in this investigation was (36 x 200mg = 7 200 mg =) 7.2 g.
(b) The researchers cut a branch from three colonies of coral each day for 12 days, so (3 x 12 =) 36 branches of coral were used in this investigation and were applied to 36 separate microarrays.

P3. Storing the tissue samples at a cold temperature will reduce the activity of microorganisms or enzymes that may digest or damage the mRNA. In addition, the rapid freezing of the sample will prevent any further transcription of genes in the tissue after the point at which they were collected, which would give invalid measurements of gene expression.

P4. The cDNA must be heated to a high temperature before it is added to the microarray so that it becomes single-stranded as its H-bonds break, and its bases are exposed. This allows the molecules to hybridise by base pairing to any cDNA molecules on the microarray with complementary sequences.

P5. Percentage increase = ((final value – initial value) ÷ initial value) x 100. Therefore, the percentage increase in gene expression of species C between day 6 and day 8 is = ((0.56 – –0.03) ÷ –0.03) x 100 = 1967 %.

P6. Evidence for this claim is that shortly after a dramatic increase in marine temperature (at day 6-7), the expression of UPR genes increases from day 6 to 8. When the temperature falls from day 8 to 9, the expression also reduces.

Evidence against this claim is that in species A, the expression appears to rise earlier than the dramatic temperature increase. Also, the gene expression does not fall to the levels seen before the rise after the temperature falls.

The study may not be completely valid: the removal of branches from the coral may result in gene expression changes, and measurement of the temperature of water 1 m away from the coral colonies may not have provided an accurate estimate of the temperature of the coral itself. In addition, an increase in gene expression (production of mRNA) does not necessarily indicate that the encoded proteins are produced (the mRNA may not be translated).

P7. The scientists should use a microarray with a wide variety of single-stranded DNA probes complementary to the base sequence of a wide variety of genes common to different species of coral. Many copies of each type of probes are placed in each spot of the microarray. They should grow coral in identical environments (e.g. temperature, pH and nutrient content of the marine environment should be the same). At the same time, the scientists should expose both coral colonies to a 'heat shock' by increasing the temperature of their environment, and then at the same time for both colonies later, remove an equal mass of coral tissue from both coral colonies and undertake reverse transcription of the mRNA in each sample using fluorescent nucleotides to produce cDNA. It is important that the cDNA prepared from each sample is labelled with a different fluorescent nucleotide (e.g. red from one species and green from the other). After heating both cDNA samples to make the DNA molecules single-stranded, they are then applied to the same microarray. Unbound sample DNA is then washed off, leaving behind only the hybridised sample DNA. A laser/ UV light is then used to detect the presence of the fluorescent hybridised probes. A digital image is collected for analysis. The scientists will be able to identify which genes are expressed in only one species of coral by identifying spots on the microarray that appear red or green, and will be able to identify genes expressed in equal amounts by both corals by spots that appear yellow.

Experimental skills 19.2

P1. After isolating a mutant (non-functional) allele for phenylalanine hydroxylase, either by using a restriction enzyme to excise it from the chromosome, synthesising the gene by automated polynucleotide synthesis or reverse transcription from mRNA, this is inserted into a vector such as a plasmid or virus. This is then used to insert the gene into a zygote or early embryo, from which a mouse lacking the ability to produce this enzyme would be produced.

P2. A gRNA molecule is produced that has a region of complementary sequence to the gene encoding phenylalanine hydroxylase. This is added with recombinant Cas9 enzyme to a zygote or early embryo, from which a mouse lacking the ability to produce this enzyme would be produced.

P3. There is a drop in blood levels of phenylalanine in first $\frac{7}{21}$ days, but afterwards it levels out and stays low for the rest of the investigation time period.

P4. The introduction of a non-mutated gene into the mouse cells results in the production of normal, active phenylalanine hydroxylase enzyme. This enzyme then proceeds to convert phenylalanine into tyrosine, which reduces the concentration of phenylalanine in the blood.

P5. Reasons for the variation in the concentration of phenylalanine over the course of this study could include:
- the diet of the mice is changed.
- the availability of food for the mice is changed.
- the mice change in size or energy requirements.

In all cases, this may cause phenylalanine intake to change during the course of the study.

P6. water / saline / virus (only) / (empty) liposomes / vector (only) / use of placebo.

Chapter review

1. (a) Hydrolysis occurs during stage P (the removal of the mRNA by hydrolysis) and during stage R (the opening of the plasmid by restriction enzymes).
   (b) Condensation occurs during stage P (the formation of a single cDNA strand from DNA nucleotides), during stage Q (the formation of a second cDNA strand from DNA nucleotides), and during stage S (the ligation of the donor gene and the cut plasmid). In all cases, phosphodiester bonds form.
   (c) Polymerisation occurs during stages P and Q, when cDNA is formed from DNA nucleotides.
   (d) Hybridisation occurs during all stages, which involve the formation of base pairs between complementary nucleotides.

2. A suggested table is as follows:

Polymerase chain reaction (PCR)	DNA replication in the cell cycle
The DNA template is separated into two strands by heating (94 °C).	The DNA template is separated into two strands by the enzyme, DNA helicase.
Primers are required and only a defined region of the DNA molecule is replicated.	Primers are not required. The entire DNA molecule is replicated.
The enzyme Taq DNA polymerase has an optimum temperature of 72 °C	Human DNA polymerase has an optimum temperature of 37 °C

3. Databases in genetic technology can be used to:
   - Store the vast quantities of data produced by the sequencing of DNA and protein molecules.
   - Compare the sequences of DNA and protein molecules between species and between individuals of a species.
   - Predict the effect of changing DNA and protein sequences on the function of the organism, and hence predict the structure of medicinal drugs that could be used to treat diseases.

4. (b) Selective breeding is the most obvious odd-one-out. Although all three processes result in a change to the genetic makeup of a species over time, this process takes many hundreds or even thousands of generations (see Chapter 17).
   (c) Factor VIII is the most obvious odd-one-out. Although the absence of all three proteins cause disease in individuals, the gene that encodes factor VIII is on the X chromosome, and therefore haemophilia is much more common in males than females.
   (d) Breast cancer susceptibility is the most obvious odd-one-out. This is because it is caused by alleles of a number of genes (e.g. *BRCA1* and *BRCA2*), and in any case the emergence of the disease in an individual is due to a combination of inheritance and environmental factors that mean it is less than certain. Whether or not an individual suffers from HD or CF is entirely due to a mutation in an allele of a very specific gene (*HTT* or *CFTR*).

5. A suggested table is as follows:

Ethical problems	Other problems
The donor DNA may insert into the genome of the host cell's genome and can disrupt its function (the cell may become cancerous).	It is difficult to put the donor DNA into target cells.
Germline gene therapy may offer a permanent cure for individuals who are born with a genetic disease, but this involves the permanent change to the genome, and the inheritance of this change by future generations.	Target cells may not express the donor DNA.

6. (a) Including a promoter next to the gene for *Bt* toxin will result in the cell activating transcription of the target gene. Promoters are regions of DNA to which transcription factors can bind, allowing RNA polymerase to begin transcription of the downstream gene.

(b) DNA ligase is able to catalyse the formation of phosphodiester bonds in condensation reactions. This is required to join together a plasmid, which has been cut using restriction enzymes, with a section of donor DNA. This forms a recombinant plasmid.

(c) In areas that grow *Bt* cotton plants, very few insect pests can feed as they will die after consumption of the GM plants. Therefore, if an insect has an ability to withstand this insecticide, due to the possession of an allele, this individual will be at a competitive advantage over others. This individual will therefore be able to survive, reproduce and pass on this trait, which will become more common in the population over time (see Chapter 17).

7. Examples of accurate analogies include:
   - The ruler. Electrophoresis allows gene technologists to measure the length of DNA molecules.
   - The scissors. Restriction enzymes cut DNA and can isolate a fragment of DNA that contains a gene.
   - The glue. DNA ligase enzyme joins DNA molecules together, e.g. an isolated gene and a cut plasmid to form a recombinant plasmid.
   - The light bulbs. Marker genes, such as the gene that encodes green fluorescent protein, will inform the gene technologist that a recombinant plasmid has been successfully absorbed by a cell and is being expressed.
   - Nuts and bolts. These nucleotides are used in genetic engineering to build RNA and DNA molecules.
   - Notebook. Databases are used to store and analyse sequence data from DNA, RNA and proteins, and can be used to make predictions.
   - Mirror. Reverse transcriptase can be used to make a strand of cDNA from a strand of mRNA, which is the complementary sequence ('mirror image').
   - Elastic bands. Plasmids are used as vectors to introduce donor DNA into a host cell.
   - Magnifying glass. A microarray is used to spot differences between gene expression patterns.
   - Equipment for gene editing: CRISPR-Cas9 can be used to cut at very specific places in a DNA molecule (using the knife), to make deletions (using the eraser), or insert extra sequence (using the pencil).
   - Equipment for PCR: A thermal cycler (blowtorch) heats DNA molecule and uses primers to join together nucleotides using enzymes such as *Taq* DNA polymerase (hammer/ screwdriver).

8. (a) The genome is the entirety of the DNA found in a cell.
   (b) The mean number of bases in one gene of *C. ethensis-2.0* is found by dividing the size of the genome (785 701) by the total number of genes (680). This is equal to (785 701 ÷ 680) = 1155 bases.
   (c) The scientists may have failed to include regulatory sequences that are required for some of these genes to be expressed.
   (d) Gene editing could be used to change the genome sequences of wild bacteria in order to reduce their ability to cause disease, or to produce useful products such as medicinal drugs or health supplements.

Practice exam-style questions

E1. A [1]

E2. (a) Restriction endonuclease. [1]
   (b) GFP is produced in high quantities in a jellyfish cell, indicating that there are large numbers of mRNA molecules coding for GFP that has been transcribed from this gene. [1] There is only one copy of the GFP gene in any jellyfish cell, however, among many other genes. [1]
   (c) Scientists can identify, and isolate, regions of DNA that contain a promoter sequence that binds to a transcription factor present in a specific type of cell. [1] Placing this region of DNA next to the donor gene that is to be expressed will mean that the recombinant protein is only expressed in that specific tissue, because RNA polymerase can bind via transcription factors. [1]
   (d) Previously, marker genes conferred resistance to antibiotics on the bacterial cells that absorbed and expressed them. [1] The use of such marker genes has now been discouraged, because it may contribute to the evolution of antibiotic-resistant strains of bacteria (there is a small possibility that the GM bacteria could transfer the plasmid to wild bacteria). [1] It is also true that the detection of fluorescent marker gene products is much quicker than conducting experiments to detect antibiotic resistance.
   (e) Products of marker genes must be: easily identifiable, [1] non-toxic to the host cell [1] and must be expressed in combination with the donor gene of interest. [1]
   (f) A virus [1] is used to deliver the GFP gene into an embryo of the zebrafish. Retroviruses can be used, which insert the genetic material into the genome, [1] resulting in all cells of the fish expressing the recombinant protein. [1]
   (g) Glo-Fish® are not intended for human consumption, and some people are concerned that GM salmon may cause long-term and unpredictable health effects on those who eat them. [1] GM salmon are grown in semi-captivity, compared with Glo-Fish®, and therefore there is a greater risk that they will escape and breed with wild fish. [1]

E3. (a) During the time period indicated by X, the DNA template is heated to around 94 °C. [1] At this high temperature, the hydrogen bonds between the two polynucleotide strands are broken so that the two strands are separated and the bases are exposed to produce template strands for complementary copying. [1]
   (b) Exact answer will depend on artwork (expected: 50 seconds).
   (c) *Taq* DNA polymerase is an enzyme, so it is not changed by the reaction and can catalyse condensation reactions in successive PCR cycles. [1] The enzyme is not denatured by the high temperatures used in PCR because it is thermostable. [1] The high temperatures used during PCR mean that the process is more efficient (faster) than if normal (e.g. human) DNA polymerase was used. [1]
   (d) The independent variable in this investigation is the ratio of G/C to A/T in the primer dimer. [1] The dependent variable is the melting temperature (T_m). [1]
   (e) The investigators standardised the following factors in an attempt to ensure that the investigation was valid [max. 3]:
      - pH [1]
      - Each primer was 18 nucleotides long. [1]
      - The same volume/ concentration of primer solution was used in each test. [1]
      - The primer solution was heated slowly from an initial temperature of room temperature. [1]
   (f) The results of this investigation were considered reliable because the scientists repeated the tests 10 times for each primer and then calculated a mean value. [1]
   (g) 1 cm^3 of the original stock solution of 0.1 mmol dm^{-3} was added to 9 cm^3 of distilled water. [1] This is then thoroughly

mixed, before 1 cm³ of this solution is added to 9 cm³ of distilled water. [1] The process is repeated a third time, which will give 10 cm³ of a solution of 0.1 μmol cm⁻³. [1]

(h) The Pearson's correlation coefficient should be calculated for this data. [1] This is because the data is normally distributed and continuous. [1]

(i) As the proportion of G/C in the primer dimers increases, the melting temperature increases. This suggests that there are more hydrogen bonds that form between the two primers/ stronger forces of attraction between them, [1] which results in a requirement for greater heat energy to separate the two strands. [1]

E4. (a) PCR is required to amplify DNA. [1] This means that a large amount of DNA, for screening purposes, can be obtained from a very small sample, such as a few blood or cheek cells. [1]

(b) During gel electrophoresis, DNA molecules, which have a negative charge due to their phosphate groups, move towards the positive electrode (anode) through a gel that provides resistance to their movement. [1] Longer/ more massive fragments of DNA will move more slowly than shorter/ lighter fragments in a given time, [1] and therefore will move a lesser distance. [1]

(c) Person B tested positive for Huntington's disease [1] This is because they have a band that has travelled the least distance, and therefore is larger than all the others. [1]

(d) The scientists could separate a mixture of DNA fragments of known length at the same time as the samples of DNA from the individuals. [1] Comparing the position of the bands from the individuals with the position of the bands in this mixture will allow the scientists to estimate the number of CAG repeats in the fragments corresponding to the bands on the gel. [1]

(e) Person C has two alleles for *HTT* that have exactly the same number of CAG repeats, [1] or Person C has one allele with very few CAG repeats and this has run far down the gel and is not shown. [1]

(f) Suggestions may include:

Advantages: screening for faulty *HTT* alleles can provide potentially affected individuals with the knowledge they need to make an informed choice whether or not to have children, or prepare for their upcoming disease by making a lifestyle change. [1]

Disadvantages: a negative test for HD will relieve worry and will have a positive implication for life insurance, whereas a positive test for a faulty HTT allele could result in great distress as there is no cure for HD and an individual will know that they will develop the disease with complete certainty. [1]

(g) Gene editing involves the cutting of DNA at a very specific position in the genome, determined by the sequence of a guide RNA (gRNA). Scientists could design a gRNA that is complementary to the region of sequence around the CAG repeats, [1] and delete a number of CAG repeats from the faulty *HTT* allele (to fewer than 40 copies). [1] However, this would need to be done in the germline cells (gametes or very early embryo), to ensure that all cells of an individual have the edited allele. [1]

E5. (a) Amplification involves the production of many copies of DNA molecules from a template molecule, which can happen in a process called PCR. [1]

(b) The primers produced by the researchers should be short (15–20 nucleotide) lengths of single-stranded DNA that

have a base sequence complementary to sequences of DNA either side of the STR regions in the pangolin genome. [1] PCR using these primers will therefore amplify the STR regions, which differ in length between different individuals. [1]

(c) The researchers should amplify a number of different STRs in the DNA taken from the seized scales, and also amplify the same STRs in the DNA of pangolins native to Africa as well as those native to Asia. [1] The result will be a mixture of fragments of DNA of different lengths, the size of which is dependent on the number of repeats in the STRs. Next, the researchers should subject the three samples of DNA fragments to gel electrophoresis. [1] Comparing the positions of the bands in each of the three gel lanes will give an indication of the degree of genetic similarity between the organisms from which the samples were obtained. If the position of bands corresponding to the seized scales is more similar to the pattern of the bands corresponding to the African pangolin DNA sample than the Asian sample, then it is likely that the seized scales were African in origin. [1]

A sketch may look something like this. In this example, five STRs have been amplified, with samples A and B (seized scales and African pangolins respectively) having more similarity than either do with C (Asian pangolins):

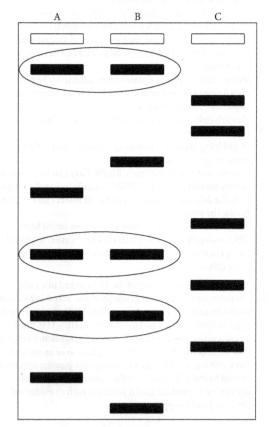

(identical bands are circled).

(d) PCR is a very sensitive technique, because any small quantities of DNA present in the reaction vessel will be amplified at the same time as the majority of the template molecules. [1] Packing the scales close together may mean that small sections of biological material are mixed, reducing the likelihood that the scientists will be able to estimate the relative proportions of scales that originated from Africa and Asia. [1]

(e) It is important in captive breeding programmes to avoid breeding individuals that are very closely related: this will increase the likelihood of the offspring genetic diseases affecting the offspring. [1mark] Genetic fingerprinting (DNA profiling) of potential parents can provide an individuation of the degree of relatedness between them, and captive breeders can select individuals that have fewer bands in common. [1]

Index

Acknowledgements

The publishers wish to thank the following for permission to reproduce photographs. Every effort has been made to trace copyright holders and to obtain their permission for the use of copyright materials. The publishers will gladly receive any information enabling them to rectify any error or omission at the first opportunity.

(t = top, c = centre, b = bottom, r = right, l = left)

This textbook includes images that are licensed under an Open Government Licence v3.0, Attribution-ShareAlike 4.0 International Licence and Creative Commons Attribution-NonCommercial 4.0 International Licence where indicated.

p 1 Heiti Paves/Shutterstock, p 2l Lebendkulturen.de /Shutterstock, p 2c Science History Images / Alamy Stock Photo, p 2r DAVID SCHARF / SCIENCE PHOTO LIBRARY, p 5t Videologia/Shutterstock, p 5b Science History Images / Alamy Stock Photo, p 6 D. Kucharski K. Kucharska/ Shutterstock, p 8 BIOPHOTO ASSOCIATES / SCIENCE PHOTO LIBRARY, p 9t Dr. Norbert Lange/Shutterstock, p 9b Asier Romero/Shutterstock, p 10 BIOPHOTO ASSOCIATES / SCIENCE PHOTO LIBRARY, p 11 BIOPHOTO ASSOCIATES / SCIENCE PHOTO LIBRARY, p 12 Science History Images / Alamy Stock Photo, p 13 Science History Images / Alamy Stock Photo, p 14t KEITH R. PORTER / SCIENCE PHOTO LIBRARY, p 14b Lebendkulturen.de /Shutterstock, p 15 DR DAVID FURNESS, KEELE UNIVERSITY / SCIENCE PHOTO LIBRARY, p 16 BIOPHOTO ASSOCIATES / SCIENCE PHOTO LIBRARY, p 17 Science History Images / Alamy Stock Photo, p 22 EYE OF SCIENCE / SCIENCE PHOTO LIBRARY, p 23 Jagoush/ Shutterstock, p 25 ANDREW LAMBERT PHOTOGRAPHY / SCIENCE PHOTO LIBRARY, p 26t David Martindill, p 26bl David Martindill, p 26bc David Martindill, p 26br David Martindill, p 27l ANDREW LAMBERT PHOTOGRAPHY / SCIENCE PHOTO LIBRARY, p 27cl ANDREW LAMBERT PHOTOGRAPHY / SCIENCE PHOTO LIBRARY, p 27cr ANDREW LAMBERT PHOTOGRAPHY / SCIENCE PHOTO LIBRARY, p 27r David Martindill, p 29 Science History Images / Alamy Stock Photo, p 35 nehophoto/Shutterstock, p 41 Jose Luis Calvo/Shutterstock, p 51t Francois Loubser /Shutterstock, p 51b SCOTT CAMAZINE / SCIENCE PHOTO LIBRARY, p 54 PopTika/Shutterstock, p 55 adapted from Office of National Statistics licensed under the Open Government Licence v3.0 (http://www.nationalarchives.gov. uk/doc/open-government-licence/version/3/), p 61 Rich Carey/Shutterstock, p 63l David Martindill, p 63r Universal Images Group North America LLC / Alamy Stock Photo, p 66 J.C. REVY, ISM / SCIENCE PHOTO LIBRARY, p 71 David Martindill, p 87 Vadim Sadovski/Shutterstock, p 94 Africa Studio/Shutterstock, p 95 DON W. FAWCETT / SCIENCE PHOTO LIBRARY, p 96 ESB Basic/Shutterstock, p 102 SergeyIT/Shutterstock, p 103 David Martindill, p 108 John Wollwerth/Shutterstock, p 114 Rattiya Thongdumhyu /Shutterstock, p 116 Science Photo Library / Alamy Stock Photo, p 117 David Martindill, p 119 kudla/Shutterstock, p 121 mahirart/Shutterstock, p 126 background GL Archive / Alamy Stock Photo, p 126 inset T.W./Shutterstock, p 127l Kateryna Kon/Shutterstock, p 127r Nathan Devery /Shutterstock, p 129l Telomere length is paternally inherited and is associated with parental lifespan, Omer T. Njajou, et.al., PNAS July 17, 2007 104 (29) 12135-12139, Copyright (2007) National Academy of Sciences, U.S.A., p 129r Beth Harvey /Shutterstock, p 130 royaltystockphoto.com/Shutterstock, p 134t HERVE CONGE, ISM / SCIENCE PHOTO LIBRARY, p 134ct HERVE CONGE, ISM / SCIENCE PHOTO LIBRARY, p 134cb HERVE CONGE, ISM / SCIENCE PHOTO LIBRARY, p 134b HERVE CONGE, ISM / SCIENCE PHOTO LIBRARY, p 140 D. Kucharski K. Kucharska/Shutterstock, p 141 Rattiya Thongdumhyu/Shutterstock, p 142 Dimarion/Shutterstock, p 143 All for you friend/Shutterstock, p 149 Galyna Andrushko/Shutterstock, p 150 WILL & DENI MCINTYRE / SCIENCE SOURCE / SCIENCE PHOTO LIBRARY, p 153 A. BARRINGTON BROWN, © GONVILLE & CAIUS COLLEGE / SCIENCE PHOTO LIBRARY, p 164 with thanks to Prof. Tsuneko Okazaki, p 173 DR ELENA KISELEVA / SCIENCE PHOTO LIBRARY, p 183 David Lade/Shutterstock, p 184l YSurasit/Shutterstock, p 184r Aleksey Stemmer /Shutterstock, p 185 Fotofermer/Shutterstock, p 186t Ernest Cooper/Shutterstock, p 186b John Boud / Alamy Stock Photo, p 187 DR DAVID FURNESS, KEELE UNIVERSITY / SCIENCE PHOTO LIBRARY, p 188 J.C. REVY, ISM / SCIENCE PHOTO LIBRARY, p 189l BIOPHOTO ASSOCIATES / SCIENCE PHOTO LIBRARY, p 189r J.C. REVY, ISM / SCIENCE PHOTO LIBRARY, p 190l Claudio Divizia/Shutterstock, p 190r Mike Rosecope /Shutterstock, p 191 CLAUDE NURIDSANY & MARIE PERENNOU / SCIENCE PHOTO LIBRARY, p 196t ADA_ photo/Shutterstock, p 196bl Henrik Larsson/Shutterstock, p 196br POWER AND SYRED / SCIENCE PHOTO LIBRARY, p 203 ZUMA Press, Inc. / Alamy Stock Photo, p 208t National Geographic Image Collection / Alamy Stock Photo, p 208c DENNIS KUNKEL MICROSCOPY / SCIENCE PHOTO LIBRARY, p 208b Jose Luis Calvo/Shutterstock, p 210t somersault1824/Shutterstock, p 210c somersault1824 /Shutterstock, p 210b somersault1824/Shutterstock, p 211t SIRIKWAN DOKUTA/Shutterstock, p 211b Jarun Ontakrai /Shutterstock, p 217 Africa Studio/Shutterstock, p 226 first vector trend/Shutterstock, p 228t HERVE CONGE, ISM / SCIENCE PHOTO LIBRARY, p 228b ALVIN TELSER / SCIENCE PHOTO LIBRARY, p 229 BIOPHOTO ASSOCIATES / SCIENCE PHOTO LIBRARY, p 230r STEVE GSCHMEISSNER / SCIENCE PHOTO LIBRARY, p 230l PROF. P. MOTTA/DEPT. OF ANATOMY/UNIVERSITY "LA SAPIENZA", ROME / SCIENCE PHOTO LIBRARY, p 231 ASTRID & HANNS-FRIEDER MICHLER / SCIENCE PHOTO LIBRARY, p 232 MICROSCAPE / SCIENCE PHOTO LIBRARY, p 236 adapted from Mortality in relation to smoking: 50 years' observations on male British doctors" by Doll R., Peto R., Boreham J, Sutherland I. BMJ, June 2004; 328(7455), p.1519. Epub 2004 Jun 22. https://www.ncbi.nlm. nih.gov/pubmed/15213107, data courtesy of Dr Hongchao Pan, p 239 TUM2282/Shutterstock, p 240 BSIP SA / Alamy Stock Photo, p 242t Kletr/Shutterstock, p 242c age fotostock / Alamy Stock Photo, p 242b tantzi13/Shutterstock,

Acknowledgements

p 243 Phanie / Alamy Stock Photo, p 246 Science Photo Library / Alamy Stock Photo, p 248 Max Roser. Published online at OurWorldInData.org under CC-BY-SA Licence (https://creativecommons.org/licenses/by-sa/4.0/), p 250 Kallayanee Naloka/Shutterstock, p 257 2Ban/Shutterstock, p 258 STEVE GSCHMEISSNER / SCIENCE PHOTO LIBRARY, p 261 reproduced with permission of the © ERS 2019: European Respiratory Journal 35 (5) 1039-1047; DOI: 10.1183/09031936.00036709 Published 1 May 2010, p 264 STEVE GSCHMEISSNER / SCIENCE PHOTO LIBRARY, p 272 HANK MORGAN / SCIENCE PHOTO LIBRARY, p 274t Nordroden/Shutterstock, p 274b ProStockStudio /Shutterstock, p 278 Africa Studio/Shutterstock, p 280 NYPL / SCIENCE SOURCE / SCIENCE PHOTO LIBRARY, p 282 Ravenash/Shutterstock, p 286 STEVE GSCHMEISSNER / SCIENCE PHOTO LIBRARY, p 288 Josep Suria /Shutterstock, p 289 J.C. REVY, ISM / SCIENCE PHOTO LIBRARY, p 291 Hurst Photo/Shutterstock, p 300 Science History Images / Alamy Stock Photo, p 302 Jason Lindsey / Alamy Stock Photo, p 303 Jubal Harshaw/Shutterstock, p 308 JHVEPhoto/Shutterstock, p 310 Science History Images / Alamy Stock Photo, p 318 Rudmer Zwerver/Shutterstock, p 327 Hriana/Shutterstock, p 329l Jose Luis Calvo /Shutterstock, p 329r CNRI / SCIENCE PHOTO LIBRARY PHOTO LIBRARY, p 334 Ondrej83/Shutterstock, p 337 Ozgur Coskun/Shutterstock, p 339t PROF. P. MOTTA / DEPT. OF ANATOMY / UNIVERSITY "LA SAPIENZA", ROME / SCIENCE PHOTO LIBRARY, p 339b J.L. KEMENY / ISM / SCIENCE PHOTO LIBRARY, p 342 Choksawatdikorn/ Shutterstock, p 343 Kateryna Kon/Shutterstock, p 351 Kallayanee Naloka/Shutterstock, p 357 Jacob Lund /Shutterstock, p 361 CNRI / SCIENCE PHOTO LIBRARY, p 362 Jose Luis Calvo/Shutterstock, p 369 Science History Images / Alamy Stock Photo, p 374 Jose Luis Calvo /Shutterstock, p 378 Chrispo/Shutterstock, p 379 Martin Shields / Alamy Stock Photo, p 386 petarg/Shutterstock, p 387 Marmaduke St. John / Alamy Stock Photo, p 390l DR KEITH WHEELER / SCIENCE PHOTO LIBRARY, p 390r agefotostock / Alamy Stock Photo, p 395 Martin Shields / Alamy Stock Photo, p 397 claire norman/Shutterstock, p 398 Friedrich Stark / Alamy Stock Photo, p 408 Colin Woods /Shutterstock, p 430 Henri Koskinen/Shutterstock, p 431 Joe McDonald/Shutterstock, p 439t Luis_Martinez/Shutterstock, p 439cl melissamn/Shutterstock, p 439cr COULANGES /Shutterstock, p 439b Everett Historical/Shutterstock, p 442t Henrik Larsson/Shutterstock, p 442b Steve McWilliam /Shutterstock, p 443 Matthew Ball/Shutterstock, p 447 Animal Photography / Alamy Stock Photo, p 449t Photographer253 /Shutterstock, p 449b WIRACHAIPHOTO/Shutterstock, p 452t Andrew Parkinson / Alamy Stock Photo, p 452bl Olga_i/Shutterstock, p 452bc Geza Farkas/Shutterstock,

p 452br Juniors Bildarchiv GmbH / Alamy Stock Photo, p 454 Randy Bjorklund/Shutterstock, p 457 GUDKOV ANDREY /Shutterstock, p 458tl National Geographic Image Collection / Alamy Stock Photo, p 458tc J Hopwood/Shutterstock, p 458tr Michael Wiegmann/Shutterstock, p 458bl Nolichuckyjake/Shutterstock, p 458br Vladimir Vasiltvich /Shutterstock, p 462 Fotokostic/Shutterstock, p 463 Lebendkulturen.de/Shutterstock, p 464t Bildagentur Zoonar GmbH/Shutterstock, p 464c Ryzhkov Sergey/Shutterstock, p 464b Menno Schaefer/Shutterstock, p 466 Ethan Daniels/ Shutterstock, p 469t Tisha 85/Shutterstock, p 469b SARAWUT KUNDEJ/Shutterstock, p 481 Brian Clifford/Shutterstock, p 484 Jono Photography/Shutterstock, p 485 Gina Fellendorf /Shutterstock, p 491 ETH Zurich / Jonathan Venetz, p 496 DR GOPAL MURTI / SCIENCE PHOTO LIBRARY, p 499 Newscom / Alamy Stock Photo, p 501 Hemis / Alamy Stock Photo, p 502 Kallayanee Naloka/Shutterstock, p 505 David Martindill, p 506 science photo/Shutterstock, p 508 reprinted with permission of AAAS. Modified from "Tidal heat pulses on a reef trigger a fine-tuned transcriptional response in corals to maintain homeostasis" by Lupita J. Ruiz-Jones and Stephen R. Palumbi, Science Advances, 08 Mar 2017, Vol. 3, no. 3, e1601298. Fig 1 C, D, https://advances.sciencemag.org/ content/3/3/e1601298?TB_iframe=true&width=370.8&height= 658.8. © The Authors, some rights reserved; exclusive licensee American Association for the Advancement of Science. Distributed under a Creative Commons Attribution NonCommercial License 4.0 (CC BY-NC) (http:// creativecommons.org/licenses /by-nc/4.0/), p 515 Danita Delmont/Shutterstock, p 517t konstantinks/Shutterstock, p 517tc Creative Stall/Shutterstock, p 517c Visual Generation/ Shutterstock, p 517bc Panda Vector /Shutterstock, p 517b Anatolir/Shutterstock, p 526 STEVE GSCHMEISSNER / SCIENCE PHOTO LIBRARY, p 527 Miroslav Hlavko/ Shutterstock, p 534t Rattiya Thongdumhyu/Shutterstock, p 534b D. Kucharski K. Kucharska/Shutterstock, p 539t Science Photo Library / Alamy Stock Photo, p 539b adapted from Asychterz18 (https://commons.wikimedia.org/wiki/ File:Diffrences_in_Stomata_Opening_Throughout_the_Day_ for_C3_plants_and_CAM_plants_(1).svg) under Attribution-ShareAlike 4.0 International (https://creativecommons.org/ licenses/by-sa/4.0/deed.en), p 544 STEVE GSCHMEISSNER / SCIENCE PHOTO LIBRARY, p 545 based on WHO, Global Tuberculosis Report 2012, p 547 TRossJones/Shutterstock, p 551 Science History Images / Alamy Stock Photo, p 552 reproduced from Climate Change and World Agriculture (RLE, Pollution, Climate & Change), 1st Edition by Martin L., Parry, published by Routledge. © Routledge, 2019, reproduced by arrangement with Taylor & Francis Books UK, p 553 Mps197/Shutterstock, p 564 Guido Vermeulen-Perdaen/ Shutterstock, p 566 Sirilak S/Shutterstock.